Sustainable Design Through Process Integration

Sustainable Design Through Process Integration

Fundamentals and Applications to Industrial Pollution Prevention, Resource Conservation, and Profitability Enhancement

MAHMOUD M. EL-HALWAGI
The Artie McFerrin Department of Chemical Engineering
Texas A&M University
College Station, Texas, USA

AMSTERDAM • BOSTON • HEIDELBERG • LONDON • NEW YORK • OXFORD • PARIS • SAN DIEGO
SAN FRANCISCO • SINGAPORE • SYDNEY • TOKYO
Butterworth-Heinemann is an imprint of Elsevier

Butterworth-Heinemann is an imprint of Elsevier
225 Wyman Street, Waltham, MA 02451, USA
The Boulevard, Langford Lane, Kidlington, Oxford, OX5 1GB, UK

Copyright © 2012 Elsevier Inc. All rights reserved

No part of this publication may be reproduced or transmitted in any form or by any means, electronic or mechanical, including photocopying, recording, or any information storage and retrieval system, without permission in writing from the publisher. Details on how to seek permission, further information about the Publisher's permissions policies and our arrangements with organizations such as the Copyright Clearance Center and the Copyright Licensing Agency, can be found at our website: www.elsevier.com/permissions.

This book and the individual contributions contained in it are protected under copyright by the Publisher (other than as may be noted herein).

Notices
Knowledge and best practice in this field are constantly changing. As new research and experience broaden our understanding, changes in research methods, professional practices, or medical treatment may become necessary.

Practitioners and researchers must always rely on their own experience and knowledge in evaluating and using any information, methods, compounds, or experiments described herein. In using such information or methods they should be mindful of their own safety and the safety of others, including parties for whom they have a professional responsibility.

To the fullest extent of the law, neither the Publisher nor the authors, contributors, or editors, assume any liability for any injury and/or damage to persons or property as a matter of products liability, negligence or otherwise, or from any use or operation of any methods, products, instructions, or ideas contained in the material herein.

Library of Congress Cataloging-in-Publication Data
El-Halwagi, Mahmoud M., 1962-
  Sustainable design through process integration : fundamentals and applications to industrial pollution prevention, resource conservation, and profitability enhancement / Mahmoud M. El-Halwagi.
   p. cm.
 Includes bibliographical references.
 ISBN 978-1-85617-744-3 (hardback)
 1. Chemical processes. 2. Sustainable engineering. I. Title.
 TP155.7.E476 2012
 660'.28—dc23
                                                                2011020300

British Library Cataloguing-in-Publication Data
A catalogue record for this book is available from the British Library

For information on all Butterworth-Heinemann publications
visit our Web site at www.elsevierdirect.com

Printed in the United States of America

11 12 13 14 10 9 8 7 6

Working together to grow
libraries in developing countries

www.elsevier.com | www.bookaid.org | www.sabre.org

ELSEVIER   BOOK AID International   Sabre Foundation

To my parents, my wife, and my sons, with love and gratitude

CONTENTS

Preface ix

1. Introduction to Sustainability, Sustainable Design, and Process Integration 1
2. Overview of Process Economics 15
3. Benchmarking Process Performance Through Overall Mass Targeting 63
4. Direct-Recycle Networks: A Graphical Approach 89
5. Synthesis of Mass-Exchange Networks: A Graphical Approach 111
6. Combining Mass-Integration Strategies 133
7. Heat Integration 147
8. Integration of Combined Heat and Power Systems 165
9. Property Integration 201
10. Direct-Recycle Networks: An Algebraic Approach 223
11. Synthesis of Mass-Exchange Networks: An Algebraic Approach 231
12. Synthesis of Heat-Induced Separation Networks for Condensation of Volatile Organic Compounds 237
13. Design of Membrane-Separation Systems 243
14. Overview of Optimization 255
15. An optimization approach to direct recycle 287
16. Synthesis of mass-exchange networks: A mathematical programming approach 299
17. Synthesis of reactive mass-exchange networks 315
18. Mathematical Optimization Techniques for Mass Integration 327
19. Mathematical Techniques for the Synthesis of Heat-Exchange Networks 345
20. Synthesis of Combined Heat and Reactive Mass-Exchange Networks 357
21. Design of Integrated Biorefineries 365
22. Macroscopic approaches of process integration 375
23. Concluding Thoughts: Launching Successful Process-Integration Initiatives and Applications 393

Appendix I Conversion Relationships for Concentrations and Conversion Factors for Units 397
Appendix II Modeling of Mass-Exchange Units for Environmental Applications 401

Index 415

Preface

One of the most important challenges facing humanity is the need for a sustainable development that accommodates the escalating demands for natural resources while leaving future generations with the opportunities to realize their potential. This challenge is especially important for the chemical process industries that are characterized by the enormous usage of natural resources. To effectively address this challenge, it is inevitable for industry to embrace the concept of sustainable design, which involves process-design activities that lead to economic growth, environmental protection, and social progress for the current generation without compromising the potential of future generations to have an ecosystem that meets their needs. Consequently, a growing number of industries are launching sustainable-design initiatives that are geared toward enhancing the corporate stewardship of the environment. Although these initiatives are typically clear in their strategic goals, they are very difficult for technical managers and process engineers to transform into viable actions. A sustainable design should endeavor to conserve natural resources (mass and energy), recycle and reuse materials, prevent pollution, enhance yield, improve quality, advance inherent safety, and increase profitability. The question is how to achieve and reconcile these objectives? Processing facilities are complex systems of unit operations and streams. Designing these facilities or improving their performance typically entails the screening of numerous alternatives. Because of the enormous number of design alternatives, laborious conventional engineering approaches that are based on generating and testing case studies are unlikely to provide effective work processes or reach optimal solutions. Indeed, what is needed is a systematic framework and associated concepts and tools that methodically guide designers to the global insights of the process, identify root causes of the problems or key areas of opportunities, benchmark the performance of the process, and develop a set of design recommendations that can attain the true potential of the process.

Over the past three decades, significant advances have been made in treating chemical processes as integrated systems and developing systematic tools to determine practically achievable benchmarks. This framework is referred to as process integration and is defined as a holistic approach to design and operation that emphasizes the unity of the process. Process integration can be used to systematically enhance and reconcile various process objectives, such as cost effectiveness, yield enhancement, energy efficiency, and pollution prevention. Many archival papers have been published on different aspects of process integration. Because of the specialized nature of these papers, readership has been mostly confined to academic researchers in the field. On the other hand, many industrial projects have been successfully implemented on specific aspects of process integration. Because of the confidential nature of most of these projects, details have not been widely available in the public domain. This book was motivated by the need to reach out to a much wider base of readers who are interested in systematically developing sustainable designs through process integration. The book is appropriate for senior-level undergraduate or first-year graduate courses on process design, sustainability, or process synthesis and integration. It is also tailored to serve as a self-study textbook for process engineers and technical managers involved in process innovation, development, design and improvement, pollution prevention, and energy conservation. A key feature of the book is the emphasis on benchmarking the performance of a process or subprocess and then methodically detailing the steps needed to attain these performance targets in a cost-effective manner.

The approach of this book is to first explain the problem statement and scope of applications, followed by the generic concepts, procedures, and tools that can be used to solved the problem. Next, case studies and numerical examples are given to demonstrate the applicability of the tools and procedures. Chapter 1 introduces the key concepts of sustainability, sustainable design, and process integration. Motivating examples are given on the development and integration of sustainable design alternatives. The chapter also describes the learning outcomes of the books. Chapter 2 provides a detailed coverage of process economics including cost types and estimation, depreciation, break-even analysis, time value of money, and profitability analysis. Applications involve a broad range of conventional and contemporary problems in the process industries. Because of the extensive nature of the chapter, it can be used in senior-level process design and economics courses. Chapter 3 introduces the concept of overall benchmarking (targeting) and focuses on the identification of performance targets for the consumption of fresh materials, the discharge of waste materials, and the production of maximum yield. Chapters 4 through 9 present graphical techniques (pinch diagrams) for the targeting of direct-recycle systems, mass-exchange networks, overall processes, heat-exchange networks, combined heat and power systems, and property integration. Chapters 10 through 13 are based on algebraic procedures for the design of direct-recycle networks, mass-exchange networks, heat-induced separators, and membrane-separation networks. Chapter 14 covers the basic approaches to the formulation of optimization problems as mathematical programs and the different types of formulations. Examples are given on transforming tasks and concepts into optimization formulations. Also, the use of the software LINGO is described. Chapters 15 through 20 are devoted to the solution of sustainable design problems through optimization. Several classes of problems are addressed, including direct-recycle networks, mass-exchange networks, heat-exchange networks, and combined heat and reactive mass-exchange networks. Chapter 21 covers the conceptual design and techno-economic assessment of integrated biorefineries. The focus is on top-level and quick synthesis and screening of alternative designs. Macroscopic process integration approaches are addressed in Chapter 22, with several applications such as eco-industrial parks, material flow analysis, environmental impact assessment, and life cycle analysis. The book culminates in Chapter 23, which offers a discussion on commercial applicability of process integration for sustainable design, track record and pitfalls in implementing process integration, and starting and sustaining process integration initiatives and projects.

Various individuals have positively impacted my path of learning about and contributing to sustainable design through process integration. I very much appreciate the professional associates and leaders of the process systems engineering and the sustainability communities whose contributions have made a paradigm shift in the understanding and tackling of sustainable design problems. I am especially grateful to Dr. Dennis Spriggs (president of Matrix Process Integration) who has mentored me in numerous industrial applications and has consistently shown the power of the "science of the big picture" in tackling complex industrial challenges in a smooth and insightful manner. I am also thankful to the academic partners with whom I had the honor of collaborating. Specifically, I would like to thank the following professors and their students: Drs. Ahmed Abdel-Wahab (Texas A&M University-Qatar), Mert Atilhan (Qatar University), Mario Eden (Auburn University), Nimir Elbashir (Texas A&M University-Qatar), Amro El-Baz (Zagazig University), Fadwa Eljack (Qatar University), Xiao Feng (China University of Petroleum), Dominic C. Y. Foo (University of Nottingham, Malaysia Campus), Arturo Jiménez-Gutiérrez (Instituto Tecnológico de Celaya), Ken Hall (Texas A&M University), Mark lottzapple (Texas A&M University), Viatcheslav Kafarov (Universidad Industrial de Santander), B. J. Kim (Soongsil University), Patrick Linke (Texas A&M University-Qatar), Vladimir Mahalec (McMaster University), Sam Mannan (Texas A&M University), Pedro Medellín Milán (Universidad Autónoma de San Luis Potosí), Denny Ng (University of Nottingham, Malaysia Campus), Martín Picón-Núñez (Universidad de Guanajuato), José María Ponce-Ortega (Universidad Michoacana de San Nicolás de Hidalgo), Abeer Shoaib (Suez Canal University), Paul Stuart (Ecole Polytechnique de Montréal), and Raymond Tan (De La Salle University).

I am very grateful to the numerous undergraduate students at Texas A&M University and Auburn University as well as attendees of my industrial workshops, short courses, and seminars whose invaluable feedback and input was instrumental in developing and refining the book.

I am indebted to my former and current graduate students. I have learned much from this distinguished group of scholars, which includes: Nesreen Ahmed (Suez Canal University), Nasser Al-Azri (Sultan Qaboos University), Hassan Alfadala (Barwa), Eid Al-Mutairi (King Fahd University of Petroleum and Minerals), Abdul-Aziz Almutlaq (King Saud University), Meteab Al-Otaibi (SABIC), Saad Al-Sobhi (Qatar University), Musaed Al-Thubaiti (Aramco), Selma Atilhan (Texas A&M University-Qatar), Srinivas "B.K." Bagepalli (Danaher), Buping Bao, Abdullah Bin Mahfouz (SABIC), Ian Bowling (Chevron), Ming-Hao Chiou, Benjamin Cormier (BP), Eric Crabtree (Parsons), Alec Dobson (Solutia), Russell Dunn (Vanderbilt University), Brent Ellison (LightRidge Resources), René Elms (Bryan Research and Engineering), Fred Gabriel (Honeywell), Kerron Gabriel, Walker Garrison (Valero), Adam Georgeson (Bryan Research and Engineering), Ian Glasgow (Mustang Engineering), Murali Gopalakrishnan (ExxonMobil), Daniel Grooms (Invista), Ahmad Hamad (Marathon Oil Company), Natalie Hamad, Dustin Harell (Intel), Rasha Hasaneen (GE Corporate Initiatives), Ronnie Hassanen (GE Energy), Ana Carolina Hortua (KBR), Vasiliki Kazantzi (Technological Educational Institute of Larissa), Houssein Kheireddine, Eva Lovelady (Mustang Engineering), Rubayat Mahmud (Intel), Tanya Mohan (Air Products), Lay Myint (Shell Global Solutions), Bahy Noureldin (Aramco), Mohamed Noureldin, Madhav Nyapathi (Shell), Gautham "P. G." Parthasarathy (Solutia), Eric Pennaz, Grace Pokoo-Aikins (Cambridge Environmental Technologies), Viet Pham (Dow), Xiaoyun Qin (Baker Engineering and Risk), Jagdish Rao (Shell), Arwa Rabie (Dow), Andrea Richburg (3M), Brandon Shaw (Foster Wheeler), Mark Shelley (Andrews Kurth LLP), Chris Soileau (Veritech), Carol Stanley (Walter Energy), Lakeshia Stewart (Honeywell), Eman Tora (National Research Center), Ragavan Vaidyanathan (Jacobs Engineering), Ting Wang (KBR), Anthony Warren (General Electric), Key Warren (Southern Company), Matt Wolf (Honeywell), José Zavala (Universidad de Guanajuato), and Mingjie Zhu (AtoFina).

The financial support of my process integration research by various federal, state, industrial, and international sponsors is gratefully acknowledged. I am also indebted to Mr. Artie McFerrin for his generous endowment and enthusiastic support that allowed me to pursue exciting and exploratory research and to transfer the findings to the classroom.

I would like to thank the editing and production team at Elsevier, especially Ms. Fiona Geraghty, Mr. Jonathan Simpson, and Ms. Heather Tighe, as well as Ms. Teresa Christie and her team at MPS North America for their excellent work on the various phases of production.

I am very grateful to my mother for being a constant source of love, inner peace, and support throughout my life. I am truly indebted to my father, Dr. Mokhtar El-Halwagi for being my most profound mentor and role model, introducing me to the fascinating world of chemical and environmental engineering, and teaching me the most valuable lessons in the profession and in life. I am also grateful to my grandfather, the late Dr. Mohamed El-Halwagi, for instilling in me a deep love for chemical engineering and a passion to seek knowledge and to pass it on. I am thankful to my brother, Dr. Baher El-Halwagi, for his constant support and encouragement. I owe a great debt of gratitude to my wife, Amal, for her unconditional love, unstinting understanding, unlimited encouragement, and unwavering support. With her impressive engineering skills, she has always been my first reader and my most constructive critic, and with her superb human qualities, she has constantly been my sustained source of comfort, compassion, wisdom, and inspiration. Finally, I am grateful to my sons, Omar and Ali, for being the sunshine of my life, for their warmth and love, for their remarkable achievements, and for their genuine care about humanity, which gives me great hope for a more sustainable world with a better tomorrow.

Mahmoud M. El-Halwagi
College Station, Texas

CHAPTER 1

INTRODUCTION TO SUSTAINABILITY, SUSTAINABLE DESIGN, AND PROCESS INTEGRATION

Industrial processes exert some of the most profound impacts on the ecosystem. These impacts are attributed to several factors including the significant usage of natural resources, the environmental discharges associated with the processing, and the ecological effects of the products. The process industries span a wide range of commodities including chemical, petroleum, gas, petrochemical, pharmaceutical, biofuel, food, microelectronics, metal, textile, and forestry feedstocks and products. The objectives of preventing pollution, conserving resources, increasing productivity, and enhancing profitability are among the top priorities of the process industries. Process engineers and managers who are routinely charged with tasks of achieving these objectives face the following primary challenges:

- How to systematically evolve solutions and innovative designs
- How to efficiently assess and screen process alternatives
- How to navigate through the complexities of industrial processes and develop an insightful understanding of the process, its limitations, and its opportunities
- How to reconcile the different objectives of the process (for example, economic, technical, environmental)
- How to continue process development and improvement in ways that can be sustained

The foregoing challenges raise the issues of what constitutes a sustainable improvement of the process and how to methodically and efficiently address these challenges. The following sections provide a brief discussion on sustainability and the role of process integration as a powerful and effective framework for sustainable design and for addressing the aforementioned process-engineering challenges.

WHAT IS SUSTAINABILITY?

Although there are several definitions of sustainability, the most commonly quoted definition derives from the definition of *sustainable development* in the "Brundtland Report" of the 1987 World Commission on Environment and Development (WCED, 1987), in which *sustainable development* means "*meeting the needs of the present without compromising the ability of future generations to meet their own needs.*" A group of professionals at the U.S. Environmental Protection Agency proposed the following definition: "*sustainability occurs when we maintain or improve the material and social conditions for human health and the environment over time without exceeding the ecological capabilities that support them*" (Sikdar, 2003). There is a growing interest in sustainability because of:

- Increasing population, industrialization, and standards of living
- Dwindling natural resources (for example, fossil fuels) and increase in the consumptions of the nonrenewable resources
- Global climatic changes
- Risk to biodiversity and ecosystems

Sustainability is based on balancing three principal objectives: environmental protection, economic growth, and societal equity (see Fig. 1.1). These are sometimes referred to as the "*triple bottom line: people, planet, and profit*" (Elkington, 1994).

Metrics and *indicators* are used to assess the sustainability performance of a process or a system, to evaluate the progress toward enhancing sustainability, and to assist decision makers in evaluating alternatives. The terms *metrics* and *indicators* are typically used interchangeably to provide a measure of sustainability. However, a metric usually gives a quantitative characterization or an index value, whereas indicators provide a narrative description in addition to the quantitative characterization (Tanzil and Beloff, 2005) and may include one or more metrics. There are numerous sustainability metrics, indicators, and approaches published by researchers (for example, Sikdar, 2011, 2009a, 2009b; Fan and Zhang, 2011; Jiménez-González and Constable, 2011; Powell, 2010; Uhlman and Saling, 2010; Ukidwe and Bakshi, 2008; Piluso et al., 2008; Cabezas et al., 2007; Martins et al., 2007; Beloff et al., 2005; Tsoka et al., 2004; Schwartz et al., 2002; Pennington et al., 2001) and professional organizations such as the Institution of Chemical Engineers (IChemE, 2002) and the American Institute of Chemical Engineers (AIChE) (for example, Cobb et al., 2009). One way of categorizing sustainability indicators is to classify them based on the economic, environmental, and social dimensions of sustainability as one-, two-, and three-dimensional metrics (Sikdar, 2003) as follows:

- **One-dimensional metrics** are based on only one of the economic, environmental, and social dimensions. Examples of one-dimensional economic metrics include capital investment, operating cost, return on investment, and payback period. Examples of one-dimensional environmental metrics include toxicity, biological

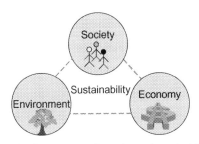

FIGURE 1-1 The three primary dimensions of sustainability.

Sustainable Design Through Process Integration.
© 2012 Elsevier Inc. All rights reserved.

TABLE 1-1 Examples of GWP of Some GHGs over Two Time Horizons

GHG	GWP (20-Year Time Horizon)	GWP (100-Year Time Horizon)
Carbon dioxide (CO_2)	1	1
Methane (CH_4)	56	21
Nitrous oxide (N_2O)	280	310
Refrigerant HFC-23 (CHF_3)	9,100	11,700

oxygen demand (BOD) of wastewater, chemical oxygen demand (COD) of wastewater, ozone depletion in the stratosphere, acidification of the atmosphere and aquatic ecosystems (resulting from the emission of acidifying chemicals such as sulfur and nitrogen oxides), and aquatic eutrophication (which involves excessive growth of biomass that can be exacerbated by the discharge of mineral nutrients such as nitrogen and phosphorus compounds into water bodies). Another one-dimensional environmental metric is the *global warming potential* (GWP) introduced by the United Nations Intergovernmental Panel on Climate Change (UN IPCC) (for example, Houghton et al., 1992, 1990). The GWP is intended to account for the impact of emissions of greenhouse gases (GHGs) on global warming. Specifically, GWP is a measure of the relative radiative effects of the emissions of several GHGs. Each GHG is given a GWP relative to CO_2 (which is taken as the basis with a GWP being 1). Therefore, the GWP is expressed in units of CO_2 equivalent (for example, tonne CO_2 equivalent). Values of the GWP for different GHGs are estimated for a specific time horizon over which the impact of such GHGs is tracked and integrated. Examples of the GWP values for two time horizons are shown in Table 1.1. The global warming index (GWI) is defined as follows:

$$[1.1] \qquad GWI = \sum_i m_i * GWP_i$$

where m_i is the mass of GHG i emitted over a certain period.

- **Two-dimensional metrics** are based on the simultaneous assessment of two out of the three sustainability dimensions. This category includes economic-environmental, socioeconomic, and socioenvironmental indicators. In this context, a particularly useful philosophy is *eco-efficiency* proposed by the World Business Council for Sustainable Development (WBCSD, 2000) as the following: "Eco-efficiency is achieved by the delivery of competitively priced goods and services that satisfy human needs and bring quality of life, while progressively reducing ecological impacts and resource intensity throughout the life cycle to a level at least in line with the earth's estimated carrying capacity. In short, it is concerned with creating more value with less impact." Specific application of eco-efficiency to the process industries involves the assessment and enhancement of metrics associated with the following aspects (Uhlman and Saling, 2010; Tanzil and Beloff, 2005):

- Material consumption: The use of feedstocks, water, and material utilities has a major impact on the depletion of nonrenewable resources and the discharge of wastes. Inefficient material use negatively affects the economic and the environmental dimensions of sustainability. An example of material consumption metric is the mass intensity index that may be defined as:

$$[1.2] \qquad \text{Mass intensity index} = \frac{\text{Mass of raw materials} - \text{Mass of products}}{\text{Mass of products}}$$

In the case of water, the index may be defined as:

$$[1.3] \qquad \text{Water intensity index} = \frac{\text{Mass of fresh water used}}{\text{Mass of products}}$$

- Energy consumption: Energy is a major driving force for operating industrial processes. Excessive usage of energy leads to economic losses and negative environmental impact (for example, emission of GHGs, contribution to ozone depletion and atmospheric acidification). One way of measuring energy efficiency is the energy intensity index that may be defined as:

$$[1.4] \qquad \text{Energy intensity index} = \frac{\text{Net energy used in the process}}{\text{Mass of products}}$$

- Environmental discharges: The release of hazardous and toxic pollutants causes harmful (and sometimes irreversible) effects on the environment. It also has negative economic consequences either because of the required cost of treatment or because of the financial liability to the industrial sources of these discharges.
- Land use: When land is used for an industrial purpose (directly as in the case of installing facilities or indirectly as in the case of planting biomass for the production of biofuels), there are important ecological and societal consequences. For instance, substituting one type of a crop for another (to provide a feedstock to biorefineries) affects the use of water resources, involves the use and discharge of different chemicals, changes the sequestration of carbon dioxide during photosynthesis, and impacts the communities around the farmed areas.

It is worth noting that metrics such as the mass, water, and energy intensity indices can be used to compare different projects and processes. Furthermore, the sustainability impact of process modifications can be assessed through the concept of an *incremental return on sustainability* (IROS) (Spriggs et al., 2009). For instance, consider an additional project for a process to reduce GHG emissions. The project leads to an improvement in the environmental impact but requires additional energy consumption. In this case,

$$[1.5] \quad \text{IROS} = \frac{\text{Change in environmental impact}}{\text{Change in net energy usage}}$$

- Therefore, for an energy-reduction project to be acceptable, it must meet a minimum value of the IROS, which guarantees a basic level of environmental performance. For instance, a minimum limit may be the best in class (for example, kilogram [kg] CO_2 equivalent emission per kilojoule [kJ]).
- **Three-dimensional metrics** assess sustainability by integrating the economic, environmental, and social aspects.

The foregoing discussion highlights the importance of enhancing productivity, conserving resources, and abating pollution ("getting more for less") in the process industries and describes several methods for assessing the sustainability of various industrial processes. A central question is not just how to assess sustainability of an industrial process but *how to achieve* a sustainable performance and enhance it. The next section introduces sustainable design through process integration as an enabling tool to attain sustainability in a methodical, effective, and generally applicable way.

WHAT IS SUSTAINABLE DESIGN THROUGH PROCESS INTEGRATION

Sustainable design of industrial processes may be defined as the design activities that lead to economic growth, environmental protection, and social progress for the current generation without compromising the potential of future generations to have an ecosystem that meets their needs. The following are the principal objectives of a sustainable design:

- Resource (mass and energy) conservation
- Recycle/reuse
- Pollution prevention
- Profitability enhancement
- Yield improvement
- Capital–productivity increase and debottlenecking
- Quality control, assurance, and enhancement
- Process safety

These objectives are closely related to the seven themes Keller and Bryan (2000) identified as the key drivers for process-engineering research, development, and changes in the primary chemical process industries. These themes are:

- Reduction in raw material cost
- Reduction in capital investment
- Reduction in energy use
- Increase in process flexibility and reduction in inventory
- Ever-greater emphasis on process safety
- Increased attention to quality
- Better environmental performance

Again, the question is *how to methodically and effectively achieve the objectives of a sustainable design*. The answer is ***process integration!***

A chemical process is an integrated system of interconnected units and streams. Proper understanding and solution of process problems should not be limited to symptoms of the problems but should identify the root causes of these problems by treating the process as a whole. Effective improvement and synthesis of the process must account for this integrated nature. Therefore, integrating process resources is a critical element in designing and operating cost-effective and sustainable processes. **Process integration** *is a holistic approach to process design, retrofitting, and operation that emphasizes the unity of the process* (El-Halwagi, 1997). In light of the strong interaction among process units, resources, streams, and objectives, process integration offers a unique framework along with an effective set of methodologies and enabling tools for sustainable design. The strength and attractiveness of process integration stem from its ability to systematically offer the following:

- Fundamental understanding of the global insights of a process and the root causes of performance limitations
- Ability to benchmark the performance of various objectives for the process ahead of detailed design through targeting techniques
- Effective generation and screening of solution alternatives to achieve the best-in-class design and operation strategies

Process integration involves the following activities (El-Halwagi, 2006):

1. **Task identification:** The first step in synthesis is to explicitly express the goal we are aiming to achieve and describe it as an *actionable* task. The actionable task should be defined in such a way so as to capture the essence of the original goal. For instance, pollution prevention may be described as a task of reducing certain discharges of the process to a certain extent, while quality enhancement may be described as a task to reach a specific composition or certain properties of a product.
2. **Targeting:** The concept of targeting is one of the most powerful contributions of process integration. **Targeting** refers to the identification of performance benchmarks ahead of detailed design. This is critical in the process integration motto of "*big picture first, details later.*" In a way, you can find the ultimate answer without having to specify how it may be reached. Targeting allows us to determine how far we can push the process performance and sheds useful insights on the exact potential and realizable opportunities for the process. Even if we elect not to reach the target, it is still useful to benchmark current performance versus the true potential of the process. This is particularly useful in comparing the sustainability metrics with the ultimate performance of the process.
3. **Generation of alternatives (synthesis):** Given the enormous number of possible solutions to reach the target (or the defined task), it is necessary to use a framework that is rich enough to embed all

configurations of interest and represent alternatives that aid in answering questions such as: How should streams be rerouted? What are the needed transformations (for example, separation, reaction, heating, and so on)? Should we use separations to clean up wastewater for reuse? To remove what? How much? From which streams? What technologies should be employed (for instance, should we use extraction, stripping, ion exchange, or a combination)? Where should they be used? Which solvents? What type of columns? Should we change operating conditions of some units? Which units and which operating conditions? The right level of representation for generating alternatives is critically needed to capture the appropriate design space. Westerberg (2004) underscores this point by stating that "It is crucial to get the representation right. The right representation can enhance insights. It can aid innovation." The generation of such design alternatives and representations is effectively handled through process synthesis, which involves putting together separate elements into a connected or a coherent whole. The term *process synthesis* dates back to the early 1970s, and gained much attention with the seminal book of Rudd et al. (1973). **Process synthesis** may be defined as "the discrete decision-making activities of conjecturing (1) which of the many available component parts one should use, and (2) how they should be interconnected to structure the optimal solution to a given design problem" (Westerberg, 1987). Process synthesis is concerned with the activities in which the various process elements are combined and the flow sheet of the system is generated so as to meet certain objectives. Therefore, the aim of process synthesis is "to optimize the logical structure of a chemical process, specifically the sequence of steps (reaction, distillation, extraction, etc.), the choice of chemical employed (including extraction agents), and the source and destination of recycle streams" (Johns, 2001). Hence, in process synthesis, we know process inputs and outputs and are required to revise the structure and parameters of the flow sheet (for retrofitting design of an existing plant) or create a new flow sheet (for grassroot design of a new plant). This is shown in Fig. 1.2.

Reviews of process synthesis techniques are available in literature (for example, Foo et al., 2011; Diwekar and Shastri, 2011; Majozi, 2010; Foo, 2009; Turton et al., 2009; Kemp, 2009; Seider et al., 2008; Towler and Sinnott, 2008; Smith, 2005; Westerberg, 2004; Dunn and El-Halwagi, 2003; Furman and Sahinidis, 2002; Bagajewicz, 2000; El-Halwagi and Spriggs, 1998; Biegler et al., 1997).

4. **Selection of alternative(s) (synthesis):** Once the search space has been generated to embed the appropriate alternatives, it is necessary to extract the optimum solution from among the possible alternatives. This step is typically guided by some performance metrics that assist in ranking and selecting the optimum alternative. Graphical, algebraic, and mathematical optimization techniques may be used to select the optimum alternative(s). It is worth noting that the generation and selection of alternatives are *process synthesis* activities.

5. **Analysis of selected alternative(s):** Although synthesis is aimed at combining the process elements into a coherent whole, analysis involves the decomposition of the whole into its constituent elements for individual study of performance. Hence, process analysis can be contrasted (and complemented) with process synthesis. Once an alternative is generated or a process is synthesized, its detailed characteristics (for example, flow rates, compositions, temperature, and pressure) are predicted using analysis techniques. These techniques include mathematical models, empirical correlations, computer-aided process simulation tools, evaluation of sustainability metrics, techno-economic analysis, safety review, and environmental impact assessment. In addition, process analysis may involve predicting and validating performance using experiments at the lab and pilot-plant scales, and even actual runs of existing facilities. Thus, in *process analysis* problems, we know the process inputs along with the process structure and parameters while we seek to determine the process outputs (see Fig. 1.3).

Therefore, process synthesis and analysis serve as the two primary pillars for sustainable design through process integration with synthesis generating alternatives and analysis evaluating the generated alternatives. Figure 1.4 is a schematic representation of such interaction.

Over the past three decades, numerous contributions have been made in the field of process integration. These contributions may be classified in different ways. One method of classification is based on the three primary areas of integration: mass, energy, and properties. **Mass integration** is a systematic methodology that provides a fundamental understanding of the global flow of mass within the process and employs this understanding in identifying performance targets and optimizing the generation and routing of species throughout the process. On the other hand, **energy integration** is a systematic methodology that provides a fundamental understanding of energy utilization within the process and employs this understanding in identifying energy targets and optimizing heat-recovery and energy-utility systems. Finally, **property integration** is a functionality-based, holistic approach to the allocation

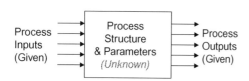

FIGURE 1-2 Process synthesis problems.

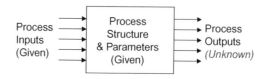

FIGURE 1-3 Process analysis problems.

CHAPTER 1 Introduction to Sustainability, Sustainable Design, and Process Integration

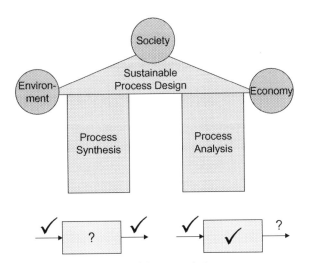

FIGURE 1-4 Pillars of sustainable process design.

and manipulation of streams and processing units, which is based on the tracking, adjustment, assignment, and matching of functionalities throughout the process. The fundamentals and applications of mass, energy, and property integration have been reviewed in literature (for example, Foo et al., 2011; Majozi, 2010; Rossiter, 2010; Foo, 2009; Kemp, 2007; El-Halwagi, 2006; Smith, 2005; El-Halwagi et al., 2004; Dunn and El-Halwagi, 2003; Hallale, 2001; El-Halwagi and Spriggs, 1998; El-Halwagi, 1997; and Shenoy, 1995).

MOTIVATING EXAMPLES ON THE GENERATION AND INTEGRATION OF SUSTAINABLE DESIGN ALTERNATIVES

Consider the acrylonitrile (AN) process shown in Fig. 1.5a (El-Halwagi, 2006, 1997). The main reaction in the process involves the vapor phase catalytic reaction of propylene, ammonia, and oxygen at 842°F (450°C) and 2 atmosphere (atm). To produce acrylonitrile (C_3H_3N) and water, for example:

$$C_3H_6 + NH_3 + 1.5O_2 \xrightarrow{catalyst} C_3H_3N + 3H_2O.$$

The reaction products are quenched in an indirect-contact cooler/condenser, which condenses a portion of the reactor off-gas. The remaining off-gas is scrubbed with water, and then decanted into an aqueous layer and an organic layer. The organic layer is fractionated in a distillation column under slight vacuum, which is induced by a steam-jet ejector. Steam is generated by heated boiler feed water (BFW). Wastewater is collected from four process streams: off-gas condensate, aqueous layer of decanter, distillation bottoms, and jet-ejector condensate. The wastewater stream is fed to the biotreatment facility. At present, the biotreatment facility is operating at full hydraulic capacity and, consequently, it constitutes a bottleneck for the plant. The plant has a sold-out profitable product and wishes to expand. Our task is to eliminate the bottleneck from the process.

The intuitive response to debottlenecking the process is to construct an expansion to the biotreatment facility (or install another one). This solution focuses on the symptom of the problem: The biotreatment is filling up; therefore, we must expand its capacity. A legitimate question is whether other solutions, probably superior ones, will address the problem by making in-plant process modifications as opposed to an "end-of-pipe" solution. Invariably, the answer in this case and most other process design problems is "yes." If so, how do we determine the root causes of the problem (not just the symptoms), and how can we generate superior solutions? Where do we start and how do we address the problem?

For now, let us start with a conventional engineering approach involving a brainstorming session among a group

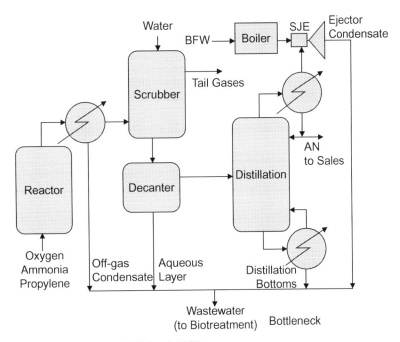

FIGURE 1-5a Process for AN manufacturing. *Source: El-Halwagi (1997).*

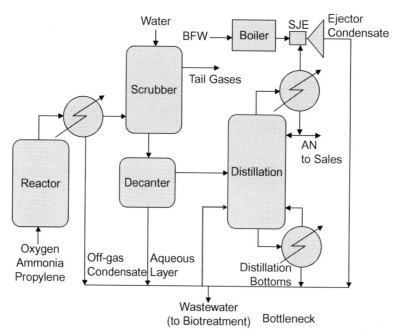

FIGURE 1-5b Recycle to the distillation column. *Source: El-Halwagi (2006).*

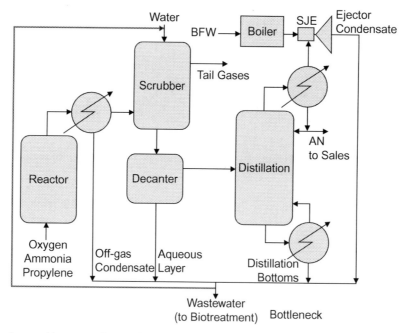

FIGURE 1-5c Recycle to replace scrubber water. *Source: El-Halwagi (2006).*

of process engineers who will generate a number of ideas and evaluate them. Because the objective is to debottleneck the biotreatment facility, an effective approach may be based on reducing the influent wastewater flow rate into biotreatment. One way of reducing wastewater flow rate is to adopt a wastewater recycle strategy in which it is desired to recycle some (or all) of the wastewater. For instance, the plant could recycle some of the wastewater to the distillation column (see Fig. 1.5b). After analyzing this solution, it does not seem to be effective. The fresh water to

the process is still the same, water generated by the main AN-producing reaction is the same, and therefore the wastewater leaving the plant will remain the same. So, the plant could employ a recycling strategy that replaces fresh water with wastewater. This way, the fresh water use in the process is reduced and, consequently, the wastewater leaving the process will be reduced as well. One option is to recycle the wastewater to the scrubber (see Fig. 1.5c), assuming that it is feasible to process the wastewater in the scrubber without negatively impacting the process

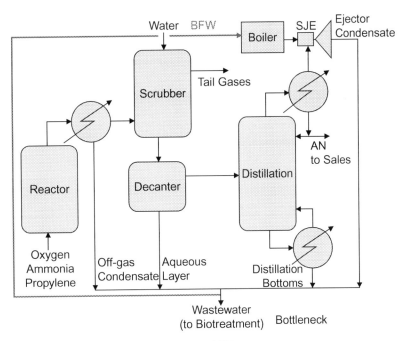

FIGURE 1-5d Recycle to substitute boiler feed water. *Source: El-Halwagi (2006).*

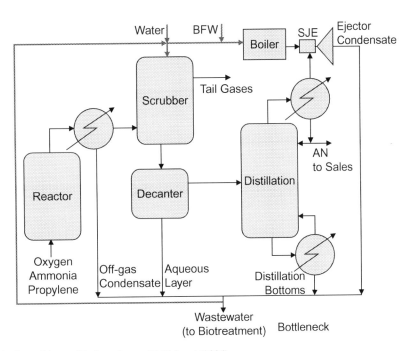

FIGURE 1-5e Recycle to both scrubber and boiler. *Source: El-Halwagi (2006).*

performance. In such cases, both fresh water and wastewater will be reduced. Alternatively, it may be possible to recycle the wastewater to the boiler (see Fig. 1.5d). Along the same lines, the wastewater may be recycled to both the scrubber and the boiler (see Fig. 1.5e). However, how should the wastewater be distributed between the two units? One can foresee many possibilities for distribution (50/50, 51/49, 60/40, 99/1, and so on). Another alternative is to consider segregating (avoiding the mixing of) the wastewater streams. Segregation would prevent some wastewater streams from mixing with the more polluted streams, thereby enhancing their likelihood for recycling. For instance, the off-gas condensate and the decanter aqueous layer may be segregated from the two other wastewater streams and recycled to the scrubber and the boiler (see Fig. 1.5f). Clearly, there are many alternatives for segregation and recycle. To safeguard against the accumulation of impurities or the detrimental effects of replacing fresh water with waste streams, it may be necessary to consider the use of separation technologies to clean up

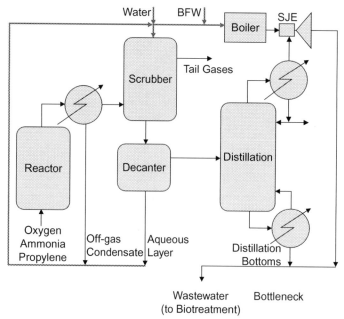

FIGURE 1-5f Segregation of wastewater and recycle of two segregated streams. *Source: El-Halwagi (2006).*

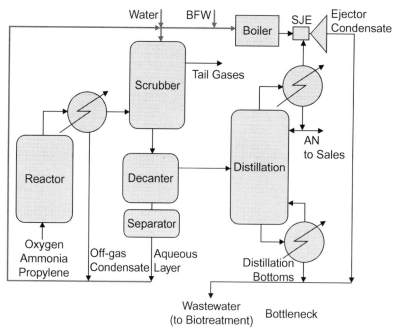

FIGURE 1-5g Combined separation and recycle. *Source: El-Halwagi (2006).*

the streams and render them in a condition acceptable for recycling. For example, a separator may be installed to treat the decanter wastewater (see Fig. 1.5g). But, what separation technologies should be used? To remove what? From which streams? Figures 1.5h through 1.5j are just three possibilities (out of numerous alternatives) for the type and allocation of separation technologies. And so on! Clearly, there are *infinite numbers of alternatives* to solve this problem. So many decisions have to be made on the rerouting of streams, the distribution of streams, the changes to be made in the process (including design and operating variables), the substitution of materials and reaction pathways, and the replacement or addition of units.

Notwithstanding the numerous design alternatives, process integration can determine the performance target and synthesize the optimal solution without enumeration. As will be shown by the overall mass targeting tools described in Chapter 3, the benchmarks for water usage

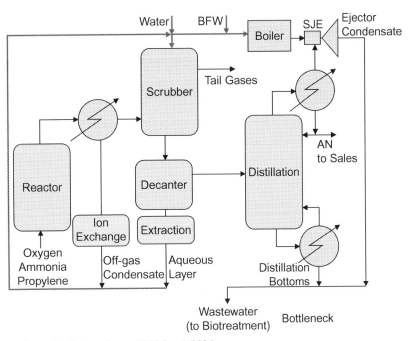

FIGURE 1-5h Defining separation technologies. *Source: El-Halwagi (2006).*

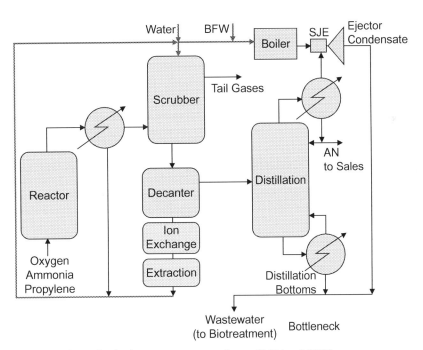

FIGURE 1-5i Hybrid separation technologies for the decanter wastewater. *Source: El-Halwagi (2006).*

and discharge can be first determined before detailed design and without the need to create alternative configurations (similar to the ones shown by Figs. 1.5b through 1.5j). Figure 1.6 shows the values of these targets. Next, the optimal solution (shown by Fig. 1.7) is systematically synthesized using the mass-integration techniques described in Chapters 4 through 6.

The following observations may be inferred from the foregoing discussion:

- There are typically numerous alternatives that can solve a typical sustainable design problem.
- The optimum solution may not be intuitively obvious.

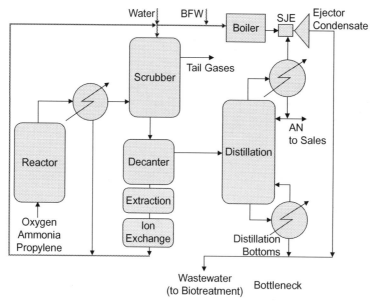

FIGURE 1-5j Switching the order of separation technologies. *Source: El-Halwagi (2006).*

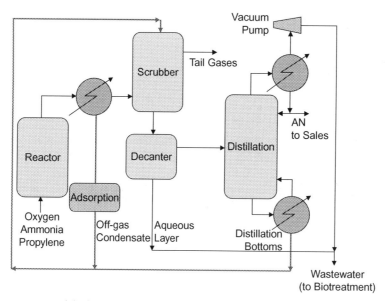

FIGURE 1-6 Benchmarking water usage and discharge for the AN example before detailed design.

- One should not focus on the symptoms of the process problems. Instead, one should identify the root causes of the process deficiencies.
- It is necessary to understand and treat the process as an integrated system.
- There is a critical need to systematically extract the optimum solution from among the numerous alternatives without enumeration.

Until recently, there were three primary conventional engineering approaches to addressing sustainable design problems:

- **Brainstorming and solution through scenarios**: A select few of the engineers and scientists most familiar with the process work together to suggest and synthesize several conceptual design scenarios (typically three to five). For instance, the foregoing exercise of generating alternatives for the AN case study falls under this category. Each generated scenario is then assessed (for example, through simulation, techno-economic analysis, and so on) to examine its feasibility and to evaluate some performance metrics (for example, cost, safety, reliability, flexibility, operability, environmental impact, and so on). These metrics are used to rank the generated scenarios and to select a recommended solution. This recommended solution may be inaccurately referred to as the "optimum solution" when in fact it is only optimum out of the few generated

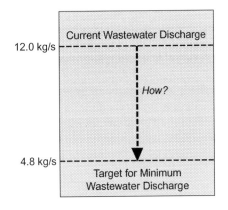
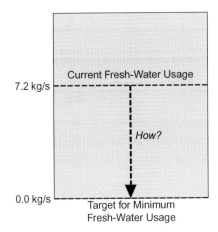

FIGURE 1-7 Optimal solution to AN case study. *Source: El-Halwagi (2006).*

alternatives. Indeed, it may be far away for the true optimum solution.
- **Adopting/evolving earlier designs:** In this approach, a related problem that has been solved earlier is identified. The problem may be at the same plant or another plant. Then, its solution is either copied, adopted, or evolved to suit the problem at hand and to aid in the generation of a similar solution.
- **Heuristics:** Over the years, process engineers have discovered that certain design problems may be categorized into groups or regions, each having a recommended way of solution. Heuristics is the application of experience-derived knowledge and rules of thumb to a certain class of problems. It is derived from the Greek word *heuriskein*, which means "to discover." Heuristics have been used extensively in industrial applications (for example, Harmsen, 2004).

Over the years, these approaches have provided valuable solutions to industrial problems and are commonly used. Notwithstanding the usefulness of these approaches in providing solutions that typically work, they have several serious limitations (Sikdar and El-Halwagi, 2001):

- They cannot enumerate the infinite alternatives: Because these approaches are based on brainstorming few alternatives or evolving an existing design, the generated alternatives are limited.
- They are not guaranteed to come close to optimum solutions: Without the ability to extract the optimum from the infinite alternatives, these approaches may not provide effective solutions (except for very simple cases, extreme luck, or near-exhaustive effort). Just because a solution works and is affordable does not mean that it is a good solution. Additionally, when a solution is selected from few alternatives, it should not be called an optimum solution. It is only optimum with respect to the few generated alternatives.
- They are time and money intensive: Because each generated alternative should be assessed (at least from a techno-economic perspective), significant efforts and expenses are involved in generating and analyzing the enumerated solutions.
- They have limited range of applicability: Heuristics and rules of thumb are most effective when the problem at hand is closely related to the class of problems and design region for which the rules have been derived. However, they must be used with extreme care. Even subtle differences from one process to another may render the design rules invalid.
- They do not shed light on global insights and key characteristics of the process: In addition to solving the problem, it is beneficial to understand the underlying phenomena, root causes of the problem, and insightful criteria of the process. Trial and error as well as heuristic rules rarely provide these aspects.
- They severely limit groundbreaking and novel ideas: If the generated solutions are derived from the last design that was implemented or based exclusively on the experience of similar projects, what will drive the "out-of-the-box" thinking that leads to process innovation.

The good news is that recent advances in process integration have led to the development of systematic, fundamental, and generally applicable techniques that can be learned and applied to overcome the aforementioned limitations and methodically address process-improvement problems. Problems such as debottlenecking and water conservation described in the AN example can be readily and methodically solved to identify the optimal solution.

Next, let us consider the pharmaceutical processing facility shown by Fig. 1.8. The process has an adiabatic reactor. The feedstock entering the reactor (C_1) is preheated from 300 Kelvin (K) to 550 K. The gaseous stream leaving the reactor (H_1) at 520 K is cooled to 330 K, then sent to a recovery unit. The product stream leaving the bottom of the reactor is fed to a washing and purification network. The top stream leaving the separation unit (H_2) is cooled from 380 K to 300 K prior to storage. The solvent used in washing (C_2) is heated from 320 K to 380 K before entering the washing unit.

At present, the process uses 4,870 kilowatts (kW) of an external heating utility and 2,300 kW of an external cooling utility. Because there are process hot streams to be cooled and process cold streams to be heated, it is beneficial to integrate the heat exchange between the hot and the cold streams before using external utilities. There are numerous alternatives for transferring heat from the hot

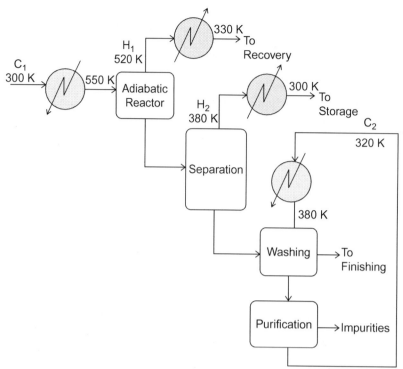

FIGURE 1-8 Simplified flow sheet of a pharmaceutical process.

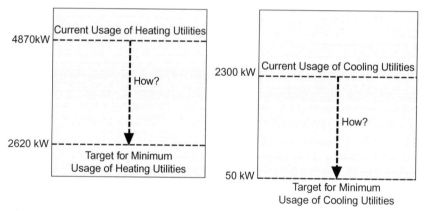

FIGURE 1-9 Benchmarking heating and cooling utilities before detailed design.

to the cold streams. These alternatives differ in the order of matches and the extent of heat transferred from the hot to the cold streams. Instead of enumerating and comparing these many alternatives, heat-integration techniques (described in Chapters 7 and 19) offer a methodical way to first determine the targets for minimum heat and cooling utilities and then aid the designer in generating a network of heat exchangers that reaches the target. Figure 1.9 shows the values of the benchmarks for minimum heating and cooling utility targets. Figure 1.10 shows a network of heat exchanger reaching the utility targets. Each heat exchanger is represented by an ellipse with the heat-transfer rate noted inside it. Temperatures are placed next to each stream. The reduction in heating and cooling utilities is attributed to energy conservation through proper exchange of heat between the hot and the cold streams, namely heat integration.

The previously mentioned examples of mass integration (water recycle in this case) and heat integration are just samples of the types of problems that can be systematically addressed using generally applicable process integration techniques.

STRUCTURE AND LEARNING OUTCOMES OF THE BOOK

This book presents the fundamentals and applications of process integration and how they can be used for generating best-in-class sustainable designs. Holistic approaches, methodical techniques, and step-by-step procedures are presented and illustrated by a wide variety of case studies. Visualization, algebraic, and mathematical programming techniques are used to explain and address process integration problems. Chapter 2 gives an overview of process

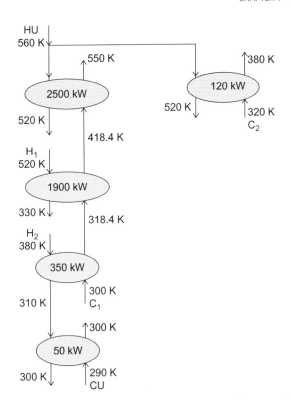

FIGURE 1-10 A network of heat exchangers attaining the targets for minimum heating and cooling utilities.

economics and the assessment of economic criteria pertaining to sustainability. Chapters 3 through 6 focus on graphical approaches for mass integration. Chapters 7 and 8 give graphical and algebraic tools for heat integration and combined heat and power systems. Chapter 9 presents graphical techniques for property integration. Chapters 10 and 11 describe algebraic tools for mass integration. Chapters 12 and 13 cover energy-induced separations such as condensation and membrane systems. The rest of the book introduces mathematical programming techniques, starting with a tutorial on how to formulate optimization problems, solve them using a software (LINGO), and continuing to cover various classes of problems for mass, energy, and property integration. The scope of problems ranges from identification of overall performance targets to integration of separation systems, recycle networks, heat-exchange networks, energy conservation, design of biorefineries, and macroscopic integrated systems. Numerous case studies are used to illustrate the theories, concepts, and tools. Table 1.2 summarizes the key learning outcomes of the book and the associated chapters.

In addition to the methodologies, concepts, and tools described in this book, as well as the specific learning outcomes summarized in Table 1.2, the book presents an overarching message: Process integration provides holistic, systematic, generally applicable approaches to benchmark performance ahead of detailed design and then generates best-in-class sustainable design alternatives that reach the desired targets. If you do not already have a working experience in process integration, this book will present you with a paradigm shift that is intended to systematize and enhance the way you target the performance of processing plants and devise effective and innovative strategies for attaining the sustainability objectives of a process. It is hoped that you will use process integration as a powerful vehicle in the journey of inducing continuous process improvement and improving the sustainability metrics of the process. Welcome aboard!

TABLE 1-2 Primary Learning Outcomes of the Book and Associated Chapters

Learning Outcomes (You will be able to)	Associated Chapters (When you study Chapter #)
Apply different techniques to estimate fixed cost, operating cost, and total capital investment	2
Conduct break-even analysis and use it to determine minimum selling price or subsidy for environmental purposes	2
Evaluate profitability criteria of a project and screen alternative projects based on economic criteria	2
Determine overall mass targets (performance benchmarks) for a process with objectives such as minimum usage of raw materials, minimum discharge of waste, and maximum process yield	3
Determine targets for no-/low-cost strategies for maximum recycle of process streams, minimum purchase of the fresh materials, and minimum discharge of waste	4, 9, 10, 15
Screen different mass-exchange technologies and design the optimal network of separation units to achieve a set task	5, 11, 16, 17
Develop mass-integration strategies for a process	6, 18
Determine benchmarks for minimum heating and cooling utilities of a process and design a network of heat exchangers to attain the utility targets	7, 19
Design combined heat and power (cogeneration) systems	8
Optimize the process performance based on properties and functionalities	9
Design energy-induced separation systems (for example, recovery of volatile organic compounds via condensation, membrane separation)	12, 13
Formulate sustainable design tasks as a mathematical program (optimization problems) and solve them using graphical and computer-aided software	14
Solve mass- and energy-integration problems using mathematical programming	15, 16, 17, 18, 19, 20
Design integrated biorefineries	21
Combine macroscopic analyses (for example, life cycle analysis, environmental impact assessment) with process integration strategies	22

REFERENCES

Bagajewicz M: A review of recent design procedures for water networks in refineries and process plants, *Comp Chem Eng* 24(9–10):2093–2113, 2000.

Beloff M, Lines M, Tanzil D, editors: *Transforming sustainability into action: the chemical industry*, Hoboken, New Jersey, 2005, John Wiley & Sons, Inc..

Biegler LT, Grossmann IE, Westerberg AW: Systematic methods of chemical process design, New Jersey, 1997, Prentice Hall.

Cabezas H, Whitmore HW, Pawlowski CW, Mayer AL: On the sustainability of an integrated model system with industrial, ecological, and macroeconomic components, *Resour Conserv Recycl* 50(2):122–129, 2007.

Cobb C, Schuster D, Beloff B, Tanzil D: The AIChE Sustainability Index: the factors in detail, *Chem Eng Prog* January:60–63, 2009.

Diwekar U, Shastri Y: Design for environment: a state-of-the-art review, *Clean Techn Environ Plicy* 13:227–240, 2011.

Dunn RF, El-Halwagi MM: Process integration technology review: background and applications in the chemical process industry, *J Chem Tech Biotech* 78:1011–1021, 2003.

El-Halwagi MM: *Pollution prevention through process integration: systematic design tools*, San Diego, 1997, Academic Press.

El-Halwagi MM: *Process integration*, Amsterdam, 2006, Elsevier.

El-Halwagi MM, Glasgow IM, Eden MR, Qin X: Property integration: componentless design techniques and visualization tools, *AIChE J* 50(8):1854–1869, 2004.

El-Halwagi MM, Spriggs HD: Solve design puzzles with mass integration, *Chem Eng Prog* 94(August):25–44, 1998.

Elkington J: Towards the sustainable corporation: win-win-win business strategies for sustainable development, *Calif Manage Rev* 36(2):90–100, 1994.

Fan LT, Zhang T: Estimation of energy dissipation and cost: the foundation for sustainability assessment in process design. In Foo DCY, El-Halwagi MM, Tan RR, editors: *Recent advances in sustainable process design and optimization*, "Advances in process systems engineering," vol 3, 2011, World Scientific Publishing Company.

Foo DCY: State-of-the-art review of pinch analysis techniques for water network synthesis, *Ind Eng Chem Res* 48:5125–5159, 2009.

Foo DCY, El-Halwagi MM, Tan RR, editors: *Recent advances in sustainable process design and optimization*, Series on Advances in Process Systems Engineering, 2011, World Scientific Publishing Company.

Furman KC, Sahinidis NV: A critical review and annotated bibliography for heat exchanger network synthesis in the 20th century, *Ind Eng Chem Res* 41(10):2335–2370, 2002.

Hallale N: Burning bright: trends in process integration, *Chem Eng Prog* 97(7):30–41, 2001.

Harmsen GJ: Industrial best practices of conceptual process design, *Chem Eng Processing* 43:677–681, 2004.

Houghton JT, Callander BA, Varney SK, editors: Climate change: the supplement report to IPCC scientific assessment, Cambridge, UK, 1992, Cambridge University Press.

Houghton JT, Callander BA, Varney SK, editors: Climate change: the IPCC scientific assessment, Cambridge, UK, 1990, Cambridge University Press.

IChemE (Institution of Chemical Engineers): The sustainability metrics: sustainable development progress metrics recommended for use in the process industries, IChemE, 2002, Rugby.

Jiménez-González C, Constable DJC: *Green chemistry and engineering: a practical design approach*, Hoboken, New Jersey, 2011, John Wiley and Sons, Inc.

Johns WR: Process synthesis poised for a wider role, *Chem Eng Prog* 97(4):59–65, 2001.

Keller GE, II, Bryan PF: Process engineering: moving in new directions, *Comp Chem Eng* 96(1):41–50, 2000.

Kemp I: *Pinch analysis and process integration: a user guide on process integration for the efficient use of energy*, ed 2, 2007, Butterworth-Heinemann.

Majozi T: *Batch chemical process integration: analysis, synthesis, and optimization*, Heidelberg, 2010, Springer.

Martins AA, Mata TM, Costa CAV, Sikdar SK: Framework for sustainability metrics, *Ind Eng Chem Res* 46:2962–2973, 2007.

Pennington DW, Norris G, Hoagland T, Bare JC: An introduction to metrics for the environmental comparison of process and product alternatives. In Sikdar SK, El-Halwagi MM, editors: *Process design tools for the environment*, 2001, Taylor and Francis, pp 65–102.

Piluso C, Huang YL, Lou HH: Ecological input-output analysis based sustainability analysis of industrial systems, *Ind Eng Chem Res* 47(6):1955–1966, 2008.

Powell JB: Sustainability metrics, indicators, and indices for the process industries. In Harmsen J, Powell JB, editors: *Sustainable development in the process industries: cases and impact*, Hoboken, New Jersey, 2010, John Wiley and Sons, Inc.

Rossiter AP: Improve energy efficiency via heat integration, *Chem Eng Prog* 106(12):33–42, 2010.

Rudd DF, Powers GJ, Siirola JJ: *Process synthesis*, New Jersey, 1973, Prentice Hall.

Schwartz J, Beloff B, Beaver E: Use sustainability metrics to guide decision-making, *Chem Eng Prog* 98(7):58–63, 2002.

Seider WD, Seader JD, Lewin DR, Widagdo S: *Product and process design principles: synthesis, analysis, and design*, ed 3, New York, 2008, Wiley.

Shenoy UV: *Heat exchange network synthesis: process optimization by energy and resource analysis*, Houston, Texas, 1995, Gulf Pub. Co.

Sikdar SK: Quo vadis energy sustainability? *Clean Technol Environ Policy* 11(4):367–369, 2009.

Sikdar SK: On aggregating multiple indicators into a single metric for sustainability, *Clean Technol Environ Policy* 11(2):157–161, 2009.

Sikdar SK: What about industrial water sustainability? *Clean Technol Environ Policy* 13(1):1, 2011.

Sikdar SK: Sustainable development and sustainability metrics, *AIChE J* 49(8):1928–1932, 2003.

Sikdar SK, El-Halwagi MM, editors: *Process design tools for the environment*, Taylor and Francis, 2001.

Smith R: *Chemical process design and integration*, New York, 2005, Wiley.

Spriggs HD, Brant B, Rudnick D, Hall KR, El-Halwagi MM: Sustainability metrics for highly integrated biofuel production facilities. In El-Halwagi MM, Linninger AA, editors: *Design for energy and the environment: proceedings of the 7th international conference on the Foundations of Computer-Aided Process Design (FOCAPD)*, Taylor & Francis, 2009, CRC Press, pp 399–404.

Tanzil D, Beloff B: Implementing sustainable development: decision-support approaches and tools. In Beloff M, Lines M, Tanzil D, editors: *Transforming sustainability into action: the chemical industry*, Hoboken, New Jersey, 2005, John Wiley & Sons, Inc., pp 199–328.

Towler G, Sinnott R: *Chemical engineering design: principles, practice and economics of plant and process design*, Amsterdam, 2008, Elsevier.

Tsoka C, Johns WR, Linke P, Kokossis A: Towards sustainability and green chemical engineering: tools and technology requirements, *Green Chem* 6:401–406, 2004.

Turton R, Bailie C, Wallace B, Whiting B, Shaeiwitz JA: *Analysis, synthesis, and design of chemical processes*, ed 3, 2009, Prentice Hall.

Uhlman BW, Saling P: Measuring and communicating sustainability through eco-efficiency analysis, *Chem Eng Prog* December:17–26, 2010.

Ukidwe MU, Bakshi BR: Resource intensities of chemical industry sectors in the United States via input-output network models, *Comp Chem Eng* 32(9):2050–2064, 2008.

WBCSD (World Business Council for Sustainable Development) *Eco-efficiency: creating more value with less impact*, 2000, World Council for Sustainable Development.

WCED (World Commission on Environment and Development) *Our common future* (The Brundtland Report), New York, 1987, Oxford University Press.

Westerberg AW: Process synthesis: a morphological view. In Liu YA, McGee HA, Jr., Epperly WR, editors: *Recent developments in chemical process and plant design*, New York, 1987, Wiley, pp 127–145.

Westerberg AW: A retrospective on design and process synthesis, *Comp Chem Eng* 28:447–458, 2004.

CHAPTER 2
OVERVIEW OF PROCESS ECONOMICS

An essential component of any sustainable design is the economic aspect. In designing a new process or retrofitting an existing one, many of the technical and environmental decisions are strongly impacted by the economic factors. Therefore, it is critical for engineers to study process economics to be able to do the following:

- Evaluate feasibility of new projects
- Improve performance of existing processes
- Make design and operating decisions
- Compare alternatives
- Decide strategic directions for the company
- Establish sound policies for process and product objectives (for example, matters pertaining to the environment, safety, and quality)

This chapter provides an overview of basic concepts and tools in cost estimation and economics of chemical processes to answer the following questions:

- What are the cost items involved in installing and operating a process?
- What types of cost estimation can be carried out and to what level of accuracy?
- How do we estimate the cost of building a plant or implementing a project?
- How do we account for the changes in market conditions and for the time value of money?
- How do we estimate the recurring costs associated with running the plant?
- How do we assess the economic viability of a project?
- How do we screen competing projects and select the most attractive alternative(s)?

To address these questions, this chapter focuses on the following topics:

- Cost types and estimation
- Depreciation
- Break-even analysis
- Time value of money
- Profitability analysis

COST TYPES AND ESTIMATION

Two primary costs should be evaluated: capital investment and operating costs. The **total capital investment** (TCI) or *capital cost* is the money needed to purchase and install the plant and all its ancillaries and to provide for the necessary expenses required to start the process operation. Once the plant is in production mode, then the continuous expenses needed to run the plant are referred to as **operating costs**. The following sections provide the basis for how these costs are estimated, what the key elements in each cost are, and how these costs can be extrapolated and updated. Specifically, the following costs are discussed:

- Capital (fixed, working, and total)
- Equipment
- Operating
- Production (total annualized cost)

CAPITAL COST ESTIMATION

The TCI of a process is classified into two types of expenditures: *fixed capital investment* and *working capital investment*:

[2.1] $$TCI = FCI + WCI$$

The **fixed capital investment** (FCI) is the money required to pay for the processing equipment and the ancillary units, acquiring and preparing the land, civil structures, facilities, and control systems. In turn, the FCI is further classified into two components: manufacturing (or direct) and nonmanufacturing. The *manufacturing FCI* involves the fixed-cost items that are directly associated with production such as the processing equipment, installation, piping, pumping/compression, process instrumentation, process utility facilities and distribution, process waste treatment systems, and all the civil work associated with the production units. The *nonmanufacturing FCI* includes the fixed-cost items that are not directly tied to production such as land, analytical laboratories, storage areas, nonprocess utilities and waste treatment, engineering centers, research and development laboratories, administrative offices, cafeterias and restaurants, and recreational facilities. On the other hand, the **working capital investment** (WCI) is the money needed to pay for the operating expenditures up to the time when the product is sold as well as the expenses required to pay for stockpiling raw materials before production (typically one to two months of raw materials are stockpiled prior to production). An important characteristic of the WCI is that it is recoverable at the end of the project. For instance, if the plant stockpiles a two-month reserve of raw materials prior to starting the operation and maintains a two-month stock throughout the operation, then there is no need to buy raw materials during the last two months of the project, and the value of the WCI can indeed be fully recovered. Typically, the WCI ranges between 10 percent and 25 percent of the TCI. Figure 2.1 is a summary of the main components constituting the TCI.

As the word *estimation* indicates, there is a level of uncertainty in most cost estimation studies. Depending on the objective of the cost estimation, there are different desired levels of accuracy. The Association for the Advancement of Cost Engineering (AACE) International (www.aacei.org) defines several types of cost estimation studies along with their accuracy ranges. Other classifications are also available in the literature. Table 2.1 provides a brief description of some types of estimates along with their accuracy levels and needed information. An important perspective in conducting a specific type of cost estimation study is to attain the "right level of details." If there is a need to quickly generate ballpark cost estimates of a potential project to decide on whether or not further assessment is needed, then an order-of-magnitude estimate or a study estimate may be all that is required at this stage. Little effort is spent and very rough estimates are obtained. However, these estimates may be exactly what is needed to decide on whether or not to consider the project any further and to charter more detailed studies where the time and effort are then justified.

Several approaches are used for capital cost estimation. The following are some of the most commonly used methods:

1. Manufacture's quotation
2. Computer-aided tools
3. Capacity ratio with exponent
4. Updates using cost indices
5. Factors based on equipment cost
6. Empirical correlations
7. Turnover ratio

The following describes basic concepts and applications of these methods.

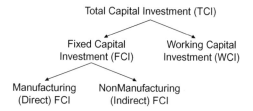

FIGURE 2-1 Main components of TCI.

1. **Manufacturer's quotation**

   Capital cost estimates may be obtained directly from specialized manufacturers and engineering firms. Depending on the level of details of the planned project, availability of resources, time limitations, and objective of the cost estimate, different levels of accuracy can be obtained.

2. **Computer-aided tools**

   Various computer-aided resources may be used to evaluate the FCI. Of particular importance are tools that are linked to process simulators because they are integrated with equipment performance and sizing as evaluated by simulation. Examples include ICARUS (www.aspentech.com) and SuperPro Designer (http://www.intelligen.com).

3. **Capacity ratio with exponent**

   When there are two identical (or at least very similar) plants of different capacities with the FCI known for one of them, then the FCI of the other plant may be estimated as follows:

   $$[2.2] \qquad FCI_B = FCI_A \left(\frac{Capacity_B}{Capacity_A}\right)^x$$

   Where FCI_B and FCI_A are the fixed capital investments of plants B and A, and $Capacity_B$ and $Capacity_A$ are the capacities (for example, flow rate of main product) of plants B and A, the exponent x is usually less than 1.0 and should be evaluated based on the actual data of the specific process. If no such data are available, x may be assumed to be 0.6 to 0.7 (hence, the name *sixth-tenths-factor rule*) for order-of-magnitude estimates. Equation 2.2 is an expression of the economy of scale, which indicates that as the plant capacity increases, the FCI per unit production decreases. For instance, when $x = 0.6$, doubling the plant capacity does not double the FCI but instead leads to 52 percent increase in FCI. Table 2.2 shows values of x for different pieces of equipment.

TABLE 2-1 Classification of Cost Estimation Studies

Type/Objective of Cost Estimation	Accuracy Level	Level of Project Definition (expressed as % of project completely defined)	Type of Needed Information
Order-of-magnitude estimate or concept screening	−50/+100%	0 to 2%	Experience or cost data of a similar plant or basic information on sold product and capacity
Study estimate or preliminary feasibility	−30/+50%	1 to 20%	Preliminary description of the process flow sheet and duty data of the main equipment
Preliminary estimate or budget authorization	−20/+30%	10 to 50%	Equipment sizing and basic simulation
Definitive estimate or project control estimate	−15/+25%	40 to 80%	Detailed equipment data (e.g., sizing, simulation, design specifications, drawings)
Contractor's estimate or detailed estimate	−5/+10%	75 to 100%	Detailed simulation, complete engineering drawings, mechanical and electrical datasheets, design specifications, process layout, site survey

Source: Adapted and revised from the AACE International; Dysert (2001); Christensen and Dysert (2005); and Coker (2007).

TABLE 2-2 Typical Plant Scaling Exponents for Selected Processes

Product	Process	Size Range	Exponent x	Reference
Acetaldehyde	Ethylene conversion	25,000 to 100,000 tonne/yr	0.70	Remer and Chai (1990a)
Acetic acid	methanol conversion	3000 to 75,000 tonne/yr	0.59	Garrett (1989)
Ammonia	Natural gas reforming	365,000 to 550,000 tonne/yr	0.66	Gerrard (2000)
Adipic acid	Cyclohexanol conversion	7000 to 330,000 tonne/yr	0.64	Garrett (1989)
Ethylene	Cracking of ethane	500 to 2000 million (MM) lb/yr	0.60	Towler and Sinnott (2008)
Ethylene oxide	Direct oxidation of ethylene	20,000 to 200,000 tonne/yr	0.78	Remer and Chai (1990a); Dysert (2001)
Hydrogen	Steam reforming of methane	10 to 150 MM standard cubic feet (SCF)/day	0.79	Towler and Sinnott (2008)
Polyethylene	High-pressure polymerization of ethylene	40,000 tonne/yr	0.69	Salem (1981)
Polyvinyl chloride	Polymerization of vinyl chloride	20,000 tonne/yr	0.60	Salem (1981)

Example 2-1 Using capacity ratios with exponents for estimating FCI

A processing facility is designed to convert waste cooking oil and vegetable oil to biodiesel. The FCI of the process producing 40 MM gal/yr is estimated to be $23 MM (Elms and El-Halwagi, 2009). Estimate the FCI of a similar process producing 20 MM gal/yr. Also, conduct a sensitivity analysis on the effect of production rate on the FCI per annual gallon (that is, FCI per gal/yr).

SOLUTION

Assuming a capacity exponent of 0.6, we have:

$$\text{FCI of 20 MM gal/yr process} = \text{FCI of 40 MM gal/yr process} * \left(\frac{20\,\text{MM gal/yr}}{40\,\text{MM gla/yr}}\right)^{0.6}$$
$$= \$15\,\text{MM}$$

The following two economy-of-scale observations are worth noting:

- When the capacity of the plant is doubled from 20 to 40 MM gal/yr, the FCI is not doubled. Instead, it increases by about 50 percent (from $15 to 23 MM).
- When two of the 20 MM gal/yr plants are built, they will cost $30 MM, which is 30 percent more expensive than building a single 40 MM gal/yr process.

Next, the FCI per annual gallon is calculated. For the 40 MM gal/yr process:

$$\text{The FCI per annual gallon} = \frac{23*10^6}{40*10^6}$$
$$= \$0.58/\text{annualgal}$$

Similarly, for the 20 MM gal/yr process,

$$\text{the FCI per annual gallon} = \frac{15*10^6}{20*10^6}$$
$$= \$0.75/\text{annualgal}$$

The same calculation is repeated for production rates ranging from 10 MM to 50 MM gal/yr; Fig. 2.2 shows the results. Again, the impact of economy of scale is illustrated by the reduction in the fixed cost per annual gallon with the increase in the plant size.

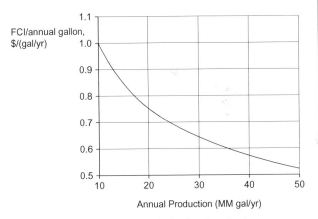

FIGURE 2-2 Sensitivity analysis for biodiesel production.

4. **Updates using cost indices**

Cost estimates are made and reported for a given time. With inflation and price fluctuation, it is necessary to account for the FCI as a function of time. Cost indices are very useful in adjusted cost estimates based on time. A basis value (for example, 100) of the cost index is set for a certain year. The values of the cost index are published regularly (for example, monthly or quarterly). Suppose that the cost of a given plant at time t1 (referred to as FCI_{t1}) is known. It is desired to get the cost of an identical plant at time t2 (designated as FCI_{t2}). The cost index is used as follows:

$$[2.3] \qquad FCI_{t2} = FCI_{t1}\left(\frac{\text{Cost index at time } t2}{\text{Cost index at time } t1}\right)$$

There are several useful cost indices. These include the Chemical Engineering Plant Cost Index (published monthly at Chemical Engineering, www.ChE.com), the Nelson-Farrar Refinery Construction Index (published monthly at Oil and Gas Journal, www.ogjonline.com), and the Engineering News Record Construction Index (published weekly at Engineering News Record, http://enr.construction.com). In using cost indices, it is usually advisable to keep the updating period within 20 years. Table 2.3 provides a listing of recent values of the Chemical Engineering Plant Cost Index.

5. **Ratio factors based on delivered equipment cost**

This approach is based on assigning all the items in the FCI and the TCI a ratio to the delivered equipment

> **Example 2-2 Updating the FCI using a cost index**
>
> The FCI of a 1500-ton/day ammonia plant in 2002 is estimated to be $120 MM (Couper, 2003). Estimate the FCI of a similar plant producing 2000-ton ammonia/day in 2009.
>
> **SOLUTION**
>
> First, the cost of 1500-ton/day plant needs to be updated to 2009; then it should be scaled up to 2000 ton/day. From Table 2.3, the values of the Chemical Engineering Plant Cost Index for 2002 and 2009 are 395.6 and 521.9, respectively. Therefore,
>
> $$\text{FCI of the 1500-ton/day plant in 2009} = \text{FCI of the 1500-ton/day plant in 2002} * \frac{521.9}{395.6}$$
> $$= \$158.3\text{MM}$$
>
> Table 2.2 gives a Capacity cost exponent of 0.66 for the ammonia plant. Hence,
>
> $$\text{FCI of the 2000-ton/day plant in 2009} = \text{FCI of 1500-ton/day plant in 2009} * \left(\frac{2000}{1500}\right)^{0.66}$$
> $$= \$191.4\text{MM}$$

TABLE 2-3 Recent Values of the Chemical Engineering Plant Cost Index

Year	Chemical Engineering Plant Cost Index*
2001	394.3
2002	395.6
2003	402.0
2004	444.2
2005	468.2
2006	499.6
2007	525.4
2008	575.4
2009	521.9
2010	550.8

*(Basis: Between 1957 and 1959, the value of index = 100.)
Source: Chemical Engineering, www.ChE.com/PCI.

TABLE 2-4 Values of the FCI Lang Factor

Type of Plant	FCI Lang Factor
Solid	3.10
Solid-Fluid	3.63
Fluid	4.74

Source: Lang (1948).

cost. In his seminal work, Lang (1948) proposed a method for estimating the FCI of a chemical plant based on the cost of the equipment delivered (but not installed) to the plant. Upon surveying the detailed costs of 14 plants, Lang developed factors for predicting the various cost items such as installation, instrumentation, piping, electrical systems, facilities, engineering, and so on. Lumped together, these factors are referred to as the Lang factor, which is used in estimating the FCI as follows:

[2.4a]
$$\text{FCI} = \text{FCI Lang Factor} * \text{Delivered Equipment Cost}$$

Hence,

[2.4b]
$$\text{FCI} = \text{FCI Lang Factor} * \sum_{q=1}^{N_{Equipment}} C_q^{Delivered}$$

where q is the index for the process equipment, $N_{Equipment}$ is the total number of equipment in the process, and $C_q^{Delivered}$ is the cost of equipment q delivered to the plant site.

The value of the Lang factor depends on the type of the materials processed in the facility (for example, solid, solid-fluid, fluid) as shown by Table 2.4. To utilize the Lang factor method, the cost of the different pieces of equipment must first be determined. The estimation of equipment cost is discussed later in this chapter.

Peters et al. (2003) revised the Lang factor by developing an itemized list of the various components involved in estimating the FCI. Assuming a ratio of 15:85 for the WCI:FCI, Peters et al. (2003) also provided an estimation for evaluating the TCI using the following expression:

[2.5]
$$\text{TCI} = \text{TCI Lang Factor} * \text{Delivered Equipment Cost}$$

The revised Lang factors are given in Table 2.5, and the details of the different items contributing to these factors are given in Table 2.6. For each $100 spent on purchased equipment cost (delivered to the gates of the plant), Table 2.6 gives the corresponding expenditures for the different cost items. As can also be seen from Table 2.6, Peters et al. (2003) broke down the FCI into two categories: *total direct cost*, which is also referred to as *total installed equipment cost* (which accounts for purchased equipment cost, installation, instrumentation, piping, electrical work, and so on) and *total indirect cost* (which accounts for the nonmanufacturing FCI items such as engineering, legal expenses, contractor's fees, contingency, and so on).

Another modification of the Lang approach is the *Hand method* (Hand, 1958), which assigns different cost factors depending on the type of equipment. Hand classified the process equipment into eight categories. Each category has a value of its multiplier depending on the specific requirements for civil work, installation, piping, insulation, and so on. Table 2.7 includes the equipment categories and the values of the associated factors to be used in the following expression:

[2.6]
$$\text{FCI} = \sum_{q=1}^{N_{Equipment}} f_q^{Hand} C_q^{Delivered}$$

TABLE 2-5 Values of the Revised Lang Factor

Type of Plant	FCI Lang Factor	TCI Lang Factor
Solid	4.0	4.7
Solid-Fluid	4.3	5.0
Fluid	5.0	5.9

Source: Peters et al. (2003).

TABLE 2-6 Values of the Items Contributing to the Revised Lang Factor

Item	Solid Processing Plant	Solid-Fluid Processing Plant	Fluid Processing Plant
Direct Costs			
Purchased equipment delivered to the plant	100	100	100
Purchased equipment installation	45	39	47
Instrumentation and controls (installed)	18	26	36
Piping (installed)	16	31	68
Electrical systems (installed)	10	10	11
Buildings	25	29	18
Yard improvements	15	12	10
Service facilities	40	55	70
Total Direct Plant Cost	269	302	360
Indirect Costs			
Engineering and supervision	33	32	33
Construction expenses	39	34	41
Legal expenses	4	4	4
Contractor's fees	17	19	22
Contingency	35	37	44
Total Indirect Plant Cost	128	126	144
Fixed Capital Investment	397	428	504
Working Capital Investment (15/85 of FCI)	70	75	89
Total Capital Investment	467	503	593

Source: Peters et al. (2003).

where f_q^{Hand} is the Hand factor for equipment q and $C_q^{Delivered}$ is the delivered cost of equipment q.

The rationale for the different factors is that the installation expenses differ depending on the type of equipment. For instance, fired heaters and compressors cost less to install than distillation columns. Another observation is that the instruments are treated as a separate category of equipment and should be added to the cost estimate of the other units.

It is worth noting that companies may have their own experience-based factors that are appropriately developed for specific applications. Whenever available, these factors should be used in favor of the generic multipliers.

TABLE 2-7 Values of the Hand Factors for Different Equipment Categories

Equipment Type	Hand Factor (f_q^{Hand})
Compressors	2.5
Distillation columns	4.0
Fired heaters	2.0
Heat exchangers	3.5
Instruments	4.0
Miscellaneous equipment	2.5
Pressure vessels/tanks	4.0
Pumps	4.0

6. **Empirical correlations**

There are useful expressions to estimate the cost of specific industries or general classes of processes. Depending on the basis for these expressions, their general applicability and accuracy vary significantly. The following are examples of simplified correlations that should be cautiously used for rough and quick estimates of FCI (expressed in 2010 US $), depending on the predominant phase nature of the process:

For gas-phase plants (adapted with modification from Timm [1980], Gerrard [2000], and Coker [2007]):

[2.7] $$FCI = 36,000*N*F^{0.62}$$

where N is the number of functional units (these are major processing steps such as separation, reaction, and finishing but not heat exchange or compression/pumping unless they have substantial duties such as refrigeration and liquefaction), and F is the process throughput tonne/yr.

For liquid- and/or solid-phase plants (adapted with modification from Bridgwater and Mumford [1979], Gerrard [2000], and Coker [2007]):

[2.8a] $$FCI = 7000*N*F^{0.68} \quad \text{For } F \geq 60,000 \text{ tonne/yr}$$

and

[2.8b] $$FCI = 458,000*N*F^{0.30} \quad \text{For } F < 60,000 \text{ tonne/yr}$$

where N is the number of functional units and F is the process throughput tonne/yr.

Example 2-3 FCI estimation based on factors of delivered equipment cost

Table 2.8 gives the type and cost of the units to be used in a retrofitting project in a fluid-processing facility. Additionally, the instrumentation and control systems for this project are estimated to have a delivered cost of $3.6 MM. Estimate the FCI using the Lang factors as revised by Peters et al. (2003) and the Hand method.

SOLUTION

For a fluid-processing plant, the Lang factor (revised by Peters et al., 2003) for estimating the FCI from delivered equipment cost is 5.0. Hence,

$$\text{FCI (Lang method)} = 5.0*(2.0 + 2.5 + 4.0 + 1.5)$$
$$= \$50.0\text{MM}$$

On the other hand, the Hand method assigns different installation multipliers for the different types of units and also considers the instrumentation system to be a separate unit. Table 2.9 is the worksheet showing the estimation of the FCI. Therefore, using the Hand method, we get:

$$\text{FCI (Hand method)} = 8.0 + 5.0 + 14.0 + 14.4 + 6.0$$
$$= \$47.4\text{MM}$$

TABLE 2-8 Units and Costs of the Retrofitting Project of Example 2.3

Unit	Delivered Equipment Cost ($ MM)
Distillation columns	2.0
Fired heater	2.5
Heat exchangers	4.0
Tanks	1.5

TABLE 2-9 Using the Hand Method for Example 2.3

Equipment	Delivered Equipment Cost ($ MM)	Hand Factor	Installed Equipment Cost ($ MM)
Distillation columns	2.0	4.0	8.0
Fired heaters	2.5	2.0	5.0
Heat exchangers	4.0	3.5	14.0
Instruments	3.6	4.0	14.4
Pressure vessels/tanks	1.5	4.0	6.0

Example 2-4 Using an empirical correlation to estimate the FCI

Ethanol may be produced from lignocellulosic wastes. One of the largest sources of lignocellulosic wastes in the United States is corn stover, which is the material left on the surface of the cornfield after harvesting. It contains the stalks, the leaves, the husks, and the cobs. A stover-to-ethanol plant produces 71,000 tonne/yr of ethanol and contains eight functional units: feedstock handling, pretreatment, simultaneous saccharification/fermentation distillation, solid/syrup separation, wastewater treatment, boilers, and turbogeneration/utilities (McAloon et al., 2000). Estimate the FCI of the plant.

SOLUTION

Using Eq. 2.8a,

$$\text{FCI} = 7000*8*(71{,}000)^{0.68}$$
$$= \$111\text{MM}$$

The 2010 updated FCI reported by McAloon et al. (2000) is $191 MM. Although the difference is −42 percent, it is important to recall that such a result is usually acceptable for order-of-magnitude estimates and that reaching the result involved very little effort. If there is further interest in the process, then a more detailed study (such as the one conducted by McAloon et al., 2000) is carried out. Clearly, in other applications, the level of accuracy may vary substantially.

7. **Turnover ratio**

For a plant producing multiple products, let p be an index for the products and $N_{products}$ be the total number of salable products. Therefore, the annual sales of the plant (assuming that all produced products are sold) are calculated as follows:

$$[2.9] \quad \text{Annual sales} = \sum_{p=1}^{N_{Products}} \text{Annual production rate of product } p * \text{Selling price of product } p$$

The *turnover ratio* (sometimes known as the *capital ratio* or the *asset turnover ratio*) of a plant is defined as:

$$[2.10a] \quad \text{Turnover ratio} = \frac{\text{Annual Sales}}{\text{FCI}}$$

The **turnover ratio** provides an indication on how efficiently the process equipment is used to make sales.

When the turnover ratio is known for a specific company at a given production level, the FCI may be estimated using the turnover ratio and the annual sales by rearranging Eq. 2.10a:

$$[2.10b] \quad \text{FCI} = \frac{\text{Annual Sales}}{\text{Turnover Ratio}}$$

The turnover ratio is also useful in predicting the FCI (for a new plant of a given production rate and selling price of the product) when no or little design or economic data are available. In such cases, assuming a "reasonable" value of the turnover ratio can be very useful in estimating the FCI. The challenge in such cases is the selection of that "reasonable" value when very little is known about the process. The value of the turnover ratio varies significantly depending on the process technology, plant size, and market conditions. Garrett (1989) proposed the use of a turnover ratio of 0.5 to estimate the FCI from the annual sales. Garrett based this assumption on various data for the turnover ratios and the observation that for

> **Example 2-5 Using the turnover ratio to estimate the FCI**
>
> A gas-processing facility produces a mixture of ethylene and propylene (production rates are 545,000 and 273,000 tonne/yr, respectively). If the selling prices of ethylene and propylene are $800 and $1100/tonne, respectively, estimate the FCI of the process using the turnover ratio.
>
> **SOLUTION**
>
> The annual sales of the plant can be calculated as follows:
>
> $$\text{Annual sales} = 545{,}000*800 + 273{,}000*1{,}100$$
> $$= \$736\text{MM}$$
>
> For a quick and rough estimate, let us assume a turnover ratio of 2.0. Therefore,
>
> $$\text{FCI} = \frac{736}{2}$$
> $$= \$368\text{MM}.$$
>
> The cost of this plant as reported by Seider et al. (2009) and updated to 2010 is $440 MM.

many companies the annual revenues are roughly equal to the company's assets (which are assumed to be 50 percent depreciated).

To assess the typical values and variability of turnover ratios, recent and updated data on the FCI and selling prices have been compiled for key industries as shown by Table 2.10. The resulting turnover ratios demonstrate a great deal of variability, ranging from 0.4 to 21.8, with many values lying between 0.5 and 3.5. For the results shown in Table 2.8, a 0.5 value of the turnover ratio seems to be too low. Instead, a value of 2.0 appears to be more reasonable. Therefore, when very little is known about the process and when there is a need for a very rapid estimate, it is proposed to use (with great caution) *a value of 2.0 for the turnover ratio.* Knowing the plant production rate of the main product(s) and the selling price of the product(s), dividing the annual sales by a turnover ratio of 2.0 gives a very rough estimate of the FCI. Notwithstanding the potential large inaccuracies associated with this method and the fact that it does not account for the economy of scale, the turnover ratio is very convenient in getting quick estimates with very limited data.

EQUIPMENT COST ESTIMATION

The cost of the processing equipment is the most important element in evaluating the capital investment of the plant. As mentioned earlier, the equipment cost is also a good basis for estimating the fixed capital investment of a process such as in the case of the Lang factor method. Therefore, a key aspect in most cost estimation studies is the evaluation of the cost of individual pieces of equipment. It is worth distinguishing between three types of equipment cost:

- Free on board (FOB): This is the cost of the equipment at the manufacturer's loading docks, shipping trucks, rail cars, or barges at the vendor's fabrication facility. The purchaser still has to pay for equipment freight, installation, insulation, instrumentation, electric work, piping, engineering work, and construction. When there is a reference to purchased equipment cost, it typically corresponds to the FOB basis.
- Delivered equipment cost: This term corresponds to the equipment cost delivered to the buyer. It is the sum of the FOB and the delivery costs (for example, freight, transportation insurance, importation taxes). Clearly, the latter costs depend on the distance between the supplier and the buyer, the type and size of equipment, the mode of transportation, and the tax laws of the countries of origin and destination.
- Installed equipment cost: This is the sum of the delivered equipment cost plus the installation costs (for example, labor, civil structure, and foundation work). As shown by Table 2.4, the installation costs are typically in the range of 40 to 50 percent of the delivered equipment cost.

The following methods may be employed to estimate the equipment cost:

1. Manufacture's quotation
2. Computer-aided tools
3. Capacity ratio with exponent
4. Updates using cost indices
5. Cost charts

A description of these methods is given in the following sections:

1. Manufacturer's quotation

When sufficient details are available on the size or duty of the equipment, quotations can be obtained from the proper manufacturers. Usually, an initial estimate is first obtained based on a preliminary design or task of the equipment. This helps in the screening of multiple alternatives and the comparison of performance and cost. Later, as formal procurement procedures are followed, more accurate cost estimates are obtained from the vendor. If the equipment cost is relatively high or if competitive offers are required, a bidding process followed by a techno-economic analysis is used to select the equipment and to determine the price.

2. Computer-aided tools

Computer-aided tools may be used to estimate the FOB cost of equipment. As mentioned earlier, tools that are linked to process simulators are particularly useful because they are integrated with equipment performance and sizing as evaluated by simulation. Examples include ICARUS (www.aspentech.com) and SuperPro Designer (http://www.intelligen.com). Web-based resources may also be used for order-of-magnitude estimates (for example, http://www.matche.com/EquipCost/, provided courtesy of *Matches*, and http://highered.mcgraw-hill.com/

TABLE 2-10 Estimated* Values of Turnover Ratios

Main Product of the Process	Production Rate of Main Product (10^3 tonne/yr)	Selling Prices of Product $/tonne	Annual Sales ($ MM/yr)	FCI ($ MM)	Turnover Ratio (yr^{-1})	Reference for FCI
Ammonia	300	360	108	66	1.6	Garrett (1989)
Ammonium nitrate	360	190	68	28	2.4	Seider et al. (2009)
Acetic acid	18	880	16	15	1.1	Garrett (1989)
Acetic acid	200	880	176	136	1.3	Hydrocarbon Processing (2003)
Adipic acid	295	1700	502	195	2.6	Towler and Sinnott (2008)
Alkyl benzene (linear)	73	1030	75	75	1.0	Hydrocarbon Processing (2003)
Benzene	141	1025	145	27	5.4	Hydrocarbon Processing (2003)
Biodiesel	133	1130	150	23	6.5	Elms and El-Halwagi (2009)
Bio-gasoline/Bio-jet fuel	86	1100	95	131	0.7	Pham et al. (2010)
Butene-1	18	748	13	12	1.1	Hydrocarbon Processing (2003)
Cumene	273	1300	355	43	8.3	Seider et al. (2009)
Cumene	300	1300	390	31	12.6	Hydrocarbon Processing (2003)
Cyclohexane	182	750	137	11	12.5	Hydrocarbon Processing (2003)
Diesel (from gas-to-liquid)	5674	800	4539	9210	0.5	Bao et al. (2010)
Ethanol (from corn)	71	550	39	39	1.0	McAloon et al. (2000)
Ethanol (from corn stover)	71	1130	80	191	0.4	McAloon et al. (2000)
Ethanol (from corn stover)	159	1130	180	358	0.5	Kazi et al. (2010)
Ethylene dichloride	455	400	182	114	1.6	Seider et al. (2009)
Ethylbenzene	1273	1200	1528	114	13.4	Seider et al. (2009)
Ethylbenzene	455	1200	546	25	21.8	Hydrocarbon Processing (2003)
Ethylene	568	800	625	691	0.7	Towler and Sinnott (2008)
Ethylene	83	800	66	116	0.6	Hydrocarbon Processing (2003)
Methanol	5	300	1500	558	2.7	Hydrocarbon Processing (2003)
Methanol	300	300	90	59	1.5	Garrett (1989)
Nitric acid	636	237	151	71	2.1	Seider et al. (2009)
Paraxylene	750	1540	1155	609	1.9	Hydrocarbon Processing (2003)
Phenol	182	1320	240	255	0.9	Towler and Sinnott (2008)
Phosphoric acid	1455	500	728	71	10.3	Seider et al. (2009)
Phosphoric acid	18	500	9	9	1.0	Garrett (1989)
Propylene	164	1190	195	28	7.0	Hydrocarbon Processing (2003)
Styrene	1136	1430	1624	284	5.7	Seider et al. (2009)
Styrene	500	1430	715	113	6.3	Hydrocarbon Processing (2003)
Styrene	25	1430	36	29	1.2	Hydrocarbon Processing (2003)
Sulfuric acid	1818	80	145	43	3.4	Seider et al. (2009)
Sulfuric acid	300	80	24	45	0.5	Garrett (1989)

*The FCI values were updated to mid-2010 using the Chemical Engineering Plant Cost Index. The selling prices were obtained from several sources including vendors, trade journals (primarily ICIS Chemical Business, vol. 277, July 12–18 [2010]), and historical data from www.icis.com. Judgment calls were made to reconcile and revise the data.

TABLE 2-11 Typical Scaling Exponents for Selected Pieces of Equipment

Equipment	Size Range/Sizing Criterion	Exponent x	Reference
Blowers (centrifugal)	0.5 to 4.7 m^3/s	0.59	Peters et al. (2003)
Compressor (reciprocating)	150 to 750 kW	0.80	Garrett (1989)
Cooling tower	5000 to 30,000 gpm of water flow	0.77	Brown (2007)
Ejectors (steam jet)	Steam flow rate, kg/s	0.52	Axtell and Robertson (1986)
Furnaces	Heat duty, kW	0.78 to 0.80	Axtell and Robertson (1986); Towler and Sinnott (2008)
Heat exchangers (shell-and-tube, floating head)	10 to 900 m^2	0.60	Peters et al. (2003); Ulrich and Vasudevan (2004)
Heat exchangers (shell-and-tube, fixed sheet)	10 to 40 m^2	0.44	Peters et al. (2003)
Jacketed vessel	1 to 800 m^3	0.60	Ulrich and Vasudevan (2004)
Refrigeration units	5 to 10,000 kW	0.60 to 0.70	Chauvel (1981); Ulrich and Vasudevan (2004)
Tank (floating roof)	200 to 70,000 m^3	0.60	Ulrich and Vasudevan (2004)
Tank (spherical 0 to 5 barg)	100 to 10,000 m^3	0.60 to 0.70	Ulrich and Vasudevan (2004); Towler and Sinnott (2008)
Trays (sieve)	1 to 3 m diameter	0.86	Peters et al. (2003)
Wastewater treatment	400 to 400,000 m^3/d	0.64	Seider et al. (2009)

sites/0072392665/student_view0/cost_estimator.html, provided courtesy of Peters et al. [2003]).

3. **Capacity ratio with exponent**

When there are two identical (or at least very similar) pieces of equipment of different capacities with the FOB equipment cost known for one of them, then the FOB of the other piece of equipment may be estimated as follows:

$$[2.11] \quad FOB_B = FOB_A \left(\frac{Size_B}{Size_A} \right)^x$$

Where FOB_B and FOB_A are the FOB costs of equipment B and A, and $Size_B$ and $Size_A$ are the sizes (for example, heat-transfer area of a heat exchanger, number of stages of a separation column, horsepower of a pump) of equipment B and A, the exponent x is the economy-of-scale factor and is typically less than 1.0. Its exact value varies from one type of equipment to another. If no data are available for x, it may be assumed to be 0.6 to 0.7 (hence, the name *sixth-tenths-factor rule*) for order-of-magnitude estimates. Equation 2.11 for equipment cost is consistent with Eq. 2.2 for the FCI because these two costs are usually related through a multiplier (for example, the Lang factor). Remer and Chai (1990) compiled an extensive list of the scaling exponents for numerous pieces of equipment. For illustration purposes, Table 2.11 gives values of x for different pieces of equipment.

It is worth noting that there are cases when the economy of scale is not applicable. One example is the case of modular units where larger capacities require the usage of multiple identical units. An example is reverse-osmosis units that are manufactured with specific membrane areas and are capable of handling a narrow range of flow rate. Therefore, doubling the flow rate implies doubling the number of reverse-osmosis modules and (almost) doubling the cost of the modules. Another exception to the economy of scale is when the size of the unit exceeds the applicability range; then parallel processing is recommended. For instance, if the flow rate to a tray column requires a diameter bigger than the practical norm for sieve trays, the flow rate can be split, and two (or more) columns of practical diameters are used.

4. **Updates using cost indices**

In some cases, the cost of a piece of equipment is available from a previous study and it is desired to evaluate its present cost. Because of inflation and other economic changes, it is necessary to correlate equipment cost as a function of time. In this regard, cost indices are useful tools. A cost index is an indicator of how equipment cost varies over time. The ratio of cost indices at two different times provides an estimate for the extent of equipment cost inflation between these two times. Hence,

$$[2.12] \quad FOB_{t2} = FOB_{t1} \left(\frac{Cost\ index\ at\ time\ t2}{Cost\ index\ at\ time\ t1} \right)$$

where FOB_{t1} and FOB_{t2} are the purchased equipment costs at times t1 and t2, respectively.

Various cost indices are published regularly. A commonly used index is the Marshall and Swift (*M&S*) equipment cost index published in the monthly magazine *Chemical Engineering* (www.ChE.com). Table 2.12 gives recent values of the M&S index. For atmospheric pollution control equipment, the Vatavuk cost index may be used (Vatavuk, 1995). It is not recommended to use cost indices if the updating period exceeds 10 years.

Example 2-6 Estimating cost of a heat exchanger

A shell-and-tube heat exchanger has a surface area of 100 m². Its cost in 2003 was $92,000. What was the cost of a similar heat exchanger with double the surface area in 2009?

SOLUTION

Using the M&S cost index,

$$\text{Cost of the 100-m}^2 \text{ heat exchanger in 2009} = 92{,}000 * \left(\frac{1468.6}{1{,}123.6}\right)$$
$$= \$120{,}248$$

Assuming a capacity exponent of 0.6, we get

$$\text{Cost of the 200-m}^2 \text{ heat exchanger in 2009} = 120{,}248 * \left(\frac{200}{100}\right)^{0.6}$$
$$= \$182{,}262$$

TABLE 2-12 Recent Values of the M&S Equipment Cost Index

Year	M&S Equipment Cost Index[#]
2001	1093.9
2002	1104.2
2003	1123.6
2004	1178.5
2005	1244.5
2006	1302.3
2007	1373.3
2008	1449.3
2009	1468.6
2010	1457.4

Source: Chemical Engineering, www.ChE.com/PCI.
[#]*(Basis: In 1926, value of index = 100.)*

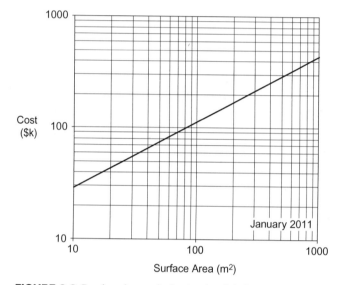

FIGURE 2-3 Purchased cost of a floating-head shell-and-tube heat exchanger (carbon steel, 1035 kPa pressure rating).

5. Cost charts

Equipment cost may be related to one or more basic sizing criteria. Examples of these sizing criteria include diameter for mass exchange trays, diameter and height for packed columns, heat transfer area for heat exchangers, horsepower for pumps and compressors, volumetric flow rate for fans and blowers, and weight or volume for storage tanks. Extensive compilations of charts correlating equipment cost to sizing criteria have been published in books, journals, and vendor catalogs. Figures 2.3 through 2.6 are sample charts intended for rough estimates and illustration purposes. All costs are FOB basis. For more equipment cost charts and data, readers are referred to Brown (2007), Coker (2007), Couper (2003), Garrett (1989), Gerrard (2000), Peters et al. (2003), Seider et al. (2009), Towler and Sinnott (2008), Turton et al. (2009), and Ulrich and Vasudevan (2004).

It is also worth noting that, unless otherwise specified, most cost data for equipment are based on near-atmospheric or some base-case conditions and the use of carbon steel as a material of construction. When the process involves harsh conditions that require special materials of construction or high pressures/vacuum, two corrections must be made for materials of construction and pressure using the following factors: [2.13]

$$\text{Materials factor for material M} = \frac{\text{Purchased cost of equipment in material of construction M}}{\text{Purchased cost of equipment in carbon steel}}$$

The values of the materials factor vary significantly depending on the market conditions, the prices of metals, and the difficulty of equipment manufacturing using the specific material of construction. For preliminary estimates, Table 2.13 provides typical values of the materials factor. Actual values should be obtained from equipment vendors. The basis value of 1.0 is given to carbon steel because it is usually the material of choice for the process industries when corrosive materials are not involved.

If the cost of materials becomes excessive, the application of a lining layer to carbon steel is used to provide protection without incurring the full cost of the expensive

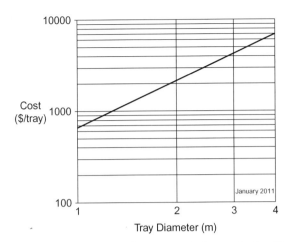

FIGURE 2-4 Purchased cost of a sieve tray (carbon steel, when 10 or more trays are purchased).

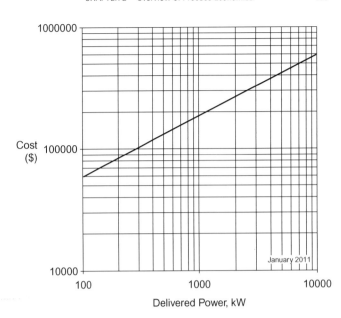

FIGURE 2-6 Purchased cost of a steam turbine.

TABLE 2-13 Typical Values of Material Factors

Material of Construction	Materials Factor
Carbon steel	1.0
Cast steel	1.2
Aluminum	1.6
Bronze	1.6
304 stainless steel	1.8
316 stainless steel	2.1
Copper	2.3
Hastelloy C	2.4
Monel	3.2
Nickel	4.5
Inconel	4.7
Titanium	8.0

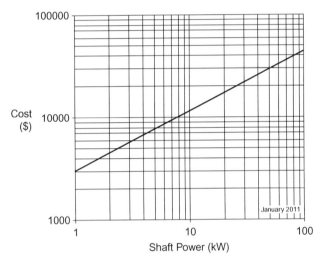

FIGURE 2-5 Purchased cost of a centrifugal pump (carbon steel).

Example 2-7 Including the materials factor in estimating the cost of a heat exchanger

Estimate the January 2011 purchased cost of a 60-m² floating-head shell-and-tube heat exchanger made of titanium.

SOLUTION

From Fig. 2.3, the purchased cost of the 60-m² carbon-steel exchanger is $80,000. To adjust for the materials of construction, the materials factor of 8.0 for titanium is used. Hence,

Purchased cost of titanium exchanger = 8.0*80,000 = $640,000.

---

material of construction. On the other hand, accounting for the pressure, the following factor is used:

[2.14]

$$\text{Pressure factor for pressure P} = \frac{\text{Purchased cost of equipment under pressure P}}{\text{Purchased cost of equipment under base-case pressure}}$$

Values of the pressure factor vary depending on the type of equipment and the handled materials. Such values are typically included in cost charts or can be obtained as part of the vendor's quotation.

OPERATING COST ESTIMATION

Once the plant is ready for production, the continuous/recurring expenses required to run the process are

TABLE 2-14 Typical Prices of Selected Utilities (2010)

Energy Utility	Cost
Natural gas	$4 to 8/MM Btu (or per 1000 SCF)
Petroleum	$12 to 15/MM Btu
Steam	$4 to 15/ton
Electricity	$0.05 to 0.10/kWh
Cooling tower water	$0.03 to 0.10/m^3
Process water	$0.50 to 1.50/m^3
Refrigeration	$20 to 50/MM Btu

referred to as the operating cost. Items such as raw materials, material utilities, energy utilities, waste handling, labor, maintenance, lab charges, research and development, royalties, and overheads are among the key expenses for the operating cost. The following is a discussion of the major items and how they can be estimated:

- **Raw materials:** Based on the material balance information for the process, the needed quantities of the fresh purchases of raw materials are determined. The unit prices of these raw materials are typically obtained through direct quotations or contractual agreements with the suppliers. Estimates of the chemical prices may be obtained from specialized journals such as ICIS *Chemical Business* (www.icis.com), which was formerly known as the *Chemical Marketing Reporter* and is published weekly. Historical data on the prices of key chemicals may also be obtained online at http://www.icis.com/StaticPages/a-e.htm.
- **Material utilities:** Although raw materials are transformed into the main products and byproducts of the process, other ingredients that are used for various processing objectives (for example, mass separating agents, solvents, catalysts, and so on). Even if these material utilities are regenerated and reused, there is the cost of regeneration and the cost of the fresh makeup to compensate for the losses throughout the process. The unit prices of the material utilities are determined similar to the methods described earlier for the cost of raw materials.
- **Energy utilities:** Various forms of energy utilities are used to drive reactions, affect separation, provide the necessary heating and cooling, and convey materials. These energy utilities include fuel, electricity, steam, water, refrigerants, air, and so on. Energy balance calculations quantify the usage of the energy utilities. The unit prices vary depending on quantity, time, and location. Table 2.14 gives typical prices of selected utilities. Web resources are also useful in getting updated values. A particularly useful site for the prices of natural gas and oil is provided courtesy of the Energy Information Administration (EIA) of the U.S. Department of Energy (http://tonto.eia.doe.gov/oog/info/ngw/ngupdate.asp).
- **Labor:** The labor cost depends on several factors including the number and skills of the employees, the number of shifts per day, and the extent of process automation. The prevailing wages for the different types of employees at the targeted region should also be considered. A useful web resource is provided courtesy of the U.S. Bureau of Labor Statistics (http://www.bls.gov/).
- **Maintenance:** There are two major types of maintenance: preventive (scheduled regularly to maintain smooth and safe operation) and responsive (carried out to address operational problems). Maintenance costs include labor and materials. They vary depending on the process, schedules, and plant conditions. A typical range for annual maintenance and repairs cost is 5 to 10 percent of the FCI.

PRODUCTION COST ESTIMATION

The production (or manufacturing) cost accounts for all expenditures involved in the operating cost needed to run the process (for example, raw materials, utilities, labor) as well as the cost of the capital investment needed for the process equipment and ancillary systems (for example, buildings and structures). Before developing an expression for the manufacturing cost, it is useful to consider the following points:

- Recalling the definition of total capital investment, it is the sum of FCI and WCI. In principle, the WCI and the land portion of the FCI are recoverable at the end of the project. Therefore, in accounting for the cost of capital investment impacting the production cost, attention should be paid to the irrecoverable portion of the FCI.
- Although the FCI is a one-time investment (and therefore has the units of money), the operating cost is incurred on a continuous basis (and hence has the units of cost per time, for example, $/yr). Hence, there is a need to annualize the FCI to be able to add and reconcile the FCI and the operating cost. In this context, the concept of *depreciation* is particularly useful and is discussed in the following section.

A more detailed discussion on the production cost will be presented later, starting with the break-even analysis.

DEPRECIATION

The process equipment and associated hardware deteriorate and lose value over their useful life period. The causes are physical (for example, wear and tear), chemical (corrosion), safety (resulting from an accident or to prevent one), and functional (technological or product obsolescence). **Depreciation** is an annual income tax deduction that is intended to allow the company to recover the cost of property (for example, process equipment) over a certain recovery period. When the depreciation charges are shielded from the company's income taxes, the saved money can be used for the replacement of the process assets or for capital recovery of the initial value of the asset. Land and the WCI cannot be depreciated because they are recoverable in principle. Depreciation does not involve an actual transfer of cash. It is an accounting tool. However, it has a major impact on the company's taxes and, therefore, on the cost of the product and the profit of the company. There are several methods to calculate depreciation. The following methods will be covered:

- Linear (straight-line) method
- Declining-balance method
- Modified accelerated cost recovery system

LINEAR DEPRECIATION (STRAIGHT-LINE METHOD)

The simplest and most commonly used method for determining depreciation is referred to as the **straight-line method** in which

$$[2.15] \quad d = \frac{V_0 - V_s}{N}$$

where d is the annual depreciation, V_0 is the *initial value* of the property, N is the *recovery period* over which depreciation is made (sometimes referred to as the *useful life period* or the *service life* of the property) in years, and V_s is the *salvage value*, which corresponds to the worth of the property at the end of the recovery period. If the property can no longer be used for its original purpose by the end of the Nth year, it may still be sold for its value of metal (or materials). In such cases, V_s is called the scrap value of the property. The limits on annual depreciation are set by federal tax regulations. For most processing equipment, N ranges from 5 to 20 years. For many chemical process units, a value of 10 years may be used in the linear depreciation method.

The book value of the property at the end of the year n ($V_{b,n}$) is defined as the difference between the initial value of the property and the sum of the depreciation charges over the n years; that is,

$$[2.16a] \quad V_{b,n} = V_0 - \sum_{t=1}^{n} d_t$$

where t is the index for time in years, n is a specific year for which the book value is calculated, and d_t is the depreciation charge in the tth year. When the linear depreciation method is used, the annual depreciation charge is constant every year ($d_t = d$) and the book value is given by:

$$[2.16b] \quad V_{b,n} = V_0 - n*d$$

It is worth noting that, at the end of the recovery period (useful life period of the project), $n = N$ and, by substituting for d from Eq. 2.15 and 2.16b, we get:

$$[2.17] \quad V_{b,N} = V_0 - N*\left(\frac{V_0 - V_s}{N}\right) = V_s$$

which indicates that the book value of the property at the end of the recovery period is the salvage value and that the property is fully paid off or written off. Hence, no more depreciation is allowed beyond the salvage value.

The depreciation scheme can be applied to the depreciable FCI (essentially the FCI minus the cost of the land). Because the FCI is paid for at the beginning of the project (even if it is financed), depreciation serves as an equivalent, distributed, and tax-deductible expense to the company's production cost that accounts for the distribution of the depreciable FCI over the useful life period of the plant. In such cases, depreciation is referred to as the *annualized fixed cost* (AFC) and is given by:

$$[2.18] \quad AFC = \frac{FCI_0 - FCI_s}{N}$$

where FCI_0 is the initial value of the depreciable FCI, FCI_s is the salvage or scrap value of the FCI at the end of the service life, and N is the service life of the property in years.

The annual *fixed charges* of the plant include the AFC, property taxes, property insurance, salaries that are independent of production size, and other constant charges. They are characterized by their constant values regardless of the production level of the plant. For many conceptual design studies of retrofitting projects, when the AFC is significantly larger than the other fixed charges, the AFC is used to represent the annual fixed charges of the plant. Consequently, the production cost (also known as the *total annualized cost* "TAC") is given by:

$$[2.19] \quad TAC = AFC + AOC$$

where AOC is the annual operating cost of the process.

Example 2-8 Estimation of total annualized cost of a scrubber

A gas-cleaning project requires the installation of two scrubbers whose FCI is $1.2 MM. Recovery period of the heat exchangers is taken to be 10 years. The salvage value of the exchangers is $0.2 MM. The scrubbing process uses $1.5*10^5$ kilogram (kg)/yr of fresh solvent purchased at $0.4/kg. The cost of the fresh solvent is considered to be the dominant operating cost of this scrubbing system. What is the total annualized cost of the scrubbing system?

SOLUTION

The AFC is obtained by calculating the annual depreciation charge through Eq. 2.18:

$$AFC = \frac{1.2*10^6 - 0.2*10^6}{10}$$
$$= \$100,000/yr$$

The annual operating cost is obtained as follows:

$$AOC = \$0.4/kg * 1.5*10^5 \, kg/hr$$
$$= \$60,000/yr$$

Substituting into Eq. 2.19, we get:

$$TAC = 100,000 + 60,000$$
$$= \$160,000/yr$$

DECLINING-BALANCE METHOD

Because of the risks involved in projects and the decline in the value of money over time, it is desirable to accelerate depreciation by setting aside higher depreciation charges in the earlier years of the project. The **declining-balance (fixed-percentage) method** is an accelerated depreciation scheme in which the annual depreciation charge is taken as a fixed fraction of the book value at the end of the previous year. Although this fixed fraction remains constant throughout the life of the project, the depreciation charges are set aside at higher values in the earlier years of the project.

Let:

α = fixed depreciation fraction
d_n = depreciation charge for the nth year
n = a specific year in the life of the project
N = the recovery period (or the last year in the recovery period)
V_0 = initial value of the property
V_n = book value of the property at the end of year n
V_s = Salvage value of the property at the end of the recovery period (N years)

Hence,

[2.20] $$d_n = \alpha V_{n-1}$$

In the linear depreciation method, the rate of recovering the capital via depreciation (that is, $d/(V_0 - V_s)$) is $1/N$. The tax laws in the United States allow higher rates of capital recovery via depreciation, provided that such rates do not exceed twice the rate of the linear method. Therefore,

$$\frac{1}{N} \leq \alpha \leq \frac{2}{N}$$

Two values of α are commonly used: 1.5 (in the *150 percent declining-balance method*) and 2.0 (in the *double declining-balance [DDB] method*). Let us construct the following scheme for evaluating the depreciation charges and the book values over the recovery period of the project.

Year	Book Value at Beginning of the Year	Depreciation Charge	Book Value at End of the Year
1	V_0	αV_0	$(1-\alpha)V_0$
2	$(1-\alpha)V_0$	$(1-\alpha)\alpha V_0$	$(1-\alpha)^2 V_0$
.			
.			
n	$(1-\alpha)^{n-1}V_0$	$(1-\alpha)^{n-1}\alpha V_0$	$(1-\alpha)^n V_0$

For a selected value of α, there is no guarantee that the book value at the end of its useful life period will be the salvage value. Furthermore, the book value is not allowed at the end of any year to drop below the salvage value. Therefore, a common practice is to use the *combined depreciation method*, which is a hybrid between the declining-balance and the straight-line methods. In the combined depreciation method, the declining-balance depreciation is used until the last year of the project in which the last depreciation charge is calculated using the linear method. The book value at the end of the recovery period is the salvage value; that is,

[2.21] $$V_N = V_s$$

Therefore, the depreciation charge for the last year:

[2.22] $$d_N = V_{N-1} - V_s$$

But what if the declining-balance method yielded a book value at the end of a certain year n, which is less than the salvage value? In such cases, the combined depreciation method requires that the depreciation scheme be switched from the declining-balance method to the linear method, starting from the end of year n-1 to the end of the project. Nonetheless, the book value at the end of year n-1 (beginning of year n) is V_{n-1}, and the remaining period from the end of year n-1 to the end of the project is $N - (n-1)$. Therefore, using the linear method, the annual depreciation charge for the year n and for the ensuing years until the end of the project is given by:

[2.23] $$d_n = \frac{V_{n-1} - V_s}{N - (n-1)}$$

and

[2.24] $$d_n = d_{n+1} = \ldots = d_N$$

MODIFIED ACCELERATED COST RECOVERY SYSTEM (MACRS)

Although the linear depreciation method is widely used in conceptual design studies, the **Modified Accelerated Cost Recovery System (MACRS)** depreciation method is used in the United States for preparing corporate tax forms and calculating federal income taxes. The method is described by the U.S. Internal Revenue Service (www.IRS.gov), especially the IRS publication 946 (IRS, 2009). Here are some key highlights of a MACRS category referred to as the MACRS general depreciation system (GDS):

- MACRS uses a combination of declining-balance and linear depreciation methods, starting with the declining-balance method and switching to the linear method when the annual depreciation charge for the remainder of the depreciable capital using the linear method is higher than that calculated by the declining-balance method.
- The IRS defines a "class life" for each type of equipment/industry. For each class life, a certain recovery period (over which the depreciation calculations are made) is allowed. For many pieces of equipment involved in the chemical process industries, the class life is 10 years and the recovery period is 7 years.
- For equipment with class lives less than or equal to 10 years, the DDB method is used, and for those with class lives greater than 10 years, the 150 percent declining-balance method is used.

Example 2-9 Depreciation using the double declining-balance method

A new process equipment has an initial cost of $900,000 and a projected salvage value of $90,000. The recovery period for the equipment is taken as 10 years. It is desired to perform the following depreciation-related calculations.
1. Using the double declining-balance (DDB) method, calculate the annual depreciation charges and the book values over the equipment life.
2. Compare the book value of the equipment over the useful life period using the DDB method (with linear adjustment in the last year using the combined depreciation method) versus the straight-line method.
3. If the salvage value of the equipment is projected to be $130,000, recalculate the depreciation charges using the DDB method.

SOLUTION
1. First, let us calculate the depreciation charges and book value using only the DDB method. The fixed depreciation fraction is $2/10 = 0.2$. The first depreciation charge is $0.2*900,000 = \$180,000$. The book value at the end of the first year is $900,000 - 180,000 = \$720,000$. The same calculations are repeated for the rest of the 10 years, and the results are shown by Table 2.15. The book value at the end of the 10th year is $96,637, which does not match the salvage value. Therefore, the combined depreciation method is used by switching to the linear method in the last year for which the depreciation charge is calculated from Eq. 2.22:

$$d_{10} = 120,796 - 90,000 = \$30,796$$

Hence, Table 2.15 is revised to show the change in the last year, as illustrated by Table 2.16.

2. For the linear depreciation method, the depreciation charge is:

$$d = \frac{900,000 - 90,000}{10} = \$81,000/\text{yr}$$

Therefore, the plot of the book value versus time is a straight line with a slope of $81,000/yr. Figure 2.7 shows the results of the linear and the combined (DDB and linear in last year) methods. As the plots show, the book value using the DDB method is lower than that of the linear method, which confirms that in the accelerated depreciation schemes, higher depreciation charges are set aside in the earlier years of the project. Eventually, both methods reach the salvage value at the end of the useful life period.

3. Let us now consider the case when the salvage value is $130,000. As can be seen from Table 2.17, if the DDB method is used, the book value at the end of the 9th year is 120,976, which is less than the salvage value. Therefore, the combined depreciation method is used by applying linear depreciation to the last two years with an annual depreciation charge (according to Eq. 2.23) being:

$$d_9 = d_{10} = \frac{150,995 - 130,000}{10 - 8} = \$10,498/\text{yr}$$

The book values are adjusted for the last two years, and Table 2.17 shows the results.

TABLE 2-16 Results of the DDB Method (with Linear Adjustment Using the Combined Depreciation Method) for Example 2.9

Recovery Year	Annual Depreciation Charge ($)	Book Value at End of Year ($)
0		900,000
1	180,000	720,000
2	144,000	576,000
3	115,200	460,800
4	92,160	368,640
5	73,728	294,912
6	58,982	235,930
7	47,186	188,744
8	37,749	150,995
9	30,199	120,796
10	30,796	90,000

TABLE 2-15 DDB Results for Example 2.9

Recovery Year	Annual Depreciation Charge ($)	Book Value at End of Year ($)
0		900,000
1	180,000	720,000
2	144,000	576,000
3	115,200	460,800
4	92,160	368,640
5	73,728	294,912
6	58,982	235,930
7	47,186	188,744
8	37,749	150,995
9	30,199	120,796
10	24,159	96,637

TABLE 2-17 Results of the DDB Method (with Linear Adjustment Using the Combined Depreciation Method) for Example 2.9, When the Salvage Value Is $130,00

Recovery Year	Annual Depreciation Charge ($)	Book Value at End of Year ($)
0		900,000
1	180,000	720,000
2	144,000	576,000
3	115,200	460,800
4	92,160	368,640
5	73,728	294,912
6	58,982	235,930
7	47,186	188,744
8	37,749	150,995
9	10,498	140,498
10	10,498	130,000

(Continued)

Example 2-9 Depreciation using the double declining-balance method (*Continued*)

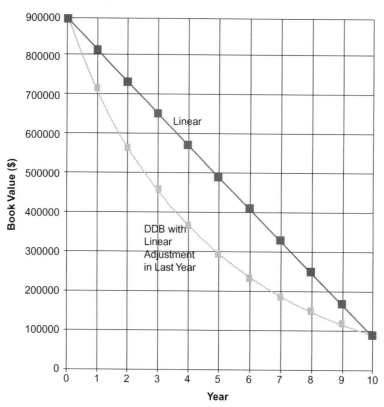

FIGURE 2-7 The book value versus recovery year using two depreciation methods (DDB and linear) for Example 2.9.

Example 2-10 Calculations of the MACRS method

To illustrate the MACRS method, let us follow the calculations for a unit with an initial cost of $100.00 and a recovery period of 7 years, for which the DDB method is used. In the first year, the DDB method gives an annual depreciation charge of $(2/7)*100 = \$28.57$. However, the MACRS method permits only the depreciation of half of that first year. Therefore, the first annual depreciation charge using the DDB method is $14.29. If the linear depreciation is used, the depreciation charge is $(100.00/7) = \$14.29$ for the full first year. Therefore, the half-year DBB depreciation charge in the first year is used. The book value at the end of the first year is $100.00 - 14.29 = \$85.71$. For the second year, the DDB method gives an annual depreciation charge of $(2/7)*85.71 = \$24.49$. Because depreciation in the first year was carried out over 0.5 year, then for the linear method, the remaining time is 6.5 years. The annual depreciation charge for the book value is $(85.71/6.5) = \$11.30$. Because the DDB method yields a larger depreciation charge, it is selected. Hence, the book value at the end of the second year is $85.71 - 24.49 = \$61.22$. In the third year, the depreciation charge using the DDB method is $(2/7)*61.22 = \$17.49$, while the linear depreciation of the book value for the remaining 5.5 years is $(61.22/5.5) = \$11.13$. The DDB method still yields a larger depreciation charge than the linear method. Therefore, the book value at the end of the third year is $61.22 - 17.49 = \$43.73$. In the fourth year, the DDB-based depreciation charge is $(2/7)*43.73 = \$12.49$, while the linear-depreciation charge is $(43.73/4.5) = \$9.72$. Therefore, the book value at the end of the fourth year is $43.73 - 12.49 = \$31.24$. In the fifth year, the depreciation charge calculated through the DDB method is $(2/7)*31.24 = \$8.93$ and the linear depreciation gives the same annual charge of $(31.24/3.5) = \$8.93$. Hence, the book value at the end of the fifth year is $31.24 - 8.93 = \$22.31$. In the sixth year, the DDB-based depreciation charge is $(2/7)*22.31 = \$6.37$, while the linear-depreciation charge is $(22.31/2.5) = \$8.92$. Therefore, the switch is made from the DDB method to the linear depreciation method. The book value at the end of the sixth year is $22.31 - 8.92 = \$13.39$. The linear depreciation for the seventh year is $(13.39/1.5) = \$8.93$, and the book value at the end of the seventh year is $13.39 - 8.93 = 4.46$. There is still 0.5 year remaining in the recovery period. Because MACRS reaches full depreciation of the capital by the end of the recovery period, then the annual depreciation charge in the last year is $4.46 (equal to the book value at the end of the seventh year). For general applicability, Table 2.18 lists the results of this example on a percentage basis of the annual depreciation charge to the initial value of the property.

TABLE 2-18 Percentage of Capital Depreciated Annually over a 7-Year Recovery Period Using the MACRS Method

Recovery Year n	Depreciation Rate $\frac{d_n}{V_0}*100\%$
1 (0.5 year)	14.29
2	24.49
3	17.49
4	12.49
5	8.93
6	8.92
7	8.93
8 (0.5 year)	4.46

- A half-year convention is used initially and toward the end of class life for the declining-balance calculations. This means that, in the first year of the recovery period, when the declining-balance method is used, only 50 percent of the depreciation charges can be made (assuming that the operation was started in the middle of the tax year or that the property was half productive in the first year). Because only half of the first year was used, the half-year convention leads to the extension of the depreciation period by a half year toward the end of the calculation. For instance, if a seven-year recovery period is used, then half a year is used initially followed by six full years then half a year toward the end. Therefore, although the recovery period is seven years, the depreciation calculations are spread over eight years.
- Depreciation is made for the full initial value of the property while disregarding the salvage value.

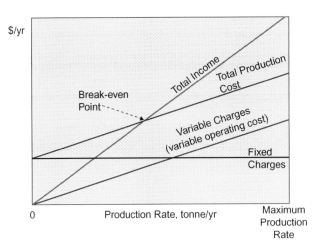

FIGURE 2-8 Break-even analysis.

BREAK-EVEN ANALYSIS

As mentioned earlier, the total production cost includes two expenses:

- Fixed charges: These are expenditures that are independent of the production capacity or the plant activity. Examples include depreciation (or annualized fixed cost "AFC") and fixed operating costs (for example, salaries, wages, property taxes, property insurance) that are independent of the production levels. Hence,

[2.25]
$$\text{Annual fixed charges} = \text{AFC (or annual depreciation charges)} + \text{fixed annual operating costs}$$
(for example, salaries, wages, property taxes, insurance)

- Variable charges: These are costs that are dependent on the extent of production such as raw materials, labor, material utilities, energy utilities, waste handling, royalties, and so on. Such charges can be conveniently represented by the variable annual operating cost (AOC). Typically, the variable AOC is linearly proportional to the production capacity.

On the other hand, the project will result in revenues or savings for the plant. In a production project, the revenues may correspond to the sales of the products and byproducts. In resource conservation projects, the savings are calculated based on the reduction in the consumption of material and energy utilities of the process. Some environmental project may garner revenue as a result of conservation of natural resources or subsidies (for example, credit for the reduction in greenhouse gas emissions).

The **break-even analysis** is intended to determine the conditions under which the total production cost is equal to the process revenues. Figure 2.8 shows a graphical representation of the process expenditures and revenue versus the production flow rate. First, the annual fixed charges are represented as a horizontal line because they are essentially independent of the production level. Next, the variable charges are plotted over the production range. In many cases, the variable charges are linearly proportional to the production level, starting with zero when there is no production and reaching its maximum value when the production rate is maximum. The slope of the variable charges is the operating cost per unit production (for example, $/tonne product). The fixed and variable charges are added through superposition by moving the variable charges line parallel to itself and starting it from the value of the fixed charges. The result is the line representing the total production cost of the process. On the other hand, the total income is represented by the line starting from the origin and reaching the maximum income when the production rate is at its maximum rate. The slope of the total income line is the revenue per unit production (for example, selling price $/tonne of product). If there are other sources of income (for example, subsidies), then:

[2.26]
$$\text{Annual income} = \text{annual sales} + \text{other revenues}$$

The annual income is the sum of annual sales and the other revenues. The intersection of the total production cost and the total income is referred to as the **break-even point** (BEP). Therefore, at the BEP:

[2.27]
Annual fixed charges + Variable AOC = Annual income

Below the BEP, the total production cost is greater than the total income and the process loses money. Conversely, above the BEP, the process makes a profit. Clearly, a company would like to have a BEP production rate to be as low as possible. Additionally, the company should operate higher than the BEP and to endeavor to operate at production rates that are as high as possible provided that there is a market for the product. The reality, however, is that it is unlikely for a process to operate at maximum production capacity. Operational problems as well as preventive and responsive maintenance activities result in *turnaround periods* during which the process is shut down

Example 2-11 Break-even production rate

A process has an FCI of $310 MM. The recovery period for depreciating the FCI is taken to be 10 years. The salvage value of the process is $10 MM. In addition to the AFC (depreciation), other fixed charges include $20 MM/yr of fixed operating costs for the process (property taxes, insurance, production-independent salaries, and so on). The variable operating cost of the process is $100/tonne. Maximum production capacity of the process is 200,000 tonne/yr. The selling price of the product is $485/tonne. What is the break-even production rate (tonne/yr)?

SOLUTION

The AFC is calculated by determining the annual depreciation charge via Eq. 2.18:

$$AFC = \frac{300*10^6 - 10*10^6}{10}$$
$$= \$30\text{MM/yr}$$

Therefore,

The annual fixed charges of the process = AFC + Fixed operating cost

$$= 30*10^6 + 20*10^6$$
$$= \$50\text{MM/yr}$$

The variable annual operating cost at maximum production capacity is obtained as follows:

$$AOC^{max} = \$100/\text{tonne}*200,000\text{ tonne/yr}$$
$$= \$20\text{MM/yr}$$

The AOC is represented versus the production rate as a straight line extending from an annual cost of zero when there is no production to $20 MM/yr when the production rate is at 200,000 tonne/yr. To add the AOC to the annual fixed charges of the process, superposition is used by moving the AOC line up to start from the fixed charges of $50 MM/yr.

The value of the annual sales at maximum production capacity is calculated as follows:

$$AS^{max} = \$485/\text{tonne}*200,000\text{ tonne/yr}$$
$$= \$97\text{MM/yr}$$

Figure 2.9 represents the break-even analysis and identifies the BEP as the point of intersection between annual production costs and annual sales at a production rate of approximately 130,000 tonne/yr.

The BEP can also be determined algebraically. Let the annual production rate at the BEP be X_{BEP} tonne/yr. The AOC and the annual sales at the BEP can be expressed in terms of X_{BEP} as $100*X_{BEP}$ and $485*X_{BEP}$ respectively. At the BEP, the annual expenses are equal to the annual sales. Therefore,

$$50*10^6 + 100*X_{BEP} = 485*X_{BEP}$$

Solving for X_{BEP}, we get

$$X_{BEP} = 129,870 \text{ tonne/yr}$$

Therefore, the on-stream efficiency at the BEP is $(129,870/200,000)*100\% = 65\%$.

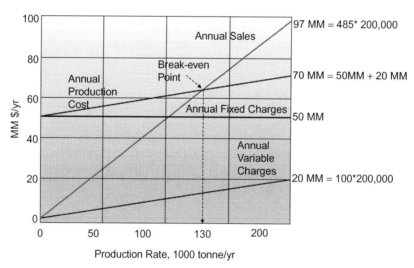

FIGURE 2-9 Locating the BEP for Example 2.11.

partially or completely. Furthermore, market conditions may necessitate temporary reduction in the production rates to maintain a certain selling price or to stay within the demand level. These factors lead to the concept of *on-stream efficiency* defined as:

[2.28]
$$\text{On-stream efficiency} = \left(\frac{\text{Actual operating hours of the process per year}}{8760}\right)*100\%$$

where 8760 is the total number of hours per year (24*365). Alternatively, the following definition may be used:

[2.29]
$$\text{On-stream efficiency} = \left(\frac{\text{Actual annual production}}{\text{maximum production}}\right)*100\%$$

An on-stream efficiency of 95 percent or higher is typically considered to represent operational excellence, although lower efficiencies (for example, 85 percent or higher) are more realistically targeted.

Example 2-12 Using the break-even analysis for product pricing

A company is launching a new product P. The depreciable FCI of the process is $450 MM and the salvage value is $50 MM. A 10-year linear depreciation scheme is to be used to calculate the AFC. Other fixed charges for the process (property taxes, insurance, and so on) are $5 MM/yr. The maximum production capacity of the process is 100,000 tonne/yr of product P. The operating cost of the process is $110/tonne. It is desired to break even at a production rate of 40 percent of the maximum process capacity. What should be the selling price of the product?

SOLUTION

Using Eq. 2.18, the AFC is calculated:

$$AFC = \frac{450*10^6 - 50*10^6}{10}$$
$$= \$40 MM/yr$$

The annual fixed charges of the process

$$= 40*10^6 + 5*10^6$$
$$= \$45 MM/yr$$

At the BEP, the production rate is 0.4*100,000 = 40,000 tonne/yr. Therefore,

$$AOC \text{ at the BEP} = \$110/\text{tonne} * 40,000 \text{ tonne/yr}$$
$$= \$4.4 MM/yr$$

Let C^P be the selling price ($/tonne) of product P. Therefore, the annual sales at the BEP is $40,000*C^P$.

At the BEP,

$$45,000,000 + 4,400,000 = 40,000*C^P$$

Hence,

$$C^P = \$1235/\text{tonne}$$

Example 2-13 Sensitivity analysis via the break-even analysis

In the previous example, it is desired to study the impact of potential fluctuations in the selling price of product P on the on-stream efficiency at the BEP. Develop a plot showing this sensitivity analysis for a product selling in the price range of $600 to $1600/tonne.

SOLUTION

Designating the annual production rate at the BEP by X_{BEP} tonne/yr, we can write the break-even analysis as:

$$45*10^6 + 110*X_{BEP} = C^P * X_{BEP}$$

For each value of C^P, a corresponding X_{BEP} is obtained. The on-stream efficiency at the BEP is calculated by dividing X_{BEP} by the maximum production rate of 100,000 tonne/yr. Figure 2.10 shows these results. It is worth pointing out that as the selling price decreases to $600/tonne, the process must have an on-stream efficiency of 92 percent to break even. As will be discussed later in the chapter, higher production will be needed to make an acceptable level of profitability. Therefore, a selling price of $600/tonne is unacceptable. On the other hand, as the selling price increases to $1600/tonne, the process has a comfortable on-stream efficiency of 30 percent.

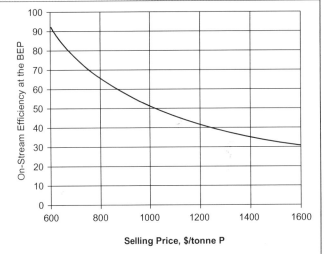

FIGURE 2-10 Sensitivity analysis of on-stream efficiency versus selling price for Example 2.13.

The break-even analysis is relatively simple to conduct and provides valuable insights. In addition to determining the production rate below which the operation loses money, it can also be used to determine the sensitivity of the process to variations (for example, in production rate, in unit selling price, in unit operating cost, and so on). The analysis can also be used to determine appropriate selling prices of new products so as to stay above the BEP.

TIME-VALUE OF MONEY

Over time, there is always a change in the cost of goods and services. As has been mentioned earlier, cost indices can be used to account for the time-based variation in the cost of equipment and plants. There are also more general cost indices that address the broader categories of goods and services. A particularly useful example is the Consumer Price Index (CPI), which is published regularly by the U.S. Department of Labor, Bureau of Labor Statistics. The CPI describes the variations in the prices of a cross-section of goods and services purchased by urban households such as foods, beverages, housing, apparel, transportation, medical care, recreation, education, communications, and government-charged fees for water, sewage, and other services. The CPI has a value of 100 for a 1982–1984 reference base. The ratio of CPI at

Example 2-14 Accounting for plant size in determining the minimum product price via the break-even analysis

Hydrogen is to be produced from biomass. The process involves several units including gasification to syngas (containing primarily CO, H_2, CO_2, hydrocarbons, and other gases such as NH_3 and H_2S) and tars, steam reforming ($C_nH_m + nH_2O = (n + m/2)H_2 + nCO$) and water-gas shift ($CO + H_2O = CO_2 + H_2$), gas cleanup and conditioning, and hydrogen separation. The following data are available for a 2000 dry tonne biomass/day (mostly adapted with revisions and updates from Spath et al., 2005):

Hydrogen yield = 70 kg H_2 produced from the process/tonne dry biomass fed to the process

Delivered equipment cost = $60 MM
Fixed annual operating costs:
 Salaries and wages = $2.1 MM/yr
 Plant overheads = $2.0 MM/yr
 Maintenance = $3.1 MM/yr
 Insurance and taxes = $3.1 MM/yr
Variable operating costs (expressed as $/kg of produced H_2):
 Biomass feedstock = $0.75/kg of produced H_2
 Other raw materials = $0.11/kg of produced H_2
 Catalysts = $0.05/kg of produced H_2
 Waste disposal = $0.03/kg of produced H_2
 Electricity = $0.08/kg of produced H_2

Assuming a 100 percent on-stream efficiency and a 10-year linear depreciation with no salvage value,

1. Calculate the minimum selling price of hydrogen needed to break even for the 2000-tonne biomass/day plant when it runs for full capacity with a 100 percent on-stream efficiency.
2. If the plant size is unknown, develop a relationship between the plant size and the minimum selling price of hydrogen. Consider varying the plant size from 2000 tonne/day to 20,000 tonne/day. For simplicity, assume that:
   - the FCI is related to capacity through the six-tenths-factor rule.
   - the fixed operating costs (expressed on an annual basis as $/yr) will remain roughly the same regardless of the plant size.
   - the variable operating costs (expressed as $/kg of produced hydrogen) will remain the same (which means that the variable annual operating cost will change linearly with the production rate).
3. As the plant size increases, biomass will have to be transported from farther area. In such cases, the transportation cost should be included. The first 2000 tonne/day of biomass are locally available and, therefore, the feedstock cost ($0.75/kg of produced H_2) includes the transportation cost. Above 2000 tonne/day, the biomass will have to be hauled from a long distance. The higher the biomass demand, the larger the cost of transportation per tonne of biomass. For flow rates greater than 2000 tonne/day, the transportation cost is given by:

Annual transportation cost of biomass ($/yr) = 0.4*(Flow rate of biomass in tonne/day − 2000)*(Flow rate of biomass in tonne/day)

Resolve part (b) while accounting for the transportation cost.

SOLUTION

1. For the plant processing 2000 tonne dry biomass/day:
Using the revised Lang factor from Table 2.5 for a solid-fluid process:

$$FCI = 4.3*60 = \$258 MM$$

Applying a linear depreciation scheme over 10 years, we get:

$$\text{Annual depreciation charges} = \frac{258}{10} = \$25.8 MM$$

From the given data,

$$\text{Fixed annual operating cost} = 2.1 + 2.0 + 3.1 + 3.1 = \$10.3 MM/yr$$

Because the variable operating costs are given per kg of produced H_2, we need to calculate the annual production rate of H_2. For a hydrogen yield of 70 kg produced H_2/tonne of dry biomass fed to process:

Annual production rate of H_2 = 70 kg H_2/tonne biomass*2000 tonne biomass/day*365 days/yr
 = 51.1 MM kg H_2/yr

But,

Variable operating costs for the process = 0.75 + 0.11 + 0.05 + 0.03 + 0.08
 = $1.02/kg of produced H_2

Therefore, at full production rate of 51.1 MM kg H_2/yr:

Variable annual operating cost = $1.02/kg of produced H_2 * 51.1 MM kg H_2/yr
 = $52.12 MM/yr

Let $C^{hydrogen\ min}$ be the minimum selling price of hydrogen ($/kg) needed to break even at full production rate. Therefore, applying the break-even equation:

$$C^{hydrogen\ min} * 51.1*10^6 = (25.8*10^6 + 10.3*10^6) + 52.12*10^6$$

that is,

$$C^{hydrogen\ min} = \$1.73/kg\ H_2$$

Alternatively, the minimum selling price can be calculated via the TAC:

$$TAC = AFC + AOC$$

But,

AOC = Fixed annual operating cost + Variable annual operating cost

Therefore, at full production capacity:

$$AOC = 10.30 + 51.12 = \$61.42 MM/yr$$

and

$$TAC = 25.80 + 61.42 = \$87.22/yr$$

The value of the TAC represents that total annual production cost of 51.01 MM kg H_2 per year.
Thus,

$$C^{hydrogen\ min} = \frac{88.22*10^6}{51.1*10^6} = \$1.73/kg\ H_2$$

2. Let us write the FCI as a function of plant size expressed in terms of flow rate of biomass feedstock. Using the sixth-tenths-factor rule:

$$FCI (\$) = 258*10^6 *(\text{Flow rate of biomass in tonne per day}/2000)^{0.6}$$
$$= 2.698*10^6 *(\text{Flow rate of biomass in tonne/day})^{0.6}$$

Hence,

Annual depreciation ($/yr) = $0.2698*10^6$ *(Flow rate of biomass in tonne/day)$^{0.6}$

and

Variable annual operating cost ($/yr) = 1.02*(Product flow rate in kg of produced H_2 per year)

Because the hydrogen yield is 70 kg H_2 produced from the process/tonne dry biomass fed to the process, then

Variable annual operating cost ($/yr) = 1.02*(70 kg H_2/tonne biomass)*(Flow rate of biomass tonne biomass/day)*(365 day/yr)
 = 26,061*Flow rate of biomass (tonne/day)

(Continued)

Example 2-14 Accounting for plant size in determining the minimum product price via the break-even analysis (Continued)

Also, for a minimum selling price of hydrogen:

Annual sales ($/yr) = $C^{hydrogen\,min}$ ($/kg H_2)*Product flow rate in kg of produced H_2 per year

= $C^{hydrogen\,min}$ *(70kg H_2/tonne biomass)*(Flow rate of biomass tonne biomass/day)*(365 day/yr)

= 25,550*$C^{hydrogen\,min}$*Flow rate of biomass (tonne/day)

Applying the break-even analysis, we get:

25,550*$C^{hydrogen\,min}$*Flow rate of biomass (tonne/day) = 0.2698*10^6*(Flow rate of biomass in tonne/day)$^{0.6}$ + 26,061*Flow rate of biomass (tonne/day) + 10.3*10^6

For a given value of the biomass flow rate, the minimum selling price of hydrogen can be determined. For instance, when the flow rate of the biomass feedstock is 2000 tonne/day, we get:

25,550*$C^{hydrogen\,min}$*2000 = 0.2698*10^6*(2000)$^{0.6}$ + 26,061*2000 + 10.3*10^6

that is,

$C^{hydrogen\,min}$ = $1.73/kg H_2

which is the same answer obtained in part (a). When the flow rate of the biomass feedstock is varied from 2000 to 20,000 tonne/yr, the resulting sensitivity analysis can be plotted as shown by Fig. 2.11. The reduction in the hydrogen selling price per kg with the increase in the plant size is attributed to the economy of scale.

3. When the transportation cost is added, the break-even equation becomes:

25,550*$C^{hydrogen\,min}$*Flow rate of biomass (tonne/day) = 0.2698*10^6*(Flow rate of biomass in tonne/day)$^{0.6}$ + 26,061*Flow rate of biomass (tonne/day) + 0.4*(Flow rate of biomass in tonne/day − 2000)*(Flow rate of biomass in tonne/day)

Several values of the biomass flow rate (between 2000 and 20,000 tonne/day) are used, and the break-even equation is solved to get the minimum selling price of hydrogen. The sensitivity analysis is shown by Fig. 2.12. Although the increase in the plant capacity leads to a reduction in the depreciation charges per kg of product (due to the economy of scale), the transportation cost per kg of product increases (because the material has to be hauled from farther areas). This tradeoff causes the nonmonotonic behavior shown by Fig. 2.12, leading to a minimum selling price of $1.45/kg H_2 when the biomass flow rate is 9581 tonne/day.

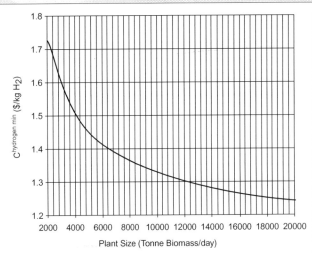

FIGURE 2-11 Minimum selling price of hydrogen as a function of plant size (without transportation cost).

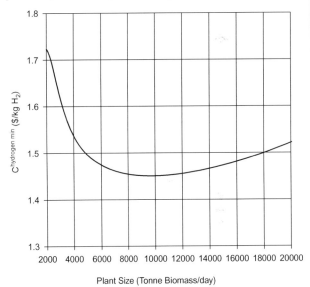

FIGURE 2-12 Minimum selling price of hydrogen as a function of plant size (with transportation cost).

two different times provides an estimate for the extent of relative change in price of goods and services. Hence,

[2.30] $$Price_{t2} = Price_{t1}\left(\frac{CPI\ at\ time\ t2}{CPI\ at\ time\ t1}\right)$$

where $Price_{t1}$ and $Price_{t2}$ are the purchased prices of goods and services at times t1 and t2, respectively. For instance, for each $100 spent on goods and services during the period between 1982 and 1984 (with a CPI value of 100), the average price for the same goods and services in 2009 (when the CPI value was 214.5) is $214.5. This rise in price reflects a decline in the purchasing power of the money. Such an increase in the average prices of goods and services over a period of time is referred to as **inflation**. Conversely, deflation corresponds to a decrease in the average prices of goods and services over a period of time. A useful measure of inflation (or deflation) is the annual **inflation rate**, which is defined as annualized percentage change in the value of CPI over two consecutive years:

[2.31] $$\text{Inflation rate in year n} = \left(\frac{CPI\ in\ year\ n - CPI\ in\ year\ n-1}{CPI\ in\ year\ n-1}\right)*100\%$$

For example, for the years 2008 and 2007, when the CPIs were 215.303 and 207.342, the inflation rate for 2008 is (215.303 − 207.342)/207.342*100% = 3.8%.

A negative inflation rate corresponds to deflation. Figures 2.13 and 2.14 illustrate the values of the U.S. CPI and annual inflation rates over the period between 1954 and 2009. As shown by these figures, the general trend is an increase in the value of the CPI, which designates inflation. Nonetheless, in a few cases, the annual inflation is negative (for example, −0.4% in both 1955 and 2009), which corresponds to deflation.

The U.S. Department of Labor, Bureau of Labor Statistics, publishes another useful index called the Producer Prices Index for Chemicals and Allied Products. It reflects the time-based changes in the average production cost of various chemicals. Figure 2.15 represents recent data for this index.

The foregoing discussion highlights the importance of accounting for the time-value of money in cost estimation studies. The next sections offer an overview of the following basic concepts in time-value of money involved in industrial economic studies:

1. Compound interest
2. Cash flow diagrams
3. Annuities

COMPOUND INTEREST OF A SINGLE PAYMENT

Consider an amount of money deposited as a single payment in a bank for a number of years or a loan to be paid off over a number of years. Let:

P = Principal (original amount of the deposit or initial amount of the loan)

N = Number of interest periods (for example, years)

i = interest rate based on a unit interest period (for example, annual interest rate)

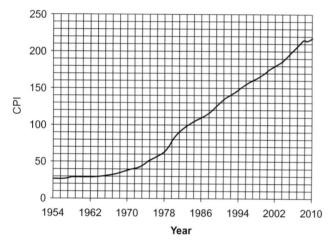

FIGURE 2-13 Time-based variation in the U.S. Consumer Price Index. *Source: Based on data published by the U.S. Department of Labor, Bureau of Labor Statistics, All Urban Consumers, All Items, ftp://ftp.bls.gov/pub/special.requests/cpi/cpiai.txt.*

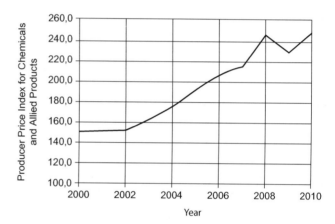

FIGURE 2-15 Time-based variation in the U.S. Producer Price Index for Chemicals and Allied Products. *Source: Based on data published by the U.S. Department of Labor, Bureau of Labor Statistics, http://data.bls.gov/PDQ/servlet/SurveyOutputServlet?series_id 5 WPU06&data_tool 5 XGtable.*

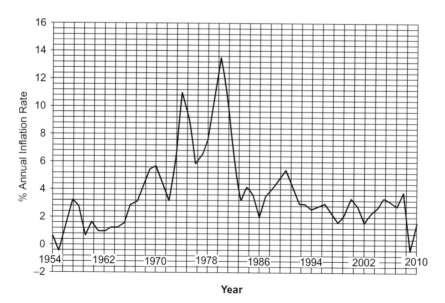

FIGURE 2-14 Time-based variation in the U.S. annual inflation rate. *Source: Based on data published by the U.S. Department of Labor, Bureau of Labor Statistics, ftp://ftp.bls.gov/pub/special.requests/cpi/cpiai.txt.*

Example 2-15 Future value of a single payment

A graduating senior has deposited her first-month salary of $5000 in a bank at an annual interest rate of 5 percent. How much will she have in the bank after 50 years?

SOLUTION

P = $5000
i = 0.05
N = 50 years

Substituting into Eq. 2.32, we have:

$$F = 5000*(1 + 0.05)^{50}$$
$$= \$57,337$$

Example 2-16 Compound interest with a monthly interest rate

A credit card company charges an interest rate of 1.5 percent compounded monthly. What is the equivalent annual interest rate?

SOLUTION

Let us take a basis of P = $1.

$$i = 0.015 \text{ (compounded monthly)}$$

For one year, N = 12 months.
Therefore, after one year:

$$F = (1 + 0.015)^{12}$$
$$= 1.1956$$

But,

$$F = 1 + i_{equivalent\ annual}$$

that is,

$$i_{equivalent\ annual} = 0.1956$$

Therefore, the equivalent annual interest rate is 19.56 percent (not 1.5*12 = 18 percent).

Example 2-17 Principal of a single payment

A graduating senior wishes to have $50,000 in her bank account after 50 years by making a single deposit now and leaving it in her account for 50 years while earning an annual interest rate of 5 percent. How much does she have to deposit now?

SOLUTION

F = $50,000
i = 0.05
N = 50 years

Substituting into Eq. 2.33, we have:

$$P = 50,000*(1 + 0.05)^{-50}$$
$$= \$4360$$

The **compound interest** corresponds to the transaction where an interest rate is charged to the principal, then for each interest period, the interest is charged to the sum of the principal and the accumulated interest charges. The following scheme describes the compound interest calculations where the future value (future worth) of the compounded amount, F, at the end of each period is the sum of the compounded interest at the end of the previous period plus the interest earned during that period.

Period	Interest earned during the period	Future Value at the end of each period
1	Pi	$P + Pi = P(1 + i)$
2	$P(1 + i)i$	$P(1 + i) + P(1 + i)i = P(1 + i)^2$
3	$P(1 + i)^2 i$	$P(1 + i)^2 + P(1 + i)^2 i = P(1 + i)^3$
N	$P(i + i)^{N-1} i$	$P(1 + i)^{N-1} + P(1 + i)^{N-1} i = P(1 + i)^N$

Therefore, the general formula for the future value after N periods is:

$$[2.32] \qquad F = P(1 + i)^N$$

It is worth noting that when i is expressed as an annual percentage rate, then N should be in years. Otherwise, if i is given based on a different time period, then N should have the same units of that time period.

By rearranging Eq. 2.32, we get:

$$[2.33] \qquad P = \frac{F}{(1 + i)^N}$$

which may be used to calculate:

- The principal (P) that needs to be deposited as a single payment to accumulate a future value F after N periods at an interest rate i

Example 2-18 Present worth of a future value

As mentioned earlier, the value of money tends to erode over time because of inflation. Suppose that after 50 years, you will have a fortune of $1 MM. Assuming that the average annual inflation rate will be 3 percent for the next 50 years, how much is the million dollars worth in terms of today's dollars?

SOLUTION

$F = \$1.0*10^6$
$i = 0.03$
$N = 50$ years

Substituting into Eq. 2.33, we have:

$$P = 10^6 * (1 + 0.03)^{-50}$$
$$= \$228,107$$

(By today's standards, you will not be a millionaire but you may be called "almost a quarter millionaire"!)

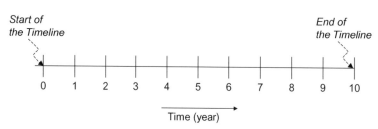

FIGURE 2-16 Constructing the timeline of a cash flow diagram.

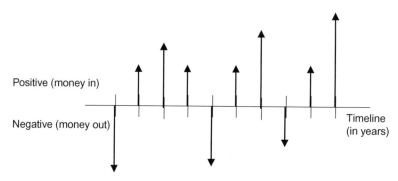

FIGURE 2-17 A typical cash flow diagram.

or

- The present value or present worth (P) of a future value (F) that will be available after N years with an anticipated average annual inflation rate i for the next N years. As such, Eq. 2.33 can be used to bring future values *back to the present*.

CASH FLOW DIAGRAM

The cash flow diagram is a useful tool in representing financial transactions of a period of time and in accounting for the time-value of money. First, a horizontal axis is drawn to represent the timeline of the financial transactions (see Fig. 2.16). The increments on the timeline scale correspond to the targeted periods (for example, years or months). Then, vertical arrows are drawn at different times to account for cash flows (financial inputs and outputs) using the following convention:

- Downward-pointing arrows below the timeline represent costs/expenditures/withdrawals (money out "−")
- Upward-pointing arrows above the timeline represent incomes/savings/deposits (money in "+")

Figure 2.17 is a schematic representation of a typical cash flow diagram.

Any cash flow F_N at a future time, N, can be converted to present value using Eq. 2.33.

[2.33] $$P = \frac{F}{(1 + i)^N}$$

The cash flow diagram is particularly useful in evaluating the different revenues and expenditures over the life of the project.

ANNUITIES

An **annuity** is a series of constant payments or withdrawals made at equal time intervals. It is commonly used in the payment of FCI over a period of time, home mortgages, savings as part of a retirement plan, life insurance, and so on. Let:

A = Uniform annuity installment (for example, constant deposits or withdrawals) made at the end of each period as shown by Fig. 2.18.
N = Number of periods (for example, years)

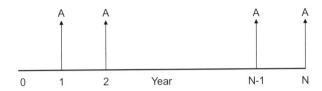

a. An Annuity Scheme with Deposits

b. An Annuity Scheme with Withdrawals

FIGURE 2-18 Cash flow diagrams for annuity schemes involving uniform (a) deposits and (b) withdrawals at the end of each year.

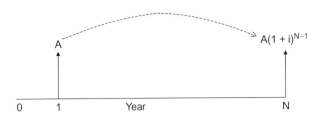

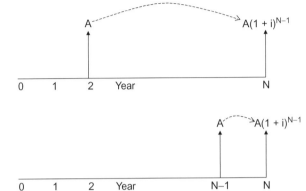

FIGURE 2-19 Conversion of annuity installments to future worths.

i = Interest rate based on a unit interest period (for example, annual interest rate)

Let us consider the first annuity payment, A, made at the end of the first year. At end of Nth year, this deposit will have stayed for $N - 1$ years, earning compound interest. Therefore, after N years, the future value of the annuity deposited at end of first year will be $A(1 + i)^{N-1}$. Similarly, the second annuity deposit will have a future value of $A(1 + i)^{N-2}$ by the end of the Nth year. The same approach is adopted for the remaining years as shown by Fig. 2.19. The deposit made by the end of year $N - 1$ will accrue interest for one year and will have a future value of $A(1 + i)$ at the end of the Nth year. Finally, the last deposit made at the end of the Nth year will have a future value of A. Summing up all these terms, we get the *future sum of the annuity*, F, as:

$$F = A(1 + i)^{N-1} + A(1 + i)^{N-2} \\ + A(1 + i)^{N-3} + \ldots A(1 + i)^2 + A(1 + i) + A$$

Multiplying both sides by $(1 + i)$, we get

$$F(1 + i) = A(1 + i)^N + A(1 + i)^{N-1} \\ + A(1 + i)^{N-2} + \ldots + A(1 + i)^2 + A(1 + i)$$

Subtracting the two equations, we obtain

$$Fi = A(1 + i)^N - A$$

Factoring A and dividing both sides by i, we have

[2.34] $$F = A\left(\frac{(1 + i)^N - 1}{i}\right)$$

which gives the *future sum of the annuity* after N period as a function of the uniform annuity payments, the interest rate, and the N time periods.

It is also beneficial to calculate how much the sum of the annuity is worth in terms of today's dollars. Recalling Eq. 2.33 for converting future values to present values, we get the *present sum of the annuity*, P, as:

[2.35] $$P = A\left(\frac{(1 + i)^N - 1}{i(1 + i)^N}\right)$$

For compounding and discounting expressions, the American National Standards Institute (www.ANSI.org) proposed the following shorthand notations:

For compound interest for N:

(F/P, i, N), which represents $F/P = (1 + i)^N$ and is used to calculate a future value F for a single payment P at an interest rate i after N years.

(P/F, i, N), which corresponds to $P/F = (1 + i)^{-N}$ and is the discounting of a single amount F to calculate the present value P at an interest/discount rate i after N years.

(F/A, i, N), which represents $F/A = \left(\frac{(1 + i)^N - 1}{i}\right)$

and is used to obtain the future worth of multiple amounts of an annuity A recurring N times over N years at a discount rate i

(P/A, i, N), which designates $P/A = \left(\frac{(1 + i)^N - 1}{i(1 + i)^N}\right)$

and is used to obtain the present worth P of multiple amounts of an annuity A recurring N times over N years at a discount rate i

(A/P, i, N), which designates $A/P = \left(\frac{i(1 + i)^N}{(1 + i)^N - 1}\right)$

and is used to calculate the annuity installment A, which is to be repeated over N years to provide a present worth P over N years at a discount rate i.

Example 2-19 Future sum of annuity

Some companies offer to provide contributions matching the employee's investment and invest both in a tax-deferred annuity fund (for example, 401K plans in the United States). A young engineer intends to deposit $12,000 per year in a tax-deferred annuity fund (half of which will be deducted from the employee's salary and the other half will be contributed by the employer). The fund will earn an annual interest rate of 5 percent. How much will the engineer have in the annuity fund after 30 years?

SOLUTION

$A = \$12,000/yr$
$i = 0.05$
$N = 30$ years

Substituting in Eq. 2.34, we get

$$F = 12,000 * \left(\frac{(1 + 0.05)^{30} - 1}{0.05} \right)$$
$$= \$797,266$$

It is worth comparing this result with that of annuity deposits not earning any interest ($30*12 = \$360,000$).

Example 2-20 Present sum of annuity

In the previous example, what is the present sum of the annuity?

SOLUTION

Using Eq. 2.35, we get:

$$P = 12,000 * \left(\frac{(1 + 0.05)^{30} - 1}{0.05(1 + 0.05)^{30}} \right)$$
$$= \$184,469$$

Example 2-21 Using an annuity scheme to calculate AFC

A project has a depreciable FCI of $10 MM. Using an annuity scheme over 10 years with an interest rate of 10 percent, calculate the AFC.

SOLUTION

Using Eq. 2.36,

$$AFC = 10*10^6 \left(\frac{0.1(1 + 0.1)^{10}}{(1 + 0.1)^{10} - 1} \right)$$
$$= \$1,627,454/yr$$

It is interesting to note that if a 10-year linear depreciation scheme were used, the AFC would have been $1 MM/yr. The additional $627,454/yr is the cost of capital because of the interest involved in borrowing. Alternatively, the 10-year annuity calculation of the AFC gives almost the same result as a 6-year linear depreciation scheme. In other words, the cost of capital in this example prolongs paying off the FCI by about four years.

The formula for the present sum of the annuity can also be used to annualize the FCI. As shown earlier, if interest is not included, a linear depreciation scheme may be used to calculate the AFC. When interest is included as the cost of capital and when P represents the FCI of the plant, then annualizing the FCI over N years is equivalent to calculating the uniform annuity installments; that is,

$$[2.36] \quad AFC = FCI \left(\frac{i(1 + i)^N}{(1 + i)^N - 1} \right)$$

In such cases, the ratio of AFC/FCI = is called the *capital recovery factor* or the *annual capital charge ratio*. Hence,

$$[2.37] \quad \text{Capital recovery factor} = \left(\frac{i(1 + i)^N}{(1 + i)^N - 1} \right)$$

PROFITABILTY ANALYSIS

Consider a project that requires a TCI of $50 MM and will produce an annual profit of $10 MM. Is this project profitable enough? Just because a project makes profit does not mean that it is profitable enough. Now consider another project proposed to the same company with an estimated TCI of $70 MM and it will generate $15 MM/yr of profit. Which project should be selected? Although each company employs a number of economic metrics for the economic assessment of proposed projects, there are several profitability criteria that are broadly used in industry. The next sections will answer the questions of assessing the profitability of a project and screening competing alternatives. The following topics will be covered:

- Profitability criteria without the time-value of money:
    * Return on investment
    * Payback period

Example 2-22 Sensitivity of capital recovery ratio

Develop a graph showing the dependence of the capital recovery ratio on the interest rate and the number of years used in annualization of the FCI over the following ranges:

$$0.05 \leq i \leq 0.20$$
$$10 \leq N \leq 20$$

SOLUTION

Equation 2.37 is used to generate the curves shown on Fig. 2.20.

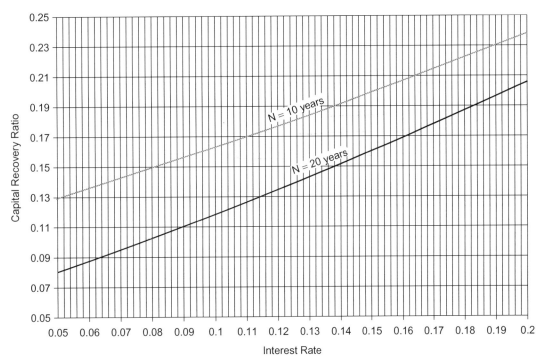

FIGURE 2-20 Sensitivity analysis for the capital recovery ratio of Example 2.22.

- Profitability criteria with time-value of money
  - Net present value
  - Discounted cash flow return on investment
  - Discounted cash flow payback period
- Comparison of alternatives
  - Net present value
  - Annual cost/revenue
  - Total annualized cost
  - Incremental return on investment

PROFITABILITY CRITERIA WITHOUT THE TIME-VALUE OF MONEY

Two criteria are commonly used for assessing the profitability of a project without including interest or the time-value of money: return on investment and payback period.

Return on investment (ROI) is defined as:

$$[2.38] \quad ROI = \frac{\text{Annual Profit}}{\text{TCI}} * 100\%$$

Typically, the ROI uses the annual net (after-tax) profit (sometimes referred to as the net income per year or the annual after-tax cash flow); that is,

$$[2.39] \quad ROI = \frac{\text{Annual Net (After-Tax) Profit}}{\text{TCI}} * 100\%$$

where

[2.40]
$$\text{Annual net (after-tax) profit} = \text{Annual gross profit} - \text{Annual income taxes}$$

and

[2.41] Annual gross profit = Annual income − Annual operating cost

where

[2.42] Annual income = Annual sales + other revenues (for example, subsidies)

In calculating the income taxes, it is important to recall that the annual depreciation charges are shielded from the taxes; that is,

[2.43]
Taxable annual gross profit = Annual gross profit − Depreciation

> **Example 2-23 Calculation of ROI**
>
> A proposed project requires a TCI of $4 MM and provides an annual net (after-tax) profit of $20,000/yr. What is the ROI of the project? Should the company implement it?
>
> **SOLUTION**
>
> $$ROI = (20,000/4,000,000)*100\% = 0.5\%$$
>
> Because of the small ROI, the project is not recommended. The company is better off saving the money in a bank or other financial investments that do not carry the same risk of implementing this industrial project while getting higher returns.

> **Example 2-24 ROI for a mass integration project**
>
> A mass integration project requires piping for recycling and two mass exchangers. The project results in $3.5 MM/yr of annual savings. The project requires a fixed capital investment of $2.0 MM. The land is already available and will not be charged. The working capital investment is taken as 15/85 of the fixed capital investment. The annual operating cost of the project is $0.3 MM/yr. Depreciation is calculated over 10 years with no salvage value. The corporate tax rate for the project is 30 percent of the annual taxable gross profit. What is the ROI of the project?
>
> **SOLUTION**
>
> Annual FCI depreciation (annualized fixed cost) = (2,000,000 − 0.0)/10
> = $200,000/year
>
> Thus,
>
> Total annualized cost = Annualized fixed cost + Annual operating cost
> = 200,000 + 300,000 = $500,000/year
>
> But,
>
> Annual net (after-tax) profit
> = (Annual income − Total annualized cost)*(1 − Tax rate) + Depreciation
>
> Hence,
>
> Annual net (after-tax) profit = (3,500,000 − 500,000)*(1 − 0.30) + 200,000
> = $2,300,000/yr
>
> Total capital investment = Fixed capital investment + Working capital investment
> = 2,000,000 + (15/85)*2,000,000
> = 2,000,000*(100/85) = $2,353,000
>
> Therefore,
>
> ROI = (2,300,000/2,353,000)*100%
> = 98%
>
> This is a very attractive project.

Hence,

[2.44] Annual income taxes = Taxable annual gross profit*Tax rate

Therefore,

[2.45] Annual net (after-tax) profit = Annual gross profit − (Annual gross profit − Depreciation)* Tax rate

By subtracting and adding depreciation, we get

[2.46]

Annual net (after-tax) profit = (Annual gross profit − Depreciation) + Depreciation − (Annual gross profit − Depreciation)*Tax rate
= (Annual gross profit − Depreciation)*(1 − Tax rate) + Depreciation

Hence,

[2.47] Annual net (after-tax) profit = (Annual income − Annual operating cost − Depreciation)*(1 − Tax rate) + Depreciation

[2.48] = (Annual income − Total annualized cost)*(1 − Tax rate) + Depreciation

The ROI has the units of percentage per year. It is analogous (and should be compared) to interest rates from banks and return on investment from investments in the financial markets. Clearly, the higher the ROI, the more desirable the project. Unless a project is mandated with no profit requirement (for example, to meet environmental regulations or safety objectives), companies insist on a *minimum acceptable value of the ROI* (also referred to as the *hurdle rate*) before approving a project. In many cases, a hurdle rate of 10 to 15 percent for the ROI is required. As the risk in the project increases, the choice of the minimum acceptable ROI may also significantly increase.

Both the ROI and the PBP calculations are simple and provide useful insights. Nonetheless, they do not account for interest and the time-value of money. The following section introduces different metrics to address profitability while accounting for interest and the time-value of money.

PROFITABILITY CRITERIA WITH THE TIME-VALUE OF MONEY

When the time-value of money is considered, the following profitability criteria may be used:

- Net present value
- Discounted cash flow return on investment
- Discounted payback period

Example 2-25 Using the ROI for determining environmental subsidies

A group of engineers is considering the formation of a company to be called Aggies Renewable Engineering, which specializes in the introduction of sustainable sources of energy. A potential project for the company is the manufacture of biodiesel from several renewable sources including algae and waste cooking oil. The growth of algae requires carbon dioxide as one of the main nutrients. Therefore, CO_2 may be sequestered from the flue gases of a power plant or an industrial source to be used for algae cultivation. Then, the feedstocks including the algal oil and the waste cooking oil are pretreated and processed to produce biodiesel. Substitution of the petro-diesel with the more sustainable biodiesel offers a reduction in the net emission of greenhouse gases including CO_2. On the other hand, a techno-economic study reveals that biodiesel from algae and other renewable sources is more expensive that petro-diesel. Therefore, to launch this new process, the company is requesting a government subsidy in the form of a tax credit. The following data are available for a biodiesel plant with a maximum production capacity of 40 MM gal/yr (Pokoo-Aikins et al., 2010; Elms and El-Halwagi, 2010); Myint and El-Halwagi, 2009):
- FCI of the plant is $25 MM
- WCI of the plant is $5 MM
- Operating cost 5 $2.87/gal biodiesel produced
- Selling price of biodiesel 5 $3.00/gal
- Tax rate 5 35%
- A linear depreciation scheme is to be used over 5 years with no salvage value (that is, annual depreciation = $5 MM/yr)
- When biodiesel is used in lieu of petro-diesel, the reduction in greenhouse gas emissions is taken to be of 8 kg CO_2 equivalent/gal biodiesel

The company desires to make a 30 percent ROI at the maximum production capacity. What should be the value of the tax credit (or a tax-exempted subsidy) expressed as $/gal biodiesel and $/tonne CO_2 equivalent? Examine the impact of carbon credit on the company's ROI.

SOLUTION

Let us add the tax credit ($/yr) to the annual income in Eq. 2.47, while noting that the tax credit is not taxable. Hence,

Annual net (after-tax) profit
= (Annual income − Annual operating cost − Depreciation)*
  (1 − Tax rate) + Tax credit + Depreciation

Therefore,

Annual net profit = (40MMgal/yr*$3.00/gal − 40MMgal/yr*$2.87/gal − 5.0*10^6)
(1 − 0.35) + Tax credit + 5.0*10^6 = 5.13*10^6 + Tax credit

Therefore,

$$ROI = \frac{5.13*10^6 + \text{Tax credit}}{30*10^6} *100\%$$

when the ROI is 30 percent, the tax credit should be $3.87 MM/yr.

Because the maximum production is 40 MM gal/yr, then

Tax credit per gallon of biodiesel = 3.87*10^6/40.00*10^6 = $0.097/gal

And, in terms of CO_2 credit,

Tax credit per tonne of CO_2 mitigated = ($0.097/gal)/(0.008 tonne CO_2 equivalent/gal)
= $12.1/tonne CO_2 equivalent

Figure 2.21 illustrates the impact of varying the carbon credit on the company's ROI.

Payback Period:

The **payback period** (PBP), which is sometimes referred to as the *simple payback period* or the *payout period*, is defined as:

$$[2.49] \qquad PBP = \frac{\text{Depreciable FCI}}{\text{Annual Net (After-Tax) Profit}}$$

It is an indication on how fast the depreciable FCI can be recovered. Therefore, the shorter the PBP, the more attractive the project. In many cases, a maximum PBP of 5 to 10 years is allowed. In some of the risky projects, payout periods of less than three years or even less than one year have been selected. On the other hand, for some of the long-term strategic projects, 20 years (or even longer) for the PBP have been accepted. Notice that the annual cash flow excludes depreciation (to avoid double counting because payback period is aimed at recovering back the FCI, which is also the aim of depreciation). Also, notice that the investment to be recovered is the depreciable FCI. It excludes land that is recoverable in principle. It also uses FCI, not TCI, because the working capital can be fully recovered at the end of the project.

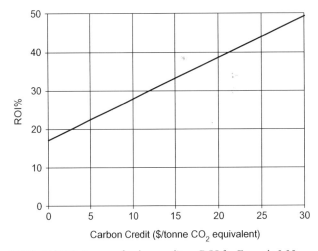

FIGURE 2-21 Impact of carbon credit on ROI for Example 2.25.

Net Present Value (NPV) or Net Present Worth (NPW)

Over the useful life of the project, expenditures and revenues are made over different times. Typically, the expenses start with the purchase of the land, equipment, installation, working capital, and start-up expenses. In large projects, the procurement and installation of the equipment may take multiple years. Then, the project begins to yield revenue. These expenditures and revenues constitute cash flows over the duration of the project. Such cash flows should be brought forward and backward to a reference time referred to as the *"present"* time. The net present value (NPV) is the cumulative value (revenues − expenses) adjusted to the reference time. The "present" time may be taken as the beginning of expenditures or the

Example 2-26 Payback period for a mass integration project

Let us revisit the mass integration project of Example 2.24. Calculate the simple payback period of the project.

SOLUTION

From the previous example,

$$FCI = \$2,000,000$$
$$\text{Annual net (after-tax) profit} = \$2,300,000/yr$$

Therefore,

$$PBP = (2,000,000/2,300,000)$$
$$= 0.87 \, yr$$

which is indeed quite attractive.

Example 2-27 Sensitivity analysis for the PBP as a function of operating cost

Example 2.25 for the production of biodiesel is revisited. The following are the key data:
- FCI of the plant is $25 MM
- WCI of the plant is $5 MM
- Operating cost 5 $2.87/gal biodiesel produced
- Selling price of biodiesel 5 $3.00/gal
- Tax rate 5 35%
- A linear depreciation scheme is to be used over 5 years with no salvage value (that is, annual depreciation = $5 MM/yr)

The operating cost of the plant is dominated by the cost of raw materials and may vary from $2.60 to $3.20/gal. No subsidy is given in this example. Conduct a sensitivity analysis to show the impact of the operating cost on the PBP.

SOLUTION

Let C^{op} be the operating cost per gallon of biodiesel. Therefore,

$$\text{Annual net profit} = (40 MMgal/yr * \$3.00/gal - 40 MMgal/yr * \$C^{op}/gal - 5.0*10^6)$$
$$(1 - 0.35) + 5.0*10^6 = 79.75*10^6 - 26.00*10^6 C^{op}$$

It is important to note that this expression is valid as long as the taxable income is positive; that is,

$$40 MMgal/yr * \$3.00/gal - 40 MMgal/yr * \$C^{op}/gal - 5.0*10^6 > 0$$

or

$$C^{op} < \$2.88/gal$$

Higher than this operating cost, the taxable income is not positive and no taxes are collected. Substituting for the tax rate to be zero, we get:

$$\text{Annual net profit} = 40 MMgal/yr * \$3.00/gal - 40 MMgal/yr * \$C^{op}/gal$$
$$2.88 \leq C^{op} \leq 3.00$$

Finally, when the operating cost reaches the selling price, the gross profit becomes zero and the FCI will not be recovered (the PBP tends to infinity as C^{op} approaches $3.00/gal). It is worth noting that between the 35 percent tax rate and the no-tax situation, there are usually income brackets that will gradually decrease the tax rate as the gross income decreases. For simplicity in this example, the tax rate is assumed to drop immediately from 35 percent to 0 percent when no profit is made. Hence, the following are the expressions for the PBP over the different regions of operating cost:

$$PBP = \frac{25.0*10^6}{79.75*10^6 - 26.00*10^6 C^{op}} \quad C^{op} < \$2.88/gal$$

$$PBP = \frac{25.0*10^6}{120.00*10^6 - 40.00*10^6 C^{op}} \quad 2.88 \leq C^{op} < 3.00$$

$$PBP \to \infty \quad \text{as} \quad C^{op} \to 3.00$$

Figure 2.22 shows the results of the sensitivity analysis. An important observation is how sensitive this plant is to modest changes in the operating cost. This is the unfortunate reality for those biofuel processes whose economics are dominated by the cost of raw materials. It is particularly true for cases when there is a strong competition for the feedstock between fuel and food (for example, soybean for biodiesel and corn for ethanol). The fuel–food feud typically raises the price for both. Modest increases in the cost of feedstocks can mean the difference between making attractive profit and filing for bankruptcy (or seeking government bailout!).

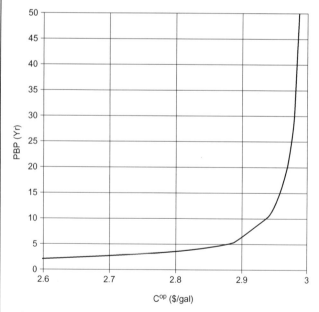

FIGURE 2-22 Sensitivity analysis for the payback period when the operating cost of a biodiesel plant is varied.

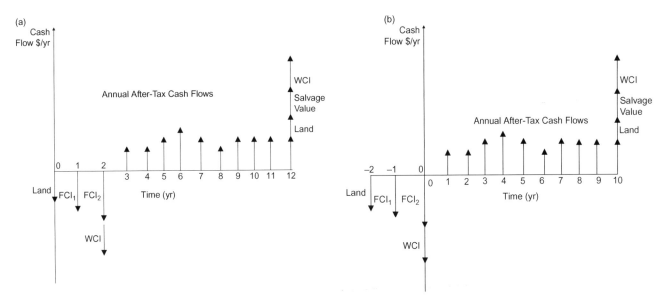

FIGURE 2-23 a. Cash flow diagram with "present" time taken as the beginning of expenditures. b. Cash flow diagram with "present" time taken as the beginning of operation.

start of operation. Figure 2.23 shows both scenarios. The choice of the present time does not alter the conclusions of the profitability analysis. It only impacts the value of NPV because it assigns it to a specific time. In the rest of this chapter, the present (zero) time will be taken as the start of expenditure. The FCI is spent over two installments (FCI_1 and FCI_2). At the end of the useful life of the project, the value of land, salvage value of the depreciable FCI, and the WCI are recoverable. Although different methods can be used to predict the values of land and WCI at the end of the project by accounting for the time-value of money and potential market conditions, in many cases their values are simply taken equal to their initial values without adjusting for the time-value of money or the potential market conditions.

To calculate the NPV, the cash flows (inflows or expenditures) for any year N is brought backward to the zero time using Eq. 2.33 to get the *discounted cash flows*:

$$P = F(1 + i)^{-N}$$

Then, the discounted cash flows are summed up to give the NPV; that is,

$$(2.50) \quad NPV = \sum_{N=0}^{N_{Final}} ACF_N (1 + i)^{-N}$$

Where ACF_N is the annual cash flow for year N (which may be negative in the case of outflows or positive in the case of inflows). N_{Final} is the final year of the project. The term $(1 + i)^{-N}$ is referred to as the *discount factor*, and i is called the *discount rate*. In the NPV calculations, many companies take i to be the company's minimum acceptable ROI (the hurdle rate), which is usually more conservative than the interest rate or the inflation rate. This choice is reasonable given that the company has the choice to invest in this project or in other projects while expecting to exceed the minimum acceptable ROI. Therefore, the value of money should be discounted at the hurdle rate. Otherwise, a different project should be pursued.

It is desired to have a positive NPV, which means that after paying all the expenses of the project and absorbing an annual decline in the value of money set to the discount rate, the revenues of the project will give net value of the NPV expressed in present. In other words, when the NPV is positive, the project over its life period has earned an ROI more than the discount rate. Therefore, the higher the value of the NPV, the more attractive the project is. A negative NPV is a sign of a project that should not be recommended because the ROI of the project is less than the discount rate.

It is worth noting that there are several consecutive uniform annual after-tax cash flows of $800,000/yr. Instead of converting each one of them to present value using the discount factor, it is easier to use the present sum of annuity formula:

$$P = A\left(\frac{(1 + i)^N - 1}{i(1 + i)^N}\right)$$

with A = $800,000/yr, i = 0.15, and N = 5. Hence,

$$P = 800,000\left(\frac{(1 + 0.15)^5 - 1}{0.15(1 + 0.15)^5}\right)$$
$$= \$2,681,724$$

Example 2-28 NPV of a project

A new process will be installed. Land is first purchased for $20 MM. Then, the equipment will be procured and installed over two years. The corresponding FCI costs for years 1 and 2 are $120 MM and $180 MM, respectively. The WCI for the project is $45 MM and will be spent at the end of the second year. Operation is started at the beginning of the third year. The process is continued for 11 years from the start of operation, during which revenue is made on an annual basis. The annual after-tax cash flows may vary from one year to another. For instance, in year 3 (which corresponds to the first year of operation), the plant does not operate in full capacity and incurs some start-up costs. Therefore, its annual after-tax cash flow is less than that in subsequent years. Also, major maintenance and retrofitting activities in year 8 cause a reduction in the annual after-tax cash flow. At the end of the useful life period of the plant, the land and WCI are estimated to be recovered for $20 MM and $45 MM, respectively, and the salvage value is estimated to be $30 MM. Table 2.19a shows a summary of the annual after-tax cash flows of the project. Using a discount rate of 10 percent, calculate the NPV of the project.

SOLUTION

Table 2.19b shows the worksheet for the NPV calculations by converting the nondiscounted cash flows to discounted values, and then summing them up. The cumulative discounted cash flow diagram is shown by Fig. 2.24. As the results indicate, the project yields $44.99 MM of NPV. This means that after paying all the expenses of the project and absorbing a 10 percent annual erosion in the value of money, the revenues of the project will give net value of $44.99 MM in present dollars (where present here corresponds to the start of expenditure).

TABLE 2-19a Annual Cash Flows for Example 2.28

End of Year	Annual (Nondiscounted) Cash Flow (in $ MM)
0	−20 (Land)
1	−120 (FCI_1)
2	−225 (−185 for FCI_2 and −45 for WCI)
3	40
4	65
5	65
6	70
7	75
8	55
9	60
10	60
11	75
12	75
13	165 (70 from annual after-tax cash flow, 20 from land, 45 from WCI, 30 from salvage value)

TABLE 2-19b Worksheet for the NPV Calculation of Example 2.28

End of Year	Annual (Nondiscounted) Cash Flow (in $ MM)	Discount Factor $(1+i)^{-n}$	Discounted Cash Flow (in $ MM)	Cumulative Discounted Cash Flow (in $ MM)
0	−20 (Land)	1.0000	−20.000	−20.000
1	−120 (FCI_1)	0.9091	−109.092	−129.092
2	−225 (−185 for FCI_2 and −45 for WCI)	0.8264	−185.940	−315.032
3	40	0.7513	30.052	−284.980
4	65	0.6830	44.395	−240.585
5	65	0.6209	40.359	−200.226
6	70	0.5645	39.515	−160.711
7	75	0.5132	38.490	−122.221
8	55	0.4665	25.658	−96.563
9	60	0.4241	25.446	−71.117
10	60	0.3855	23.130	−47.987
11	75	0.3505	26.288	−21.699
12	75	0.3186	23.895	2.196
13	165 (70 from annual after-tax cash flow, 20 from land, 45 from WCI, 30 from salvage value)	0.2897	47.794	49.990

(Continued)

The WCI and salvage value recovered at the end of the fifth year should also be brought backward to the present through the discount factor; that is,

$$P = F(1+i)^{-N}$$

with $F = 200{,}000$, $i = 0.15$, and $N = 5$.

Thus,

$$P = 200{,}000\,(1 + 0.15)^{-5}$$
$$= \$99{,}435$$

Now, we are in a position to add the discounted cash flows to get the NPV:

Example 2-28 NPV of a project (Continued)

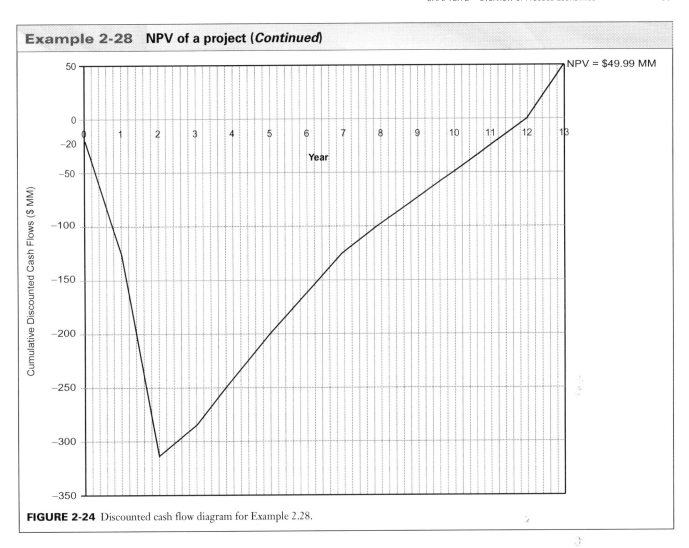

FIGURE 2-24 Discounted cash flow diagram for Example 2.28.

Example 2-29 NPV with uniform annual after-tax cash flows

Consider a project with the annual cash flows shown in Table 2.20. Using an annual discounted rate of 15 percent, calculate the NPV of the project.

TABLE 2-20 Nondiscounted Annual After-Tax Cash Flows for Example 2.29

End of Depreciation Year	Annual After-Tax Cash Flow ($)
0	−2,200,000 (for TCI)
1	800,000
2	800,000
3	800,000
4	800,000
5	1,000,000 (of which $200,000 is for recovered WCI and salvage value)

SOLUTION

First, let us represent the cash flow diagram for the project as shown by Fig. 2.25a. The discounted cash flows are calculated in Table 2.21, and the resulting NPV is $581,159. The discounted cash flow diagram is depicted by Fig. 2.25b.

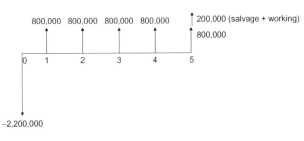

FIGURE 2-25a Nondiscounted cash flow diagram for Example 2.29.

(Continued)

Example 2-29 NPV with uniform annual after-tax cash flows (*Continued*)

TABLE 2-21 Worksheet for the NPV Calculation of Example 2.29

End of Year	Annual After-Tax Cash (Nondiscounted) Flow ($)	Discount Factor $(1 + i)^{-n}$	Discounted Cash Flow ($)	Cumulative Discounted Cash Flow ($)
0	−2,200,000 (for TCI)	1.000000	−2,200,000	−2,200,000
1	800,000	0.869565	695,652	−1,504,348
2	800,000	0.756144	604,915	−899,433
3	800,000	0.657516	526,013	−373,420
4	800,000	0.571753	457,402	83,982
5	1,000,000 (of which $200,000 is for recovered WCI and salvage value)	0.497177	497,177	581,159

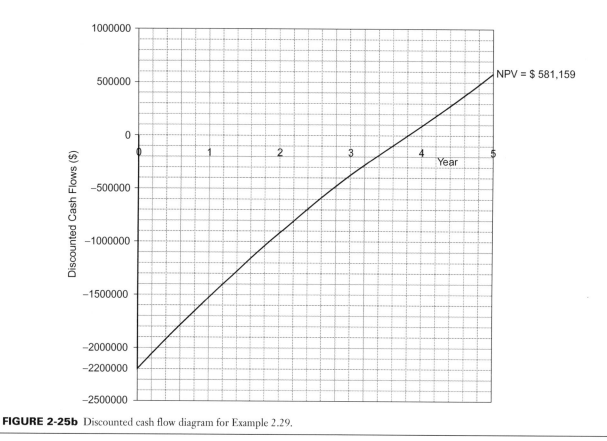

FIGURE 2-25b Discounted cash flow diagram for Example 2.29.

$$NPV = -2,200,000 + 2,681,724 + 99,435$$
$$= \$581,159$$

which is the same answer obtained earlier using the worksheet shown in Table 2.21.

Discounted Cash Flow Return on Investment (Internal Rate of Return)

The *discounted cash flow return on investment (DCF ROI)*, also known as the *internal rate of return (IRR)*, is the value of i that renders the NPV to be zero. As such, the DCF ROI provides the ROI for the project when the time-value of money is considered and when all expenses and revenues are accounted for over the life period of the project. The higher the value of the DCF ROI, the more attractive the project is. Mathematically, the DCF ROI can be determined by expressing the NPV in terms of an unknown i and setting the NPV value to be zero. This will require the solution of a nonlinear algebraic equation in one unknown, which can be solved using trial and error or other numerical techniques. If the calculated DCF ROI is greater than (or equal to) the minimum acceptable ROI set by the company, then the project is recommended because it has exceeded (or met) the company's hurdle rate of an acceptable ROI. Otherwise, if the DCF ROI is less than the minimum acceptable ROI, the project is not recommended.

Example 2-30 Calculation of the DCF ROI

Determine the DCF ROI (or IRR) for the project with the data given in Table 2.22. If the company has a minimum acceptable ROI of 10 percent, should this project be recommended? Compare the DCF with the nondiscounted ROI.

SOLUTION

The five installments of $1,000,000 of annual after-tax cash flows can be brought back to the present using the present sum of annuity formula, and the $400,000 of recovered WCI and salvage value can be converted to present value using the discount factor. Therefore, the NPV is given by:

$$NPV = -3{,}1500{,}000 + 1{,}000{,}000\left(\frac{(1+i)^5 - 1}{i(1+i)^5}\right) + \frac{400{,}000}{(1+i)^5}$$

Setting the NPV = 0 and solving for i, we get the DCF ROI to be 0.20. The same result can be obtained using trial and error for several values of i and plotting the resulting NPV versus i to graphically determine which value of i gives a zero value of NPV. Figure 2.26 shows the results of the trial and error. Similarly, the worksheet of the discounted cash flow diagram can be used by setting different values of i and iterating until the NPV becomes zero. Because the DCF ROI (or IRR) is greater than the 10 percent minimum acceptable ROI, it is recommended for positive consideration toward implementation.

The value of the nondiscounted ROI may be calculated using Eq. 2.38:

$$ROI = \frac{\text{Annual Net (After-Tax) Profit}}{TCI} * 100\%$$
$$= \frac{1{,}000{,}000}{3{,}150{,}000} * 100\%$$
$$= 31.7\%$$

TABLE 2-22 Annual After-Tax Cash Flows for Example 2.30

End of Year	Annual After-Tax Cash Flow ($)
0	−3,150,000 (for TCI)
1	1,000,000
2	1,000,000
3	1,000,000
4	1,000,000
5	1,400,000 (of which $400,000 is for recovered WCI and salvage value)

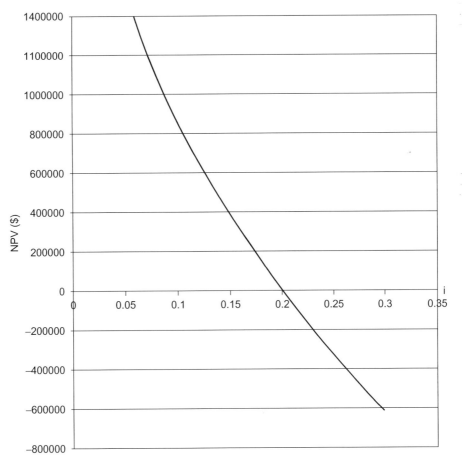

FIGURE 2-26 Trial-and-error plot to determine the DCF ROI for Example 2.30.

Example 2-31 Using the DCF ROI to determine minimum selling price

We have seen how the break-even analysis can be used to estimate a minimum selling price of a product when the time-value of money is not considered. Now, we show how the DCF ROI along with the NPV can be used to calculate a minimum selling price when the time-value of money and the cash flow diagram are considered. The case study in this example is an ethanol plant that uses corn stover as a feedstock. The following data (mostly extracted with revisions and updates from Kazi et al., 2010) are available:

- The plant capacity is 2000 tonnes of corn stover/day
- The plant produces 54 MM gal/yr of ethanol as the primary product and sells electricity power ($12 MM/yr) as a co-product
- Depreciable FCI = $358 MM (spent 50 percent then 50 percent after one and two years, respectively, from purchasing the land)
- Land cost = $25 MM
- WCI = $53 MM
- Operation starts at the end of the second year and continues for 20 years from the beginning of operation
- Feedstock cost = $58 MM/yr
- Cost of other raw materials, enzymes, and catalysts = $55 MM/yr
- Waste disposal = $8 MM/yr
- Fixed annual operating costs (salaries, property taxes, insurance) = $10 MM/yr
- Tax rate = 35 percent of taxable income
- Depreciation using MACRS with a 7-year recovery period is to be used
- The plant has a negligible salvage value

TABLE 2-23 Calculation of the Annual Depreciation of the $358 MM of FCI of the Ethanol Plant Over a 7-Year Recovery Period Using the MACRS Method

Recovery Year n	Depreciation Rate $\frac{d_n}{V_0}*100\%$	Depreciation Charge ($MM/yr)
1 (0.5 year) which is in the third year of the project because operation starts after two years from purchasing the land	14.29	51.16
2	24.49	87.67
3	17.49	62.61
4	12.49	44.71
5	8.93	31.97
6	8.92	31.93
7	8.93	31.97
8 (0.5 year)	4.46	15.98

TABLE 2-24 Worksheet for Annual After-Tax (Nondiscounted) Cash Flows for Example 2.31

End of Year	Annual Sales from Ethanol and Electricity ($ MM)	Annual Operating Cost ($ MM)	Annual Depreciation ($ MM)	FCI and WCI ($ MM)	Annual After-Tax (Nondiscounted) Cash Flow ($ MM)
0	0	0	0	−25.00 (Land)	−25.00
1	0	0	0	−179.00 (FCI$_1$)	−179.00
2	0	0	0	−232.00 (−179.00 for FCI$_2$ and −53.00 for WCI)	−232.00
3	54*C^{EtOH} + 12	131.00	51.16	0	(54*C^{EtOH} + 12.00 − 131.00−51.16)*(1.00−0.35) + 51.16 = 35.1*C^{EtOH} −59.44
4	54*C^{EtOH} + 12	131.00	87.67	0	35.1*C^{EtOH} −46.66
5	54*C^{EtOH} + 12	131.00	62.61	0	35.1*C^{EtOH} −55.44
6	54*C^{EtOH} + 12	131.00	44.71	0	35.1*C^{EtOH} −61.70
7	54*C^{EtOH} + 12	131.00	31.97	0	35.1*C^{EtOH} −66.16
8	54*C^{EtOH} + 12	131.00	31.93	0	35.1*C^{EtOH} −66.17
9	54*C^{EtOH} + 12	131.00	31.97	0	35.1*C^{EtOH} −66.16
10	54*C^{EtOH} + 12	131.00	15.98	0	35.1*C^{EtOH} −71.76
11	54*C^{EtOH} + 12	131.00	0	0	(54*C^{EtOH} + 12.00 − 131.00)*(1.00 − 0.35) = 35.1*C^{EtOH} −77.35
12	54*C^{EtOH} + 12	131.00	0	0	35.1*C^{EtOH} −77.35
13	54*C^{EtOH} + 12	131.00	0	0	35.1*C^{EtOH} −77.35
14	54*C^{EtOH} + 12	131.00	0	0	35.1*C^{EtOH} −77.35
15	54*C^{EtOH} + 12	131.00	0	0	35.1*C^{EtOH} −77.35
16	54*C^{EtOH} + 12	131.00	0	0	35.1*C^{EtOH} −77.35
17	54*C^{EtOH} + 12	131.00	0	0	35.1*C^{EtOH} −77.35
18	54*C^{EtOH} + 12	131.00	0	0	35.1*C^{EtOH} −77.35
19	54*C^{EtOH} + 12	131.00	0	0	35.1*C^{EtOH} −77.35
20	54*C^{EtOH} + 12	131.00	0	0	35.1*C^{EtOH} −77.35
21	54*C^{EtOH} + 12	131.00	0	0	35.1*C^{EtOH} −77.35
22	54*C^{EtOH} + 12	131.00	0	78.00 (25.00 from sold land and 53.00 from recovered WCI)	35.1*C^{EtOH} −26.65

(Continued)

Example 2-31 Using the DCF ROI to determine minimum selling price (Continued)

What is the selling price of ethanol that will yield a zero value of the NPV of the plant over 20 years of operation using a 10 percent IRR?

SOLUTION

Let us assume that the annual operating cost will remain constant throughout the 20 years of the project with:

AOC = Variable annual operating cost + Fixed annual operating cost
= (58 + 55 + 8) + 10 = $131MM/yr

Depreciation is carried out over 7 years. Using the MACRS scheme, the depreciation charges are shown in Table 2.23. It is important to recall the half-year convention that spreads the depreciation over 8 years. It is also worth noting that the recovery period for depreciation (7 years) is different from the useful life period of the project (20 years) over which the NPV calculation will be done. Beyond the recovery period (years 9 to 20), there are no depreciation charges. Let C^{EtOH} be the selling price of ethanol ($/gal) and assume it to be constant over the life period of the plant. Therefore, the annual sales ($ MM/yr) of ethanol can be expressed as 54 MM gal/yr*$C^{EtOH}$$/gal. Table 2.24 is a worksheet for calculating the annual after-tax (nondiscounted) cash flows. Using the discount factor, Table 2.25 gives the discounted cash flows for the project. When these discounted cash flows are summed, the result is the NPV of the project, which is given by:

$$NPV = 247.11 \cdot C^{EtOH} - 842.77$$

For the minimum selling price of ethanol, the NPV is set to zero and we get:

$$C^{EtOH} = \$3.41/gal$$

TABLE 2-25 Worksheet for the NPV Calculation of Example 2.31

End of Year	Annual After-Tax (Nondiscounted) Cash Flow ($ MM)	Discount Factor $(1 + i)^{-n}$	Annual After-Tax (Discounted) Cash Flow ($ MM)
0	−25.00	1.0000	−25.00
1	−179.00	0.9091	−162.73
2	−232.00	0.8264	−191.74
3	$(54 \cdot C^{EtOH} + 12.00 - 131.00 - 51.16) \cdot (1.00 - 0.35) + 51.16 = 35.1 \cdot C^{EtOH} - 59.44$	0.7513	$26.37 \cdot C^{EtOH} - 44.66$
4	$35.1 \cdot C^{EtOH} - 46.66$	0.6830	$23.97 \cdot C^{EtOH} - 31.87$
5	$35.1 \cdot C^{EtOH} - 55.44$	0.6209	$21.97 \cdot C^{EtOH} - 34.42$
6	$35.1 \cdot C^{EtOH} - 61.70$	0.5645	$19.81 \cdot C^{EtOH} - 34.83$
7	$35.1 \cdot C^{EtOH} - 66.16$	0.5132	$18.01 \cdot C^{EtOH} - 33.95$
8	$35.1 \cdot C^{EtOH} - 66.17$	0.4665	$16.37 \cdot C^{EtOH} - 30.87$
9	$35.1 \cdot C^{EtOH} - 66.16$	0.4241	$14.89 \cdot C^{EtOH} - 28.06$
10	$35.1 \cdot C^{EtOH} - 71.76$	0.3855	$13.53 \cdot C^{EtOH} - 27.67$
11	$(54 \cdot C^{EtOH} + 12.00 - 131.00) \cdot (1.00 - 0.35) = 35.1 \cdot C^{EtOH} - 77.35$	0.3504	$12.30 \cdot C^{EtOH} - 27.11$
12	$35.1 \cdot C^{EtOH} - 77.35$	0.3186	$11.18 \cdot C^{EtOH} - 24.65$
13	$35.1 \cdot C^{EtOH} - 77.35$	0.2897	$10.17 \cdot C^{EtOH} - 22.41$
14	$35.1 \cdot C^{EtOH} - 77.35$	0.2633	$9.24 \cdot C^{EtOH} - 20.37$
15	$35.1 \cdot C^{EtOH} - 77.35$	0.2394	$8.40 \cdot C^{EtOH} - 18.52$
16	$35.1 \cdot C^{EtOH} - 77.35$	0.2176	$7.64 \cdot C^{EtOH} - 16.83$
17	$35.1 \cdot C^{EtOH} - 77.35$	0.1978	$6.94 \cdot C^{EtOH} - 15.30$
18	$35.1 \cdot C^{EtOH} - 77.35$	0.1799	$6.31 \cdot C^{EtOH} - 13.91$
19	$35.1 \cdot C^{EtOH} - 77.35$	0.1635	$5.74 \cdot C^{EtOH} - 12.65$
20	$35.1 \cdot C^{EtOH} - 77.35$	0.1486	$5.22 \cdot C^{EtOH} - 11.50$
21	$35.1 \cdot C^{EtOH} - 77.35$	0.1351	$4.74 \cdot C^{EtOH} - 10.45$
22	$35.1 \cdot C^{EtOH} - 26.65$	0.1228	$4.31 \cdot C^{EtOH} - 3.27$
			Summation (NPV) = $247.11 \cdot C^{EtOH} - 842.77$

An advantage of the DCF ROI is that it gives a more realistic estimate of the returns of the project compared to the simple ROI calculation, which does not take into account the time-value of money or the changes in the financial transactions of the project over time. On the other hand, compared to the NPV, the DCF ROI has the disadvantage of not providing insights on the size of the project. It just gives the discount rate that renders the NPV to be zero without characterizing how much money was actually made or lost. As mentioned before, the NPV describes the present value of the net size of the investment after all expenditures and revenues have been accounted for and adjusted using discount factors.

The discounted ROI gives a more optimistic estimate than that of the DCF ROI because it does not account for the time-value of money. As such, it should be used with caution.

Example 2-32 Calculation of the DPBP

What is the DPBP for Example 2.28? Compare the result with the simple PBP.

SOLUTION

The land cost is $20 MM at zero time, and the WCI is $45 MM at the end of year 2. Using a discount rate of 10 percent,

$$\text{The discounted non-FCI expenses} = -20 - \frac{45}{(1+0.1)^2}$$
$$= \$ -57.190 \text{MM}$$

Next, the DPBP is located by identifying the time at which the cumulative discounted cash flow of the project is -$57.190 MM. As demonstrated by Fig. 2.27, the DPBP is found to be 9.6 years from the beginning of expenditure (that is, 7.6 years from the start of operation). Alternatively, the DPBP may be found via interpolation from the worksheet given by Table 2.19. The value of $-\$57.190$ MM lies between years 9 and 10, with cumulative cash flows of $-\$71.117$ M and $-\$47.987$ MM, respectively. Using interpolation, we get:

$$\text{DPBP} = 9 + \frac{-57.190 - (-71.117)}{-47.987 - (-71.117)}$$
$$= 9.6 \text{ years}$$

To evaluate the simple (nondiscounted) PBP, let us recall the definition:

$$\text{PBP} = \frac{\text{Depreciable FCI}}{\text{Annual Net (After-Tax) Profit}}$$

For the simple PBP, the depreciable FCI is not discounted; that is,

$$\text{Depreciable FCI} = 120 + 185 = \$305 \text{MM}$$

Also, the annual after-tax profit is not discounted and is taken as the average of the operating years of the project; that is,

$$\text{Average annual after-tax profit} = \frac{40 + 65 + 65 + 70 + 75 + 55 + 60 + 60 + 75 + 75 + 70}{11}$$
$$= \$64.55 \text{MM/yr}$$

Therefore,

$$\text{PBP} = \frac{305}{64.55}$$
$$= 4.7 \text{ years}$$

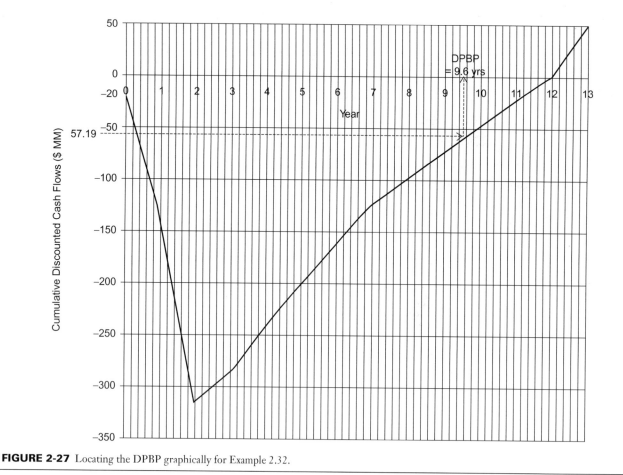

FIGURE 2-27 Locating the DPBP graphically for Example 2.32.

It is also possible to solve the problem using trial and error. An iterative value of the selling price of ethanol is assumed and the NPV is calculated for a DCF ROI of 10 percent. The procedure is repeated until the calculated NPV is zero.

Discounted Payback Period

The simple payback period defined by Eq. 2.49 does not account for the time-value of money or for the cash flow variations over the life period of the project. To overcome these limitations, the notion of a discounted payback

Example 2-33 Comparison of different profitability criteria

Consider a process for the production of a specialty chemical. The land is purchased for $30 MM. The FCI is spent over the next two years for a total of $540 MM. The WCI ($60 MM) is spent toward the end of the second year. Operation starts at the beginning of the third year. Table 2.26 shows the estimated annual production, selling price of the product, and annual operating cost (which accounts for raw materials, utilities, labor, maintenance, and so on). The project has an anticipated useful life period of 10 years from the start of operation. The straight-line depreciation method is used with a 10-year recovery period. At the end of the project, the WCI and salvage value are estimated to $60 MM and $40 MM, respectively. A tax rate of 35 percent is applied. The company uses a discount rate of 10 percent. Calculate the NPV, DCF ROI, and discounted PBP for the project.

TABLE 2-26 Data for Example 2.33

End of Year	Estimated Annual Production (MM tonne/yr)	Estimated Selling Price of Product ($/tonne)	Estimated Annual Operating Cost ($ MM/yr)
3	1.0	200	130
4	2.0	210	260
5	2.0	220	260
6	2.0	230	270
7	2.2	240	300
8	1.5	250	260
9	2.2	260	320
10	2.2	270	330
11	2.0	280	310
12	1.5	290	280

SOLUTION

Table 2.27 is a worksheet to evaluate the different items leading to the calculation of the annual after-tax cash flow. The estimated annual sales can be obtained simply by multiplying the production times the unit selling price. For depreciation, a linear scheme is used to get an annual depreciation charge of $50 MM/yr. During the years of production, the annual after-tax cash flow is calculated using Eq. 2.47:

Annual net (after-tax) profit = (Annual income − Annual operating cost − Depreciation)*(1 − Tax rate) + Depreciation

The annual income for years 3 through 11 is the annual sales. For year 12, the annual income is the sum of the annual sales, the recovered WCI, and the salvage value.

TABLE 2-27 Worksheet to Evaluate Annual After-Tax Cash Flows

End of Year	Land ($ MM)	FCI ($ MM)	WCI ($ MM)	Estimated Annual Production (MM tonne/yr)	Estimated Selling Price of Product ($/tonne)	Estimated Annual Sales ($MM/yr)	Estimated Annual Operating Cost ($ MM/yr)	Depreciation ($ MM/yr)	Annual After-Tax Cash Flow ($ MM/yr)
0	−30								−30.0
1		−180							−180.0
2		−360	−60						−420.0
3				1.0	200	200	130	50	63.0
4				2.0	210	420	260	50	121.5
5				2.0	220	440	260	50	134.5
6				2.0	230	460	270	50	141.0
7				2.2	240	528	300	50	165.7
8				1.5	250	375	260	50	92.3
9				2.2	260	572	320	50	181.3
10				2.2	270	594	330	50	189.1
11				2.0	280	560	310	50	180.0
12		40	60	1.5	290	435	280	50	183.3

period (DPBP) or discounted cash flow payback period (DCF PBP) is used. Because the WCI and the land are recoverable in principle, the DPBP is the time needed to recover the depreciable FCI. The following steps are used to determine the DPBP:

- Develop the discounted cash flow diagram for the project.
- Calculate the discounted values of the WCI and the land brought back to time zero (present time).

Let us call this value the discounted non-FCI expenses. Because it is an expenditure, it is assigned a negative sign.

- On the discounted cash flow diagram, the time at which the cumulative discounted cash flow equals the discounted non-FCI expenses defines the DPBP. At the DPBP, enough revenues (discounted to time zero) have been generated to pay for all the TCI expenses (discounted to time zero) except the discounted values of the WCI and the land (which will ultimately be recovered at the end of the project). This means that sufficient profit has been made to pay off the depreciable FCI with the values of both the profit and the FCI discounted to the present time to have a consistent basis. It is worth pointing out that for projects featuring a positive NPV, the value of the discounted non-FCI expenses will be encountered twice: once during the expenditure phase when the TCI is being spent (the downswing portion of the cash flow curve) and once during the revenue phase (the upswing portion of the cash flow curve). The DPBP is located on the upswing portion of the curve.

Although the DPBP requires more effort and information than the PBP, it offers a more accurate estimation of the time needed to recover the depreciable FCI from the accrued cash flows of the project. A key disadvantage for the DPBP is that it is not impacted by the cash flows following the payout period and, therefore, does not shed light on the profitability of the project following the DPBP.

The simple PBP is determined from the start of operation. Therefore, the 4.7-year simple PBP should be compared with the 7.6-year discounted PBP. The difference between the values of the DPBP and the simple PBP underscores the importance of accounting for the time-value of money.

Next, the cumulative discounted cash flows are calculated as shown in the worksheet of Table 2.28.

Based on the results of the worksheet, the discounted cash flow diagram is plotted in Fig. 2.28. The NPV is $152.057 MM.

The land cost is $30 MM at zero time and the WCI is $60 MM at the end of year 2. Using a discount rate of 10 percent,

$$\text{The discounted non-FCI expenses} = -30 - \frac{60}{(1 + 0.1)^2} = \$-79.587\,\text{MM}$$

From Fig. 2.28, the DPBP is 8.5 years from the beginning of expenditure, which corresponds to 6.5 years from the start of production.

TABLE 2-28 Worksheet for the NPV Calculation of Example 2.33

End of Year	Annual (Non-Discounted) Cash Flow (in $ MM)	Discount Factor $(1 + i)^{-N}$	Discounted Cash Flow (in $ MM)	Cumulative Discounted Cash Flow (in $ MM)
0	−30.0	1.0000	−30.000	−30.000
1	−180.0	0.9091	−163.636	−193.636
2	−420.0	0.8264	−347.107	−540.744
3	63.0	0.7513	47.332	−493.412
4	121.5	0.6830	82.986	−410.426
5	134.5	0.6209	83.514	−326.912
6	141.0	0.5645	79.591	−247.321
7	165.7	0.5132	85.030	−162.291
8	92.3	0.4665	43.059	−119.232
9	181.3	0.4241	76.889	−42.343
10	189.1	0.3855	72.906	30.563
11	180.0	0.3505	63.089	93.652
12	183.3	0.3186	58.405	152.057

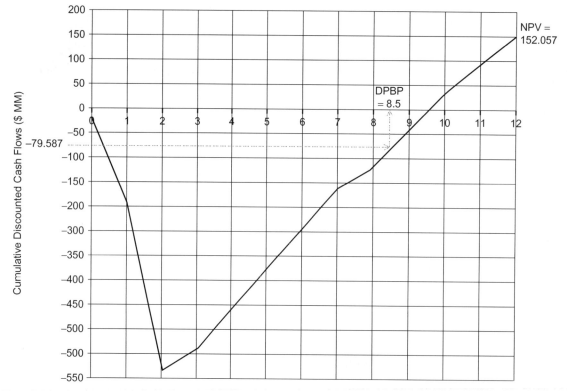

FIGURE 2-28 Determining the NPV and the DPBP for Example 2.33.

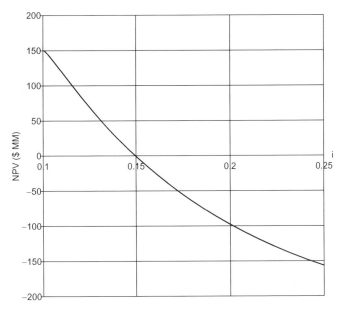

FIGURE 2-29 Determination of the DCF ROI for Example 2.33.

The NPV is calculated for several values of the discount rate, i. The results are plotted in Fig. 2.29. The DCF ROI is obtained when the NPV is zero. As the plot shows, the DCF ROI is 15 percent.

The NPV of the project is good for the size of the project. All the project expenditures will be paid off while allowing for a 10 percent annual deterioration of the value of money, and the project still makes $152.057 MM of net profit (expressed as present dollars). The DPBP of 6.5 years from the start of operation is also good. Although many companies prefer a PBP of 3 to 5 years, 6.5 years is acceptable when there is a limited risk in the process technology, the supply and cost of the raw materials, and the market for the product. Finally, the DCF ROI of 15 percent exceeds the company's hurdle rate of 10 percent. Therefore, the three indicators point to the direction of recommending the project.

COMPARISON OF ALTERNATIVES

The previous profitability criteria dealt with the assessment of individual projects. In many cases, there are several alternatives. These alternatives may involve options for the company to invest in different options, various implementations to deliver the same performance, or modifications of a base-case project to achieve several metrics. Although different criteria and financial analysis metrics should be used in screening alternatives and finalizing decisions, the following criteria are particularly useful in the comparison of alternatives:

- Net present value
- Annual cost/revenue
- Total annualized cost
- Incremental return on investment

Net Present Value (Net Present Worth)

As discussed earlier, the NPV is one of the most insightful profitability criteria. In addition to assessing individual projects, it can also be used to screen competing projects. The NPV accounts for:

- the time value of money.
- the size of investment and the inflows and outflows of cash over the life of the project.
- different durations of project alternatives.

When competing projects are considered, the project with the highest NPV (within the company's constraints and financial guidelines) wins.

Annual Cost/Revenue

The annual cost (or revenue) of a project is a uniform annual number that distributes the NPV of the project over a given period. The term **annual cost** is used when the NPV of the project is negative, whereas the term **annual revenue** is used when the NPV of the project is positive. A negative NPV for a project is acceptable in cases when there are mandates to carry out the project (for example, enhancing safety, complying with environmental regulations, and so on). The following are some observations for the annual cost/revenue:

- It converts all the costs/revenues of the project into equal annual installments (annuities). Therefore, this annual value can be readily included in the company's financial calculations.
- It is used when the TCI is financed via an annuity scheme.
- It accounts for the time-value of money.
- It accounts for different durations of project alternatives.
- In the case of revenue-making projects, the project with the highest annual revenue (within the company's constraints and financial guidelines) wins.
- In the case of net-cost projects, the project with the least annual cost (within the company's constraints and financial guidelines) wins.

The annual cost/revenue is closely tied to the NPV via the following expression:

[2.51a]
$$NPV = \text{Annual Cost/Revenue} * \left(\frac{(1+i)^N - 1}{i(1+i)^N} \right)$$

or

[2.51b]
$$\text{Annual Cost/Revenue} = NPV \left(\frac{i(1+i)^N}{(1+i)^N - 1} \right)$$

Total Annualized Cost

The total annualized cost has been defined earlier by Eq. 2.19:

[2.19] $$TAC = AFC + AOC$$

The TAC can be used in break-even calculations and in trading off fixed cost versus operating cost. It also has a

Example 2-34 Comparison of alternatives based on NPV

A company is considering the investment in one of two projects: A and B. The nondiscounted cash flows for both projects are given in Tables 2.29 and 2.30. The company uses a discount rate of 15 percent. Which project should be recommended?

SOLUTION

Tables 2.31 and 2.32 show the worksheets for calculating the NPVs for projects A and B. Because project B has a uniform cash flow of $700,000/yr for five years, an annuity calculation may be used in lieu of Table 2.32, as follows:

$$P = A\left[\frac{(1+i)^N - 1}{i(1+i)^N}\right]$$

with $A = \$700,000/\text{yr}$, $i = 0.15$, and $N = 5$. Hence,

$$P = 700,000\left[\frac{(1+0.15)^5 - 1}{0.15(1+0.15)^5}\right]$$
$$= \$2,346,509$$

The WCI and salvage value recovered at the end of the fifth year should also be discounted to the present; that is,

$$P = F(1+i)^{-N}$$

with $F = 200,000$, $i = 0.15$, and $N = 5$. Thus,

$$P = 200,000(1+0.15)^{-5}$$
$$= \$99,435$$

Therefore, the NPV for project B is:

$$NPV = -1,600,000 + 2,346,509 + 99,435$$
$$= \$845,944$$

which is the same answer obtained earlier using the worksheet shown in Table 2.32.

Comparing the NPV of project B ($845,944) to the NPV of project A ($310,066), project B should be recommended.

TABLE 2-29 Data for Project A of Example 2.34

Year	Nondiscounted Annual After-Tax Cash Flow ($)
0	−1,400,000 (TCI)
1	400,000
2	800,000
3	−300,000
4	800,000
5	1,000,000

TABLE 2-30 Data for Project B of Example 2.34

Year	Nondiscounted Annual After-Tax Cash Flow ($)
0	−1,600,000 (TCI)
1	700,000
2	700,000
3	700,000
4	700,000
5	900,000 (of which $200,000 is recovered WCI and salvage value)

TABLE 2-31 Worksheet for Calculating the NPV of Project A for Example 2.34

End of Year	Annual (Non-discounted) Cash Flow ($)	Discount Factor $(1+i)^{-N}$	Discounted Cash Flow ($)	Cumulative Discounted Cash Flow ($)
0	−1,400,000	1.0000	−1,400,000	−1,400,000
1	400,000	0.8696	347,826	−1,052,174
2	800,000	0.7561	604,915	−447,259
3	−300,000	0.6575	−197,255	−644,514
4	800,000	0.5718	457,403	−187,111
5	1,000,000	0.4972	497,177	310,066

TABLE 2-32 Worksheet for Calculating the NPV of Project B for Example 2.34

End of Year	Annual (Non-discounted) Cash Flow ($)	Discount Factor $(1+i)^{-N}$	Discounted Cash Flow ($)	Cumulative Discounted Cash Flow ($)
0	−1,600,000	1.0000	−1,600,000	−1,600,000
1	700,000	0.8696	608,696	−991,304
2	700,000	0.7561	529,301	−462,004
3	700,000	0.6575	460,261	−1742
4	700,000	0.5718	400,227	398,485
5	900,000	0.4972	447,459	845,944

useful application in comparing alternatives, especially the ones that are not expected to make revenue such as some of the environmental compliance projects and safety-enhancement modifications. In such cases, the competing alternatives are mutually exclusive and are designed to meet the same performance metrics. The alternative with the lowest TAC (subject to the company's constraints) is favored.

The TAC is a special case of the annual cost described in the previous section. The key difference is that the TAC does not account for the time-value of money and assumes that the AFC and the AOC will remain constant throughout the useful life period of the project.

Example 2-35 Using annual cost/revenue to compare alternatives

Consider the data given in the previous example for projects A and B. Calculate the annual revenue for both projects.

SOLUTION

Using Eq. 2.51b for the NPVs of projects A and B ($310,066 and 845,944), we have:
For project A:

$$\text{Annual revenue} = 310{,}066\left[\frac{0.15(1+0.15)^5}{(1+0.15)^5 - 1}\right]$$
$$= \$92{,}498/\text{yr}$$

For project B:

$$\text{Annual revenue} = 845{,}944\left[\frac{0.15(1+0.15)^5}{(1+0.15)^5 - 1}\right]$$
$$= \$252{,}358/\text{yr}$$

As expected from the previous example, project B should be recommended.

Example 2-36 Using the total annualized cost to compare alternatives

An environmental compliance project involves the screening of three technologies to remove 90 percent of a pollutant from a 150,000 kg/hr wastewater stream containing 10 ppm of the pollutant (Gabriel and El-Halwagi, 2005). The cost data for the three competing technologies (stripping, ion exchange, and adsorption) to remove 90 percent of the pollutant are given in Table 2.33. Using a 10-year linear depreciation with no salvage value, recommend one of the technologies.

TABLE 2-33 Cost Data for the Wastewater Treatment Example

Technology	FCI ($)	AOC ($/yr)
Stripping	249,700	13,900
Ion exchange	151,100	31,500
Adsorption	241,800	26,300

SOLUTION

Table 2.34 provides the worksheet for calculating the TAC of the three technologies. The amount of removed pollutant is the same for the three technologies and, assuming 8760 hrs/yr of operation, can be calculated through a component material balance on the pollutant:

Annual load of pollutant removed = 150,000 kg wastewater/hr*10*10^{-6}
(kg pollutant/kg wastewater)*8760 hr/yr = 13,140 kg pollutant/yr

Therefore, the TAC can be normalized on a per kg pollutant basis by dividing the TAC by the annual load of the pollutant to be removed. The results are given in the last column of Table 2.34. Based on the TAC results ($/yr or $/kg pollutant), stripping has the lowest TAC and should, therefore, be recommended.

TABLE 2-34 Worksheet for Evaluating the TAC of the Three Wastewater Treatment Technologies

Technology	FCI ($)	AFC ($/yr)	AOC ($/yr)	TAC ($/yr)	TAC/kg pollutant removed ($/kg removed)
Stripping	249,700	24,970	13,900	38,870	2.96
Ion exchange	151,100	15,110	31,500	46,610	3.55
Adsorption	241,800	24,180	26,300	50,480	3.84

Incremental Return on Investment (IROI)

When a base-case design is developed, there are usually different "add-on" alternatives or modifications to enhance the performance of the project. The question is whether each or any of them is worth the additional investment. In such cases, the incremental return on investment (IROI) is particularly useful. It is defined as:

[2.52]
$$\text{IROI} = \frac{\text{Incremental annual net (after-tax) profit of add-on project}}{\text{Incremental TCI of add-on project}} *100\%$$

where

[2.53]
Incremental annual net profit of add-on project = Annual net profit of the combined (add-on and base-case) project − annual net profit of the base-case project

and

[2.54]
Incremental TCI of add-on project = TCI the combined (add-on and base-case) project − TCI of the base-case project

The IROI must meet the company's minimum acceptable ROI for the add-on project to be recommended.

Example 2-37 Using IROI to screen alternatives

A mass integration study on the conservation of a fresh solvent has generated a base case that is a direct-recycle project. The total capital investment is $1 MM, and the project provides an annual net profit of $400,000/yr.

Two add-on alternatives are proposed as possible modifications to the base case:

Alternative 1: The addition of an absorption system to the direct recycle project. The absorption system will cost an additional $3 MM of TCI and will result in a total annual net profit of $900,000/yr for the combined based case and the absorption system.

Alternative 2: The addition of an extraction system to the direct recycle project coupled with the absorption system. The extraction system will cost an additional $2 MM of TCI of the base case and the absorption system. It will result in a total annual net profit of $1,100,000/yr for the combined base case, absorption, and extraction.

The company has a 15 percent minimum hurdle rate. Which projects should you recommend?

SOLUTION

First, the ROI for the base-case project is calculated:

$$ROI_{Direct\ Recycle} = \frac{400,000}{1,000,000}*100\% = 40\%$$

Therefore, the direct-recycle project should be recommended. Next, let us calculate the IROI for adding the absorption system:

$$IROI_{Absorption} = \frac{900,000 - 400,000}{3,000,000}*100\% = 16.7\%$$

and should, therefore, be recommended.

Next, we check the IROI for adding the extraction system to the direct-recycle project and the absorption system. Because the extraction is added to the combined (direct-recycle and absorption project), the base case for comparison becomes the direct-recycle and absorption cost and profit; that is,

$$IROI_{Extraction} = \frac{1,10,000 - 900,000}{2,000,000}*100\% = 10\%$$

which is less than the company's hurdle rate of 15 percent. Hence, the extraction project should not be recommended. Therefore, the final recommended project is to implement the direct-recycle project and to install the absorption system.

It is interesting to check the ROI of the cumulative project of direct recycle, absorption, and extraction. The annual net profit is $1.1 MM/yr and the TCI is 1.0 + 3.0 + 2.0 = $6.0 MM. Thus,

$$ROI_{Direct\ Recycle+absorption+extraction} = \frac{1,100,000}{6,000,000}*100\% = 18.3\%$$

which exceeds the company's minimum acceptable ROI of 15 percent. Therefore, if the cumulative project of direct recycle, absorption, and extraction would have been evaluated as a single project, the wrong conclusion would have been to recommend it. This discussion underscores the importance of conducting the IROI analysis for add-on projects. The ROI analysis alone for the cumulative project would have been misleading. Any additional capital expenditure beyond the base-case project must be justified by an additional profit that meets the company's ROI. In this example, the extraction project does not.

HOMEWORK PROBLEMS

2.1. A gas-to-liquid (GTL) process that produces 140,000 barrels (bbl)/day has a fixed capital investment of $12 billion. Estimate the fixed capital investment of a similar GTL plant producing 110,000 bbl/day.

2.2. A process produces ethanol from a cellulosic waste. The flow rate of the cellulosic waste feedstock is 1800 tons/day. The process yield is 90 gallons of ethanol per ton of cellulosic waste. In 2007, the fixed capital investment of the plant was estimated to be $3.10/gal of annual ethanol production.
   a. What was the FCI of the plant in 2007?
   b. What is the FCI of the plant now?

2.3. For the previous problem, use the turnover ratio to estimate the fixed cost of the plant.

2.4. The installed equipment cost in problem 2 is $40 MM. Using the Lang factor (as revised by Peters et al., 2003), what is the FCI of the plant?

2.5. A fluid-processing retrofitting project involves the addition of several units. Table 2.35 gives the type and cost of these units. Additionally, the instrumentation and control systems for this project are estimated to have a delivered cost of $4.0 MM. Estimate the FCI using the Lang factors as revised by Peters et al. (2003) and the Hand method.

TABLE 2-35 Units and Costs of the Retrofitting Project

Unit	Delivered Equipment Cost ($ MM)
Distillation columns	5.0
Fired heater	4.0
Heat exchangers	2.0
Tanks	1.0

2.6. A stover-to-ethanol plant produces 40,000 tonne/yr of ethanol and contains eight functional units: feedstock handling, pretreatment, simultaneous saccharification/fermentation distillation, solid/syrup separation, wastewater treatment, boilers, and turbo-generation/utilities (McAloon et al., 2000). Estimate the FCI of the plant using the empirical correlation based on the work of Bridgwater and Mumford (1979).

2.7. A new plant is to produce 500,000 tonne/yr of ethylene. The process has six functional units and requires a delivered equipment cost of $110 MM. The FCI of a similar plant with a production rate of 600,000 tonne/yr is $700 MM. Estimate the FCI of

the 500,000 tonne/yr ethylene plant using the following methods:
a. The Lang factor (as revised by Peters et al., 2003)
b. Sixth-tenths factor rule
c. Empirical equation based on the work of Timm (1980)
d. Turnover ratio (assume a value of 2.0)
Comment on your results.

2.8. A heat integration project requires the use of 41.2 MM Btu/hr of heating utility in the form of a medium pressure steam. Natural gas is burned in a boiler to produce the medium pressure steam, which is used to deliver heat to the process heaters. The thermal efficiency of the boiler is 65 percent and is defined as:

$$\eta_{thermal} = \frac{\text{Heat delivered by steam in heaters}}{\text{Heat generated from combustion of natural gas in the boiler}}$$

What is the annual operating cost of the heating utility?
Hint: Please use information from the Energy Information Administration website (www.eia.doe.gov) and look for the Natural Gas Weekly Update.

2.9. In the previous problem, if natural gas is substituted with crude oil and thermal efficiency is maintained at the same level, what is the annual heating cost?

2.10. A process uses 63,400 SCF/hr of natural gas. What is the annual cost of natural gas used in the process? Hint: One SCF approximately produces 1000 Btu of thermal energy.

2.11. A fixed-sheet shell-and-tube heat exchanger has a heat-transfer area of 100 m2 and an installed cost of $100,000. What is the cost of a similar heat exchanger that has a heat-transfer area of 50 m2?

2.12. Using a cost chart, estimate the current purchased cost of a cast-steel centrifugal pump that requires 100 kW of power. What is the cost if the pump is made of 316 stainless steel?

2.13. A mass integration project requires four separation units whose FCI is $50.0 MM. The useful life period of the units is taken to be 10 years. The salvage value of the units is 10 percent of the FCI. What is the annual depreciation charge using the straight-line method?

2.14. A retrofitting project has an initial FCI of $28 MM and a projected salvage value $3 MM. The recovery period for the equipment is taken as 10 years.
a. Using the DDB method, calculate the annual depreciation charges and the book values over the equipment life.
b. Compare the book value of the equipment over the useful life period using the DDB method (with linear adjustment in the last year using the combined depreciation method) versus the straight-line method.

2.15. Using the MACRS method, calculate the annual depreciation charges for a unit with an initial cost of $100.00 and a recovery period of:
a. 5 years for which the DDB method is used.
b. 10 years for which the DDB method is used.
Comment on your results.

2.16. A process has a fixed capital investment of $500 MM. The useful life period of the process is taken to be 10 years. The salvage value of the process is $50 MM. Other fixed charges for the process (property taxes, insurance, salaries, and so on) are $30 MM/yr. The operating cost of the process is $250/ton. Maximum production capacity of the process is 300,000 ton/yr. The selling price of the product is $1000/ton. What is the break-even production rate (ton/yr)?

2.17. A process has been designed to produce a new product P. Based on your design calculations, the fixed capital investment of the process is estimated to be $500 MM. The useful life period of the process is taken to be 10 years. The salvage value of the process is assumed to be 10 percent of the fixed capital investment. Other fixed charges for the process (property taxes, insurance, salaries, and so on) are $30 MM/yr. The maximum production capacity of the process is 300,000 tons/yr of product P. The operating cost of the process is $250/ton. It is desired to break even at a production rate of 33.3 percent of the maximum process capacity. What should be the selling price of the product?

2.18. Consider a $10,000 deposit in a bank account that earns 6 percent of annual interest rate. How much money will be available in that account after 10 years?

2.19. Consider a $10,000 deposit in a bank account that earns 0.5 percent of interest rate compounded monthly. How much money will be available in that account after 10 years? Comment on the results of this problem compared with those of the previous problem.

2.20. A chemical engineer would like to buy a house using an annuity scheme paid at the end of each year for the next 20 years at an 8 percent annual interest rate. The engineer can pay $20,000 today as a down payment and is willing to pay an annuity of up to $24,000 per year (paid at the end of each year). What is the maximum price of the house the engineer can afford?

2.21. A proposed project requires a total capital investment of $40 MM and provides an annual net (after-tax) profit of $10 MM/yr. What is the ROI of the project? Assuming that the working capital investment of the project is 15 percent of the total capital investment of the project, what is the payback period of the project?

2.22. A heat integration project results in saving 5 MM Btu/hr of heating utility and 14 MM Btu/hr of cooling utility. The prices of heating and cooling utilities are $8/MM Btu and $10/MM Btu, respectively. The process operates for 8000 hrs per year. The project requires the installation of three heat exchangers, pumps, and pipeline. The fixed capital investment of the project is $4.0 MM. The working capital investment is taken as 15/85 of the fixed capital investment. The annual operating cost of the project (for pumping the integrated streams) is $0.5 MM/yr.

Depreciation is calculated over 10 years with no salvage value. The corporate tax rate for the project is 25 percent of the annual taxable gross profit. What is the payback period of the project?

2.23. A solvent recovery project has the following characteristics:
- Fixed capital investment: $4.0 MM
- Working capital investment: $400,000 (fully recoverable at $400,000 at the end of the useful life of the project)
- Useful life period: 10 years
- Salvage value: $500,000
- Recovered solvent: 5 MM lb/yr
- Value of the solvent: $0.2/lb
- Annual operating cost: $200,000/yr
- The corporate tax rate for the project is 30 percent of the annual taxable gross profit

Calculate the NPV of the project for two values of the annual discount rate: 10 percent and 15 percent. Comment on your results.

2.24. A debottlenecking project has the following characteristics:
- Total capital investment: $3.6 MM
- Annual after-tax cash flow = $1 MM/yr
- Useful life period of the project = 10 years
- Salvage value and working capital at the end of the project = $400,000

Calculate the discounted cash flow return on investment for the project. If the company has a 15 percent minimum acceptable return on investment, what would you recommend for the company?

2.25. Consider a process for the production of a specialty chemical. The land is purchased for $40 MM. The FCI is spent over the next two years for a total of $660 MM. The WCI ($70 MM) is spent toward the end of the second year. Operation starts at the beginning of the third year. Table 2.36 shows the estimated annual production, selling price of the product, and annual operating cost (which accounts for raw materials, utilities, labor, maintenance, and so on). The project has an anticipated useful life period of 10 years from the start of operation. A linear depreciation over 10 years is used. At the end of the project, the WCI and salvage value are estimated to $70 MM and $60 MM, respectively. A tax rate of 35 percent is applied. The company uses a discount rate of 15 percent. Calculate the NPV, DCF ROI, and discounted PBP for the project.

2.26. An energy-conservation study has produced a base case, which is a heat-recovery project. The total capital investment is $5.0 MM and the project provides an annual net profit of $1.0 MM/yr. Two add-on alternatives are proposed as possible modifications to the base case:
- Alternative 1: The addition of an additional heat exchanger that will cost an additional $2.0 MM of TCI and will result in a total annual net profit of $1.8 MM/yr for the combined base case and the additional heat exchanger.
- Alternative 2: The addition of a steam turbine for cogeneration of heat and power. The turbine coupled with the heat-recovery network and the additional heat exchanger. The turbine will cost $4.0 MM in addition to the TCI of the base-case heat-recovery network and the additional heat exchanger. It will result in a total annual net profit of $2.2 MM/yr for the combined base case, additional exchanger, and turbine.

The company has a 15 percent minimum hurdle rate. Which projects should you recommend?

REFERENCES

Axtell O, Robertson JM: *Economic evaluation in the chemical process industries*, New York, 1986, John Wiley.

Bao B, El-Halwagi MM, Elbashir NO: Simulation, integration, and economic analysis of gas-to-liquid processes, *Fuel Process Technol* 91(7):703–713, 2010.

Bridgwater AV, Mumford CJ: *Waste recycling and pollution control handbook*, Chapter 20 UK, 1979, George Godwin.

Brown T: *Engineering economics and economic design for process engineers*, Boca Raton, Florida, 2007, CRC Press.

Chauvel A: *Manual of economic analysis of chemical processes*, New York, 1981, McGraw Hill.

Christensen P, Dysert LR: *Cost estimate classification system: as applied in engineering, procurement, and construction for the process industries*, The Association for the Advancement of Cost Engineering (AACE) International Recommended Practice No. 18R-97, Morgantown, West Virginia, 2005, AACE Publications.

Coker AK: *Ludwig's applied process design for chemical and petrochemical plants*, ed 4, vol 1, Oxford, UK, 2007, Gulf Professional Publishing, an imprint of Elsevier.

Couper JR: *Process engineering economics*, New York, 2003, Marcel Dekker.

Dysert L: Sharpen your capital cost estimation skills, *Chem Eng*:70, Oct. 2001.

Elms RD, El-Halwagi MM: Optimal scheduling and operation of biodiesel plants with multiple feedstocks, *Int J Process Syst Eng* 1(1):1–28, 2009.

Elms RD, El-Halwagi MM: The effect of greenhouse gas policy on the design and scheduling of biodiesel plants with multiple feedstocks, *Clean Technol Environ Policy* in press.

Gabriel F, El-Halwagi MM: Simultaneous synthesis of waste interception and material reuse networks: problem reformulation for global optimization, *Environ Prog* 24(2):171–180, July 2005.

Garrett DE: *Chemical engineering economics*, New York, 1989, Van Nostrand Reinhold.

Gerrard AM: *Guide to capital cost estimating*, ed 4, UK, 2000, Rugby. IChemE.

Hand WE: From flow sheet to cost estimate. *Petroleum Refiner*, September 1958, pp 331–334.

Hydrocarbon Processing: Petrochemical processes, *Handbook*. Gulf Publishing Company, Houston, USA, March 2003.

Kazi FK, Fortman J, Anex R, Kothandaraman G, Hsu D, Aden A, et al.: Techno-economic analysis of biochemical scenarios for production of cellulosic ethanol, NREL Technical Report NREL/TP-6A2-46588, Golden, Colorado, 2010, National Renewable Energy Laboratory.

Lang HJ: Simplified approach to preliminary cost estimates, *Chem Eng* June: 112–113, 1948.

McAloon A, Taylor F, Wee W, Ibsen K, Wooley R: *Determining the cost of producing ethanol from corn starch and lignocellulosic feedstocks*. Technical Report NREL/

TABLE 2-36 Data for Problem 2.25

End of Year	Estimated Annual Production (MM tonne/yr)	Estimated Selling Price of Product ($/tonne)	Estimated Annual Operating Cost ($ MM/yr)
3	1.5	400	230
4	2.5	410	380
5	2.5	410	380
6	2.5	410	380
7	2.5	440	390
8	2.5	440	390
9	2.0	460	390
10	2.0	470	390
11	2.0	480	390
12	1.5	500	400

TP-580-28893, Golden, Colorado, 2000, National Renewable Energy Laboratory.

Myint LL, El-Halwagi MM: Process analysis and optimization of biodiesel production from soybean oil, *J Clean Technol Environ Policies* 11(3):263–276, 2009.

Peters MS, Timmerhaus KD, West RE: *Plant design and economics for chemical engineers*, ed 5, New York, 2003, McGraw Hill.

Pham V, Holtzapple MT, El-Halwagi MM: Techno-economic analysis of biomass to fuel via the MixAlco process, *J Ind Microbiol Biotechnol* 37(11), 1157–1168, 2010.

Pokoo-Aikins G, Nadim A, Mahalec V, El-Halwagi MM: Design and analysis of biodiesel production from algae grown through carbon sequestration, *Clean Technol Environ Policy* 12:239–254, 2010.

Remer DS, Chai LH: Estimate costs of scaled-up process plants, *Chem Eng* April:138–175, 1990.

Remer DS, Chai LH: Design cost factors for scaling-up engineering equipment, *Chem Eng Prog* August:77–82, 1990.

Salem AB: Estimate capital costs fast, *Hydrocarbon Process* September:199–201, 1981.

Seider WD, Seader JD, Lewin DR, Widagdo S: *Product and process design principles: synthesis, analysis, and evaluation*, ed 3, New York, 2009, John Wiley and Sons Inc.

Spath P, Aden A, Eggeman T, Ringer M, Wallace B, Jechura J: *Biomass to hydrogen production detailed design and economics utilizing the Battelle Columbus Laboratory indirectly-heated gasifier*, Technical Report #TP 510-37408 Golden, Colorado, 2005, National Renewable Energy Laboratory. NREL.

Timms SRM, Phil M: Thesis, UK, 1980, Aston University.

Towler G, Sinnott R: *Chemical Engineering Design: Principles, Practice and Economics of Plant and Process Design*, Oxford, UK, 2008, Butterworth-Heinemann/Elsevier.

Turton R, Bailie RC, Whiting WB, Shaeiwitz JA: Analysis, synthesis, and design of chemical processes, ed 3, Prentice Hall, Upper Saddle River, New Jersey, USA, 2009.

Ulrich GD, Vasudevan PT: *Chemical engineering: process design and economics, a practical guide*, ed 2, Durham, NH, 2004, Process Publishing.

United States Internal Revenue Service: IRS Publication 946, How to Depreciate Property, 2009.

Vatavuk WM: A potpourri of equipment prices, *Chem Eng* August:68–73, 1995.

CHAPTER 3: BENCHMARKING PROCESS PERFORMANCE THROUGH OVERALL MASS TARGETING

Benchmarking the performance of an industrial facility refers to the determination of a standard of excellence to which the process performance can be compared. This is an important step in assessing the prospects for improving the performance of a process and in evaluating indicators and metrics for process efficiency and sustainability. In this context, the concept of targeting is very useful. **Targeting** refers to the identification of performance benchmarks that can be determined ahead of carrying out a detailed design (for new processes) or without conducting an in-depth analysis (for existing processes). The overarching philosophy in targeting is *"big picture first, details later,"* and the emphasis is on using minimum data and calculations to identify performance limits. Overall mass targeting is intended to evaluate benchmarks for the flows of streams and chemical species for the whole process or a section of the process. Examples of overall mass targets include:

- Maximum yield of desired products or by-products
- Minimum usage of raw materials
- Minimum usage of material utilities (for example, solvents, water)
- Minimum discharge of pollutants and waste streams

This chapter presents systematic tools for determining overall mass targets. The focus is on evaluating these targets without deciding which technologies should be selected, what design changes must be made, and how the detailed economic analysis impacts the targets. In other words, what is the best possible performance of the process irrespective of the type of available technologies or the cost involved in undertaking process changes? Subsequent chapters will provide the details on how these targets can be attained, how to synthesize and screen design alternatives, and how the cost-benefit analysis is used to generate cost-effective designs.

This chapter discusses two categories of overall mass targeting that are based on:

- Stoichiometric calculations, which are used for when there are very limited data and information for the process
- Mass integration, which is used for existing processes or process designs with sufficient details

STOICHIOMETRY-BASED TARGETING

Suppose that there is interest in a chemical pathway that is known to convert certain feedstocks (raw materials) into products. Before designing an industrial process based on this chemical pathway and assessing its technical and economic feasibility, it is desired to estimate the flows of the key feedstocks and products. For targeting purposes, it is reasonable to focus on two key systems of the process: reaction and separation as shown by Fig. 3.1. Each system may be composed of several units. The feedstocks are converted in the reaction system into a number of products, by-products, and waste materials. These chemicals are separated into the main product(s) and by-products to be sold, the waste materials to be treated and discharged, and the unreacted raw materials to be recycled to the reaction system.

The ensuing sections discuss two types of targets based on stoichiometry:

1. Stoichiometric targeting
2. Stoichiometric-economic targeting

STOICHIOMETRIC TARGETING

At the initial stages of assessing the process, very limited data are available. To perform an overall mass targeting of the product, stoichiometric calculations provide a convenient tool for benchmarking. Depending on the type of available data, three levels of stoichiometric targets may be determined:

1. Theoretical stoichiometric targets with full product recovery (see Fig. 3.2a): When only the process chemistry is available in the form of an overall reaction, the stoichiometric calculations are carried out assuming maximum reaction yield and full recovery of the product.
2. Actual stoichiometric targets without product losses (see Fig. 3.2b): When the process chemistry

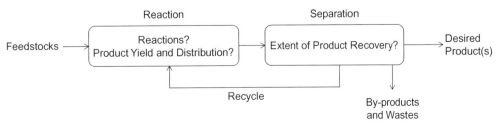

FIGURE 3-1 A generic process with reaction and separation systems.

Sustainable Design Through Process Integration.
© 2012 Elsevier Inc. All rights reserved.

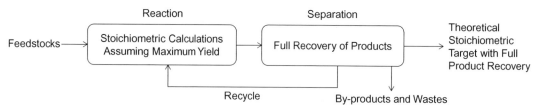

FIGURE 3-2a Theoretical stoichiometric targeting.

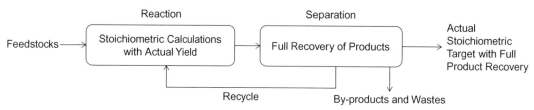

FIGURE 3-2b Actual stoichiometric targeting.

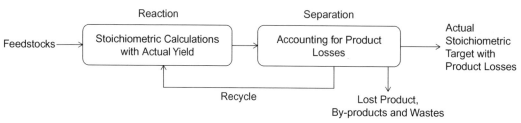

FIGURE 3-2c Accounting for product losses in separation.

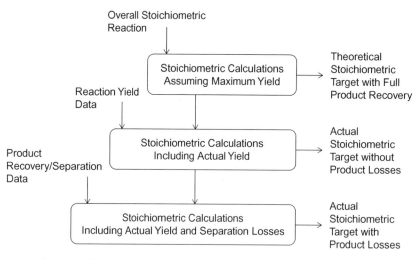

FIGURE 3-3 Data, calculations, and outputs of the three levels of stoichiometric targeting.

is available in the form of an overall reaction along with the actual yield data for the product (from experiments, thermodynamic-equilibrium models, or reaction models), the stoichiometric calculations are carried out using the actual reaction yield and full recovery of the product.

3. Actual stoichiometric targets with product losses (see Fig. 3.2c): When the process chemistry is available in the form of an overall reaction along with data on the actual yield of the product and its expected fractional recovery in the separation systems, the stoichiometric calculations are carried out using the actual reaction yield while accounting for the expected losses of the product.

Figure 3.3 summarizes the type of data, calculations, and outcomes of the three levels of stoichiometric targets.

STOICHIOMETRIC-ECONOMIC "STOICHIO-NOMIC" TARGETING

The overall mass targets based on stoichiometric calculations determined in the previous section provide benchmarks for the flows of reactants and products. These targets can be coupled with chemical-price data to quickly determine whether or not the proposed process has a potential for economic viability. This form of stoichiometric-economic targeting (which may be abbreviated as stoichio-nomic targeting) is very useful in making initial decisions on a proposed process when very limited data are available for the process. Two stoichio-nomic targets can be calculated. The first target is called the **economic gross potential** (EGP) of the process (Crabtree and El-Halwagi, 1995) and is defined as:

[3.1]
$$EGP = \sum_{p=1}^{N_{Products}} \text{Annual production rate of product p*Selling price of product p} - \sum_{r=1}^{N_{Reactants}} \text{Annual feed rate of reactant r*Purchased price of reactant r}$$

where $N_{products}$ and $N_{Reactants}$ are the number of products and reactants, respectively.

By using stoichiometric targets for the maximum possible flow rate of the products and by accounting only for the chemical price of the reactants (without considering the rest of the operating costs or the fixed cost of the process), the value of EGP gives an upper bound on the revenue of the process. If the EGP of the process is negative or slightly positive, then the process is not economically viable (unless there is some form of subsidy) and, therefore, there is no need to spend more time and effort on detailing the design or analyzing the economics of the process. Therefore,

[3.2a]

EGP > 0   The process may be considered for a more detailed analysis; higher values of EGP are more desirable

[3.2b]

EGP ≤ 0   The process is not economically viable

It is worth noting that for a negative value of EGP, the process cannot be profitable unless there is a special revenue stream such as government subsidies. Furthermore, positive values of EGP do not necessarily guarantee profitability. This is attributed to the possible reduction in the value of EGP due to the factors unaccounted for in the targeting such as reaction and separation losses, the nonfeedstock operating costs (for example, utilities, labor, waste treatment), and the fixed capital investment of the process.

Another stoichio-nomic indicator is an analogous way of representing the data used in calculating the value of EGP for the process. It is referred to as the **Metric for Inspecting Sales and Reactants** (MISR) and is defined as:

[3.3]
$$MISR = \frac{\sum_{p=1}^{N_{Products}} \text{Annual production rate of product p* Selling price of product p}}{\sum_{r=1}^{N_{Reactants}} \text{Annual feed rate of reactant r* Purchased price of reactant r}}$$

High values of MISR are desirable with a minimum value of MISR = 1 required to further consider the process for a more detailed analysis. Noting that positive values of EGP correspond to MISR values greater than one, the values of MISR can be used for assessing the economic viability of the process as follows:

[3.4a]

MISR > 1   The process may be considered for a more detailed analysis; higher values of MISR are more desirable

---

Example 3-1   Stoichiometric targeting of ethanol production from glucose

A new process is to be designed for the conversion of 150 MM kg/yr of sugar to ethanol. The sugar is taken to be in the form of glucose ($C_6H_{12}O_6$) and is converted to ethanol (C_2H_5OH) through the following overall fermentation reaction:

[3.5] $C_6H_{12}O_6 \rightarrow 2C_2H_5OH + 2CO_2$

1. Calculate the maximum theoretical stoichiometric target for ethanol.
2. Available experimental data (Krishnan et al., 1999) show that the actual reaction yield that can be obtained is 0.46 kg ethanol/kg glucose. Determine the actual stoichiometric target for ethanol.
3. In separating ethanol from the reaction mixture, it is expected to lose 5 percent of ethanol with the wastewater stream. What is the actual stoichiometric target for ethanol when the separation losses are accounted for?

SOLUTION

1. To evaluate the theoretical target for ethanol, let us assume full conversion of glucose according to the overall stoichiometric reaction:

[3.5] $C_6H_{12}O_6 \rightarrow 2C_2H_5OH + 2CO_2$

Molecular weights: 180   2*46 + 2*44

Therefore,

[3.6a]   The theoretical yield of ethanol from glucose $= \dfrac{2*46}{180}$
$= 0.51$ kg ethanol/kg glucose

For a feed rate of 150 MM kg/yr of glucose, then the theoretical stoichiometric target of ethanol =

[3.6b]   kg ethanol/kg glucose*150 MM kg glucose/yr = 76.5 MM kg ethanol/yr

2. For the reported experimental yield, The actual stoichiometric target of ethanol =

[3.7]   0.46 kg ethanol/kg glucose*150 MM kg glucose/yr = 69.0 MM kg ethanol/yr

which is 10 percent less than the theoretical stoichiometric target.

3. By accounting for 5 percent loss of ethanol in separation, The actual stoichiometric target of ethanol with product losses

[3.8] $= 0.95*69.0 = 65.6$ MM kg ethanol/yr

> ### Example 3-2 Stoichiometric targeting for a methanol plant
>
> A company is contemplating a process for the conversion of natural gas to methanol based on the following chemical pathway (Olah et al., 2006):
>
> First, steam reforming of methane is used to produce syngas (for simplicity, natural gas is considered to be methane):
>
> [3.9] $$CH_4 + H_2O \rightarrow CO + 3H_2$$
>
> The methanol synthesis reaction is:
>
> [3.10] $$CO + 2H_2 \rightarrow CH_3OH$$
>
> Because methanol synthesis requires a 1:2 stoichiometric ratio of CO to H_2, there is excess hydrogen produced from the steam-reforming reaction of methane. To take advantage of such hydrogen surplus, CO_2 is injected into the methanol-synthesis reactor where it reacts with methane through the "dry reforming" reaction:
>
> [3.11] $$CH_4 + CO_2 \rightarrow 2CO + 2H_2$$
>
> It is desired to produce 1.0 MM ton/yr of methanol. What is the target for minimum flow rate of natural gas to be fed to the process? Express the target in standard cubic foot (SCF) per year with the standard conditions taken as of 60°F and 1 atmosphere pressure (atm).
>
> **SOLUTION**
>
> For targeting purposes, assume that the three reactions are to be held in three reactors. Therefore, it is possible to adjust the flow rates to each reactor so as to meet desired ratios of chemical species. The overall reaction for the process can be obtained by combining the individual reactions each multiplied by a coefficient that ultimately yields the desired overall reaction. The chemical reactions for steam and dry reforming of methane (see Equations 3.9 and 3.11) should be combined in a way that produces the 1:2 stoichiometric ratio of CO to H_2 required for methanol synthesis. Hence, Eq. 3.9 is multiplied times two then added to Eq. 3.11 to get:
>
> [3.12] $$3CH_4 + 2H_2O + CO_2 \rightarrow 4CO + 8H_2$$
>
> Recalling the methanol-synthesis reaction (Eq. 3.10), we have:
>
> [3.13a] $$3CH_4 + 2H_2O + CO_2 \rightarrow 4CO + 8H_2 \rightarrow 4CH_3OH$$
>
> or shortly,
>
> [3.13b] $$3CH_4 + 2H_2O + CO_2 \rightarrow 4CH_3OH$$
>
> Another method for deriving the overall reaction (Eq. 3.13b) without invoking the individual steps is to use atomic balances for the postulated stoichiometry. Let us give the product a stoichiometric coefficient of one while assigning unknown coefficients a, b, and c for CH_4, $2H_2O$, and CO_2, respectively, that is,
>
> [3.14] $$aCH_4 + bH_2O + cCO_2 \rightarrow CH_3OH$$
>
> The atomic balances for carbon, hydrogen, and oxygen are respectively given by:
>
> [3.15a] $$a + c = 1$$
>
> [3.15b] $$4a + 2b = 4$$
>
> [3.15c] $$b + 2c = 1$$
>
> Multiplying Eq. 3.15a by 4 and subtracting from Eq. 3.15b, we get
>
> [3.15d] $$2b - 4c = 0 \text{ or } b = 2c$$
>
> Substituting from Eq. 3.15d into Eq. 3.15c, we obtain
>
> [3.15e] $$2c + 2c = 1 \text{ or } c = 0.25$$
>
> Combining the results from Equations 3.15d and 3.15e,
>
> [3.15f] $$b = 2*0.25 = 0.5$$
>
> which can be substituted into Eq. 3.15a to give:
>
> [3.15g] $$a = 1 - 0.25 = 0.75$$
>
> Substituting from Equations 3.15e through 3.15g into Eq. 3.14, we get
>
> [3.16a] $$0.75\,CH_4 + 0.50\,H_2O + 0.25\,CO_2 \rightarrow CH_3OH$$
>
> or
>
> [3.16b] $$3CH_4 + 2H_2O + CO_2 \rightarrow 4CH_3OH$$
>
> which matches Eq. 3.13b. Hence, for producing 4 moles of CH_3OH, the minimum target for methane is 3 moles. The desired production rate of methanol is 1.0 MM ton/yr, which corresponds to:
>
> [3.17]
> $$\text{Production rate of methanol in kilomole (kmol)/yr} = 1.0 \text{ MM ton/yr} * \frac{2000 \text{ lb}}{\text{ton}} * \frac{\text{lb mol}}{32 \text{ lb}}$$
> $$= 62.50 \text{ MM pound mole (lb mol) methanol/yr}$$
>
> Therefore,
>
> [3.18a]
> $$\text{Target for minimum flow rate of methane fed to the process} = \frac{3 \text{ lb mol methane}}{4 \text{ lb mol methanol}} * 62.50 \text{ MM lb mol methane/yr}$$
> $$= 46.88 \text{ MM lb mol methane/yr}$$
>
> At standard conditions of 60°F and 1 atm, one lb mol of gas occupies a volume of 379.5 SCF. Hence,
>
> [3.18b]
> $$\text{Target for minimum flow rate of methane fed to the process} = 46.88 \text{ MM lb mol methane/yr} * \frac{379.5 \text{ SCF}}{\text{lb mol}}$$
> $$= 17.791*10^9 \text{ SCF/yr}$$

[3.4b]

$$MISR \leq 1 \quad \text{The process is not economically viable}$$

As a rule of thumb, when alternative chemical pathways are considered, the pathway with the highest value of EGP or MISR should be prioritized for detailed analysis, followed by the other alternatives in a descending order of the value of the EGP or MISR.

MASS-INTEGRATION TARGETING

Stoichiometric targeting is used when limited data and information are available for the process. We now turn

> **Example 3-3 Stoichio-nomic targeting for CO_2 methanation**
>
> Carbon dioxide is one of the primary greenhouse gases (GHG) resulting from industrial processes. Laboratory experiments have shown that CO_2 can be converted to methane by hydrogenation over a composite catalyst via the following methanation reaction:
>
> [3.19] $$CO_2 + 4H_2 \rightarrow CH_4 + 2H_2O$$
>
> A new catalyst has been recently developed to induce high conversion of CO_2 and selectivity to CH_4 under reasonably mild conditions. An investment group, interested in reducing GHG emissions while making a profit, is considering the use of this methanation approach to convert CO_2 from industrial emissions to methane. Because CO_2 will be extracted from an industrial waste stream, it will be supplied free of charge. Hydrogen is available at \$2.00/kg. The value of methane is \$5.00/1000 SCF (at 60°F and 1 atm). The value of produced water is negligible compared to the value of methane.
>
> a. How would you advise the group of investors?
> b. If you were advising the government to offer a GHG-reduction incentive for this technology, what would you recommend as the minimum acceptable subsidy (\$/tonne CO_2)?
>
> **SOLUTION**
>
> a. Let us take a basis of 1000 SCF of methane. At standard conditions of 60°F and 1 atm, one lbmol of methane occupies a volume of 379.5 SCF. Therefore,
>
> [3.20] $$\text{Basis of 1000 SCF of methane} = 1000\,\text{SCF} * \frac{1\,\text{lb mol}}{379.5\,\text{SCF}} * \frac{1\,\text{kmol}}{2.2\,\text{lb mol}} * \frac{16\,\text{kg}}{\text{kmol}}$$
> $$= 19.2\,\text{kg methane}$$
>
> First, let us identify the theoretical target for hydrogen:
>
> [3.19] $$\begin{array}{c} CO_2 + 4H_2 \rightarrow CH_4 + 2H_2O \\ \text{Molecular weights:}\quad 44\quad 4*2\quad 16\quad 2*18 \end{array}$$
>
> For a basis of 1000 SCF (19.2 kg) of CH_4,
>
> [3.21] $$\text{Minimum target for hydrogen} = \frac{4*2\,\text{kg hydrogen}}{16\,\text{kg methane}} * 19.2\,\text{kg methane}$$
> $$= 9.6\,\text{kg hydrogen}$$
>
> [3.22] $$\text{Cost of hydrogen} = 9.6\,\text{kg hydrogen} * \$2.00/\text{kg hydrogen} = \$19.2$$
>
> As indicated in the problem statement, the selling value of 1000 SCF of methane is \$5.00, the value of produced water is negligible, and CO_2 in the feedstock is provided for free. Therefore, according to Eq. 3.1,
>
> [3.23] $$\text{EGP of the process on a basis of 1000 SCF of methane} = 5.00 - 19.2$$
> $$= \$-14.2$$
>
> Alternatively, according to Eq. 3.3,
>
> [3.24] $$\text{MISR} = \frac{5.00}{19.20} = 0.26$$
>
> Because the EGP of the process is negative or the MISR value less than one, the process should not be pursued. If the technology is not viable under stoichiometric targeting, then economic loss of the actual plant will be further exacerbated by virtue of the reaction and separation losses and the additional operating and fixed costs. This big-picture targeting provides valuable advice to the investors prior to the unnecessary pursuit of an unattractive technology. It is worth mentioning that methanation is useful in other applications such as the removal of trace amounts of CO_2 (or CO) from gaseous mixture to avoid catalyst poisoning of catalyst. In such cases, methanation is used for operational purposes and not for the value of produced methane. It is also noteworthy that the reverse reaction of methanation is the steam reforming of methane. The large difference between the value of hydrogen and the cost of methane provides an attractive EGP (\$13.3 for a basis of 1000 SCF of methane or slightly less if the cost of steam is considered), which is the main reason why steam reforming of methane is a principal route for the production of hydrogen.
>
> b. The methanation technology contributes to the reduction of CO_2 emissions. Let the minimum government subsidy be S_{CO_2} (\$/tonne CO_2). To achieve a zero value of the EGP (or MISR = 1) of the process for the basis of 1000 SCF of methane,
>
> $$5.00 + S_{CO_2} * \frac{44\,\text{kg } CO_2}{16\,\text{kg methane}} * 19.2\,\text{kg methane} * \frac{\text{tonne } CO_2}{1000\,\text{kg } CO_2} - 19.2 = 0$$
>
> Hence,
>
> [3.25] $$S_{CO_2} = \$269/\text{tonne } CO_2$$
>
> This is an extremely high subsidy. In reality, the subsidy will have to be even higher given the aforementioned reasons associated with the assumptions involved in stoichio-nomic targeting. The government will be well advised to invest in other GHG-reduction technologies.

our attention to the benchmarking of the performance of existing processes or process designs with sufficient details (for example, flow sheet, mass balances, process models). In this context, mass integration is a very effective framework for establishing targeting tools and the strategies to reach the target. **Mass integration** is a systematic methodology that provides a fundamental understanding of the global flow of mass within the process and employs this understanding in identifying performance targets and optimizing the generation and routing of species throughout the process. In this section, we address the overall mass targeting for existing processes (or process designs with sufficient details) using big-picture approaches of mass integration. The following cases will be presented:

1. Minimum waste discharge
2. Minimum fresh usage
3. Maximum product yield

TARGETING FOR MINIMUM WASTE DISCHARGE

Consider the case when it is desired to determine the target for minimizing the load of a targeted species discharged out of the process (for example, pollutants in effluents). The following discussion is taken from the approach proposed by El-Halwagi (2006). Three sets of data for that species are first collected: fresh usage, terminal discharge, and generation/depletion. The fresh usage (F) refers to the amount of the targeted species in the streams entering the process (the waste stream may have entered the process as a fresh feedstock or a material utility). The terminal discharge (T) corresponds to the load of the targeted species in streams designated as waste streams or point sources for pollution. Generation (G) refers to the net amount of the targeted species, which is produced through chemical reaction. Depletion (D) may

Example 3-4 Stoichio-nomic targeting for a corn-to-ethanol plant

A company is considering the design and installation of a new corn-to-ethanol plant. The plant is to process 10 MM bushels of corn per year. The composition of corn is given in Table 3.1. Hydrolysis and fermentation are used to convert the corn content of starch, cellulose, sugars, and hemicellulose to ethanol. For targeting purposes, the following molecular formulas are assumed for the building blocks for the species to be converted to ethanol: $C_6H_{10}O_5$ (for starch and cellulose), $C_6H_{12}O_6$ (for sugar), and $C_5H_{10}O_5$ (for hemicellulose). It is worth noting that although starch and cellulose are polymeric carbohydrates (polysaccharides) with the general formula $(C_6H_{10}O_5)_n$, considering the building block $C_6H_{10}O_5$ is appropriate for targeting purposes because the ratio of the product to the reactant is the same regardless of the value of n. The same argument is also valid for hemicellulose, which is composed of several polymerized sugars.

a. What is the theoretical stoichiometric target for ethanol production?
b. If the actual ethanol yield of the process is expected to be 80 percent of the theoretical target, corn is purchased at $5.50 per bushel (weighing 25.4 kg), and ethanol is sold for $2.65 per gallon, perform a stoichio-nomic targeting of the process.

SOLUTION

a. First, the flows are converted to MM kg/yr.

[3.26a] $\quad$ The feed flow rate of corn = 10 MM bushels/yr * $\frac{25.4 \text{ kg}}{\text{bushel}}$
$$= 254 \text{ MM kg/yr}$$

Hence, using the compositions reported in Table 3.1, we have:

[3.26b] $\quad$ The feed flow rate of starch and cellulose = (0.62 + 0.04)*254 MM
$$= 167.64 \text{ MM kg/yr}$$

[3.26c] $\quad$ The feed flow rate of sugar = 0.02*254 MM
$$= 5.08 \text{ MM kg/yr}$$

[3.26d] $\quad$ The feed flow rate of hemicellulose = 0.06*254 MM
$$= 15.24 \text{ MM kg/yr}$$

Let us develop the stoichiometric reactions and theoretical yields. For starch and cellulose, the stoichiometric equation is hydrolysis with one mole of water (to give glucose), which then becomes similar to the fermentation reaction given by Eq. 3.5, that is,

[3.27] $\quad C_6H_{10}O_5 + H_2O \rightarrow C_6H_{12}O_6 \rightarrow 2C_2H_5OH + 2CO_2$

Molecular weights: $\quad$ 162 $\quad$ + 18 $\quad\quad$ 180 $\quad\quad$ 2*46 $\quad$ + 2*44

Therefore,

[3.28] $\quad$ The theoretical yield of ethanol from starch and cellulose = $\frac{2*46}{162}$
$$= 0.57 \text{ kg ethanol/kg starch or cellulose}$$

TABLE 3-1 Composition of Corn

Ingredient	Weight Percentage (wt%)
Starch	62
Cellulose	4
Sugars	2
Hemicellulose	6
Protein	6
Oil	4
Ash	1
Water	15

For sugar, the stoichiometric equation and target have been developed in Example 3.1:

[3.5] $\quad C_6H_{12}O_6 \rightarrow 2C_2H_5OH + 2CO_2$

Molecular weights: $\quad$ 180 $\quad\quad$ 2*46 $\quad$ + 2*44

Therefore,

[3.6a] $\quad$ The theoretical yield of ethanol from sugar = $\frac{2*46}{180}$
$$= 0.51 \text{ kg ethanol/kg sugar}$$

For hemicellulose, let us give the hemicellulose a stoichiometric coefficient of one while assigning unknown coefficients a, b, and c for H_2O, C_2H_5OH, and CO_2, respectively; that is,

[3.29] $\quad C_5H_{10}O_5 + aH_2O \rightarrow bC_2H_5OH + cCO_2$

The atomic balances for carbon, hydrogen, and oxygen are respectively given by:

[3.30a] $\quad 5 = 2b + c$

[3.30b] $\quad 10 + 2a = 6b$ or $5 = -a + 3b$

[3.30c] $\quad 5 + a = b + 2c$

Rearranging Eq. 3.30a as:

[3.30d] $\quad c = 5 - 2b$

then substituting into Eq. 3.30c, we obtain

[3.30e] $\quad 5 + a = b + 10 - 4b$ or $5 = a + 3b$

Solving Equations 3.30b and 3.30e simultaneously, we get:

[3.30f] $\quad a = 0$

and

[3.30g] $\quad b = \frac{5}{3}$

Substituting from Eq. 3.30g into Eq. 3.30d, we obtain

[3.30h] $\quad c = \frac{5}{3}$

Therefore, the hemicellulose-to-ethanol stoichiometric equation can be represented by:

[3.31] $\quad C_5H_{10}O_5 \rightarrow \frac{5}{3}C_2H_5OH + \frac{5}{3}CO_2$

Molecular weights: $\quad$ 150 $\quad\quad \frac{5}{3}*46 + \frac{5}{3}*44$

Therefore,

[3.32] $\quad$ The theoretical yield of ethanol from hemicellulose = $\frac{5*46}{3*150}$
$$= 0.51 \text{ kg ethanol/kg hemicellulose}$$

(Continued)

Example 3-4 Stoichio-nomic targeting for a corn-to-ethanol plant (*Continued*)

Hence,
The theoretical stoichiometric yield of ethanol

[3.33]
$$\begin{aligned}
&= 167.64 \text{ MM kg starch and cellulose/yr}*0.57 \text{ kg ethanol/kg starch or cellulose} \\
&\quad + 5.08 \text{ MMkg sugar/yr}*0.51 \text{ kg ethanol/kg sugar} \\
&\quad + 15.24 \text{ MM kg hemicellulose/yr}*0.51 \text{ kg ethanol/kg hemicellulose} \\
&= 105.92 \text{ MM kg ethanol/yr}
\end{aligned}$$

The ethanol density is 0.79 kg/lit (where lit stands for liter) or 0.79 kg/lit*3.7854 lit/gal (where gal stands for gallon) = 2.99 kg/gal. Hence,

[3.34a]
$$\begin{aligned}
\text{The theoretical stoichiometric yield of ethanol} &= \frac{105.92 \text{ MM kg ethanol/yr}}{2.99 \text{ kg/gal}} \\
&= 35.42 \text{ MM gal ethanol/yr}
\end{aligned}$$

It is useful to report this theoretical target as a ratio to the used corn; that is,

[3.34b]
$$\begin{aligned}
\text{The theoretical stoichiometric yield of ethanol} &= \frac{35.42 \text{ MM gal ethanol/yr}}{254,000 \text{ tonne corn/yr}} \\
&= 139 \text{ gal ethanol/tonne corn (wet basis)}
\end{aligned}$$

or

[3.34c]
$$\begin{aligned}
\text{The theoretical stoichiometric yield of ethanol} &= \frac{35.42 \text{ MM gal ethanol/yr}}{10 \text{ MM bushels corn/yr}} \\
&= 3.54 \text{ gal ethanol/bushel corn}
\end{aligned}$$

It is also interesting to assess the land usage needed to produce ethanol. Suppose that 150 bushels are produced per acre per year, therefore

[3.35]
$$\begin{aligned}
\text{The theoretical stoichiometric yield of ethanol} &= 3.54 \text{ gal ethanol/bushel corn}*150 \text{ bushels/(acre.yr)} \\
&= 531 \text{ gal ethanol/(acre.yr)}
\end{aligned}$$

b. Because the actual target is 80 percent of the theoretical yield, then

[3.36a] The actual target of ethanol = 0.80*35.42 MM = 28.34 MM gal ethanol/yr

[3.36b]
$$\begin{aligned}
\text{Selling price of produced ethanol} &= 28.34 \text{ MM gal ethanol/yr}*\$2.65/\text{gal ethanol} \\
&= \$75.10 \text{ MM/yr}
\end{aligned}$$

The cost of water reacted in the process is negligible compared to the cost of corn. Therefore, the cost of feedstocks is primarily that of corn, which is calculated as:

[3.37]
$$\begin{aligned}
\text{Cost of corn fed to the process} &= 10 \text{ MM bushels corn/yr}*\$5.50/\text{bushel} \\
&= \$55.00 \text{ MM/yr}
\end{aligned}$$

Therefore, according to Eq. 3.1,

[3.38a]
$$\begin{aligned}
\text{EGP of the process} &= 75.10 \text{ MM/yr} - 55.00 \text{ MM/yr} \\
&= \$20.10 \text{ MM/yr}
\end{aligned}$$

Alternatively, according to Eq. 3.3,

[3.38b] $$\text{MISR} = \frac{75.10}{55.00} = 1.37$$

It is important to recall that the stoichio-nomic targeting using EGP or MISR only accounts for the selling price of the product and the purchased cost of the reactants. The actual process will incur the additional operating costs and the fixed cost of the plant. Therefore, the economic potential of this process is not very attractive. Furthermore, if the price of corn increases as a result of the competition between the food and the fuel, the economic potential of the process will further deteriorate. Figure 3.4 shows the effect of corn price on EGP of the process.

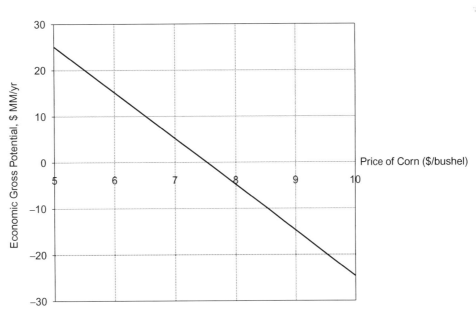

FIGURE 3-4 Effect of corn price on EGP of the ethanol process.

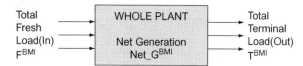

FIGURE 3-5 Overall material balance for the targeted species before mass integration. *Source: El-Halwagi (2006).*

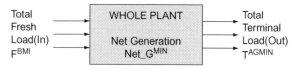

FIGURE 3-6 Overall material balance for the targeted species after minimizing the net generation. *Source: El-Halwagi (2006).*

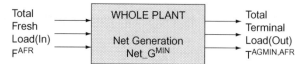

FIGURE 3-7 Overall material balance of the targeted species after fresh reduction and net-generation minimization. *Source: El-Halwagi (2006).*

take place through chemical reactions but it may also be attributed to leaks, fugitive emissions, and other losses that are not explicitly accounted for. The net generation (Net_G) of a targeted species is defined as the difference between generation and depletion; that is,

[3.39] $\quad Net\_G = G - D$

Figure 3.5 shows an overview of a typical process with inputs, generation/depletion, and terminal discharges. The overall mass balance for the targeted species is given by:

[3.40] $\quad T^{BMI} = F^{BMI} + Net\_G^{BMI}$

where T^{BMI}, F^{BMI}, and Net_G^{BMI}, respectively, refer to the terminal load, fresh load, and net generation of the targeted species before any mass integration changes.

When the objective is to reduce the terminal discharge of the targeted species, the two terms on the right-hand side of Eq. 3.40 are to be reduced. Noureldin and El-Halwagi (1999) describe the general solution strategy. However, there are several special cases for which simplified targeting methods can be developed. An example is when the net generation is independent of stream recycle and adjustments in fresh feed. In such cases, the following two-step shortcut method can be used to determine the minimum-discharge target. In the first step, the net generation of the targeted species is minimized. The net generation can be described in terms of the process design and operating variables that are allowed to be modified. The value of the minimum net generation of the targeted species is referred to as Net_G^{MIN}. As shown by Fig. 3.6, once the net generation has been minimized, the terminal discharged load after generation minimization (T^{AGMIN}) can be calculated through an overall material balance on the targeted species:

[3.41] $\quad T^{AGMIN} = F^{BMI} + Net\_G^{MIN}$

Next, the fresh (external) usage of the targeted species is minimized. Toward this objective, it is necessary to consider all the external streams entering the process that contain the targeted species. First, the appropriate design and operating variables of the process are manipulated to reduce (to the extent feasible) the fresh supply of the targeted species. The appropriate design and operating variables are the ones that influence the fresh usage of the targeted species and are allowed to be manipulated. Using process simulation, actual process data, or modeling (first principles, semi-empirical, or empirical), the dependence of the fresh usage on these manipulated variables may be expressed as:

[3.42] $\quad F = f(\text{manipulated design variables, manipulated operating variables})$

This expression can be minimized to identify the minimum load of the targeted species in the fresh feeds after reduction (designated by F^{AFR}). As can be seen from Fig. 3.7, the terminal load after minimization of net generation and reduction in fresh feed is calculated through the following mass balance:

[3.43] $\quad T^{AGMIN, AFR} = F^{AFR} + Net\_G^{MIN}$

Additional reduction in the fresh usage can be achieved through recovery and recycle. In principle, given sufficient financial resources and technologies, it is possible to recover almost all of the targeted species from the terminal streams or from paths leading to the terminal streams and render the recovered species in a condition that enables its use in lieu of the targeted species in the fresh feed. Clearly, the higher the recovery and recycle to replace the fresh load, the lower the net fresh use and the lower the terminal discharge. Consequently, to minimize the terminal discharge of the targeted species, we should recycle the maximum amount from terminal streams (or paths leading to terminal streams) to replace fresh feed. The maximum recyclable load of the targeted species is the lower of the two loads: the fresh and the terminal; that is,

[3.44] $\quad R^{MAX} = \text{argmin} \{F^{AFR}, T^{AGMIN, AFR}\}$

where argmin refers to the lowest value in the set of loads (in this case, the fresh and terminal loads).

When the recycle is implemented, the revised material balance for the targeted components after mass integration becomes:

[3.45] $\quad F^{AMI} = F^{AFR} - R^{MAX}$

This result is shown by Fig. 3.8. Therefore, the target for minimum discharge of the targeted species after

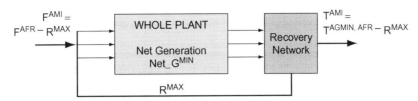

FIGURE 3-8 Targeting for minimum discharge of targeted species. *Source: El-Halwagi (2006).*

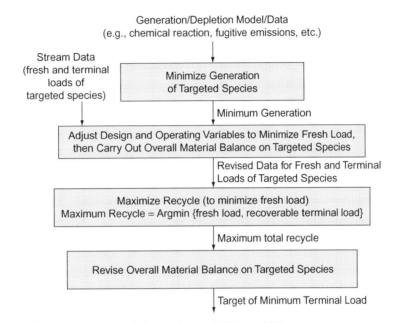

FIGURE 3-9 Procedure for identifying minimum waste discharge. *Source: El-Halwagi (2006).*

mass integration (T^{AMI}) can be calculated through the following overall material balance after mass integration:

$$[3.46] \qquad T^{AMI} = F^{AMI} + Net\_G^{MIN}$$

Alternatively, the target for minimum discharge can be calculated from a material balance around the recovery and recycle system:

$$[3.47] \qquad T^{AMI} = T^{AGMIN, AFR} - R^{MAX}$$

It is important to emphasize that, at this stage, the objective is to determine an overall mass target for the specific species. The determination of the technical and economic details of implementing the targets (for example, the recovery units) is not necessary at this targeting level. Indeed, the spirit of such benchmarking is to determine the targets without getting entangled in the details of the implementation. The ensuing chapters will cover the tools that can be used to methodically synthesize and screen the alternatives that will reach the desired target at minimum cost.

The foregoing procedure can be summarized by the flowchart shown in Fig. 3.9. First, generation and depletion information are gathered as data or models. These include information on the depletion or generation of the targeted species through chemical reactions, nonpoint loss of the species through fugitive emissions or leaks, and so on. Data are also collected on the amount of the targeted species in fresh feeds entering the process or terminal streams leaving the process. The net generation is first minimized. Then, the fresh feed of the targeted species is minimized by adjusting design and operating variables. Next, a recovery system is placed to recover maximum recyclable load (the lower of fresh and terminal loads). Hence, the fresh feed is minimized as maximum recycle is used to replace fresh feed. Finally, an overall material balance is used to calculate the target for minimum waste discharge.

TARGETING FOR MINIMUM PURCHASE OF FRESH MATERIAL UTILITIES

We now move to the case of targeting minimum fresh usage of material utilities (for example, water, solvents, additives, and so on). The overall material balance can be written as:

$$[3.48] \qquad F^{BMI} = T^{BMI} - Net\_G^{BMI}$$

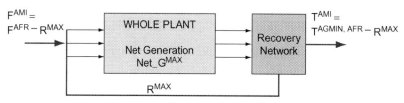

FIGURE 3-10 Targeting for minimum usage of material utilities. *Source: El-Halwagi (2006).*

For the special case when the net generation is independent of stream recycle and adjustments in fresh feed, we can use a procedure similar to the one described for minimum waste discharge. The key difference is that the net generation of the targeted species is maximized (as opposed to being minimized in the case of waste discharge). This can be deduced from the overall material balance given by Eq. 3.48, which entails maximizing the net generation to minimize the fresh load. The targeting scheme is shown in Fig. 3.10. It is worth pointing out that when the net generation is not to be altered, the target procedures for minimizing waste discharge and fresh usage become identical.

TARGETING FOR MAXIMUM PRODUCT YIELD

Maximization of the yield of a desired product is an important process objective. The following is a systematic procedure for benchmarking the process yield. It is a simplified version of the more general approach developed by Al-Otaibi and El-Halwagi (2006). First, it is important to distinguish between the reactor yield and the process yield. The actual yield of a reactor is the amount of the desired product actually obtained from the limiting reactant; that is,

[3.49]
$$\text{Actual yield of a reactor} = \frac{\text{Amount of desired product generated in the reactor}}{\text{Amount of limiting reactant fed to the reactor}}$$

where the limiting reactant is the chemical that is provided in the least amount compared to the stoichiometric requirement and will therefore be the first reactant to run out as the reaction proceeds. Suppose that a reaction produces an amount "b" of the desired product from an amount "a" of the limiting reactant. Therefore,

[3.50] Actual yield of a reactor "$\text{Yield}_{reactor}$" $= \dfrac{b}{a}$

As discussed earlier in the stoichiometric targeting, the actual yield may not attain the theoretical value of the yield for various reasons including the conversion of the limiting reactant to by-products and wastes through competing reactions (series, parallel), reverse reactions, and the selection of the design and operating conditions.

For overall mass targeting, it is important to consider the yield of the whole process and not just the reactor. The overall process yield is the amount of desired product going to sales obtained from a limiting fresh feed entering the process; that is,

[3.51]
$$\text{Process yield} = \frac{\text{Amount of desired product leaving the process to sales}}{\text{Amount of limiting fresh raw material entering the process}}$$

As an illustration, suppose that the amount of the limiting fresh material purchased and fed to the process is "A," which produces an amount "B" of the desired product that is to be sold. Hence,

[3.52] $$\text{Process yield} = \frac{B}{A}$$

Figure 3.11 is a schematic representation of a process, including the quantities involved in defining the yields of a reactor and of the whole process.

Although the reactor yield has a considerable effect on the overall process yield and plant economics, the overall process yield directly and considerably affects the process economics and capital productivity. Additionally, enhancing the process yield is closely tied to the conservation of raw materials and the minimization of waste discharge.

Before describing the targeting procedure, it is important to outline the following key causes of loss in the overall yield of the process:

- Loss in the allocation of raw materials to the reaction system
- Reaction yield not reaching its maximum
- Loss of product leaving the process in terminal streams other than the desired outlet stream to sales
- Insufficient recovery and recycle of unreacted raw materials

Therefore, the yield-maximization process is based on the following steps:

Step 1: Maximize routing of targeted raw material to the reaction system. Consider the process shown

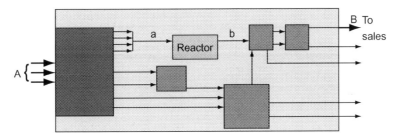

FIGURE 3-11 A generic process showing the quantities involved in evaluating the yields of the reactor and the process. *Source: Al-Otaibi and El-Halwagi (2006).*

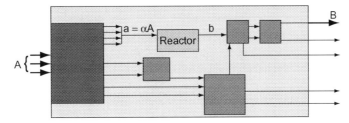

FIGURE 3-12a Evaluating feed to reactor. *Source: Al-Otaibi and El-Halwagi (2006).*

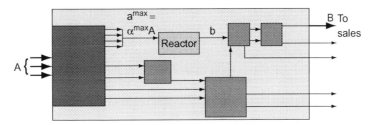

FIGURE 3-12b Maximizing routing of raw material to reactor. *Source: Al-Otaibi and El-Halwagi (2006).*

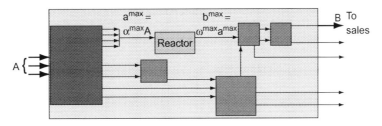

FIGURE 3-13 Maximization of the reaction yield. *Source: Al-Otaibi and El-Halwagi (2006).*

by Fig. 3.12a. Of the fresh feed amount "A," a fraction "α" is allocated to the reactor. Therefore, the objective of this step is to maximize α by using recovery units (revising the operation of existing ones if no additional capital investment is allowed or adding new units if additional capital investment is permitted) and adjusting design and operating condition to reroute the maximum possible amount of the limiting reactant to the reactor (Fig. 3.12b); that is,

[3.53] $$a^{max} = \alpha^{max} A$$

Step 2: Maximizing reactor yield. In this step, the design and operating variables of the reactor as well as the feed conditions are adjusted so as to maximize the yield of the desired product in the reactor (see Fig. 3.13); that is,

[3.53] $$b^{max} = \omega^{max} a^{max}$$

Step 3: Rerouting the product from undesirable outlets to desirable outlets. After the reactor, the

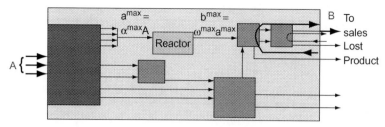

FIGURE 3-14 Routing product from undesirable outlets to desirable outlets. *Source: Al-Otaibi and El-Halwagi (2006).*

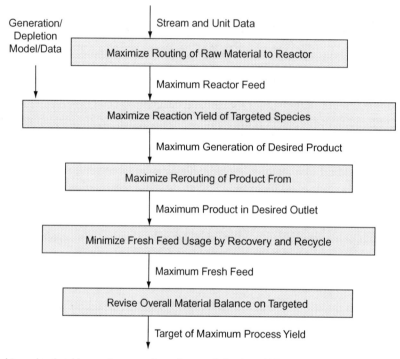

FIGURE 3-15 Flowchart for hierarchical yield-targeting procedure. *Source: Al-Otaibi and El-Halwagi (2006).*

generated product is processed through several separation and finishing units. Because of "sloppy" separations and design/operational decisions, a certain amount of the product is lost with terminal streams leaving in undesirable outlets (that is, streams other than the main product stream going to sales). Therefore, the objective of this step is to manipulate the separation and finishing units so as to minimize the losses and reroute them to the desirable outlet (main product stream). Figure 3.14 illustrates this rerouting.

Step 4: Minimizing fresh feed usage through recycle: As discussed before, recycle can be used to minimize the fresh usage. The limiting reactant can be recovered from the terminal streams (or from streams along the pathway to the terminals) and recycled to reduce the fresh usage. The maximum recyclable load of the limiting reactant is the lower of the terminal load and the fresh feed requirement of the process; that is,

[3.54]

Maximum recyclable load of raw material = argmin
  (Load of limiting reactant in terminal streams,
   fresh limiting-reactant feed requirement of the
   process)

The key steps involved in the aforementioned targeting procedure are summarized by Figure 3.15.

Example 3-5 Minimizing fresh water usage in a pulping mill

The Kraft pulping process is a common technology for converting wood chips to pulp that is subsequently made into paper. Figure 3.16 is a schematic representation of a Kraft process (Lovelady et al., 2007). In this process, 6000 tonne per day (tpd) of wood chips are cooked in a caustic solution referred to as the white liquor (composed primarily of NaOH and Na_2S). The reaction takes place in a digester. The produced pulp is washed in brown-stock washers then fed or sold to a papermaking plant. After the brown-stock washing, the spent cooking solution is referred to as weak black liquor. This liquor is concentrated by a number of multiple-effect evaporators (MEE) and a concentrator to provide a strong black liquor. The vapor streams from the MEE and the concentrator are discharged as wastewater. The strong black liquor is burned in a recovery boiler to generate heat and to chemicals (referred to as smelt). The recovery furnace is connected to an electrostatic precipitator (ESP) to remove the ashes from the off-gas prior to discharge. The smelt from the recovery furnace is composed primarily of Na_2S, Na_2CO_3, and some Na_2SO_4. The smelt is dissolved to form green liquor that is clarified to remove any undissolved materials (known collectively as dregs). The clarified green liquor is fed to a slaker where lime and water react to form slaked lime:

[3.55] $\quad\quad\quad\quad CaO + H_2O \rightarrow Ca(OH)_2$
$\quad\quad\quad$ Molecular weights $\quad$ 56 $\quad$ 18 $\quad\quad$ 74

Slaking is the primary reaction involving water in the process. The slaking reaction consumes 256 tpd of calcium oxide along with the stoichiometrically equivalent quantity of water. Therefore,

[3.56]
$$\text{The amount of water depleted in the pulping process due to the slaking reaction:} = 256 \text{ tpd CaO} * \frac{18 \text{ tonne H}_2\text{O}}{56 \text{ tonne CaO}}$$
$$= 168 \text{ tpd}$$

Slaking is followed by a causticizing reaction between the slaked lime and sodium carbonate to yield white liquor:

[3.57] $\quad\quad\quad\quad Ca(OH)_2 + Na_2CO_3 \rightarrow 2NaOH + CaCO_3$

The white liquor from the white-liquor clarifier (stream S3) is fed back to the digester for cooking the wood chips, and the chemical cycle for the white liquor is closed. Meanwhile, calcium carbonate from the causticizing reaction is processed through the wasters/filters and fed to the lime kiln where the following thermal decomposition reaction takes place:

[3.58] $\quad\quad\quad\quad CaCO_3 \rightarrow CaO + CO_2$

The produced calcium oxide is recycled back to the slaker for the reaction given by Eq. 3.55, and the calcium oxide cycle is closed.

Water usage and discharge is one of the largest environmental problems associated with the Kraft pulping process. Figure 3.17 shows a block flow diagram of the process. The process streams are designated by S, and the water flows

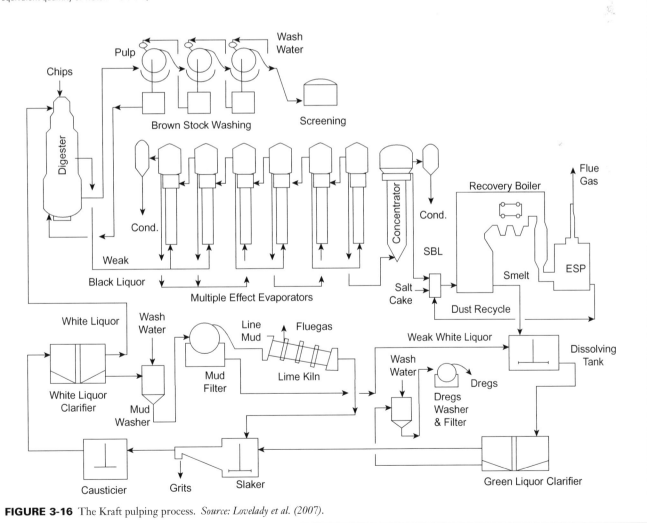

FIGURE 3-16 The Kraft pulping process. *Source: Lovelady et al. (2007).*

Example 3-5 Minimizing fresh water usage in a pulping mill (Continued)

(in tpd) are designated by W. Water enters the process as moisture in the incoming wood chips (stream S1) as well as in the freshwater streams used in the washer, screening, and washers/filters (streams S2, S6, and S24, respectively). Water leaves the system with the wet pulp (S7), from the screening (S8), from the MEE condenser (S10), from the concentrator condenser (S12), in the ESP off-gas (S15), in the filter reject (S23), in the kiln off-gas (S26), and in the slaker off-gas (S28). The objective of this example is to determine the target for minimum water consumption. The three wastewater streams (S8, S10, and S12) may be considered for recycle. On the other hand, water in the off-gas streams (S15, S26, and S28) and in the filter reject (which is a wet cake) are not considered for recycle because of the prohibitive cost of recovering recyclable water from these streams. Also, water with the pulp leaving the process (S7) is not a candidate for recycle because of the wetness requirement for the subsequent papermaking machines. Determine the targets for minimum freshwater usage and wastewater discharge in the pulping process.

SOLUTION

First, the "big-picture" data needed for targeting are gathered and represented as shown by Fig. 3.18. These include the water input to the process, the depletion of water by chemical reaction (slaking), and the terminal water outputs. Before mass integration,

[3.59]
$$\text{Current usage of fresh water} = 13{,}995 + 1450 + 5762 = 21{,}207 \text{ tpd}$$

[3.60]
$$\text{Current discharge of wastewater (in S8, S10, and S12)} = 1450 + 8901 + 1024 = 11{,}375 \text{ tpd}$$

It is worth noting that, the overall mass balance given by Eq. 3.40 is indeed satisfied for water; that is,

[3.61]
$$3000 + 13{,}995 + 1450 = 10{,}995 + 1450 + 8901 + 1024 + 1202 + 5762 - 168 \qquad + 4 + 423 + 40$$

Based on Equations 3.59 and 3.60, the fresh water usage of the process is 21,207 tpd and the flow rate of the recyclable wastewater streams is 11,375 tpd. Therefore, according to Eq. 3.44, the maximum recycle corresponds to the lower of the two flows (11,375 tpd). For targeting purposes, a recovery network is placed to purify the recycle stream and render it in a condition that enables the replacement of an equal amount of the fresh water. As Fig. 3.19 shows, the three wastewater streams can be fully recycled and the target for minimum usage of fresh water is 9832 tpd.

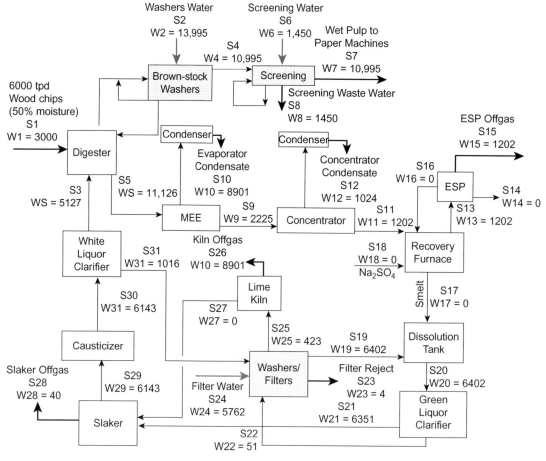

FIGURE 3-17 Block flow diagram of the Kraft pulping process with stream numbers (designated by S) and water flows in TPD (designated by W). *Source: Lovelady et al. (2007).*

(*Continued*)

Example 3-5 Minimizing fresh water usage in a pulping mill (*Continued*)

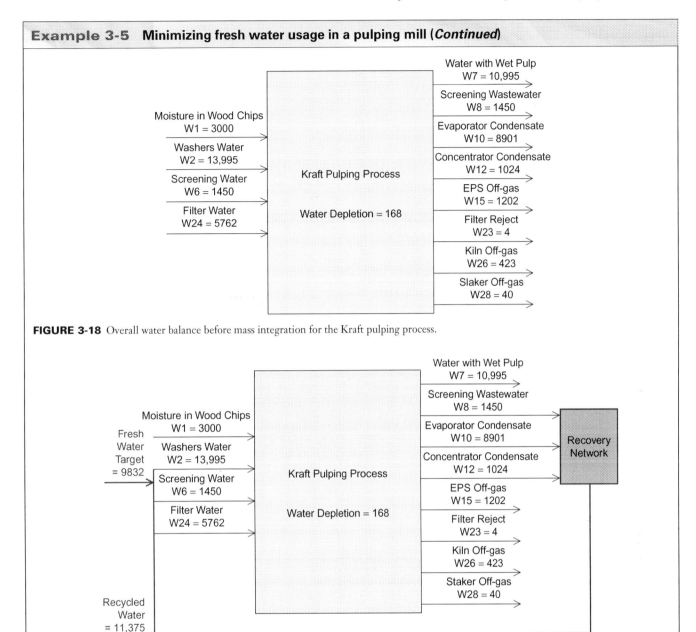

FIGURE 3-18 Overall water balance before mass integration for the Kraft pulping process.

FIGURE 3-19 Water targeting for the Kraft pulping process.

Example 3-6 Reduction of discharge in a tire-to-fuel plant

This case study is taken from El-Halwagi (2006) and adapted from El-Halwagi (1997) and Noureldin and El-Halwagi (2000). It involves a processing facility that converts scrap tires into fuel via pyrolysis. Figure 3.20 is a simplified flow sheet of the process. The discarded tires are fed to a high-temperature reactor where hydrocarbon content of the tires is broken down into oils and gaseous fuels. The oils are further processed and separated to yield transportation fuels. As a result of the pyrolysis reactions, water is formed. The amount of generated water is a function of the reaction temperature, T_{rxn}, through the following correlation:

[3.62] $\quad W_{rxn} = 0.152 + (5.37 - 7.84 \times 10^{-3} T_{rxn}) e^{(27.4 - 0.04 T_{rxn})}$

Where W_{rxn} is in kg/s and T_{rxn} is in Kelvin (K). At present, the reactor is operated at 690 K, which leads to the generation of 0.12 kg water/s. To maintain acceptable product quality, the reaction temperature should be maintained within the following range:

[3.63] $\quad 690 \leq T_{rxn}(K) \leq 740$

The gases leaving the reactor are passed through a cooling/condensation system to recover some of the light oils. To separate the oils, a decanter is used to separate the mixture into two layers: aqueous and organic. The aqueous layer is a wastewater stream whose flow rate is designated as W_1, and it contains phenol as the primary pollutant. The organic layer is mixed with the liquid products of the reactor and fed to finishing. As a result of fuel finishing, a gaseous waste is produced and flared. As a safety precaution to prevent the back propagation of fire from the flare, a seal pot (or a water valve) is placed before the flare to provide a

(*Continued*)

Example 3-6 Reduction of discharge in a tire-to-fuel plant (*Continued*)

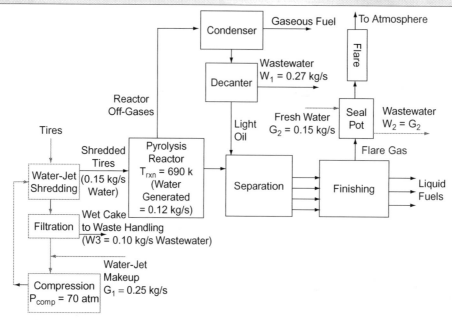

FIGURE 3-20 Simplified flow sheet of tire-to-fuel plant. *Source: El-Halwagi (2006).*

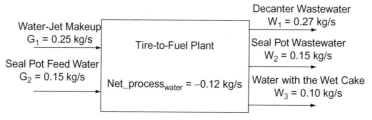

FIGURE 3-21 Overall water balance for the tire-to-fuel process. *Source: El-Halwagi (2006).*

buffer zone between the fire and the flare gas. The flow rate of the water stream passing through the seal pot is referred to as G_2, and an equivalent flow rate of wastewater stream, $W_2 = G_2$, is withdrawn from the seal pot.

Tire shredding is achieved by using high-pressure water jets. The shredded tires are fed to the process while the spent water is filtered. The wet cake collected from the filtration system is forwarded to solid waste handling. The filtrate is mixed with fresh water–jet makeup "G_1" to compensate for water losses with the wet cake "W_3" and the shredded tires. The mixture of filtrate and water makeup is fed to a high-pressure compression station for recycle to the shredding unit. The flow rate of water-jet makeup depends on the applied pressure coming out of the compression stage "P_{comp}" via the following expression:

$$[3.64] \qquad G_1 = 0.47\, e^{-0.009\, P_{comp}}$$

where G_1 is in kg/s and P_{comp} is in atm. To achieve acceptable shredding, the jet pressure may be varied within the following range:

$$[3.65] \qquad 70 \leq P_{comp}(atm) \leq 90$$

At present, P_{comp} is 70 atm, which requires a water-jet makeup flow rate of 0.25 kg/s.

The water lost in the cake is related to the mass flow rate of the water-jet makeup through:

$$[3.66] \qquad W_3 = 0.4\, G_1$$

In addition to the water in the wet cake, the plant has two primary sources for wastewater: from the decanter (W_1) and from the seal pot (W_2). At present, the values of W_1, W_2, and W_3 are 0.27, 0.15, and 0.10 kg/s, respectively. The wastewater from the decanter contains about 500 ppm of phenol. Within the range of allowable operating changes, this concentration can be assumed to remain constant. At present, the wastewater from the seal pot contains no phenol. The plant has been shipping the wastewater streams W_1 and W_2 for off-site treatment. The cost of wastewater transportation and treatment is $0.10/kg, leading to a wastewater treatment cost of approximately $1.33 MM/yr. W_3 has been processed on site. Because of the characteristics of W_3, the plant does not allow its recycle back to the process even after processing waste. The plant wishes to reduce off-site treatment of wastewater streams W_1 and W_2 to avoid cost of off-site treatment and alleviate legal liability concerns in case of transportation accidents or inadequate treatment of the wastewater. The objective of this problem is to determine a target for reducing the flow rate of terminal discharges W_1 and W_2.

SOLUTION

Figure 3.21 shows an overall water balance for the process before mass integration.

The first step in the analysis is to reduce the terminal discharge by minimizing the net generation of water. Figure 3.22 is a graphical representation of Eq. 3.62, illustrating the net generation of water through chemical reaction as a function of the reaction temperature. As can be seen from this graph, the minimum generation of water is 0.08 kg/s and is attained at a reaction temperature of 710 K.

(*Continued*)

Example 3-6 Reduction of discharge in a tire-to-fuel plant (*Continued*)

Next, we adjust design and operating parameters so as to minimize fresh water consumption. As mentioned earlier, the fresh water used in shredding is a function of pressure as given by Eq. 3.64:

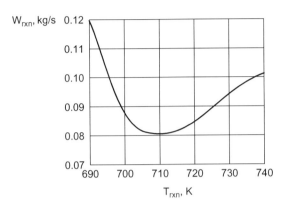

FIGURE 3-22 Rate of water consumption as a function of reaction temperature. *Source: El-Halwagi (2006).*

$$[3.64] \quad G_1 = 0.47\, e^{-0.009\, P_{comp}}$$

where P_{comp} should be maintained within a permissible interval of from 70 to 95 atm. Therefore, to minimize the fresh water needed for shredding, the value of P_{comp} should be set to its maximum limit of 95 atm. Consequently, G_1 is reduced to 0.20 kg/s. According to Eq. 3.76, the new value of W_3 is given by:

$$[3.67] \quad W_3 = 0.4*0.20 = 0.08 \text{ kg/s}$$

With the new values of G_1 and W_3 and with the water generation minimized to 0.08 kg/s, an overall water balance provides the value of W_2, to be 0.15 kg/s. These results are shown in Fig. 3.23, and represent the overall water balance after sink/generator manipulation with existing units and current process configuration. Next, we calculate the target for water usage and discharge using interception (cleaning up of recycled water) and recycle. This targeting analysis is shown in Fig. 3.24, and it yields a target of zero fresh water and 0.08 kg/s for wastewater discharge.

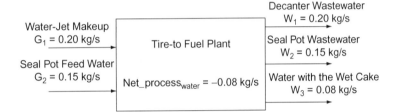

FIGURE 3-23 Overall water balance after sink/generator manipulation. *Source: El-Halwagi (2006).*

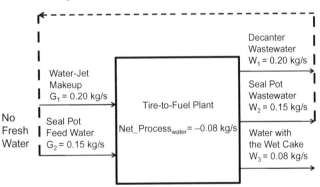

FIGURE 3-24 Targeting for minimum water usage and discharge. *Source: El-Halwagi (2006).*

Example 3-7 Yield targeting in acetaldehyde production through ethanol oxidation

Consider the process of producing acetaldehyde via ethanol oxidation shown by Fig. 3.25. The process description and data are based on the case study reported by Al-Otaibi and El-Halwagi (2006). In this case study, the following symbols are used: "S," "A," and "E," to respectively refer to a process stream, the flow of acetaldehyde (in tonne per year or tpy), and the flow of ethanol (in tpy). Ethanol feedstock (50 percent ethanol) is vaporized in a flash column. Some ethanol is lost in the bottom stream of the flash drum. The ratio of ethanol in the bottom stream (E2) to ethanol in the feed (E1) of the flash is given as a function of the flash temperature:

$$[3.68a] \quad \alpha = \frac{E2}{E1} = 10.5122 - 0.0274*T_{flash}$$

where T_{flash} is the temperature of the flash drum in K.

Equivalently, using the component mass balance for ethanol around the flash column (E1 = E2 + E3), Eq. 3.68a may be written in terms of the ratio of the ethanol in the top stream (E3) and the feed (E1) to the flash drum:

$$[3.68b] \quad (1 - \alpha) = \frac{E3}{E1}$$

The acceptable range for T_{flash} is given by the following bounds:

$$[3.69] \quad 380.0 \leq T_{flash}(K) \leq 383.5$$

At present, the flash temperature is 380 K.

(*Continued*)

Example 3-7 Yield targeting in acetaldehyde production through ethanol oxidation (*Continued*)

The vapor stream leaving the flash is mixed with air and fed to an oxidation reactor where acetaldehyde and water are produced according to the following chemical reaction:

[3.70]
$$CH_3CH_2OH + \frac{1}{2}O_2 \rightarrow CH_3CHO + H_2O$$

Assuming that the aforementioned chemical reaction is only a reaction taking place in the reactor, the ethanol consumed in the reactor can be related to acetaldehyde formed in the reactor through stoichiometry and molecular weights. Therefore,

[3.71]   Ethanol consumed in the reactor = (46/44)*acetaldehyde formed in the reactor

The reactor yield is defined as:

[3.72a] $$Y_{reactor} = \frac{\text{Mass rate of acetaldehyde formed in the reactor}}{\text{Mass flow rate of ethanol fed to the reactor}}$$

Because there is no acetaldehyde entering the reactor, therefore, the mass rate of acetaldehyde formed in the reactor is equal to A5. Hence, Eq. 3.72a may be rewritten as:

[3.72b] $$Y_{reactor} = \frac{A5}{E3}$$

The dependence of the reactor yield on the reaction temperature is given by:

[3.73a] $$Y_{reactor} = 0.33 - 4.2*10^{-6}*(T_{rxn} - 580)^2$$

where T_{rxn} is the reactor temperature (K), which is constrained by the following range of acceptable operating temperature:

[3.74] $$300 \leq T_{rxn} (K) \leq 860$$

At present, the reactor is operated at 442 K.

Scrubbing in two absorption columns is carried out using water and ethanol to recover the product and unreacted materials. The liquid leaving the bottom of the first absorption column is separated in three distillation columns. Acetaldehyde is recovered from the top of the first column. The acetaldehyde recovered in the first distillation column is a function of the heat duty of the reboiler of that column. The relationship is given by:

[3.75] $$A14 = \beta*A9$$

where

[3.76] $$\beta = 0.14*Q_R + 0.89$$

where Q_R is the reboiler heat duty (heat flow rate) in MW. The range of the reboiler duty is:

[3.77] $$0.55 \leq Q_R (MW) \leq 0.76$$

For the base case, the reboiler duty (heat flow rate) is 0.55 MW.

Organic wastes are collected from the top of the second column. The third column produces a distillate that contains ethanol, water, and minor components. The bottom product is primarily an aqueous waste and is fed to the biotreatment

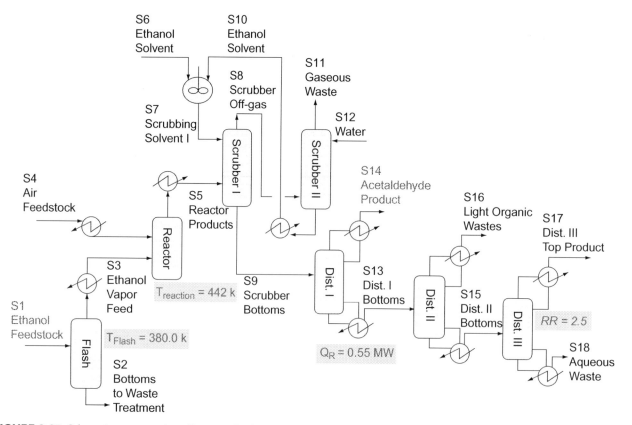

FIGURE 3-25 Schematic representation of base case for the acetaldehyde process. *Source: Al-Otaibi and El-Halwagi (2006).*

(*Continued*)

Example 3-7 Yield targeting in acetaldehyde production through ethanol oxidation (*Continued*)

facility. To reduce ethanol losses (with the aqueous waste going to biotreatment), the reflux ratio of the third distillation column may be manipulated. The following relations may be used:

[3.78] $$E17 = \gamma * E15$$

where

[3.79] $$\gamma = 0.653 * e^{(0.085 * RR)}$$

where RR is the reflux ratio in the third distillation column. Currently, the reflux ratio for the column is 2.5, and the working range for the reflux ratio is:

[3.80] $$2.5 \leq RR \leq 5.0$$

The objective of this case study is to maximize the overall process yield without adding new process equipment. The overall process yield is defined as:

[3.81a]
$$\text{Process yield} = \frac{\text{Mass flow rate of acetaldehyde in final product stream}}{\text{Mass flow rate of fresh ethanol fed to process as feedstock}}$$

The use of the phrase "ethanol fed to the process as a feedstock" excludes the use of ethanol for nonreactive purposes such as solvents (in stream S6). Therefore, the process yield is given by:

[3.81b] $$\text{Process yield} = \frac{A14}{E_{Fresh}}$$

where A14 is the mass flow rate of acetaldehyde in the final acetaldehyde product (in stream S14), and E_{Fresh} is the mass flow rate of ethanol in the fresh feedstock to the process (in stream S1 in the current process configuration).

The following flows assumptions are used throughout the case study (even after process changes):
- E6 = 400 tpy of ethanol
- Acetaldehyde leaves the process in two streams: S14 and S16
- Ethanol leaves the process in two streams: S17 and S18
- Direct recycle of ethanol is allowed only from the top of the third distillation column to the flash column

At present, the plant produces 100,000 tpy of acetaldehyde (that is, A14 = 100,000). It is desired to explore increasing the process yield using existing units and without adding new pieces of equipment. As can be seen from Eq. 3.91b, enhancing the process yield involves increasing the A14 and/or decreasing E_{Fresh}. In this case study, the desired production level of acetaldehyde is to be maintained at 100,000 tpy. What is the target for maximum overall process yield and what are the necessary process changes to reach the yield target?

SOLUTION

The first step is to calculate the current "base-case" process yield. Let us start with the acetaldehyde product stream and model the process backward until the value of the base-case fresh ethanol feed is calculated. For the 100,000 tpy production rate of acetaldehyde and a reboiler duty of the third distillation column of 0.55 MW, Equations 3.86 and 3.85 give:

[3.82] $$\beta = 0.14 * 0.55 + 0.89 = 0.967$$

and

[3.83] $$A9 = 100,000/0.967 = 103,413 \text{ tpy}$$

At present, the reactor is operated at 442 K. Using Eq. 3.73a, we get

[3.73b] $$Y_{reactor} = 0.33 - 4.2*10^{-6} * (422 - 580)^2$$
$$= 0.25 \text{ kg acetaldehyde/kg ethanol}$$

Noting that,

[3.84] $$A9 = A5$$

and combining with the definition of the reactor yield as given by Eq. 3.72b, we have:

$$E3 = \frac{103,413}{0.25} = 413,652 \text{ tpy}$$

Next, we move to the flash column. In the base case, the flash temperature is operated at 380 K. According to Eq. 3.68a,

[3.85] $$\alpha = 10.5122 - 0.0274 * 380 = 0.1002$$

Hence, according to Eq. 3.68b,

[3.86] $$E1 = \frac{413,652}{(1.0000 - 0.1002)} = 459,715 \text{ tpy}$$

[3.87] Current (base-case) process yield before mass integration $= \frac{100,000}{459,715}$
$$= 0.218$$

The objective of this case study is to identify the maximum yield of the process without adding new units and while maintaining an acetaldehyde production flow rate of 100,000 tpy. As described by Fig. 3.15, the following steps can be used to identify the yield target:

Step 1: Maximize routing of targeted raw material to the reaction system
Step 2: Maximizing reactor yield
Step 3: Rerouting the product from undesirable outlets to desirable outlets
Step 4: Minimizing fresh feed usage through recycle

Based on Step 1, the ethanol losses from the bottom of the flash should be minimized so as to maximize the routing of ethanol to the reactor. Because,

[3.88a] $$\alpha = \frac{E2}{E1} = 10.5122 - 0.0274 * T_{flash}$$

and

[3.69] $$380.0 \leq T_{flash} (K) \leq 383.5$$

Therefore, the value of α is minimized when the flash is operated at 383.5 K. Thus,

[3.88b] $$\alpha^{min} = 10.5122 - 0.0274 * 383.5$$
$$= 0.0043$$

But,

[3.68b] $$(1 - \alpha) = \frac{E3}{E1}$$

Because the maximum allocation of ethanol to the reactor corresponds to the minimum losses of ethanol in the flash column, then,

[3.89] $$E3^{max} = (1 - 0.0043) * E1$$
$$= 0.9957 * E1$$

Step 2 is aimed at maximizing the reactor yield, which is given as a function of the reaction temperature:

[3.73] $$Y_{reactor} = 0.33 - 4.2*10^{-6} * (T_{rxn} - 580)^2$$

which is maximized when T_{rxn} is 580 K; that is,

[3.90] $$Y_{reactor}^{max} = 0.33$$

(*Continued*)

Example 3-7 Yield targeting in acetaldehyde production through ethanol oxidation (*Continued*)

Using the definition of the reactor yield,

[3.72b] $$Y_{reactor} = \frac{A5}{E3}$$

Substituting from Eq. 3.90 into Eq. 3.72b, we obtain:

[3.91a] $$A5^{max} = 0.33*E3^{max}$$

which can be combined with Eq. 3.89 to give:

[3.91b] $$A5^{max} = 0.33*0.9957*E1 = 0.3286*E1$$

Step 3 is intended to maximize the rerouting of acetaldehyde from undesirable outlets to desirable outlets. This is equivalent to minimizing the acetaldehyde losses from the bottom of the first distillation column, which wind up as waste from the top of the second distillation column. The model for the first distillation column is summarized by:

[3.75] $$A14 = \beta*A9$$

where

[3.76] $$\beta = 0.14*Q_R + 0.89$$

and

[3.77] $$0.55 \leq Q_R \text{ (MW)} \leq 0.76$$

Because Eq. 3.76 is monotonically increasing as a function of the reboiler heat duty, the selection of the value of Q_R to be 0.76 provides the maximum allocation of acetaldehyde in the desired product stream; that is,

[3.92] $$\beta^{max} = 0.14*0.76 + 0.89 = 0.9964$$

Combining Equations 3.75 and 3.92, we get:

[3.93a] $$A14^{max} = 0.9964*A9^{max}$$

But,

[3.84] $$A9 = A5$$

Thus,

[3.93b] $$A14^{max} = 0.9964*A5^{max}$$

Because the desired production level of acetaldehyde is 100,000 tpy in this case study, Eq. 3.93b results in:

[3.94] $$A5^{max} = \frac{100,000}{0.9964} = 100,361 \text{ tpy}$$

Based on Eq. 3.91a,

[3.95] $$E3^{max} = \frac{100,361}{0.33} = 304,124$$

And using Eq. 3.89, we obtain:

[3.96] $$E1 = \frac{304,124}{0.9957} = 305,438 \text{ tpy}$$

Finally, in Step 4, focus is given to minimizing fresh feed usage through recycle from E17 to replace some of the purchased fresh ethanol used to supply E1. To determine the value of E17, the ethanol leaving the reactor must first be calculated. According to the reaction stoichiometry and molecular weights:

[3.71] Ethanol consumed in the reactor = (46/44)*acetaldehyde formed in the reactor

Substituting from Eq. 3.94 in Eq. 3.71, we get:

[3.97] Ethanol consumed in the reactor = (46/44)*100,361 = 104,923 tpy

A component material balance on ethanol around the reactor gives:

[3.98a] $$E5 = E3 - \text{Ethanol consumed in the reactor}$$

or

[3.98b] $$E5 = 304,124 - 104,923 = 199,201 \text{ tpy}$$

Because ethanol leaves the process in streams S17 and S18,

[3.99] $$E5 + E6 = E17 + E18$$

But, the ethanol balance around the third distillation column gives:

[3.100] $$E15 = E17 + E18$$

Therefore,

[3.101a] $$E15 = E5 + E6$$

Using the value of E5 from Eq. 3.98b into Eq. 3.100a, we obtain

[3.101b] $$E15 = 199,201 + 400 = 199,601$$

Because the ethanol recycle comes from stream S17, maximizing the ethanol implies maximizing E17. Because

[3.78] $$E17 = \gamma*E15$$

where

[3.79] $$\gamma = 0.653*e^{(0.085*RR)}$$

and

[3.80] $$2.5 \leq RR \leq 5.0$$

To maximize the value of E17, the reflux ratio should be set to 5.0. Hence,

[3.102] $$\gamma^{max} = 0.653*e^{(0.085*5.0)} = 0.9988$$

Based on Equations 3.78, 3.101b, and 3.102, the following result is obtained:

[3.103] $$E17^{max} = 0.9988*199,601 = 199,361 \text{ tpy}$$

The recycle of ethanol from the top of the third distillation column to the inlet of the flash column to substitute some of the fresh ethanol gives:

[3.104a] $$E_{Fresh} = E1 - E17$$

Therefore,

[3.104b] $$E_{Fresh}^{min} = E1 - E17^{max}$$

Substitution from Equations 3.91 and 3.103 into Eq. 3.104a gives:

[3.104c] $$E_{Fresh}^{min} = 305,438 - 199,361 = 106,077 \text{ tpy}$$

Using the definition of process yield in Eq. 3.82b, we get:

[3.105] $$\text{Maximum process yield} = \frac{100,000}{106,077} = 0.943$$

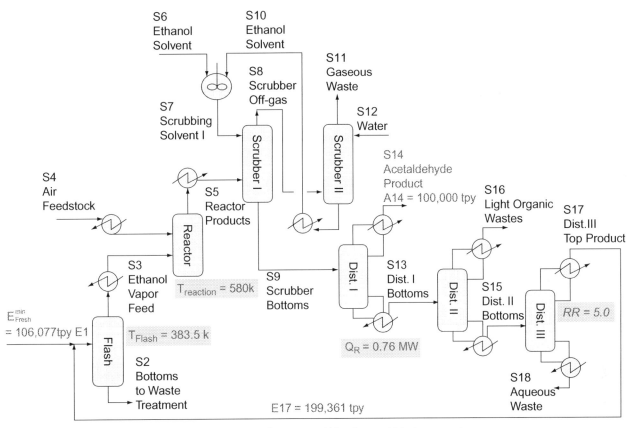

FIGURE 3-26 Process changes to achieve the maximum overall process yield for the acetaldehyde case study.

Figure 3.26 shows the process with the changes needed to reach the maximum overall process yield. There are several useful observations from the case study. Substantial yield enhancement has been achieved without adding new units. It is also emphasize that the reactor yield should not be confused with the overall process yield. Although the reactor yield is important, it is only one factor in maximizing the overall process yield. The other factors of feedstock losses, product losses, and feedstock recycle contribute significantly to the enhancement of the overall process yield. This underscores the importance of mass integration as a holistic approach to optimizing the performance of the process by considering the root causes of inefficiencies, evaluating performance benchmarks, and identifying the necessary changes to reach the desired target.

MASS INTEGRATION STRATEGIES FOR ATTAINING THE TARGETS

The overall mass targets of the process are typically determined without specifying the detailed design changes that are needed to reach the targets. Several mass integration strategies can be used to develop cost-effective implementations. These strategies include stream segregation/mixing, recycle, interception using separation devices, changes in design and operating conditions of units, materials substitution, and technology changes including the use of alternate chemical pathways. These strategies can be classified into a hierarchy of three categories

- No-/low-cost changes
- Moderate cost modifications, and
- New technologies

(El-Halwagi, 1999):

Three main factors can be used in describing these strategies: economics, impact, and acceptability. The economic dimension can be assessed by a variety of criteria such as capital cost, return on investment, net present worth, and payback period. Impact is a measure of the effectiveness of the proposed solution in reducing negative ecological and hazard consequences of the process, such as reduction in emissions and effluents from the plant. Acceptability is a measure of the likelihood of a proposed strategy to be accepted and implemented by the plant. In addition to cost, acceptability depends upon several factors, including corporate culture, dependability, safety, and operability. Figure 3.27 is a schematic representation of the typical hierarchy of mass-integration strategies. These strategies are typically in ascending order of cost and impact and in descending order of acceptability. The ensuing chapters describe the systematic mass-integration tools associated with identifying such strategies.

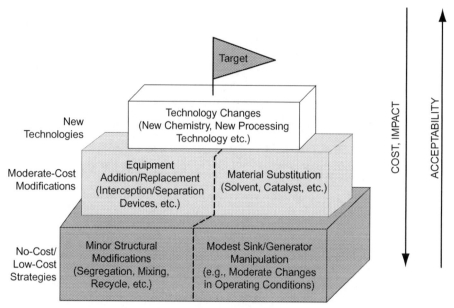

FIGURE 3-27 Hierarchy of mass-integration strategies. *Source: El-Halwagi (1999).*

HOMEWORK PROBLEMS

3.1. Hydrogen is produced through the steam reforming of natural gas. If the prices of steam and natural gas are $6.00/tonne and $5.0/1,000 SCF, respectively, comment on the potential economic viability of the process.

3.2. Urea is produced through the following two-step reactions (Tjioe and Tinge, 2010):

[3.106]

$2NH_3 + CO_2 \rightarrow NH_2COONH_4$
(ammonium carbamate)

[3.107]

$NH_2COONH_4 \rightarrow H_2O + NH_2CONH_2 (urea)$

A process uses 300,000 tonne/yr of ammonia as a feedstock.
1. What is the theoretical stoichiometric target for the maximum production of urea?
2. What is the value of the stoichio-nomic indicators of the process?
3. All the produced urea is used to produce melamine ([structure] or $C_3N_6H_6$), via a two-step process in which urea is first decomposed to isocyanic acid and ammonia, and then isocyanic acid is decomposed to melamine and carbon dioxide. What is the theoretical target for the produced melamine?

3.3. Polyethylene is produced through the polymerization of ethylene. What is the maximum theoretical yield of the process (kg polyethylene/kg ethylene)?

3.4. A specialty chemical process uses feedstock S to produce two primary products (A and M) according to the following stoichiometric reaction:

[3.108] $S \rightarrow A + M$

In the same reactor, a side reaction also takes place in parallel and produces a terminally unrecyclable (TU) component according to the following reaction:

[3.109] $S \rightarrow TU$

The conversion of S in the aforementioned reactions is virtually complete. The ratio of the rates of reaction of component S in the first and the second reaction is dependent on the temperature according to the following expression:

[3.110]

$$\frac{r_1}{r_2} = -2.778*10^{-4}*T^2 + 0.3111*T - 80.0979$$

where r_1 and r_2 are the reaction rates of the first and second reactions expressed as $\frac{kmols\ reacted}{m^3 \cdot s}$ and T is the reaction temperature (K), which is bounded by the permissible operating range:

[3.111] $500 \leq T \leq 600$

What is the stoichiometric target for the maximum molar ratio of $\frac{A\ \&\ M}{TU}$?

3.5. Ethylene oxide is formed through the oxidation of ethylene:

[3.112] $C_2H_4 + 0.5 O_2 \rightarrow C_2H_4O$

A competing combustion reaction also takes place in parallel:

[3.113] $C_2H_4 + 3O_2 \rightarrow 2CO_2 + 2H_2O$

With the use of a special catalyst and an inhibiting agent, the ratio of the rate of consuming ethylene in the combustion reaction to the rate of consuming ethylene in the oxidation reaction can be kept to less than or equal to 1:19. What is the maximum target for producing ethylene oxide (expressed as kg ethylene oxide/kg ethylene)?

3.6. The Lurgi coal gasification process (Probstein and Hicks, 2006) yields the syngas shown in Table 3.2. Two reactions are used to produce methanol:

[3.114] $CO + 2H_2 \rightarrow CH_3OH$

[3.115] $CO_2 + 3H_2 \rightarrow CH_3OH + H_2O$

Data show that the fractional conversions of CO and CO_2 to methanol are 98 percent and 50 percent, respectively. What is the expected stoichiometric target for methanol?

3.7. Rice hulls constitute an agricultural waste that can be used to produce hydrogen and other salable gases. The following experimental data are available for the oxygen gasification of rice hulls (El-Halwagi, 1985):
 - The chemical formula of rice hulls (on a dry, ash-free basis) is $C_{3.14}H_{5.23}O_{2.29}$.
 - The weight ratio of oxygen to rice hulls is 0.28.
 - At 1000 K, 70 percent of the rice hulls are converted to gases (the rest is unreacted rice hulls and solid residue). All the gasifying oxygen is consumed in forming the gases.
 - Table 3.3 gives the product distribution in the resulting gases.
Assess the economic viability of the process.

Hint: If 100 kg of rice hulls are gasified with 28 kg of oxygen, the amount of produced gas is 0.7*100 + 28 = 98 kg. The mol% composition of the gases should be converted to mass% to calculate the produced amount of each product.

TABLE 3-2 Syngas Derived by Lurgi Gasification

Component	Flow Rate, kmol/s
H_2	2.105
CO	0.763
CO_2	0.172
CH_4	0.564
C2+	0.008
N_2	0.044

Source: Probstein and Hicks (2006).

TABLE 3-3 Gaseous Products from the Oxygen Gasification of Rice Hulls

Component	mol%
H_2	23
CO	45
CO_2	21
CH_4	8
C_2H_4	3

Source: El-Halwagi (1985).

3.8. A new process is to be considered for the conversion of wood to methanol. First, wood is converted to syngas in an oxygen gasifier (Bain, 1992):

[3.116] $CH_{1.4}O_{0.6} + 0.2O_2 \rightarrow CO + 0.7H_2$

The ratio of H_2 to CO in the syngas is less than the 2:1 needed for the methanol-synthesis reaction:

[3.117] $2H_2 + CO \rightarrow CH_3OH$

Therefore, the CO produced from the oxygen gasifier is catalytically reacted with water through the shift reaction:

[3.118] $CO + H_2O \rightarrow CO_2 + H_2$

Estimate the maximum theoretical target for methanol expressed as kg methanol/kg dry wood.

3.9. It is desired to carry out a top-level stoichio-nomic targeting for a proposed process that converts a woody biomass into gasoline through fast pyrolysis followed by hydrogenation. The overall pyrolysis reaction is given by (Venderbosch and Prins, 2010):

[3.119]

$CH_{1.60}O_{0.68}$ (woody biomass) + $0.136H_2O$ (moisture in woody biomass) + $0.230O_2$ (oxygen in air) $\rightarrow$ $0.624CH_{2.59}O_{1.05}$ (pyrolysis oil) + $0.0735CH_{0.54}O_{1.30}$ (pyrolysis gas) + $0.185CH_{0.72}O_{0.22}$ (char) + $(0.131CO_2 + 0.0250H_2O + 0.095O_2)$ (flue gases)

The pyrolysis oil (or py-oil) is subsequently reacted with hydrogen to produce gasoline (simplified by the molecular formula C_8H_{18}).
The following prices may be assumed:
Woody biomass: $80/tonne
Hydrogen: $1500/tonne
Gasoline: $3.00/gal
1. Perform a stoichio-economic targeting for the process.
2. If the non-feedstock operating cost of the process is $0.85/gal of gasoline and the fixed capital investment of the process is $12.00/annual gal of gasoline, comment on the economic viability of the process. Assume a 10-year linear depreciation scheme with no salvage value.

3.10. Vinyl acetate monomer "VAM" is manufactured by reacting acetic acid with oxygen and ethylene according to the following chemical reaction:

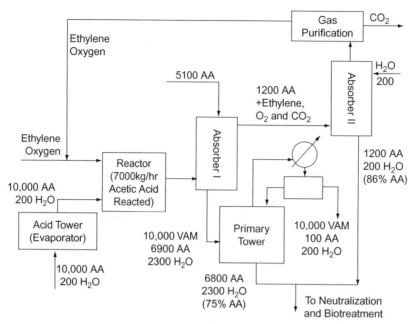

FIGURE 3-28 VAM manufacturing process. *Source: El-Halwagi (2006).*

[3.120]

$$C_2H_4 + 0.5\,O_2 + C_2H_4O_2 = C_4H_6O_2 + H_2O$$

Ethylene oxygen acetic acid VAM water

Consider the process described by El-Halwagi (2006) and shown in Fig. 3.28. A fresh feed of 10,000 kg/hr of acetic acid "AA" along with 200 kg/hr of water are evaporated in an acid tower. The vapor is fed with oxygen and ethylene to the reactor where 7000 kg/hr of acetic acid are reacted and 10,000 kg/hr of VAM are formed. The reactor off-gas is cooled and fed to first absorber, where AA (5100 kg/hr) is used as a solvent. Almost all the gases leave from the top of the first absorption column, together with 1200 kg/hr of AA. This stream is fed to the second absorption column where water (200 kg/hr) is used to scrub acetic acid. The bottom product of the first absorption column is fed to the primary distillation tower, where VAM is recovered as a top product (10,000 kg/hr) along with 200 kg/hr of water and a small amount of AA (100 kg/hr), which is not economically justifiable to recover. This stream is sent to final finishing. The bottom product of the primary tower (6800 kg/hr of AA and 2300 kg/hr of water) is mixed with the bottom product of the second absorption column (1200 kg/hr of AA and 200 kg/hr of water). The mixed waste is fed to a neutralization system followed by biotreatment. In this example, let us consider the case when no changes are made to the consumption by chemical reaction and there are no adjustments in design or operating conditions to reduce fresh AA consumption. What is the target for minimum fresh usage and minimum terminal losses of AA?

3.11. Consider the VAM process described in the previous problem. A new reaction pathway has been developed and will to be used for the production of VAM. This new reaction does not involve acetic acid. The rest of the process remains virtually unchanged, and the AA losses with the product are 100 kg/hr. What are the targets for minimum fresh usage and discharge/losses of AA?

3.12. Consider the magnetic tape manufacturing process shown by Fig. 3.29. In this process (Dunn et al., 1995), coating ingredients are dissolved in 0.09 kg/s of organic solvent and mixed to form a slurry. The slurry is suspended with resin binders and special additives. Next, the coating slurry is deposited on a base film. Nitrogen gas is used to induce evaporation rate of solvent that is proper for deposition. In the coating chamber, 0.011 kg/s of solvent are decomposed into other organic species. The decomposed organics are separated from the exhaust gas in a membrane unit. The retentate stream leaving the membrane unit has a flow rate of 3.0 kg/s and is primarily composed of nitrogen that is laden with 1.9 wt/wt% of the organic solvent. The coated film is passed to a dryer where nitrogen gas is employed to evaporate the remaining solvent. The exhaust gas leaving the dryer has a flow rate of 5.5 kg/s and contains 0.4 wt/wt% solvent. The two exhaust gases are mixed and disposed of.

In addition to the environmental problem, the facility is concerned about the waste of resources, primarily in the form of used solvent (0.09 kg/s) that costs about $2.3 million/yr. It is desired to undertake a mass-integration analysis to optimize solvent usage, recovery, and losses. Determine the target for minimizing fresh solvent usage in the process.

3.13. Consider a specialty chemical process where product P is produced from reactant A. Two parallel reactions take place:

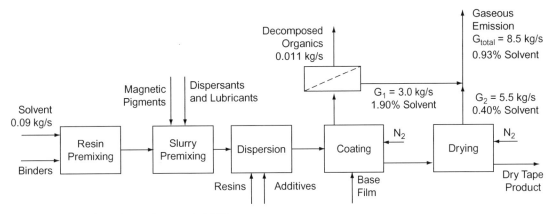

FIGURE 3-29 Magnetic tape plant. *Source: Dunn et al. (1995).*

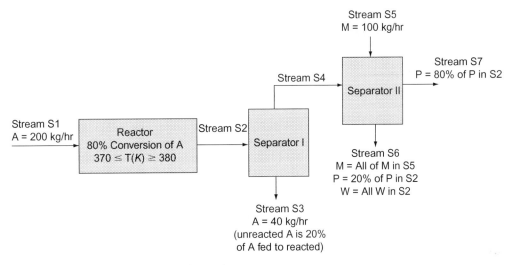

FIGURE 3-30 Schematic representation of the specialty chemical process.

[3.121]

A → P and A → W where W is a waste material (undesirable by-product).

The conversion of A in the reactor is 80 percent (that is, 80 percent of A fed to the reactor is converted to P and W). The rates of formation of P and W are dependent on the reaction temperature according to the following expressions:

[3.122]

Product P formed in the reactor (kg/hr)
$= 0.8*$ (A in S1) $- 2*(375 - T)^2$

[3.123]

Waste W formed in the reactor (kg/hr)
$= 2*(375 - T)^2$

where T is the reactor temperature in K. The reactor temperature is allowed to vary in the following range:

[3.124] $370 \leq T(K) \leq 380$

Currently, the reactor is operated at 370 K. The stream leaving the reactor (S2) is fed to the first separator, which removes all of the unreacted A as a bottom product. The top product from the first separator is fed to the second separator, which uses 100 kg/hr of solvent M to remove the waste compound W as a bottom stream. Twenty percent of the product P formed in the reactor is lost in stream S6. The top product of the second separator (S7) is composed of the remaining 80 percent of P formed in the reactor.

Figure 3.30 provides a schematic representation of the process.

Table 3.4 provides the mass balance of the streams when the reactor is operated at 370 K.

One of the metrics the company uses to assess the efficiency of the production process is referred to as mass efficiency (ME) and is defined as:

[3.125]

$$ME = \frac{\text{Output Product P in Stream S7}}{\text{Sum of Fresh Inputs for Components A and M Fed to the Process}} * 100\%$$

TABLE 3-4 Material Balance (kg/hr) When the Reaction Temperature Is 370 K

Stream\Component	A	P	W	M	Total
S1	200	0	0	0	200
S2	40	110	50	0	200
S3	40	0	0	0	40
S4	0	110	50	0	160
S5	0	0	0	100	100
S6	0	22	50	100	172
S7	0	88	0	0	88

1. What is the current value of the mass efficiency?
2. If the only change allowed in the process is to alter the reaction temperature and it is desired to maximize the amount of P in stream S7, what is the optimal temperature of the reactor (T in K)?
3. If the only change allowed in the process is to alter the reaction temperature, what is the target for maximum flow rate of product P in stream S7?
4. In addition to changing the reaction temperature, you may also use recovery (with new units) and rerouting of stream without any limitations on the cost or technology. In this case, what is the maximum flow rate of product P in stream S7?
5. In addition to changing the reaction temperature, you may also use recovery (with new units) and rerouting of stream without any limitations on the cost or technology. In this case, what is the maximum target for the mass efficiency of the process?

REFERENCES

Al-Otaibi M, El-Halwagi MM: Targeting techniques for enhancing process yield, trans. IChemE, Part A, *Chem Eng Res Des* 84(A10):943–951, 2006.

Bain RL: *Material and energy balances for methanol from biomass using biomass gasifiers*, NREL Report TP-510-17098, Golden, Colorado, National Renewable Energy Lab, 1992.

Crabtree EW, El-Halwagi MM: Synthesis of environmentally acceptable reactions, *AIChE Symp Ser* 90(303):117–127, 1995. New York, AIChE.

Dunn RF, El-Halwagi MM, Lakin J, Serageldin M: *Selection of organic solvent blends for environmental compliance in the coating industries*, Houston, Texas, Proceedings of the First International Plant Operations and Design Conference, vol 3, pp 83–107, 1995.

El-Halwagi MM: Oxygen gasification of rice hulls. In Veziroglu TN, editor: *Alternative energy sources*, vol 2, New York, 1985, Hemisphere Pub. Co., pp. 471–480.

El-Halwagi MM: Sustainable pollution prevention through mass integration. In Sikdar S, Diwekar U, editors: *Tools and methods for pollution prevention*, 1999, Kluwer, pp 233–275.

El-Halwagi MM: *Pollution prevention through process integration: systematic design tools*, San Diego, 1997, Academic Press.

El-Halwagi MM: *Process integration*, Amsterdam, 2006, Elsevier.

Krishnan MS, Ho NWY, Tsao GT: Fermentation kinetics of ethanol production from glucose and xylose by recombinant saccaromyces 1400 (pLNH33), *Appl Biochem Biotechnol* 78(1–3):373–388, 1999.

Lovelady EM, El-Halwagi MM, Krishnagopalan G: An integrated approach to the optimization of water usage and discharge in pulp and paper plants, *Int J Environ Pollut (IJEP)* 29(1–3):274–307, 2007.

Noureldin MB, El-Halwagi MM: Interval-based targeting for pollution prevention via mass integration, *Comp. Chem. Eng.* 23:1527–1543, 1999.

Olah GA, Goeppert A, Prakash GKS: *Beyond oil and gas: the methanol economy*, 2006, Wiley-VCH.

Probstein RF, Hicks RE: *Synthetic fuels*, New York, 2006, Dover Publications Inc.

Tjioe TT, Tinge JT: Integrated urea-melamine process at DSM: sustainable product development. In Harmsen J, Powell JB, editors: *Sustainable development in the process industries: cases and impact*, New Jersey, 2010, John Wiley and Sons, pp 199–208.

Venderbosch RH, Prins W: Fast pyrolysis of biomass for energy and chemicals: technologies at various scales. In Harmsen J, Powell JB, editors: *Sustainable development in the process industries: cases and impact*, New Jersey, 2010, John Wiley and Sons, pp 109–155.

CHAPTER 4

DIRECT-RECYCLE NETWORKS: A GRAPHICAL APPROACH

Chapter 3 described systematic techniques for the overall targeting of mass. These targets can be attained through several strategies such as rerouting of streams, manipulation of design and operating variables, addition of new process units, substitution of solvents and chemicals, and alteration of technologies. These strategies provide different levels of cost and benefit. In this chapter, focus is given to direct-recycle networks. **Recycle** refers to the utilization of a process stream (for example, a waste or a low-value stream) in a process unit (a sink). Although reuse is distinguished from recycle by emphasis that reuse corresponds to the reapplication of the stream for the original intent, we will use the term *recycle* in a general sense that includes reuse. Specifically, this chapter will focus on *direct recycle*, where the streams are rerouted without the addition of new pieces of equipment. As such, direct recycle is a no-/low-cost strategy because it essentially involves pumping and piping and in some cases may even be achieved without the need for additional pumping or piping. In this chapter, a holistic approach is presented for the synthesis of direct-recycle networks with the objective of determining rigorous targets for minimum usage of fresh materials, maximum recycle of process streams, and minimum discharge of waste materials. First, the design problem is stated. Then, a graphical approach (the material recycle pinch diagram) is described to identify the targets. Next, the source-sink mapping diagram is used to identify the detailed rerouting implementations that achieve the targets.

PROBLEM STATEMENT FOR THE DESIGN OF DIRECT-RECYCLE NETWORKS

Consider a process with a number of process sources ($N_{Sources}$) (for example, process streams, wastes) that can be considered for possible recycle and replacement of the fresh material and/or reduction of waste discharge. Each source, i, has a given flow rate, W_i, and a given composition of a targeted species, y_i. Available for service is a fresh (external) resource that can be purchased to supplement the use of process sources in a number of process sinks (N_{sinks}). The sinks are process units (for example, reactors, separators, and so on) that can accept recycled streams. Each sink, j, requires a feed whose flow rate, G_j^{in}, and an inlet composition of a targeted species, z_j^{in}, must satisfy certain bounds on their values. These bounds are described by the following constraints:

[4.1] $$G_j^{min} \leq G_j^{in} \leq G_j^{max} \qquad \text{where } j = 1, 2, \ldots, N_{sinks}$$

where G_j^{min} and G_j^{max} are given lower and upper bounds on admissible flow rate to unit j.

[4.2a] $$z_j^{min} \leq z_j^{in} \leq z_j^{max} \qquad \text{where } j = 1, 2, \ldots, N_{sinks}$$

where z_j^{min} and z_j^{max} are given lower and upper bounds on admissible compositions to unit j.

A particularly useful case is when pure fresh resources are used in the process sinks. In such cases, the value of is zero and the constraint on the bounds for admissible composition becomes:

[4.2b] $$0 \leq z_j^{in} \leq z_j^{max} \qquad \text{where } j = 1, 2, \ldots, N_{sinks}$$

The objective is to use a graphical procedure that determines the target for minimum usage of the fresh resource, maximum material reuse, and minimum discharge to waste.

Figure 4.1 schematically represents the problem, where there are process sources and fresh streams to be allocated to process sinks. Each stream (source) has the option of splitting in any proportion then allocated to a process sink. It is also possible to mix portions of different streams and then assign the mixtures to a process sink. Therefore, the design questions to be answered include:

- Should a stream (source) be segregated and split? To how many fractions? What should be the flow rate of each split?

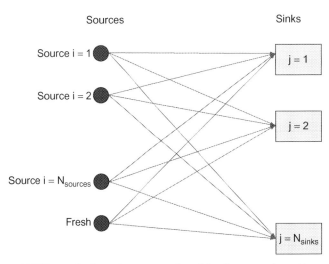

FIGURE 4-1 A schematic representation of the direct-recycle problem for matching sources and sinks.

Sustainable Design Through Process Integration.
© 2012 Elsevier Inc. All rights reserved.

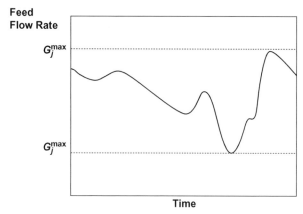

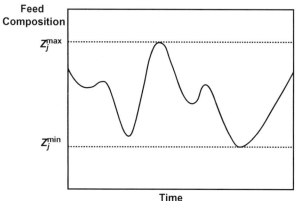

FIGURE 4-2 (a) Bounding feed flow rate based on monitored data. (b) Bounding feed composition based on monitored data. *Source: El-Halwagi (2006).*

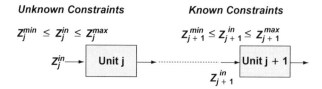

FIGURE 4-3 Constraint propagation to determine composition bounds. *Source: El-Halwagi (2006).*

- Should streams or splits of streams be mixed? To what extent?
- What should be the optimum feed entering each sink? What should be its composition?
- What is the minimum amount of fresh resource to be used?
- What is the minimum discharge of unused process sources?

The flow rate and composition bounds for each sink can be determined based on several of the following considerations (El-Halwagi, 2006):

- Technical (for example, manufacturer's specifications for operable composition ranges to avoid scaling, corrosion buildup, and so on, and operable flow rate ranges such as weeping/flooding flow rates).
- Safety (for example, to stay away from explosion regions).
- Physical (for example, saturation limits).
- Monitoring. These bounds can also be determined from historical data of operating the unit, which are typically available through the process information monitoring system). Figure 4.2 illustrates the bounding of feed flow rate and composition for sink j based on monitored data for which the sink performed acceptably.
- Constraint propagation. In some cases (see Fig. 4.3), the constraints on a sink (j) are based on critical constraints for another unit (j + 1). Using a process model to relate the inlets of units j and j + 1, we can derive the constraints for unit j based on those for unit j + 1. For instance, suppose that the constraints for unit j + 1 are given by:

[4.3a] $$0.24 \leq z^{in}_{j+1} \leq 0.28$$

and the process model relating the inlet compositions to units j and j + 1 can be expressed as:

[4.3b] $$z^{in}_{j+1} = 4 z^{in}_j$$

Therefore, the bounds for unit j are calculated to be:

[4.3c] $$0.06 \leq z^{in}_j \leq 0.07$$

SELECTION OF SOURCES, SINKS, AND RECYCLE ROUTES

In principle, it is possible to replace any fresh source of the targeted species with an equivalent amount of recycle from a terminal or an in-process stream. If the flow rate and composition of the recycled stream meet the constraints illustrated in Equations 4.1 and 4.2 for units employing fresh sources, then we can undertake direct recycle from those terminal streams to those units employing fresh sources. On the other hand, if flow rate and/or composition constraints are not met, then the terminal streams must be intercepted to render them in a condition that allows replacement of fresh sources. Here, we focus on opportunities for direct recycle, particularly for cases when the net generation of the targeted species is independent of the recycle strategy. In such cases, it is important to note that these *recycle activities should be directed to rerouting process streams to units that employ fresh resources so as to replace fresh resources with recycled process streams*. To illustrate this observation, let us consider the process shown in Fig. 4.4. In this process, three fresh streams carry the targeted species to three process sinks (j = 1 – 3). The index k is used to represent the targeted species (component). The required input load of the kth targeted species in these streams is denoted by $Fresh\_Load_{k,j}$. The targeted species leaves the process in four terminal streams, two of which (i = 1, 2) are recyclable (with or without interception), and the other two (i = 3, 4) are forbidden from being recycled. The total load from the four terminal streams is given by $Terminal\_Load_{k,1} + Terminal\_Load_{k,2} + Terminal\_Load_{k,3} + Terminal\_Load_{k,4}$.

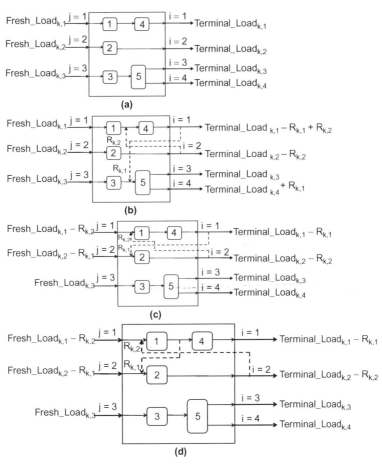

FIGURE 4-4 (a) A generic process before recycle. (b) The process after recycle to poor sinks (total terminal load of targeted species remains unchanged). (c) The process after recycle to proper sinks to replace fresh loads of targeted species (total terminal load of targeted species is reduced). (d) Recycle to replace fresh sources using in-plant and terminal streams. *Source: Noureldin and El-Halwagi (1999).*

Let us first consider recycle from terminal streams to units that do not employ fresh resources. For instance, as shown by Fig. 4.4b, let us recycle a load of $R_{k,1}$ from $i = 1$ to the inlet of unit #5 and a load of $R_{k,2}$ from $i = 2$ to the inlet of unit #4. Because we are dealing with the case where recycle activities have no effect on Net_process$_k$, the loads in the individual terminal streams are simply redistributed with the total terminal load remaining the same (Terminal_Load$_{k,1}$ + Terminal_Load$_{k,2}$ + Terminal_Load$_{k,3}$ + Terminal_Load$_{k,4}$). Therefore, in this case, sinks that do not employ fresh sources of the targeted species are poor destinations for recycle.

Next, we consider recycles that reduce fresh loads. For instance, let us examine the effect of recycling a load of $R_{k,1}$ from $i = 1$ to the inlet of unit #2 and a load of $R_{k,2}$ from $i = 2$ to the inlet of unit #1. This is shown by Fig. 4.4c. The result of the fresh source replacement is a net reduction of $R_{k,1} + R_{k,2}$ from the fresh load of species k, and consequently, the total terminal loads are reduced by $R_{k,1} + R_{k,2}$. It is worth noting that these appropriate recycles are not limited to terminal streams. Instead, what is needed is the replacement of fresh loads with recycled loads from an in-plant or a terminal source. For example, the same effect shown in Fig. 4.4c can be accomplished by recycling (with or without interception) from in-plant sources (for example, $i = 5$) as shown in Fig. 4.4d.

The foregoing discussion illustrates that, in the case where the net generation/depletion of the targeted species is independent of stream-rerouting activities, recycle should be allocated to sinks that consume a fresh resources. The recyclable sources may be terminal streams or in-plant streams on the path to terminal streams. The result of such selection of sources and sinks leads to a reduction in both fresh consumption and waste discharge.

DIRECT-RECYCLE TARGETS THROUGH MATERIAL RECYCLE PINCH DIAGRAM

As mentioned in the problem statement, the key design task is to determine how the process streams (sources) should be segregated, split, mixed, and allocated to process units (sinks). To determine the target for minimum fresh usage and minimum waste discharge, it is important to simultaneously design all the direct-recycle strategies based on a holistic approach that considers all the sources and the sinks in the process. A particularly attractive targeting approach is the *material recycle pinch diagram* developed by El-Halwagi et al. (2003). To illustrate the targeting procedure, let us first consider the case of a *pure fresh* resource that is to be replaced by process sources. Later, this restriction of using pure fresh will be relaxed. Constraint 4.2b is given by:

[4.2b] $\quad 0 \leq z_j^{in} \leq z_j^{max} \qquad$ where $j = 1, 2, \ldots, N_{sinks}$

where z_j^{in} is the composition of the impurities in the stream entering the jth sink. This stream may result from mixing various portions of different sources including the fresh. Furthermore, we consider the case when the value of the flow rate to be fed to each sink, G_j, is given (for example, the lower bound of Constraint 4.1). The load of impurities entering the jth sink is expressed as:

[4.4] $$M_j^{Sink} = G_j z_j^{in}$$

Therefore, the composition constraint of the sink (4.2b) may be replaced by the following constraint on load:

[4.5] $$0 \leq M_j^{Sink} \leq M_j^{max} \quad \text{where } j = 1, 2, \ldots, N_{sinks}$$

where

[4.6] $$M_j^{max} = G_j z_j^{max}$$

The optimality criteria are derived and explained by El-Halwagi et al. (2003). Here, the results of two key optimality criteria are just described. In addition to the proof given by El-Halwagi et al. (2003), a justification of the two rules will be given later in the chapter using the source-sink mapping diagram. The two rules can be stated as follows:

> *The source-prioritization rule*: To minimize the usage of the fresh resource, recycle of the process sources should be prioritized in order of their composition of impurities, starting with the source with the lowest concentration of impurities.
>
> *The sink-load rule*: If the sink requires the use of fresh source, its inlet impurities load should be maximized; that is,

[4.7] $$M_j^{in,optimum} = M_j^{max} \quad \text{where } j = 1, 2, \ldots, N_{sinks}$$

unless no fresh resource is to be used in this sink (in which case, the inlet load of the sink is that of the recycled/reused sources).

The **source prioritization rule** ensures that the process sources are recycled in order of their purities, starting with the "cleanest" recyclable stream and continuing as needed in an ascending order of the concentration of impurities. The rationale is that if a cleaner source is not used, there is no need to consider "dirtier" streams for recycle. The **sink-load rule** allows the maximum extent of recycle of the sources. The higher the extent of recycle replacing the fresh, the lower the usage of the fresh and the discharge of the waste. The source prioritization and sink-load rules constitute the basis for the following graphical procedure referred to as the **material recycle pinch diagram** (El-Halwagi et al., 2003):

1. Rank the sinks in ascending order of maximum admissible composition of impurities,

$$z_1^{max} \leq z_2^{max} \leq \ldots z_j^{max} \ldots \leq z_{N_{Sinks}}^{max}$$

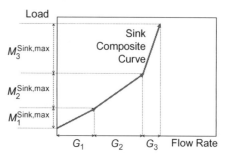

FIGURE 4-5 Developing the sink composite curve.

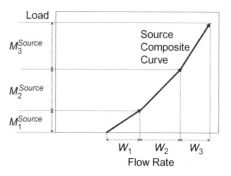

FIGURE 4-6 Developing the source composite curve.

2. Rank sources in ascending order of impurities composition; that is,

$$y_1 < y_2 < \ldots y_i \ldots < y_{N_{Sources}}$$

3. Plot the maximum admissible load of impurities in each sink ($M_j^{Sink,max} = G_j z_j^{max}$) versus its flow rate. Therefore, each sink is represented by an arrow whose vertical distance is $M_j^{Sink,max} = G_j z_j^{max}$, horizontal distance is flow rate, and slope is z_j^{max}. Start with the first sink (which has the lowest z_j^{max}). From the arrowhead of this sink, plot the second sink. Proceed to plot the rest of the sinks using superposition of the sinks' arrows in ascending order. The resulting curve is referred to as the **sink composite curve** (see Fig. 4.5). The sink composite is a cumulative representation of all the sinks and corresponds to the upper bound on their feasibility region.

4. Represent each source as an arrow by plotting the load of each source versus its flow rate. The load of the ith source is calculated through:

[4.8] $$M_i^{Source} = W_i y_i$$

Start with the source having the least composition of impurities and place its arrow tail anywhere on the horizontal axis. As will be shown later, it is irrelevant where this arrow tail is placed. Continue with the other sources and use superposition as shown in Fig. 4.6 to create a source composite curve. The **source composite curve** is a cumulative representation of all process streams considered for recycle. Now, we

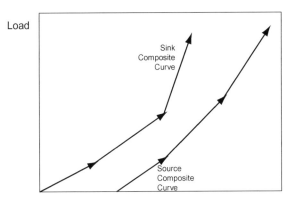

FIGURE 4-7 The sink and source composite curves.

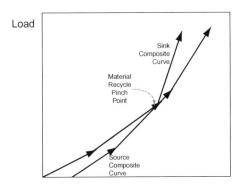

FIGURE 4-8 The material recycle pinch diagram. *Source: El-Halwagi et al. (2003).*

have the two composite curves on the same diagram (see Fig. 4.7).

5. Move the source composite stream horizontally until it touches the sink composite stream with the source composite below the sink composite in the overlapped region. The point where they touch is the material recycle pinch point (see Fig. 4.8). The flow rate of sinks below which there are no sources is the target for minimum fresh usage. The flow rate in the overlapped region of process sinks and sources represents the directly recycled flow rate. Finally, the flow rate of the sources above which there are no sinks is the target for minimum waste discharge. Those targets are shown on Fig. 4.9.

DESIGN RULES FROM THE MATERIAL RECYCLE PINCH DIAGRAM

A key insight can be observed from the material recycle pinch diagram. The pinch point distinguishes two zones. Below that point, fresh resource is used in the sinks, while unused process sources are discharged above that point. The primary characteristic for the pinch point is based on the following observation: The pinch point is the point where the load of recycled/reused sources match that of the sink. Hence, it corresponds to the most constrained point in the recycle system. If the two composite curves are not touched

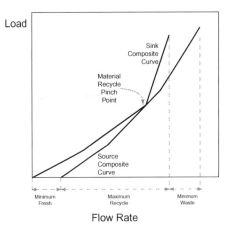

FIGURE 4-9 Identifying targets for minimum fresh usage, maximum direct recycle, and minimum waste discharge. *Source: El-Halwagi et al. (2003).*

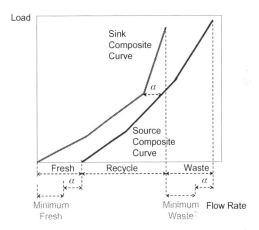

FIGURE 4-10 Passing flow rate (α) through the pinch point leads to less integration. *Source: El-Halwagi (2006).*

at the pinch (for example, by moving the source composite to the left, thereby passing a flow rate of α through the pinch), the fresh usage and waste discharge are both increased by the same magnitude of the flow rate passed through the pinch (α). Additionally, the extent of recycled flow rate is also reduced by the same magnitude (α) as shown by Fig. 4.10. On the other hand, if we move the source composite to the left of the pinch, a portion of the source composite will lie above the sink composite, thereby leading to the violation of Constraint 4.5. This situation is shown in Fig. 4.11.

The previous discussion indicates that, to achieve the minimum usage of fresh resources, maximum reuse of the process sources, and minimum discharge of waste, the following three design rules are needed:

- No flow rate should be passed through the pinch (that is, the two composites must touch).
- No waste should be discharged from sources below the pinch.
- No fresh resources should be used in any sink above the pinch.

The targeting procedure identifies the targets for fresh, waste, and material reuse without commitment to the

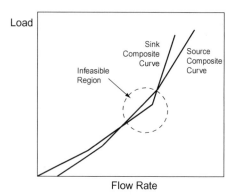

FIGURE 4-11 Violation of the pinch point leads to infeasibility. *Source: El-Halwagi (2006).*

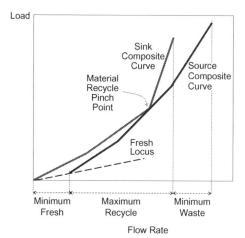

FIGURE 4-12 The material recycle pinch diagram when the fresh resource is impure. *Source: El-Halwagi (2006).*

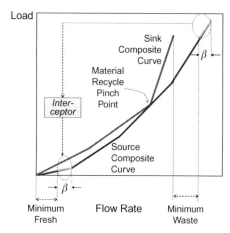

FIGURE 4-13 Insights for process modification: moving a source across the pinch. *Source: El-Halwagi (2006).*

detailed design of the network matching the sources and sinks. In detailing the solution, there can be more than one solution satisfying the identified targets. Those solutions can be identified using the source-sink mapping diagram. To compare the multiple solutions having the same target of fresh usage and waste discharge, other objectives should be used (for example capital investment, safety, flexibility, operability, and so on).

EXTENSION TO THE CASE OF IMPURE FRESH

The same targeting procedure can be extended for cases when the fresh resource is impure. In the case of pure fresh, the source composite curve was slid on the horizontal axis. The reason for this is that, with no impurities in the fresh, it does not contribute the load of impurities regardless of how much flow rate of fresh is used. Consequently, the horizontal axis serves as a locus for the fresh resource. When the fresh is impure but cleaner than the rest of the sources, its locus becomes a straight line emanating from the origin and having a slope of y_{Fresh} (composition of impurities in the fresh). Hence, the source composite curve is slid on the fresh locus until it touches the sink composite while lying below it in the overlapped region. This case is shown by Fig. 4.12. If the fresh resource was not the source with the least composition of impurities, the same procedure can be adopted by ranking the sources in ascending order of composition and placing the locus of the fresh at its proper rank.

INSIGHTS FOR PROCESS MODIFICATIONS

The material recycle pinch diagram and the associated design rules can be used to guide the engineer in making process modifications to enhance material reuse; for instance, the observation that below the pinch there is deficiency of recyclable sources, whereas above the pinch there is a surplus of sources. Therefore, sinks can be moved from below the pinch to above the pinch and sources can be moved from above the pinch to below the pinch to reduce the usage of fresh resources and the discharge of waste. Moving a sink from below the pinch to above the pinch can be achieved by increasing the upper bound on the composition constraint of the sink given by inequality (see Constraint 4.2). Conversely, moving a source from above the pinch to below the pinch can be accomplished by reducing its composition through changes in operating conditions or by adding an "interception" device (for example, separator, reactor, and so on) that can lower the composition of impurities. The design of interception networks will be handled in several chapters in this book. Figure 4.13 illustrates an example when a flow rate β is intercepted to reduce its content of impurities down to a composition similar to that of the fresh. Therefore, this flow rate is moved from above the pinch to below the pinch. Compared to the nominal case without interception, two benefits accrue as a result of this movement across the pinch: Both the usage of fresh resource and the discharge of waste are reduced by β.

Another alternative is to reduce the load of a source below the pinch (again by altering operating conditions or adding an interception device). Consequently, the cumulative load of the source composite decreases and allows an additional recycle of process sources with the result of decreasing both the fresh consumption and waste

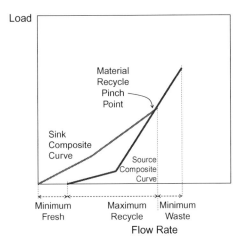

FIGURE 4-14 Example of a material recycle pinch diagram before interception. *Source: El-Halwagi (2006).*

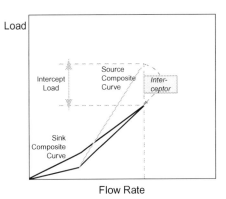

FIGURE 4-15 Insights for process modification: lowering the slope of second source through interception. *Source: El-Halwagi (2006).*

discharge. Figure 4.14 illustrates the material recycle pinch diagram before interception. Then, the second source is intercepted to remove the load protruding above the pinch. As the intercepted load is removed, the slope of the second source decreases. The new slope is the composition of the intercepted source. Consequently, the source composite curve can be slid to the left to reduce (or in this case to eliminate) the use of fresh resource as shown in Fig. 4.15.

Example 4-1 Water recycle in a formic acid plant

Consider the process of manufacturing formic acid (HCOOH), shown by Fig. 4.16. Biomass is gasified to produce a syngas mixture. A solid–gas separator is used to remove the particulates referred to as the fly ash. To avoid the problem of solid-waste disposal of fly ash while adding value, water is mixed with the fly ash to yield a paste that can be used in the manufacture of construction materials (for example, in making bricks or providing a filler for asphalt). The hot gases leaving the solid-gas separator are to be separated to different streams including a carbon monoxide stream that is used as a reactant in making formic acid. Water quenching is used to cool the hot gases leaving the solid-gas separator. The carbon monoxide stream leaving the gas-separation system is reacted with methanol to give methyl formate (HCOOCH$_3$) through the following carbonylation reaction over a strong-base catalyst:

[4.9] $$CH_3OH + CO \rightarrow HCOOCH_3$$

Methyl formate is recovered using a distillation column. Next, water is used to hydrolyze methyl formate to formic acid and methanol as follows:

[4.10] $$HCOOCH_3 + H_2O \rightarrow HCOOH + CH_3OH$$

A separation network is used to recover formic acid and to produce a stream of distilled water and another stream of organics.

There are three process sinks that use fresh water: the fly ash stabilizer, the quenching column, and the hydrolysis reactor. At present, a total of 9500 kg/hr of fresh water is used in quenching and hydrolysis. Two process streams (sources) are considered for recycle: the quenching effluent and distilled water. A total of 7000 kg/hr of wastewater is discharged as quenching effluent and distilled water. It is desired to use direct recycle to reduce fresh water usage and wastewater discharge. The following constraints define the acceptable ranges for flow rates and concentrations fed to the two process sinks.

FLY ASH STABILIZER:

[4.11] 500 ≤ Flow Rate of Stabilizing Water Fed to the Fly Ash Stabilizer (kg/hr) ≤ 600

[4.12] 0.0 ≤ Concentration of the Impurities (wt%) ≤ 10.0

QUENCHING COLUMN:

[4.13] 3000 ≤ Flow Rate of the Quenching Water Fed to the Column (kg/hr) ≤ 3500

[4.14] 0.0 ≤ Concentration of the Impurities (wt.%) ≤ 3.0

HYDROLYSIS REACTOR:

[4.15] 6000 ≤ Flow Rate of Hydrolysis Water Fed to the Reactor (kg/hr) ≤ 6500

[4.16] 0.0 ≤ Concentration of the Impurities (wt.%) ≤ 1.0

Table 4.1 summarizes the data for the process sinks. As mentioned earlier in the procedure for constructing the material recycle pinch diagram, the sinks

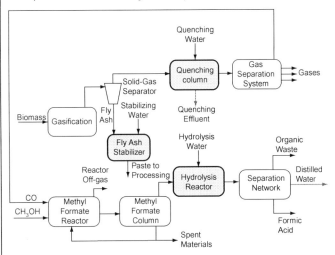

FIGURE 4-16 A block flow diagram of the formic acid process.

(Continued)

Example 4-1 Water recycle in a formic acid plant (Continued)

TABLE 4-1 Sink Data for the Formic Acid Example

Sink	Flow Rate kg/hr	Maximum Inlet Mass Fraction of Impurities	Maximum Inlet Load of Impurities, kg/hr
Hydrolysis Reactor	6000	0.0100	60
Quenching Column	3000	0.0300	90
Fly Ash Stabilizer	500	0.1000	50

TABLE 4-2 Source Data for the Formic Acid Example

Source	Flow Rate, kg/hr	Mass Fraction of Impurities	Load of Impurities, kg/hr
Distilled Water	4000	0.0175	70
Quenching Effluent	3000	0.0600	180

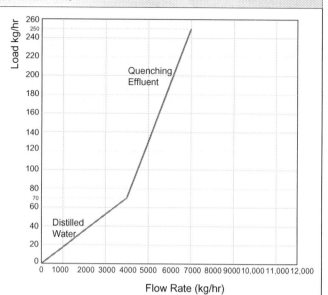

FIGURE 4-18 The source composite curve for the formic acid example.

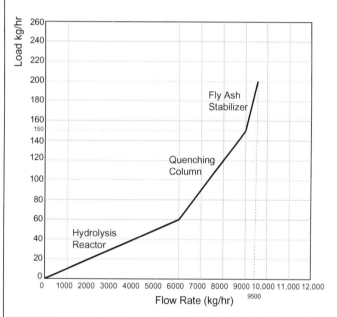

FIGURE 4-17 The sink composite curve for the formic acid example.

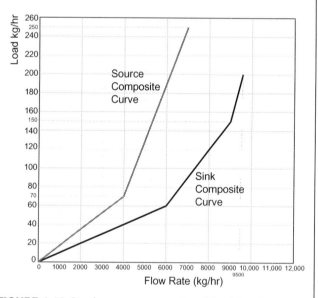

FIGURE 4-19 Simultaneous representation of the sink and source composites for the formic acid example.

are listed in ascending order of the maximum allowable inlet mass fraction of impurities. Furthermore, because the objective is to minimize the fresh usage, the lower bound of the acceptable flow rate of each sink is selected as the required flow rate. Table 4.2 provides the data for the process sources arranged in ascending order of concentration of impurities.

SOLUTION

The sink and source composite curves are constructed as shown in Figures 4.17 and 4.18. The two composite curves are shown simultaneously on Fig. 4.19. Next, the source composite curve is slid horizontally to the right until it touches the sink composite curve while the source composite lies below the sink composite at every value of flow rate. The material reuse pinch diagram is shown by Fig. 4.20. The targets for minimum fresh water usage and minimum wastewater discharge are found to be 3667 and 1167 kg/hr, respectively.

To attain the targets, it is necessary to determine how the sources should be assigned to the sinks. Here, a visualization approach is presented based on the work of El-Halwagi et al. (2003). Later, a graphical-algebraic approach (called the source-sink mapping diagram) will be presented.

An important insight from the material recycle pinch diagram is that the sources and the sinks below the pinch are balanced both in terms of flow rate and load of impurities. It is also important to recall the sink-load rule, which indicates that the load of impurities entering the sink should be maximized to minimize the usage of the fresh. Therefore, an optimum combination of sources fed to a sink below the pinch should match the flow rate and load of the sink. Furthermore, only sources below the pinch should be assigned to sinks below the pinch. For instance, one possible implementation for the feed to the hydrolysis reactor is to mix fresh water with a portion of the distilled water. The requirement is that both the flow rate and load of impurities for the mixed sources match the flow rate and maximum permissible load of the sink. Because the fresh in this case is pure, it does not contribute any load of impurities to the feed entering the sink. Hence, the required flow rate of the distilled water can be graphically determined by taking the portion of

(Continued)

Example 4-1 Water recycle in a formic acid plant (*Continued*)

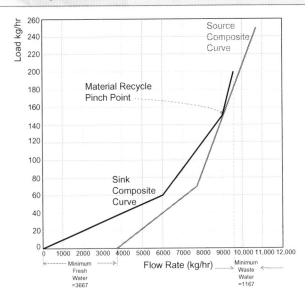

FIGURE 4-20 The material recycle pinch diagram for the formic acid example.

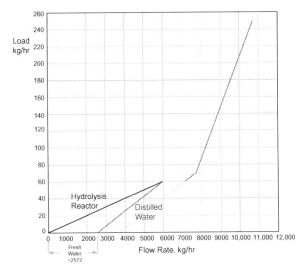

FIGURE 4-21 Assigning fresh water and distilled water the hydrolysis reactor.

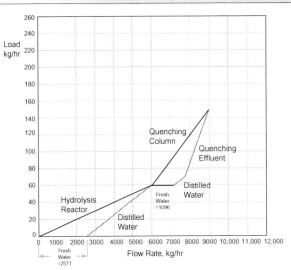

FIGURE 4-22 Assigning fresh water, distilled water, and quenching effluent to the quenching column.

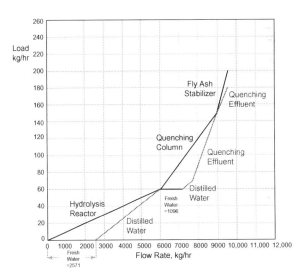

FIGURE 4-23 Assigning the quenching effluent to the fly ash stabilizer.

the distilled water that contains an impurities load of 60 kg/hr. As can be seen from Fig. 4.21, this portion is 3429 kg/hr. When moved to the left until it touches the hydrolysis reactor while lying completely below it, the rest of the flow rate (2571 kg/hr) should come from the fresh. Next, the unused portion of distilled water along with the quenching water below the pinch are mixed and fed to the quenching column. The remaining flow rate required by the quenching column (1096 kg/hr) is provided by fresh water. This is shown in Fig. 4.22. It is worth noting that the total fresh requirement (2571 + 1096 = 3667 kg/hr) matches the target identified from the material recycle pinch diagram.

The problem of matching sources and sinks above the pinch is easier than that below the pinch. Because no fresh water is used above the pinch, assigning sources to sinks involves matching flow rates without exceeding the maximum permissible load of impurities. Therefore, for the fly ash stabilizer, 500 kg/hr of the quenching effluent should be used as the feed stream. The rest of the quenching effluent (1267 kg/hr) is discharged as waste. Figure 4.23 shows the assignment of a portion of the quenching effluent to the fly ash stabilizer along with the previously shown matches below the pinch. Figure 4.24 is a schematic representation of the allocation of the sources to sinks implementing the matches shown by Fig. 4.23.

The matches shown in Fig. 4.24 illustrate that the targets for fresh water and waste discharge are consistent with the targeting values determined from the material recycle pinch diagram. But is this the only possible implementation of the targets? Let us examine another arrangement. Suppose that we assign all of the fresh target (3667 kg/hr) to the hydrolysis reactor. The rest of the feed to the hydrolysis reactor (6000 − 3667 = 2333 kg/hr) should come from a combination of the distilled water and quenching effluent such that the total load of impurities is 60 kg/hr. This is achieved by sliding the quenching effluent on the distilled water line until the total load is 60 kg/hr. The result is 1882 kg/hr of distilled water and 451 kg/hr of quenching effluent. Alternatively, flow and load balances can be solved simultaneously as follows:

[4.17] Flow rate of distilled water fed to the hydrolysis reactor + Flow rate of quenching effluent fed to the hydrolysis reactor = 2333

(*Continued*)

Example 4-1 Water recycle in a formic acid plant (*Continued*)

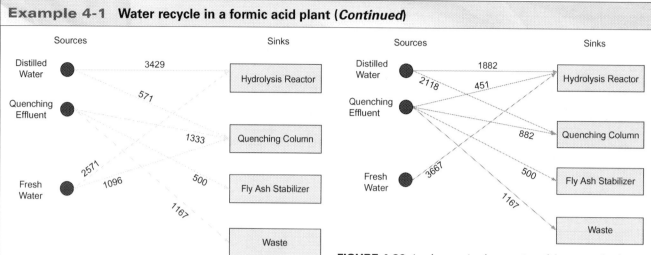

FIGURE 4-24 An implementation of the targets for the formic acid process (all numbers represent assigned flow rate from source to sink in kg/hr).

FIGURE 4-26 An alternate implementation of the targets for the formic acid process (all numbers represent assigned flow rate from source to sink in kg/hr).

[4.18]

Flow rate of distilled water fed to the hydrolysis reactor*0.0175 + Flow rate of quenching effluent fed to the hydrolysis reactor*0.06 = 60

The solution of the two balance equations gives 1882 kg/hr of distilled water and 451 kg/hr of quenching effluent. Next, the remaining portions of distilled water and quenching effluent below the pinch are assigned to the quenching column as shown by Fig. 4.25. Above the pinch, there is only one option for assigning 500 kg/hr of the quenching effluent to the fly ash stabilizer. Figure 4.26 is a schematic representation of the allocation of the sources to sinks implementing the matches shown by Fig. 4.25.

The previous discussion demonstrates that despite the unique targets for minimum fresh usage and minimum waste discharge, there can be numerous (even infinite) implementations that will attain the target. Such richness in design enables the consideration of additional criteria. For instance, the designer may include other design criteria (for example, length or cost of piping, safety, and so on) to select the implementation from among the possible alternatives that satisfy the targets for minimum fresh water usage and minimum waste water discharge.

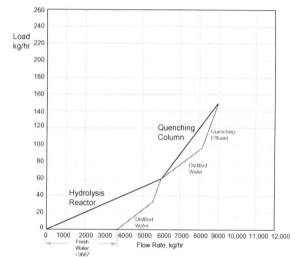

FIGURE 4-25 An alternate assignment of sources to sinks below the pinch.

THE SOURCE-SINK MAPPING DIAGRAM FOR MATCHING SOURCES AND SINKS

Example 4.1 describes a graphical method for assigning sources to sinks. This section provides an alternate graphical-algebraic method for allocating sources to sinks. This method is called the **source-sink mapping diagram** (El-Halwagi and Spriggs, 1996). For each species, a diagram is constructed by plotting the flow rate versus composition. This is shown by Fig. 4.33. On the source-sink mapping diagram, sources are represented by shaded circles, and sinks are represented by hollow circles. Constraints 4.1 and 4.2, characterizing the feasibility region for each sink, are represented by two bands for flow rate and composition. The intersection of these two bands provides a zone of acceptable feed to the sink. If a source (for example, source a) lies within this zone, it can be directly recycled to the sink (for example, sink S). Moreover, sources b and c can be mixed using the lever arm principle to create a mixed stream that can be recycled to sink S.

A particularly useful tool for application in the source-sink mapping diagram is the lever arm rule, which can be described based on component material balance. Consider the mixing of two sources, a and b, shown

Example 4-2 Water recovery in a food process (El-Halwagi, 2006)

In addition to determining direct-recycle targets, the material recycle pinch diagram can be used to assess and critique existing or proposed designs. Consider the food processing plant shown in the simplified flow sheet of Fig. 4.27. The process has many units, including two sinks that use fresh water: a feedstock washer and an off-gas scrubber. Table 4.3 provides the data for these two sinks. The process results in two aqueous streams that are sent to a biotreatment facility. These two streams are condensate I from the evaporator and condensate II from the stripper. Both streams may be recycled to partially or completely replace the fresh water used in the two sinks. The data for the two process sources are given in Table 4.4.

Because of the large consumption of fresh water and discharge of wastewater, the plant has decided to adopt direct-recycle strategies. An engineer has proposed that Condensate I be recycled to the scrubber (see Fig. 4.28). The result of this project is to eliminate the need for fresh water in the scrubber, reduce overall fresh water consumption to 8000 kg/hr, and reduce wastewater discharge (Condensate II) to 9000 kg/hr. Critique this proposed project.

SOLUTION

First, it is useful to use the material recycle pinch diagram to determine the direct-recycle targets for minimum water usage and minimum wastewater discharge. These benchmarks are obtained from the material recycle pinch diagram shown in Fig. 4.29. As can be seen from the pinch diagram, the target for minimum fresh water usage is 2000 kg/hr, whereas the target for minimum wastewater discharge is 3000 kg/hr. Because these targets are significantly superior to the results of the proposed solution (8000 kg/hr of fresh water and 9000 kg/hr of wastewater), it is important to methodically assess the suggested recycle strategy.

Next, we represent the proposed solution on the material recycle pinch diagram as shown in Fig. 4.30. Because Condensate I is matched with the scrubber, it is plotted directly below the scrubber. This is a feasible solution because the flow rate is satisfied and the load of Condensate I is less than the maximum admissible load for the scrubber. However, we notice that this solution results in passing 6000 kg/hr through the pinch. Therefore, we expect to see a 6000 kg/hr increase in fresh usage and waste discharge. Indeed, this is the case (fresh water usage is 8000 kg/hr compared to the target of 2000 kg/hr and waste discharge of 9000 kg/hr compared to the target of 3000 kg/hr). Additionally, we notice that proposed recycle does not follow the sink composition rule that calls for the maximization of the inlet composition to the sink (the proposed solution has the inlet to the scrubber being 2 percent as compared to 5 percent).

The previous discussion indicates that the proposed project should not be implemented. Instead, an integrated solution consistent with the identified targets should be recommended (see Fig. 4.31). Nonetheless, if the engineer insists on implementing the proposed recycle project, there is still an opportunity to improve upon this situation. The fact that the concentration of impurities in Condensate II is higher than the admissible concentration to the washer does not prevent us from pursuing partial recycle. Figure 4.31 is the pinch diagram for the remaining source

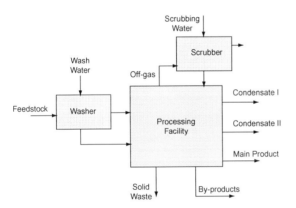

FIGURE 4-27 A simplified flow sheet of the food processing plant. *Source: El-Halwagi (2006).*

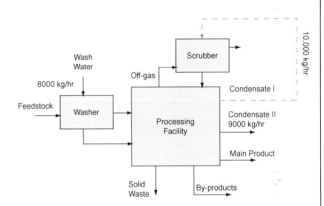

FIGURE 4-28 Proposed recycle project. *Source: El-Halwagi (2006).*

TABLE 4-3 Sink Data for the Food Processing Example

Sink	Flow Rate kg/hr	Maximum Inlet Mass Fraction	Maximum Inlet Load, kg/hr
Washer	8000	0.03	240
Scrubber	10,000	0.05	500

Source: El-Halwagi (2006).

TABLE 4-4 Source Data for the Food Processing Example

Source	Flow Rate kg/hr	Maximum Inlet Mass Fraction	Maximum Inlet Load, kg/hr
Condensate I	10,000	0.02	200
Condensate II	9000	0.09	810

Source: El-Halwagi (2006).

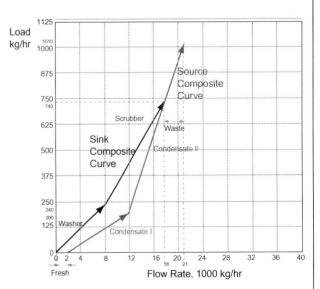

FIGURE 4-29 Pinch diagram for food processing example: targets for fresh and waste are 2000 and 3000 kg/hr, respectively. *Source: El-Halwagi (2006).*

(Continued)

Example 4-2 Water recovery in a food process (El-Halwagi, 2006) (*Continued*)

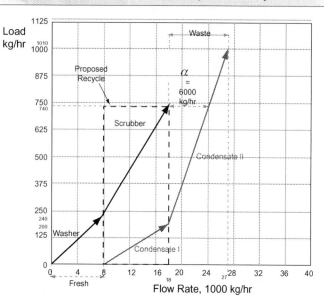

FIGURE 4-30 Representation of proposed recycle on pinch diagram which results in passing 6000 kg/hr through the pinch. *Source: El-Halwagi (2006).*

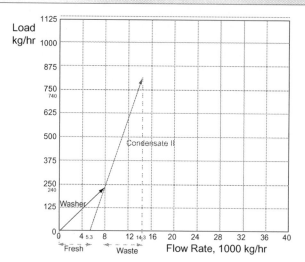

FIGURE 4-31 Pinch diagram if condensate I and the scrubber are taken out of the problem which leads to targets of 5300 and 6300 kg/hr for fresh and waste, respectively. *Source: El-Halwagi (2006).*

and sink in the problem. As can be seen from the diagram, if Condensate I is used in lieu of the scrubber water, then the target for the remaining source (Condensate II) and sink (Washer) is 5300 kg/hr in fresh water usage and 6300 kg/hr in wastewater discharge. The foregoing discussion underscores the importance of ensuring that a short-term project be compatible with an overall integrated strategy. An individual project that is seemingly flawless may be detrimental to the overall performance of the whole process and may prevent the process from ever reaching its true target. Additionally, a big-picture approach such as the material recycle pinch diagram yields insights that may not be seen by detailed engineering focusing on individual units and streams. Finally, ill-conceived (although well-intentioned) projects that are not rooted in proper understanding of the fundamentals and tools of process integration may lead to counterproductive results.

Example 4.3 Dealing with impurities in the fresh feed: Hydrogen recycle in a refinery

There is a growing demand for hydrogen usage in various industries including refineries, petrochemical plants, and biorefineries. This example deals with the recycle of hydrogen in a petroleum refinery, based on the case study reported by Alves and Towler (2002). The refinery has four sinks that use fresh hydrogen and six sources that contain hydrogen and may be recycled to reduce the usage of fresh hydrogen. The data for the process sinks and sources are summarized in Tables 4.5 and 4.6.

The available fresh hydrogen contains 0.05 mole fraction of impurities. It is desired to use the material recycle pinch diagram to determine the direct-recycle targets for minimum fresh usage of hydrogen and waste discharge.

SOLUTION

First, draw the sink composite curve. Next, in constructing the sources composite curve, draw the locus for fresh hydrogen as a locus with a slope of the impurities concentration (mole fraction of 0.05). Then, slide the source composite curve on the fresh locus until it touches the sink composite while lying completely below it at every value of the flow rate. Figure 4.32 identifies the targets of 269 and 103 mol/s for usage and discharge of hydrogen, respectively.

TABLE 4-5 Sink Data for Hydrogen Problem

Sinks	Flow (mol/s)	Maximum Inlet Impurity Concentration (mol %)	Load (mol/s)
1	2495	19.39	483.8
2	180.2	21.15	38.1
3	554.4	22.43	124.4
4	720.7	24.86	179.2

Source: Alves and Towler (2002).

TABLE 4-6 Source Data for the Hydrogen Problem

Sources	Flow (mol/s)	Impurity Concentration (mol %)	Load (mol/s)
1	623.8	7	43.7
2	415.8	20	83.2
3	1801.9	25	450.5
4	138.6	25	34.7
5	346.5	27	93.6
6	457.4	30	137.2

Source: Alves and Towler (2002).

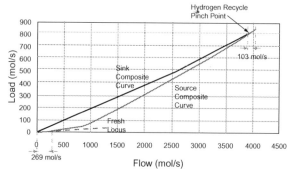

FIGURE 4-32 Hydrogen reuse pinch diagram. *Source: El-Halwagi et al. (2003).*

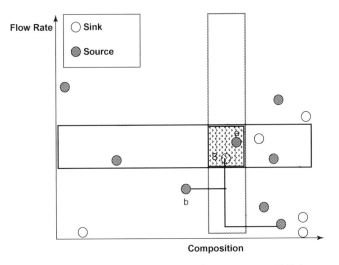

FIGURE 4-33 The source-sink mapping diagram. *Source: El-Halwagi and Spriggs (1996).*

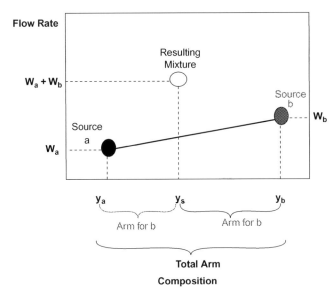

FIGURE 4-35 The lever arm rule for mixing. *Source: El-Halwagi (2006).*

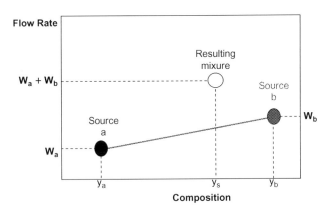

FIGURE 4-34 Mixing of sources a and b. *Source: El-Halwagi (2006).*

in Fig. 4.34. Let us designate the flow rates of the two sources by W_a and W_b and the compositions by y_a and y_b. When the two streams are mixed, the resulting flow rate is $W_a + W_b$ and the composition is y_s, which can be determined according to the following component material balance:

[4.19a] $$y_s(W_a + W_b) = y_a W_a + y_b W_b$$

which can be rearranged to:

[4.19b] $$\frac{W_a}{W_b} = \frac{y_b - y_s}{y_s - y_a}$$

or

$$\frac{W_a}{W_b} = \frac{Arm\ for\ a}{Arm\ for\ b}$$

where

[4.19d] $$Arm\ for\ a = y_b - y_s$$

and

[4.19e] $$Arm\ for\ b = y_s - y_a$$

Similarly,

[4.20a] $$\frac{W_a}{W_a + W_b} = \frac{Arm\ for\ a}{Total\ arm}$$

where the total arm is the sum of arm for a and arm for b. Hence,

[4.20b] $$Total\ arm = y_b - y_a$$

The lever arms are shown by Fig. 4.35.

An important application of the lever arm rule is to illustrate the sink-load and source prioritization rules mentioned earlier in developing the material recycle pinch diagram. Consider the case of sink j shown by Fig. 4.36, with constraints on flow rate and composition. A recyclable stream (source a) is to be mixed with the fresh to reduce the fresh usage in the sink. Depending on the flow rate proportions for mixing the fresh and source a, different compositions can be obtained. As long as the mixture lies within the feasibility region of the sink, it is acceptable. But what is the ratio of fresh to source (a) that minimizes the usage of the fresh? Alternatively, what should be the composition of the feed entering the sink? Should it be z_j^{min}, z_j^{max}, or some intermediate value, z_j^{avg}? According to the lever arm rule described by Eq. 4.20a, we get:

[4.21a] $$\frac{Fresh\ flow\ rate\ used\ in\ sink}{Total\ flow\ rate\ fed\ to\ sink} = \frac{Fresh\ arm}{Total\ arm}$$

That is,

[4.21b] $$\frac{Fresh\ flow\ rate\ used\ in\ sink}{Total\ flow\ rate\ fed\ to\ sink} = \frac{y_a - z_{Feed\ to\ sink}}{y_a - y_F}$$

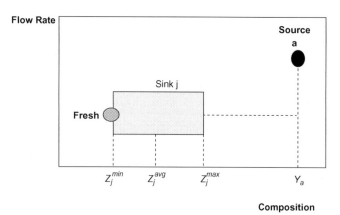

FIGURE 4-36 Selection of feed composition to sink. *Source: El-Halwagi (2006).*

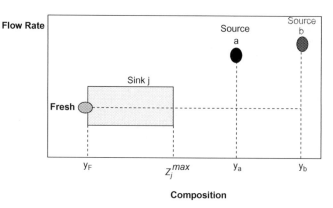

FIGURE 4-37 The source-prioritization rule. *Source: El-Halwagi (2006).*

For a given requirement of the total flow rate fed to the sink, the flow rate of the fresh is minimized when $z_{\text{Feed to sink}}$ is maximized. Hence, the composition of the feed entering the sink should be set to $z_j^{\max}$. This analysis explains the previously mentioned optimality criterion for the sink, which can be described by the **sink composition rule**, when a fresh resource is mixed with process source(s), the composition of the mixture entering the sink should be set to a value that minimizes the fresh arm. For instance, when the fresh resource is a pure substance that can be mixed with a source that contains an impurity, the composition of the mixture should be set to the maximum allowable value.

The other optimality criterion deals with the order of recycling the sources. For instance, suppose that the process sink shown by Fig. 4.37 currently uses a fresh resource, and it is desired to reduce this fresh usage by mixing the fresh stream with a process source. Two process sources (a and b) are considered for recycle. Both sources have sufficient flow rate to satisfy the sink, but their compositions exceed the maximum admissible composition to the sink. Nonetheless, upon mixing with the fresh resource in proper proportions, both process sources can be fed to the sink. The question is which source should be used: a or b?

According to Fig. 4.37 and Eq.4.21b:

[4.22a]   Fresh arm when source a is used $= y_a - z_j^{\max}$

where $z_j^{\max}$ is set as the composition of the mixture as described in the sink composition rule. Similarly,

[4.22b]   Fresh arm when source b is used $= y_b - z_j^{\max}$

The arms given by Equations 4.22a and 4.22b are referred to as the absolute fresh arms when sources a and b are used. Clearly, the absolute fresh arm for a is shorter than the fresh arm for b. However, Eq. 4.21 indicates that, to minimize the usage of the fresh resource, the relative fresh arm (which is the ratio of the absolute fresh arm to the total arm of the mixture) should be minimized. It can be shown that minimizing the absolute fresh arm for a source corresponds to minimizing the relative fresh arm

for the source. To prove this observation, we compare the relative fresh arms for sources a and b:

[4.23]   Relative fresh arm for a $= \dfrac{y_a - z_j^{\max}}{y_a - y_F}$

and

[4.24]   Relative fresh arm for b $= \dfrac{y_b - z_j^{\max}}{y_b - y_F}$

Let us now evaluate the ratio of the two relative fresh arms:

[4.25]

$$\frac{\text{Relative fresh arm for a}}{\text{Relative fresh arm for b}} = \frac{\dfrac{y_a - z_j^{\max}}{y_a - y_F}}{\dfrac{y_b - z_j^{\max}}{y_b - y_F}}$$

$$= \frac{\dfrac{y_a - z_j^{\max}}{y_a - y_F}}{\dfrac{(y_b - y_a) + (y_a - z_j^{\max})}{(y_b - y_a) - (y_a - y_F)}} = \frac{\dfrac{(y_b - y_a)}{(y_a - y_F)} + 1}{\dfrac{(y_b - y_a)}{(y_a - z_j^{\max})} + 1}$$

But, as can be seen from Fig. 4.37,

[4.26a] $y_a - y_F > y_a - z_j^{\max}$

Hence,

[4.26b] $\dfrac{y_b - y_a}{y_a - y_F} < \dfrac{y_b - y_a}{y_a - z_j^{\max}}$

Combining Eq. 4.25 and inequality (4.26b), we obtain:

[4.27] Relative fresh arm for a $<$ Relative fresh arm for b

Inequality (show in Eq. 4.27) allows us to look at the absolute fresh arms instead of calculating the relative fresh arms. This leads to a rule previously mentioned as one of the optimality criteria used in constructing the material recycle pinch diagram: the **source-prioritization rule**, a rule stating that, to minimize the usage of the fresh resource, recycle of the process sources should be prioritized in order of their fresh arms, starting with the source having the shortest fresh arm. For instance, in Fig. 4.37, source a should be recycled before source b is considered. Therefore, the recyclable source with the shortest relative fresh arm should be used first until it is completely used in recycle, and then the source with the next-to-shortest fresh arm is recycled, and so on.

Because the source- and sink-composite curves touch at the pinch point, the following conditions are observed:

[4.28] Total flow rate of the sources (including the fresh) below the pinch = Total flow rate demand of the sinks below the pinch

and

[4.29] Cumulative load of impurities in the sources (including the fresh) below the pinch = Cumulative maximum permissible load of impurities for the sinks below the pinch

To meet the balances given by Equations 4.28 and 4.29 while complying with the sink composition optimality rule, the following conditions should be satisfied:

- The sources below the pinch must be assigned to sinks below the pinch.

- Below the pinch, the total load of impurities of the sources fed to a sink must equal the maximum load of impurities allowable for that sink.

Therefore, the procedure for using the source-sink mapping diagram for allocating sources to sinks is given by the following steps:

Below the Pinch:

1. Start with any sink. Use the source prioritization rule by assigning the sources in ascending order of the fresh arm and using the fresh as needed. The total load of impurities of the sources fed to the sink must equal the maximum allowable load of impurities for that sink. The flow rate must also balance.
2. When a process source is completely recycled, the next source (in terms of the length of the fresh arm) is used and so on.
3. The same procedure is repeated for the rest of the process sinks until all the demands for the sinks are met using the sources (including the fresh) below the sink.

Above the Pinch:

1. Start with the sink immediately above the pinch. Use the sources closest to the sink to satisfy the flow rate while ensuring that the load of impurities in the sources is less than or equal to the maximum allowable load of impurities for the sink.
2. Repeat the previous step for the next sink and so on.
3. The flow rate of the sources above the pinch that have not been assigned to the sinks is discharged as waste.

Example 4-4 Matching sources and sinks for the formic acid example

Let us revisit Example 4.1, which dealt with water recycle in a formic acid plant. As a reminder, the following constraints define the acceptable ranges for flow rates and concentrations fed to the two process sinks.

Fly Ash Stabilizer:

[4.11] $500 \leq$ Flow Rate of Stabilizing Water Fed to the Fly Ash Stabilizer (kg/hr) ≤ 600

[4.12] $0.0 \leq$ Concentration of the Impurities (wt%) ≤ 10.0

Quenching Column:

[4.13] $3000 \leq$ Flow Rate of the Quenching Water Fed to the Column (kg/hr) ≤ 3500

[4.14] $0.0 \leq$ Concentration of the Impurities (wt%) ≤ 3.0

Hydrolysis Reactor:

[4.15] $6000 \leq$ Flow Rate of Hydrolysis Water Fed to the Reactor (kg/hr) ≤ 6500

[4.16] $0.0 \leq$ Concentration of the Impurities (wt%) ≤ 1.0

These constraints are represented on the source-sink mapping diagram as shown by the boxes depicting the feasibility regions for the sinks shown by Fig. 4.38. Additionally, the data for the sources given by Table 4.2 are represented by the circles.

From the material recycle pinch diagram of Fig. 4.20, two sinks lie below the pinch: hydrolysis reactor and the quenching column. Therefore, according to the design rules associated with the pinch analysis, fresh usage may be used in one or both of these sinks. As mentioned earlier, there may be multiple (even infinite) implementations for the unique targets identified by the material recycle pinch diagram. We can start with any sink below the pinch. Let us start with an implementation that begins with matching sources to the hydrolysis reactor. To substitute some of the fresh used in this sink, recycle from distilled water is considered because it is the source closest to the hydrolysis reactor sink (shortest fresh arm). According to the lever arm principle (Eq. 4.21b),

[4.30a] $\dfrac{\text{Fresh water in the hydrolysis reactor}}{6000} = \dfrac{0.00175 - 0.0100}{0.0175 - 0.0000}$

Hence,

[4.30b]   Fresh water used in the hydrolysis reactor $= 2571$ kg/hr

Based on the overall material balance around the hydrolysis reactor, we have:

[4.31]   Flow rate of distilled water fed to the hydrolysis reactor $= 6000 - 2571$
$= 3429$ kg/hr

Alternatively, component (impurities) material balance for the feed of the hydrolysis reactor yields

[4.32a]   Flow rate of distilled water fed to the hydrolysis reactor*$0.0175 +$ Fresh water to the hydrolysis reactor*$0.0000 = 6000*0.0100$

(Continued)

Example 4-4 Matching sources and sinks for the formic acid example (*Continued*)

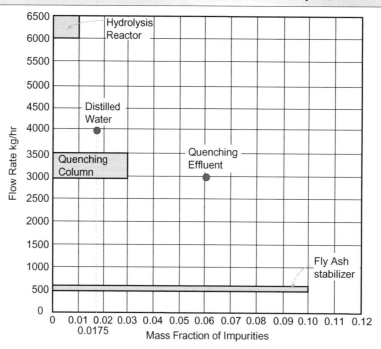

FIGURE 4-38 The source-sink mapping diagram for the formic acid problem.

That is,

[4.32b] Flow rate of distilled water fed to the hydrolysis reactor = 3429 kg/hr

which is the same result obtained from the application of the lever arm rule.

[4.33] The remaining flow rate of distilled water = 4000 − 3429
= 571 kg/hr

Because the distilled water has not been fully utilized, its remaining flow rate should be considered for recycle to the quenching column as it provides a shorter fresh arm than the quenching effluent. Because the composition of the distilled water is less than the maximum allowable composition for the quenching column, an additional source will have to be used (according to the sink composition rule below the pinch). Therefore, the remaining amount of the distilled water is recycled to the quenching column along with fresh water and some amount of the quenching effluent. A component material balance around the quenching column gives:

[4.34] 571*0.0175 + Flow rate of quenching effluent fed to the quenching column*0.0600 + Fresh water to the quenching column*0.0000 = 3000*0.0300

That is,

Flow rate of quenching effluent fed to the quenching column = 1333 g/hr

An overall material balance around the quenching column is given by:

[4.35] Fresh water to the quenching column = 3000 − 571 − 1333
= 1096 kg/hr

Combining Equations 4.30b and 4.35, we get:

[4.36] Total fresh water used in the process after recycle = 2571 + 1096
= 3667 kg/hr

which is the same target determined by the material recycle pinch diagram (see Fig. 4.20).

Finally, 500 kg/hr of the quenching effluent should be allocated to the fly ash stabilizer. Therefore,

[4.37] Remaining flow rate of quenching = 3000 − 1333 − 500
effluent discharged as wastewater = 1167 kg/hr

which matches the target identified from the material recycle pinch diagram. The results of this solution are consistent with the representation given by Fig. 4.24.

As mentioned earlier, the matching procedure can start with any sink below the pinch. The previous implementation started with the hydrolysis reactor. Let us identify another set of matches starting with the quenching column. Distilled water is the source with the shortest fresh arm to the quenching column. Nonetheless, its composition is less than the maximum allowable composition of impurities. Therefore, distilled water is to be mixed with the quenching effluent to get a flow rate of 3000 kg/hr and a mass fraction of impurities of 0.0300; that is,

[4.38] Flow rate of distilled water fed to the quenching column + Flow rate of quenching effluent fed to the quenching column = 3000 kg/hr

and

[4.39]
Flow rate of distilled water fed to the quenching column*0.0175 + Flow rate of quenching effluent fed to the quenching column*0.0600 = 3000*0.0300

Solving Equations 4.38 and 4.39 simultaneously, we get:

[4.40a] Flow rate of quenching effluent fed to the quenching column = 882 kg/hr

[4.40b] Flow rate of distilled water fed to the quenching column = 2118 kg/hr

[4.41] The remaining flow rate of distilled water = 4000 − 2118
= 1882 kg/hr

Noting from Fig. 4.20 that the flow rate at the pinch point is 9000, then

[4.42] The remaining flow rate of quenching = 9000 − 4000 − 882 − 3667
effluent below the pinch = 451 kg/hr

(*Continued*)

Example 4-4 Matching sources and sinks for the formic acid example (Continued)

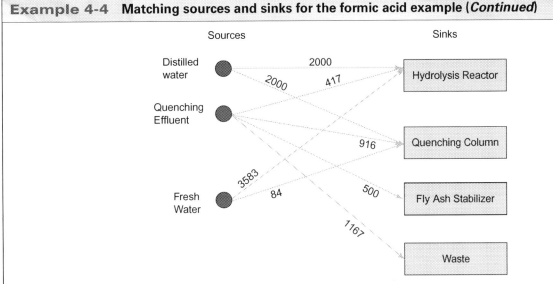

FIGURE 4-39 An alternate implementation for the targets of the formic acid example.

Because no fresh was used in the hydrolysis reactor, then all the fresh water has to be used in the quenching column. Therefore, the quenching column should receive 1882 kg/hr from distilled water, 451 kg/hr from quenching effluent, and 3667 kg/hr from fresh water. It is worth noting that the mass fraction of impurities in this mix fed to the quenching column can be calculated through balance on the impurities:

(4.43)

$$\text{Mass fraction of impurities in the mix fed to the quenching column} = \frac{1882*0.0175 + 451*0.0600 + 3667*0.0000}{6000} = 0.01$$

which, as expected, is the maximum mass fraction of impurities allowed into the hydrolysis reactor.

Above the pinch, 500 kg/hr should be assigned from the quenching effluent to the fly ash stabilizer. The implementation of these allocations matches the structure shown by Fig. 4.26. Other implementations are also possible. For instance, if we assign 2000 kg/hr of distilled water to the hydrolysis reactor along with unknown portions of the quenching effluent and fresh water, then the balance on impurities in the streams entering the hydrolysis reactor is given by:

[4.44a] $2000*0.0175 +$ Flow rate of quenching effluent assigned to the hydrolysis reactor$*0.0600 +$ Flow rate of fresh water assigned to the hydrolysis reactor$*0.0000 = 6000*0.0100$

Hence,

[4.44b] Flow rate of quenching effluent assigned to the hydrolysis reactor $= 417$ kg/hr

and,

[4.45] Flow rate of fresh water assigned to the hydrolysis reactor $= 6000 - 2000 - 417 = 3583$ kg/hr

The remaining flow rates below the pinch are allocated to the quenching column. These flow rates are $(4000 - 2000) = 2000$ kg/hr for the distilled water, $(1333 - 417) = 916$ kg/hr for the quenching water, and $(3667 - 3583) = 84$ kg/hr for the fresh water. This implementation is shown by Fig. 4.39.

Other implementations can be similarly generated. In this case study, an infinite number of alternatives can be synthesized to meet the targets determined by the materials recycle pinch diagram. Any alternative meeting the target must satisfy the aforementioned optimality criteria. The infinite number of implementations satisfying the target offers richness in design that enables the designer to add another objective to select the alternative of choice. Such objectives include proximity of sources and sinks, cost of piping, safety considerations, operational preferences, and so on. It is also worth noting that having an infinite number of alternatives satisfying the target does not mean that any implementation will meet the target. There are infinite implementations that will not meet the target. There are also infinite infeasible implementations. Finally, any implementation meeting the target must satisfy the aforementioned optimality criteria.

MULTICOMPONENT SOURCE-SINK MAPPING DIAGRAM

The source-sink diagram can be extended to represent ternary (three-component) systems. Figure 4.40 shows an equilateral triangle that represents the ternary diagram for components A, B, and C. Each apex represents 100 percent of a component. The opposite base corresponds to 0 percent of that component. Each line parallel to the base is a locus of a constant percentage of that component. Figure 4.40 shows the loci for various percentages of component A. The horizontal base of the triangle is the locus of any stream containing 0 percent of A. The other horizontal lines show the loci for compositions ranging from 10 to 90 percent A. Figure 4.41 illustrates the representation of a source containing 20 percent A, 30 percent B, and 50 percent C. Clearly, it is sufficient to represent

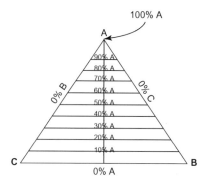

FIGURE 4-40 Ternary composition representation.

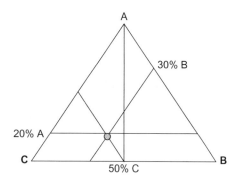

FIGURE 4-41 Representation of a ternary mixture.

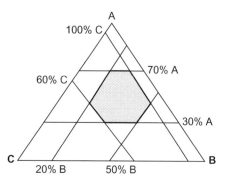

FIGURE 4-42 Representation of a sink with ternary constraints. *Source: El-Halwagi (2006).*

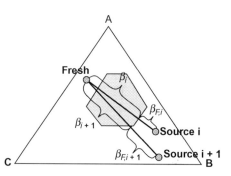

FIGURE 4-43 Level arm rule for a ternary source-sink diagram. *Source: El-Halwagi (2006).*

the loci of two compositions (for example, 20 percent A and 30 percent B) and locate the source at their intersection. The third composition is determined by passing a line parallel to the 0 percent C base through the source.

Composition constraints in the sinks can be represented on the ternary diagram. For instance, consider a sink with the following constraints:

[4.46a] $0.30 \leq$ mass fraction of A in stream entering the sink ≤ 0.70

[4.46b] $0.20 \leq$ mass fraction of B in stream entering the sink ≤ 0.50

[4.46c] $0.10 \leq$ mass fraction of C in stream entering the sink ≤ 0.60

The sink constraints are represented by Fig. 4.42. Any point within the shaded region is a feasible feed to the sink. The ternary source-sink diagram has the same rules as the sink optimality and source prioritization (for example, Parthasarathy and El-Halwagi, 2000). For instance, when two process sources (i and i + 1) are considered for recycle by mixing with a fresh resource, the lever arms should be first calculated. As can be seen from Fig. 4.43, the lever arm rules can be expressed as follows:

[4.47]
$$\frac{\text{Fresh usage when source } i \text{ is recycled}}{\text{Feed flow rate of sink}} = \frac{\beta_{F,i}}{\beta_{F,i} + \beta_i}$$

and

[4.48]
$$\frac{\text{Fresh usage when source } i+1 \text{ is recycled}}{\text{Feed flow rate of sink}} = \frac{\beta_{F,i+1}}{\beta_{F,i+1} + \beta_{i+1}}$$

Because source i has a shorter relative arm than that of i + 1, it should be used first. Source i + 1 is only used after source i has been completely recycled. This is the same concept used in source prioritization previously described in the single-component source-sink representation.

HOMEWORK PROBLEMS

4.1. Consider the magnetic tape manufacturing process shown in Fig. 4.44. A description of the process is given by Dunn et al. (1995). Coating ingredients are dissolved in 0.09 kg/s of organic solvent and mixed to form a slurry. The slurry is suspended with resin binders and special additives. Next, the coating slurry is deposited on a base film. Nitrogen gas is used to induce evaporation rate of solvent that is proper for deposition. In the coating chamber, 0.011 kg/s of solvent are decomposed into other organic species. The decomposed organics are separated from the exhaust gas in a membrane unit. The retentate stream leaving the membrane unit has a flow rate of 3.0 kg/s and is primarily composed of nitrogen that is laden with 1.9 wt/wt% of the organic solvent. The coated film is passed to a dryer where nitrogen gas is employed to evaporate the remaining solvent. The exhaust gas

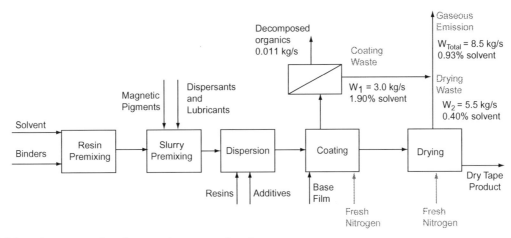

FIGURE 4-44 Schematic representation of a magnetic tape manufacturing process.

leaving the dryer has a flow rate of 5.5 kg/s and contains 0.4 wt/wt% solvent. The two exhaust gases are mixed and disposed off.

It is desired to undertake a direct-recycle initiative to use solvent-laden nitrogen (gaseous wastes) in lieu of fresh nitrogen gas in the coating and drying chambers. The following constraints on the gaseous feed to these two units should be observed:

Coating:

[4.49]

$$3.0 \leq \text{flow rate of gaseous feed (kg/s)} \leq 3.2$$

[4.50]

$$0.0 \leq wt\% \text{ of solvent} \leq 0.2$$

Dryer:

[4.51]

$$5.5 \leq \text{flow rate of gaseous feed (kg/s)} \leq 6.0$$

[4.52]

$$0.0 \leq wt\% \text{ of solvent} \leq 0.1$$

It may be assumed that outlet gas composition from the coating and the dryer chambers are independent of the entering gas compositions.

Using segregation, mixing, and direct recycle, what is minimum consumption of nitrogen gas that should be used in the process? Show one implementation for assigning the sources to the sinks while meeting the target for minimum nitrogen usage.

4.2. Vinyl acetate monomer "VAM" is manufactured by reacting acetic acid with oxygen and ethylene according to the following chemical reaction:

$$C_2H_4 + 0.5O_2 + C_2H_4O_2 = C_4H_6O_2 + H_2O$$
(ethylene oxygen acetic acid VAM water)

Consider the process shown in Fig. 4.45. In an acid tower, 10,000 kg/hr of acetic acid "AA" along with 200 kg/hr of water are evaporated. The vapor is fed with oxygen and ethylene to the reactor where 7000 kg/hr of acetic acid are reacted and 10,000 kg/hr of VAM are formed. The reactor off-gas is cooled and fed to the first absorber, where AA (5100 kg/hr) is used as a solvent. Almost all the gases leave from the top of the first absorption column together with 1200 kg/hr of AA. This stream is fed to the second absorption column where water (200 kg/hr) is used to scrub acetic acid. The bottom product of the first absorption column is fed to the primary distillation tower where VAM is recovered as a top product (10,000 kg/hr), along with small amount of AA (100 kg/hr), which is not worth recovering, and water (200 kg/hr). This stream is sent to final finishing. The bottom product of the primary tower (6800 kg/hr of AA and 2300 kg/hr of water) is mixed with the bottom product of the second absorption column (1200 kg/hr of AA and 200 kg/hr of water). The mixed waste is fed to a neutralization system followed by biotreatment.

The following technical constraints should be observed in any proposed solution.

Neutralization:

[4.53]

$$0 \leq \text{Flow Rate of Feed to Neutrazation (kg/hr)} \leq 11{,}000$$

[4.54]

$$0 \leq \text{AA in Feed to Neutralization (wt\%)} \leq 85\%$$

Acid Tower:

[4.55]

$$10{,}200 \leq \text{Flow Rate of Feed to Acid Tower (kg/hr)} \leq 11{,}200$$

[4.56]

$$0.0 \leq \text{Water in Feed to Acid Tower (wt\%)} \leq 10.0$$

First Absorber:

[4.57]

$$5{,}100 \leq \text{Flow Rate of Feed to Absorber I (kg/hr)} \leq 6000$$

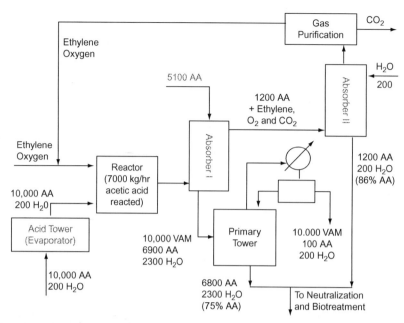

FIGURE 4-45 Schematic flow sheet of VAM process with all numbers in kg/hr. *Source: El-Halwagi (2006).*

TABLE 4-7 Sink Information for Problem 4.3

Sink	Flow (tonnes/hr)	Maximum Inlet Concentration (ppm)	Load (kg/hr)
1	120	0	0
2	80	50	4
3	80	50	4
4	140	140	19.6
5	80	170	13.6
6	195	240	46.8

Source: Sorin and Bedard (1999).

TABLE 4-8 Source Information for Problem 4.3

Sources	Flow (tonnes/hr)	Concentration (ppm)	Load (kg/hr)
1	120	100	12
2	80	140	11.2
3	140	180	25.2
4	80	230	18.4
5	195	250	48.75

Source: Sorin and Bedard (1999).

[4.58]
$$0.0 \leq \text{Water in Feed to Absorber I (wt\%)} \leq 5.0$$

It can be assumed that the process performance will not significantly change as a result of direct-recycle activities.

What is the target for minimum usage (kg/hr) of fresh acetic acid in the process if segregation, mixing, and direct recycle are used? Develop three implementations to reach the target.

4.3. Consider a process with six sinks and five sources (Sorin and Bedard, 1999). The process data are given in Tables 4.7 and 4.8. Fresh water is used in the sinks, and it is desired to replace as much fresh water as possible using direct recycle of process sources. Determine the target for minimum usage of fresh water and waste discharge after direct recycle. How many material recycle pinch points are there? What is their significance?

4.4. Consider the wastewater minimization problem Polley and Polley (2000) described. The process has four sources and four sinks, and information concerning them can be seen in Tables 4.9 and 4.10.

TABLE 4-9 Sink Data for Problem 4.4

Sinks	Flow (tonnes/hr)	Maximum Inlet Concentration (ppm)	Load (kg/hr)
1	50	20	1
2	100	50	5
3	80	100	8
4	70	200	14

Source: Polley and Polley (2000).

TABLE 4-10 Source Data for Problem 4.4

Sources	Flow (tonnes/hr)	Concentration (ppm)	Load (kg/hr)
1	50	50	2.5
2	100	100	10
3	70	150	10.5
4	60	250	15

Source: Polley and Polley (2000).

TABLE 4-11 Sink Data for the Pulp and Newsprint Problem

Sink	Flow (L/min)	Max. Allowable Fines Concentration (%)	Load (L/min)
1	200	1.000	2.0
2	400	1.000	4.0
3	355	0.020	0.1
4	150	1.000	1.5
5	13,000	1.000	130.0
6	4250	1.000	42.5
7	2800	1.000	28.0
8	4580	1.000	45.8
9	1950	1.000	19.5
10	500	1.000	5.0
11	1000	1.000	10.0
12	3000	1.000	30.0
13	435	1.000	4.4
14	310	1.000	3.1
15	60	1.000	0.6
16	1880	1.000	18.8
17	4290	1.000	42.9
18	9470	1.000	94.7
19	6500	1.000	65.0
20	620	1.000	6.2
21	55	1.000	0.6
22	70	1.000	0.7
23	320	1.000	3.2
24	1050	1.000	10.5
25	73,000	1.000	730.0
26	1765	1.000	17.7
27	235	1.000	2.4
28	95	1.000	1.0
29	20	1.000	0.2
30	180	0.000	0.0
31	160	0.018	0.0
32	30	0.018	0.0
33	20	0.018	0.0
34	315	0.000	0.0
35	315	0.000	0.0
36	930	0.018	0.2
37	460	0.018	0.1
38	30	0.018	0.0
39	30	0.018	0.0
40	315	0.000	0.0
41	315	0.000	0.0
42	110	0.018	0.0
43	110	0.018	0.0
44	190	0.000	0.0
45	190	0.000	0.0
46	100	0.000	0.0
47	20	0.000	0.0
48	15	0.000	0.0
49	60	0.018	0.0
50	30	0.018	0.0
51	100	0.000	0.0
52	20	0.000	0.0
53	100	0.000	0.0
54	20	0.000	0.0

Source: Jacob et al. (2002).

TABLE 4-12 Source Data for the Pulp and Newsprint Mill

Source	Flow (L/min)	Fines Concentration (%)	Load (L/min)
TMP clear water	25,000	0.07	17.5
TMP cloudy water	39,000	0.13	50.7
Inclined screen water	5980	0.50	29.9
Press header water	2840	0.49	13.9
Save-all clear water	6840	0.08	5.5
Save-all clear water	3720	0.1	3.7
Silo water	73,000	0.39	284.7
Machine chest white water	8585	0.34	29.2
Vacuum pump overflow	2570	0.00	0.0
Residual showers	1940	0.13	2.5

Source: Jacob et al. (2002).

It is desired to:
a. Find the targets for minimum fresh water and minimum wastewater discharge after direct recycle.
b. Develop a recycle strategy that attains the identified targets.

4.5. A thermomechanical pulp and newsprint mill has 54 sinks and 10 sources of water (Jacob et al., 2002). The source and sink data can be seen in Tables 4.11 and 4.12.

Identify a target for minimum water usage and minimum waste discharge.

Hint: To simplify the problem, notice that there are only four concentration levels of interest for the sinks. Therefore, the sinks can be lumped into four sinks with fine concentrations (percentages) of 0.000, 0.018, 0.020, and 1.000.

REFERENCES

Alves JJ, Towler GP: Analysis of refinery hydrogen distribution systems, *Ind Eng Chem Res* 41:5759–5769, 2002.
Dunn RF, El-Halwagi, MM, Lakin, J, Serageldin, M: *Selection of organic solvent blends for environmental compliance in the coating industries.* From Griffith ED, Kahn H, Cousins MC, editors: Proceedings of the First International Plant Operations and Design Conference, New York, AIChE, vol III, pp 83–107, 1995.
El-Halwagi MM: *Process integration*, Amsterdam, 2006, Academic Press/Elsevier.
El-Halwagi MM, Gabriel F, Harell D: Rigorous graphical targeting for resource conservation via material recycle/reuse networks, *Ind Eng Chem Res* 42:4319–4328, 2003.
El-Halwagi MM, Spriggs HD: *An integrated approach to cost and energy-efficient pollution prevention*, San Diego, Proceedings of Fifth World Congr. of Chem. Eng., vol III, pp 344–349, 1996.
Jacob J, Kaipe F, Couderc F, Paris J: Water network analysis in pulp and paper processes by pinch and linear programming techniques, *Chem. Eng. Communications* 189:184–206, 2002.
Noureldin MB, El-Halwagi MM: Interval-based targeting for pollution prevention via mass integration, *Comp Chem Eng* 23:1527–1543, 1999.
Noureldin MB, El-Halwagi MM: Pollution-prevention targets through integrated design and operation, *Comp Chem Eng* 24:1445–1453, 2000.
Parthasarathy G, El-Halwagi MM: Optimum mass integration strategies for condensation and allocation of multicomponent VOCs, *Chem Eng Sci* 55:881–895, 2000.
Polley GT, Polley HL: Design better water networks, *Chem Eng Prog* 96:47–52, 2000.
Sorin M, Bedard S: The global pinch point in water reuse networks, *Trans Inst Chem Eng* 77:305–308, 1999.

CHAPTER 5
SYNTHESIS OF MASS-EXCHANGE NETWORKS: A GRAPHICAL APPROACH

Mass-exchange units are ubiquitously used in industrial facilities. A mass exchanger is any direct-contact mass-transfer unit that employs a **mass-separating agent** (MSA) or a lean phase (for example, a solvent, an adsorbent, a stripping agent, an ion-exchange resin) to selectively remove certain components (for example, pollutants, products, by-products) from a rich phase (a waste, a product). Examples of mass-exchange operations include absorption, adsorption, stripping, solvent extraction, leaching, and ion exchange. Appendix II provides an overview of the principles of modeling, designing, and operating individual mass-exchange units.

In many industrial situations, the problem of selecting, designing, and operating a mass-exchange system should not be confined to assessing the performance of individual mass exchangers. Instead, multiple mass-exchange operations are involved, and several candidate MSAs are considered. In such situations, mass has to be removed from several rich streams (sources), and MSAs can be used for removing the targeted species. Therefore, adopting a systemic network approach can provide significant technical and economic benefits. In this approach, a mass-exchange system is selected, and is designed by *simultaneously* screening all candidate mass-exchange operations to identify the optimum system. This chapter defines the problem of synthesizing **mass-exchange networks** (MENs), and discusses its challenging aspects and provides a graphical approach for the synthesis of MENs.

MASS-EXCHANGE NETWORK SYNTHESIS TASK

Motivated by the need to simultaneously screen candidate MSAs and competing mass-exchange technologies while incorporating thermodynamic and economic considerations, El-Halwagi and Manousiouthakis (1989) introduced the novel problem and conceptual framework of synthesizing MENs and developed systematic techniques for their optimal design. The MEN-synthesis problem can be stated as follows: Given a number N_R of rich streams (sources) and a number N_S of MSAs (lean streams), it is desired to synthesize a cost-effective network of mass exchangers that can preferentially transfer certain undesirable species from the rich streams to the MSAs. Given also are the flow rate of each rich stream, G_i, its supply (inlet) composition y_i^s, and its target (outlet) composition y_i^t, where $i = 1, 2, ..., N_R$. In addition, the supply and target compositions, x_j^s and x_j^t, are given for each MSA, where $j = 1, 2, ..., N_S$. The flow rate of each MSA is unknown, and is to be determined so as to minimize the network cost. Figure 5.1 is a schematic representation of the problem statement.

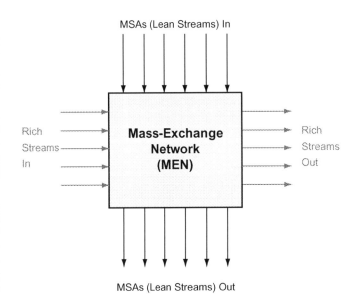

FIGURE 5-1 Schematic representation of the MEN synthesis problem. *Source: El-Halwagi and Manousiouthakis (1989).*

The candidate lean streams can be classified into N_{SP} process MSAs and N_{SE} external MSAs (where $N_{SP} + N_{SE} = N_S$). The process MSAs already exist on plant site and can be used for the removal of the undesirable species at a very low cost (virtually free). The flow rate of each process MSA that can be used for mass exchange is bounded by its availability in the plant; that is,

[5.1] $\qquad L_j \leq L_j^c \quad j = 1, 2, ..., N_{SP},$

where L_j^c is the flow rate of the jth MSA that is available in the plant. On the other hand, the external MSAs can be purchased from the market. Their flow rates are to be determined according to the overall economic considerations of the MEN.

When the rich streams are in the form of rich streams, the target composition of the undesirable species in each rich stream corresponds to the environmental regulations or the required specification prior to terminal-rich treatment. On the other hand, if the rich stream is in the form of a process stream, it is intercepted using the MEN to adjust its composition before it gets assigned or recycled to a process sink. Its target composition should satisfy the constraints imposed by the process sink.

Designers assign the target composition of the undesirable species in each MSA based on the specific circumstances of the application. The nature of such circumstances may be physical (for example, maximum solubility of the pollutant in the MSA), technical (to avoid

Sustainable Design Through Process Integration.
© 2012 Elsevier Inc. All rights reserved.

excessive corrosion, viscosity, or fouling), environmental (to comply with environmental regulations), safety (to stay away from flammability limits), or economic (to optimize the cost of subsequent regeneration of the MSA).

As can be inferred from the problem statement, the MEN synthesis task attempts to provide cost-effective solutions to the following design questions:

- Which mass-exchange operations should be used (for example, absorption, adsorption)?
- Which MSAs should be selected (for example, which solvents, adsorbents)?
- What is the optimal flow rate of each MSA?
- How should these MSAs be matched with the rich streams (that is, stream pairings)?
- What is the optimal system configuration (for example, how should these mass exchangers be arranged? Is there any stream splitting and mixing?)?

The foregoing design questions are highly combinatorial so any exhaustive enumeration technique would be hopelessly complicated. A hit-and-miss or trial-and-error approach to conjecturing the solution is likely to fail because it cannot consider the overwhelming number of decisions to be made. Hence, the designer needs practical tools to systematically address the MEN synthesis task. This chapter provides a graphical targeting approach referred to as the mass-exchange pinch diagram.

THE MEN-TARGETING APPROACH

The MEN-targeting approach is based on the identification of performance targets ahead of design and without prior commitment to the final network configuration. In the context of synthesizing MENs, two useful targets can be established:

1. **Minimum cost of MSAs.** By integrating the thermodynamic aspects of the problem with the cost data of the MSAs, one can indeed identify the minimum cost of MSAs and the flow rate of each MSA required to undertake the assigned mass-exchange duty. This can be accomplished without actually designing the network. Because the cost of MSAs is typically the dominant operating expense, this target is aimed at minimizing the operating cost of the MEN. Any design featuring the minimum cost of MSAs will be referred to as a minimum operating cost (MOC) solution.

2. **Minimum number of mass-exchanger units.** Combinatorics determines the minimum number of mass-exchanger units required in the network. This objective attempts to minimize indirectly the fixed cost of the network, because the cost of each mass exchanger is usually a concave function of the unit size. Furthermore, in a practical context, it is desirable to minimize the number of separators so as to reduce pipe work, foundations, maintenance, and instrumentation. Normally, the minimum number of units is related to the total number of streams by the following expression (El-Halwagi and Manousiouthakis, 1989):

$$[5.2] \qquad U = N_R + N_S - N_i$$

where N_i is the number of independent synthesis subproblems into which the original synthesis problem can be subdivided. In most cases, there is only one independent synthesis subproblem.

In general, these two targets are incompatible. Later, systematic techniques will be presented to identify an MOC solution and then minimize the number of exchangers satisfying the MOC.

THE CORRESPONDING COMPOSITION SCALES

A particularly useful concept in synthesizing MENs is the notion of "corresponding composition scales." It is a tool for incorporating thermodynamic constraints of mass exchange by establishing a one-to-one correspondence among the compositions of all streams for which mass transfer is thermodynamically feasible. This concept is based on a generalization of the notion of a "minimum allowable composition difference." To demonstrate this concept, let us consider a mass exchanger for which the equilibrium relation governing the transfer of a certain component from a rich stream, i, to the MSA, j, is given by the following linear expression:

$$[5.3] \qquad y = m_j x_j^* + b_j,$$

which indicates that, for a rich stream composition of y_i, the maximum theoretically attainable composition of the MSA is x_j^*. By employing a minimum allowable composition difference of ε_j, one can draw a "practical feasibility line" that is parallel to the equilibrium line but offset to its left by a distance, ε_j (see Fig. 5.2). For an operating

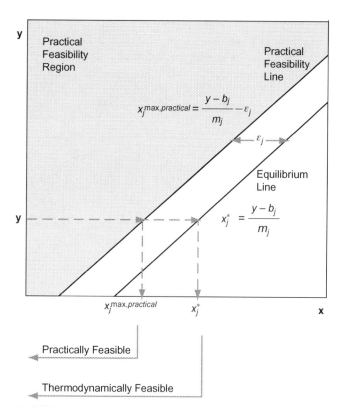

FIGURE 5-2 Establishing corresponding composition scales.

line to be practically feasible, it must lie in the region to the left of the practical feasibility line. Hence, for any pair $(y, x_j^{max,practical})$ lying on the practical feasibility line, two statements can be made. For a given y, the value $x_j^{max,practical}$ corresponds to the maximum composition of the transferrable component that is practically achievable in the MSA. Therefore, the transferrable component can be practically transferred from the rich composition (y) to the lean-stream composition ($x_j^{max,practical}$) or any lower composition (any composition to its left). Conversely, for a given $x_j^{max,practical}$, the value y corresponds to the minimum composition of the transferrable component in the rich stream that is needed to practically transfer the transferrable component from the rich stream to the MSA.

It is important to derive the mathematical expression relating y and $x_j^{max,practical}$ on the practical-feasibility line. For a given y, the values of $x_j^{max,practical}$ can be obtained by evaluating x_j^* that is in equilibrium with y and then subtracting ε_j; that is,

[5.4a] $$x_j^{max,practical} = x_j^* - \varepsilon_j$$

or

[5.4b] $$x_j^* = x_j^{max,practical} + \varepsilon_j.$$

Substituting from Eq. 5.4b into Eq. 5.3, one obtains

[5.5a] $$y = m_j(x_j^{max,practical} + \varepsilon_j) + b_j$$

or

[5.5b] $$x_j^{max,practical} = \frac{y - b_j}{m_j} - \varepsilon_j$$

Equation 5.5 can be used to establish a one-to-one correspondence among all composition scales for which mass exchange is feasible. Because most environmental applications involve dilute systems, one can assume that these systems behave ideally. Hence, the transfer of the pollutant or the targeted component is indifferent to the existence of other species in the rich stream. In other words, even if two rich streams contain species that are not identical, but share the same composition of a particular targeted component, the equilibrium composition of the targeted component in an MSA will be the same for both rich streams. Hence, a single composition scale, y, can be used to represent the concentration of the targeted component in any rich stream. Next, Eq. 5.5b can be employed to generate N_S scales for the MSAs. For a given set of corresponding composition scales $(y, x_1, x_2, ..., x_j, ..., x_{NS})$, it is thermodynamically and practically feasible to transfer the transferrable components from any rich stream to any MSA. In addition, it is also feasible to transfer the transferrable from any rich stream of a composition y to any MSA that has a composition less than the value of x_j obtained from Eq. 5.5b.

THE MASS-EXCHANGE PINCH DIAGRAM

To minimize the cost of MSAs, it is necessary to make maximum use of process MSAs before considering the application of external MSAs. In assessing the applicability of the process MSAs to remove the transferrable component, one must consider the thermodynamic limitations of mass exchange. El-Halwagi and Manousiouthakis (1989) introduced the mass-exchange pinch diagram as a holistic graphical approach for targeting and synthesizing MENs. The initial step in constructing the pinch diagram is creating a global representation for all the rich streams. This representation is accomplished by first plotting the mass exchanged by each rich stream versus its composition. Hence, each rich stream is represented as an arrow whose tail corresponds to its supply composition and its head to its target composition. The slope of each arrow is equal to the stream flow rate. The vertical distance between the tail and the head of each arrow represents the mass of targeted component that is lost by that rich stream according to

[5.6] Mass lost from the ith rich stream
$$MR_i = G_i(y_i^s - y_i^t), \quad I = 1,2,...,N_R$$

It is worth noting that the vertical scale is only relative. Any stream can be moved up or down while preserving the same vertical distance between the arrow head and tail and maintaining the same supply and target compositions. A convenient way of vertically placing each arrow is to stack the rich streams on top of one another, starting with the rich stream having the lowest target composition (see Fig. 5.3 for an illustration of two rich streams).

Having represented the individual rich streams, we are now in a position to construct the rich composite stream. A rich composite stream represents the cumulative mass lost by all the rich streams. It can be readily obtained by using the "diagonal rule" for superposition to add up mass in the overlapped regions of streams. Hence, the rich composite stream is obtained by applying linear superposition to all the rich streams. Figure 5.4 illustrates this concept for two rich streams.

Next, a global representation of all process lean streams is developed as a lean composite stream. First, we establish N_{SP} lean composition scales (one for each process MSA) that are in one-to-one correspondence with the rich scale according to the method outlined in the section titled "The Corresponding Composition Scales." Next, the mass of transferrable component that can be gained by each process MSA is plotted versus the composition scale of that MSA. Hence, each process MSA is represented as

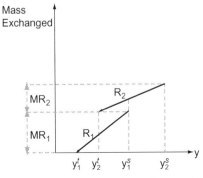

FIGURE 5-3 Representation of mass exchanged by two rich streams.

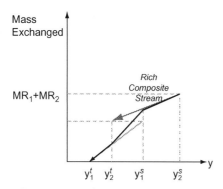

FIGURE 5-4 Constructing a rich composite stream using superposition.

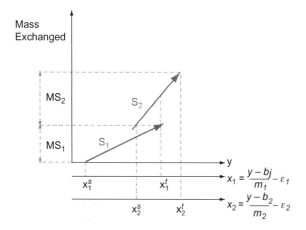

FIGURE 5-5 Representation of mass exchanged by two process MSAs.

an arrow extending between supply and target compositions (see Fig. 5.5 for a two-MSA example). The vertical distance between the arrow head and tail is given by

$$[5.7] \quad \text{Mass that can be gained by the jth process MSA} \\ MS_j = L_j^c(x_j^t - x_j^s) \qquad j = 1, 2, \ldots, N_{SP}$$

Once again, the vertical scale is only relative, and any stream can be moved up or down on the diagram. A convenient way of vertically placing each arrow is to stack the process MSAs on top of one another, starting with the MSA having the lowest supply composition (see Fig. 5.6). Hence, a lean composite stream representing the cumulative mass of the targeted component gained by all the MSAs is obtained by using the diagonal rule for superposition.

Next, both composite streams are plotted on the same diagram (see Fig. 5.7). On this diagram, thermodynamic feasibility of mass exchange is guaranteed when the lean composite stream is always above the rich composite stream. This is equivalent to ensuring that, at any mass-exchange level (which corresponds to a horizontal line), the composition of the lean composite stream is located to the left of the rich composite stream, asserting thermodynamic feasibility. Therefore, the lean composite stream can be slid down until it touches the rich composite stream. The point where the two composite streams touch is called the "mass-exchange pinch point," hence the name "pinch diagram" (see Fig. 5.7).

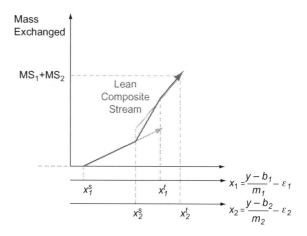

FIGURE 5-6 Construction of the lean composite stream using superposition.

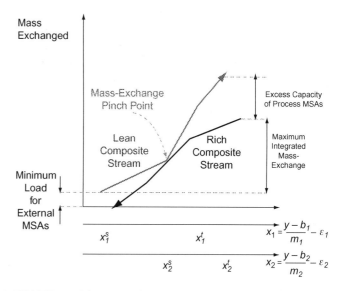

FIGURE 5-7 The mass-exchange pinch diagram. *Source: El-Halwagi and Manousiouthakis (1989).*

On the pinch diagram, the vertical overlap between the two composite streams represents the maximum amount of the transferrable component that can be transferred from the rich streams to the process MSAs. It is referred to as the "integrated mass exchange." The vertical distance of the lean composite stream that lies above the upper end of the rich composite stream is referred to as "excess process MSAs." It corresponds to that capacity of the process MSAs to remove mass that cannot be used because of thermodynamic infeasibility. According to the designer's preference or to the specific circumstances of the process, such excess can be eliminated from service by lowering the flow rate and/or the outlet composition of one or more of the process MSAs. Finally, the vertical distance of the rich composite stream that lies below the lower end of the lean composite stream corresponds to the mass to be removed by external MSAs.

As can be seen from Fig. 5.7, the pinch decomposes the synthesis problem into two regions: a rich end and a lean end. The rich end comprises all streams or parts of streams richer than the pinch composition. Similarly, the lean end includes all the streams or parts of streams leaner than the pinch composition. Above the pinch, exchange between the rich and the lean process streams takes place. External MSAs are not required. Using an external MSA above the pinch will incur a penalty, eliminating an equivalent amount of process lean streams from service. On the other hand, below the pinch, both the process and the external lean streams should be used. Furthermore, Fig. 5.7 indicates that if any mass is transferred across the pinch, the composite lean stream will move upward and, consequently, external MSAs in excess of the minimum requirement will be used. Therefore, to minimize the cost of external MSAs, mass should not be transferred across the pinch.

The previous discussion indicates that, to achieve the targets for maximum integration of mass exchange from the rich stream to process MSAs and minimum load to be removed by the external MSAs, the following three design rules are needed:

- No mass should be passed through the pinch (that is, the two composites must touch).
- No excess capacity should be removed from MSAs below the pinch.
- No external MSAs should be used above the pinch.

It is worth pointing out that these observations are valid only for the class of MEN problems covered in this chapter. When the assumptions employed in this chapter are relaxed, more general conclusions can be made. For instance, it will be shown later that the pinch analysis can still be undertaken even when a plant have no process MSAs.

Example 5-1 Recovery of benzene from the gaseous emission of a polymer production facility

Figure 5.8 shows a simplified flow sheet of a copolymerization plant. The copolymer is produced via a two-stage reaction. The monomers are first dissolved in a benzene-based solvent. The monomer mixture is fed to the first stage of reaction where a catalytic solution is added. Several additives (extending oil, inhibitors, and special additives) are mixed in a mechanically stirred column. The resulting solution is fed to the second-stage reactor, where the copolymer properties are adjusted. The stream leaving the second-stage reactor is passed to a separation system that produces four fractions: copolymer, unreacted monomers, benzene, and gaseous waste. The copolymer is fed to a coagulation and finishing section. The unreacted monomers are recycled to the first-stage reactor, and the recovered benzene is returned to the monomer-mixing tank. The gaseous waste, R_1, contains benzene as the primary targeted component that should be recovered. The stream data for R_1 are given in Table 5.1.

Two process MSAs and one external MSA are considered for recovering benzene from the gaseous waste. The two process MSAs are the additives, S_1, and the liquid catalytic solution, S_2. They can be used for benzene recovery at virtually no operating cost. In addition to its positive environmental impact, the recovery of benzene by these two MSAs offers an economic incentive because it reduces the benzene makeup needed to compensate for the processing losses. Furthermore, the additives mixing column can be used as an absorption column by bubbling the gaseous waste into the additives. The mixing pattern and speed of the mechanical stirrer can be adjusted to achieve a wide variety of mass-transfer tasks. The stream data for S_1 and S_2 are given in Table 5.2. The equilibrium data for benzene in the two process MSAs are given by:

$$[5.8] \quad y_1 = 0.25 x_1$$

and

$$[5.9] \quad y_1 = 0.50 x_2$$

where y_1, x_1, and x_2 are the mole fractions of benzene in the gaseous waste, S_1 and S_2 respectively. For control purposes, the minimum allowable composition difference for S_1 and S_2 should not be less than 0.001.

The external MSA, S_3, is an organic oil that can be regenerated using flash separation. The operating cost of the oil (including pumping, makeup, and regeneration) is $0.05/kg mol of recirculating oil. The equilibrium relation for transferring benzene from the gaseous waste to the oil is given by

$$[5.10] \quad y_1 = 0.10 x_3$$

The data for S_3 are given in Table 5.3. The absorber sizing equations and fixed cost were given in Example II.2 in Appendix II. Using the graphical pinch approach, synthesize a cost-effective MEN that can be used to remove benzene from the gaseous waste (see Fig. 5.9).

SOLUTION

Constructing the Pinch Diagram

As has been described earlier, the rich composite stream is first plotted as shown in Fig. 5.10. Next, the lean composite stream is constructed for the two process MSAs. Equation 5.5 is employed to generate the correspondence among the composition scales y, x_1, and x_2. The least permissible values of the minimum allowable composition difference are used ($\varepsilon_1 = \varepsilon_2 = 0.001$). Later, it will be shown that these values are optimum for an MOC solution. Next, the mass exchangeable by each of the two process lean streams is represented as an arrow versus its respective composition scale (see Fig. 5.11a). As demonstrated by Fig. 5.11b, the lean composite stream is obtained by applying superposition to the two lean arrows. Finally, the pinch diagram is constructed by combining Figures 5.10 and 5.11b. The lean composite stream is slid vertically until it is completely above the rich composite stream.

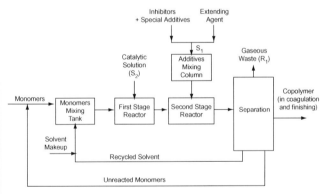

FIGURE 5-8 A simplified flow sheet of a copolymerization process.

TABLE 5-1 Data of Rich (Waste) Stream for the Benzene Removal Example

Stream	Description	Flow Rate G_i kg mol/s	Supply Composition (mole fraction) y_i^s	Target Composition (mole fraction) y_i^t
R_1	Off-gas from product separation	0.2	0.0020	0.0001

(Continued)

Example 5-1 Recovery of benzene from the gaseous emission of a polymer production facility (Continued)

TABLE 5-2 Data of Process Lean Streams for the Benzene Removal Example

Stream	Description	Upper Bound on Flow Rate L_j^c kg mol/s	Supply Composition of Benzene (mole fraction) x_j^s	Target Composition of Benzene (mole fraction) x_j^t
S_1	Additives	0.08	0.003	0.006
S_2	Catalytic solution	0.05	0.002	0.004

TABLE 5-3 Data for the External MSA for the Benzene Removal Example

Stream	Description	Upper Bound on Flow rate L_j^c kg mole/s	Supply Composition of Benzene (mole fraction) x_j^s	Target Composition of Benzene (mole fraction) x_j^t
S_3	Organic oil	∞	0.0008	0.0100

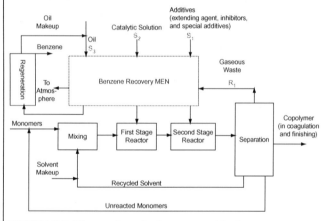

FIGURE 5-9 The copolymerization process with a benzene recovery MEN.

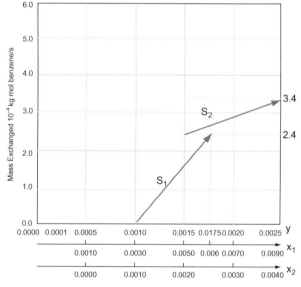

FIGURE 5-11A Representation of the two process MSAs for the benzene recovery example.

Interpreting Results of the Pinch Diagram

As can be seen from Fig. 5.12, the pinch is located at the corresponding mole fractions (y, x_1, and x_2) = (0.0010, 0.0030, and 0.0010). The excess capacity of the process MSAs is 1.4×10^{-4} kg mol benzene/s and cannot be used because of thermodynamic and practical feasibility limitations. This excess can be eliminated by reducing the outlet compositions and/or flow rates of the process MSAs. Because the inlet composition of S_2 corresponds to a mole fraction of 0.0015 on the y scale, the rich load immediately above the pinch (from $y = 0.0010$ to $y = 0.0015$) cannot be removed by S_2. Therefore, S_1 must be included in a MOC solution. Indeed, S_1 alone can be used to remove all the rich load above the pinch (2×10^{-4} kg mol benzene/s). To reduce the fixed cost by minimizing the number of mass exchangers, it is preferable to use a single solvent above the pinch rather than two solvents. This is particularly attractive given the availability of the mechanically stirred additives-mixing column for absorption. Hence, the excess capacity of the process MSAs is eliminated by avoiding the use of S_2 and reducing the flow rate and/or outlet composition of S_1. There are infinite combinations of L_1 and x_1^{out} that can be used to remove the excess capacity of S_1 according to the following material balance:

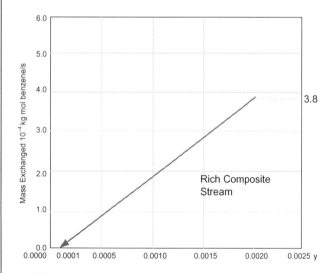

FIGURE 5-10 Rich composite stream for the benzene recovery example.

(Continued)

Example 5-1 Recovery of benzene from the gaseous emission of a polymer production facility (Continued)

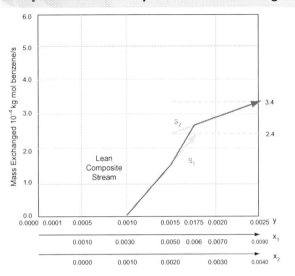

FIGURE 5-11B Construction of the lean composite stream for the two process MSAs of the benzene recovery example.

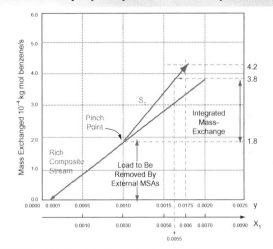

FIGURE 5-13 Graphical identification of x_1^{out}.

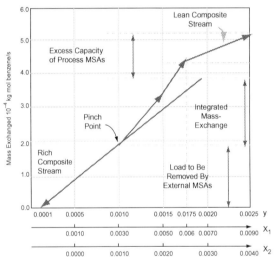

FIGURE 5-12 The pinch diagram for the benzene recovery example ($\varepsilon_1 = \varepsilon_2 = 0.001$).

[5.11a] Benzene load above the pinch to be removed by $S_1 = L_1(x_1^{out} - x_1^s)$ that is,

[5.11b] $\quad\quad\quad\quad 2 \times 10^{-4} = L_1(x_1^{out} - 0.003)$

Nonetheless, because the additives-mixing column will be used for absorption, the whole flow rate of S_1 (0.08 kg mol/s) should be fed to the column. Hence, according to Eq. 5.11b, the outlet composition of S_1 is 0.0055. The same result can be obtained graphically as shown in Fig. 5.13. It is worth recalling that the target composition of an MSA is only an upper bound on the actual value of the outlet composition. As was shown in this example, the outlet composition of an MSA is typically selected so as to optimize the cost of the system.

Selection of the Optimal Value of ε_1

Because S_1 is a process MSA with almost no operating cost and because it is to be used in process equipment (the mechanically stirred column) that does not require additional capital investment for utilization as an absorption column,

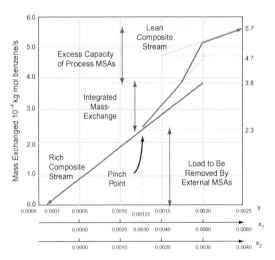

FIGURE 5-14 The pinch diagram when ε_1 is increased to 0.002.

S_1 should be utilized to its maximum practically feasible capacity for absorbing benzene. The remaining benzene load (below the pinch) is to be removed using the external MSA. The higher the benzene load below the pinch, the higher the operating and fixed costs. Therefore, in this example, it is desired to maximize the integrated mass exchanged above the pinch. As can be seen on the pinch diagram when ε_1 increases, the x_1 axis moves to the right relative to the y axis and, consequently, the extent of integrated mass exchange decreases, leading to a higher cost of external MSAs. For instance, Fig. 5.14 demonstrates the pinch diagram when ε_1 is increased to 0.002. The increase of ε_1 to 0.002 results in a load of 2.3×10^{-4} kg mol benzene/s to be removed by external MSAs (compared to 1.8×10^{-4} kg mol benzene/s for $\varepsilon_1 = 0.001$), an integrated mass exchange of 1.5×10^{-4} kg mol benzene/s (compared to 2.0×10^{-4} kg mol benzene/s for $\varepsilon_1 = 0.001$), and an excess capacity of process MSAs of 1.9×10^{-4} kg mol benzene/s (compared to 1.4×10^{-4} kg mol benzene/s for $\varepsilon_1 = 0.001$). Thus, the optimum ε_1 in this example is the smallest permissible value given in the problem statement to be 0.001.

It is worth noting that there is no need to optimize over ε_2. As previously shown, when ε_2 was set equal to its lowest permissible value (0.001), S_1 was selected as the optimal process MSA above the pinch. On the pinch diagram, as ε_2 increases, S_2 moves to the right, and the same arguments for selecting S_1 over S_2 remain valid.

(Continued)

Example 5-1 Recovery of benzene from the gaseous emission of a polymer production facility (*Continued*)

Optimizing the Use of the External MSA

The pinch diagram (Fig. 5.12) demonstrates that below the pinch, the load of the rich stream has to be removed by the external MSA, S_3. This renders the remainder of this example identical to Example II.2 solved in Appendix II. Therefore, the optimal flow rate of S_3 is 0.0234 kg mol/s, and the optimal outlet composition of S_3 is 0.0085. Furthermore, the minimum total annualized cost of the benzene recovery system is $41,560/yr (see Fig. II.13 in Appendix II).

Constructing the Synthesized Network

The previous analysis shows that the MEN comprises two units: one above the pinch in which R_1 is matched with S_1, and one below the pinch in which the remainder load of R_1 is removed using S_3. Figure 5.15 illustrates the network configuration.

It is worth noting the MEN–synthesis problems may require the screening of multiple external MSAs. The next example addresses this issue.

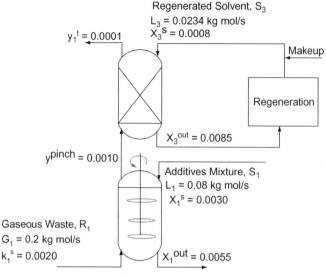

FIGURE 5-15 Optimal MEN for the benzene recovery example.

Example 5-2 Dephenolization of aqueous wastes

Consider the oil recycling plant shown in Fig. 5.16. In this plant, two types of waste oil are handled: gas oil and lube oil. The two streams are first de-ashed and demetallized. Next, atmospheric distillation is used to obtain light gases, gas oil, and a heavy product. The heavy product is distilled under vacuum to yield lube oil. Both the gas oil and the lube oil should be further processed to attain desired properties. The gas oil is steam stripped to remove light and sulfur impurities, then hydrotreated. The lube oil is dewaxed/de-asphalted using solvent extraction followed by steam stripping.

The process has two main sources of wastewater. These are the condensate streams from the steam strippers. The principal pollutant in both wastewater streams is phenol. Phenol is of concern primarily because of its toxicity, oxygen depletion, and turbidity. In addition, phenol can cause objectionable taste and odor in fish flesh and potable water.

Several techniques can be used to separate phenol. Solvent extraction using gas oil or lube oil (process MSAs S_1 and S_2, respectively) is a potential option. Besides the purification of wastewater, the transfer of phenol to gas oil and lube oil is a useful process for the oils. Phenol tends to act as an oxidation inhibitor and serves to improve color stability and reduce sediment formation. The data for the waste streams and the process MSAs are given in Tables 5.4 and 5.5, respectively.

Three external technologies are also considered for the removal of phenol. These processes include adsorption using activated carbon, S_3, ion exchange using a polymeric resin, S_4, and stripping using air, S_5. The equilibrium data for the transfer of phenol to the jth lean stream is given by $y = m_j x_j$, where the values of m_j are 2.00, 1.53, 0.02, 0.09, and 0.04 for S_1, S_2, S_3, S_4, and S_5, respectively. Throughout this example, a minimum allowable composition difference, ε_j, of 0.001(kg phenol)/(kg MSA) will be used.

The Pinch Diagrams

The pinch diagram is constructed as shown in Fig. 5.17. The pinch is located at the corresponding mass fraction (y, x_1, x_2) = (0.0168, 0.0074, 0.0100). The excess capacity of the process MSAs is 0.0184 kg phenol/s. This excess can be eliminated by reducing the outlet compositions and/or flow rates of the process MSAs. For instance, if the designer elects to remove this excess by lowering the flow rate of S_2, the actual flow rate of S_2 can then be calculated as follows:

$$[5.12] \qquad L_2 = 3 - \frac{0.0184}{0.03 - 0.01} = 2.08 \text{ kg/s}$$

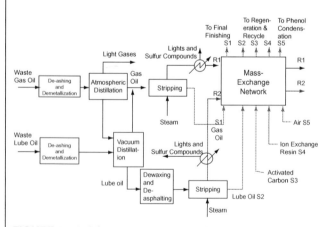

FIGURE 5-16 Schematic representation of an oil recycling plant.

TABLE 5-4 Data of Waste Streams for the Dephenolization Example

Stream	Description	Flow Rate G_i, kg/s	Supply Composition y_i^s	Target Composition y_i^t
R_1	Condensate from first stripper	2	0.050	0.010
R_2	Condensate from second stripper	1	0.030	0.006

(*Continued*)

Example 5-2 Dephenolization of aqueous wastes (*Continued*)

TABLE 5-5 Data of Process MSAs for the Dephenolization Example

Stream	Description	Upper Bound on Flow Rate, L_j^c, kg/s	Supply Composition, x_j^s	Target Composition, x_j^t
S_1	Gas oil	5	0.005	0.015
S_2	Lube oil	3	0.010	0.030

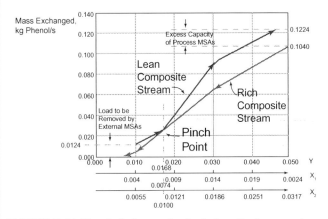

FIGURE 5-17 The pinch diagram for the dephenolization example.

Figure 5.17 also indicates that 0.0124 kg phenol/s are to be removed using external MSAs. The following section discusses screening the three MSAs to select the one that yields a MOC.

Cost Estimation and Screening of External MSAs

To determine which external MSA should be used to remove this load, it is necessary to determine the supply and target compositions as well as unit cost data for each MSA. Toward this end, one ought to consider the various processes each MSA undergoes. For instance, activated carbon, S_3, has an equilibrium relation (adsorption isotherm) for adsorbing phenol that is linear up to a lean-phase mass fraction of 0.11, after which activated carbon is quickly saturated and the adsorption isotherm levels off. Hence, x_3^t is taken as 0.11. It is also necessary to check the thermodynamic feasibility of this composition. Equation 5.5a can be used to calculate the corresponding composition on the y scales:

[5.13] $\qquad y = 0.02(0.11 + 0.001) = 0.0022$

This value is less than the supply compositions of R_1 and R_2. Hence, it is possible to transfer phenol from both waste streams to S_3. Furthermore, this value is less than the composition at the tail of the lean composite stream (0.01). Therefore, S_3 will not eliminate any phenol that can be removed by the process MSAs (S_1 and S_2). After adsorbing phenol, activated carbon is fed to a steam regeneration column. In this unit, a mass ratio 2:1 of steam to adsorbed phenol is used to remove almost all adsorbed phenol. The regenerated activated carbon is recycled back to the MEN. During regeneration, 5 percent of the activated carbon is lost and has to be replenished by fresh makeup activated carbon. The steam leaving the regeneration column is condensed. The phenol in the condensate is separated via decantation followed by distillation. The cost of decantation and distillation is almost equal to the value of the recovered phenol. Hence, the operating cost of activated carbon (C_3, $/kg of recirculating activated carbon) can be obtained as follows:

C_3 = cost of makeup + cost of regeneration
= 0.05 × unit cost of fresh activated carbon $/kg +
2 × cost of steam $/kg × amount of phenol adsorbed
(kg phenol/kg activated carbon)

Considering the unit costs of activated carbon and steam to be 1.60 and 5×10^{-3} $/kg, respectively, we get

[5.14] $\qquad C_3 = 0.05 \times 1.60 + 2 \times 5 \times 10^{-3} x_3^t$

But, the mass of activated carbon needed to remove 1 kg of phenol from the waste streams can be evaluated through material balance,

[5.15a] $\qquad$ 1 kg of phenol = kg activated carbon $(x_3^t - 0)$

that is,

[5.15b] $\qquad$ kg of activated carbon needed to remove 1 kg of phenol from waste streams = $\dfrac{1}{x_3^t}$

Hence, one can multiply Equations 5.14 and 5.15b to obtain the cost of removing 1 kg of phenol from waste streams using activated carbon, C_3' ($/kg phenol removed) as

[5.16] $\qquad C_3' = \dfrac{0.08}{x_3^t} + 0.01$

Substituting the value of x_3^t (0.11 kg phenol/kg activated carbon) into Equations 5.14 and 5.16, one obtains

[5.17] $\qquad C_3 = \$0.081$/kg recirculating activated carbon

and

[5.18] $\qquad C_3' = \$0.737$/kg of removed phenol

If the ion-exchange resin is used for removing phenol, it is regenerated by employing caustic soda to convert phenol into sodium phenoxide (a salable compound) according to the following reaction:

$$C_6H_5OH + NaOH \rightarrow C_6H_5ONa + H_2O$$

The reaction is thermodynamically favorable and proceeds to convert almost all the phenol to sodium phenoxide. Next, sodium phenoxide is separated from the caustic solution via distillation. The distillation cost is almost equal to the value of generated sodium phenoxide. Hence, the main operating costs of the process are the amount of sodium hydroxide that reacts with phenol and the cost of makeup resin. The molecular weights of phenol and sodium hydroxide are 94 and 40, respectively, so one can readily see that the amount of reacted sodium hydroxide is 0.426 (kg NaOH)/(kg phenol). The rate of ion-exchange resin deactivation is taken as 5 percent. Hence, the flow rate of resin makeup is a 0.05 fraction that of recirculating resin. Thus, the operating cost of the resin (C_4, $/kg of recirculating resin) is determined through

C_4 = cost of makeup + cost of regeneration
= 0.05 × unit cost of fresh resin $/kg + 0.426 ×
cost of sodium hydroxide $/kg ×
amount of phenol adsorbed (kg phenol/kg resin)

(*Continued*)

Example 5-2 Dephenolization of aqueous wastes (*Continued*)

Considering the unit costs of resin and sodium hydroxide to be 3.80 and 0.30 $/kg, respectively, one obtains

[5.19]
$$C_4 = 0.05 \times 3.80 + 0.426 \times 0.30 x_4^t$$
$$= 0.19 + 0.128 x_4^t$$

The mass of the resin needed to remove one kg of phenol from the waste streams can be determined from a material balance,

[5.20] 1 kg of phenol removed from waste streams = mass of resin $(x_4^t - 0)$.

Therefore, the cost of removing 1 kg of phenol from waste streams using ion-exchange resin C_4' ($/kg phenol removed), is determined by combining Equations 5.19 and 5.20 to yield

[5.21]
$$C_4' = \frac{0.19}{x_4^t} + 0.128$$

As Eq. 5.21 indicates, the higher the value of x_4^t, the lower the removal cost. So, what is the highest possible value of x_4^t that should be used? Because no mass is transferred across the pinch, material balance above the pinch is completely satisfied by using the process MSAs. By extending the target composition of S_4 to be above the pinch, one transfers a load that can be removed by process MSAs (for free) to S_4. Hence, when the operating cost of the network is considered, there is no economic incentive to pass S_4 above the pinch point. Therefore, the optimum target composition of S_4 is that corresponding to the pinch. Because the pinch composition on the waste scale is 0.0168, one can readily evaluate the corresponding composition for S_4 by applying Eq. 5.5b,

[5.22]
$$x_4^t = \frac{0.0168}{0.09} - 0.001 = 0.186$$

By using this value, we can now invoke Equations 5.19 and 5.21 to get

[5.23] $C_4 = \$0.214$/kg of recirculating resin

and

[5.24] $C_4' = \$1.150$/kg of removed phenol

When air stripping is used to remove phenol from a waste stream, the gaseous stream leaving the MEN cannot be discharged to the atmosphere owing to air-quality regulations. Hence, the air leaving the MEN is fed to a phenol-recovery unit in which a refrigerant is used to condense phenol. Based on the cooling duty, the cost of handling air in the condensation unit is

[5.25] $C_5 = \$0.06$/kg air

The composition of phenol in the air leaving the MEN should be below the lower flammability limit (LFL). But, the LFL for phenol in air is 5.8 w/w%. An operating composition less than 50 percent of the LFL is typically suggested. Hence,

[5.26] $x_5^t = 0.5 \times 0.058 = 0.029$

This value corresponds to a corresponding y-scale composition of 0.0012, which is less than the supply mass fraction of phenol in either waste stream as well as the pinch composition. Hence, it is thermodynamically feasible for S_5 to recover phenol from R_1 and R_2. In addition, any amount of phenol removed by S_5 does not overlap with the load handled by the process MSAs. Therefore, the operating cost of air needed to remove 1 kg of phenol can be evaluated as follows:

[5.27] $C_5' = 0.06/x_5^t = \$2.069$/kg of removed phenol

Because it is thermodynamically feasible for any of the three external MSAs to remove the remaining phenolic load (0.0124 kg phenol/s), one should select the MSA that minimizes the operating cost. This is achieved by comparing the values of C_j' reported by Equations 5.18, 5.24, and 5.27 to be 0.737, 1.150, and 2.069 $/kg of removed phenol by using activated carbon, ion-exchange resin, and air, respectively. Therefore, activated carbon is the optimum external MSA. In addition, it is possible to evaluate the flow rate and operating cost of activated carbon as

[5.28] $L_3 = 0.0124/(0.11 - 0) = 0.1127$ kg/s

Also,

[5.29a]
$$\text{Annual operating cost} = \left(\frac{\$0.081}{\text{kg activated carbon}}\right)\left(0.1127 \frac{\text{kg activated carbon}}{\text{s}}\right)$$
$$\left(\frac{3600\text{s}}{\text{hr}}\right)\left(\frac{8760\text{hr}}{\text{yr}}\right)$$
$$= \$288 \times 10^3 /\text{yr}.$$

Equivalently, the same result can be obtained through the following equation:

[5.29b]
$$\text{Annual operating cost} = \left(\frac{\$0.737}{\text{kg phenol removed}}\right)(0.0124)\left(\frac{\text{kg phenol removed}}{\text{s}}\right)$$
$$\left(\frac{3600\text{s}}{\text{hr}}\right)\left(\frac{8760\text{hr}}{\text{yr}}\right)$$
$$= \$288 \times 10^3 /\text{yr}.$$

Once again, this MOC solution is a design target that has been identified even before determining the network configuration and equipment design. If this target is acceptable to the company, more detailed design efforts are warranted to scrutinize equilibrium as well as cost data, configure the system, and develop the detailed internal design of the units. This approach of "breadth first, depth later" is a key element in the overall design philosophy presented in this book.

CONSTRUCTING PINCH DIAGRAMS WITHOUT PROCESS MSAS

So far, an MOC solution has been identified through a two-stage process. First, the use of process MSAs is maximized by constructing the pinch diagram with the lean composite stream composed of process MSAs only. In the second stage, the external MSAs are screened to remove the remaining load at minimum cost. Suppose that the process does not have any process MSAs. How can a lean composite line be developed? The following shortcut method can be employed to construct the pinch diagram for external MSAs.

First, the rich composite line is plotted. Then, Eq. 5.5 is employed to generate the correspondence among the rich composition scale, y, and the lean composition scales for all external MSAs. Each external MSA is then represented versus its composition scale as a horizontal arrow extending between its supply and target compositions (see Fig. 5.18a). Several useful insights can be gained from this diagram. Let us consider three MSAs: S_1, S_2, and S_3, whose costs ($/kg of recirculating MSA) are c_1, c_2, and c_3, respectively. These costs can be converted into $/kg of removed species, c_j^r, as follows:

[5.30] $\quad c_j^r = \dfrac{c_j}{x_j^{out} - x_j^s} \quad$ where j = 1, 2, 3.

If there are no thermodynamic limitations, the value of the outlet composition of the lean stream is the target composition.

If arrow S_2 lies completely to the left of arrow S_1 and c_2^r is less than c_1^r, one can eliminate S_1 from the problem because it is thermodynamically and economically inferior to S_2. On the other hand, if arrow S_3 lies completely to the left of arrow S_2 but c_3^r is greater than c_2^r, one should retain both MSAs. To minimize the operating cost of the network, separation should be staged to use the cheapest MSA where it is feasible. Hence, S_2 should be used to remove all the rich load to its left while the remaining rich load is removed by S_3 (see Fig. 5.18a). The flow rates of S_2 and S_3 are calculated by simply dividing the rich load removed by the composition difference for the MSA. Now that the MSAs have been screened and their optimal flow rates have been determined, one can construct the pinch diagram as shown in Fig. 5.18b.

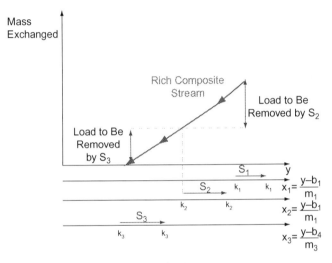

FIGURE 5-18A Screening external MSAs.

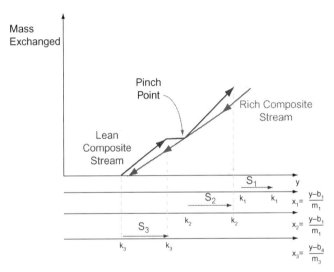

FIGURE 5-18B Constructing the pinch diagram for external MSAs.

Example 5-3 Toluene removal from wastewater

Toluene is to be removed from a wastewater stream. The flow rate of the waste stream is 10 kg/s and its inlet composition of toluene is 500 parts per million by weight (ppmw). It is desired to reduce the toluene composition in water to 20 ppmw. Three external MSAs are considered: air (S_2) for stripping, activated carbon (S_2) for adsorption, and a solvent extractant (S_3). The data for the candidate MSAs are given in Table 5.6. The equilibrium data for the transfer of the pollutant from the waste stream to the jth MSA is given by

[5.31] $\quad\quad\quad\quad\quad\quad y_1 = m_j x_j$

where y_1 and x_j are the mass fractions of the toluene in the wastewater and the jth MSA, respectively.

Use the pinch diagram to determine the minimum operating cost of the MEN.

SOLUTION

Based on the given data and Eq. 5.30, we can calculate c_j' to be 0.38, 5.53, and 43.90 $/kg of toluene removed for air, activated carbon, and the extractant, respectively. Because air is the least expensive, it will be used to remove all the load to its right (0.0045 kg toluene/s, as can be seen from Fig. 5.19a). Therefore, the flow rate of air can be calculated as

[5.32] $\quad$ Flow rate of air $= \dfrac{0.0045}{19{,}000 \times 10^{-6} - 0} = 0.237$ kg/s

Because activated carbon lies to the left of the extractant and has a lower c_j', it will be used to remove the remaining load (0.0003 kg toluene/s, as shown by Fig. 5.19a). The flow rate of activated carbon is 0.015 kg/s. For 8000 operating hours per year, the annual operating cost of the system is $96,700/year. The annualized fixed cost can be calculated after equipment sizing (as shown in Example 2.1).

TABLE 5-6 Data for MSAs of Toluene Removal Example

Stream	Upper Bound on Flow Rate L_j^c kg/s	Supply Composition (ppmw) x_j^s	Target Composition (ppmw) x_j^t	m_j	ε_j ppmw	C_j $/kg MSA
S_1	∞	0	19,000	0.0084	6000	7.2×10^{-3}
S_2	∞	100	20,000	0.0012	15,000	0.11
S_3	∞	50	2100	0.0040	10,000	0.09

(Continued)

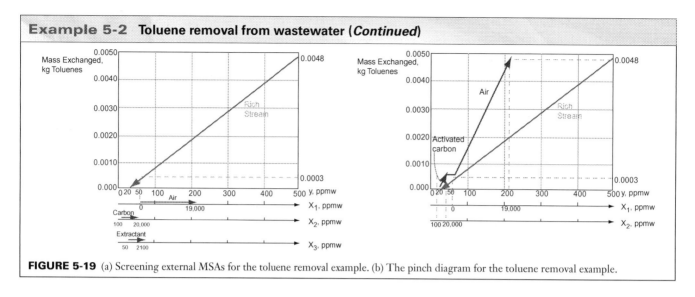

FIGURE 5-19 (a) Screening external MSAs for the toluene removal example. (b) The pinch diagram for the toluene removal example.

CONSTRUCTION OF THE MEN CONFIGURATIOVN WITH MINIMUM NUMBER OF EXCHANGERS

As mentioned in the section entitled "The Men-Targeting Approach," the targeting approach adopted for synthesizing MENs attempts first to minimize the cost of MSAs by identifying the flow rates and outlet compositions of MSAs that yield minimum operating cost (MOC). This target has been tackled in the sections on "The Mass-Exchange Pinch Diagram" and "Constructing Pinch Diagrams without Process MSAS" in this chapter. The second step in the synthesis procedure is to minimize the number of exchangers that can realize the MOC solution.

A convenient shortcut method is to match streams on the mass-exchange pinch diagram by passing mass horizontally from rich to lean streams. When two streams overlap horizontally on the pinch diagram, the mass transferred from rich to lean is the lower of the two mass-exchange loads. This shortcut method was used in constructing the network for Example 5.1. In this section, a more general procedure is presented.

Owing to the existence of a pinch that decomposes the problem into two subproblems, one above and one below the pinch, the minimum number of mass exchangers compatible with a MOC solution, U_{MOC}, can be obtained by the following expression (El-Halwagi and Manousiouthakis, 1989):

[5.33a]
$$U_{MOC} = U_{MOC, above\ pinch} + U_{MOC, below\ pinch}$$

where

[5.33b]
$$U_{MOC, above\ pinch} = N_{R, above\ pinch} + N_{S, above\ pinch} - N_{i, above\ pinch}$$

and

[5.33c]
$$U_{MOC, below\ pinch} = N_{R, below\ pinch} + N_{S, below\ pinch} - N_{i, below\ pinch}$$

where $U_{MOC, above\ pinch}$ is the number of MOC units above the pinch, $N_{R, above\ pinch}$ is the number of rich streams above the pinch, $N_{S, above\ pinch}$ is the number of lean streams below the pinch, $N_{i, above\ pinch}$ is the number of independent problems above the pinch, $U_{MOC, below\ pinch}$ is the number of MOC units below the pinch, $N_{R, below\ pinch}$ is the number of rich streams below the pinch, $N_{S, below\ pinch}$ is the number of lean streams below the pinch, and $N_{i, below\ pinch}$ is the number of independent problems below the pinch. The number of independent problems (above or below the pinch) includes cases such as when a rich and a lean stream have exactly the same load and, therefore, only one mass exchanger is needed to match the two streams. Having determined U_{MOC}, we should then proceed to match the pairs of rich and lean streams. In the following, it will be shown that matching has to start from the pinch and must satisfy a number of feasibility criteria.

FEASIBILITY CRITERIA AT THE PINCH

To guarantee the minimum cost of MSAs, no mass should be transferred across the pinch. By starting stream matching at the pinch, designers avoid any situation that may result in such a transfer. Moreover, at the pinch, all matches feature a driving force (between operating and equilibrium lines) equal to the minimum allowable composition difference, e_j. Because the pinch represents the most thermodynamically constrained region for design, the number of feasible matches in this region is severely limited. Thus, when synthesis is started at the pinch, the freedom of design choices at later steps will not be prejudiced.

This discussion clearly shows that the synthesis of a MEN should start at the pinch and proceed in two directions separately: the rich and the lean ends. To facilitate the design procedure both above and below the pinch, we next establish feasibility criteria for the characterization of stream matches at the pinch. These criteria identify the essential matches or topology options at the pinch. Such matches will be referred to as "pinch matches" or "pinch exchangers." The feasibility criteria will also inform designers whether or not stream splitting is required at the pinch. Once away from the pinch, these feasibility criteria do not have to be considered, and matching the rich and the lean streams becomes a relatively simple task. To identify the feasible pinch topologies, the following two feasibility criteria will be applied to the stream data: stream population and operating line versus equilibrium line.

Stream Population

Let us first consider the case above the pinch. In a MOC design, any mass exchanger immediately above the pinch will operate with the minimum allowable composition difference at the pinch side. Because all rich streams at the pinch must be brought to their pinch composition, any rich stream leaving a pinch exchanger can operate only against a lean stream at its pinch composition. Therefore, for each pinch match, at least one lean stream (or branch) has to exist per each rich stream. In other words, for an MOC design, the following inequality must apply at the rich end of the pinch

$$[5.34a] \qquad N_{ra} \leq N_{la},$$

where N_{ra} is the number of rich streams or branches immediately above the pinch, and N_{la} is the number of lean streams or branches immediately above the pinch. If this inequality does not hold for the stream data, one or more of the lean streams will have to be split.

Conversely, immediately below the pinch, each lean stream must be brought to its pinch composition. At this composition, any lean stream can operate only against a rich stream at its pinch composition or higher. Because an MOC design does not permit the transfer of mass across the pinch, each lean stream immediately below the pinch will require the existence of at least one rich stream (or branch) at the pinch composition. Therefore, immediately below the pinch, the following criterion must be satisfied:

$$[5.34b] \qquad N_{lb} \leq N_{rb}$$

where N_{lb} is the number of lean streams or branches immediately below the pinch, and N_{rb} is the number of rich streams or branches immediately below the pinch. Again, splitting of one or more of the rich streams may be necessary to realize this inequality.

OPERATING LINE VERSUS EQUILIBRIUM LINE

Consider the mass exchanger shown in Fig. 5.20. The lean end of this exchanger is immediately above the pinch. A component material balance for the pollutant around the exchanger can be written as

$$[5.35] \qquad G_i(y_i^{in} - y_i^{pinch}) = L_j(x_j^{out} - x_j^{pinch})$$

but, at the pinch,

$$[5.36] \qquad y_i^{pinch} = m_j(x_j^{pinch} + \varepsilon_j) + b_j$$

To ensure thermodynamic feasibility at the rich end of the exchanger, the following inequality must hold:

$$[5.37] \qquad y_i^{in} \geq m_j(x_j^{out} + \varepsilon_j) + b_j$$

Substituting from Equations 5.36 and 5.37 into Eq. 5.35, one gets

$$G_i[m_j(x_j^{out} + \varepsilon_j) + b_j - m_j(x_j^{pinch} + \varepsilon_j) - b_j]$$
$$\leq L_j(x_j^{out} - x_j^{pinch})$$

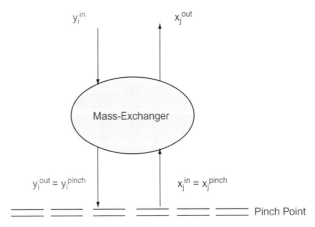

FIGURE 5-20 A mass exchanger immediately above the pinch.

and hence

$$[5.38a] \qquad \frac{L_j}{m_j} \geq G_i$$

This is the feasibility criterion for matching a pair of streams (i,j) immediately above the pinch. By recalling that the slope of the operating line is L_j/G_i, one can phrase Eq. 5.38a as follows: For a match immediately above the pinch to be feasible, the slope of the operating line should be greater than or equal to the slope of the equilibrium line. On the other hand, one can similarly show that the feasibility criterion for matching a pair of streams (i, j) immediately below the pinch is given by

$$[5.38b] \qquad \frac{L_j}{m_j} \leq G_i$$

Once again, stream splitting may be required to guarantee that inequality (Eq. 5.38a or 5.38b) is realized for each pinch match. It should also be emphasized that the feasibility criteria (Equations 5.34 and 5.38) should be fulfilled only at the pinch. Once the pinch matches are identified, it generally becomes a simple task to complete the network design. Moreover, designers always have the freedom to violate these feasibility criteria at the expense of increasing the cost of external MSAs beyond the MOC requirement.

NETWORK SYNTHESIS

The feasibility criteria described by Equations 5.34 and 5.38 can be employed to synthesize an MEN possessing the minimum number of exchangers that satisfy the MOC solution. The following representation will be used to illustrate the network structure graphically. Rich streams are represented by vertical arrows running at the left of the diagram. Compositions (expressed as weight ratios of the key component in each stream) are placed next to each arrow. A match between two streams is indicated by placing a circle on each of the streams and connecting them by a line. Mass-transfer loads of the key component for each exchanger are noted in appropriate units (for example, kg pollutant/s) inside the circles. The pinch is represented by two horizontal dotted lines.

To demonstrate the applicability of these feasibility criteria, we now revisit the dephenolization case study (Example 5.2). As discussed earlier, the synthesis ought to start at the pinch and proceed in two directions, toward the rich and the lean ends. We begin with the rich end of the problem. Above the pinch, we have two rich streams and two MSAs. Hence, the minimum number of exchangers above the pinch can be calculated according to Eq. 5.34b as

[5.39] $U_{MOC,above\ pinch} = 2 + 2 - 1 = 3$ exchangers.

Immediately above the pinch, the number of rich streams is equal to the number of MSAs; thus, the feasibility criterion given by Eq. 5.34a is satisfied. The second feasibility criterion (Eq. 5.38a) can be checked through Fig. 5.21. By comparing the values of L_j/m_j with G_i for each potential pinch match, one can readily deduce that it is feasible to match S_1 with either R_1 or R_2 immediately above the pinch. Nonetheless, although it is possible to match S_2 with R_2, it is infeasible to pair S_2 with R_1 immediately above the pinch. Therefore, one can match S_1 with R_1 and S_2 with R_2 as rich-end pinch exchangers.

When two streams are paired, the exchangeable mass is the lower of the two loads of the streams. For instance, the mass-exchange loads of R_1 and S_1 are 0.0664 and 0.0380 kg/s, respectively. Hence, the mass exchanged from R_1 to S_1 is 0.0380 kg/s. Owing to this match, the capacity of S_1 above the pinch has been completely exhausted and S_1 may now be eliminated from further consideration in the rich-end subproblem. Similarly, 0.0132 kg/s of phenol will be transferred from R_2 to S_2, thereby fulfilling the required mass-exchange duty for R_2 above the pinch. We are now left with only two streams above the pinch (R_1 and S_2). The remaining load for R_1 is $0.0664 - 0.0380 = 0.0284$ kg/s. Also, the removal capacity left in S_2 is $0.0416 - 0.0132 = 0.0284$ kg/s. As expected, the remaining loads of R_1 and S_2 above the pinch are equal, because no mass is passed through the pinch. Hence, material balance must be satisfied above the pinch. The two streams (R_1 and S_2) are therefore matched, and the synthesis subproblem above the pinch is completed. This rich-end design is shown in Fig. 5.22b. The intermediate compositions can be calculated through component material balances. For instance, the composition of S_2 leaving its match with R_2 and entering is match with R_1, $x_2^{intermediate}$, can be calculated via a material balance around the R_2-S_2 exchanger; that is,

[5.40a] $x_2^{intermediate} = 0.0100 + \dfrac{0.0132}{2.08} = 0.0164$

or a material balance around the R_1-S_2 exchanger:

[5.40b] $x_2^{intermediate} = 0.0300 - \dfrac{0.0284}{2.08} = 0.0164$

Having completed the design above the pinch, we can now move to the problem below the pinch. Figure 5.23 illustrates the streams below the pinch. It is worth noting

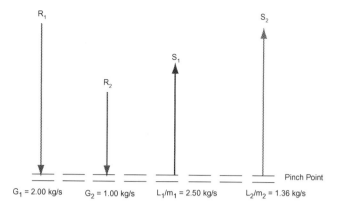

FIGURE 5-21 Feasibility criteria above the pinch for the dephenolization example.

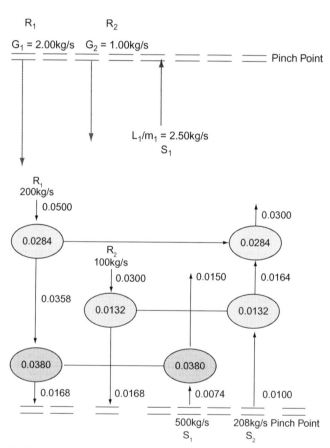

FIGURE 5-22 (a) Feasibility criteria below the pinch for the dephenolization example. (b) Rich-end design for the dephenolization example.

that immediately below the pinch, only streams R_1, R_2, and S_1 exist. Stream S_3 does not reach the pinch point and so is not considered when the feasibility criteria of matching streams at the pinch are applied. Because N_{rb} is 2 and N_{lb} is 1, inequality (Eq. 5.34b) is satisfied. Next, the second feasibility criterion (Eq. 5.38b) is examined. As can be seen in Fig. 5.22, S_1 cannot be matched with either R_1 or R_2, because L_1/m_1 is greater than G_1 and G_2. Hence, S_1

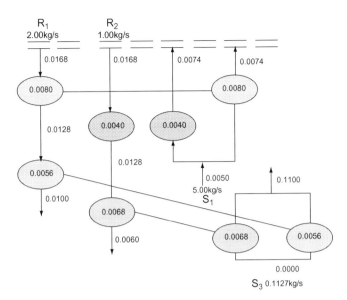

FIGURE 5-23 Lean-end design for the dephenolization example.

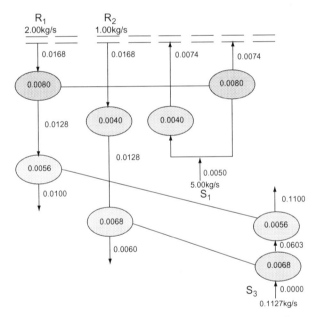

FIGURE 5-24 Lean-end design for the dephenolization example.

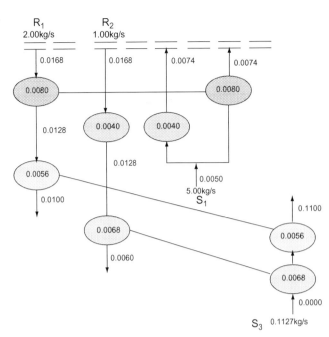

FIGURE 5-25 Lean-end design for the dephenolization example.

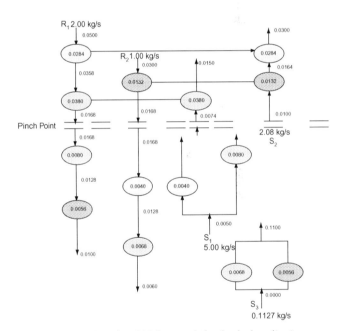

FIGURE 5-26 Complete MOC network for the dephenolization example.

must be split into two branches: one to be matched with R_1 and the other to be paired with R_2. There are infinite number of ways to split L_1 so as to satisfy Eq. 5.38b. Let us arbitrarily split L_1 in the same ratio of G_1 to G_2, that is, to 3.33 and 1.67 kg/s. This split realizes the inequality (Eq. 5.38b), because $3.33/2 < 2$ and $1.67/2 < 1$.

The remaining loads of R_1 and R_2 can now be eliminated by S_3 (activated carbon). Several configurations can be envisioned for S_3. These structures include a split design (Fig. 5.23), a serial design in which S_3 is first matched with R_1 (Fig. 5.24), and a serial design in which S_3 is first matched with R_2 (Fig. 5.25). The number of exchangers below the pinch is four, which is one more than $U_{MOC,\,below\,the\,pinch}$. Once again, $U_{MOC,\,below\,the\,pinch}$ is just a lower bound on the number of exchangers and need not be realized exactly.

It is now possible to generate a MOC network by combining the rich-end design with a lean-end design. For instance, if Figs. 5.21b and 5.23 are merged, the MOC MEN shown in Fig. 5.26 is obtained.

TRADING OFF FIXED COST VERSUS OPERATING COST

As has been mentioned in the section titled "The MEN-Targeting Approach," two design targets can be determined for an MEN: MOC and minimum number of units (aimed at minimizing fixed cost). The section on

"Construction of the Men Configuration with Minimum Number of Exchangers" presented a method that is aimed at constructing MENs with minimum number of exchangers satisfying the MOC target. In general, the two targets of fixed and operating costs are not necessarily compatible; the designer should trade off both costs. Three principal approaches may be used in conjunction with the graphical approaches for trading off fixed and operating costs:

1. Varying the mass-exchange driving forces
2. Mixing rich streams
3. Using mass-load paths

TRADING OFF FIXED AND OPERATING COSTS BY VARYING THE MASS-EXCHANGE DRIVING FORCES

As discussed in Section II.5 in Appendix II, the minimum allowable composition differences can be used to trade off fixed versus operating costs. Typically, an increase in ϵ_j leads to an increase in the MOC of the network (see Figs. 5.12 and 5.14) and a decrease in the fixed cost of the system. Hence, the minimum allowable composition differences can be iteratively varied until the total annualized cost of the system is attained (see Fig. II.13 in Appendix II, which is shown below as Fig. 5.27).

TRADING OFF FIXED AND OPERATING COSTS BY MIXING RICH STREAMS

The other common approach for trading off fixed costs against operating costs is mixing of rich streams. An important class of problems involves the mixing of rich streams. In some cases, the plant operation and the environmental regulations allow such mixing, which typically decreases the number of mass exchangers and, consequently, the fixed cost. On the other hand, mixing various rich streams normally increases the MOC of the system. This can be explained on a pinch diagram because mixing results in a right shift of the rich composite stream. Such a shift may make a process MSA or a low-cost external MSA infeasible, increasing the operating cost of the MSAs. There are also cases when mixing rich streams does not affect the MOC solution. For instance, consider the previous dephenolization example. Suppose that process considerations allow mixing of the two rich streams. Furthermore, assume that the rich recovery task is described in terms of retaining an 80 percent phenol recovery from the mixed stream. The data for the mixed rich stream are given in Table 5.7. A pinch analysis (see Fig. 5.28) demonstrates that the MOC solution is still the same. In this case, mixing the rich streams is recommended because it reduces the number of mass exchangers without affecting the MOC solution.

TRADING OFF FIXED AND OPERATING COSTS USING MASS-LOAD PATHS

A third method for trading off fixed versus operating costs is the use of *mass-load paths* (El-Halwagi and Manousiouthakis, 1989). A **mass-load path** is a continuous connection that starts with an external MSA and concludes with a process MSA. By shifting the loads along a path, one can add an excess of external MSAs to replace an equivalent amount of process MSA. This procedure eliminates exchangers at the expense of additional operating cost. For instance, in Fig. 5.26, the path

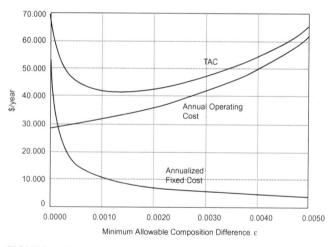

FIGURE 5-27 Using mass transfer driving force to trade off fixed cost versus operating cost (for Example 5.1).

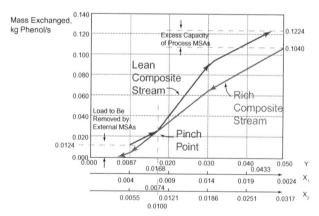

FIGURE 5-28 The pinch diagram for the dephenolization example with mixing of rich streams.

TABLE 5-7 Data of Rich Streams for Dephenolization Example with Mixing of Rich Streams

Stream	Description	Flow Rate G_i, kg/s	Supply Composition (Mass Fraction) y_i^s	Target Composition (Mass Fraction) y_i^t
R_{mixed}	Mixed R_1 and R_2	3	0.0433	0.0087

S_3-R_2-S_1 can be used to shift a load of 0.0040 kg/s from S_1 to S_3. This load shift will lead to the elimination of the exchanger, pairing R_2 with S_1 below the pinch. Therefore, the flow rate and/or outlet composition of S_1 should be reduced to remove the unused capacity of 0.004 kg phenol/s. For instance, if the outlet composition of S_2 is kept unchanged, the flow rate of S_2 should be adjusted to 4.60 kg/s. An additional benefit of load shifting is that the two exchangers matching S_1 with R_1 above and below the pinch can now be combined into a single exchanger. However, the flow rate of S_3 will have to be increased by $0.0040/(0.1100 - 0) = 0.0364$ kg/s. This step incurs an additional operating cost, which can be calculated as

[5.41]
$$\text{Additional annual operating cost due to increase in flow rate of } S_3$$
$$= \left(0.0364 \frac{\text{kg activated carbon}}{\text{s}}\right)\left(\frac{\$0.081}{\text{kg activated carbon}}\right)$$
$$\left(3600 \frac{\text{s}}{\text{hr}}\right)\left(\frac{8760 \text{ hr}}{\text{yr}}\right)$$
$$\approx \$93 \times 10^3/\text{yr}$$

or

$$\text{Additional annual operating cost due to increase in flow rate of } S_3$$
$$= \left(0.004 \frac{\text{kg phenol}}{\text{s}}\right)\left(\frac{\$0.737}{\text{kg phenol removed}}\right)$$
$$\left(3600 \frac{\text{s}}{\text{hr}}\right)\left(\frac{8760 \text{ hr}}{\text{yr}}\right)$$
$$\approx \$93 \times 10^3/\text{yr}$$

Designers should compare this additional cost with the fixed-cost savings accomplished by eliminating two exchangers (the exchanger matching R_2 with S_1 below the pinch is removed, and the two exchangers matching R_1 with S_1 are combined into a single unit). This comparison provides the basis for determining whether or not the mass-load path should be employed. Figure 5.29 illustrates the network configuration after the mass-load path is used to eliminate two exchangers.

Three additional mass-load paths (S_3-R_2-S_2, S_3-R_1-S_1, S_3-R_1-S_2) can be employed to shift removal duties from the process MSAs to the external MSA until all the rich load (0.104 kg phenol/s) is removed by the external MSA (see Fig. 5.30). In this case, the operating cost of the system is

[5.42]
$$\text{Annual operating cost}$$
$$= \left(0.104 \frac{\text{kg phenol}}{\text{s}}\right)\left(\frac{\$0.737}{\text{kg phenol}}\right)\left(3600 \frac{\text{s}}{\text{hr}}\right)\left(\frac{8760 \text{ hr}}{\text{yr}}\right)$$
$$\approx \$2.417 \text{ million/yr}.$$

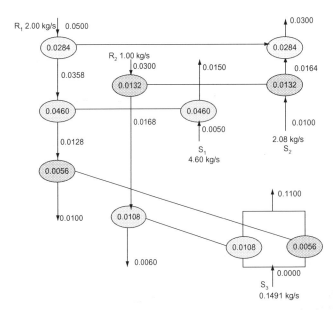

FIGURE 5-29 Reduction of number of exchangers using a mass-load path.

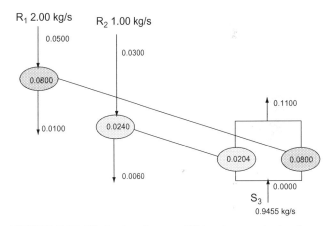

FIGURE 5-30 Elimination of process MSAs using mass-load paths.

By comparing Equations 5.29 and 5.42, we notice that the application of the mass-load paths has resulted in an operating cost increase of \$2.129 MM/yr. On the other hand, the number of mass exchangers has dropped from seven to two. Economic criteria (for example, total annualized cost or incremental return on investment) can be used to determine the extent to which the fixed cost should be traded off with operating cost.

HOMEWORK PROBLEMS

5.1. A processing facility converts scrap tires into fuel via pyrolysis. Figure 5.31 is a simplified block flow diagram of the process. The discarded tires are fed to a high-temperature reactor where heat breaks down the hydrocarbon content of the tires into oils and gaseous fuels. The oils are further processed and separated to yield transportation fuels. The reactor off-gases are cooled to condense light oils. The condensate is decanted into two layers: organic and aqueous. The organic layer is mixed with the liquid products of the reactor. The aqueous layer is a wastewater stream whose organic content must be reduced prior to discharge. The primary pollutant in the wastewater is a heavy hydrocarbon. The data for the wastewater stream are given in Table 5.8.

A process lean stream and three external MSAs are considered for removing the pollutant. The process lean stream is a flare gas (a gaseous stream fed to the flare) that can be used as a process stripping agent. To prevent the back-propagation of fire from the flare, a seal pot is used. An aqueous stream is passed through the seal pot to form a buffer zone between the fire and the source of the flare gas. Therefore, the seal pot can be used as a stripping column in which the flare gas strips the organic pollutant off the wastewater while the wastewater stream constitutes a buffer solution for preventing back-propagation of fire.

Three external MSAs are considered: a solvent extractant (S_2), an adsorbent (S_3), and a stripping agent (S_4). The data for the candidate MSAs are given in Table 5.9. The equilibrium data for the transfer of the pollutant from the waste stream to the jth MSA is given by

$$[5.33] \quad y_1 = m_j x_j$$

where y_1 and x_j are the mass fractions of the organic pollutant in the wastewater and the jth MSA, respectively.

For the given data, use the pinch diagram to determine the minimum operating cost of the MEN. Construct a network of mass exchangers that satisfies the MOC target with the minimum number of exchangers.

5.2. If the fixed cost is disregarded in the previous problem, what is the lowest target for operating cost of the MEN? Hint: Set all the ε_js equal to zero.

5.3. Consider the coke oven gas (COG)-sweetening process shown in Fig. 5.32. The basic objective of COG sweetening is the removal of acidic impurities, primarily hydrogen sulfide, from COG (a mixture of H_2, CH_4, CO, N_2, NH_3, CO_2, and H_2S). Hydrogen sulfide is an undesirable impurity, because it is corrosive and contributes to SO_2 emission when the COG is burnt. The existence of ammonia in COG and the selectivity of aqueous ammonia in absorbing H_2S suggests that aqueous ammonia is a candidate solvent

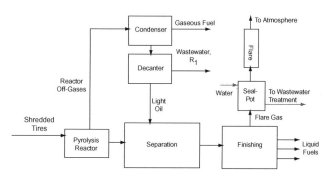

FIGURE 5-31 A simplified block flow diagram of a tire-to-fuel process.

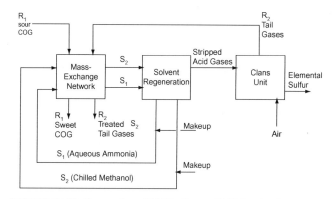

FIGURE 5-32 Sweetening of COG. *Source: El-Halwagi and Manousiouthakis (1989).*

TABLE 5-8 Data for the Wastewater Stream of Tire Pyrolysis Plant

Stream	Description	Flow Rate G_i, kg/s	Supply Composition (ppmw) y_i^s	Target Composition (ppmw) y_i^t
R_1	Aqueous layer from decanter	0.2	500	50

TABLE 5-9 Data for the MSAs of the Tire Pyrolysis Problem

Stream	Upper Bound on Flow Rate L_j^c kg/s	Supply Composition (ppmw) x_j^s	Target Composition (ppmw) x_j^t	m_j	ε_j ppmw	C_j $/kg MSA
S_1	0.15	200	900	0.5	200	—
S_2	∞	300	1000	1.0	100	0.001
S_3	∞	10	200	0.8	50	0.020
S_4	∞	20	600	0.2	50	0.040

(process lean stream, S_1). It is desirable that the ammonia recovered from the sour gas compensate for a large portion of the ammonia losses throughout the system and, thus, reduce the need for ammonia makeup. Besides ammonia, an external MSA (chilled methanol, S_2) is also available for service to supplement the aqueous ammonia solution as needed.

The purification of the COG involves washing the sour COG, R_1, with sufficient aqueous ammonia and/or chilled methanol to absorb the required amounts of hydrogen sulfide. The acid gases are subsequently stripped from the solvents and the regenerated MSAs are recirculated. The stripped acid gases are fed to a "Claus unit" where elemental sulfur is recovered from hydrogen sulfide. In view of air pollution control regulations, the tail gases leaving the Claus unit, R_2, should be treated for partial removal of the unconverted hydrogen sulfide. Table 5.10 summarizes the stream data.

Using the pinch diagram with $\varepsilon_1 = \varepsilon_2 = 0.0001$, find the minimum cost of MSAs required to handle the desulfurization of R_1 and R_2. Where is the pinch located?

5.4. Figure 5.33 is a simplified flow diagram of an oil refinery. The process generates two major sources of phenolic wastewater: one from the catalytic cracking unit and the other from the visbreaking system. Two technologies can be used to remove phenol from R_1 and R_2: solvent extraction using light gas oil S_1 (a process MSA) and adsorption using activated carbon S_2 (an external MSA). Table 5.11 provides data for the streams. A minimum allowable composition difference, e_j, of 0.01 can be used for the two MSAs.

By constructing a pinch diagram for the problem, find the minimum cost of MSAs needed to remove phenol from R_1 and R_2. How do you characterize the point at which both composite streams touch? Is it a true pinch point? Synthesize a network of mass exchangers that can achieve the MOC solution using minimum number of mass exchangers.

5.5. Fig. 5.34 shows the process flow sheet for an ethylene/ethylbenzene plant (Stanley and El-Halwagi, 1995). Gas oil is cracked with steam in a pyrolysis furnace to form ethylene, low BTU gases, hexane, heptane, and heavier hydrocarbons. The ethylene is then reacted with benzene to form ethylbenzene. Two wastewater streams are formed: R_1, which is the quench water recycle for the cooling tower, and R_2, which is the wastewater from the ethylbenzene portion of the plant. The primary pollutant present in the two wastewater streams is benzene. Benzene must be removed from stream R_1 down to a concentration of 200 ppm before R_1 can be recycled back to the cooling tower. Benzene must also be removed from stream R_2 down to a concentration of 360 ppm before R_2 can be sent to biotreatment. The data for streams R_1 and R_2 are shown in Table 5.12.

There are two process MSAs available to remove benzene from the wastewater streams. These process MSAs are hexane (S_1) and heptane (S_2). Hexane is

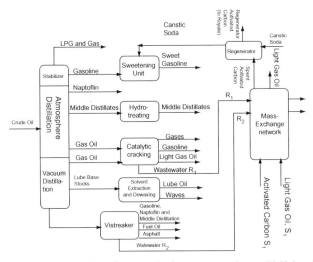

FIGURE 5-33 Dephenolization of refinery wastes. *Source: El-Halwagi et al. (1992).*

TABLE 5-10 Stream Data for the COG-Sweetening Problem

	Rich stream			MSAs					
Stream	G_i (kg/s)	y_i^s	y_i^t	Stream	L_j^c kg/s	x_j^s	x_j^t	m_j	b_j	c_j \$/kg
R_1	0.90	0.0700	0.0003	S_1	2.3	0.0006	0.0310	1.45	0.000	0.00
R_2	0.10	0.0510	0.0001	S_2	∞	0.0002	0.0035	0.26	0.000	0.10

TABLE 5-11 Stream Data for Refinery Problem

	Rich stream			MSAs					
Stream	G_i (kg/s)	y_i^s	y_i^t	Stream	L_j^c (kg/s)	x_j^s	x_j^t	m_j	b_j	c_j \$/kg
R_1	8.00	0.10	0.01	S_1	10.00	0.01	0.02	2.00	0.00	0.00
R_2	6.00	0.08	0.01	S_2	∞	0.00	0.11	0.02	0.00	0.08

available at a flow rate of 0.8 kg/s and supply composition of 10 ppmw whereas heptane is available at a flow rate of 0.3 kg/s and supply composition of 15 ppmw. The target compositions for hexane and heptane are unknown and should be determined by the engineer designing the MEN. The mass-transfer driving forces, ε_1 and ε_2, should be at least 30,000 and 20,000 ppmw, respectively. The equilibrium data for benzene transfer from wastewater to hexane and heptane are:

[5.34] $\qquad y = 0.011 x_1$

and

[5.35] $\qquad y = 0.008 x_2,$

where y, x_1, and x_2 are given in mass fractions.

Two external MSAs are considered for removing benzene: air (S_3) and activated carbon (S_4). Air is compressed to 3 atm before stripping. Following stripping, benzene is separated from air using condensation. Henry's law can be used to predict equilibrium for the stripping process. Activated carbon is continuously regenerated using steam in the ratio of 1.5 kg steam: 1 kg of benzene adsorbed on activated carbon. Makeup at the rate of 1 percent of recirculating activated carbon is needed to compensate for losses due to regeneration and deactivation. Over the operating range, the equilibrium relation for the transfer of benzene from wastewater onto activated carbon can be described by:

[5.36] $\qquad y = 7.0 \times 10^{-4} x_4$

1. Using the pinch diagram, determine the pinch location, minimum load of benzene to be removed by external MSAs, and excess capacity of process MSAs. How do you remove this excess capacity?
2. Considering the four candidate MSAs, what is the MOC needed to remove benzene?

5.6. Consider the magnetic tape manufacturing process (Dunn et al., 1995) shown in Fig. 5.35. First, coating ingredients are dissolved in 0.09 kg/s of organic solvent and mixed to form a slurry. The slurry is suspended with resin binders and special additives. Next, the coating slurry is deposited on a base film. Nitrogen gas is used to induce evaporation rate of solvent that is proper for deposition. In the coating chamber, 0.011 kg/s of solvent are decomposed into other organic species. The decomposed organics are separated from the exhaust gas in a membrane unit. The retentate stream leaving the membrane unit has a flow rate of 3.0 kg/s and is primarily composed of nitrogen that is laden with 1.9 wt/wt% of the organic solvent. The coated film is passed to a dryer where nitrogen gas is employed to evaporate the remaining solvent. The exhaust gas leaving the dryer has a flow rate of 5.5 kg/s and contains 0.4 wt/wt% solvent. The two exhaust gases are mixed and disposed of.

Due to environmental regulations, it is required to reduced the total solvent emission to 0.06 kg/s (by removing 25 percent of current emission). Three MSAs can be used to remove the solvent from the gaseous emission. The equilibrium data for the transfer of the organic solvent to the jth lean stream

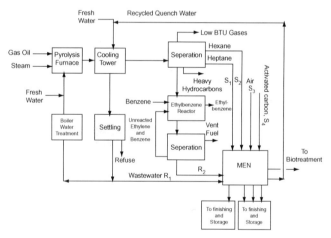

FIGURE 5-34 Process flow sheet for an ethylene/ethylbenzene plant. *Source: Stanley and El-Halwagi (1995).*

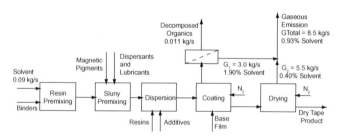

FIGURE 5-35 Schematic representation of a magnetic-tape manufacturing process.

TABLE 5-12 Data for the Waste Streams of the Ethylbenzene Plant

Stream	Description	Flow Rate G_i kg/s	Supply Composition (ppmw) y_i^s	Target Composition (ppmw) y_i^t
R_1	Wastewater from settling	100	1000	200
R_2	Wastewater from ethylbenzene separation	50	1800	360

TABLE 5-13 Data for the MSAs

Stream	Upper Bound on Flow Rate L_j^c kg/s	Supply Composition (Mass Fraction) x_j^s	Target Composition (Mass Fraction) x_j^t	m_j	ε_j (Mass Fraction)	C_j $/kg MSA
S_1	∞	0.014	0.040	0.4	0.001	0.002
S_2	∞	0.020	0.080	1.5	0.001	0.001
S_3	∞	0.001	0.010	0.1	0.001	0.002

is given by $y = m_j x_j$, where the values of m_j are given in Table 5.13. Throughout this problem, a minimum allowable composition difference, ε_j, of 0.001(kg organic solvent)/(kg MSA) is to be used. The data for the MSAs are given in Table 5.13.

1. Using the pinch diagram, determine which solvent(s) should be employed to remove the solvent? What is the MOC for the solvent removal task? Construct a network of mass exchangers that uses the minimum number of exchangers to satisfy the MOC target. Hint: Consider segregating the two waste streams and removing solvent from one of them. The annualized fixed cost of a mass exchanger, $/yr, may be approximated by 18,000 (Gas flow rate, kg/s)$^{0.65}$.

2. The value of the recovered solvent is $0.80/kg of organic solvent. What is the annual gross revenue (annual value of recovered solvent − total annualized cost of solvent recovery system)?

NOMENCLATURE

b_j	intercept of equilibrium line for the jth MSA
C_j	unit cost of the jth MSA including regeneration and makeup, $/unit flow rate of recirculating MSA
C_j^r	unit cost of the jth MSA required to remove a unit mass/mole of the key pollutant
G_i	flow rate of the ith waste stream
i	index for rich streams
j	index for MSAs
L_j	flow rate of the jth MSA
L_j^C	upper bound on the available flow rate of the jth MSA
m_j	slope of equilibrium line for the jth MSA
MR_i	mass/moles of pollutant lost from the ith rich stream as defined by Eq. 5.6
MS_j	mass/moles of pollutant gained by the jth MSA as defined by Eq. 5.7
N_i	number of independent synthesis problems
N_S	number of MSAs
N_{SE}	number of external MSAs
N_{SP}	number of process MSAs
N_R	number of rich (waste) streams
R_i	the ith rich stream
S_j	the jth MSA
U	minimum number of mass-exchange units, defined by Eq. 5.2
x_j	composition of key component in the jth MSA
x_j^{in}	inlet composition of key component in the jth MSA
$x_j^{in,max}$	maximum practically feasible inlet composition of key component in the jth MSA
$x_j^{in,*}$	maximum thermodynamically feasible inlet composition of key component in the jth MSA
$x_j^{max,practical}$	maximum practically feasible composition of the key component in the jth MSA
x_j^{out}	outlet composition of key component in the jth MSA
x_j^s	supply composition of the key component in the jth MSA
x_j^t	target composition of the key component in the jth MSA
x_j^*	composition of key component in the jth MSA that is in equilibrium with y_i
y	composition scale for the key component in any rich stream
y_i	composition of key component in the ith rich stream
y_i^s	supply composition of key component in the ith rich stream
y_i^t	target composition of key component in the ith rich stream

GREEK LETTER

ε_j	minimum allowable composition difference for the jth MSA

REFERENCES

Dunn RF, El-Halwagi MM, Lakin J, Serageldin M: Selection of organic solvent blends for environmental compliance in the coating industries. In Griffith ED, Kahn H, Cousins MC, editors: *Proceedings of the first international plant operations and design conference*, vol III, New York, AIChE, pp 83–107, 1995.

El-Halwagi MM, El-Halwagi AM, Manousiouthakis V: Optimal design of dephenolization networks for petroleum-refinery wastes, *Trans Inst Chem Eng* 70(B):131–139, August 1992.

El-Halwagi MM, Manousiouthakis V: Synthesis of mass exchange networks, *AIChE J* 35(8):1233–1244, 1989.

Stanley C, El-Halwagi MM: Synthesis of mass exchange networks using linear programming techniques. In Rossiter AP, editor: *Waste minimization through process design*, New York, McGraw Hill, pp 209–224, 1995.

CHAPTER 6
COMBINING MASS-INTEGRATION STRATEGIES

The previous three chapters set the stage for developing mass-integration strategies. Chapter 3 presented several approaches to the determination of overall mass targets depending on the type and extent of available data. These targets are identified ahead of detailed design. As mentioned in Chapter 3, these targets can be attained by a combination of strategies as shown by Fig. 6.1, where impact is a measure of the effectiveness of the proposed solution in partially reaching the target. The cumulative impact of all the strategies yields the desired target. Acceptability is a measure of the likelihood of a proposed strategy to be accepted and implemented by the plant. Chapters 4 and 5 gave important classes of systematic mass-integration tools associated with identifying such strategies. Direct recycle was covered by Chapter 4 as a key element in no-/low-cost strategies involving segregation, mixing, and rerouting of streams without the addition of new equipment. The synthesis of mass-exchange networks was described in Chapter 5 as an illustration of the addition of new units (interception) and the selection or substitution of solvents to achieve a separation task. This chapter combines the tools presented in the previous three chapters. First, the process representation from a species viewpoint is presented. Then, the combination of mass-integration tools is demonstrated through the applicability to a case study on the production of acrylonitrile.

PROCESS REPRESENTATION FROM A MASS-INTEGRATION SPECIES PERSPECTIVE

Once an overall mass target is determined, it is necessary to develop cost-effective strategies to reach the target. For a given target, there are numerous design decisions that must be judiciously made. These include addressing the following challenging questions:

- What are the optimum stream-rerouting strategies?
- Should any streams be segregated, mixed, or rerouted? Which ones?
- Which streams should be recycled/reused? To which units?
- Should design variable and/or operating conditions of existing units be altered? Which ones? To what extent?
- Is there a need to add/replace units? Which ones? Where to add/replace?
- Should interception (for example, separation) devices be added? Which streams should be intercepted? To remove what? To what extent?
- Which separating agents should be selected for interception?
- What is the optimal flow rate of each separating agent?
- How should these separating agents be matched with the rich streams (that is, stream pairings)?

Because of the prohibitively large number of alternatives involved in answering these questions, a systematic procedure is needed to extract the optimum solution(s) without enumerating them. This is the role provided by mass integration. Mass integration is a holistic and systematic methodology that provides a fundamental understanding of the global flow of mass within the process and employs this understanding in identifying performance targets and optimizing the allocation, separation, and generation of streams and species. Mass integration is based on fundamental principles of chemical engineering combined with system analysis using graphical

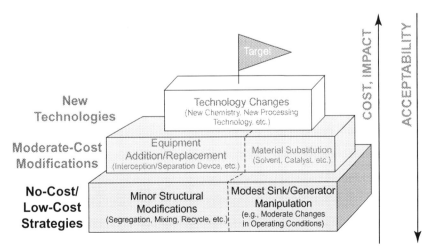

FIGURE 6-1 Hierarchy of mass-integration strategies. *Source: El-Halwagi (1999).*

Sustainable Design Through Process Integration.
© 2012 Elsevier Inc. All rights reserved.

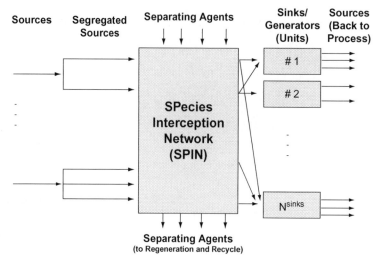

FIGURE 6-2 A process from a species perspective. *Source: El-Halwagi et al. (1996).*

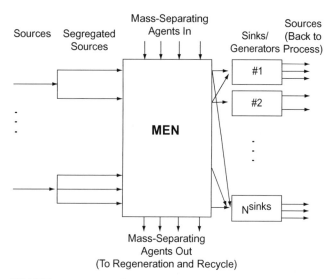

FIGURE 6-3 A process from a species viewpoint when MSAs are used for interception.

and discharge media. Streams leaving the sinks become, in turn, sources. Therefore, sinks are also generators of the targeted species. Each sink/generator may be manipulated via design and/or operating changes to affect the flow rate and composition of what each sink/generator accepts and discharges.

Stream characteristics (for example, flow rate, composition, pressure, temperature, and so on) can be modified by adding new units that intercept the streams prior to being fed to the process sinks and condition their properties to the desired values. This is performed by a species-interception network (SPIN) that may use mass- and energy-separating agents. **Interception** denotes the utilization of new unit operations to adjust the composition, flow rate, and other properties of certain process streams to make them acceptable for existing process sinks. A particularly important class of interception devices is separation systems. These separations may be induced by the use of mass-separating agents (MSAs) and/or energy-separating agents (ESAs). A systematic technique is needed to screen the multitude of separating agents and separation technologies to find the optimal separation system. The synthesis of MSA-induced physical-separation systems has been covered in Chapter 5 through the synthesis of mass-exchange networks (MENs). Other interception systems are covered throughout the book. The MEN can be used to prepare sources for recycle. When MSAs are used as the only type of separating agents, the SPIN is replaced with a MEN and Fig. 6.2 is revised as shown in Fig. 6.3.

and optimization-based tools. To develop detailed mass-integration strategies, let us represent the process flow sheet from a species viewpoint (for example, El-Halwagi and Spriggs, 1998; El-Halwagi et al., 1996), as shown in Fig. 6.2. For each targeted species, there are sources (streams that carry the species) and process sinks (units that can accept the species). Process sinks include reactors, separators, heaters/coolers, biotreatment facilities,

---

Example 6-1 Application of mass integration to debottleneck an acrylonitrile process and reduce water usage and discharge (El-Halwagi, 1997)

Acrylonitrile (AN, C_3H_3N) is manufactured via the vapor-phase ammoxidation of propylene:

$$C_3H_6 + NH_3 + 1.5O_2 \xrightarrow{catalyst} C_3H_3N + 3H_2O$$

The reaction takes place in a fluidized-bed reactor in which propylene, ammonia, and oxygen are catalytically reacted at 450°C and 2 atm. The reaction is a single pass with almost complete conversion of propylene. The reaction products are cooled using an indirect-contact heat exchanger that condenses a fraction of the reactor off-gas. The remaining off-gas is scrubbed with water, and then decanted into an aqueous layer and an organic layer. The organic layer is fractionated in a distillation column under slight vacuum, which is induced by a steam-jet ejector. Figure 6.4 shows the process flow sheet along with pertinent material balance data.

(Continued)

Example 6-1 Application of mass integration to debottleneck an acrylonitrile process and reduce water usage and discharge (El-Halwagi, 1997) (*Continued*)

The wastewater stream of the plant is composed of the off-gas condensate, the aqueous layer of the decanter, the bottom product of the distillation column, and the condensate from the steam-jet ejector. This wastewater stream is fed to the biotreatment facility. Because the biotreatment facility is currently operating at full hydraulic capacity, it constitutes a bottleneck for the plant. Plans for expanding production of acrylonitrile (AN) are contingent upon debottlenecking the biotreatment facility by reducing its influent or installing an additional treatment unit. The new biotreatment facility will cost about $4 million in capital investment and $360,000/yr (where yr stands for year) in annual operating cost, leading to a TAC of $760,000/yr with a 10-year linear depreciation. The objective of this case study is to use mass-integration techniques to devise cost-effective strategies to debottleneck the biotreatment facility.

The following technical constraints should be observed in any proposed solution:

Scrubber

[6.1] $\quad 5.8 \leq$ flow rate of wash feed (kg/s) ≤ 6.2

[6.2] $\quad 0.0 \leq$ ammonia content of wash feed (ppm NH_3) ≤ 10.0

Boiler Feed Water (BFW)

[6.3] $\quad$ Ammonia content of BFW (ppm NH_3) = 0.0

[6.4] $\quad$ AN content of BFW (ppm AN) = 0.0

Decanter

[6.5] $\quad 10.6 \leq$ flow rate of feed (kg/s) ≤ 11.1

Distillation Column

[6.6] $\quad 5.2 \leq$ flow rate of feed (kg/s) ≤ 5.7

[6.7] $\quad 0.0 \leq$ Ammonia content of feed (ppm NH_3) ≤ 30.0

[6.8] $\quad 80.0 \leq$ AN content of feed (wt% AN) ≤ 100.0

Furthermore, for quality and operability objectives, the plant does not wish to recycle the AN product stream (top of distillation column), the feed to the distillation column, and the feed to the decanter.

Three external MSAs are considered for removing ammonia from water: air (S_1), activated carbon (S_2), and an adsorbing resin (S_3). The data for the candidate MSAs are given in Table 6.1. The equilibrium data for the transfer of the pollutant from the waste stream to the jth MSA is given by

[6.9] $\quad y_1 = m_j x_j$,

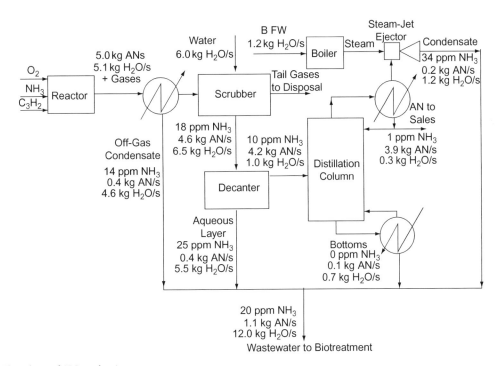

FIGURE 6-4 Flow sheet of AN production.

TABLE 6-1 Data for MSAs of the AN Problem

Stream	Upper Bound on Flow Rate L_j^c	Supply Composition (ppmw) x_j^s	Target Composition (ppmw) x_j^t	m_j	ε_j ppmw	C_j $/kg MSA	C_j' $/kg NH_3 removed
S_1	∞	0	6	1.4	2	0.004	667
S_2	∞	10	400	0.02	5	0.070	180
S_3	∞	3	1100	0.01	5	0.100	91

(*Continued*)

Example 6-1 Application of mass integration to debottleneck an acrylonitrile process and reduce water usage and discharge (El-Halwagi, 1997) (*Continued*)

where y_1 and x_j are weight-based parts per million of ammonia in the wastewater and the jth MSA, respectively.

SOLUTION

The first step in the analysis is to identify the target for debottlenecking the biotreatment facility. An overall water balance for the plant (Fig. 6.5a) can be written as follows:

Water in + Water generated by chemical reaction = Wastewater out + Water losses

Because the wastewater discharge is larger than fresh-water flow rate, it is possible, in principle, to bring wastewater to a quality that can substitute fresh water using segregation, mixing, recycle, and interception. Furthermore, sink/generator manipulation can be employed to reduce flow rate of fresh water. Hence, fresh water usage in this example can in principle be completely eliminated, and for the same reaction conditions and water losses, the target for wastewater discharge can be calculated from the overall water balance as follows (Fig. 6.5b):

[6.10] *Target of minimum discharge to biotreatment* = 5.1 − 0.3
= 4.8 kg water/s

As can be seen from Fig. 6.5b, the corresponding target for minimum usage of fresh water is potentially zero.

Figure 6.6 illustrates the gap between current process performance and benchmarked mass targets. Having identified this target, let us now determine how to best attain the target. It is also necessary to sequence and integrate the solution strategies.

First, we develop no-/low-cost strategies starting with minor process modifications. An operating parameter that can be altered to reduce fresh-water usage (and consequently wastewater discharge) is the flow rate of the feed to the scrubber. As given by Constraint 6.1, the flow rate of fresh water fed to the first scrubber may be reduced to 5.8 kg/s. This is a minor process modification that involves setting the flow control valve. Compared to the current usage of 6.0 kg/s, the net result is a reduction of 0.2 kg/s in water usage (and discharge). These results are shown in Fig. 6.7. To track water, the wastewater discharge is described in terms of kg H_2O/s (not as total flow rate of wastewater including other species).

Next, we consider other no-/low-cost strategies including segregation, mixing, and direct-recycle opportunities. First, we identify the relevant sources and sinks. Once the streams composing the terminal wastewater are segregated, we get four sources that can be potentially recycled. Fresh water used in the scrubber and the boiler provides two more sources. To reduce wastewater discharge to biotreatment, fresh water must be reduced. Hence, we should focus our attention on recycling opportunities to sinks that employ fresh water, namely, the scrubber and the boiler. Figure 6.8 illustrates the sources and sinks involved in the analysis.

Direct-recycle strategies are based on segregating, rerouting, and mixing of sources without the use of new equipment. Hence, there is no SPIN involved in

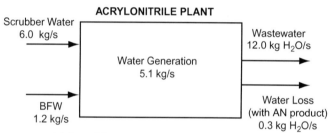

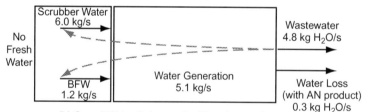

FIGURE 6-5 Establishing targets for biotreatment influent: overall water balance (a) before and (b) after mass integration.

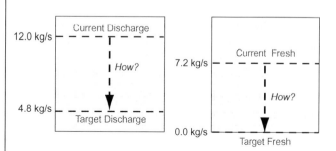

FIGURE 6-6 Benchmarking water usage and discharge.

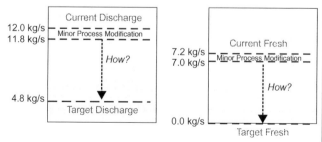

FIGURE 6-7 Reduction in water usage and discharge (expressed as kg H_2O/s) after minor process modification.

(*Continued*)

Example 6-1 Application of mass integration to debottleneck an acrylonitrile process and reduce water usage and discharge (El-Halwagi, 1997) (*Continued*)

direct recycle. Therefore, for direct recycle, Fig. 6.8 is revised by eliminating the SPIN as shown in Fig. 6.9. Because of the stringent limitation on the boiler feed water (BFW) (no ammonia or AN), no recycled stream can be used in lieu of fresh water (segregation, mixing, recycle, and interception can reduce, but not eliminate, ammonia/AN content). Consequently, in Fig. 6.9, none of the process sources are allocated to the boiler.

The boiler should not be considered as a sink for recycle (with or without interception). Instead, it should be handled at the stage of sink/generator manipulation. This leaves us with the five segregated sources and two sinks (boiler and scrubber). The data for the sources and sinks are summarized in Tables 6.2 and 6.3, respectively. Using these data, the material recycle pinch diagram is constructed by first developing the sink composite curve (Fig. 6.10), then the source composite curve (Fig. 6.11), and the material recycle pinch diagram (Fig. 6.12). As can be seen from Fig. 6.12, when direct recycle is used, the following targets can be obtained:

[6.11a]   Minimum fresh water = 2.1 kg/s

[6.11b]   Maximum direct recycle = 4.9 kg/s

[6.11c]   Minimum waste discharge = 8.0 kg/s

As mentioned earlier, to track water, it is useful to express the flow rate of waste discharge as kg H_2O/s (not as total flow including other species). This can be readily calculated by noting that the reduction in fresh-water usage is equal to the reduction in wastewater discharge (expressed as kg H_2O/s). Because,

[6.11d]   Reduction in fresh-water usage = 7.2 − 2.1 = 5.1 kg/s

We have

[6.11e]   Minimum waste discharge (expressed as kg H_2O/s) = 12.0 − 2.1 = 6.9 kg H_2O/s[1]

Figure 6.13 shows these results.

It is beneficial to detail the direct-recycle strategies to attain the identified targets. Toward that end, the source-sink mapping diagram can be used to identify which sources should be recycled to which sinks. Figure 6.14 is a source-sink mapping representation of the problem. As mentioned earlier, because of the stringent requirements on the feed to the boiler sink (0.0 composition of any

---
[1] The difference between the wastewater discharge expressed as total flow rate (8.0 kg/s) versus that expressed in terms of H_2O (6.9 kg/s) accounts for the 1.1 kg/s of collective AN losses in the wastewater.

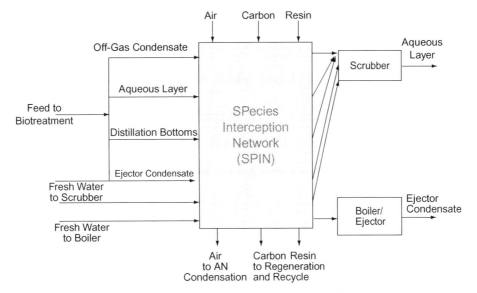

FIGURE 6-8 Segregation, mixing, interception, and recycle representation for the AN case study.

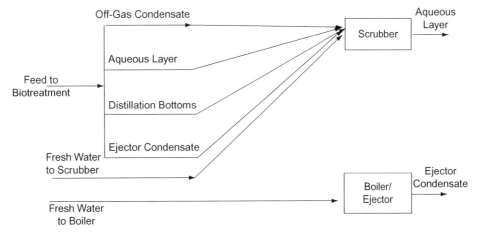

FIGURE 6-9 Schematic representation of segregation, direct recycle, and mixing for the AN example. *Source: El-Halwagi (1997).*

Example 6-1 Application of mass integration to debottleneck an acrylonitrile process and reduce water usage and discharge (El-Halwagi, 1997) (*Continued*)

TABLE 6-2 Sink Data for the AN Example

Sink	Flow Rate kg/s	Maximum Inlet Composition of NH_3 ppm	Maximum Inlet Load, 10^{-6} kg NH_3/s
BFW	1.2	0.0	0.0
Scrubber	5.8	10.0	58.0

TABLE 6-3 Source Data for the AN Example

Source	Flow Rate kg/s	Inlet Composition of NH_3 ppm	Inlet Load, 10^{-6} kg NH_3/s
Distillation Bottoms	0.8	0.0	0.0
Off-Gas Condensate	5.0	14	70.0
Aqueous Layer	5.7[a]	25	142.5
Jet-Ejector Condensate	1.4	34	47.6

[a] *Because the flow rate of the feed to the scrubber has been reduced by 0.2 kg/s as a result of minor process modification, the flow rate of the aqueous layer is assumed to decrease from 5.9 to 5.7 kg/s.*

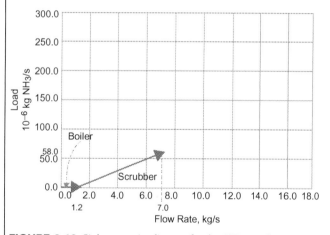

FIGURE 6-10 Sink composite diagram for the AN example.

pollutant), it is not possible to replace any of the feed to the boiler through recycle (even with interception). Consequently, we will not represent the boiler sink on the source-sink mapping diagram. Following the lever arm rules, including sink-feed conditions and source-prioritization rules described in Chapter 4, the following statements can be made:

- The composition of the feed to the boiler should be set to its maximum value (10 ppm).
- The use of the sources should be prioritized as follows: distillation bottoms, off-gas condensate, aqueous layer, then jet-ejector condensate.

Hence, we start by considering the use of the first two sources (shortest fresh arms): distillation bottoms and off-gas condensate. The flow rate resulting from combining these two sources (5.8 kg/s) is sufficient to run the scrubber. However, its ammonia composition[2] as determined by the lever-arm principle is 12.1 ppm,

---

[2] Algebraically, this composition can be calculated as follows:

$$\frac{(5.0 \text{ kg/s})(14 \text{ ppm } NH_3) + 0}{5.8 \text{ kg/s}} = 12.1 \text{ ppm } NH_3$$

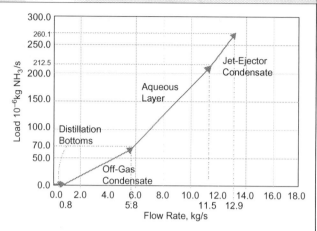

FIGURE 6-11 Source composite diagram for the AN example.

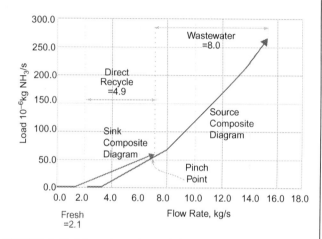

FIGURE 6-12 Material recycle pinch diagram for the AN example.

which lies outside the zone of permissible recycle to the scrubber. As shown by Fig. 6.15, the maximum flow rate of the off-gas condensate to be recycled to the scrubber[3] is determined as follows:

[6.12]

$$\frac{\text{Arm of Gas Condensate}}{\text{Total Arm}} = \frac{\text{Flow Rate of Recycled Gas Condensate}}{\text{Flow Rate of Scrubber Feed}}$$

i.e.,

$$\frac{10-0}{14-0} = \frac{\text{Flow Rate of Recycled Gas Condensate}}{5.8}$$

Hence, flow rate of recycled gas condensate = 4.1 kg/s, and the flow rate of fresh water is 0.9 kg/s (5.8 − 0.8 − 4.1). Therefore, direct recycle can reduce the fresh water consumption (and consequently the influent to biotreatment) by 5.1 kg/s. This is exactly the same target identified from the material recycle pinch analysis as described by Eq. 6.11d.

---

[3] Again, algebraically this flow rate can be calculated as follows:

$$\frac{\text{Flow rate of recycled off-gas condensate} \times 14 \text{ ppm } NH_3 + 0 + 0}{5.8 \text{ kg/s}} = 10 \text{ ppm } NH_3$$

where the numerator represents the ammonia in recycled off-gas condensate, distillation bottoms (none), and fresh water (none). Hence, flow rate of recycled off-gas condensate = 4.14 kg/s.

(*Continued*)

Example 6-1 Application of mass integration to debottleneck an acrylonitrile process and reduce water usage and discharge (El-Halwagi, 1997) (*Continued*)

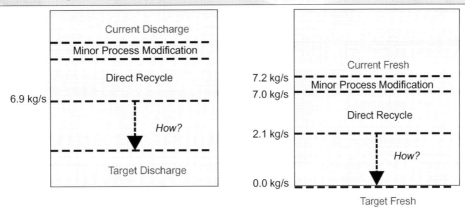

FIGURE 6-13 Reduction in water usage and discharge (expressed as kg H$_2$O/s) after minor process modification and direct recycle.

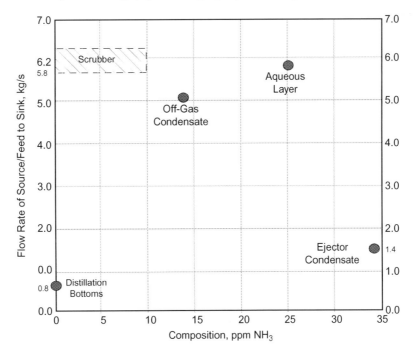

FIGURE 6-14 Source-sink mapping diagram for the AN example.

With the details of the direct-recycle strategies determined, it is possible to determine the implementation cost. The primary cost of direct recycling is pumping and piping. Assuming that the TAC for pumping and piping is $80/(m.yr) and assuming that the total length of piping is 600 m, the TAC for pumping and piping is $48,000/yr.

Because not all the off-gas condensate has been recycled, there is no need to consider recycle from any other source (because they have longer fresh arms). Therefore, there are no more direct-recycle opportunities and we have exhausted the no-/low-cost strategies. Next, we move to adding new units and we consider interception.

Before screening interception devices, it is necessary to determine the interception task. For all of the off-gas condensate to be recycled to the boiler, its ammonia content has to be reduced. As can be seen from Fig. 6.16, to fully recycle the off-gas condensate, the ammonia load to be removed is 12*10^{-6} kg/s. The composition of the intercepted off-gas condensate is the slope of the intercepted stream, which is 11.6 ppm. Therefore, the interception task is to reduce the composition of ammonia in the off-gas condensate from 14.0 ppm to 11.6 ppm. The same result can be obtained algebraically as follows:

[6.13a]

Load removed from off-gas condensate = flow rate of off-gas condensate*
(supply composition − target composition)

that is,

$$12*10^{-6} = 5.0*(14.0 - \text{target composition})$$

Therefore,

[6.13b]

Target (intercepted) composition of ammonia in off-gas condensate = 11.6 ppm

The same result may also be obtained through the source-sink mapping diagram as shown in Fig. 6.17. Alternatively, it may be calculated as follows:

$$\frac{(5.0 \text{ kg/s})y^t \text{ ppm } NH_3 + 0}{5.8 \text{ kg/s}} = 10.0 \text{ ppm } NH_3$$

(*Continued*)

Example 6-1 Application of mass integration to debottleneck an acrylonitrile process and reduce water usage and discharge (El-Halwagi, 1997) (*Continued*)

that is,

[6.13] $$y^t = 11.6 \text{ ppm}$$

To synthesize an optimal MEN for intercepting the off-gas condensate, we construct the pinch diagram as shown in Fig. 6.18. Because the three MSAs lie completely to the left of the rich stream, they are all thermodynamically feasible.

Hence, we choose the one with the least cost ($/kg NH$_3$ removed), namely the resin. The annual operating cost for removing ammonia using the resin is:

[6.14] $$5 \frac{\text{kg Liquid}}{\text{s}} * (14.0*10^{-6} - 11.6*10^{-6}) \frac{\text{kg } NH_3}{\text{kg Liquid}} * 91 \frac{\$}{\text{kg } NH_3} *$$
$$3600*8760 \frac{\text{s}}{\text{yr}} = \$34,437/\text{yr}.$$

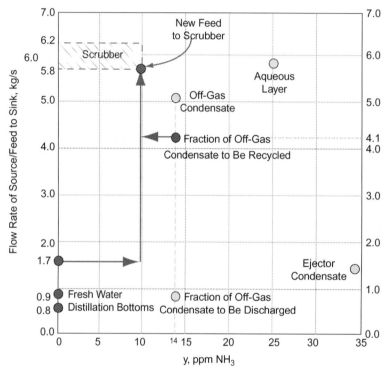

FIGURE 6-15 Direct-recycle strategies for the AN example.

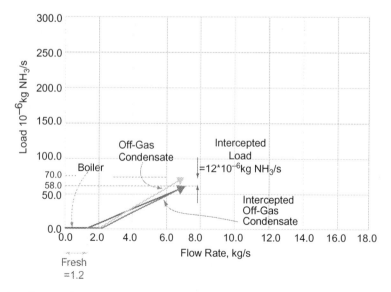

FIGURE 6-16 Intercepting the off-gas condensate.

(*Continued*)

Example 6-1   Application of mass integration to debottleneck an acrylonitrile process and reduce water usage and discharge (El-Halwagi, 1997) (*Continued*)

The annualized fixed cost of the adsorption column along with its ancillary equipment (for example, regeneration, materials handling, and so on) is estimated to be about $85,000/yr. Therefore, the TAC for the interception system is $119,437/yr.

As a result of minor process modification, segregation, interception, and recycle, we have eliminated the use of fresh water in the scrubber, leading to a reduction in fresh-water consumption (and influent to biotreatment) by 6.0 kg/s. Therefore, the target for segregation, interception, and recycle has been realized. Figure 6.19 shows these results.

Next, we focus our attention on sink/generator manipulation to remove fresh-water consumption in the steam-jet ejector. The challenge here is to alter the design and/or operation of the boiler, the ejector, or the distillation column to reduce or eliminate the use of steam. Several solutions may be proposed including:

- Replacing the steam-jet ejector with a vacuum pump. The distillation operation will not be affected. The operating cost of the ejector and the vacuum pump are comparable. However, a capital investment of $75,000 is needed to purchase the pump. For a five-year linear depreciation with negligible salvage value, the annualized fixed cost of the pump is $15,000/year.
- Operating the column under atmospheric pressure, thereby eliminating the need for the vacuum pump. Here a simulation study is needed to examine the effect of pressure change.
- Relaxing the requirement on BFW quality to a few parts per million of ammonia and AN. In this case, recycle and interception techniques can be used to significantly reduce the fresh-water feed to the boiler and, consequently, the net wastewater generated.

Figure 6.20 illustrates the revised flow sheet with segregation, interception, recycle, and sink/generator manipulation. As can be seen from the figure, the flow rate of the terminal wastewater stream has been reduced to 4.8 kg H_2O/s. This is exactly the same *target* predicted in Fig. 6.5b. To refine the material balance throughout the plant, a simulation study is needed. This is an effective use of simulation in coordination with process synthesis and integration.

Figures 6.21 and 6.22 are impact diagrams (sometimes referred to as Pareto charts) for the reduction in wastewater and the associated TAC. These figures illustrate the cumulative impact of the identified strategies on biotreatment influent and cost.

We are now in a position to discuss the merits of the identified solutions. As can be inferred from Fig. 6.20, the following benefits can be achieved:

- Acrylonitrile production has increased from 3.9 kg/s to 4.6 kg/s, which corresponds to an 18 percent yield enhancement for the plant. This production increase is a result of better allocation of process streams: the essence of mass integration. For a selling value of $0.6/kg of AN, the additional production of 0.7 kg AN/s can provide an annual revenue of $13.3 million/yr!

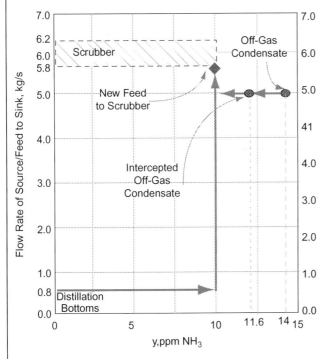

FIGURE 6-17 Determination of interception task for the off-gas condensate.

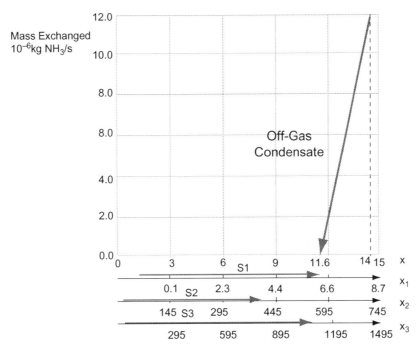

FIGURE 6-18 The mass-exchange pinch diagram for the AN case study.

(*Continued*)

Example 6-1 Application of mass integration to debottleneck an acrylonitrile process and reduce water usage and discharge (El-Halwagi, 1997) (*Continued*)

- Fresh-water usage and influent to the biotreatment facility are decreased by 7.2 kg/s. The value of fresh water and the avoidance of treatment cost are additional benefits.
- Influent to biotreatment is reduced to 40 percent of current level. Therefore, the plant production can be expanded 2.5 times the current capacity before the biotreatment facility is debottlenecked again.

Clearly, this is a superior solution to the installation of an additional biotreatment facility.

It is instructive to draw some conclusions from this case study and emphasize the design philosophy of mass integration. First, the target for debottlenecking the biotreatment facility was determined ahead of design. Then, systematic tools were used to generate optimal solutions that realized the target. Next, an analysis study for performance was needed to refine the results. This is an efficient approach to understanding the global insights of the process, setting performance targets, realizing these targets, and saving time and effort by focusing on the big picture first and dealing with the details later. This is a fundamentally different approach than using the designer's subjective decisions to alter the process and check the consequences using detailed analysis. It is also different from using simple end-of-pipe treatment solutions. Instead, the various species are optimally allocated throughout the process. Therefore, objectives such as yield enhancement, pollution prevention, and cost savings can be simultaneously addressed. Indeed, pollution prevention (when undertaken with the proper techniques) can be a source of profit for the company, not an economic burden.

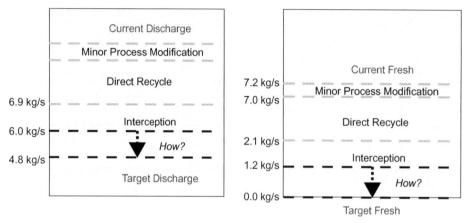

FIGURE 6-19 Reduction in water usage and discharge (expressed as kg H_2O/s) after minor process modification, direct recycle, and interception.

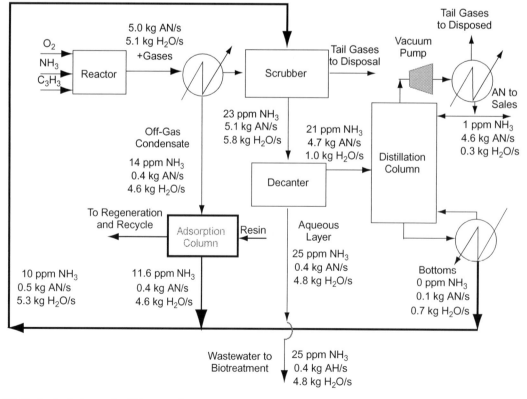

FIGURE 6-20 Optimal solution to the AN case study.

(*Continued*)

Example 6-1 Application of mass integration to debottleneck an acrylonitrile process and reduce water usage and discharge (El-Halwagi, 1997) (*Continued*)

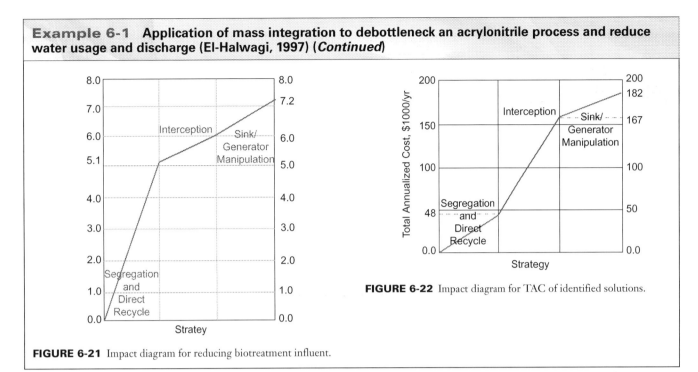

FIGURE 6-21 Impact diagram for reducing biotreatment influent.

FIGURE 6-22 Impact diagram for TAC of identified solutions.

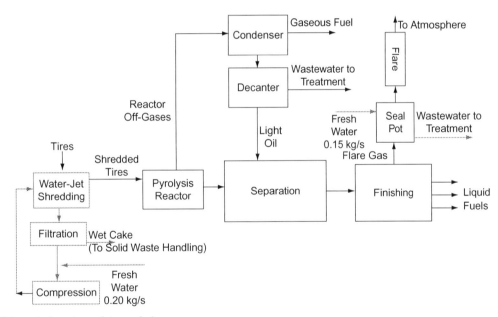

FIGURE 6-23 Schematic flow sheet of tire-to-fuel process.

HOMEWORK PROBLEMS

6.1. Let us revisit the tire-to-fuel process described in Problem 5.1. Figure 6.23 is a more detailed flow sheet. Tire shredding is achieved by using high-pressure water jets. The shredded tires are fed to the process while the spent water is filtered. The wet cake collected from the filtration system is forwarded to solid-waste handling. The filtrate is mixed with 0.20 kg/s of fresh-water makeup to compensate for water losses with the wet cake (0.08 kg water/s) and the shredded tires (0.12 kg water/s). The mixture of filtrate and water makeup is fed to a high-pressure compression station for recycle to the shredding unit. Due to the pyrolysis reactions, 0.08 kg water/s are generated.

The plant has two primary sources for wastewater: the decanter (0.20 kg water/s) and the seal pot (0.15 kg/s). The plant has been shipping the wastewater for off-site treatment. The cost of wastewater transportation and treatment is $0.01/kg leading to a wastewater treatment cost of approximately $110,000/yr. The plant wishes to stop off-site treatment of wastewater to avoid cost of off-site treatment ($110,000/yr) and alleviate legal-liability concerns in case of transportation accidents or

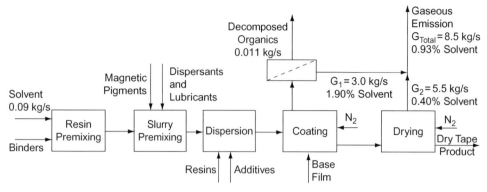

FIGURE 6-24 Schematic representation of a magnetic tape manufacturing process. *Source: Dunn et al. (1995).*

TABLE 6-4 Data for the MSAs

Stream	Upper Bound on Flow Rate L_j^c kg/s	Supply Composition (Mass Fraction) x_j^s	Target Composition (Mass Fraction) x_j^t	m_j	ε_j (Mass Fraction)	C_j $/kg MSA
S_1	∞	0.014	0.040	0.4	0.001	0.002
S_2	∞	0.020	0.080	1.5	0.001	0.001
S_3	∞	0.001	0.010	0.1	0.001	0.002

inadequate treatment of the wastewater. The objective of this problem is to eliminate or reduce off-site wastewater treatment to the extent feasible. For capital budget authorization, the plant has the following economic criterion:

[6.15]
$$\text{Payback period} = \frac{\text{Fixed capital investment}}{\text{Annual savings}} \leq 3 \text{ years}$$

where

Annual savings = Annual avoided cost of off-site treatment − Annual operating cost of on-site system

In addition to the information provided by Problem 3.1, the following data are available:

Economic Data

- Fixed cost of extraction system associated with S_2, $ = 120,000$ (Flow rate of wastewater, kg/s)$^{0.58}$
- Fixed cost of adsorption system associated with S_3, $ = 790,000$ (Flow rate of wastewater, kg/s)$^{0.70}$
- Fixed cost of stripping system associated with S_4, $ = 270,000$ (Flow rate of wastewater, kg/s)$^{0.65}$
- A biotreatment facility that can handle 0.35 kg/s wastewater has a fixed cost of $240,000 and an annual operating cost of $60,000/yr.

Technical Data

Water may be recycled to two sinks: the seal pot and the water-jet compression station. The following constraints on flow rate and composition of the pollutant (heavy organic) should be satisfied:

Seal Pot

- $0.10 \leq$ Flow rate of feed water (kg/s) ≤ 0.20
- $0 \leq$ Pollutant content of feed water (ppmw) ≤ 500

Makeup to Water-Jet Compression Station

- $0.18 \leq$ Flow rate of makeup water (kg/s) ≤ 0.20
- $0 \leq$ Pollutant content of makeup water (ppmw) ≤ 50

6.2. Consider the magnetic-tape manufacturing process previously described by Problem 4.7. In this process (Dunn et al., 1995), several binders used in coating are dissolved in 0.09 kg/s of organic solvent. The mixture is additionally mixed and suspended with magnetic pigments, dispersants, and lubricants. A base file is coated with the suspension. Nitrogen gas is used for blanketing and to aid in evaporation of the solvent. As a result of thermal decomposition in the coating chamber, 0.011 kg/s of solvent are decomposed into other organic species. The decomposed organics are separated from the exhaust gas in a membrane unit. The reject stream leaving the membrane unit has a flow rate of 3.0 kg/s and is primarily composed of nitrogen, which contains 1.9 wt/wt% of the organic solvent. The coated film is passed to a dryer where nitrogen gas is employed to evaporate the remaining solvent. The exhaust gas leaving the dryer has a flow rate of 5.5 kg/s and contains 0.4 wt/wt% solvent. The two exhaust gases are mixed and disposed of. Figure 6.24 is a schematic representation of the process.

To recover the solvent from the exhaust gases, there are three candidate MSAs. The equilibrium data for the transfer of the organic solvent to the jth lean stream is given by $y = m_j x_j$, where the values of m_j are given in Table 6.4. A minimum allowable composition difference of 0.001(kg organic solvent)/(kg MSA) is to be used.

The annualized fixed cost of a mass exchanger, \$/yr, maybe approximated by 18,000 (gas flow rate, kg/s)$^{0.65}$. The value of the recovered solvent is \$0.80/kg of organic solvent.

In addition to the environmental problem, there is also an economic incentive to recover the solvent from the exhaust gases. The value of 0.09 kg/s of solvent is approximately \$2.3 million/yr.

The objective of this problem is to develop a minimum-cost solution that minimizes the usage of fresh solvent in the process. The solution strategies may include segregation, mixing, recycle, and interception. To assess the economic potential of the solution, evaluate the payback period for your solution.

REFERENCES

Dunn RF, El-Halwagi MM, Lakin J, Serageldin M: Selection of organic solvent blends for environmental compliance in the coating industries. In Griffith ED, Kahn H, Cousins MC, editors: *Proceedings of the first international plant operations and design conference, AIChE*, vol III, New York, pp 83–107, 1995.

El-Halwagi MM: *Pollution prevention through process integration*, San Diego, Academic Press, 1997.

El-Halwagi MM: Sustainable pollution prevention through mass integration. In Sikdar S, Diwekar U, editors: *Tools and methods for pollution prevention*, Kluwer pub, pp 233–275, 1999.

El-Halwagi MM, Hamad AA, Garrison GW: Synthesis of waste interception and allocation networks, *AIChE J* 42(11):3087–3101, 1996.

El-Halwagi MM, Spriggs HD: Solve design puzzles with mass integration, *Chem. Eng. Prog.* vol. 94(August):25–44, 1998.

CHAPTER 7
HEAT INTEGRATION

Heating and cooling are among the most common operations in the process industries. Significant quantities of external heating and cooling utilities are used to drive reaction, induce separation, and render units and streams in desirable states of operation. To satisfy the need for heating and cooling, industrial operations require the extensive use of different forms of energy. For instance, fossil fuels are widely used as feedstocks of boilers to generate heating media (for example, steam, heating oil) or to drive power cycles that drive refrigeration systems. The excessive usage of external heating utilities incurs a substantial economic burden in the form of operating costs, depletes unsustainable energy resources (for example, fossil fuels), and generates large quantities of greenhouse gas emissions. Consequently, considerable attention has been given to the problem of conserving energy through heat integration and the synthesis of heat-exchange networks (HENs). The basic idea for heat integration is that there are process streams and units that need to be heated and other process streams and units that need to be cooled. Before using external utilities to provide the necessary heating and cooling, heat integration seeks to transfer the heat from the process hot streams and units to the process cold streams and units. The remaining heating and cooling tasks are then fulfilled using the external heating and cooling utilities. This chapter presents the key tools used in identifying rigorous targets for minimum heating and cooling utilities, selecting utilities, and synthesizing HENs that attain the desired targets.

HEN-SYNTHESIS PROBLEM STATEMENT

In defining the HEN-synthesis problem, it is useful to discuss the necessary information required to set up the heat-integration problem. The following data should be extracted from the process:

- Which streams and units are to be heated (the cold streams and heat sinks)? What is the heating duty required for each one? What is the temperature change associated with each cold stream and unit?
- Which streams and units are to be cooled (the hot streams and heat sources)? What is the cooling duty required for each one? What is the temperature change associated with each hot stream and unit?
- What are the available types of external heating and cooling utilities? What are their characteristics?

One version of the HEN-synthesis problem can be stated as follows:

Given a number N_H of process hot streams (to be cooled) and a number N_C of process cold streams (to be

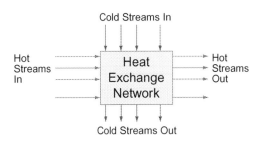

FIGURE 7-1 Synthesis of HENs.

heated), it is desired to synthesize a cost-effective network of heat exchangers that can transfer heat from the hot streams to the cold streams. Given also are the heat capacity (flow rate x specific heat) of each process hot stream, $FC_{P,u}$; its supply (inlet) temperature, T_u^s; and its target (outlet) temperature, T_u^t, where $u = 1, 2,..., N_H$. In addition, the heat capacity, $fc_{P,v}$, and supply and target temperatures, t_v^s and t_v^t, are given for each process cold stream, where $v = 1, 2,..., N_C$.[1] Available for service are N_{HU} heating utilities and N_{CU} cooling utilities whose supply and target temperatures (but not flow rates) are known. In this version of the HEN-synthesis problem statement, focus is given to hot and cold streams with changes in sensible heat. Figure 7.1 is a schematic representation of the HEN problem statement.

For a given system, the synthesis of HENs entails answering several questions:

- Which heating/cooling utilities should be employed?
- What is the optimal heat load to be removed/added by each utility?
- How should the hot and cold streams be matched (that is, stream pairings)?
- What is the optimal system configuration (sequencing of stream and heat exchangers)?

Numerous methods have been developed for the synthesis of HENs. These methods have been reviewed by Foo et al. (2011), Rossiter (2010), Majozi (2010), Kemp (2009), Smith (2005), Furman and Sahinidis (2002), Shenoy (1995), Linnhoff (1993), Gundersen and Naess (1988), Douglas (1988), Linnhoff and Hindmarsh (1983), and Linnhoff et al. (1982). One of the most powerful aspects in synthesizing HENs is the identification of minimum utility targets ahead of

---

[1] In this version of the HEN-synthesis problem, focus is given to the exchange of sensible heat among process streams. Heat sources and sinks can be represented through surrogate streams with the flow rate*heat capacity calculated by dividing the heat duty of the unit by the difference in temperature. For cases when there is no change in temperature (for example, latent heat for streams or isothermal operation of units), an approximation may be made to allow for a small change in the temperature (for example, one-degree difference between supply and target temperatures).

TABLE 7-1 Analogy between MENs and HENs

MENs	HENs
Transferred commodity: Mass	Transferred commodity: Heat
Donors: Rich streams	Donors: Hot streams
Recipient: Lean streams	Recipient: Cold streams
Rich composition: y	Hot temperature: T
Lean composition: x	Cold temperature: t
Slope of equilibrium: m	Slope of equilibrium: 1
Intercept of equilibrium: b	Intercept of equilibrium: 0
Driving force: ε	Driving force: ΔT^{min}

Source: El-Halwagi (2006).

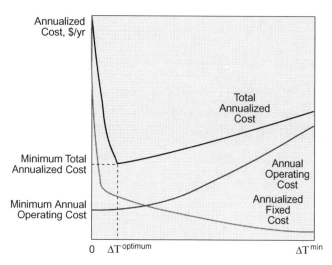

FIGURE 7-2 Role of minimum approach temperature.

designing the network. The following sections present graphical and algebraic methods of determining minimum-utility targets.

MINIMUM UTILITY TARGETS VIA THE THERMAL PINCH DIAGRAM

Consider a heat-exchange operation that transfers heat from a hot stream to a cold stream. An important boundary is reaching thermal equilibrium, which is simply attained when the temperatures of the hot and the cold stream are equal; that is,

$$[7.1] \qquad T = t$$

where T is the temperature of the hot stream and t is the temperature of the cold stream. Thermal equilibrium requires an infinitely large heat-transfer area. For the heat exchanger to be practically feasible, a minimum heat-exchange driving force (referred to as ΔT^{min}) should be employed. This minimum driving force can be used to establish a one-to-one correspondence between the temperatures of the hot and the cold streams for which heat transfer is feasible; that is,

$$[7.2] \qquad T = t + \Delta T^{min}$$

This expression ensures that the heat-transfer considerations of the second law of thermodynamics are satisfied. For a given pair of corresponding temperatures (T, t), it is thermodynamically and practically feasible to transfer heat from any hot stream whose temperature is greater than or equal to T for any cold stream whose temperature is less than or equal to t. It is worth noting the analogy between Equations 7.2 and 3.5. Thermal equilibrium is a special case of mass-exchange equilibrium with T, t, and ΔT^{min} corresponding to y_i, x_j, and ε_j, respectively, whereas the values of m_j and b_j are one and zero, respectively. Table 7.1 summarizes the analogous terms in the synthesis of mass- and heat-exchange networks. Similar to the role of ε_j in cost optimization of MENs, ΔT^{min} can be used to trade off capital versus operating costs in HENs as shown in Fig. 7.2. When ΔT^{min} approaches zero, the heat-transfer area approaches infinity and, consequently, the fixed cost tends to infinity. As the heat-transfer driving force increases, the fixed cost decreases but the operating cost increases. Such effects lead to the nonmonotonic behavior shown by Fig. 7.2, which can be used to determine an optimal driving force. More elaborate techniques can be used by assigning a different driving force for each stream or pair of streams to trade off capital versus operating costs.

To accomplish the minimum usage of heating and cooling utilities, it is necessary to maximize the heat exchange among process streams. In this context, one can use a very useful graphical technique referred to as the "thermal pinch diagram." This technique is primarily based on the work of Linnhoff and colleagues (for example, Linnhoff and Hindmarsh, 1983) Umeda et al. (1979) and Hohmann (1971). The first step in constructing the thermal pinch diagram is creating a global representation for all the hot streams by plotting the enthalpy exchanged by each process hot stream versus its temperature.[2] Hence, a hot stream losing sensible heat[3] is represented as an arrow whose tail to its supply temperature and its head corresponds to its target temperature. Assuming constant heat capacity over the operating range, the slope of each arrow is equal to $F_u C_{p,u}$. The vertical distance between the tail and the head of each arrow represents the enthalpy lost by that hot stream according to the following expression:

Heat lost from the uth hot stream
$$HH_u = F_u C_{p,u}(T_u^s - T_u^t)$$

where

$$[7.3] \qquad u = 1, 2, \ldots, N_H$$

---

[2] In most HEN literature, the temperature is plotted versus the enthalpy. However, in this chapter, enthalpy is plotted versus temperature to draw the analogy with MEN synthesis. Furthermore, when there is a strong interaction between mass and energy objectives, the enthalpy expressions become nonlinear functions of temperature. In such cases, it is easier to represent enthalpy as a function of temperature.

[3] Whenever there is a change in phase, the latent heat should also be included.

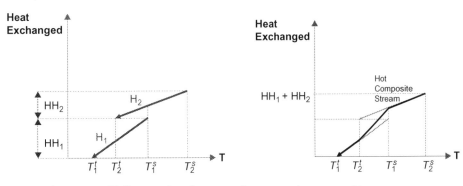

FIGURE 7-3 (a) Representing hot streams. (b) Constructing a hot composite stream using superposition.

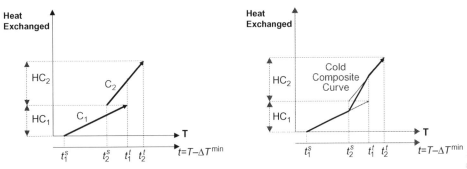

FIGURE 7.4 (a) Representing cold streams. (b) Constructing a cold composite stream using superposition (dashed line represents composite line).

Note that any stream can be moved up or down while preserving the same vertical distance between the arrow head and tail and maintaining the same supply and target temperatures. Similar to the graphical superposition described in Chapter 4, one can create a hot composite stream using the diagonal rule. Figures 7.3a, b illustrate this concept for two hot streams (dashed line represents composite line).

Next, a cold-temperature scale, t, is created in one-to-one correspondence with the hot temperature scale, T, using Eq. 7.2. The enthalpy of each cold stream is plotted versus the cold temperature scale, t. The vertical distance between the arrow head and tail for a cold stream is given by

Heat gained by the vth cold stream
$$HC_v = f_v c_{p,v}(t_v^t - t_v^s)$$

where

[7.4] $\qquad v = 1, 2, \ldots, N_C$

In a similar manner to constructing the hot-composite line, a cold composite stream is plotted (see Fig. 7.4a, b for an example with two cold streams).

Next, both composite streams are plotted on the same diagram (Fig. 7.5). On this diagram, thermodynamic feasibility of heat exchange is guaranteed when, at any heat-exchange level (which corresponds to a horizontal line), the temperature of the cold composite stream is located to the left of the hot composite stream (that is, temperature of the hot is higher than or equal to the cold temperature plus the minimum approach temperature). Hence, for a given set of corresponding temperatures, it

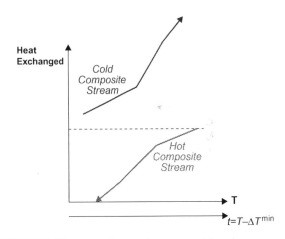

FIGURE 7-5 Placement of composite streams with no heat integration. *Source: El-Halwagi (2006).*

is thermodynamically and practically feasible to transfer heat from any hot stream to any cold stream.

The cold composite stream can be moved up and down, which implies different heat-exchange decisions. For instance, if we move the cold composite stream upward in a way that leaves no horizontal overlap with the hot composite stream, then there is no integrated heat exchange between the hot composite stream and the cold composite stream as seen in Fig. 7.5. When the cold composite stream is moved downward so as to provide some horizontal overlap, some integrated heat exchange can be achieved (Fig. 7.6). However, if the cold composite stream is moved downward such that a portion of the cold is placed to the right of the hot composite stream, heat transfer from hot

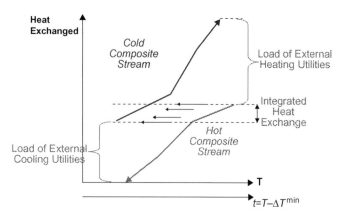

FIGURE 7-6 Partial heat integration. *Source: El-Halwagi (2006).*

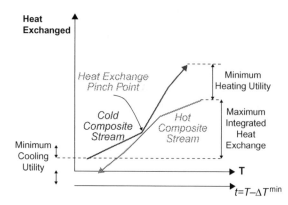

FIGURE 7-8 The thermal pinch diagram.

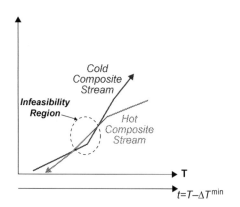

FIGURE 7-7 Infeasible heat integration. *Source: El-Halwagi (2006).*

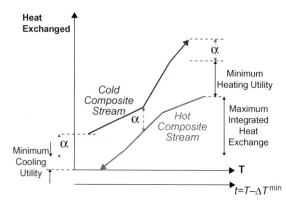

FIGURE 7-9 Penalties associated with passing heat through the pinch. *Source: El-Halwagi (2006).*

to cold becomes infeasible (Fig. 7.7). Therefore, the optimal situation is constructed when the cold composite stream is slid vertically until it touches the rich composite stream while lying completely to the left of the hot composite stream at any horizontal level. Therefore, the cold composite stream can be slid down until it touches the hot composite stream. The point where the two composite streams touch is called the "thermal pinch point." As Fig. 7.8 shows, one can use the pinch diagram to determine the minimum heating and cooling utility requirements. Again, the cold composite line cannot be slid down any further; otherwise, portions of the cold composite stream would be to the right of the hot composite stream, causing thermodynamic infeasibility. On the other hand, if the cold composite stream is moved up (that is, passing heat through the pinch), less heat integration is possible, and consequently, additional heating and cooling utilities are required. Therefore, for a minimum utility usage, the following design rules must be observed:

- No heat should be passed through the pinch.
- Above the pinch, no cooling utilities should be used.
- Below the pinch, no heating utilities should be used.

Figure 7.9 illustrates the first rule. The passage of a heat flow through the pinch (α) results in a double penalty: an increase of α in both heating utility and cooling utility. The second and third rules can be explained by noting that above the pinch there is a surplus of cooling capacity. Adding a cooling utility above the pinch will replace a load that can be removed (virtually for no operating cost) by a process cold stream. A similar argument can be made against using a heating utility below the pinch.

Example 7-1 Pharmaceutical facility

Consider the pharmaceutical processing facility illustrated in Fig. 7.10. The feed mixture (C_1) is first heated to 550 K, and then fed to an adiabatic reactor where an endothermic reaction takes place. The off-gases leaving the reactor (H_1) at 520 K are cooled to 330 K prior to being forwarded to the recovery unit. The mixture leaving the bottom of the reactor is separated into a vapor fraction and a slurry fraction. The vapor fraction (H_2) exits the separation unit at 380 K and is to be cooled to 300 K prior to storage. The slurry fraction is washed with a hot immiscible liquid at 380 K. The wash liquid is purified and recycled to the washing unit. During purification, the temperature drops to 320 K. Therefore, the recycled liquid (C_2) is heated to 380 K. Two utilities are available for service: HU_1 and CU_1. The cost of the heating and cooling utilities (in $/10^6$ kJ) are 6 and 8, respectively. Stream data are given in Table 7.2.

In the current operation, the heat exchange duties of H_1, H_2, C_1, and C_2 are fulfilled using the cooling and heating utilities. Therefore, the current annual operating cost of utilities is

(Continued)

Example 7-1 Pharmaceutical facility (*Continued*)

$$[(4750 + 120)kW \times 6 \times 10^{-6}\frac{\$}{kJ} + (1900 + 400)kW \times$$
$$8 \times 10^{-6}\frac{\$}{kJ}]3600 \times 8760\frac{s}{yr} = \$1,501,744/yr.$$

The objective of this case study is to use heat integration via the pinch diagram to reduce this operating cost. A value of $\Delta T^{min} = 10\,K$ is used.

SOLUTION

Figures 7.11 through 7.13 illustrate the hot composite stream, the cold composite stream, and the pinch diagram, respectively. As can be seen from Fig. 7.13, the two composite streams touch at 310 K on the hot scale (300 K on the cold scale). The minimum heating and cooling utilities are 2620 kW and 50 kW, respectively, leading to an annual operating cost of

$$(2620\,kW \times 6 \times 10^{-6}\frac{\$}{kJ} + 50\,kW \times 8 \times 10^{-6}\frac{\$}{kJ})3600 \times$$
$$8760\frac{s}{yr} = \$508,360/yr$$

This is only 34 percent of the operating cost prior to heat integration. Once the minimum operating cost is determined, a network of heat exchangers can be synthesized. The network synthesis will be described later in this chapter. The trade-off between capital and operating costs can be established by iteratively varying ΔT^{min} until the minimum total annualized cost is attained.

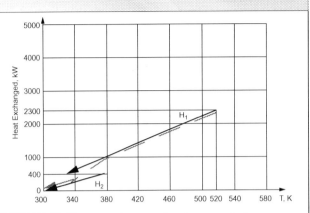

FIGURE 7-11 Hot composite stream for pharmaceutical process.

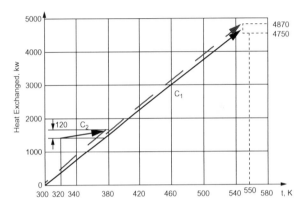

FIGURE 7-12 Cold composite stream for pharmaceutical process.

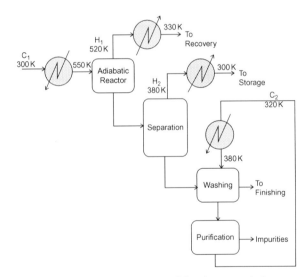

FIGURE 7-10 Simplified flow sheet of the pharmaceutical process.

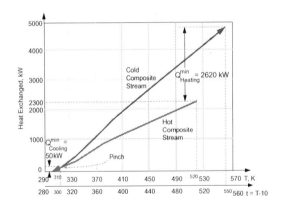

FIGURE 7-13 Thermal pinch diagram for the pharmaceutical process.

TABLE 7-2 Stream Data for Pharmaceutical Process

Stream	Flow Rate × Specific Heat, kW/K	Supply Temperature, K	Target Temperature, K	Enthalpy Change, kW
H_1	10	520	330	−1900
H_2	5	380	300	−400
HU_1	?	560	520	?
C_1	19	300	550	4750
C_2	2	320	380	120
CU_1	?	290	300	?

MINIMUM UTILITY TARGETS USING THE ALGEBRAIC CASCADE DIAGRAM

Notwithstanding the insights obtained by the graphical thermal pinch diagram, in some cases, it is desirable to use an algebraic approach. Examples include problems with numerous streams with much overlapping that can render the construction of the composite curves cumbersome. Another example is the case with temperature ranges that are vastly different to the extent that they may skew the representation of certain streams. This section presents an algebraic approach referred to as the **cascade diagram**, which is based on the *problem-table algorithm* developed by Linnhoff and Flower (1978). The first step is the construction of a **temperature-interval diagram** (TID), which is a useful tool for ensuring thermodynamic feasibility of heat exchange. Two corresponding temperature scales are generated: hot and cold. The scale correspondence is determined using Eq. 7.2. Each stream is represented as a vertical arrow whose tail corresponds to its supply temperature, while its head represents its target temperature. Next, horizontal lines are drawn at the heads and tails of the arrows. These horizontal lines define a series of temperature intervals $z = 1, 2, \ldots, n_{int}$. Within any interval, it is thermodynamically feasible to transfer heat from the hot streams to the cold streams. It is also feasible to transfer heat from a hot stream in an interval z to any cold stream that lies in an interval below it.

Next, we construct a table of exchangeable heat loads (TEHL) to determine the heat-exchange loads of the process streams in each temperature interval. The exchangeable load of the uth hot stream (losing sensible heat) that passes through the zth interval is defined as

$$[7.5] \qquad HH_{u,z} = F_u C_{p,u}(T_{z-1} - T_z)$$

where T_{z-1} and T_z are the hot-scale temperatures at the top and the bottom lines defining the zth interval. On the other hand, the exchangeable capacity of the vth cold stream (gaining sensible heat) that passes through the zth interval is computed through

$$[7.6] \qquad HC_{v,z} = f_v c_{p,v}(t_{z-1} - t_z)$$

where t_{z-1} and t_z are the cold-scale temperatures at the top and the bottom lines defining the zth interval.

Having determined the individual heating loads and cooling capacities of all process streams for all temperature intervals, one can also obtain the collective loads (capacities) of the hot (cold) process streams. The collective load of hot process streams within the zth interval is calculated by summing up the individual loads of the hot process streams that pass through that interval; that is,

$$[7.7] \qquad HH_z^{Total} = \sum_{\substack{u \text{ passes through interval } z \\ \text{where } u=1,2,\ldots,N_H}} HH_{u,z}$$

Similarly, the collective cooling capacity of the cold process streams within the zth interval is evaluated as follows:

$$[7.8] \qquad HC_z^{Total} = \sum_{\substack{v \text{ passes through interval } z \\ \text{and } v=1,2,\ldots,N_C}} HC_{v,z}$$

As has been mentioned earlier, within each temperature interval, it is thermodynamically as well as technically feasible to transfer heat from a hot process stream to a cold process stream. Moreover, it is feasible to pass heat from a hot process stream in an interval to any cold process stream in a lower interval. Hence, for the zth temperature interval, one can write the following heat-balance equation:

$$[7.9] \qquad r_z = HH_z^{Total} - HC_z^{Total} + r_{z-1}$$

where r_{z-1} and r_z are the residual heats entering and leaving the zth interval. Figure 7.14 illustrates the heat balance around the zth temperature interval.

In this equation, r_0 is zero, because no process streams exist above the first interval. In addition, thermodynamic feasibility is ensured when all the r_zs are nonnegative. Hence, a negative r_z indicates that residual heat is flowing upward, which is thermodynamically infeasible. All negative residual heats can be made nonnegative if a hot load equal to the most negative r_z is added to the problem. This load is referred to as the minimum heating utility requirement, $Q_{Heating}^{min}$. Once this hot load is added, the cascade diagram is revised. A zero residual heat designates the thermal pinch location. The load leaving the last temperature interval is the minimum cooling utility requirement, $Q_{Cooling}^{min}$.

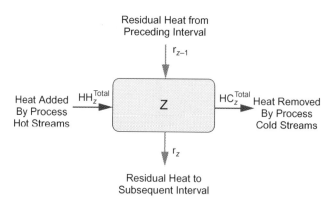

FIGURE 7-14 Heat balance around a temperature interval.

Example 7-2 Revisiting the case study on the pharmaceutical facility using the algebraic cascade diagram

We now solve the pharmaceutical case study described earlier using the algebraic cascade diagram. The first step is the construction of the TID (Fig. 7.15). Next, the TEHLs for the process hot and cold streams are developed (Tables 7.3 and 7.4). Figures 7.16 and 7.17 show the cascade diagram calculations. The results obtained from the revised cascade diagram are identical to those obtained using the graphical pinch approach.

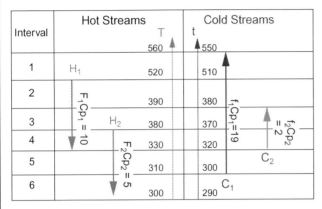

FIGURE 7-15 Temperature interval diagram for pharmaceutical case study.

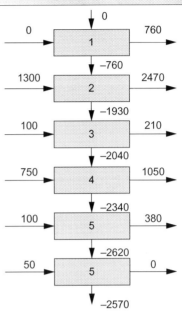

FIGURE 7-16 Cascade diagram for the pharmaceutical case study.

TABLE 7-3 TEHL for Process Hot Streams

Interval	Load of H_1 (kW)	Load of H_2 (kW)	Total Load (kW)
1	–	–	–
2	1300	–	1300
3	100	–	100
4	500	250	750
5	–	100	100
6	–	50	50

TABLE 7-4 TEHL for Process Cold Streams

Interval	Capacity of C_1 (kW)	Capacity of C_2 (kW)	Total Capacity (kW)
1	760	–	760
2	2470	–	2470
3	190	20	210
4	950	100	1050
5	380	–	380
6	–	–	–

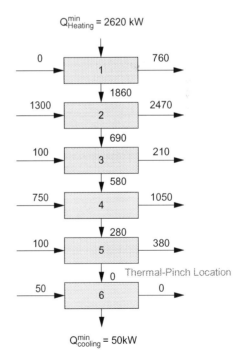

FIGURE 7-17 Revised cascade diagram for the pharmaceutical case study.

(Continued)

Example 7-2 Revisiting the case study on the pharmaceutical facility using the algebraic cascade diagram (*Continued*)

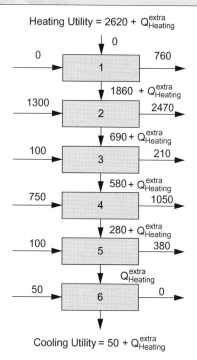

As mentioned earlier, for minimum utility usage, no heat should be passed through the pinch. Let us illustrate this point using the cascade diagram. Suppose that we use $Q_{Heating}^{extra}$ kW more than the minimum heating utility. As can be seen from Fig. 7.18, this additional heating utility passes down through the cascade diagram in the form of an increased residual heat load. At the pinch, the residual load becomes $Q_{Heating}^{extra}$. The net effect is not only an increase in the heating utility load, but also an equivalent increase in the cooling utility load.

FIGURE 7-18 Consequences of passing heat through the pinch.

SCREENING OF MULTIPLE UTILITIES USING THE GRAND COMPOSITE REPRESENTATION

The aforementioned graphical thermal pinch diagram and its numerical analog, the cascade diagram, identify the minimum heating and cooling utilities. These targets are determined without specifying the type of utilities to be used or whether several heating or cooling utilities are needed. In many cases, multiple utilities are available for service. These utilities must be screened so as to determine which one(s) should be used and the task of each utility. To minimize the cost of utilities, it may be necessary to stage the use of utilities such that at each level the use of the cheapest utility ($/kJ) is maximized while ensuring its feasibility. A convenient way of screening multiple utilities is the grand composite curve (GCC).

The GCC may be directly constructed from the cascade diagram (Linnhoff et al., 1982). To illustrate the procedure for constructing the GCC, let us consider the cascade diagram shown in Fig. 7.19a. The residual heat loads are shown leaving the temperature intervals. Suppose that r_4 is the most negative residual. As mentioned previously, this infeasibility and all other infeasibilities are removed by adding the absolute value of r_4 to the top of the cascade diagram. This value is also the minimum heating utility. The residual loads are recalculated with the load leaving the last temperature interval being the minimum cold utility as illustrated by Fig. 7.19b.

Each residual heat corresponds to a hot temperature and a cold temperature. To have a single-temperature representation, we use an adjusted temperature scale that is calculated as the arithmetic average of the hot and the cold temperature; that is,

[7.10] $$\text{Adjusted temperature} = \frac{T + t}{2}$$

Given the relationship between the hot and cold temperature (described by Eq. 7.2), we get:

[7.11a] $$\text{Adjusted temperature} = T - \frac{\Delta T^{min}}{2}$$

[7.11b] $$= t + \frac{\Delta T^{min}}{2}$$

Next, we represent the adjusted temperature versus the residual enthalpy as shown in Fig. 7.20a. This representation is the GCC. The pinch point corresponds to the zero-residual point. Additionally, the top and bottom residuals represent the minimum heating and cooling utilities. The question is how to distribute these loads over the multiple utilities. Any time, the enthalpy representation is given by a line drawn from left to right; it corresponds to a surplus of heat in that interval. Conversely,

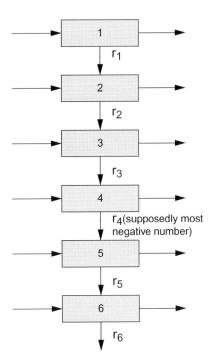

FIGURE. 7-19A Cascade diagram.

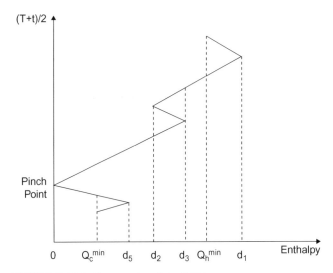

FIGURE 7-20A Construction of the GCC.

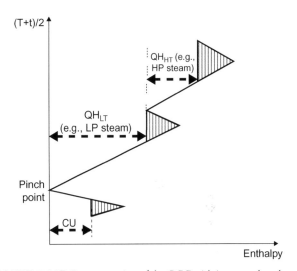

FIGURE 7-20B Representation of the GCC with integrated pockets and optimal placement of utilities.

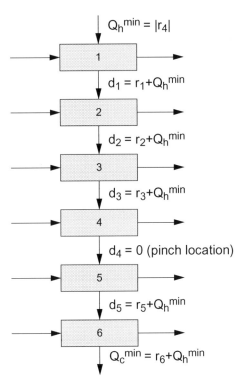

FIGURE 7-19B Revised cascade diagram.

when an enthalpy line is drawn from right to left, it corresponds to deficiency in heat in that interval. A heat surplus may be used to satisfy a heat residual below it. Therefore, the shaded regions (referred to as "pockets")

shown in Fig. 7.20b are fully integrated by transferring heat from process hot streams to process cold streams. Then, we represent each utility based on its temperature. The adjusted temperature of a heating utility is given by Eq. 7.11a whereas that for a cooling utility is given by Eq. 7.11b. We start with the cheapest utility and maximize its use by filling the enthalpy gap (deficiency) at that level. Then, we move up for heating utilities and down for cooling utilities and continue to fill the enthalpy gaps by the cheapest utility at that level. Figure 7.20b is an illustration of this concept by screening low- and high-pressure steam where the low-pressure steam is cheaper ($/kJ) than the high-pressure steam. It is worth noting that the sum of the heating loads of the low- and high-pressure steams is equal to the minimum heating utility (the value of the top heat residual).

Example 7-3 Using the GCC for utility selection (El-Halwagi, 2006)

TABLE 7-5 Stream Data for Example 7.3

Stream	Flow Rate × Specific Heat MMBtu/hr, °F	Supply Temperature, °F	Target Temperature, °F	Enthalpy Change MMBtu/hr
H_1	0.5	650	150	−250.0
H_2	2.0	550	500	−100.0
C_1	0.9	490	640	135.0
C_2	1.5	360	490	195.0

Source: El-Halwagi (2006).

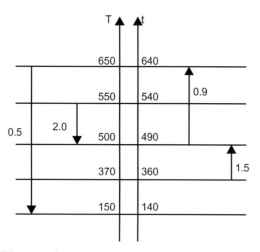

FIGURE 7-21 The temperature interval diagram for Example 7.3. *Source: El-Halwagi (2006).*

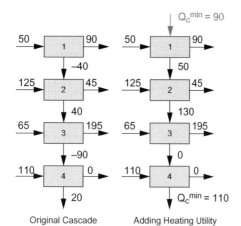

FIGURE 7-22 The cascade and revised cascade diagrams for Example 7.3. *Source: El-Halwagi (2006).*

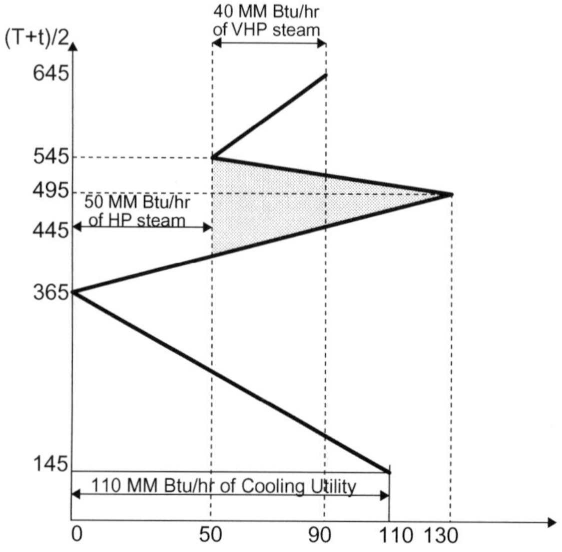

FIGURE 7-23 The GCC for Example 7.3. *Source: El-Halwagi (2006).*

Consider the stream data given in Table 7.5. Available for service are two heating utilities: a high-pressure (HP) steam and a very high-pressure (VHP) steam whose temperatures are 450°F and 660°F, respectively. The VHP steam is more expensive than the HP steam. Also available for service is a cooling utility whose temperature is 100°F. The minimum approach temperature is taken as 10°F. Figures 7.21 through 7.23 represent the temperature-interval diagram, cascade diagram, and the GCC. As can be seen from Fig. 7.22, the minimum heating requirement is 90 MM Btu/hr. To maximize the use of the HP steam, we represent the HP on the GCC (a horizontal line at 450 − (10/2) = 445°F). The deficit below this line is 50 MM Btu/hr. Therefore, the duty of the HP steam is 50 MM Btu/hr, and the rest of the heating requirement (40 MM Btu/hr) will be provided by the VHP steam.

STREAM MATCHING AND THE SYNTHESIS OF HEAT-EXCHANGE NETWORKS

This section presents a systematic method for the matching of hot and cold streams to synthesize a network of heat exchangers that satisfy the identified targets for minimum heating and cooling utilities (the minimum operating cost "MOC"). It is worth recalling the analogy between MENs and HENs, including the matching rules for MENs described in Chapter 5. For HENs featuring minimum-utility usage, there is no heat flow across the pinch. Therefore, it is possible to decompose the synthesis task into two subproblems: one above the pinch and one below the pinch. The target for the minimum number of heat exchangers satisfying the MOC is given by (Hohmann, 1971; Linnhoff et al., 1982):

[7.12] $U_{MOC} = U_{MOC,above\ pinch} + U_{MOC,below\ pinch}$

where

[7.12b] $U_{MOC,above\ pinch} = N_{H,above\ pinch} + N_{C,above\ pinch} - N_{i,above\ pinch}$

and

[7.12c] $U_{MOC,below\ pinch} = N_{H,below\ pinch} + N_{C,below\ pinch} - N_{i,below\ pinch}$.

where $U_{MOC,\ above\ pinch}$ is the number of MOC units above the pinch, $N_{H,above\ pinch}$ is the number of hot streams (including heating utilities) above the pinch, $N_{C,above\ pinch}$ is the number of cold streams (including cooling utilities) below the pinch, $N_{i,above\ pinch}$ is the number of independent problems above the pinch, $U_{MOC,below\ pinch}$ is the number of MOC units below the pinch, $N_{H,below\ pinch}$ is the number of hot streams below the pinch, $N_{C,below\ pinch}$ is the number of cold streams below the pinch, and $N_{i,below\ pinch}$ is the number of independent problems below the pinch. The number of independent problems (above or below the pinch) includes cases such as when a hot and a cold stream have exactly the same load and, therefore, only one heat exchanger is needed to match the two streams. To determine the specific matches satisfying these targets, the design is started at the pinch and moved away according to the following rules (Linnhoff and Hindmarsh, 1983):

STREAM POPULATION RULES FOR MATCHING

Let us first consider the case above the pinch. In an MOC design, any heat exchanger immediately above the pinch will operate with the minimum temperature driving force ΔT^{min} (because the two composites touch at the pinch, then all pinch matches have the minimum driving force at the pinch). Therefore, for each pinch match, at least one cold stream (or branch) has to exist per each hot stream to be able to cool the hot stream down to the pinch temperature (on the hot scale). In other words, for an MOC design, the following inequality must apply at the hot end of the pinch (that is, immediately above the pinch):

[7.13a] $N_{ha} \leq N_{ca}$

where N_{ha} is the number of hot streams or branches immediately above the pinch, and N_{ca} is the number of cold streams or branches immediately above the pinch. If this inequality does not hold for the stream data, one or more of the cold streams will have to be split to maintain the branch(es) of the cold stream(s) at the pinch temperature at the cold inlet of the pinch exchangers.

Conversely, immediately below the pinch, each cold stream must be heated to reach the pinch temperature (on the cold scale). To reach this temperature, there has to be at least one hot stream (or a branch of a hot stream) at the pinch. Because an MOC design does not permit the transfer of heat across the pinch, each cold stream immediately below the pinch will require the existence of at least one hot stream (or branch) at the pinch. Therefore, immediately below the pinch, the following criterion must be satisfied:

[7.13b] $N_{cb} \leq N_{hb}$

where N_{cb} is the number of cold streams or branches immediately below the pinch, and N_{hb} is the number of hot streams or branches immediately below the pinch. Splitting of one or more of the hot streams may be necessary to realize this inequality.

FLOW RATE*SPECIFIC HEAT RULES FOR MATCHING

Consider the heat exchanger shown in Fig. 7.19. The cold end of this exchanger is immediately above the pinch. A heat balance around the exchanger can be written as:

[7.14] $F_u C_{p,u}(T_u^{in} - T^{Pinch}) = f_v c_{p,v}(t_v^{out} - t^{Pinch})$

where the symbols for the inlet and outlet temperatures are shown on Fig. 7.24. At the pinch:

[7.15] $T^{Pinch} = t^{Pinch} + \Delta T^{min}$

Substituting from Eq. 7.15 into Eq. 7.14, we get:

[7.16] $F_u C_{p,u}(T_u^{in} - t^{Pinch} - \Delta T^{min}) = f_v c_{p,v}(t_v^{out} - t^{Pinch})$

For the feasibility of heat transfer at the hot end of the exchanger, the following condition must apply to maintain a minimum temperature driving force:

[7.17] $T_u^{in} \geq t_v^{out} + \Delta T^{min}$

Combining Equations 7.16 and 7.17, we obtain the following inequality:

$$F_u C_{p,u}(t_v^{out} + \Delta T^{min} - t^{Pinch} - \Delta T^{min}) \leq f_v c_{p,v}(t_v^{out} - t^{Pinch})$$

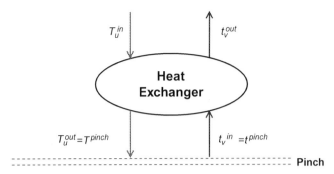

FIGURE 7-24 A heat exchanger immediately above the pinch.

or

[7.18] $F_u C_{p,u} \leq f_v c_{p,v}$ Immediately above the pinch

[7.19] $f_v c_{p,v} \leq F_u C_{p,u}$ Immediately below the pinch

Inequality 7.18 indicates that a required feasibility condition for the match of a hot stream and a cold stream immediately above the pinch is that the product of the flow rate*specific heat for the hot stream (or a branch of a hot stream) must be less than or equal to that for the matched cold stream (or branch of cold stream). If this condition is not satisfied, splitting a hot stream is used to reduce the flow rate of the matched branch of the cold stream so as to meet the feasibility criterion.

Conversely, immediately below the pinch, the following condition can be derived:

Splitting of the cold stream can be used to satisfy this feasibility criterion. Matching the streams is started at the pinch (immediately above and immediately below). Once the pinch matches are identified, it generally becomes a simple task to complete the network design. Moreover, designers always have the freedom to violate these feasibility criteria at the expense of increasing the cost of heating and cooling utilities beyond the MOC requirement. When some of the heat exchangers already exist, retrofitting techniques can be used to include these units in the synthesis procedure (for example, Al-Thubaiti et al., 2008).

Example 7-4 Network synthesis of a heat-exchanger network in a specialty chemical plant

Consider the specialty chemical processing facility shown in Fig. 7.25. The process has two adiabatic reactors. The intermediate product leaving the first reactor (C_1) is heated from 450 K to 510 K before being fed to the second reactor. The gaseous stream leaving the second reactor (H_1) at 520 K is cooled to 380 K and then sent to a disposal unit. The stream leaving the bottom of the second reactor is fed to a separation network. The product stream leaving the finishing, recovery, and separation network (H_2) is cooled from 490 K to 460 K prior to sales. Table 7.6 gives the stream data.

SOLUTION

First, the minimum utility targets are identified through the algebraic approach. The TID, the cascade diagram, and the revised cascade diagram are shown by Figures 7.26, 7.27, and 7.28, respectively. Therefore, the minimum heating and cooling utilities are 15,000 and 11,000 kW, respectively.

To synthesize the HEN that attains the identified utility targets, the problem is decomposed into two subproblems: one above the pinch and one below the pinch. Figure 7.29 shows the streams above the pinch. The number of hot streams immediately above the pinch (H_1 and H_2) is greater than the number of cold streams immediately above the pinch (C_1). Therefore, according to Inequality 7.13a, the cold stream C_1 should be split into two portions: one to be matched with H_1 and another to be matched with H_2. There are numerous ways to split the two portions of C_1 that satisfy Inequality 7.18. Figure 7.30 shows a particularly effective split where a portion of flow rate*specific heat of stream C_1 is taken as 200 kW/K to match that of hot stream (H_1) above the pinch. Hence, this portion and H_1 above the pinch have the same heat load, which can be transferred in a single heat exchanger. The remaining flow rate*specific heat of stream C_1 (400 kW/K) is matched to H_2 above the pinch to exchange 9000 kW.

The following representation will be used to illustrate the network structure graphically. Vertical arrows running at the left of the diagram represent hot streams. Temperatures are placed next to each arrow. A match between two streams is indicated by placing an ellipse on each of the streams and connecting

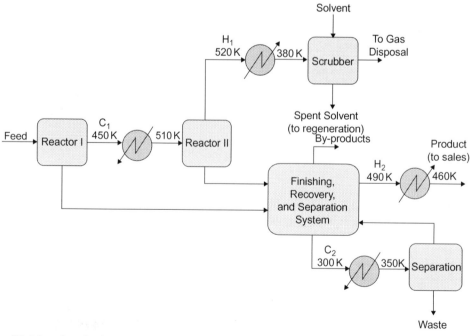

FIGURE 7-25 Simplified flow sheet for the specialty chemical processing facility.

(Continued)

Example 7-4 Network synthesis of a heat-exchanger network in a specialty chemical plant (*Continued*)

them by a line. Heat-transfer rates are noted in appropriate units (for example, kW) inside the ellipses. The pinch is represented by two horizontal dotted lines. When two streams are paired, the exchangeable rate of heat transfer is the lower of the two heat loads of the streams. For instance, the heat-exchange rate of H_2 and C_1 matched above the pinch are 300*(490 − 460) = 9000 kW and 400*(480 − 450) = 12,000 kW. Therefore, the exchangeable load between H_2 and C_1 is 9000 kW. The temperature of the C_1 branch leaving the match with H_2 is calculated through heat balance; that is,

[7.20]

$$\text{Temperature of the } C_1 \text{ branch leaving the match with } H_2 = 450 + \frac{9000}{400}$$
$$= 472.5 \, K$$

The rest of the matches are similarly carried out, and Fig. 7.31 shows the results. There are no cold streams immediately below the pinch. Therefore, the matches are relatively simple. Matching H_1 with C_2 below the pinch entails transferring the lower of the two heat-transfer rates: 200*(460 − 380) = 16,000 kW and 100*(350 − 300) = 5000 kW. Hence, 5000 kW should be transferred from H_1 to C_2. The remaining load of H_1 (16,000 − 5000 = 11,000 kW) is transferred to the cooling utility as shown by Fig. 7.32. Other configurations are also possible. For instance, H_1 below the pinch can be split into two streams in parallel and matched with C_2 and the cooling utility (compared to the serial structure shown by Fig. 7.32). It is also worth noting that the calculated heating and cooling utilities are consistent with the values determined from the targeting approach.

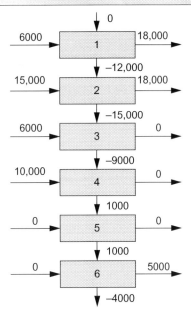

FIGURE 7-27 The cascade diagram for the specialty chemical processing facility.

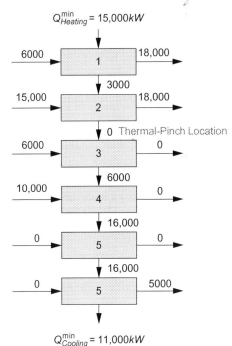

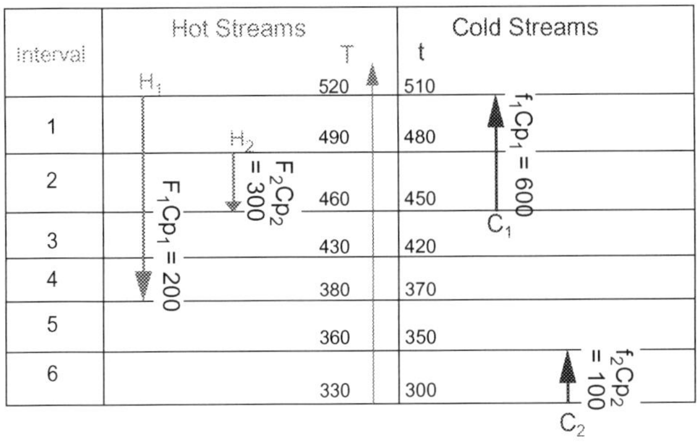

FIGURE 7-26 The TID for the specialty chemical processing facility.

FIGURE 7-28 The revised cascade diagram for the specialty chemical processing facility.

TABLE 7-6 Stream Data for the Specialty Chemical Process

Stream	Flow Rate × Specific Heat kW/K	Supply Temperature, K	Target Temperature, K	Enthalpy Change kW
H_1	200	520	380	−28,000
H_2	300	490	460	−9000
C_1	600	450	510	36,000
C_2	100	300	350	5000

(*Continued*)

Example 7-4 Network synthesis of a heat-exchanger network in a specialty chemical plant (*Continued*)

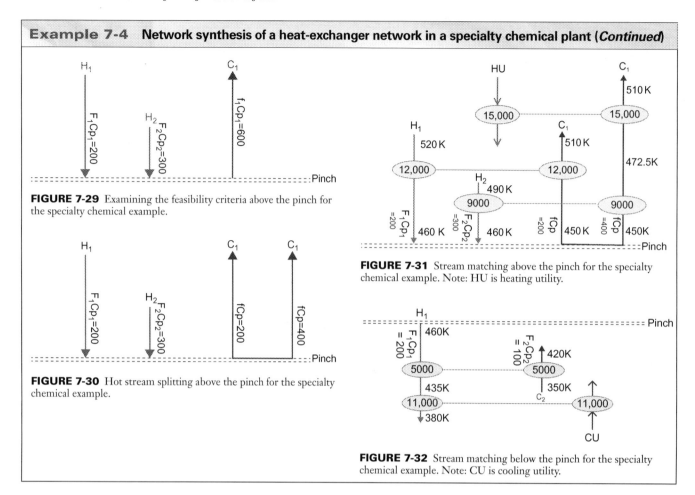

FIGURE 7-29 Examining the feasibility criteria above the pinch for the specialty chemical example.

FIGURE 7-30 Hot stream splitting above the pinch for the specialty chemical example.

FIGURE 7-31 Stream matching above the pinch for the specialty chemical example. Note: HU is heating utility.

FIGURE 7-32 Stream matching below the pinch for the specialty chemical example. Note: CU is cooling utility.

TABLE 7-7 Stream Data for Problem 7.1

Stream	Flow Rate × Specific Heat (Btu/hr °F)	Supply Temperature (°F)	Target Temperature (°F)	Enthalpy Change (10^3 Btu/hr)
H_1	1000	250	120	−130
H_2	4000	200	100	−400
HU_1	?	280	250	?
C_1	3000	90	150	180
C_2	6000	130	190	360
CU_1	?	60	80	?

Source: Douglas (1988).

HOMEWORK PROBLEMS

7.1. A plant has two process hot streams (H_1 and H_2), two process cold streams (C_1 and C_2), a heating utility (HU_1), and a cooling utility (CU_1). The problem data are given in Table 7.7. A value of $\Delta T^{min} = 10°F$ is used. Using graphical and algebraic techniques, determine the minimum heating and cooling requirements for the problem. Also, develop an HEN implementation matching the hot and the cold streams to achieve the minimum utility targets.

7.2. Consider a process that has two process hot streams (H_1 and H_2), two process cold streams (C_1 and C_2), a heating utility (HU_1, which is a saturated vapor that loses its latent heat of condensation), and a cooling utility (CU_1). The problem data are given in Table 7.8. The cost of the heating utility is \$6/10^6 kJ added, and the cost of the coolant is \$4/10^6 kJ. A value of $\Delta T^{min} = 10 K$ is used. Determine the minimum heating and cooling requirements for the process and synthesize a network of heat exchangers that meets the targets.

TABLE 7-8 Stream Data for Problem 7.2

Stream	Flow Rate × Specific Heat, kW/°C	Supply Temperature, °C	Target Temperature, °C
H_1	10.55	249	138
H_2	8.79	160	93
HU_1	?	270	270
C_1	7.62	60	160
C_2	6.08	116	260
CU_1	?	38	82

Source: Papoulias and Grossmann (1983).

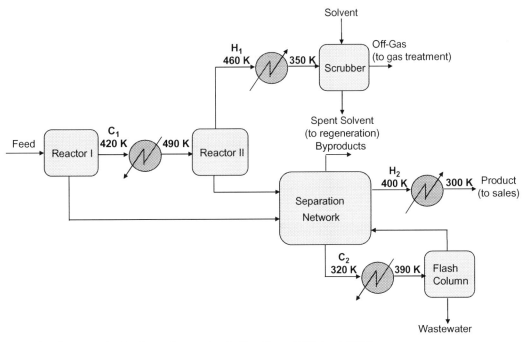

FIGURE 7-33 Simplified flow sheet for the chemical processing facility. *Source: El-Halwagi (2006).*

7.3. Consider the chemical processing facility illustrated in Fig. 7.33 (El-Halwagi, 2006). The process has two adiabatic reactors. The intermediate product leaving the first reactor (C_1) is heated from 420 K to 490 K before being fed to the second reactor. The off-gases leaving the reactor (H_1) at 460 K are cooled to 350 K prior to being forwarded to the gas-treatment unit. The product leaving the bottom of the reactor is fed to a separation network. The product stream leaving the separation network (H_2) is cooled from 400 K to 300 K prior to sales. A by-product stream (C_2) is heated from 320 K to 390 K before being fed to a flash column. Stream data are given in Table 7.9.

In the current operation, the heat exchange duties of H_1, H_2, C_1, and C_2 are fulfilled using the cooling and heating utilities. Therefore, the current usage of cooling and heating utilities are 83,000 and 56,000 kW, respectively.

The objective of this problem is to identify the target for minimum heating and cooling utilities and to synthesize a network of heat exchangers that achieves the utility targets. A value of $\Delta T^{min} = 10$ K is used.

7.4. A plant has one process hot stream (H_1) and two process cold streams (C_1 and C_2). Two heating utilities (HU_1 and HU_2) are available for service, as well as a cooling utility. The two heating utilities are condensing steams whose costs are 5 and 7 ($/10^6$ Btu), respectively. A minimum driving force of 10°F is required for any heat-exchange duty. Table 7.10 provides the data for the process. Determine the optimal usage of the two heating utilities and the cooling utility. Synthesize a network of minimum number of heat exchangers satisfying the minimum utility targets.

7.5. Consider the formic acid process shown by Fig. 7.34 (Tora and El-Halwagi, 2010). Methanol and carbon monoxide are reacted to produce methyl formate, which is later distilled. The bottom stream of the methyl formate column (C_1) enters the reboiler where it is heated to create the boil-up vapor that is returned

TABLE 7-9 Stream Data for the Chemical Process

Stream	Flow Rate × Specific Heat, kW/K	Supply Temperature, K	Target Temperature, K	Enthalpy Change, kW
H_1	300	460	350	−33,000
H_2	500	400	300	−50,000
C_1	600	420	490	42,000
C_2	200	320	390	14,000

Source: El-Halwagi (2006).

TABLE 7-10 Stream Data for the Hot and Cold Streams of Problem 7.4

Stream	Flow Rate × Specific Heat, Btu/°F · hr	Supply Temperature, °F	Target Temperature, °F
H_1	20,000	300	100
HU_1	?	300	300
HU_2	?	500	500
C_1	20,000	190	290
C_2	40,000	190	390
CU_1	?	30	40

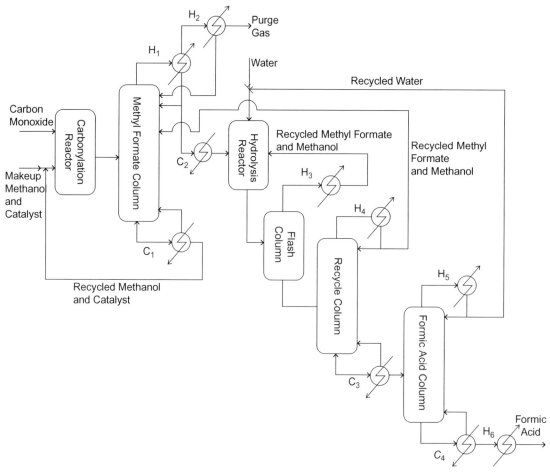

FIGURE 7-34 Formic acid process. *Source: Tora and El-Halwagi (2010).*

TABLE 7-11 Data for the Process Hot and Cold Streams for the Formic Acid Process

Stream	Supply Temperature, K	Target Temperature, K	Rate of Enthalpy Change, MW
H_1	338	333	−5.3
H_2	333	288	−0.3
H_3	330	288	−1.4
H_4	310	308	−2.7
H_5	393	308	−4.3
H_6	423	308	−0.2
C_1	368	370	5.8
C_2	333	393	0.8
C_3	368	383	2.9
C_4	423	430	4.7

Source: Tora and El-Halwagi (2010).

to the column. The vapor product (H_1) from the top of the methyl formate column is cooled and partially condensed. The remaining vapor (H_2) is further condensed using a refrigerant in a partial condenser. The noncondensed gases are purged and incinerated. Methyl formate is heated (C_2) then hydrolyzed to produce formic acid and regenerate methanol. A flash column is used to recover methyl formate and methanol, which are cooled (H_3) and recycled to the hydrolysis reactor. The liquid stream leaving the bottom of the flash unit is separate in a distillation column for further recovery of methyl formate and methanol. The overhead condenser requires cooling (H_4), and the reboiler requires heating (C_3). An additional distillation column is used to recover formic acid. The overhead condenser requires cooling (H_5), and the reboiler requires heating (C_4). The concentrated formic acid (H_6) is cooled prior to storage. The stream data are given in Table 7.11.

Two cooling utilities are available for service. Cooling water can cool the hot streams down to 308 K and a refrigeration system can be used to cool the hot streams to 288 K. A minimum driving force ($\Delta T^{\min}$) value of 10 K is required.

a. Using the GCC, determine the optimal targets for the two cooling utilities and for a heating utility.
b. Synthesize a network of heat exchangers that meets the utility targets.

NOMENCLATURE

$C_{P,u}$	specific heat of hot stream u [kJ/(kg K)]
$c_{P,v}$	specific heat of cold stream v [kJ/(kg K)]
f	flow rate of cold stream (kg/s)
F	flow rate of hot stream (kg/s)
$HC_{v,z}$	cold load in interval z
$HH_{u,z}$	hot load in interval z
MOC	minimum operating cost
N_C	number of process cold streams
N_{ca}	number of cold streams immediately above the pinch
N_{cb}	number of cold streams immediately below the pinch
N_{CU}	number of cooling utilities
N_H	number of process hot streams
N_{ha}	number of hot streams immediately above the pinch
N_{hb}	number of hot streams immediately below the pinch
N_{HU}	number of process cold streams
$Q^{\min}_{Cooling}$	minimum cooling utility
$Q^{\min}_{Heating}$	minimum heating utility
r_z	residual heat leaving interval z
t	temperature of cold stream (K)
t_v^s	supply temperature of cold stream v (K)
t_v^t	target temperature of cold stream v (K)
T_u^s	supply temperature of hot stream u (K)
T_u^t	target temperature of hot stream u (K)
T	temperature of hot stream (K)
u	index for hot streams
U_{MOC}	minimum number of heat-exchange units satisfying the minimum operating cost
v	index for cold streams
z	temperature interval

GREEK LETTER

$\Delta T^{\min}$	minimum approach temperature (K)

REFERENCES

Al-Thubaiti M, Al-Azri N, El-Halwagi M: Optimize heat transfer networks: an innovative method applies heat integration to cost effectively retrofit bottlenecks in utility systems, *Hydrocarbon Process*:109–115, March 2008.

Douglas JM: *Conceptual design of chemical processes*, New York, McGraw Hill, 1988.

El-Halwagi MM: *Process integration*, Amsterdam, Academic Press/Elsevier, 2006.

Foo DCY, El-Halwagi MM, Tan RR, editors: *Recent advances in sustainable process design and optimization*, Series on Advances in Process Systems Engineering, World Scientific Publishing Company, 2011.

Furman KC, Sahinidis NV: A critical review and annotated bibliography for heat exchanger network synthesis in the 20th century, *Ind Eng Chem Res* 41(10):2335–2370, 2002.

Gundersen T, Naess L: The synthesis of cost optimal heat exchanger networks: an industrial review of the state of the art, *Comp Chem Eng* 12(6):503–530, 1988.

Hohmann EC: *Optimum networks for heat exchange*, PhD Thesis, California, Los Angeles, University of Southern, 1971.

Kemp I: *Pinch analysis and process integration: a user guide on process integration for the efficient use of energy*, ed 2, Butterworth-Heinemann, 2009.

Linnhoff B: Pinch analysis: a state of the art overview, *Trans Inst Chem Eng* 71(A5):503–522, 1993.

Linnhoff B, Flower JR: Synthesis of heat exchanger networks: I. Systematic generation of energy optimal networks, *AIChE J* 24(4):633–642, 1978.

Linnhoff B, Hindmarsh E: The pinch design method for heat exchanger networks, *Chem Eng Sci* 38(5):745–763, 1983.

Linnhoff B, Townsend DW, Boland D, Hewitt GF, Thomas BEA, Guy AR, Marsland RH: A user guide on process integration for the efficient use of energy, UK, IChemE, 1982.

Majozi T: *Batch chemical process integration: analysis, synthesis, and optimization*, Heidelberg, Springer, 2010.

Papoulias SA, Grossmann IE: A structural optimization approach in process synthesis. II. Heat recovery networks, *Comput Chem Eng* 7(6):707–721, 1983.

Rossiter AP: Improve energy efficiency via heat integration, *Chem Eng Prog* 106(12):33–42, 2010.

Shenoy UV: *Heat exchange network synthesis: process optimization by energy and resource analysis*, Houston, Texas, Gulf Publ. Co, 1995.

Smith R: *Chemical process design and integration*, New York, Wiley, 2005.

Tora EA, El-Halwagi MM: Integration of solar energy into absorption refrigerators and industrial processes, *Chem Eng Technol* 33(9):1495–1505, 2010.

Umeda T, Itoh J, Shiroko K: A thermodynamic approach to the synthesis of heat integration systems in chemical processes, *Comput Chem Eng* 3:273–282, 1979.

CHAPTER 8
INTEGRATION OF COMBINED HEAT AND POWER SYSTEMS

The integration of thermal energy through the synthesis of heat-exchange networks (HENs) is a key element of an effective energy-management strategy. As discussed in Chapter 7, substantial savings in heating and cooling utilities can be achieved by implementing heat integration in industrial facilities. In addition to heating and cooling, other forms of energy should also be considered through the concept of energy integration. One way of defining **energy integration** (El-Halwagi, 2006) is that it is a holistic approach to the design and operation of energy systems involving the generation, allocation, transformation, and exchange of all forms of energy including heat and work (or heat rate and power when the system is studied on a per unit-time basis). The objective of this chapter is to present the principal aspects of energy integration with focus on the integrated design of heat and power systems, commonly referred to as combined heat and power (CHP). First, the chapter discusses the concepts of heat engines and heat pumps. Key thermodynamic cycles are also described and integrated with the thermal pinch analysis. Then, design rules are given for the placement of heat pumps and heat engines. Next, a systematic approach is given for the targeting of cogeneration systems dispensing process heating and power.

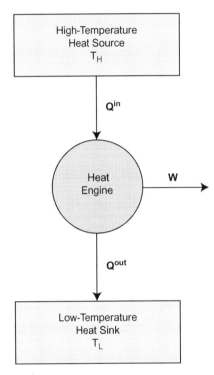

FIGURE 8-1 A heat engine.

HEAT ENGINES

A heat engine is a device that uses energy in the form of heat to provide work (for example, shaft work that can be used as is or converted to electric energy). It is the basis for many cyclic processes that are used in industry. The following sections give the basics of heat engines and the important case of a steam power plant.

Because of the second law of thermodynamics, it is impossible for a cyclic heat engine to convert all the input heat into useful work. Therefore, heat must be discharged. A heat engine has three principal components:

1. A heat source (or hot reservoir) that provides the heat input (Q^{in}) to the engine. The temperature of the heat source is referred to as T_H.
2. A heat sink (or cold reservoir) that receives the discharged heat (Q^{out}) from the engine. The temperature of the heat sink is referred to as T_L.
3. An object on which work (W) is done.

There are numerous industrial applications of heat engines. A common application is the automobile engine (internal combustion engine) where heat is generated during one part of the cycle and is used in another part of the cycle to produce useful work. Another application of heat engines is in power plants with steam turbines where fuel is burned to generate heat that is added to water to produce steam that is let down through a turbine to produce useful work. Heat engines may be modeled through thermodynamic cycles such as Brayton, Otto, Diesel, Stirling, Rankine, and Carnot.

Figure 8.1 is a schematic representation of a heat engine. The energy balance around the heat engine can be written as follows:

[8.1] $$Q^{in} = W + Q^{out}$$

The efficiency of the heat engine provides a measure of how much work can be extracted from the added heat; that is,

[8.2] $$\eta = \frac{W}{Q^{in}}$$

As mentioned earlier, the second law of thermodynamics prevents the engine from transforming all the input heat to useful work. Consequently, the efficiency of the heat engine cannot reach 100 percent. The most efficient heat engine cycle is the **Carnot cycle** described by Nicolas Carnot in 1824. The Carnot cycle involves four steps that

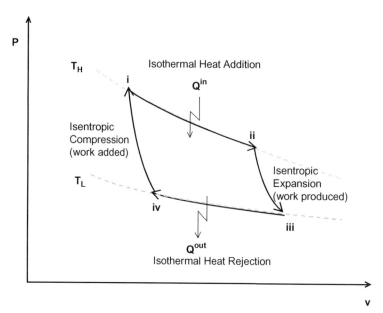

FIGURE 8-2a A pressure-volume representation of a Carnot cycle for a heat engine.

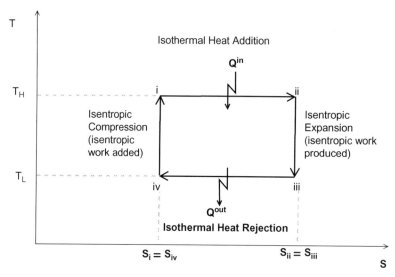

FIGURE 8-2b A temperature-entropy representation of a Carnot cycle for a heat engine.

can be represented on a pressure-volume diagram (Fig. 8.2a) or a temperature-entropy diagram (Fig. 8.2b). The cycle assumes the idealistic case of a working fluid that behaves as an ideal gas when in the gaseous phase and operates through a cycle composed of reversible processes. Although these are practically unattainable conditions, they provide the basis for rigorous bounding on the performance of a heat engine. The following is a description of these four steps that constitute the Carnot cycle:

- Isothermal expansion from state i to state ii: Heat, Q^{in}, is added to the working fluid from the high-temperature heat source at temperature T_H.
- Reversible adiabatic (isentropic) expansion of the working fluid from state ii to state iii: As a result of the expansion, the temperature is reduced from T_H to T_L. The gas expansion is used to produce work by the system.
- Isothermal compression from state iii to state iv: Heat, Q^{out}, is rejected from the working fluid to the low-temperature heat sink at temperature T_L.
- Reversible adiabatic (isentropic) compression from state iv to state i: Work is added to the working fluid to compress it isentropically. As a result of the compression, the temperature is increased from T_L to T_H and the cycle is completed.

For a reversible isothermal process, the absolute value of the heat added or removed is equal to the product of temperature times the absolute value of change in entropy. Therefore,

[8.3] $$Q^{in} = T_H * (S_{ii} - S_i)$$

and

[8.4] $$Q^{out} = T_L * (S_{iii} - S_{iv})$$

Substituting from Equations 8.3 and 8.4 into the overall energy balance of the system given by Eq. 8.1 and noting that $S_i = S_{iv}$ and $S_{ii} = S_{iii}$, the net work produced by the cycle (work produced during expansion minus work added during compression) is calculated as:

$$[8.5] \qquad W = (T_H - T_L)(S_{ii} - S_i)$$

Combining Equations 8.2, 8.3, and 8.5, we get the following expression for the efficiency of a Carnot heat engine:

$$[8.6] \qquad \eta_{Carnot} = 1 - \frac{T_L}{T_H}$$

where T_L and T_H are the absolute temperatures of the heat sink and source, respectively.

Although practically unachievable, the efficiency of a Carnot cycle provides a very useful benchmark for maximum efficiency of a cyclic heat engine. This efficiency target has the advantage that it can be determined ahead of detailed design and requires simple data (the temperatures of the heat source and sink).

STEAM TURBINES AND POWER PLANTS

A common example of a heat engine is a power plant with a steam turbine. As shown in Fig. 8.4, the power plant includes a boiler that corresponds to the high-temperature heat source where a fuel is burned to transfer heat to water, thereby producing high-pressure steam. The high-pressure steam is directed toward a turbine where the pressure energy of the steam is converted into rotational energy in the form of a shaft work. The shaft work can be converted to electric energy through an electric generator. The steam leaving the turbine is cooled/condensed, and the condensate is fed to a pump that increases the pressure of the condensate and returns the water back to the boiler to be heated, transformed into steam, and the cycle continues. Typically, the generated work from the turbine is significantly higher than the work used in the pump. Therefore, the net effect of the system is that the heat from the boiler is converted into work while the discharged heat (with the steam exiting the turbine) is transferred to the condenser that serves as the low-temperature heat sink.

A useful tool for representing the thermodynamic aspects of the steam power plant is the Mollier diagram (for example, Smith et al., 2005). There are several forms of the Mollier diagram. One form is the temperature versus entropy representation. Consider the vapor-liquid portion of the Mollier diagram shown by Fig. 8.5. The curved line to the left of the critical point is the saturated-liquid line while the curved line to its right is the saturated-vapor line. To the left of the saturated-liquid line lies the subcooled-liquid region and to the right of the saturated-vapor line lies the superheated-vapor region. The area under the curve is the two-phase liquid-vapor region. The dashed path represents heating a liquid from a subcooled state to a superheated state under constant pressure. Under a given pressure, water starts and ends evaporation at the same temperature. Therefore, the two-phase vapor-liquid segment is horizontal.

Example 8-1 Efficiency and work of a Carnot heat engine

A heat-engine cycle is to be designed to absorb 0.80 MW from a heat reservoir at 800 K and to discharge heat to a cold sink at 300 K. Before designing the engine, it is desired to determine the following:
1. Maximum theoretical efficiency of the heat engine
2. Maximum generated work rate (power)
3. Minimum rate of heat discharge to the cold sink

SOLUTION

The targets for maximum efficiency, maximum work, and minimum heat discharge are obtained by considering a Carnot cycle. According to Eq. 8.6, the maximum efficiency is calculated as follows:

$$[8.7] \qquad \eta_{Carnot} = 1 - \frac{300}{800} = 0.625$$

Generated power is determined through Eq. 8.2:

$$[8.8] \qquad W = 0.625 * 0.80 = 0.50 \, MW$$

Performing the energy balance given by Eq. 8.1, we get the rate of heat discharge at 800 K to be:

$$[8.9] \qquad Q^{out} = 0.80 - 0.50 = 0.30 \, MW$$

Figure 8.3 is a schematic representation of this Carnot heat engine.

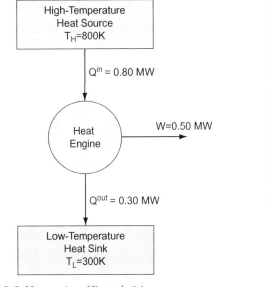

FIGURE 8-3 Heat engine of Example 8.1.

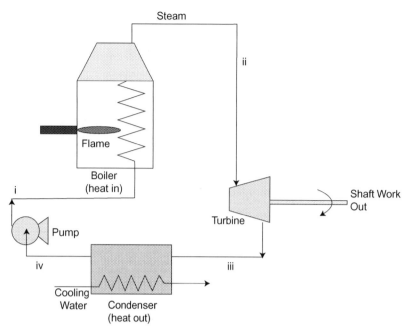

FIGURE 8-4 A schematic representation of a steam power plant.

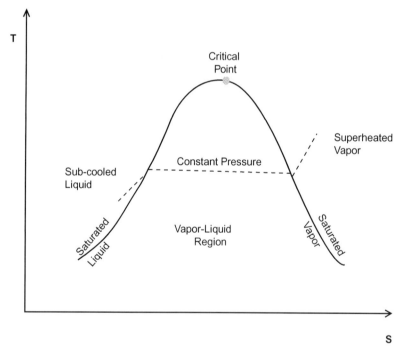

FIGURE 8-5 The temperature-entropy Mollier diagram.

If the steam power plant of Fig. 8.4 is described by a Carnot cycle, then the working fluid (water) undergoes four steps in the following four units (as shown by Fig. 8.6):

- The boiler or steam generator (which represents the high-temperature heat source): Isothermal expansion of saturated liquid (state i) to saturated steam (state ii) where heat (Q^{in}) is added in the boiler at temperature T_H.
- The turbine: Reversible adiabatic (isentropic) expansion from state ii (saturated steam) to a lower pressure at an arbitrary two-phase state iii comprised of a mixture of saturated liquid and vapor. As a result of the expansion, the temperature is reduced from T_H to T_L. The expansion of steam in the turbine is used to produce work by the system.
- The condenser (which represents the low-temperature heat sink): Partial condensation of two-phase mixture from state iii to state iv where heat (Q^{out}) is rejected from the two-phase mixture to a cooling

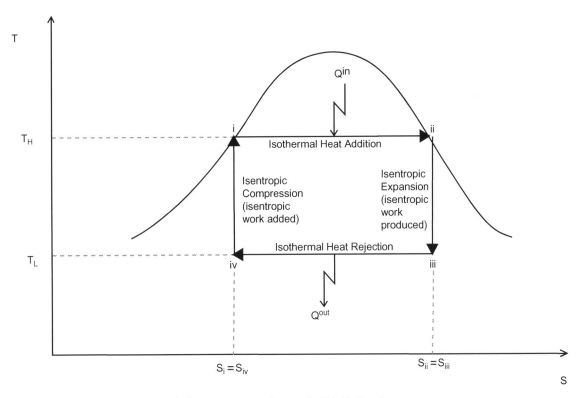

FIGURE 8-6 Representation of the Carnot cycle for a steam power plant on the T-S Mollier diagram.

medium (for example, cooling water) representing the low-temperature heat sink at temperature T_L.
- The pump: Reversible adiabatic (isentropic) compression of the two-phase mixture at state iv to a saturated liquid water at state i; work is added to the two-phase mixture to compress it isentropically. As a result of the compression, the temperature is increased from T_L to T_H and the cycle is completed.

As indicated previously, the Carnot cycle is an idealistic system because it assumes ideal gas behavior and reversible processes. In the case of a steam power plant, the Carnot cycle has additional limitations that render it difficult to implement. Expanding saturated steam to a two-phase mixture can cause mechanical damage to the turbine blades. Similarly, pumping a two-phase mixture can cause erosion of the pump internals and instabilities in the flow. Finally, it is very difficult to design a pump that will convert a two-phase feed (at state iv) to a saturated liquid (at state i). Hence, the performance of the Carnot cycle is taken as a benchmark to compare with practically implementable cycles.

A common system that can be practically implemented for steam power plants is the Rankine cycle, which has two primary differences with the Carnot cycle:

- Enough heat is added to state i so that state ii exceeds the saturated-level state and becomes a superheated steam.
- Enough heat is removed from state iii so that state iv becomes a saturated liquid.

Therefore, the four steps associated with a Rankine cycle (Fig. 8.7) are undertaken in the following four units:

- The boiler or steam generator (which represents the high-temperature heat source): Isobaric (constant-pressure at the pressure of the boiler) heating of a subcooled liquid (state i) until it reaches the saturated-liquid state at the pressure of the boiler, then evaporates to saturated liquid under constant temperature and pressure, and then is further heated to become a superheated steam (state ii).
- The turbine: Reversible adiabatic (isentropic) expansion from state ii (superheated steam) to a lower pressure at an arbitrary two-phase state iii comprised of a mixture of saturated liquid and vapor. As a result of the expansion, the temperature is reduced from T_H to T_L. The expansion of steam in the turbine is used to produce work by the system. In cases when the exhaust steam is to be used for heating the process, the outlet state iii is kept above the saturated vapor line.
- The condenser (which represents the low-temperature heat sink): Constant pressure, constant temperature, total condensation of the two-phase mixture from state iii to a saturated liquid (state iv) where heat (Q^{out}) is rejected from the two-phase mixture to a cooling medium (for example, cooling water) representing the low-temperature heat sink at temperature T_L.
- The pump: Reversible adiabatic (isentropic) compression of the two-phase mixture at state iv to a saturated liquid water at state i: Work is added to the two-phase

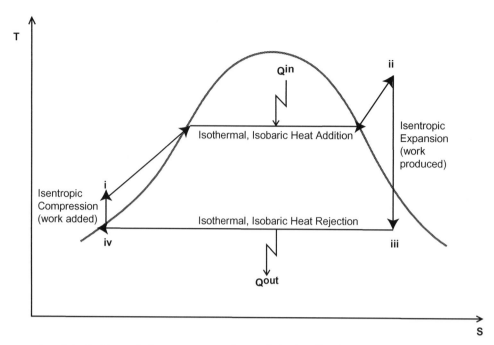

FIGURE 8-7 Representation of the Rankine cycle for a steam power plant on the T-S Mollier diagram.

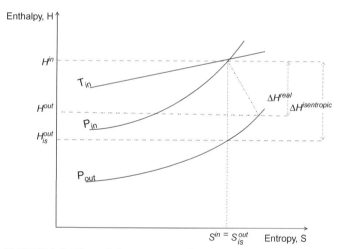

FIGURE 8-8 The enthalpy-entropy Mollier diagram for modeling a steam turbine.

mixture to compress it isentropically. As a result of the compression, the temperature is increased from T_L to T_H and the cycle is completed.

Next, the modeling of the steam turbine is developed. Let the pressure and temperature of the steam leaving the boiler and entering the turbine be designated by P_{in} and T_{in}. The corresponding enthalpy and entropy are given by the symbols H^{in} and S^{in}. When steam is let down from higher pressure to lower pressure in a steam turbine, the power generation can be obtained from the enthalpy-entropy Mollier diagram (for example, Smith et al., 2005) as illustrated in Figure 8.8. For a given pair of inlet pressure and temperature and for a given outlet pressure of the turbine, the isentropic enthalpy change in the turbine can be determined as:

[8.10] $$\Delta H^{isentropic} = H^{in} - H^{out}_{is}$$

where $\Delta H^{isentropic}$ is the specific isentropic enthalpy change in the turbine, H^{in} is the specific enthalpy of the steam at the inlet temperature and pressure of the turbine, and H^{out}_{is} is the specific isentropic enthalpy at the outlet pressure of the turbine. In reality, a turbine cannot extract all the energy generated by the isentropic expansion. Therefore, an isentropic efficiency term is defined as follows:

[8.11] $$\eta_{is} = \frac{\Delta H^{real}}{\Delta H^{isentropic}}$$

Therefore,

[8.12] $$\eta_{is} = \frac{H^{in} - H^{out}}{H^{in} - H^{out}_{is}}$$

where η_{is} is the isentropic efficiency, ΔH^{real} is the actual specific enthalpy difference across the turbine, and H^{out} is the actual specific enthalpy of the steam at the outlet temperature and pressure of the turbine. The value of η_{is} depends on the type, size, and internal design of the steam turbine and can be obtained from the manufacturer. Correlations also exist for the prediction of the isentropic efficiency (for example, Mavromatis and Kokossis, 1998; Varbanov et al., 2004).

Based on Equations 8.10, 8.11 and 8.12, the specific work, w, produced by the turbine is expressed as:

[8.13] $$w = \Delta H^{real} = \eta_{is}(H^{in} - H^{out}_{is})$$

Depending on the flow rate of steam passing through the turbine, $\dot{m}$, the power produced by the turbine, W, is given by:

$$[8.14] \quad W = \dot{m}\eta_{is}(H^{in} - H^{out}_{is})$$

In addition to the graphical approach via the Mollier diagram, the steam-turbine calculations can also be carried out using the steam tables (for example, Smith et al., 2005; Harvey and Parry, 1999).

Alternatively, the shortcut method developed by Al-Azri et al. (2009) may be used for approximate calculations in modeling a steam turbine topping system (which uses high-pressure steam to generate power in the turbine and then discharges low-pressure superheated steam to be used later for process heating). Al-Azri et al. (2009) proposed the following approximate correlations to predict steam table values for the purpose of conceptual design and targeting. The saturation temperature of the steam (T_{sat}) at pressure P is described as a function of the steam pressure and is calculated by using the following correlation:

$$[8.15] \quad T_{sat} = 112.72\, P^{0.2289}$$

where P is in psi (absolute) and T_{sat} is in °F.

For a given pressure and temperature at the inlet of the turbine, the corresponding entropy of the superheated steam at the inlet of the turbine can be calculated using the following equation:

$$[8.16] \quad S = (-0.5549\ln(T_{sat}) + 3.7876)T^{(0.1001\exp(0.0017 T_{sat}))}$$

where S is the entropy of the superheated steam in Btu/(lb R), T is the temperature of the superheated steam in °F, and T_{sat} is the saturation temperature, corresponding with the inlet pressure, in °F.

Given the entropy at the superheated-steam conditions and given the saturation temperature, the superheated-steam enthalpy is obtained from the following correlation:

$$[8.17] \quad H = 0.2029\, T_{sat}\, s^{3.647} + 817.35$$

where H is in Btu/lb, s is in Btu/(lb R), and T_{sat} is in °F.

Finally, the enthalpy of the saturated liquid (h) may be approximated by:

$$[8.18] \quad h = 1.043\, T_{sat} - 41.000$$

where h is in Btu/lb, s is in Btu/(lb R), and T_{sat} is in °F.

The remaining units in the cycle (the condenser, the pump, and the boiler) can be modeled using energy balances, as described in the following.

The Condenser

Assuming no heat losses, the cooling duty of the condenser ($Q^{Condenser}$) is the difference in the enthalpy of the condensing steam from state iii to state iv (on Fig. 8.7). Hence,

[8.19]
$$Q^{Condenser} = \text{Flow rate of steam}*(\text{Enthalpy of stream leaving the turbine} - \text{Enthalpy of saturated liquid leaving the condenser})$$

The Pump

Consider an isentropic pump that increases the pressure of the saturated liquid at state iv to a subcooled liquid at state i at the original pressure of the cycle. The isentropic power for an incompressible fluid can be approximated by the following expression, which assumes a constant density (inverse of the specific volume) of the fluid:

$$[8.20] \quad W_{Pump} = \text{Flow rate of the condensate}*\frac{(P_{pump,\,out} - P_{pump,\,in})}{\rho_{Liquid}*\eta_{Pump}}$$

Where W_{pump} is the power provided to the pump, the flow rate of the condensate is in units of mass per unit time, $P_{Pump,out}$ is the outlet pressure of the subcooled water leaving the pump, $P_{Pump,in}$ is the inlet pressure of the saturated liquid entering the pump, ρ_{Liquid} is the density of the liquid, and η_{Pump} is the pump efficiency. For instance, if the flow rate is in kg/s, the pressure is in Pa (or N/m²), and the density is in kg/m³, then the power is in N.m/s, which is J/s or Watt (W).

The Boiler

The rate of heat supplied by the boiler to the evaporating stream (Q^{boiler}) is obtained through a heat balance to isobarically move the stream from state i to state ii (Fig. 8.7):

[8.21]
$$Q^{Boiler} = \text{Flow rate of the steam}*(\text{Enthalpy of the steam leaving the boiler} - \text{Enthalpy of the water entering the boiler})$$

When the fuel is burned to provide the heat for the boiler, there are heat losses. To account for such losses and to evaluate the rate of heat that should be supplied by the fuel, we need to use the thermal efficiency of the boiler, which is defined as:

$$[8.22] \quad \eta_{Boiler} = \frac{\text{Heat received by the evaporating stream (i.e., } Q^{Boiler})}{\text{Input heat to the boiler from fuel combustion}}$$

Example 8-2 Using the steam tables to model the performance of a steam power plant in a Rankine cycle

A steam power plant is operated according to the concept of a Rankine cycle. A steam turbine receives superheated steam at 610 psi and 650°F and discharges steam at 200 psi. The turbine has a power-generation isentropic efficiency of 45 percent. The flow rate of steam is 68,200 lb/hr (8.59 kg/s). The steam leaving the turbine is fed to a condenser to produce saturated liquid at 200 psi. The condensate is pumped isentropically to a subcooled liquid at 610 psi. The pump efficiency is 70 percent. The subcooled liquid leaving the pump is subsequently fed to an isobaric boiler that produces superheated steam at 610 psi and 650°F, and the cycle continues. The thermal efficiency of the boiler is 60 percent. Using the steam tables, calculate:
1. The power generated by the steam turbine
2. The cooling duty of the condenser
3. The power to be provided to the pump
4. The heat to be generated from fuel combustion in the boiler

SOLUTION
1. **The Turbine:**
Using the steam tables (for example, Smith et al., 2005) at the inlet conditions of the turbine (610 psi and 650°F), the following data are obtained:

[8.23] $\quad H^{in} = 1321.3$ Btu/lb

[8.24] $\quad S^{in} = 1.5598$ Btu/(lb. °R)

It is also worth noting that at the inlet pressure of 610 psi, the saturation temperature is:

[8.25] $\quad T_{sat} = 487.98$ °F

For an isentropic expansion of steam, the outlet entropy of steam is 1.5598 Btu/(lb°R). This entropy and the outlet pressure of 200 psi are sufficient to define the state of outlet steam. Therefore, from the steam tables, the data for the isentropic outlet steam are:

[8.26] $\quad$ Isentropic outlet temperature $= 400$ °F

[8.27] $\quad H^{out}_{is} = 1210.1$ Btu/lb

It is useful to note that at 200 psi, the saturation temperature of steam is 381.80°F, which confirms that the isentropic outlet steam at 400°F is superheated. If the outlet steam is a two-phase mixture, the steam quality (fraction of vapor in the mixture) is calculated by equating the entropy of the mixture to the weighted average of the entropies of saturated vapor and liquid.

According to Eq. 8.14, the power produced by the turbine is:

[8.28a] $\quad W = 68,200$ lb/hr$*0.45*(1,321.3 - 1,210.1)$
$\quad\quad\quad = 3,412,728$ Btu/hr

or, in SI units:

[8.28b] $\quad W = 3,412,728 \dfrac{\text{Btu}}{\text{hr}} * \dfrac{\text{hr}}{3600\text{s}} * \dfrac{1055.1\text{J}}{\text{Btu}} * \dfrac{\text{MW.s}}{10^6\text{J}}$
$\quad\quad\quad = 1.00$ MW

To calculate the actual outlet enthalpy of steam leaving the turbine, Eq. 8.12b is used:

[8.29a] $\quad 0.45 = \dfrac{1321.3 - H^{out}}{1321.3 - 1210.1}$

Hence, the actual outlet enthalpy of steam is calculated to be:

[8.29b] $\quad H^{out} = 1271.26$ Btu/lb

2. **The Condenser:**
The isobaric condensation of steam discharges the condensate at 200 psi. From the steam tables, the temperature, entropy, and enthalpy of saturated liquid at 200 psi are:

[8.30a] $\quad t = 381.80$ °F

[8.30b] $\quad s = 0.5438$ Btu/(lb °R)

[8.30c] $\quad h = 355.31$ Btu/lb

The cooling duty of the condenser can be obtained from a heat balance around the condenser. Using the data from Equations 8.29b and 8.30c, we get:

[8.31] $\quad$ Cooling duty of the condenser $= 68,200$ lb/hr $* (1271.26 - 355.31)$ Btu/lb
$\quad\quad\quad\quad = 62,467,790$ Btu/hr

3. **The Pump:**
Next, an isentropic pump is used to increase the pressure of the condensate from 200 psi (or $200*6.895*10^3 = 1.379*10^6$ Pa) back to the original pressure of 610 psi ($4.206*10^6$ Pa). From the steam tables, the specific volume of water is $1.148*10^{-3}$ m³/kg at $1.379*10^6$ Pa and $1.260*10^{-3}$ m³/kg at $4.206*10^6$ Pa. Therefore, we assume an average specific volume value of $1.204*10^{-3}$ m³/kg, which corresponds to a density of 831 kg/m³. The power added to the pump is obtained using Eq. 8.20:

[8.32a] $\quad W_{Pump} = 8.59\dfrac{\text{kg}}{\text{s}} * \dfrac{(4.206*10^6 - 1.379*10^6)\dfrac{\text{N}}{\text{m}^2}}{831\dfrac{\text{kg}}{\text{m}^3}*0.7}$
$\quad\quad\quad = 41.7*10^3$ N.m/s
$\quad\quad\quad = 41.7*10^3$ J/s
$\quad\quad\quad = 41.7$ kW

[8.32b] $\quad = 41.7*10^3$ J/s $* \dfrac{3,600\text{ s}}{\text{hr}} * \dfrac{\text{Btu}}{1,055.1\text{ J}}$
$\quad\quad\quad = 142,280$ Btu/hr

4. **The Boiler:**
The results in Equations 8.32a and 8.32b represent the power that must be added to the pump. Because of the pump efficiency, only 70 percent of the power reaches the pumped water, that is,

[8.33a] $\quad$ Power reaching the pumped water $= 0.7*142,280$
$\quad\quad\quad\quad = 99,596$ Btu/hr

According to an energy balance around the pump, the rate of change in enthalpy in the pumped water is approximately equal to the power added to the water entering. Hence, the enthalpy of the water leaving the pump can be calculated using the results from Equations 8.30c and 8.33a as follows:

$\quad\quad 99,596 = 68,200*(\text{Enthalpy of water leaving the pump} - 355.31)$

that is,

[8.33b] $\quad$ Enthalpy of water leaving the pump $= 356.8$ Btu/lb

The rate of heating of the evaporating water in the boiler is calculated from Eq. 8.21, with data from Equations 8.23 and 8.33b:

[8.34] $\quad Q^{Boiler} = 68,200$ lb/hr$*(1321.3 - 356.8)$ Btu/lb
$\quad\quad\quad = 65,778,900$ Btu/hr

(Continued)

Example 8-2 Using the steam tables to model the performance of a steam power plant in a Rankine cycle (Continued)

Using the thermal efficiency of the boiler (60 percent) and Eq. 8.22, we get:

[8.35] Input heat to boiler from fuel combustion = 65,778,900/0.6
= 109,631,500 Btu/hr

It is useful to make some observations from the results of this example:
- That power provided to the pump (47.1 kW) is significantly smaller than the power produced by the turbine (1000 kW). This is typically the case in most steam power plants. Therefore, in calculating power to heat ratios, the pumping power is usually neglected compared to the power produced by the turbine.
- One measure for assessing the efficiency of this steam power plant is to use the cycle efficiency defined by Eq. 8.2 but adjusted to account for the real power generated by the turbine and the input heat from fuel combustion to the boiler. This ratio can be calculated from the results given by Equations 8.29 and 8.35 as

[8.36] Power to input heat ratio = 3,412,728/109,631,500 = 0.031

This poor performance is primarily due to the relatively small drop in pressure in the turbine, to the extraction of superheated steam exiting the turbine (to be used for heating the process as will be described later), and the low turbine efficiency. Order-of-magnitude higher ratios have been observed in industrial practices when high inlet pressures were used in condensing turbines operating with high-isentropic efficiency.
- Because of the energy losses associated with inefficiencies in the turbine, the pump, and the boiler, the net work of the cycle is not equal to the net heat added. Recalling results from Equations 8.28a and 8.33b,

[8.37] The net work performed by the working fluid = 3,412,728 − 142,280
= 3,270,448 Btu/hr

On the other hand, based on the results of Equations 8.35 and 8.31,

[8.38] Net heat added to the working fluid = 109,631,500 − 62,467,790
= 47,162,710 Btu/hr

Figure 8.9 summarizes the key energy results for the example.

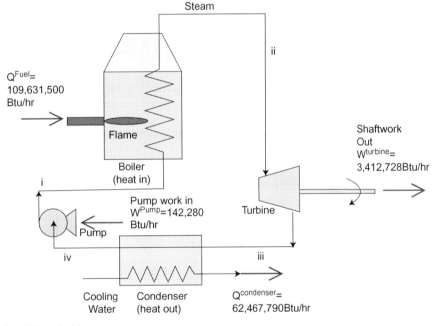

FIGURE 8-9 Key results of Example 8.2.

Example 8-3 Using approximate correlations to model the performance of a steam turbine in a Rankine cycle

Using the approximate correlations developed by Al-Azri et al. (2009), calculate the power generated by the steam turbine of Example 8.2 and compare the data and the results with those of the steam tables.

SOLUTION

Using Eq. 8.15, at the inlet pressure of 610 psi, the saturation temperature is:

[8.39] $T_{sat} = 112.72\,(610^{0.2289})$
$= 489.29\,°F$

This estimate is to be compared with the data from the steam tables listed by Eq. 8.25 as 487.98 °F.

Substituting for T_{sat} and T into Eq. 8.16, we obtain:

[8.40] $S = (-0.5549\,\ln(489.29) + 3.7876)(650^{(0.1001\exp(0.0017 \ast 489.29))})$
$= 1.5573\,Btu/hr$

and using, Eq. 8.17, we get:

[8.41] $H = 0.2029 \ast 489.29 \ast (1.5573^{3.647}) + 817.35$
$= 1316.7\,Btu/lb$

These results should be compared with the data from the steam tables:
$H = 1321.3\,Btu/lb$ and $S = 1.5598$.

(Continued)

Example 8-3 Using approximate correlations to model the performance of a steam turbine in a Rankine cycle (*Continued*)

For an isentropic expansion of steam, the outlet entropy of steam is 1.5573 Btu/(lb°R) and the pressure is 200 psi. Using Eq. 8.15, we have the saturation temperatures at the outlet conditions to be:

[8.42]
$$T_{sat} = 112.72(200^{0.2289})$$
$$= 379.06\ °F$$

Equation 8.17 is used to calculate the isentropic outlet enthalpy as follows:

[8.43]
$$H_{is}^{out} = 0.2029 * 379.06 * 1.5573^{3.647} + 817.35$$
$$= 1204.2\ Btu/lb$$

Based on Eq. 8.14, the power produced by the turbine is:

[8.44a]
$$W = 68,200\ lb/hr * 0.45 * (1316.7 - 1204.2)$$
$$= 3,452,625\ Btu/hr$$

or, in SI units:

[8.44b]
$$W = 3,452,625 \frac{Btu}{hr} * \frac{hr}{3600s} * \frac{1055.1 J}{Btu} * \frac{MW.s}{10^6 J}$$
$$= 1.01\ MW$$

This result should be compared with the turbine result of Example 8.2, using the steam tables that gave a turbine power of 1.00 MW (Eq. 8.28b). As can be seen, the correlations are convenient for numerical calculations compared to the steam tables, and they give an appropriate result for conceptual design and targeting purposes.

PLACEMENT OF HEAT ENGINES AND INTEGRATION WITH THERMAL PINCH ANALYSIS

When heat engines are used, they have a strong interaction with the heat integration of the process. Because the heat engine extracts and discharges heat, the following questions should be addressed:

- Where should the heat engine be placed on the temperature scale?
- What is the source of heat for the engine?
- Where should the heat leaving the engine be discharged?
- How should the heat engine be interfaced with the HEN?

To answer these questions, let us first consider the cascade diagram for the HEN of the process. Constructing the cascade diagram was described in Chapter 7. Suppose the cascade diagram of the HEN is represented by Fig. 8.10 and involves five temperature intervals with the pinch location corresponding to the temperature between intervals three and four (that is, lower temperature for interval three, which is upper temperature for interval four). The residual heat for the kth interval is designated by d_k. The pinch location corresponds to zero heat residual. The minimum heating and cooling utilities are designated by Q_H^{min} and Q_C^{min}.

Townsend and Linnhoff (1983a, b) identified several important observations for the placement of heat engines. To discuss these observations, let us consider Figures 8.11 through 8.13. First, Fig. 8.11 illustrates the placement of the heat engine completely above the pinch. For the HEN, the region above the pinch has a deficit in heat, which is equal to minimum heating utility. Meanwhile, the heat engine has a heat surplus in the form of heat discharge to the low-temperature sink, which is an indication of a thermodynamic waste. Therefore, it is beneficial to match the heat deficit of the HEN with the heat discharge from the engine. This can be achieved by placing the heat exchanger above the pinch. In this case, heat is extracted for the heat engine from a high-temperature heat source that is hotter than the required temperature for the heating utility, and the discharged heat is selected to match the minimum heating utility. Therefore, the process requirement for the heating utility serves as the heat sink for the heat engine. Consequently, the

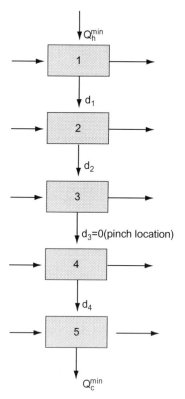

FIGURE 8-10 Cascade diagram for a HEN.

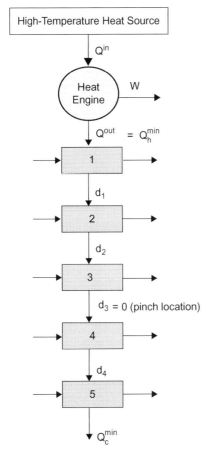

FIGURE 8-11 Placement of the heat engine above the pinch.

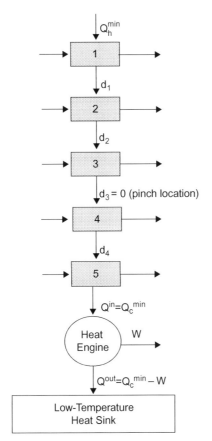

FIGURE 8-12 Placement of the heat engine below the pinch.

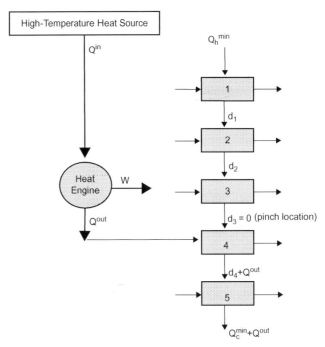

FIGURE 8-13 Placement of the heat engine across the pinch.

discharged heat from the heat engine is fully utilized to provide the heating utility for the process. This way, the heat discharged from the heat engine is no longer wasted and is effectively utilized in the HEN. So, although the efficiency of a heat engine is thermodynamically limited, such efficiency can be boosted if the heat engine is augmented with the HEN. As shown by Fig. 8.11, when the heat engine is properly placed above the pinch, the heat extracted from the hot source is transformed either to useful work or discharged as heat that is used to provide heating utility.

Figure 8.12 shows the placement of the heat engine below the pinch. For the HEN, the region below the pinch has a net surplus in heat. Therefore, a sensible proposition is to place the heat engine below the pinch such that the HEN surplus (required cooling utility) is used as the heat input to the heat engine. Consequently, two benefits accrue: The heat surplus below the pinch is removed by the heat engine, and this extracted heat is partially transformed to useful work.

Finally, let us consider placing the heat engine across the pinch where heat is extracted from above the pinch and discharged below the pinch. As can be seen from Fig. 8.13, there is no benefit from this arrangement. No reduction is achieved in heating or cooling utilities, and the heat engine performs in a similar way to the case when it is operated separately from the HEN.

The previous discussion illustrates the need to appropriately place the heat engine. As observed by Townsend and Linnhoff (1983a, b), *the heat engine should be placed above the pinch or below the pinch (but not across the pinch)*.

Example 8-4 Placement of a heat engine

Consider the chemical processing facility (El-Halwagi, 2006) illustrated in Fig. 8.14. The process has two adiabatic reactors. The intermediate product leaving the first reactor (C_1) is heated from 420 K to 490 K before being fed to the second reactor. The off-gases leaving the reactor (H_1) at 460 K are cooled to 350 K prior to being forwarded to the gas-treatment unit. The product leaving the bottom of the reactor is fed to a separation network. The product stream leaving the separation network (H_2) is cooled from 400 K to 300 K prior to sales. A by-product stream (C_2) is heated from 320 K to 390 K before being fed to a flash column. Stream data are given in Tables 8.1 and 8.2.

In the current operation, the heat exchange duties of H_1, H_2, C_1, and C_2 are fulfilled using the cooling and heating utilities. Therefore, the current usage of cooling and heating utilities for the process are 83,000 and 56,000 kW, respectively.

The power for the process is provided via a heat engine that approximates the performance of a Carnot heat engine. The heat engine draws 40,000 kW of heat from a hot reservoir at 833 K and discharges heat to a heat sink at 500 K. To assess the performance of the power and heat systems, it is desired to carry out the following tasks.

1. Determine the generated power and rate of heat discharge (cooling) for the heat engine.
2. Perform heat integration for the process to determine the minimum heating and cooling utilities. A value of $\Delta T^{min} = 10\,K$ is used.
3. Integrate the heat engine with the HEN.

SOLUTION

1. **Heat engine calculations**
   According to Eq. 8.6, the efficiency is calculated as follows:

   $$[8.45] \quad \eta_{Carnot} = 1 - \frac{500}{833} = 0.4$$

   Generated power is determined through Eq. 8.2:

   $$[8.46] \quad W = 0.4 * 40{,}000 = 16{,}000 \text{ kW}$$

   Performing the energy balance given by Eq. 8.1, we get the rate of heat discharge at 500 K to be:

   $$[8.47] \quad Q^{out} = 40{,}000 - 16{,}000 = 24{,}000 \text{ kW}$$

   Figure 8.15 is a schematic representation of this heat engine.

2. **HEN targeting**
   The TID and the cascade diagram are shown by Figures 8.16 and 8.17.

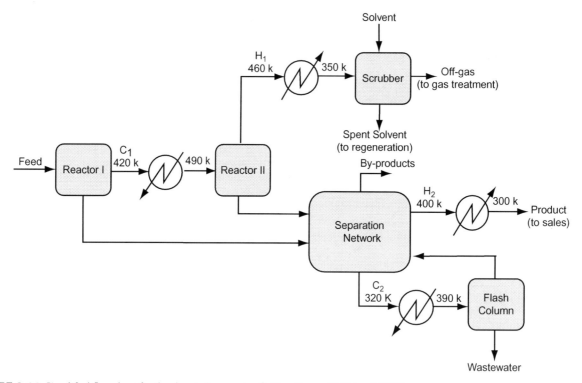

FIGURE 8-14 Simplified flow sheet for the chemical processing facility. *Source: El-Halwagi (2006).*

TABLE 8-1 Hot-Stream Data for the Chemical Process

Stream	Flow Rate × Specific Heat, kW/K	Supply Temperature, K	Target Temperature, K	Enthalpy Change, kW
H_1	300	460	350	−33,000
H_2	500	400	300	−50,000

Source: El-Halwagi (2006).

(Continued)

Example 8-4 Placement of a heat engine (*Continued*)

3. **Integration of the heat engine with the HEN**
   The heat is discharged from the heat engine at 500 K, which is hot enough to provide a portion of the needed heating utility for the HEN. Therefore, the heating utility is reduced from 33,000 kW to 9000 kW, as shown by Fig. 8.18. Additionally, the cooling utility needed to remove the 24,000 kW of heat discharged from the heat engine is completely eliminated and is now effectively used to reduce the heating utility requirement of the HEN.

TABLE 8-2 Cold-Stream Data for the Chemical Process

Stream	Flow Rate × Specific Heat, kW/K	Supply Temperature, K	Target Temperature, K	Enthalpy Change, kW
C_1	600	420	490	42,000
C_2	200	320	390	14,000

Source: El-Halwagi (2006).

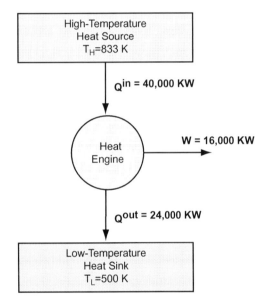

FIGURE 8-15 Heat engine of Example 8.4. *Source: El-Halwagi (2006).*

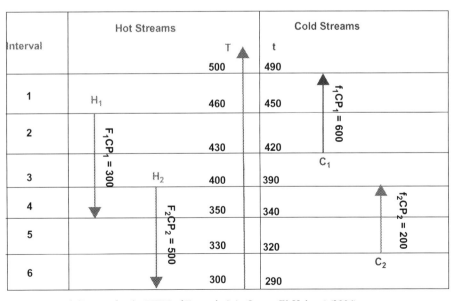

FIGURE 8-16 Temperature interval diagram for the HEN of Example 8.4. *Source: El-Halwagi (2006).*

(*Continued*)

Example 8-4 Placement of a heat engine (*Continued*)

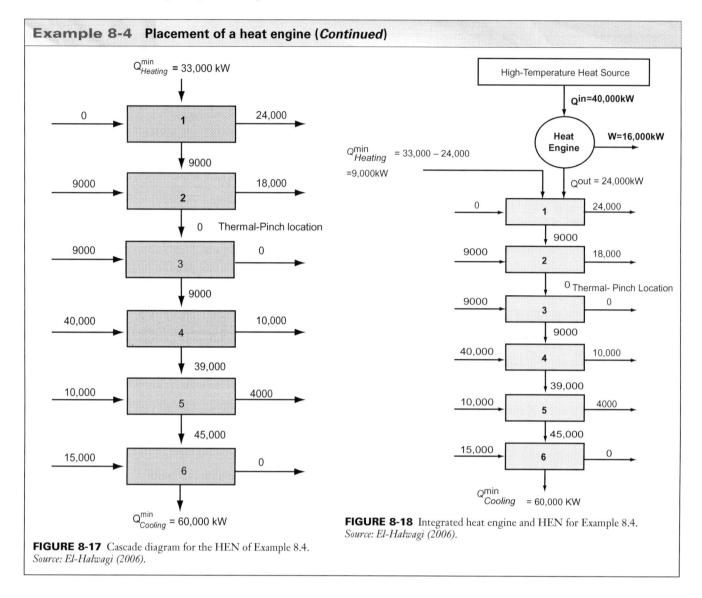

FIGURE 8-17 Cascade diagram for the HEN of Example 8.4.
Source: El-Halwagi (2006).

FIGURE 8-18 Integrated heat engine and HEN for Example 8.4.
Source: El-Halwagi (2006).

HEAT PUMPS

A **heat pump** is a device that uses external energy in the form of external work to extract heat from a low-temperature heat source to a high-temperature heat sink. Consequently, a heat pump conceptually operates opposite to a heat engine as it uses work to force the flow of heat from low to high temperature. In its simplest form, a heat pump has three elements:

1. A heat source (or cold reservoir) that provides the heat input (Q^{in}) to the heat pump. The temperature of the heat source is referred to as T_L.
2. A heat sink (or hot reservoir) that receives the discharged heat (Q^{out}) from the heat pump. The temperature of the heat pump is referred to as T_H.
3. An external source that delivers work (W).

The concept of a heat pump is widely used in many applications such as the refrigeration where heat is extracted from a low-temperature source (for example, object to be cooled) and is discharged in another part of the systems (for example, the surroundings) by exerting external work on the system. Other applications are intended for heating by "lifting" the heat from a low-temperature source using work to provide heat at a high-temperature sink.

Figure 8.19 is a schematic representation of a heat pump. The energy balance around the heat pump can be written as follows:

[8.48] $$Q^{in} + W = Q^{out}$$

The coefficient of performance (*COP*) of the heat pump provides a measure of how efficient the heat pump is in extracting heat from the cold reservoir or discharging heat to the hot reservoir relative to how much work is added. Therefore, the *COP* may be defined on a cooling or a heating basis; that is,

[8.49a] $$COP_{Cooling} = \frac{Q^{in}}{W}$$

where $COP_{Cooling}$ is the coefficient of performance of the heat pump on a cooling basis. It gives a measure of efficiency of the heat pump in removing heat from the low-temperature source relative to the work added

and

[8.49b] $$COP_{Heating} = \frac{Q^{out}}{W}$$

where $COP_{Heating}$ is the coefficient of performance of the heat pump on a heating basis. It gives a measure of efficiency of the heat pump in discharging heat to the high-temperature source relative to the work added.

There are several types and implementations of heat pumps. In the following sections, three types of heat pumps are covered:

1. Closed-cycle vapor-compression heat pumps using a separate working fluid (refrigerant)
2. Open-cycle mechanical vapor recompression using a process stream as the working fluid
3. Absorption refrigeration cycles

CLOSED-CYCLE VAPOR-COMPRESSION HEAT PUMPS USING A SEPARATE WORKING FLUID (REFRIGERANT)

The main components in such a closed-cycle vapor-compression heat pump system are two heat exchangers (a condenser and an evaporator), a compressor, and an expansion device. Figure 8.20 gives a schematic representation of the closed-cycle vapor-compression heat pump. Another important element of the vapor-compression heat pump is a volatile liquid that is referred to as the working fluid or the refrigerant and is circulated through the four components. Examples of refrigerants include hydrochlorofluorocarbons (HCFCs) and hydrofluorocarbons (HFCs). The working fluid (refrigerant) is used to remove heat from the high-temperature heat source (through a heat exchanger—an evaporator) and to deliver heat to the low-temperature heat sink (through a heat-exchanger—a condenser). A saturated or superheated vapor leaving the evaporator is fed to the compressor where an external energy (for example, mechanical, electric) is added. As a result of the compression, the pressure and temperature of the vapor increase. The hot, compressed vapor is fed to a condenser where it condenses and releases heat to a hot reservoir or heat sink. Then, the working fluid is expanded through an expansion valve and fed to the evaporator. The temperature of the working fluid fed to the evaporator is maintained below the temperature of the heat source (or cold reservoir). Therefore, heat flows from the heat source to the working fluid in the evaporator, resulting in the working fluid becoming a saturated or superheated vapor (which is the original state in which we started), and the cycle continues.

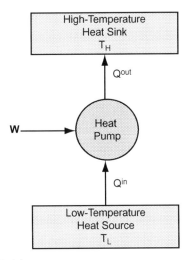

FIGURE 8-19 A heat pump.

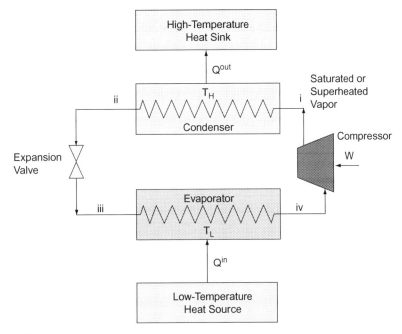

FIGURE 8-20 A vapor-compression heat pump.

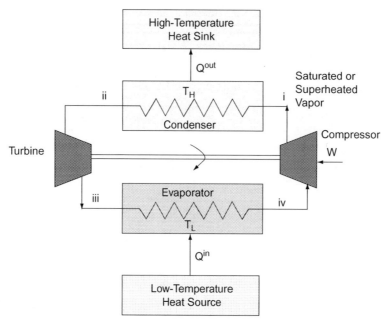

FIGURE 8-21 Brayton cycle heat pump.

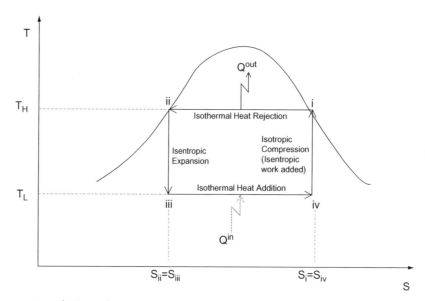

FIGURE 8-22A T-S representation of a Carnot heat pump.

A common form of the vapor-compression heat pump used in refrigeration is the Brayton cycle shown in Fig. 8.21. In this cycle, the external work is reduced by expanding the vapor in a turbine instead of an expansion valve. The work generated in the turbine is used to supplement the work exerted on the compressor.

The most efficient vapor-compression heat pump is the Carnot cycle, which involves the following four steps shown by Fig. 8.22a:

- Isothermal expansion from state i to state ii: Heat, Q^{in}, is discharged from the working fluid to the high-temperature heat source at temperature T_H.
- Reversible adiabatic (isentropic) expansion of the working fluid from state ii to state iii: As a result of the expansion, the temperature is reduced from T_H to T_L.
- Isothermal compression from state iii to state iv: Heat, Q^{out}, is removed by the working fluid from the low-temperature heat sink at temperature T_L.
- Reversible adiabatic (isentropic) compression from state iv to state i: Work is added to the working fluid to compress it isentropically. As a result of the compression, the temperature is increased from T_L to T_H, and the cycle is completed.

The Carnot cycle is an idealistic system. A more practical cycle is shown by Fig. 8.22b and involves the isentropic compression of a saturated vapor to a superheated state (i), which is condensed under constant pressure and temperature using a high-temperature sink to obtain a high-pressure saturated liquid state (ii). The saturated liquid is expanded under constant entropy to give a

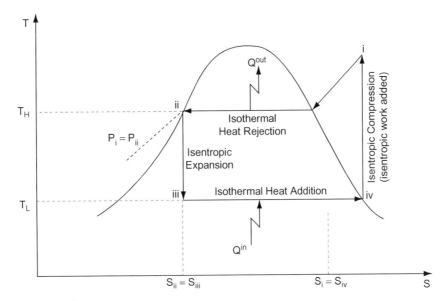

FIGURE 8-22B T-S representation of a heat pump.

low-pressure, low-temperature liquid-vapor mixture (state iii). The cold vapor-liquid mixture extracts heat from a low-temperature sink (the unit being refrigerated) under constant pressure and temperature to give a vapor at the saturated state (iv), and the cycle continues.

It is useful to model the Carnot heat pump to determine a performance target. For a reversible isothermal process, the absolute value of the heat added or removed is equal to the product of temperature times the absolute value of change in entropy. Therefore,

$$[8.50] \qquad Q^{out} = T_H^*(S_{ii} - S_i)$$

and

$$[8.51] \qquad Q^{in} = T_L^*(S_{iii} - S_{iv})$$

Let us start with the COP on a cooling basis,

$$[8.49a] \qquad COP_{Cooling} = \frac{Q^{in}}{W}$$

Using Eq. 8.48, W is substituted with the net heat from Eq. 8.48. Hence,

$$[8.52] \qquad COP_{Cooling} = \frac{Q^{in}}{Q^{out} - Q^{in}}$$

Substituting from Equations 8.50 and 8.51 into Eq. 8.52, we obtain:

$$[8.53a] \qquad COP_{Cooling}^{Carnot} = \frac{T_L}{T_H - T_L}$$

where T_L and T_H are the absolute temperatures of the cold reservoir (heat source) and hot reservoir (heat sink), respectively. If the working fluid is not in thermal equilibrium with the cold and hot reservoirs, then T_L and T_H are the absolute temperatures of the working fluid extracting heat from the cold reservoir and discharging heat to the hot reservoir, respectively.

Similarly,

$$[8.53b] \qquad COP_{Heating}^{Carnot} = \frac{T_H}{T_H - T_L}$$

The ideal limit for the COP of a heat pump is that of the Carnot cycle. The Carnot heat pump (although impractical) offers the highest value of the COP. The smaller the temperature lift $(T_H - T_L)$, the higher the COP and the better the performance. Depending on the values of T_L and T_H, the COP may exceed one. This is the reason for using the term COP instead of efficiency (which typically implies a fraction of one). Values of COP higher than one do not violate thermodynamic laws because energy is not generated in the system. It is simply moved from a lower temperature to a higher temperature. Another observation is obtained by subtracting Equations 8.53b and 8.53a to get:

$$[8.54] \qquad COP_{Heating} = COP_{Cooling} + 1$$

Therefore, the COP of a heat pump when used primarily for heating purposes is higher than when the heat pump (operating between the same T_L and T_H) is used primarily for cooling purposes. The reason is that, in the case of heating, the work added is converted to heating, which adds to the desired purpose. On the other hand, in the case of cooling, the work converted to heating does not contribute to the desired purpose of cooling. In some applications, the heat pump may be switched between dual purposes: heating and cooling. A common example is the dual-purpose heat pump for residential applications.

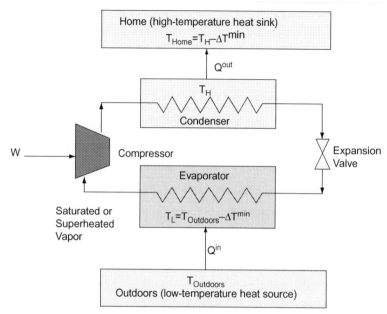

FIGURE 8-23A A heat pump in the heating mode.

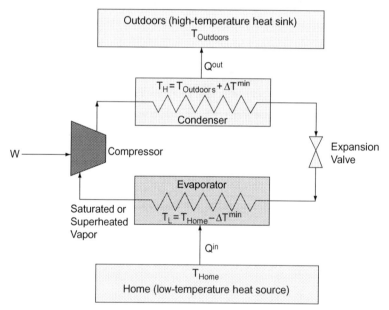

FIGURE 8-23B A heat pump in the cooling mode.

A heat exchanger (for example, coil) is placed inside the home and another one is placed in the outdoors. In the home-heating mode (for example, in winter) shown by Fig. 8.23a, the outdoor coil serves as the evaporator and extracts heat from the low-temperature heat source (the surroundings), and the coil inside the home serves as the evaporators and provides heat to the high-temperature heat sink (the home). A key characteristic of the dual-purpose heat pump is the usage of a valve that enables the reversal of the flow direction of the working fluid. Therefore, to switch from the heating mode to the cooling mode, the flow direction of the working fluid is reversed. Hence, in the home-cooling (air conditioning) mode (for example, in summer) shown by Fig. 8.23b, the coil inside the home serves as the evaporator and extracts heat from the low-temperature heat source (the home) and the outdoor coil serves as the evaporators and provides heat to the high-temperature heat sink (the surroundings). A minimum temperature difference (ΔT^{min}) is needed for heat transfer. Different values of ΔT^{min} may be used for the heat source to the evaporator coil versus the condenser coil to the heat sink.

Example 8-5 Calculating the refrigeration duty for a vapor-compression refrigeration cycle

A process unit is to be cooled using a vapor-compression heat pump that recirculates a refrigerant (tetrafluoroethane, which is commercially referred to as R-134a). The high-pressure refrigerant is throttled down to an evaporator pressure of 1 atm, which corresponds to a boiling temperature of 247 K. The refrigerant in the evaporator coil extracts heat from the process with a driving force of 5 K. Next, the refrigerant is compressed using a compressor that provides 100 kW to the refrigerant. The compressed refrigerant is fed to the condenser, which is cooled by cooling water at 300 K. The heat-transfer driving force between the refrigerant and the cooling water in the condenser is 10 K. The COP is 70 percent efficient compared to the COP of a Carnot cycle. Determine the refrigeration duty of the cycle.

SOLUTION

The refrigerant temperature in the condenser is 10 K higher than the temperature of the cooling water; that is,

[8.55] $$T_H = 300 + 10 = 310 \text{ K}$$

According to Eq. 8.53a,

[8.56] $$COP^{Carnot}_{Cooling} = \frac{247}{310 - 247} = 3.92$$

This means that for each kW of power added by the compressor, a maximum of 3.92 kW can be removed from the low-temperature heat source. Because the actual COP is 70 percent of the Carnot COP, then

[8.57] $$COP_{Cooling} = 0.7*3.92 = 2.74$$

Using Eq. 8.49a, we get the refrigeration duty to be

[8.57] $$Q^{in} = 2.74*100 = 274 \text{ kW}$$

According to Eq. 8.48,

[8.58] $$Q^{out} = 274 + 100 = 374 \text{ kW}$$

Figure 8.24 shows the key features of the refrigeration cycle.

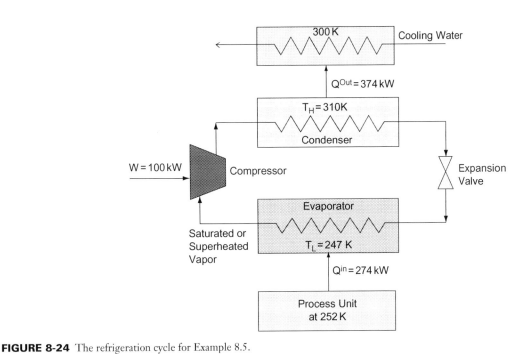

FIGURE 8-24 The refrigeration cycle for Example 8.5.

Example 8-6 Calculating the power needed for a dual-purpose vapor-compression heat pump

A heat pump is to be used to provide heating in cold weather and cooling in hot weather. The following average data are available for the cold weather:
- Required heating rate for the home = 36,000 Btu/hr
- Desired home temperature = 75°F
- Temperature of the surroundings = 55°F
- The actual COP is 65 percent of the Carnot COP for heating.
- A minimum driving force of 10°F is needed for any heat transfer.

The following average data are available for the hot weather:
- Required cooling rate for the home = 72,000 Btu/hr (or 6 tons of refrigeration where a ton of refrigeration corresponds to 12,000 Btu/hr

and the phrase is traditionally based on the refrigeration duty needed to freeze one ton of water)
- Desired home temperature = 70°F
- Temperature of the surroundings = 90°F
- The actual COP is 65 percent of the Carnot COP for cooling.
- A minimum driving force of 10°F is needed for any heat transfer.

If the compressor efficiency is 60 percent, determine the required size of the compressor expressed as the power intake in horsepower.

(*Continued*)

Example 8-6 Calculating the power needed for a dual-purpose vapor-compression heat pump (*Continued*)

SOLUTION
During Cold Weather (Heating Mode)

For the *COP* calculations, the refrigerant temperature in the condenser is 10°F higher than the temperature of the home, and the temperature has to be in absolute units (that is, Rankine). Therefore,

[8.59] $$T_H = (75 + 460) + 10 = 545\ R$$

The temperature of the refrigerant in the evaporator is 10°F lower than the surroundings. Hence,

[8.60] $$T_L = (55 + 460) - 10 = 505\ R$$

Using Eq. 8.53b,

[8.61] $$COP_{Heating}^{Carnot} = \frac{545}{545 - 505} = 13.63$$

Because the actual *COP* is 65 percent of the Carnot *COP*, then

[8.62] $$COP_{Heating} = 0.65*13.63 = 8.86$$

Using Eq. 8.49b, we get the power requirement to be

[8.63] $$W = 36{,}000/8.86 = 4063\ Btu/hr$$

According to Eq. 8.48,

[8.64] $$Q^{in} = 36{,}000 - 4063 = 31{,}937\ Btu/hr$$

During Hot Weather (Air Conditioning Mode)

The refrigerant temperature in the condenser is 10°F higher than the temperature of the surroundings; that is,

[8.65] $$T_H = (90 + 460) + 10 = 560\ R$$

The temperature of the refrigerant in the evaporator is 10°F lower than the home. Hence,

[8.66] $$T_L = (70 + 460) - 10 = 520\ R$$

Using Eq. 8.53a,

[8.67] $$COP_{Cooling}^{Carnot} = \frac{520}{560 - 520} = 13.00$$

Because the actual *COP* is 65 percent of the Carnot *COP*, then

[8.68] $$COP_{Cooling} = 0.65*13.00 = 8.45$$

Using Eq. 8.49b, we get the power requirement to be

[8.69] $$W = 72{,}000/8.45 = 8521\ Btu/hr$$

According to Eq. 8.48,

[8.70] $$Q^{out} = 72{,}000 + 8521 = 80{,}521\ Btu/hr$$

By comparing Equations 8.63 and 8.69, the compressor should be sized to deliver 8521 Btu/hr. Given that the compressor efficiency is 60 percent, the

[8.71] $$\text{Actual (brake) power provided to the compressor} = 8521/0.6 = 14{,}201\ Btu/hr$$

To convert to hp,

[8.72] $$\text{Actual (brake) horsepower provided to the compressor} = \frac{14{,}201\ Btu/hr}{2545\ Btu/hr.hp} = 5.58\ hp$$

The closest larger compressor that is commercially available should be selected (for example, 5.75 hp). The results for both seasons are shown by Figures 8.25a and b.

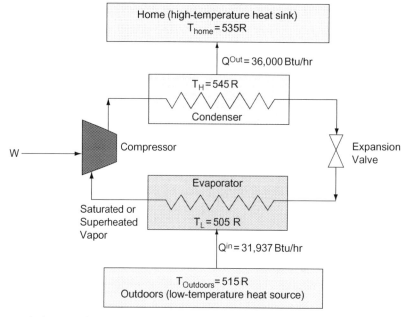

FIGURE 8-25a The heating mode for Example 8.5.

(*Continued*)

Example 8-6   Calculating the power needed for a dual-purpose vapor-compression heat pump (*Continued*)

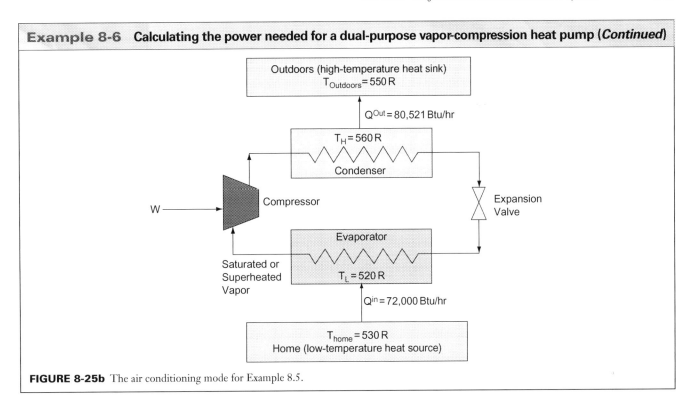

FIGURE 8-25b The air conditioning mode for Example 8.5.

VAPOR-COMPRESSION HEAT PUMPS AND THERMAL PINCH DIAGRAM

Heat pumps provide an efficient way of elevating the quality of heat. Therefore, they have the potential to be integrated with the HEN. The challenge is in proper placement of the heat pump. To illustrate the role of placing the heat pump, let us consider the following three cases for a vapor-compression heat pump:

- Placing the heat pump above the pinch: As can be seen in Fig. 8.26, this arrangement does not provide any benefit. Although the heating utility is reduced by W, an equivalent amount of work must be added to the heat pump, thereby nullifying the benefit. In fact, this situation leads to an economic loss because work is valued higher than heat.
- Placing the heat pump below the pinch: In this case, there are two penalties: Work is exerted on the heat pump without a benefit, and the cooling utility increases by a similar amount. Again, this represents an economic loss. Figure 8.27 illustrates this scenario.
- Placing the heat pump across the pinch: Figure 8.28 shows this arrangement. It offers two benefits: a reduction in heating and cooling utilities. For a work, W, exerted on the heat pump, the heating utility is reduced by $W + Q^{in}$, and the cooling utility is reduced by Q^{in}.

The previous discussion illustrates the need to appropriately place the heat pumps. As observed by Townsend and Linnhoff (1983a, b), *the heat pump should be placed across the pinch (not above the pinch or below the pinch).*

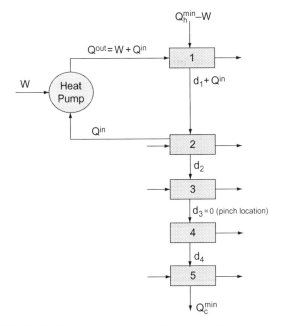

FIGURE 8-26 Placing the heat pump above the pinch.

Similar observations can be made using the grand composite curve (GCC). Also, the shape of the GCC can guide designers in determining the placement and size of a heat pump (Kemp, 2009; Smith, 2005; Benstead and Sharman, 1990; Linnhoff et al., 1982). Because the COP decreases as the lift in temperature increases, using

186 Sustainable Design Through Process Integration

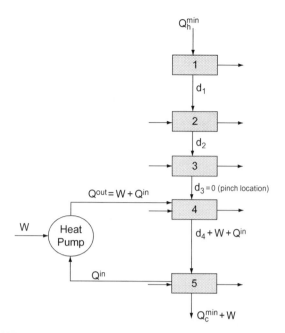

FIGURE 8-27 Placing the heat pump below the pinch.

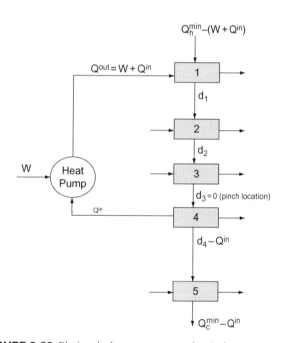

FIGURE 8-28 Placing the heat pump across the pinch.

the heat pump across the pinch should be limited to a relatively small temperature lift.

OPEN-CYCLE MECHANICAL VAPOR RECOMPRESSION USING A PROCESS STREAM AS THE WORKING FLUID

The closed-cycle vapor compression system uses a working fluid that is recirculated in a separate loop while exchanging heat with sources and sinks. In the open-cycle vapor recompression system, a process stream is used as the working fluid. This cycle has several useful

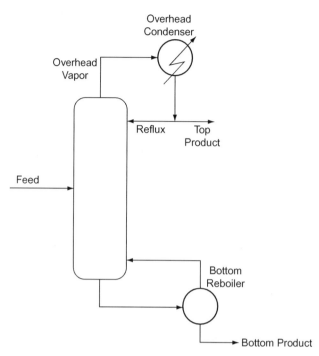

FIGURE 8-29 A distillation column.

applications. An important application is the mechanical vapor recompression (MVR) in distillation columns (for example, Shevni et al., 2011; Meili and Stuecheli, 1987; Danziger, 1979; Null, 1976) and its extension for internally heat integrated distillations columns (for example, Gadalla et al., 2005; Nakaiwa et al., 2003). To illustrate the concept of MVR, consider a distillation column with close-boiling components (Fig. 8.29). In such cases, the temperature of the vapor leaving the top of the column (and entering the overhead condenser) is slightly less than the temperature of the liquid leaving the bottom of the distillation column (and entering the reboiler). The overhead stream is to be cooled, and the bottom stream is to be heated. Therefore, if the temperature of the overhead vapor is increased above that of the bottom stream plus a minimum driving force, the two streams can be thermally integrated, and the utility requirements for the overhead condenser and the reboiler will decrease. Compression of the overhead vapor can achieve the heating task. Several implementations can be synthesized to represent the MVR concept. Figure 8.30 shows a possible configuration. The overhead vapor is compressed to a superheated state and is used to drive the reboiler in an exchanger that serves as a condenser for the overhead stream as well as a reboiler for the bottom stream. The condensed high-pressure overhead stream is expanded in a flash column to produce a liquid stream that provides the liquid top product, and the column reflux and a vapor stream that is mixed with the overhead vapor and compressed. A trim cooler may be used to adjust the heat balance and render the overhead in the desired state for flash to obtain the required flows of the top product and the reflux. Figure 8.31 shows an alternate implementation. Here an overhead exchanger is used to transfer heat from the recycled high-pressure liquid overhead to the low-pressure

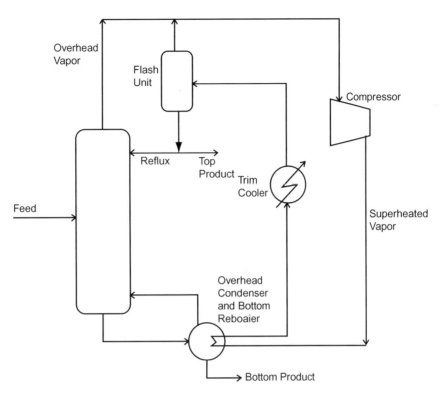

FIGURE 8-30 A distillation configuration with MVR.

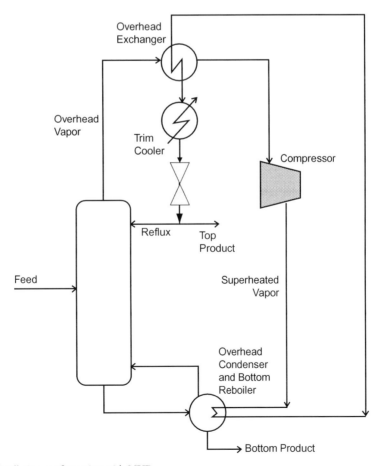

FIGURE 8-31 An alternate distillation configuration with MVR.

overhead stream leaving the top of the column. Such an exchange is beneficial for cooling the recycled overhead stream and heating the overhead stream leaving the top of the column prior to compression. A trim cooler may be used to obtain the product stream and the reflux at the desired conditions.

- Several other applications of MVR in the process industries are based on the aforementioned concept. A cost-benefit analysis should be carried out to determine the merits of installing the MVR system versus its capital and operating costs. As a general guideline, the temperature lift through compression should not be high. Otherwise the efficiency and the required power will be high. This is the reason that close-boiling distillation is a potential application for MVR. For preliminary design calculations, it is useful to use the following formulas for isentropic compression to evaluate the temperature lift and power:

Pressure-temperature expression for the isentropic compression of a gas from temperature and pressure of T_{in} and P_{in} to temperature and pressure of T_{out} and P_{out}

[8.73]
$$\frac{T_{out}}{T_{in}} = \left(\frac{P_{out}}{P_{in}}\right)^{\left(\frac{\gamma-1}{\gamma}\right)}$$

where γ is the ratio of heat capacity under constant pressure to heat capacity under constant volume. For ideal gases, the following values of γ may be used:

[8.74a] $\quad \gamma = 1.67$ for monatomic gases

[8.74b] $\quad = 1.4$ for diatomic gases

[8.74c] $\quad = 1.3$ for simple polyatomic gases

The compression power may be calculated as follows:

[8.75]
$$\text{Compression power} = \overset{\circ}{m}_{gas}\left(\frac{\gamma}{\gamma-1}\right)\left(\frac{RT_{in}}{M_{gas}\eta_{isentropic}}\right)\left[\left(\frac{P_{out}}{P_{in}}\right)^{\left(\frac{\gamma-1}{\gamma}\right)} - 1\right]$$

where $\overset{\circ}{m}_{gas}$ is the mass flow rate of the compressed gas, R is the universal gas constant, M_{gas} is the molecular weight of the gas, and $\eta_{isentropic}$ is the isentropic efficiency of the compressor. Substituting from Eq. 8.74 into Eq. 8.75, we get:

[8.76]
$$\text{Compression power} = \overset{\circ}{m}_{gas}\left(\frac{\gamma}{\gamma-1}\right)\left(\frac{R}{M_{gas}\eta_{isentropic}}\right)(T_{out} - T_{in})$$

Example 8-7 Using MVR to reduce heating and cooling utilities

A plant has a process hot stream and a process cold stream. Tables 8.3 give the stream data.

Stream H_1 is to be cooled to a temperature 360 K or lower. In the current operation, it is cooled to 360 K. Furthermore, the heat exchange duties of H_1 and C_1 are currently being fulfilled using external cooling and heating utilities. Therefore, the current values of heating and cooling utilities for the process are 2.25 and 1.00 MW, respectively (see Fig. 8.32). The flow rate of stream H_1 is 100 kg/s, its molecular weight is 28, and its pressure is 101.325 kPa (1 atm). A value of $\Delta T^{min} = 10$ K is used as the minimum driving force for any heat transfer.

As can be seen from the temperature interval diagram shown by Fig. 8.33a, the two streams do not overlap over a feasible temperature range. Therefore, under current conditions, there is no opportunity for heat integration. The objective of this example is to use MVR to induce heat integration. Perform the targeting calculations and comment on your results.

SOLUTION

For the heat pumped H_1 to be effective in transferring heat to C_1, it should be lifted to the top of interval 1 (385 K). Using Eq. 8.73, we get

$$\frac{385}{370} = \left(\frac{P_{out}}{101.325}\right)^{\frac{1.4-1}{1.4}}$$

Therefore,

[8.77] $\quad P_{out} = 116.446$ kPa

TABLE 8-3 Stream Data for the MVR Example

Stream	Flowrate × Specific Heat MW/K	Supply Temperature, K	Target Temperature, K
H_1 (Process Diatomic Gas)	0.10	370	≤360
C_1 (Liquid Hydrocarbon)	0.15	360	375

(Continued)

Example 8-7 Using MVR to reduce heating and cooling utilities (*Continued*)

And the power needed for the compressor is calculated through Eq. 8.76:

[8.78]
$$\text{Compression power} = 100\,\frac{\text{kg}}{\text{s}} * \left(\frac{1.4}{1.4-1}\right) * \left(\frac{8.314\,\frac{\text{kJ}}{\text{kmol.K}}}{28\,\frac{\text{kg}}{\text{kmol}} * 0.85}\right) * (385 - 370)$$
$$= 1834\text{ kW}$$
$$= 1.834\text{ MW}$$

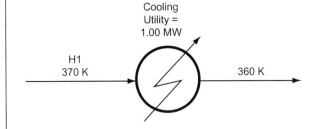

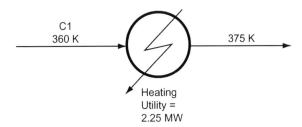

FIGURE 8-32 The unintegrated heating and cooling for Example 8.7.

The revised temperature interval diagram after heat pumping is shown by Fig. 8.33b. Now that stream H_1 is at a temperature level high enough, it can exchange heat with stream C_1 in interval 1. The exchangeable heat load is the lower of the two loads of the two streams in the intervals where they overlap. The lower of the two loads is that of H_1. Therefore,

[8.79]
$$\text{Heat transferred from } H_1 \text{ to } C_1 = 0.1*(385 - 370)$$
$$= 1.50\text{ MW}$$

Therefore,

[8.80]
$$\text{The heating utility after } H_1 - C_1 \text{ integration} = 2.25 - 1.50$$
$$= 0.75\text{ MW}$$

H_1 leaves the heat exchanger in which it is matched with C_1 at 370 K and 116.446 kPa (assuming negligible pressure drop). This stream can be throttled through an expansion valve. Assuming an isentropic expansion, Eq. 8.73 gives:

$$\frac{T_{out}}{370} = \left(\frac{101.325}{116.446}\right)^{\frac{1.4-1}{1.4}}$$

Therefore,

[8.81]
$$T_{out} = 356\text{ K}$$

Figure 8.34 illustrates the implementation of the MVR system, which results in reducing the heating and cooling utilities by 1.50 and 1.00 MW, respectively. This entails the use of 1.83 MW in the form of electric power to drive the compressor. Although the reduction in heating and cooling utilities exceeds the additional power for compression, electric power is usually valued higher than heating steam or oil and cooling water (but less valuable than refrigeration). Therefore, an economic analysis should be carried out to compare the savings in heating and cooling utilities versus the fixed cost of the MVR system and the operating cost of the electric power needed for compression. In this context, profitability criteria such as the return on investment and payback period are very useful.

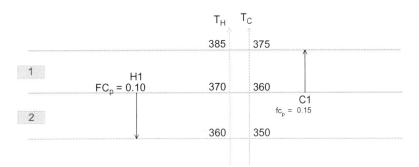

FIGURE 8-33a The temperature interval diagram for Example 8.7.

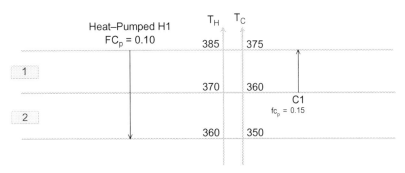

FIGURE 8-33b The revised cascade diagram for Example 8.7 after heat pumping.

(*Continued*)

Example 8-7 Using MVR to reduce heating and cooling utilities (*Continued*)

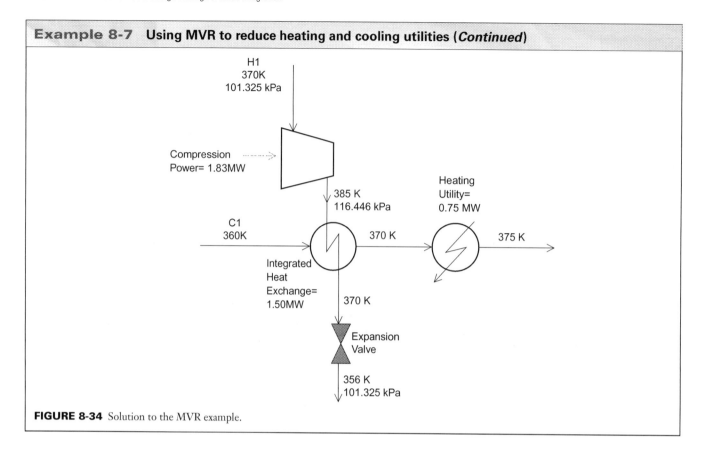

FIGURE 8-34 Solution to the MVR example.

ABSORPTION REFRIGERATION CYCLES

Another type of a heat pump is the absorption refrigeration (AR) cycle. Compared to vapor compression heat pumps where the major source of work is typically mechanical, the AR heat pumps are driven primarily by thermal energy. The source of heat may be a fossil fuel (for example, gas) or a high-pressure steam or waste heat from the process. The AR cycle heat pump uses two fluids: a working fluid and an absorbent. Although in vapor-compression heat pumps, the compressor is used to create the pressure difference to circulate the working fluid, in the absorption-cycle heat pumps, the absorbent is used to circulate the working fluid. This is done through three main steps. First, the low-pressure vapor from the evaporator is absorbed in the liquid absorbent. The dissolution of the working fluid in the absorbent is exothermic and heat must be removed from the solution. Next, the pressure of the solution is increased using a pump that consumes a modest amount of mechanical or electric energy. Then, the vapor is stripped or released in a stripper (or generator) where heat is added to the solution to induce stripping. The absorbent is returned to the absorber, the high-pressure vapor is fed to the condenser, and the rest of the cycle is similar to the vapor-compression heat pump.

The most common fluids used for the working fluid and absorbent are:

- water (working fluid) and lithium bromide (absorbent). This is normally used for applications above 273 K (for example, air conditioning).

- ammonia (working fluid) and water (absorbent). This is normally used for applications below 273 K (for example, refrigeration).

Figures 8.35a and b illustrate an overview and detailed sketches for the absorption cycle heat pumps.

Because the primary driver of the cycle is the heat added to the stripper (Q_{Str}), the *COP* for an AR cycle is typically defined as:

$$[8.82] \qquad COP_{AR} = \frac{Q^{in}}{Q_{Str}}$$

Several sources may be used to provide that heat added to the stripper. These sources include excess process heat, fossil fuels, and renewable fuels such as solar energy. To integrate the various sources of energy, some of which may fluctuate over time, heat can be delivered via a heat-transfer medium. A common example is the use of hot-water loops whereby water removes heat from the heat sources, dispatches heat to the stripper at a temperature T_{Str}^{in}, then leaves the heat exchanger associated with the stripper at a temperature T_{Str}^{out} and returns back to pick up heat and complete the hot-water loop (Fig. 8.36).

Prior to the use of external energy forms (for example, fossil, solar, biofuels), maximum usage should be made of the excess process heat. The GCC can be used to gain insights for the AR system (for example, Tora and El-Halwagi, 2010). The hot water leaving the stripper and entering the heat recovery and storage network has a temperature (T_{Str}^{out}). It is important to recall that the

CHAPTER 8 Integration of Combined Heat and Power Systems   **191**

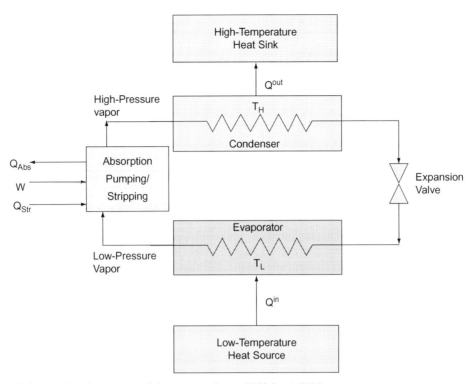

FIGURE 8-35a Overall sketch of the absorption-cycle heat pump. *Source: El-Halwagi (2006)*

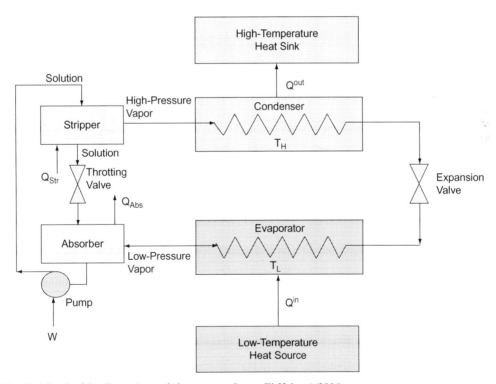

FIGURE 8-35b Detailed sketch of the absorption-cycle heat pump. *Source: El-Halwagi (2006)*

GCC represents the temperature versus enthalpy where the temperature is the arithmetic average of the temperatures of the hot and cold streams. This adds $\Delta T^{min}/2$ to the cold-stream temperature and subtracts $\Delta T^{min}/2$ from the hot-stream temperature. Hence, the temperature of the water stream leaving the heat exchanger of the stripper is represented on the GCC as $T_{Str}^{out} + \Delta T^{min}/2$. For this stream to receive heat in the heat recovery and storage network, the heat source must be at a higher temperature. Let us use the term $\Delta T^{Exch,HR}$ to designate the

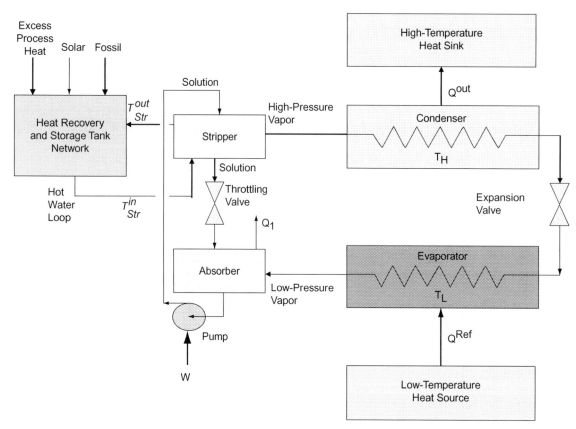

FIGURE 8-36 Heat supply to AR via a heat-recovery network and a hot-water loop. *Source: Tora and El-Halwagi (2010).*

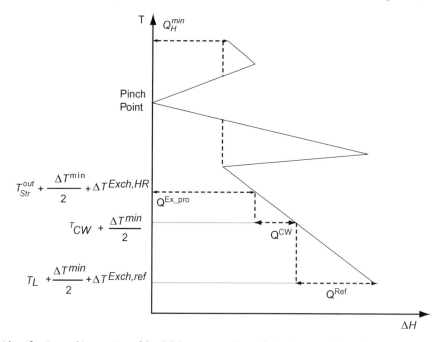

FIGURE 8-37 Target identification and integration of the GCC representation with the hot-water loop. *Source: Tora and El-Halwagi (2010).*

minimum heat-transfer driving force for a heat source transferring heat to the water in the heat recovery and storage tank. Therefore, the excess process heat must be at least at a temperature of $T_{Str}^{out} + \Delta T^{min}/2 + \Delta T^{Exch,HR}$. As shown by Fig. 8.37, the GCC can be used to determine the excess heat to be recovered for usage in AR ($Q^{Ex,Pro}$). Two benefits accrue as a result of transferring $Q^{Ex,Pro}$ from the excess process heat to the recirculating hot-water loop: reduction of the required cooling utility by $Q^{Ex,Pro}$ and supply of $Q^{Ex,Pro}$ to drive the AR system (which saves an equivalent amount of heat that would have otherwise been provided by an external energy source). The GCC also shows the utility targets for cooling water (Q^{CW}) and for the refrigerant (Q^{Ref}).

Example 8-8 Using the GCC to target AR integration

Consider the stream data for the heat-integration problem described by Tora and El-Halwagi (2010) and summarized by Table 8.4. A hot-water loop is used to extract heat from excess process heat and external energy sources, which then dispatches heat to the AR stripper. Cooling water may be used as an external utility to cool the hot streams down to 314 K. For cooling tasks below the cooling water level, an LiBr-water AR system is used to provide chilling water at 283 K. A minimum driving force (ΔT^{min}) value of 10 K is used. Develop the GCC and use it to determine how much excess process heat can be used to drive the AR system. If the COP of the AR cycle is 0.7, what is the refrigeration duty generated by the extracted process heat?

TABLE 8-4 Data for the Process Hot and Cold Streams for the AR Example

Stream	Supply Temperature, K	Target Temperature, K	Rate of Enthalpy Change, kW
H_1	398	338	−2325
H_2	358	298	−1125
H_3	318	298	−875
C_1	308	513	4100
C_2	308	328	725

Source: Tora and El-Halwagi (2010).

SOLUTION

Figure 8.38 plots the temperature interval diagram. The cascade diagram is developed to represent the heat balances around the temperature intervals (Fig. 8.39a). The cascade diagram is revised to remove the negative residuals and to determine the minimum heating and cooling utilities to be 2.5 MW and 2.0 MW, respectively (Fig. 8.39b). The residuals from the revised cascade diagram are used to plot the GCC for the process. Analysis of the GCC (see Fig. 8.40) shows that the process has a refrigeration need of 1 MW, cooling water duty of 0.25 MW, and an excess process heat of 0.75 MW that can be transferred to the hot-water loop to drive the LiBr-water AR system whose COP is 0.7. Therefore,

$$\text{Refrigeration duty generated by the excess process heat} = 0.7*0.75 = 0.525 \text{ MW} \quad [8.83]$$

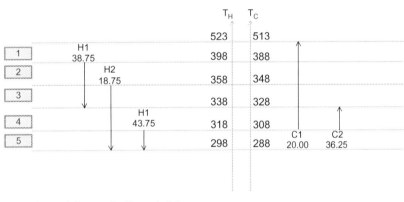

FIGURE 8-38 The temperature interval diagram for Example 8.8.

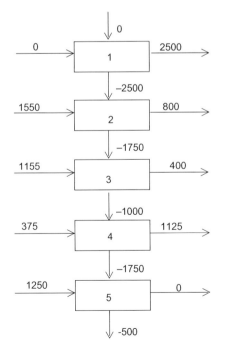

FIGURE 8-39a The cascade diagram for Example 8.8 (all numbers are in kW).

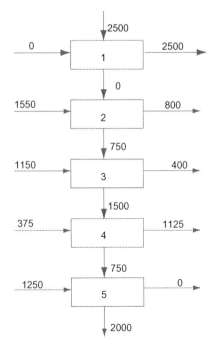

FIGURE 8-39b The revised cascade diagram for Example 8.8 (all numbers are in kW).

(Continued)

Example 8-8 Using the GCC to target AR integration (*Continued*)

The rest of the needed refrigeration (1.000−0.525 = 0.475 MW) is driven by heat extracted from external sources as follows:

[8.84] $\quad$ External heat needed to drive AR $= \dfrac{0.475}{0.7}$
$$= 0.679 \text{ MW}$$

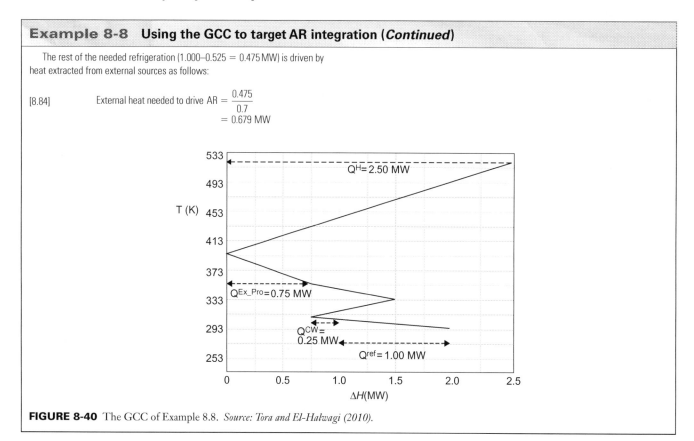

FIGURE 8-40 The GCC of Example 8.8. *Source: Tora and El-Halwagi (2010).*

COGENERATION TARGETING

The previous sections illustrate the benefits of combining heat and power. In this regard, steam plays a major role in the process. Steam is a convenient medium in capturing and delivering heat. Steam may also have nonheating uses in the process (for example, stripping agent, blanketing, and so on). Each application of steam requires a certain pressure (and/or temperature). For instance, as was seen in Chapter 7, the grand composite curve provides a method for utility selection, including steam levels. In addition to using steam for heating and nonheating purposes, it may be advantageous to generate the steam at higher pressures than the required levels and then pass the steam through turbines to generate electric power (or shaft work). This way, the steam plays a dual role: It provides the process requirements and produces power. This is often referred to as **cogeneration**, and when the process requirements are in the form of heating demands, it is referred to as combined heat and power.

This section provides a shortcut method to identifying the cogeneration target for a process. More details on the method can be found in El-Halwagi et al. (2009).

Consider a process with:

1. A number of combustible wastes and by-products;
2. Specific heating and cooling demands; and
3. Nonheating steam demands (for example, tracing, blanketing, stripping, and so on).

It is desired to identify a target for power cogeneration that makes effective use of the combustible wastes while fulfilling all process demands for heating and nonheating steam.

First, we start by considering the combustible wastes that can be burned in existing boilers and industrial furnaces that provide heat, typically in the form of steam. Depending on the specific boiler or industrial furnace, the steam is generated at a specific pressure and is fed to the appropriate steam header. The steam demands can be determined through heat and mass integration. Heat integration (for example, grand composite analysis) can be used to determine the heating steam requirements and their levels. Mass integration can be used to determine the nonheating steam demands. The result of the mass integration analysis is the identification of process supply of steam (from the combustible wastes) and the process demand of steam (for nonheating purposes). The supplies and demands are determined in terms of quantities and levels of the steam headers. Those supplies and demands are now known both in terms of quantities and pressures (or header level). A typical industrial process has several pressure levels in its steam system. Consider a process with the following headers: very high pressure (VHP), high pressure (HP), medium pressure (MP), and low pressure (LP). Depending on the supply and demand to each header, the net balance of steam for each header will be positive (surplus) or negative (deficit). For instance, Fig. 8.41 shows a case where there is a surplus of steam at the VHP and HP levels, whereas there is a deficit of steam at the MP and LP levels. To satisfy the steam deficits in the MP and LP levels, steam may be let down from the surplus headers and/or supplied from fired boilers. These decisions will be made later.

When the objective is to target cogeneration potential for a process, it is advantageous to determine this target without detailed calculations. Conventional turbine models discussed earlier in the chapter require individual turbine calculations to evaluate the outlet enthalpy at constant entropy and the flow rate of steam passing through the turbine. As such, it is useful to develop expressions for estimating turbine power using easily determined terms. Turbines are placed between steam headers with known temperature and pressures. Consequently, the specific enthalpies of these steam headers are also known. El-Halwagi et al. (2009) proposed an approximation of the turbine power to be based on the header enthalpies. They introduced the term **extractable energy**, which is based on of the header levels that the turbine actually operates between, rather than the isentropic conditions at the outlet pressure.

$$[8.85] \qquad e_{Header} = \eta_{Header} H_{Header}$$

where e_{Header} is the extractable energy for a given header, η_{Header} is an efficiency term, and H_{Header} is the specific enthalpy at a given set of conditions for the header. Let the extractable power, E_{Header}, of a header be defined as:

$$[8.86] \qquad E_{Header} = \dot{m}\eta_{Header} H_{Header} = \dot{m} e_{Header}$$

The advantage of this definition is that it does not involve detailed turbine calculations such as isentropic outlet enthalpy.

Then, the power generation expression can be rewritten as the difference between the inlet and outlet extractable power:

$$[8.87] \qquad W = E^{in} - E^{out}$$

where E^{in} is the extractable power at the header conditions feeding the inlet steam to the turbine, and E^{out} is the extractable power at the header conditions receiving the outlet steam from the turbine. Fig. 8.42b shows a representation of this form of power evaluation on a Mollier diagram. As can be seen from this figure, the header efficiency is given by:

$$[8.88] \qquad \eta_{Header} = \frac{\Delta H^{real}}{H^{in} - H^{out}}$$

Based on the concept of extractable energy, El-Halwagi et al. (2009) developed a graphical targeting procedure that identifies cogeneration targets and serves as the basis for developing feasible turbine network designs that meet the predicted target. According to this method, the

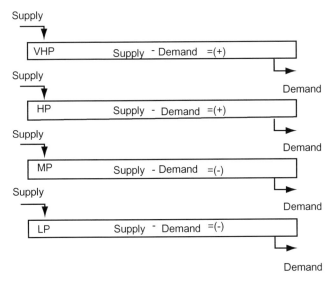

FIGURE 8-41 Header balance. *Source: El-Halwagi et al. (2009).*

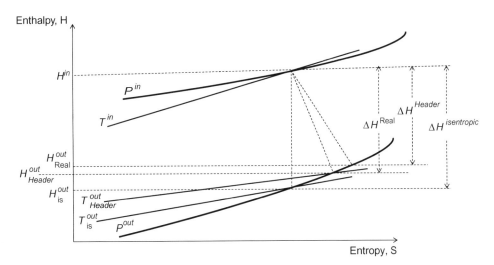

FIGURE 8-42 Approximation of turbine performance using header enthalpies. *Source: El-Halwagi et al. (2009)*

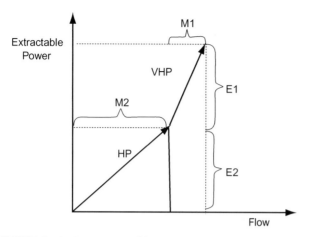

FIGURE 8-43 Construction of the extractable power surplus composite curve. *Source: El-Halwagi et al. (2009).*

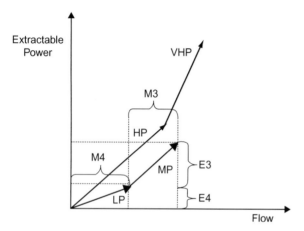

FIGURE 8-44 Construction of the extractable power surplus and deficit composite curves. *Source: El-Halwagi et al. (2009).*

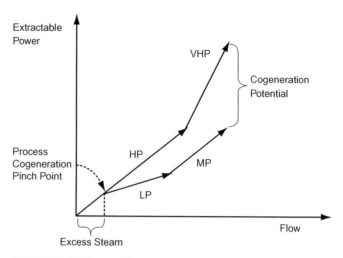

FIGURE 8-45 Extractable power cogeneration targeting pinch diagram. *Source: El-Halwagi et al. (2009).*

extractable power expressed by Eq. 8.86 is plotted for each header versus the net flow rate of the header. First, the extractable power is plotted versus the steam flow rate for each surplus header in ascending order of pressure levels. The result of this superposition is the development of a *surplus composite line*. Similarly, the extractable power for the deficit headers is plotted in ascending order, leading to the *deficit composite line*. Consider the case shown in Fig. 8.43, which is based on Fig. 8.41. There are two surplus headers (VHP and HP) and two deficit headers (MP and LP). The extractable power at the VHP, HP, MP, and LP levels are referred to as E1, E2, E3, and E4, respectively, and the steam flow rates of these headers are referred to as M1, M2, M3, and M4, respectively. The construction of the surplus composite line is shown by Fig. 8.43. The deficit composite line can be constructed in a similar manner.

The next step is to represent the two composite lines on the same diagram (Fig. 8.44). Hence, the cogeneration potential of the system can be evaluated by shifting the deficit composite line to the right and up until it is directly below the terminal point of the surplus line. As illustrated by Fig. 8.45, the vertical distance (of the "jaw") between the two terminal points of the surplus and the deficit composite lines is the *target for cogeneration potential*. This diagram is referred to as the *extractable power cogeneration targeting pinch diagram*. To guarantee that the cogeneration target is feasible, higher-pressure surplus headers must be directly above lower-pressure deficit headers. By letting down steam from the higher-pressure headers to the lower-pressure headers, the deficit is removed while power can be generated by virtue of the difference between the extractable powers. Therefore, both steam demands (heating and nonheating) are satisfied while power is cogenerated. The steam flow rate of the portion of the surplus composite line that does not overlap with the deficit composite line represents excess steam. Because there is no header demand for this excess, it can be used for power generation (not cogeneration) by letting it down through a condensing turbine, used for other process purposes, or simply vented.

Example 8-9 Cogeneration target (El-Halwagi et al., 2009)

Consider a process with four steam headers: VHP, HP, MP, and LP. The temperature and pressure of these steam headers are given in Table 8.5. For this case study, the power efficiency for any header is taken as 70 percent (that is, $\eta = 0.7$).

As a result of mass and heat integration, a number of steam streams were supplied to the headers. The results are given in Table 8.6.

By plotting the extractable power versus flow rate, the results are shown by Fig. 8.46. Next the deficit composite line is shifted to up and to the right such that its terminal point is vertically aligned with the terminal point of the surplus composite line (Fig. 8.47). As can be seen from Fig. 8.47, the cogeneration target is 21.1 MMBtu/h, and there is an excess of approximately 51,000 lb/h of steam being generated within the process. The excess steam in this case should be let down through a condensing turbine because it primarily comes from a process source. This additional power is considered to be generation (and not cogeneration) because it does not contribute to satisfying other process needs.

(Continued)

Example 8-9 Cogeneration target (El-Halwagi et al., 2009) (Continued)

TABLE 8-5 Steam Header Data

Header	Pressure (psia)	Temperature (°F)	Specific Enthalpy (Btu/lb)
VHP	600	805	1414
HP	160	600	1327
MP	80	395	1226
LP	60	300	1181

Source: El-Halwagi et al. (2009).

TABLE 8-6 Surplus/Deficit Steam Flow Rates and Extractable Power

Surplus Header	Pressure (psia)	Flow Rate (lb/h)	Net Enthalpy Difference per Hour (MMBtu/h)	Extractable Power (MMBtu/h)
VHP	600	119,519	169	118.3
HP	160	101,676	135	94.5
Deficit Headers				
MP	80	110,114	−135	94.5
LP	60	60,019	−71	49.7

Source: El-Halwagi et al. (2009).

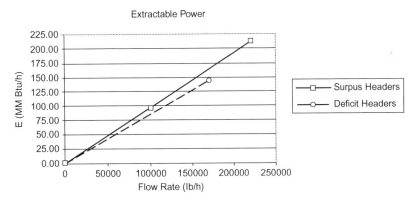

FIGURE 8-46 Unshifted extractable power versus flow rate plot. *Source: El-Halwagi et al. (2009).*

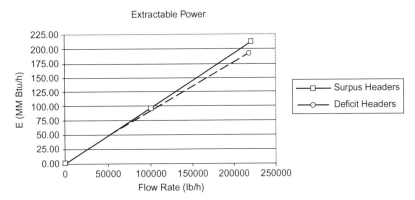

FIGURE 8-47 Shifted extractable power versus flow rate plot. *Source: El-Halwagi et al. (2009).*

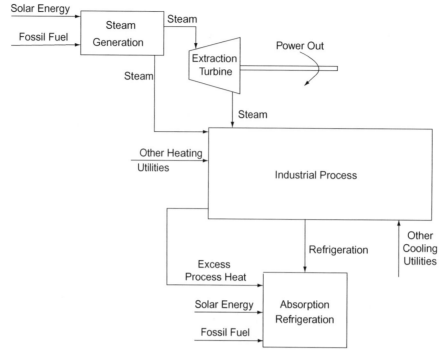

FIGURE 8-48 An integrated view of a solar-assisted trigeneration system. *Source: Tora and El-Halwagi (2011).*

ADDITIONAL READINGS

Several methods have been used to assess the cogeneration opportunities of a process. Dhole and Linnhoff (1992) proposed to use exergy analysis to estimate cogeneration opportunities. **Exergy** is a measure of the useful work available in a heat source. Raissi (1994) developed the TH-shaftwork targeting model. This model is based on an observation that the specific enthalpy at the turbine outlet minus the specific enthalpy of saturated water is relatively constant regardless of the outlet conditions. The TH-shaftwork target combines this observation with the observation that specific power can be approximated by a linear function of the outlet saturation temperature (Klemes et al., 1997). Patel et al. (2005) developed a thermodynamic framework for work integration. Mathematical-programming techniques have also been made in the area of modeling and optimization of turbine network systems. The turbine hardware model (Mavromatis, 1996; Mavromatis and Kokossis, 1998) is based on the Willans line (commonly used in turbine modeling to represent steam consumption versus rated power of the turbine) and utilizes typical maximum efficiency plots and rules of thumb to target cogeneration potential. The turbine hardware model also incorporates complex turbines by modeling them as sets of simpler turbines. Mohan and El-Halwagi (2007) developed an algebraic procedure for cogeneration targeting and included biomass as one of the energy sources. Al-Azri et al. (2009) developed a shortcut algorithmic approach for the optimization of cogeneration systems. Mahmud et al. (2011) presented a procedure for the simultaneous design of combined heat and power along with mass integration. Ponce-Ortega et al. (2011) incorporated technical, economic, and environmental criteria in the design of absorption-refrigeration systems while including the use of renewable energy.

The concept of combined heat and power (CHP) can be further extended to combined cooling, heating, and power (CCHP), which is commonly referred to as **trigeneration**. Details on trigeneration can be found in literature reviews (for example, Al-Sulaiman et al., 2010; Wu and Wang, 2006). Tora and El-Halwagi (2011) provided a conceptual design approach to trigeneration systems that integrated excess process heat (following heat integration) with external forms of energy (for example, fossil fuels and solar energy). Figure 8.48 is a schematic representation of the integrated trigeneration problem.

HOMEWORK PROBLEMS

8.1. Consider a heat engine that approximates the performance of a Carnot heat engine. The heat engine draws 3000 kW of heat from a hot reservoir at 800 K and discharges heat to a heat sink at 600 K. Determine the following:
   1. Efficiency of the heat engine
   2. Generated work rate (power)
   3. Rate of heat discharge to heat sink

8.2. Consider the heat engine of Problem 8.1. Suppose that this heat engine is used in the same process described by the pharmaceutical process of Example 7.1. What is the benefit for integrating the heat engine with the HEN?

8.3. A steam power plant is operated as a Rankine cycle. The steam turbine receives superh eated steam at 700 psi and 650°F and discharges steam at 380 psi. The turbine

TABLE 8-7 Stream Data for the MVR Example

Stream	Flow Rate × Specific Heat, MW/K	Supply Temperature, K	Target Temperature, K
H_1 (Simple Polyatomic Gas)	0.05	320	≤290
C_1	0.08	310	340

TABLE 8-8 Propylene Process Heating/Cooling Data

Stream	FC_p (MMBtu/h °F)	T_{Supply} (°F)	T_{Target} (°F)
H1	3.40	110	80
H2	1.60	50	30
H3	0.70	120	110
H4	1.00	105	80
C1	1.60	180	230
C2	1.41	130	157
C3	0.76	160	218
C4	3.98	115	120

Source: Mahmud et al. (2011).

TABLE 8-9 Process Sink Data

Sink	Flow Rate (lb/hr)	Maximum Inlet Mass Fraction (Impurities)	Maximum Inlet Load (lb/hr)
VA Process Reactor	34,000	0.20	6800

Source: Mahmud et al. (2011).

TABLE 8-10 Process Source Data

Source	Flow Rate (lb/hr)	Outlet Mass Fraction (Impurities)	Maximum Inlet Load (lb/hr)	Heating Values (Btu/lb)
De-ethanizer	10,000	0.48	4800	1000
Absorption Column	20,000	0.65	13,000	1500

Source: Mahmud et al. (2011).

TABLE 8-11 Raw Material Data

Raw Material	Cost ($/lb)	Mass Fraction (Impurities)
Fresh	0.11	0.00

Source: Mahmud et al. (2011).

TABLE 8-12 Fresh Fuel Data

Fresh Fuel	Cost ($/MMBtu)	Heating Value (Btu/lb)
Fuel	2.6	13,400

Source: Mahmud et al. (2011).

has a power-generation isentropic efficiency of 60 percent. The flow rate of steam is 50,000 lb/hr. The steam leaving the turbine is fed to a condenser to produce saturated liquid at 380 psi. The condensate is pumped isentropically to a subcooled liquid at 700 psi. The pump efficiency is 75 percent. The subcooled liquid leaving the pump is subsequently fed to an isobaric boiler that produces superheated steam at 700 psi and 650°F, and the cycle continues. The thermal efficiency of the boiler is 70 percent. Using the steam tables, calculate:
1. The power generated by the steam turbine,
2. The cooling duty of the condenser,
3. The power to be provide to the pump, and
4. The heat to be generated from fuel combustion in the boiler.

8.4. Calculate the power generated by the steam turbine of Problem 8.3 using the correlations of Al-Azri et al. (2009).

8.5. A separation unit is cooled using a vapor-compression heat pump. The throttled refrigerant boils at 260 K and extracts heat from the separation unit with a driving force of 3 K. Next, the refrigerant is compressed using a compressor that provides 300 kW to the refrigerant. The compressed refrigerant is fed to the condenser, which is cooled by cooling water at 300 K. The heat-transfer driving force between the refrigerant and the cooling water in the condenser is 10 K. The COP is 60 percent efficient compared to the COP of a Carnot cycle. Calculate the refrigeration duty of the cycle.

8.6. A production facility has a process hot stream and a process cold stream. The stream data are given by Table 8.7.

The flow rate of stream H_1 is 55 kg/s, its molecular weight is 16, and its pressure is 101.325 kPa.

A value of $\Delta T^{min} = 10$ K is used as the minimum driving force for any heat transfer. Synthesize an MVR to induce heat integration and evaluate the utility targets. Develop an economic analysis and comment on your results.

8.7. Resolve Example 8.9 when the net surplus flow rate of the VHP header is 150,000 lb/hr.

8.8. Consider the propylene manufacturing process by catalytic dehydrogenation of propane, which is described by Mahmud et al. (2011). Tables 8.8 through 8.14 summarize the data for the heating, cooling, process sinks, process sources, and external

TABLE 8-13 Process Fuel Data

Process Fuel	Flow (lb/hr)	Heating Value (Btu/lb)
Depropanizer Bottom	20,000	13,400

Source: Mahmud et al. (2011).

TABLE 8-14 Electricity Consumption Data

	Demand (MW)	Cost ($/kWh)
Electricity	12	0.06

Source: Mahmud et al. (2011).

resources for the case study. The objective of this problem is to optimize fresh and utility consumption, waste recycle, recovery of material, and thermal value from waste, external fuel, and electricity consumption.

REFERENCES

Al-Azri N, Al-Thubaiti M, El-Halwagi MM: An algorithmic approach to the optimization of process cogeneration, *J Clean Tech Env Policy* 11(3):329–338, 2009.

Al-Sulaiman FA, Jamdullaphur F, Dincer I: Trigeneration: a comprehensive review based on prime movers, *Int J Energy Res*, doi:10.1002/er.1687, 2010.

Benstead R, Sharman FW: Heat pumps and pinch technology, *Heat Recovery Systems & CHP* 10(4):387–398, 1990.

Danziger R: Distillation with vapor recompression, *Chem Eng Prog* 75:58–64, 1979.

Dhole VR, Linnhoff B: *Comp Chem Eng* 17:S101–S109, 1992.

El-Halwagi MM: *Process integration*, Amsterdam, 2006, Academic Press/Elsevier.

El-Halwagi MM, Harell D, Spriggs HD: Targeting cogeneration and waste utilization through process integration, *Appl Energy* 86(6):880–887, 2009.

Gadalla M, Olujic Z, Sun L, De Rijke A, Jansens PJ: Pinch analysis-based approach to conceptual design of internally heat-integrated distillation columns, *Trans IChemE Chem Eng Res Des* 83(8A):987–993, 2005.

Harvey AH, Parry WT: Keep your steam tables up to date, *Chem Eng Prog* 95(1):45, 1999.

Kemp I: *Pinch analysis and process integration: a user guide on process integration for the efficient use of energy*, ed 2, Butterworth-Heinemann, 2009.

Klemes J, Dhole VR, Raissi K, Perry SJ, Puigjaner L: Targeting and design methodology for reduction of fuel, power, and CO_2 on total sites, *J Appl Therm Eng* 17:993, 1997.

Linnhoff B: Pinch analysis: a state of the art overview, *Trans Inst Chem Eng Chem Eng Res Des* (71)1993.

Linnhoff B, Townsend DW, Boland D, Hewitt GF, Thomas BEA, Guy AR, et al. *User guide on process integration for the efficient use of energy*, Rugby, Warwickshire, UK, ICHemE Press, 1982.

Mahmud R, Harell D, El-Halwagi MM: A process integration framework for the optimal design of combined heat and power systems in the process industries. In Foo DCY, El-Halwagi MM, Tan RR, editors: *Recent advances in sustainable process design and optimization (advances in process systems engineering)*, World Scientific Publishing Company, (Chapter 14), 2011.

Mavromatis SP: *Conceptual design and operation of industrial steam turbine networks*, PhD thesis Manchester, UK, University of Manchester Institute of Science and Technology, 1996.

Mavromatis SP, Kokossis AC: Conceptual optimization of utility networks for operational variation-I targets and level optimization, *Chem Eng Sci* 53:1585, 1998.

Meili A, Stuecheli A: Distillation columns with direct vapor recompression, *Chem Eng* 94:133–143, 1987.

Mohan T, El-Halwagi MM: An algebraic targeting approach for effective utilization of biomass in cogeneration systems through process integration, *J Clean Tech Env Policy* 9(1):13–25, 2007.

Nakaiwa M, Huang K, Endo A, Ohmori T, Akiya T, Takamatsu T: Internally heat integrated distillation columns: a review, *Trans IChemE, Part A, Chem Eng Res Des* 81:162–177, 2003.

Null HR: Heat pumps in distillation, *Chem Eng Prog* 72:58–64, 1976.

Patel B, Hildebrandt D, Glasser D, Hausberger B: Thermodynamic analysis of processes: 1. Implications of work integration, *Ind Eng Chem Res* 44(10):3529–3537, 2005.

Ponce-Ortega JM, Tora EA, Gonzalez-Campos JB, El-Halwagi MM: Integration of renewable energy with industrial absorption refrigeration systems: systematic design and operation with technical, economic, and environmental objectives, *Ind Eng Chem Res* (in press, 2011, dx.doi.org/10.1021/ie200141j).

Raissi K: *Total site integration*, PhD thesis Manchester, UK, University of Manchester Institute of Science and Technology 1994.

Shevni AA, Herron DM, Agrawal R: Energy efficiency limitations of the conventional heat integrated distillation column for binary distillation, *Ind Eng Chem Res* 50:119–130, 2011.

Smith JM, Van Ness HC, Abbott MM: *Introduction to chemical engineering thermodynamics*, ed 7, McGraw Hills, 2005.

Smith R: *Chemical process design and integration*, New York, 2005, Wiley.

Tora EA, El-Halwagi MM: Integration of solar energy into absorption refrigerators and industrial processes, *Chem Eng Technol* 33(9):1495–1505, 2010.

Tora EA, El-Halwagi MM: Integrated conceptual design of solar-assisted trigeneration systems, *Comp Chem Eng* 35(9):1807–1814, 2011.

Townsend DW, Linnhoff B: Heat and power networks in process design. I. Criteria for placement of heat engines and heat pumps in process networks, *AIChE J* 29:742–747, 1983a.

Townsend DW, Linnhoff B: Heat and power networks in process design. II. Design procedure for equipment selection and process matching, *AIChE J* 29:747–754, 1983b.

Varbanov PS, Doyle S, Smith R: Modelling and optimization of utility systems, *Trans IChemE* 82:561–578, 2004.

Wu DW, Wang RZ: Combined cooling, heating and power: a review, *Prog Energy Combustion Sci* 32(5–6):459–495, 2006.

CHAPTER 9

PROPERTY INTEGRATION

A common feature of the previously covered mass-integration techniques is that they are "chemo-centric"; namely, they are based on tracking individual chemical species. Nonetheless, many material reuse problems are driven and governed by properties or functionalities of the streams and not by their chemical constituency. The following are some examples of property-based problems:

- The usage of material utilities (for example, solvents) relies on their characteristics such as equilibrium distribution coefficients, viscosity, and volatility without the need to chemically characterize these materials.
- Constraints on process units that can accept recycled/reused process streams and wastes are not limited to compositions of components but are also based on the properties of the feeds to processing units.
- The performance of process units depends on properties. For instance, a heat exchanger performs based on the heat capacities and heat-transfer coefficients of the matched streams. The chemical identity of the components is only useful to the extent of determining the values of heat capacities and heat-transfer coefficients. Similar examples can be given for many other units (for example, vapor pressure in condensers, specific gravity in decantation, relative volatility in distillation, Henry's coefficient in absorption, density and head in pumps, density, pressure ratio, heat-capacity ratio in compressors, and so on).
- Quantities of emissions are dependent on properties of the pollutants (for example, volatility, solubility, and so on).
- The environmental regulations involve limits on properties (for example, pH, color, toxicity, total organic carbon (TOC), biological oxygen demand (BOD), ozone-depleting ability).
- Tracking numerous chemical pollutants is prohibitively difficult (for example, complex hydrocarbons and lignocellulosic materials), whereas tracking properties is manageable.

Therefore, it is important to have a systematic design methodology that is based on properties and functionalities. In response, El-Halwagi and his coworkers introduced the paradigm of property integration. **Property integration** is a functionality-based, holistic approach to the allocation and manipulation of streams and processing units, which is based on the tracking, adjustment, assignment, and matching of functionalities throughout the process (El-Halwagi et al., 2004).

In this chapter, we focus on the problem of identifying rigorous targets for direct reuse in property-based applications through visualization techniques. The chapter also discusses the identification of interception tasks. First, the problem of direct recycle with single-property constraints is addressed through a material recycle pinch diagram similar to the one presented in Chapter 3. Then, we deal with the problem of multiple properties.

PROPERTY-BASED MATERIAL RECYCLE/REUSE PINCH DIAGRAM

The first problem we address is the property-based direct recycle when the constraints are based on one property. More details on the problem and its solution technique can be found in Kazantzi and El-Halwagi (2005). The problem can be stated as follows:

Consider a process with a number N_{sinks} of process sinks (units). Each sink, j, requires a feed with a given flow rate, G_j, and an inlet property, p_j^{in}, that satisfies the following constraint:

$$[9.1] \quad p_j^{min} \leq p_j^{in} \leq p_j^{max} \quad \text{where } j = 1, 2, \ldots, N_{sinks}$$

where p_j^{min} and p_j^{max} are the specified lower and upper bounds on admissible property to unit j.

The plant has a number $N_{sources}$ of process sources (for example, process streams, wastes) that can be considered for possible reuse and replacement of the fresh material. Each source, i, has a given flow rate, W_i, and a given property, p_i.

Available for service is a fresh (external) resource whose property value is p_{Fresh} and can be purchased to supplement the use of process sources in sinks. The objective is to develop a noniterative graphical procedure that determines the target for minimum usage of the fresh resource, maximum material reuse, and minimum discharge to waste. The problem can be schematically represented through a source-sink allocation as shown in Fig. 9.1.

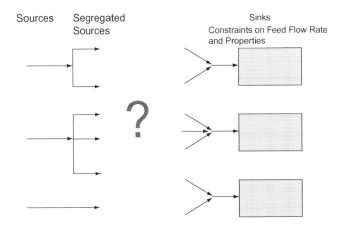

FIGURE 9-1 Schematic representation of the property-based material reuse problem.

According to this representation, each source is allowed to split and be forwarded to all sinks. In particular, the objective here is to determine the optimum flow rate from each source to each sink so as to minimize the consumption of the fresh resource.

When several sources are mixed, it is necessary to evaluate the property of the mixture as a function of the flow rate and property of each stream. Consider the following mixing rule for estimating the resulting property of the mixture:

$$\bar{F} * \psi(\bar{p}) = \sum_i F_i * \psi(p_i) \tag{9.2}$$

where $\psi(p_i)$ is the property-mixing operator and $\bar{F}$ is the total flow rate of the mixture, which is given by:

$$\bar{F} = \sum_i F_i \tag{9.3}$$

The property-mixing operators can be evaluated from first principles or estimated through empirical or semi-empirical methods. For instance, consider the mixing of two liquid sources whose flow rates are F_1 and F_2, volumetric flow rates are V_1 and V_2, and densities are ρ_1 and ρ_2. Suppose that the volumetric flow rate of the mixture is the sum of the volumetric flow rates of the two streams; that is,

$$\bar{V} = V_1 + V_2 \tag{9.4}$$

Recalling the definition of density and designating the total flow rate of the mixture by $\bar{F}$, we get

$$\frac{\bar{F}}{\bar{\rho}} = \frac{F_1}{\rho_1} + \frac{F_2}{\rho_2} \tag{9.5}$$

Comparing Equations 9.2 and 9.5, we can define the density-mixing operator as:

$$\psi(\rho_i) = \frac{1}{\rho_i} \tag{9.6}$$

Next, we define the property load of a stream as follows:

$$M_i^{Source} = W_i * \psi(p_i) \tag{9.7}$$

For source i, let

$$y_i = \psi(p_i) \tag{9.8}$$

Therefore, Eq. 9.7 can be rewritten as

$$M_i^{Source} = W_i * y_i \tag{9.9}$$

We can rewrite the sink constraints given by Eq. 9.2 in terms of the property-mixing operator as:

$$\psi_j^{min} \le \psi_j^{in} \le \psi_j^{max} \tag{9.10}$$

To develop the targeting procedure, let us first start with the following special case for which the fresh source has superior properties to all other streams and the sink constraint is given by:

$$\psi^{Fresh} \le \psi_j^{in} \le \psi_j^{max} \tag{9.11}$$

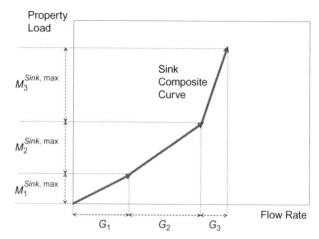

FIGURE 9-2 Developing sink composite curve.

where

$$\psi^{Fresh} = \psi(p_{Fresh}) \tag{9.12}$$

The property load of a sink is defined as the product of the flow rate times the property operator of the feed to the sink. Hence, we define the maximum property load for a sink as

$$M_j^{Sink,max} = G_j * \psi_j^{max} \tag{9.13}$$

For sink j, let

$$z_j^{max} = \psi_j^{max} \tag{9.14}$$

Therefore, Eq. 9.13 can be rewritten as

$$M_j^{Sink,max} = G_j * z_j^{max} \tag{9.15}$$

It is beneficial to observe the similarity between Equations 9.9 and 9.15, with their equivalent expressions for loads of impurities given in Chapter 4 for direct recycle using concentration constraints. Therefore, the property loads are analogous to the mass loads. Consequently, we can develop a targeting procedure similar to the one given by the material recycle pinch diagram. This tool is referred to as the *property-based material recycle (or reuse) pinch diagram* (Kazantzi and El-Halwagi, 2005) and can be constructed through the following procedure:

1. **Sink data.** Gather data on the flow rate and acceptable range of targeted property for each sink as in Constraint 9.1. Using the admissible range of property value, calculate the maximum value of the property operator (ψ_j^{max}). Then, evaluate the maximum admissible property load ($M_j^{Sink,max}$) using Eq. 9.13. Rank the sinks in ascending order of ψ_j^{max}.
2. **Sink composite.** Using the required flow rate for each sink (G_j) and the calculated values of the maximum admissible loads ($M_j^{Sink,max}$), develop a representation for each sink in ascending order. Superposition is used to create a sink composite curve. This is shown in Fig. 9.2.
3. **Fresh line.** Use Eq. 9.12 to evaluate the property operator of fresh (ψ_{Fresh}). A locus of the fresh line

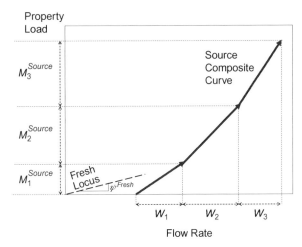

FIGURE 9-3 Source composite curve.

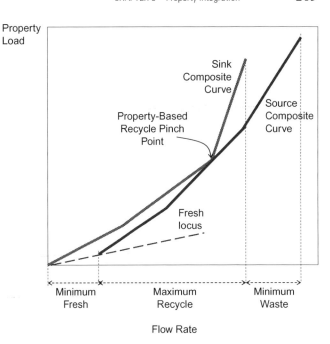

FIGURE 9-4 Property-based material recycle pinch diagram. *Source: Kazantzi and El-Halwagi (2005).*

is drawn starting from the origin with a slope of ψ_{Fresh}.

4. **Source data.** Gather data on the flow rate and property value for each process source. Using the functional form of the property operator, calculate the value of the property operator for each source (ψ_j). Rank the sources in ascending order of ψ_j. Also, calculate the property load of each source (M_i) using Eq. 9.7.
5. **Source composite.** Using the flow rate for each source (G_i) and the calculated values of the property operator (ψ_j), develop a representation for each source in ascending order. Superposition is used to create a source composite curve. As mentioned in step 3, the locus of the fresh stream is represented by plotting a line with a slope of ψ_{Fresh}. The previous steps are represented in Fig. 9.3.
6. **Pinch diagram.** The source composite curve is slid on the fresh line, pushing it to the left while keeping it below the sink composite curve until the two composites touch at the pinch point, with the source composite completely below the sink composite in the overlapped region. Determine the minimum consumption of fresh resource and the minimum discharge of the waste as shown by the pinch diagram (Fig. 9.4).

Similar to the design rules mentioned in Chapter 3, the following three design rules are needed for the minimum usage of fresh resources, maximum reuse of the process sources, and minimum discharge of waste:

- No property load should be passed through the pinch (that is, the two composites touch).
- No waste should be discharged from sources below the pinch.
- No fresh should be used in any sink above the pinch.

The targeting procedure identifies the targets for fresh, waste, and material reuse without commitment to the detailed design of the network matching the sources and sinks.

For other cases of property-based pinch diagrams (including the case when the property operators of the sources and the fresh are above those of the sinks or the case when the fresh source has a property operator that lies in the midst of property operators of process sources), readers are referred to the paper by Kazantzi and El-Halwagi (2005).

PROCESS MODIFICATION BASED ON PROPERTY-BASED PINCH DIAGRAM

The property-based material recycle/reuse pinch diagram and the aforementioned design rules can be used to guide engineers in making process modifications to enhance material reuse (for instance, the observation that below the pinch there is deficiency of reusable sources, whereas above the pinch there is a surplus of sources). Therefore, sinks can be moved from below the pinch to above the pinch, and sources can be moved from above the pinch to below the pinch to reduce the usage of fresh resources and the discharge of waste. Moving a sink from below the pinch to above the pinch can be achieved by increasing the upper bound on the property constraint of the sink given by Inequality 9.9. Conversely, moving a source from above the pinch to below the pinch can be accomplished by reducing the value of the property operator through changes in operating conditions or by adding an "interception" device (for example, heat exchanger, separator, reactor, and so on) that can lower the value of the property operator. In such cases, the problem representation shown in Fig. 9.1 can be extended to account for

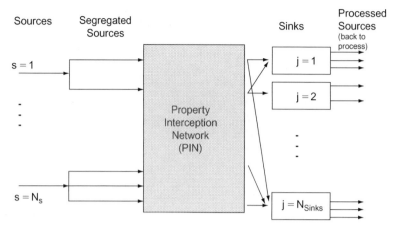

FIGURE 9-5 Schematic representation of property-based allocation and interception. *Source: El-Halwagi et al. (2004).*

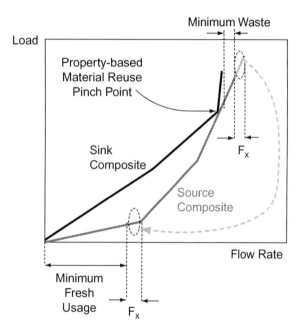

FIGURE 9-6 Insights for process modification: moving a source across the pinch. *Source: Kazantzi and El-Halwagi (2005).*

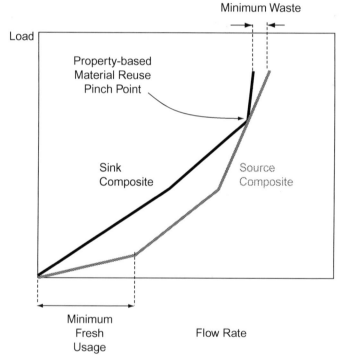

FIGURE 9-7 Insights for process modification: reducing the load of a source. *Source: Kazantzi and El-Halwagi (2005).*

the addition of new "interception" devices that serve to adjust the properties of the sources. Figure 9.5 is a schematic representation of the extended problem statement, which incorporates stream allocation, interception, and process modification. Figure 9.6 illustrates an example of the pinch-based representation when a flow rate F_x is intercepted and its value of property operator is reduced to match that of the fresh. Compared to the nominal case of Fig. 9.4, two benefits accrue as a result of this movement across the pinch: both the usage of fresh resource and the discharge of waste are reduced by F_x.

Another alternative is to reduce the load of a source below the pinch (again by altering operating conditions or adding an interception device). Consequently, the cumulative load of the source composite decreases and allows an additional reuse of process sources with the result of decreasing both the fresh consumption and waste discharge. Figure 9.7 demonstrates the revised pinch diagram for Fig. 9.4, after the load of the process source (and therefore its slope) below the pinch has been decreased. When slid on the fresh line, an additional amount of the source above the pinch can be reused, resulting in a net reduction in fresh and waste.

Example 9-1 Solvent recycle in metal degreasing with a single property (Shelley and El-Halwagi, 2000; Kazantzi and El-Halwagi, 2005)

A metal degreasing process presented in Fig. 9.8 is considered here. Currently, a fresh organic solvent is used in the degreaser and the absorber. A reactive thermal processing and solvent regeneration system is used to decompose the grease and the organic additives, and regenerate the solvent from the degreaser. The liquid product of the solvent regeneration system is reused in the degreaser, while the gaseous product is passed through a condenser, an absorber, and a flare. The process produces two condensate streams: Condensate I from the solvent regeneration unit and Condensate II from the degreaser. The two streams are currently sent to hazardous waste disposal. Because these two streams possess many desirable properties that enable their possible use in the process, it is recommended to consider their recycle/reuse. The absorber and the degreaser are the two process sinks. The two process sources satisfy many properties required for the feed of the two sinks. An additional property should be examined, namely, Reid vapor pressure (RVP), which is important in characterizing the volatility, makeup, and regeneration of the solvent.

The mixing rule for RVP d is given by the following expression:

$$\overline{RVP}^{1.44} = \sum_{i=1}^{N_s} x_i RVP_i^{1.44} \quad [9.16]$$

Moreover, the inlet flow rates of the feed streams to the degreaser and absorber, along with their constraints on the property (RVP), are given in Table 9.1.

The RVP for Condensate I is a function of the thermal regeneration temperature as follows:

$$RVP_{\text{Condensate I}} = 0.56 e^{\left(\frac{T-100}{175}\right)} \quad [9.17]$$

where $RVP_{\text{Condensate I}}$ is the RVP of condensate I in atm and T is the temperature of the thermal processing system in K. The acceptable range of this temperature is 430 K to 520 K. At present, the thermal processing system operates at 515 K, leading to an RVP of 6.0. The data for Condensate I and Condensate II are given in Table 9.2.

1. What is the target for minimum fresh usage when direct recycle is used?
2. If the process is to be modified by manipulating the temperature of the condenser, what is the target for minimum fresh usage?

SOLUTION

1. The values of property loads for the sinks and the sources are calculated as shown in Tables 9.3 and 9.4. It is worth noting that all the sources satisfy the lower bound property constraints for the sinks. Therefore, we focus on the maximum admissible values of the property constraints for the sinks. Once the property loads are calculated, we construct the sink and source composite curves in ascending order of operators and slide the source composite on the fresh line until the two composites touch at the pinch point, as shown in Fig. 9.9. Thus, we can graphically determine the fresh consumption, which is 2.38 kg/s and the waste discharge, which is 2.38 kg/s as well. The fresh solvent consumption is, therefore, reduced by 66 percent.

TABLE 9-1 Flow Rates and Bounds on Properties of Sinks

Sink	Flow Rate (kg/s)	Lower Bound on RVP (atm)	Upper Bound on RVP (atm)
Degreaser	5.0	2.0	3.0
Absorber	2.0	2.0	4.0

Source: Kazantzi and El-Halwagi (2005).

TABLE 9-2 Properties of Process Sources and Fresh

Source	Flow Rate (kg/s)	RVP (atm)
Process Condensate I	4.0	6.0
Process Condensate II	3.0	2.5
Fresh Solvent	To be determined	2.0

Source: Kazantzi and El-Halwagi (2005).

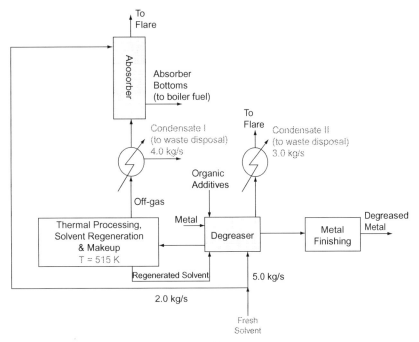

FIGURE 9-8 A degreasing plant. *Source: Kazantzi and El-Halwagi (2005).*

(Continued)

Example 9-1 Solvent recycle in metal degreasing with a single property (Shelley and El-Halwagi, 2000; Kazantzi and El-Halwagi, 2005) (Continued)

2. As mentioned earlier, reuse can be enhanced through process modifications. For instance, if it is desired to completely eliminate the fresh consumption by reducing the load of Condensate I, then we reconstruct the pinch diagram with the two process sources only. Because Condensate I lies above the absorber, it is necessary to decrease the slope of Condensate I until it touches the sink composite curve, with Condensate I completely below the sink composite curve. Fig. 9.10 shows that the property load of Condensate I has to be reduced until it becomes 26.26 kg. atm$^{1.44}$/s. Recalling the definition of property load, we get:

$$[9.18a] \quad 26.26 = 4.0 * RVP^{1.44}_{Condensate\ I}$$

TABLE 9-3 Calculating Maximum Property Loads for the Sinks

Sink	Flow Rate (kg/s)	Upper Bound on RVP (atm)	[Upper Bound on RVP]^1.44 (atm$^{1.44}$)	Maximum Property Load, (kg. atm$^{1.44}$/s)
Degreaser	5.0	3.0	4.87	24.35
Absorber	2.0	4.0	7.36	14.72

TABLE 9-4 Calculating Property Loads for the Sources

Source	Flow Rate (kg/s)	RVP (atm)	RVP^1.44 (atm$^{1.44}$)	Property Load, (kg. atm$^{1.44}$/s)
Process Condensate I	4.0	6.0	13.20	52.80
Process Condensate II	3.0	2.5	3.74	11.22

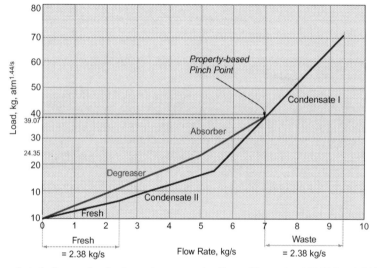

FIGURE 9-9 Property-based pinch diagram for the degreasing case study. *Source: Kazantzi and El-Halwagi (2005).*

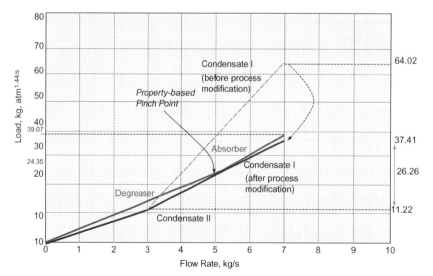

FIGURE 9-10 Identification of process modification task for the degreasing case study.

(Continued)

Example 9-1 Solvent recycle in metal degreasing with a single property (Shelley and El-Halwagi, 2000; Kazantzi and El-Halwagi, 2005) (*Continued*)

Hence,

[9.18b] $$RVP_{Condensate\ I} = 3.69\ \text{atm}$$

Substituting from Eq. 9.18b into Eq. 9.17, we have

[9.19a] $$3.69 = 0.56 e^{\left(\frac{T-100}{175}\right)}$$

that is,

[9.19b] $$T = 430\ \text{K}$$

Hence, to achieve this RVP, the temperature of the thermal processing system should be reduced to 430 K. The solution is shown by Fig. 9.11.

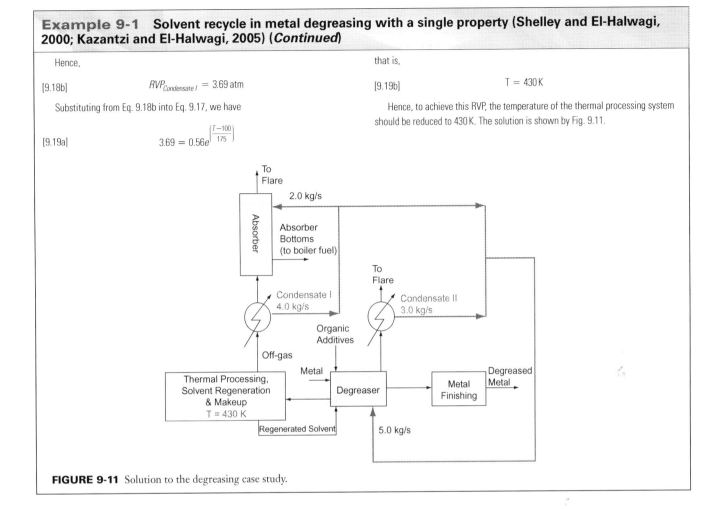

FIGURE 9-11 Solution to the degreasing case study.

CLUSTERING TECHNIQUES FOR MULTIPLE PROPERTIES

The previously presented pinch diagram is suitable for cases governed by a single dominant property. When the sink constraints and the process sources are characterized by multiple properties, it is necessary to address these properties simultaneously. Because properties (or functionalities) form the basis of performance of many units, it will be very insightful to develop design procedures based on key properties instead of key compounds. The challenge, however, is that although chemical components are conserved, properties are not. Therefore, the question is whether or not it is possible to track these functionalities instead of compositions. Shelley and El-Halwagi (2000) introduced the notion of "componentless design," which serves as the basis for property-based integration. Specifically, Shelley and El-Halwagi (2000) devised conserved quantities, called **clusters**, that act as surrogate properties and enable the conserved tracking of functionalities instead of components. Therefore, the design and integration can be carried out in the cluster domain by focusing on the relevant properties. Then, the design can be mapped back from the cluster domain to the property domain. Shelley and El-Halwagi (2000) give the basic mathematical expressions for clusters, which are summarized in the following section. Suppose that we have $N_{Sources}$ streams. Each stream, i, is characterized by N_p raw properties. The index for the properties is designated by r. Consider the class of properties whose mixing rules for each raw property, r, are given by the following equation:

[9.20] $$\psi_r(\bar{p}_r) = \sum_{i=1}^{N_{Sources}} x_i \psi_r(p_{r,i})$$

Equation 9.20 is another form of Eq. 9.2, where x_i is the fractional contribution of the ith stream into the total flow rate of the mixture, and $\psi_r(p_{r,i})$ is an operator on $p_{r,i}$, which can be normalized into a dimensionless operator by dividing by a reference value

[9.21] $$\Omega_{r,i} = \frac{\psi_r(p_{r,i})}{\psi_r^{ref}}$$

Then, an *AU*gmented *P*roperty (AUP) index for each stream, i, is defined as the summation of the dimensionless raw property operators:

[9.22] $$AUP_i = \sum_{r=1}^{N_p} \Omega_{r,i} \quad i = 1, 2, \ldots, N_{Sources}$$

The *cluster for property r in stream i*, $C_{r,i}$, is defined as follows:

[9.23] $$C_{r,i} = \frac{\Omega_{r,i}}{AUP_i}$$

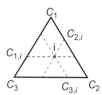

FIGURE 9-12a Ternary representation of intra-stream conservation of clusters within the same streams. *Source: Shelley and El-Halwagi (2000).*

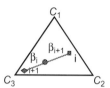

FIGURE 9-12b Lever arm addition for clusters of two streams. *Source: Shelley and El-Halwagi (2000).*

Property clusters are useful quantities that enable the conserved tracking of properties and the derivation of important design tools. In particular, property clusters possess two key characteristics (Shelley and El-Halwagi, 2000): intra- and inter-stream conservation:

Intra-stream conservation: For any stream i, the sum of clusters must be conserved, adding up to a constant (for example, unity); that is,

$$[9.24] \quad \sum_{r=1}^{N_p} C_{r,i} = 1 \quad i = 1, 2, \ldots, N_{Sources}$$

Figure 9.12a illustrates the ternary representation of Eq. 9.24 for three clusters of the ith stream.

Each of the ternary axes ranges from 0.0 to 1.0.

Inter-stream conservation: When two or more streams are mixed, the resulting individual clusters are conserved. This is represented by consistent additive rules in the form of lever arm rules, so that the mean cluster property of two or more streams with different property clusters can be easily determined graphically. The lever arm additive rule for clusters can be expressed by:

$$[9.25] \quad \bar{C}_r = \sum_{i=1}^{N_{Sources}} \beta_i C_{r,i} \quad r = 1, 2, \ldots, N_p$$

where $\bar{C}_r$ is the mean cluster for the rth property resulting from adding the individual clusters of $N_{Sources}$ streams, and β_i represents the fractional lever arm of cluster ($C_{r,i}$) of stream i. The cluster arm is given by:

$$[9.26] \quad \beta_i = \frac{x_i \, AUP_i}{\overline{AUP}}$$

where

$$[9.27] \quad \overline{AUP} = \sum_{i=1}^{N_{Sources}} x_i \, AUP_i$$

Although the mixing of the original properties may be based on nonlinear rules, the clusters are tailored to exhibit linear mixing rules in the cluster domain. When two sources (i and i + 1) are mixed, the locus of all mixtures on the cluster ternary diagram is given by the straight line connecting sources i and i + 1. Depending on the fractional contributions of the streams, the resulting mixture splits the mixing line in ratios β_i and β_{i+1}. Figure 9.12b illustrates the lever arm addition for clusters of two streams, i and i + 1. It looks similar to the lever arm mixing rules for the three-component source-sink mapping diagram presented in Chapter 3. It is worth noting that Eq. 9.25 describes a revised lever arm rule for which the arms are different from those for the mixing of masses. In addition to mass fractions, they also include ratios of the augmented properties of the mixed streams. The appearance of the augmented property index in the lever arm rule allows one-to-one mapping from raw properties to property clusters and vice versa when streams are mixed.

It is instructive to verify the inter- and inter-stream conservation rules for the clusters as well as the expressions for the lever arms and augmented properties. First, we start by summing the individual clusters in the same stream (expressed by Eq. 9.23) and substituting Eq. 9.22 to get:

$$[9.28] \quad \sum_{r=1}^{N_p} C_{i,s} = \frac{\sum_{r=1}^{N_p} \Omega_{r,i}}{AUP_i} = \frac{AUP_i}{AUP_i} = 1$$

This proves the intra-stream conservation. Furthermore, if we mix streams (with each stream possessing a cluster of $C_{r,i}$ and a fractional flow rate contribution of x_i), we get a resulting cluster of $\bar{C}_r$. According to the definition given by Eq. 9.23, we have

$$[9.29] \quad \bar{C}_r = \frac{\bar{\Omega}_r}{\overline{AUP}}$$

Dividing both sides of Eq. 9.20 by ψ_r^{ref} and employing the definition given by Eq. 9.21, we get

$$[9.30] \quad \bar{\Omega}_r = \frac{\psi_r(\bar{p}_r)}{\psi_r^{ref}} = \frac{\sum_{i=1}^{N_{Sources}} x_i \psi_r(p_{r,i})}{\psi_r^{ref}} = \sum_{i=1}^{N_{Sources}} x_i \Omega_{r,i}$$

Substituting from Eq. 9.30 into Eq. 9.29, we have:

$$[9.31] \quad \bar{C}_r = \sum_{i=1}^{N_{Sources}} \frac{x_i \Omega_{r,i}}{\overline{AUP}}$$

Substituting from Eq. 9.23 into Eq. 9.31, we obtain:

$$[9.32] \quad \bar{C}_r = \sum_{i=1}^{N_{Sources}} \frac{x_i \, AUP_i}{\overline{AUP}} C_{r,i}$$

Let us denote the cluster-mixing arm of the ith stream by

$$[9.33] \quad \beta_i = \frac{x_i \, AUP_i}{\overline{AUP}}$$

Therefore, Eq. 9.32 can be written as

$$[9.25] \quad \bar{C}_r = \sum_{i=1}^{N_{Sources}} \beta_i C_{r,i} \quad r = 1, 2, \ldots, N_p$$

which is the inter-stream conservation rule. Now that we have defined the cluster arm, let us derive the expression for $\overline{AUP}$. Equation 9.24 can be applied to the mixture stream to give:

$$[9.34] \quad \sum_{r=1}^{N_p} \bar{C}_r = 1$$

Substituting Eq. 9.25 into Eq. 9.34, we get

$$[9.35a] \quad \sum_{r=1}^{N_p} \sum_{i=1}^{N_{Sources}} \beta_i C_{r,i} = \sum_{i=1}^{N_{Sources}} \sum_{r=1}^{N_p} \beta_i C_{r,i} = 1$$

Hence,

$$[9.35b] \quad \sum_{i=1}^{N_{Sources}} \beta_i \sum_{r=1}^{N_p} C_{r,i} = 1$$

Substituting from Eq. 9.24 into Eq. 9.35b, we get

$$[9.36] \quad \sum_{i=1}^{N_{Sources}} \beta_i = 1$$

which can be combined with Eq. 9.33 to give:

$$[9.37] \quad \frac{1}{\overline{AUP}} \sum_{i=1}^{N_{Sources}} x_i \, AUP_i = 1$$

which can be rearranged to give the proof for Eq. 9.27:

$$[9.27] \quad \overline{AUP} = \sum_{i=1}^{N_{Sources}} x_i \, AUP_i$$

CLUSTER-BASED SOURCE-SINK MAPPING DIAGRAM FOR PROPERTY-BASED RECYCLE AND INTERCEPTION

With multiple properties involved, the overall problem for property integration involving the recycle, reuse, and interception of streams can be stated as follows (El-Halwagi et al., 2004): "Given a process with certain sources (streams) and sinks (units) along with their properties and constraints, it is desired to develop graphical techniques that identify optimum strategies for allocation and interception that integrate the properties of sources, sinks, and interceptors so as to optimize a desirable process objective (for example, minimum usage of fresh resources, maximum utilization of process resources, minimum cost of external streams) while satisfying the constraints on properties and flow rate for the sinks."

Our objective is to develop visualization tools that systematically optimize a certain process objective (for example, minimum usage of fresh resources, maximum utilization of process resources, minimum cost of external streams) while satisfying the constraints on properties and flow rate for the sinks. The solution strategies include a combination of allocation and interception. Allocation of sources involves the segregation and mixing of streams and their assignment to units throughout the process. Interception involves the use of processing units (typically new equipment) to adjust the properties of the various streams. If no capital investment is available for new interception devices, then a no-/low-cost solution will be based on the allocation of external sources and the recycle/reuse of internal sources to meet the constraints of the sinks.

To address the aforementioned problem, the following design decisions must be made:

- How is the geometrical shape of the feasibility region for each sink identified?
- To which sinks should the sources be allocated?
- Is there a need to use a fresh (external) source? If yes, how much and where should it be employed?
- Is there a need for segregation or mixing?
- Is interception required to modify the properties of the sources? What are the optimal interception tasks to reach process targets or objectives?

The three-property cluster source-sink mapping diagram is analogous to the ternary source-sink mapping diagram described in Chapter 3. However, constructing the boundaries of the feasibility region (BFR) for the sink is not as straightforward as in the case of composition-based case. Because the constraints are given in terms of properties whereas the cluster source-sink diagram is in the cluster domain, plotting the BFR requires derivation of construction rules. These rules have been derived by El-Halwagi et al. (2004) and are summarized as follows:

- The BFR can be accurately represented by no more than *six* linear segments.
- When extended, the linear segments of the BFR constitute three convex hulls (cones) with their heads lying on the three vertices of the ternary cluster diagram. This observation is shown by Fig. 9.13a.
- The six points defining the BFR can be determined before constructing the BFR. These six points are characterized by the following values of dimensionless operators (Ωs) for the sink constraints: ($\Omega_{i,s}^{min}$, $\Omega_{j,s}^{min}$, $\Omega_{k,s}^{max}$), ($\Omega_{i,s}^{min}$, $\Omega_{j,s}^{max}$, $\Omega_{k,s}^{max}$), ($\Omega_{i,s}^{min}$, $\Omega_{j,s}^{max}$, $\Omega_{k,s}^{min}$), ($\Omega_{i,s}^{max}$, $\Omega_{j,s}^{max}$, $\Omega_{k,s}^{min}$), ($\Omega_{i,s}^{max}$, $\Omega_{j,s}^{min}$, $\Omega_{k,s}^{min}$), and ($\Omega_{i,s}^{max}$, $\Omega_{j,s}^{min}$, $\Omega_{k,s}^{max}$). This observation is shown by Fig. 9.13b.

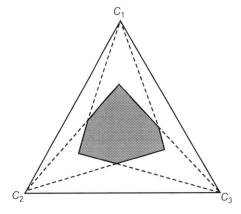

FIGURE 9-13a The BFR is bounded by three cones emanating from the cluster vertices. *Source: El-Halwagi et al. (2004).*

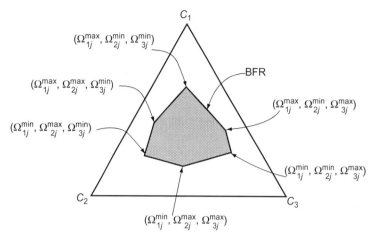

FIGURE 9-13b The six points defining the boundaries of the feasibility region of a sink j. *Source: El-Halwagi et al. (2004).*

PROPERTY-BASED DESIGN RULES FOR RECYCLE AND INTERCEPTION

With the rigorous determination of the BFR, we can now proceed and describe the design rules. In particular, the following visualization techniques and revised lever arm rules that can be systematically used for mixing points and property modification tasks. We also present the optimality conditions for selecting values of the augmented property index of a sink. For the derivation of these rules, the reader is referred to El-Halwagi et al. (2004).

Source-prioritization rule: When two sources (i and i + 1) are mixed to satisfy the property constraints of a sink with source *i being more expensive than i + 1*, minimizing the mixture cost is achieved by selecting the minimum feasible value of x_i.

It is important to note that x_i cannot be directly visualized on the ternary cluster diagram. Instead, the lever arms on the ternary cluster diagram represent another quantity, β_i. The two terms are related through the augmented property index (AUP) as described earlier:

$$[9.33] \quad \beta_i = \frac{x_i \, AUP_i}{\overline{AUP}}$$

Using Eq. 9.27, let us rewrite Eq. 9.33 in the case of mixing two sources (i and i + 1):

$$[9.38a] \quad \beta_i = \frac{x_i AUP_i}{x_i AUP_i + (1 - x_i) AUP_{i+1}}$$

Rearranging, we get

$$[9.38b] \quad x_i = \frac{\beta_i AUP_{i+1}}{\beta_i AUP_{i+1} + (1 - \beta_i) AUP_i}$$

Next, when we take the first derivative of x_i with respect to β_i, we get:

$$[9.39] \quad \frac{dx_i}{d\beta_i} = \frac{AUP_{s+1}[\beta_i AUP_{i+1} + (1 - \beta_i) AUP_i] - \beta_i AUP_{i+1}[AUP_{i+1} - AUP_i]}{[\beta_i AUP_{i+1} + (1 - \beta_i) AUP_i]^2}$$

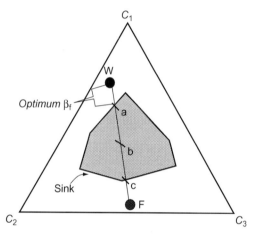

FIGURE 9-14 Determination of optimal mixing point for fresh resources and a process source. *Source: El-Halwagi et al. (2004).*

Rearranging and simplifying, we get

$$[9.40] \quad \frac{dx_i}{d\beta_i} = \frac{AUP_i AUP_{i+1}}{[\beta_i AUP_{i+1} + (1 - \beta_i) AUP_i]^2}$$

With both AUP_i and AUP_{i+1} being nonnegative, the right-hand side of Eq. 9.40 is also nonnegative. Therefore, x_i as a function of β_i is monotonically increasing. From this, we can state that following rule (El-Halwagi et al., 2004):

Lever arm source-prioritization rule: On a ternary cluster diagram, minimization of the cluster arm of a source corresponds to minimization of the flow contribution of that source. In other words, *minimum β_i corresponds to minimum x_i*.

For instance, consider the case of a fresh (external) resource (F) whose flow rate is to be minimized by recycling the maximum possible flow rate of a process stream (for example, waste W), as shown by Fig. 9.14. The straight line connecting the two sources represents the locus for any mixture of W and F. The resulting mixture splits the total mixing arm in the ratios of β_F to β_W. The intersection of the mixing line with the feasibility region

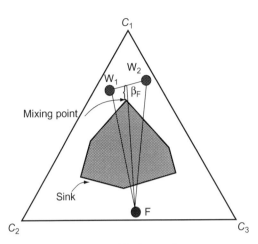

FIGURE 9-15 Determination of fresh arm when two process sources are used. *Source: El-Halwagi et al. (2004)*.

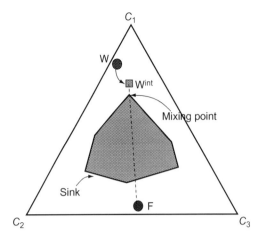

FIGURE 9-16 Task identification for the interception system. *Source: El-Halwagi et al. (2004)*.

of the sink gives the line segment representing all feasible mixtures. This is shown on Fig. 9.14 by the segment connecting points a and c. The question is: What should be the optimum mixing point (for example, a, b, or c)? Our objective is to minimize the mixture cost. Based on the lever arm and prioritization rules, the optimal mixing point corresponds to the minimum feasible β_F. This is point a as shown on Fig. 9.14. It is worth mentioning that this is a necessary condition only. For sufficiency, values of the augmented property index and flow rate should match as well. El-Halwagi et al. (2004) explain this issue in detail.

The same concept can be generalized when more than two sources are mixed. For instance, consider the mixing of a fresh resource (F) with two process streams (W_1 and W_2). As can be seen from Fig. 9.15, the mixing region is defined by the triangle connecting points F, W_1, and W_2. Any mixture of the three sources can be represented by a point in that triangle. It is worth noting that the line connecting represents all possible mixtures of W_1 and W_2. For a given mixing point (for example, the mixing point shown in Fig. 9.15), one can graphically determine the fresh arm (β_F) and use it to calculate the flow rate of the fresh source according to Eq. 9.33.

In case the target for recycling a process stream (W) is not met, more of the process stream may be recycled by adjusting its properties via an interception device (for example, separation, reaction, and so on). Our objective is to alter the properties of W such that the use of F is minimized. The logical question is: What should be the task of the interception device in changing the properties (and consequently the cluster values) of the process stream? The cluster diagram can rigorously answer this question. As shown by Fig. 9.16, for the selected mixing point and the desired value of β_F, the fresh arm can be drawn to determine the desired location of the intercepted internal stream (W^{int}). Additionally, Eq. 9.38b may be employed to determine the desired value of the augmented property index for W^{int} (because the values of the augmented property index are known for F and the mixing point of the sink). Once the ternary cluster value for W^{int} and its augmented property index are determined, Eq. 9.23, which defines

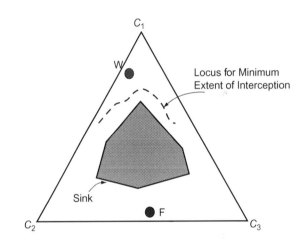

FIGURE 9-17 Development of locus for minimum interception. *Source: El-Halwagi et al. (2004)*.

the cluster equation, is solved backward to determine the dimensionless property operators for the intercepted stream. Next, Eq. 9.21 is solved to calculate the raw properties of W^{int}. This is the minimum extent of interception (as measured on the cluster domain) to achieve maximum recycle of W or minimum usage of the fresh, because additional interception will still lead to the same target of minimum usage but will result in a mixing point inside the sink and not just on the periphery of the sink. Once the task for the interception system is defined, conventional process synthesis techniques can be employed to develop the design and operating parameters for the interception system. The same procedure can be repeated for various mixing points, resulting in the task identification of the locus for minimum extent of interception (see Fig. 9.17).

DEALING WITH MULTIPLICITY OF CLUSTER-TO-PROPERTY MAPPING (EL-HALWAGI ET AL., 2004)

Consider a certain point whose coordinates on the ternary cluster diagram are given by C_1^{sink}, C_2^{sink}, C_3^{sink}. This cluster point may correspond to multiple combinations of property points[1]. In other words, as a result of the nonlinear mapping from the property domain to the cluster domain, it is possible to have multiple property points ($n_{Multiple}$) from the feasible property domain that result in the same value of clusters; that is,

[9.41]
$$(C_1^{sink}, C_2^{sink}, C_3^{sink}) \equiv (p_{1,m}, p_{2,m}, p_{3,m})$$
where $m = 1, 2, \ldots, n_{Multiple}$ and $(p_{1,m}, p_{2,m}, p_{3,m})$ are feasible for the considered sink.

For each property combination $(p_{1,m}, p_{2,m}, p_{3,m})$, the corresponding augmented property index is designated by AUP_m. The set of all of these multiple feasible values of AUP_m is defined as:

[9.42]
$$SET\_AUP_{(C_1^{sink}, C_2^{sink}, C_3^{sink})}^{Feasible} = \{AUP_m | m = 1, 2, \ldots n_{Multiple}\}.$$

Looking at the minimum and maximum values of AUP_m, the range for the values of AUP_m is given by the following interval:

[9.43]
$$INTERVAL\_AUP_m = [\text{Argmin } AUP_m, \text{Argmax } AUP_m]$$

where Argmin AUP_m and Argmax AUP_m are the lowest and highest values of $AUP_m \in SET\_AUP_{(C_1^{sink}, C_2^{sink}, C_3^{sink})}^{Feasible}$.

As a result of such multiplicity, ***three conditions must be satisfied to ensure feasibility of feeding sources (or mixtures of sources) into a sink***:

1. The cluster value for the source (or mixture of sources) must be contained within the feasibility region of the sink on the cluster ternary diagram.
2. The values of the augmented property index for the source (or mixture of sources) and the sink must match.
3. The flow rate of the source (or mixture of sources) must lie within the acceptable feed flow rate range for the sink.

Suppose that two sources i and i + 1 are mixed in a ratio of x_i to x_{i+1} such that the resulting mixture has a cluster value which matches that of a feasible sink cluster; that is,

[9.44] $(C_1^{mixture}, C_2^{mixture}, C_3^{mixture}) = (C_1^{sink}, C_2^{sink}, C_3^{sink})$

Equation 9.44 corresponds to the aforementioned second condition for ensuring feasibility of feeding sources into a sink. As stated earlier, this is a necessary but not sufficient condition for satisfying the sink constraint. In addition to satisfying the previous condition and to satisfying the flow rate (third condition), sufficiency is guaranteed when the value of the augmented property for the mixture matches a feasible value of the augment property for the sink; that is,

[9.45]
$$AUP_{mixture} = AUP_m \in SET\_AUP_{(C_1^{sink}, C_2^{sink}, C_3^{sink})}^{Feasible}$$

---

[1] Points on the BFR are exceptions. Each point on the BFR has a one-to-one mapping to the property domain. The reason for such one-to-one mapping stems from the uniqueness of the six vertices on the BFR. Mixtures of unique points also correspond to unique points.

So, which of the $n_{Multiple}$ feasible values of the augmented properties in the set $SET\_AUP_{(C_1^{sink}, C_2^{sink}, C_3^{sink})}^{Feasible}$ should be selected?

Recalling the source prioritization rule and designating source i to be more expensive than source i + 1, minimizing x_i results in minimizing cost of the mixture. Consequently, we should select an AUP_m, which minimizes x_i. This can be determined by establishing the relationship between AUP_m and x_i.

Let us denote the numerical values of the augmented properties for sources i and i + 1 as AUP_i and AUP_{i+1}, respectively. These are constants. According to Eq. 9.27, we can describe the augmented property of the mixture in terms of the individual augmented properties. Substituting into Eq. 9.45, we get

[9.46a] $\quad x_i AUP_i + (1 - x_i) AUP_{i+1} = AUP_m$

or

[9.46b] $\quad AUP_m = x_i (AUP_i - AUP_{i+1}) + AUP_{i+1}$

Hence,

[9.47]
$$x_i = \frac{AUP_m - AUP_{i+1}}{AUP_i - AUP_{i+1}}$$

Therefore,

[9.48]
$$\frac{\partial x_i}{\partial AUP_m} = \frac{1}{AUP_i - AUP_{i+1}}$$

which is monotonically increasing if $AUP_i > AUP_{i+1}$ and monotonically decreasing if $AUP_i < AUP_{i+1}$. Therefore, to minimize x_i (and consequently the cost), we should select

[9.49a] $\quad AUP_m^{optimum} = \text{Argmin } AUP_m \quad \text{if} \quad AUP_i > AUP_{i+1}$

[9.49b] $\quad AUP_m^{optimum} = \text{Argmax } AUP_m \quad \text{if} \quad AUP_i < AUP_{i+1}$

These results can be shown graphically in Figures 9.18a and b. According to Eq. 9.46b, the relationship between AUP_m and x_s is represented by a straight line whose slope is $AUP_s - AUP_{s+1}$. If $AUP_s > AUP_{s+1}$, the slope is positive (see Fig. 9.18a), which corresponds to Eq. 9.49a. On the other hand, when $AUP_s < AUP_{s+1}$, the slope is negative (see Fig. 9.18b), which corresponds to Eq. 9.49b.

As described by Eq. 9.46a, the AUP of the sink should match that of the mixture (first condition for feasibility). If no possible mixture can have an AUP matching that selected for the sink (for example, Argmax AUP_m as per Eq. 9.49a), then we systematically decrease the value of the sink's AUP, starting with Argmax AUP_m until we get the highest value of $AUP_m \in SET\_AUP_{(C_1^{sink}, C_2^{sink}, C_3^{sink})}^{Feasible}$, which matches that of the mixture. A similar procedure is adopted for the conditions of Eq. 9.49b, by systematically increasing the value of the sink's AUP, starting with Argmin AUP_m until we get the highest value of $AUP_m \in SET\_AUP_{(C_1^{sink}, C_2^{sink}, C_3^{sink})}^{Feasible}$, which matches that of the mixture.

Now that the foregoing rules and tools have been developed, it is useful to proceed to a case study to illustrate the applicability of these rules and tools.

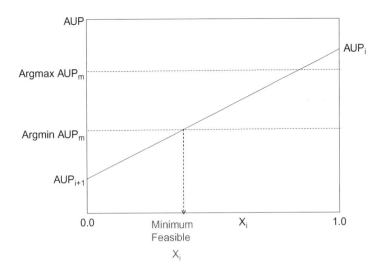

FIGURE 9-18a Selection of optimum AUP_m when $AUP_i > AUP_{i+1}$. *Source: El-Halwagi et al. (2004).*

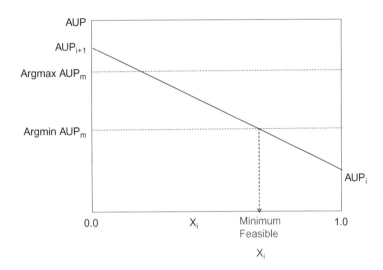

FIGURE 9-18b Selection of optimum AUP_m when $AUP_i < AUP_{i+1}$. *Source: El-Halwagi et al. (2004).*

Example 9-2 VOC recycle in a metal degreasing plant with multiple properties

Consider the metal degreasing process (Shelley and El-Halwagi, 2000) shown in Fig. 9.19 and partly described by Example 9.1 (for a single property). The process uses a fresh organic solvent in the absorption column and the degreaser. The solvent usage is based on three properties: sulfur content, density, and Reid vapor pressure (RVP). The solvent is used in the absorption column to capture lights that escape from the solvent regeneration unit. The organic solvent is also used in the degreaser to degrease the metal parts. Currently, the off-gas volatile organic compounds (VOCs) that evaporate from the degreasing process are flared, leading to economic loss and environmental pollution. In this problem, it is desired to explore the possibility of condensing and reusing the off-gas VOCs instead of flaring, thereby reducing the usage of fresh solvent.

The fresh organic solvent has the following properties:

[9.50] $\qquad S_{fresh} = 0.1\ (wt\%)$

[9.51] $\qquad \rho_{fresh} = 610\ kg/m^3$

[9.52] $\qquad RVP_{fresh} = 2.1\ atm$

The sinks have the following constraints on the properties and flow rates:

The absorber

[9.53] $\qquad 0.0 \leq S_{absorber}\ (wt\%) \leq 0.1$

[9.54] $\qquad 530 \leq \rho_{absorber}\ (kg/m^3) \leq 610$

[9.55] $\qquad 1.5 \leq RVP_{absorber}\ (atm) \leq 2.5$

[9.56] $\qquad 4.4 \leq F_{absorber}\ (kg/min) \leq 6.2$

The degreaser

[9.57] $\qquad 0.0 \leq S_{degreaser}\ (wt\%) \leq 1.0$

[9.58] $\qquad 555 \leq \rho_{degreaser}\ (kg/m^3) \leq 615$

(Continued)

Example 9-2 VOC recycle in a metal degreasing plant with multiple properties (*Continued*)

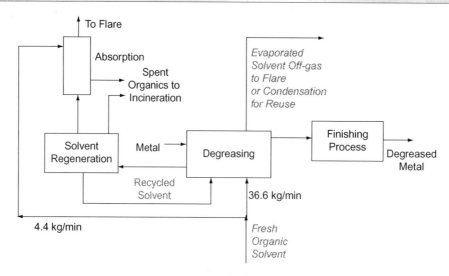

FIGURE 9-19 A degreasing plant. *Source: Shelley and El-Halwagi (2000).*

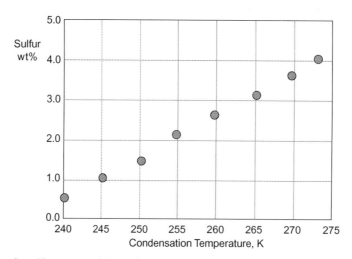

FIGURE 9-20 Experimental data for sulfur content of the condensate versus condensation temperature. *Source: Shelley and El-Halwagi (2000).*

[9.59] $\quad 2.1 \leq RVP_{degreaser}\,(atm) \leq 4.0$

[9.60] $\quad 36.6 \leq F_{degreaser}\,(kg/min) \leq 36.8$

Experimental data are available for the degreaser off-gas condensate. Samples of the off-gas were taken and then condensed at various condensation temperatures, and the three properties, as well as flow rate of the condensate, were measured as shown in Figs. 9.20 through 9.23.

The following mixing rules can be used to evaluate the properties resulting from mixing several streams (Shelley and El-Halwagi, 2000):

[9.61] $\quad \overline{S}(wt\%) = \sum_{s=1}^{N_s} x_s\, S_s(wt\%)$

[9.62] $\quad \dfrac{1}{\overline{\rho}} = \sum_{s=1}^{N_s} \dfrac{x_s}{\rho_s}$

[9.63] $\quad \overline{RVP}^{1.44} = \sum_{s=1}^{N_s} x_s\, RVP_s^{1.44}$

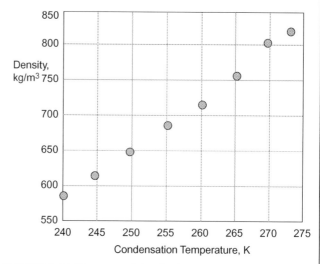

FIGURE 9-21 Experimental data for density of the condensate versus condensation temperature. *Source: Shelley and El-Halwagi (2000).*

(*Continued*)

Example 9-2 VOC recycle in a metal degreasing plant with multiple properties (*Continued*)

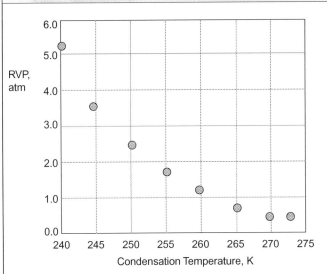

FIGURE 9-22 Experimental data for Reid vapor pressure of the condensate versus condensation temperature. *Source: Shelley and El-Halwagi (2000).*

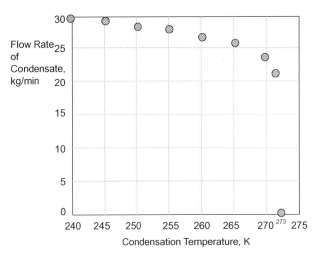

FIGURE 9-23 Experimental data for condensate flow rate versus condensation temperature. *Source: Shelley and El-Halwagi (2000).*

SOLUTION

To transform the problem from the property domain to the cluster domain, let us arbitrarily select the following reference values of the raw properties:

[9.64] $$S^{ref} = 0.005 \text{ (mass fraction)}$$

[9.65] $$\rho^{ref} = 1000 \text{ kg/m}^3$$

[9.66] $$RVP^{ref} = 1.0 \text{ atm}$$

As an illustration, consider the conversion of the fresh properties to clusters. Using the data given by Equations 9.50 through 9.52 and the operators defined by Equations 9.61 through 9.63, we get the following values of the fresh operators:

[9.67] $$\psi_S^{Fresh} = 0.001$$

[9.68] $$\psi_\rho^{Fresh} = \frac{1}{610} = 1.64*10^{-3} \text{ m}^3/\text{kg}$$

[9.69] $$\psi_{RVP}^{Fresh} = 2.1^{1.44} = 2.91 \text{ atm}^{1.44}$$

Let us also calculate the values of the property operators for the reference properties. Hence,

[9.70] $$\psi_S^{Ref} = 0.005$$

[9.71] $$\psi_\rho^{Ref} = \frac{1}{1000} = 1.0*10^{-3} \text{ m}^3/\text{kg}$$

[9.72] $$\psi_{RVP}^{Ref} = 1.00^{1.44} = 1.00 \text{ atm}^{1.44}$$

Using the definition of Eq. 9.21, we get

[9.73] $$\Omega_S^{Fresh} = \frac{0.001}{0.005} = 0.20$$

[9.74] $$\Omega_\rho^{Fresh} = \frac{1.64*10^{-3}}{1.0*10^{-3}} = 1.64$$

[9.75] $$\Omega_{RVP}^{Fresh} = \frac{2.91}{1.00} = 2.91$$

According to Eq. 9.22, we have

[9.76] $$AUP_{Fresh} = 0.20 + 1.64 + 2.91 = 4.75$$

Therefore, the cluster values for fresh solvent can be calculated as follows:

[9.77] $$C_S^{Fresh} = \frac{0.20}{4.75} = 0.04$$

[9.78] $$C_\rho^{Fresh} = \frac{1.64}{4.75} = 0.35$$

[9.79] $$C_{RVP}^{Fresh} = \frac{2.91}{4.75} = 0.61$$

Similarly, the property data for the condensate at different temperatures can be converted to clusters. For instance, as can be seen from Figs. 9.20 through 9.22, the properties of the condensate at 240 K are:

[9.80] $$S^{Condensate \text{ at } 240} = 0.005 \text{ (mass fraction)}$$

[9.81] $$\rho^{Condensate \text{ at } 240} = 580 \text{ kg/m}^3$$

[9.82] $$RVP^{Condensate \text{ at } 240} = 5.2 \text{ atm}$$

These 245 K condensate properties can be converted to the following cluster values:

[9.83] $$C_S = 0.074$$

[9.84] $$C_\rho = 0.128$$

[9.85] $$C_{RVP} = 0.798$$

and the value of the augmented property:

[9.86] $$AUP_{Condensate} = 13.465$$

Next, the sink constraints are used to calculate the cluster values for the vertices of the BFR of the absorber (Table 9.5) and the degreaser (Table 9.6).

Figure 9.24 shows the results, including the absorber (from the results of Table 9.5). A mixture between the fresh solvent and a recycled condensate lies on the straight line connecting the two sources. As can be seen from Fig. 9.24, all straight lines connecting the fresh with the condensate lie outside the absorber. Therefore, there are no feasible recycles of the condensate, and the fresh solvent must be used to provide all the required flow rate of the absorber. This is a useful insight

(*Continued*)

Example 9-2 VOC recycle in a metal degreasing plant with multiple properties (Continued)

TABLE 9-5 Vertices of the BFR for the Absorber

Characteristic Dimensionless Operators	Corresponding Values of Raw Properties	Corresponding Values of Clusters	AUP_{Sink}
$(\Omega_{S,SinkI}^{min}, \Omega_{\rho,SinkI}^{min}, \Omega_{RVP,SinkI}^{max})$	0.00	0.00	
	530	0.34	5.628
	6.2	0.66	
$(\Omega_{S,SinkI}^{min}, \Omega_{\rho,SinkI}^{max}, \Omega_{RVP,SinkI}^{max})$	0.00	0.00	
	610	0.30	5.381
	6.2	0.70	
$(\Omega_{S,SinkI}^{min}, \Omega_{\rho,SinkI}^{max}, \Omega_{RVP,SinkI}^{min})$	0.00	0.00	
	610	0.48	3.432
	4.4	0.52	
$(\Omega_{S,SinkI}^{max}, \Omega_{\rho,SinkI}^{max}, \Omega_{RVP,SinkI}^{min})$	0.01	0.06	
	610	0.45	3.632
	4.4	0.49	
$(\Omega_{S,SinkI}^{max}, \Omega_{\rho,SinkI}^{min}, \Omega_{RVP,SinkI}^{min})$	0.01	0.05	
	530	0.49	3.880
	4.4	0.46	
$(\Omega_{S,SinkI}^{max}, \Omega_{\rho,SinkI}^{min}, \Omega_{RVP,SinkI}^{max})$	0.01	0.04	
	530	0.32	5.828
	6.2	0.64	

Source: El-Halwagi (2006).

TABLE 9-6 Vertices of the BFR for the Degreaser

Characteristic Dimensionless Operators	Corresponding Values of Raw Properties	Corresponding Values of Clusters	AUP_{Sink}
$(\Omega_{S,SinkI}^{min}, \Omega_{\rho,SinkI}^{min}, \Omega_{RVP,SinkI}^{max})$	0.00	0.00	
	555	0.20	9.16
	4.0	0.80	
$(\Omega_{S,SinkI}^{min}, \Omega_{\rho,SinkI}^{max}, \Omega_{RVP,SinkI}^{max})$	0.00	0.00	
	615	0.18	8.99
	4.0	0.82	
$(\Omega_{S,SinkI}^{min}, \Omega_{\rho,SinkI}^{max}, \Omega_{RVP,SinkI}^{min})$	0.00	0.00	
	615	0.36	4.54
	2.1	0.64	
$(\Omega_{S,SinkI}^{max}, \Omega_{\rho,SinkI}^{max}, \Omega_{RVP,SinkI}^{min})$	0.01	0.31	
	615	0.25	6.54
	2.1	0.44	
$(\Omega_{S,SinkI}^{max}, \Omega_{\rho,SinkI}^{min}, \Omega_{RVP,SinkI}^{min})$	0.01	0.30	
	555	0.27	6.71
	2.1	0.43	
$(\Omega_{S,SinkI}^{max}, \Omega_{\rho,SinkI}^{min}, \Omega_{RVP,SinkI}^{max})$	0.01	0.18	
	555	0.16	11.16
	4.0	0.66	

Source: El-Halwagi (2006).

that spares designers the need to consider infeasible recycles. Additionally, the ternary cluster diagram can be used to propose changes (for example, relaxation of the sink constraints, substitution of the fresh solvent, alteration of the condensate properties) to enable feasible recycle.

Figure 9.25 illustrates the ternary cluster diagram with the degreaser (from the results of Table 9.6). This representation can be used to determine feasible blends between the fresh solvent and the condensate as well as their relative flow rates. For instance, if the condensate at 240 K is to be recycled, a straight line is drawn

(Continued)

CHAPTER 9 Property Integration

Example 9-2 VOC recycle in a metal degreasing plant with multiple properties (*Continued*)

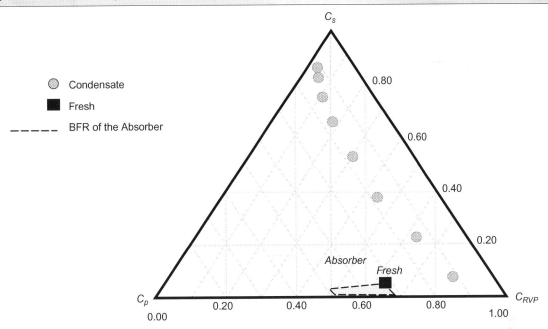

FIGURE 9-24 The ternary property-cluster diagram for the absorber.

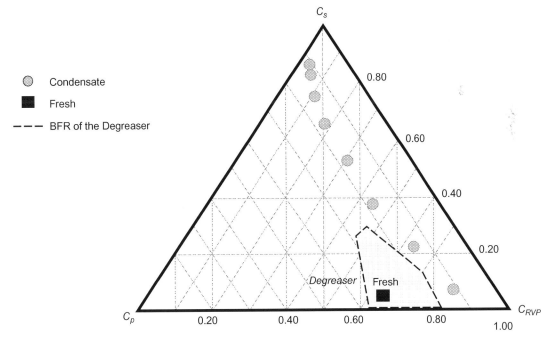

FIGURE 9-25 The ternary property-cluster diagram for the degreaser.

to connect the condensate with the fresh solvent. The relative arm of the fresh is obtained by calculating the ratio of the fresh arm to the total arm connecting the condensate and the fresh solvent. From Fig. 9.26, the fresh and the total arms are measured, and the ratio is:

[9.87] $$\beta_{Fresh} = 0.21$$

Recalling the expression for the relative arm from Eq. 9.33,

[9.33] $$\beta_i = \frac{x_i\, AUP_i}{\overline{AUP}}$$

and the form given by Eq. 9.38a, when two sources (i and 1+i) are mixed:

[9.38a] $$\beta_i = \frac{x_i\, AUP_i}{x_i\, AUP_i + (1 - x_i)AUP_{i+1}}$$

(*Continued*)

Example 9-2 VOC recycle in a metal degreasing plant with multiple properties (*Continued*)

Therefore, when the fresh and the condensate are mixed, we have:

[9.88]
$$\beta_{Fresh} = \frac{x_{Fresh} AUP_{Fresh}}{x_{Fresh} AUP_{Fresh} + (1 - x_{Fresh}) AUP_{Condensate}}$$

Substituting from Equations 9.76, 9.86, and 9.87 into Eq. 9.88, we obtain:

$$0.21 = \frac{x_{Fresh} * 4.75}{x_{Fresh} * 4.75 + (1 - x_{Fresh}) * 13.465}$$

that is,

[9.89]
$$x_{Fresh} = 0.43$$

and

[9.90]   Flow rate of the fresh = $0.43 * 36.6 = 15.74$ kg/min

[9.91]   Flow rate of the recycled condensate = $(1 - 0.43) * 36.6 = 20.86$ kg/min

As Fig. 9.23 shows, there is sufficient flow rate of the condensate (30.0 kg/min) to provide for the recyclable 20.86 kg/min. Other condensation temperatures can be similarly considered. The final selection depends on the detailed design and economics that trade off the value of the recovered solvent and saved fresh versus the cost of the condensation system.

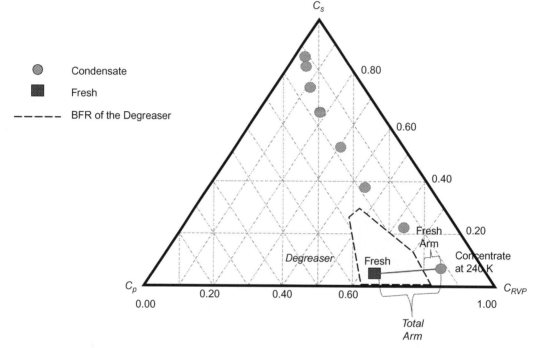

FIGURE 9-26 Determining the fresh usage via the ternary property-cluster diagram for the degreaser.

RELATIONSHIP BETWEEN CLUSTERS AND MASS FRACTIONS

Because compositions represent a special class of properties, it is useful to think of mass integration as a special case of property integration. Analogy has been shown between the material recycle pinch diagram and the property-based pinch diagram and between the source-sink ternary diagram and its cluster-based analogue (El-Halwagi, 2006). Therefore, it is useful to illustrate the relationship between mass fractions and clusters. Let us designate the mass fraction of component r in stream i by $y_{r,i}$. When a number $N_{sources}$ of streams are mixed, we get the following balance equations:

Overall material balance:

[9.3]
$$\overline{F} = \sum_{i=1}^{N_{Sources}} F_i$$

Component material balance:

[9.92]
$$\overline{F}\,\overline{y}_r = \sum_{i=1}^{N_{Sources}} F_i\, y_{r,i}$$

Diving both sides of Eq. 9.92 by $\overline{F}$, we get

[9.93]
$$\overline{y}_r = \sum_{i=1}^{N_{Sources}} x_i\, y_{r,i}$$

where

[9.94]
$$x_i = \frac{F_i}{\overline{F}}$$

Equation 9.93 is analogous with Eq. 9.20, with the mass fraction being the property of choice and the operator on the mass fraction corresponding to the mass fraction itself; that is,

[9.95]
$$p_{r,i} = y_{r,i}$$

and

$$\psi_r(p_{r,i}) = y_{r,i} \quad [9.96]$$

Let us choose a reference value of 1.0 for all mass fractions. Therefore,

$$\Omega_{r,i} = y_{r,i} \quad [9.97]$$

Substituting from Eq. 9.97 into Eq. 9.22 and recalling that the sum of mass fractions in a stream is 1.0, we get:

$$AUP_i = 1.0 \quad [9.98]$$

Substituting from Eq. 9.98 into Eq. 9.23, we get,

$$C_{r,i} = y_{r,i} \quad [9.99]$$

Therefore, when the property of choice is the mass fraction, the cluster of the property is also the mass fraction.

Additional observations may also be noted. For instance, comparing Equations 9.25 and 9.93, we get

$$\beta_i = x_i \quad [9.100]$$

Also, when Equations 9.98 and 9.100 are substituted in Eq. 9.26, we get that the AUP for a sink or a mixture is always equal to 1.0.

The foregoing observations reiterate the earlier statement that mass integration may be regarded as a special case of property integration.

ADDITIONAL READINGS

The area of property integration has recently received much attention. The concept of property-based clusters and componentless design was introduced by Shelley and El-Halwagi (2000). The overall framework for property integration and the derivation of optimality criteria for the clusters are provided by El-Halwagi et al. (2004). Property integration can be used to develop property-based process modification (Kazantzi et al., 2004a and b) as well as property-based recycle and interception (El-Halwagi et al., 2004; Kazantzi and El-Halwagi, 2005; Gabriel et al., 2003a, b; Glasgow et al., 2001). Algebraic approaches for property integration have been developed to address any number of properties (Qin et al., 2004). Furthermore, algebraic techniques have been developed by Foo et al. (2006) and extended by Ng et al. (2010) to solve the property-based pinch diagram via a cascade analysis. Property integration has been used to solve batch problems (Chen et al., 2010; Foo, 2010, Ng et al., 2008). Unsteady-state processes involving property-based scheduling and operation have been addressed by Grooms et al. (2005). Property-based clusters have been used to develop reverse problem formulations as well as integrated process and product design (Chemmangattuvalappil et al., 2010; Bommareddy et al., 2010; Solvason et al., 2009; Eljack et al., 2007, 2005; Kazantzi et al., 2007; Eden et al., 2004, 2002). Molecular clustering was used as a basis to integrate solvent and process design (Papadopoulos and Linke, 2006). Property-based modeling and its impact on process and product design has been discussed by Gani and Pistikopoulos (2002). Optimization techniques have been used to expand the applicability of property integration in tackling a broad class of problems regarding sustainable design (Kheireddine et al., 2011; Nápoles-Rivera et al., 2010a and b; Ponce-Ortega et al., 2010 and 2009; Das et al., 2009; Seingheng et al., 2007).

HOMEWORK PROBLEMS

9.1. Consider the microelectronics manufacturing facility (Gabriel et al., 2003a, Kazantzi and El-Halwagi, 2005) represented by Fig. 9.27. The wafer fabrication (Wafer Fab) section and the combined chemical and mechanical processing (CMP) section are identified as the sinks of the problem that both accept ultra pure water (UPW) as their feed. There are also two main process sources that are available for reuse to the sinks, that is, the 50 percent spent rinse and the 100 percent spent rinse. We are interested in reusing them as feed to the sinks, to reduce the ultra pure water consumption. The main characteristic that we consider here, to evaluate the reuse of the rinse streams to the sinks, is resistivity (R), which constitutes an index of the ionic content of aqueous streams.

The mixing rule for resistivity is the following (Gabriel et al., 2003a):

$$\frac{1}{\overline{R}} = \sum_{i=1}^{N_s} \frac{x_i}{R_i} \quad [9.101]$$

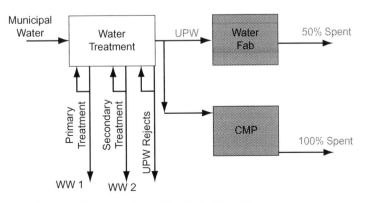

FIGURE 9-27 A microelectronics manufacturing flow sheet. *Source: Gabriel et al. (2003a).*

TABLE 9-7 Flow Rates and Bounds on Properties of Sinks

Sink	Flow Rate (gal/min)	Lower Bound on R (kΩ/cm)	Upper Bound on R (kΩ/cm)
Wafer Fab	800	16,000	20,000
CMP	700	10,000	18,000

TABLE 9-8 Properties of Process Sources and Fresh

Source	Flow Rate (gal/min)	R (kΩ/cm)
50% Spent	1000	8000
100% Spent	1000	2000
UPW (fresh)	To be determined	18,000

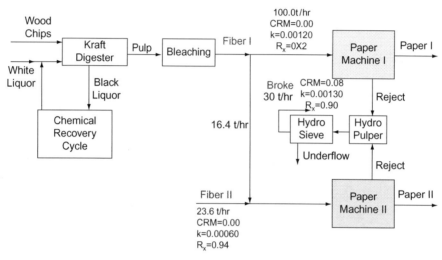

FIGURE 9-28 Schematic representation of the pulp and paper process. *Source: El-Halwagi et al. (2004).*

Moreover, the inlet flow rates of the feed streams to the wafer fabrication and the CMP sections, along with their constraints on resistivity, are given in Table 9.7, whereas the source flow rates and property values are given in Table 9.8.

Using direct recycle, what is the target for fresh (ultra pure) water usage?

9.2. Figure 9.28 is a schematic representation of the papermaking process (El-Halwagi et al., 2004). A Kraft pulping process is used to yield digested pulp using a chemical-pulping process. The digested pulp is passed to a bleaching system to produce bleached pulp (fiber I). The plant also purchases an external pulp (fiber II). Two types of paper are produced through two papermaking machines (sinks I and II). Paper machine I employs 100 tons/hr of fiber I. On the other hand, a mixture of fibers I and II (16.4 and 23.6 tons/hr, respectively) is fed to paper machine II. Because of the occasional malfunctions of the process operation, a certain amount of partly and completely manufactured paper is rejected. These waste fibers are referred to as *reject*. The reject is typically passed through a hydro-pulper and a hydro-sieve with the net result of producing an underflow, which is burnt, and an overflow (referred to as *broke*) that goes to waste treatment. It is worth noting that the broke contains fibers that may be partially recycled for papermaking.

Develop optimal solutions to the following design questions:

1. Direct recycle and reallocation: What is the optimal allocation of the three fiber sources (fiber I, fiber II, and broke) for a direct recycle/reuse situation (no new equipment)?
2. Interception of broke: To maximize the use of process resources and minimize wasteful discharge (broke), how should the properties of broke be altered so as to achieve its maximum recycle?

The performance of the paper machines and, consequently, the quality of the produced papers, rely on three primary properties (Biermann, 1996; Brandon, 1981; and Willets, 1958):

- **Objectionable Material (OM)**: This refers to the undesired species in the fiber (expressed as mass fraction).
- **Absorption Coefficient (k)**: This is an intensive property that provides a measure of absorptivity of light into the fibers (black paper has a high value of k).

Hemicellulose and cellulose have very little absorption of light in the visible region. However, lignin has a high absorbance. Therefore, light absorbance is mostly attributed to lignin. The light absorption coefficient is a very useful property in determining the opacity of the fibers.

- **Reflectivity (R_∞)**: This is defined as the reflectance of an infinitely thick material compared to an absolute standard, which is magnesium oxide (MgO).

The mixing rules for OM and k are linear (Brandon, 1981); that is,

$$\overline{OM} = \sum_{s=1}^{N_s} x_s OM_s \qquad [9.102]$$

$$\overline{k}\left(\frac{m^2}{g}\right) = \sum_{s=1}^{N_s} x_s k_s \left(\frac{m^2}{g}\right) \qquad [9.103]$$

On the other hand, a nonlinear empirical mixing rule for R_∞ is developed using data from Willets (1958):

$$\overline{R}_\infty^{5.92} = \sum_{s=1}^{N_s} x_s R_{\infty_s}^{5.92} \qquad [9.104]$$

Tables 9.9 and 9.10 describe the constraints for the two sinks, whereas Table 9.11 provides the data on the properties of the sources.

TABLE 9-9 Constraints for Paper Machine I

Property	Lower Bound	Upper Bound
OM (mass fraction)	0.00	0.02
k (m²/gm)	0.00115	0.00125
R_∞	0.80	0.90
Flow Rate (ton/hr)	100	105

Source: El-Halwagi et al. (2004).

TABLE 9-10 Constraints for Paper Machine II

Property	Lower Bound	Upper Bound
OM (mass fraction)	0.00	0.00
k (m²/gm)	0.00070	0.00125
R_∞	0.85	0.90
Flow Rate (ton/hr)	40	40

Source: El-Halwagi et al. (2004).

NOMENCLATURE

Argmax	highest value of an element in a set
Argmin	lowest value of an element in a set
AUP	augmented property index
AUP_m	augmented property index for point m in a sink
AUP_i	augmented property index for sources
$\overline{AUP}$	augmented property index of mixture
C	cluster
$C_{r,I}$	cluster of property r in source i
$\overline{C}$	cluster of mixing
F_i	flow rate of source i, tons/hr
i	index for sources
$INTERVAL\_AUP_m$	the interval bounding the range for the AUP at sink point m
j	index for sinks
k	absorption coefficient
m	a feasible point satisfying sink constraints
$n_{Multiple}$	number of multiple property points leading to the same value of the cluster
$N_{Sources}$	number of sources
p_i	ith property
$p_{i,m}$	ith property for point m in the sink
r	index for properties or clusters
R_∞	reflectivity
$SET\_AUP^{Feasible}_{(C_1^{sink}, C_2^{sink}, C_3^{sink})}$	set of all feasible AUPs for property combinations yielding the same value of clusters ($C_1^{sink}, C_2^{sink}, C_3^{sink}$)
W	internal process stream
X_i	fractional contribution of the ith stream into the total flow rate of the mixture

TABLE 9-11 Properties of Fiber Sources

Source	OM (mass fraction)	k (m²/gm)	R_∞	Maximum Available Flow Rate (ton/hr)	Cost ($/ton)
Broke	0.08	0.00130	0.90	30	0
Fiber I	0.00	0.00120	0.82	∞	210
Fiber II	0.00	0.00060	0.94	∞	400

Source: El-Halwagi et al. (2004).

SUBSCRIPTS

K	absorption coefficient
m	a feasible point satisfying sink constraints
OM	objectionable materials
R	reflectivity

SUPERSCRIPTS

Feasible	a feasible point in a sink
int	intercepted value
ref	reference value

GREEK LETTERS

β_i	mixing arm of stream i on the ternary cluster diagram
ψ_r	operator used in the mixing formula for the rth property
$\Omega_{r,i}$	normalized, dimensionless operator for the rth property of the ith source

REFERENCES

Biermann CJ: *Handbook of pulping and papermaking*, San Diego, California, Academic Press, 1996.

Bommareddy S, Chemmangattuvalappil NG, Solvason CC, Eden MR: Simultaneous solution of process and molecular design problems using an algebraic approach, *Comp Chem Eng* 34:1481–1486, 2010.

Brandon CE: Properties of paper. In Casey JP, editor: *Pulp and paper chemistry and chemical technology*, ed 3, vol III, New York, New York, John Wiley & Sons, 1981.

Chemmangattuvalappil NG, Solvason CC, Bommareddy S, Eden MR: Reverse problem formulation approach to molecular design using property operators based on signature descriptors, *Comp Chem Eng* 34:2062–2071, 2010.

Chen CL, Lee JY, Ng DKS, Foo DCY: A unified model for property integration for batch and continuous processes, *AIChE J* 56(7):1845–1858, 2010.

Das AK, Shenoy UV, Bandyopadhyay S: Evolution of resource allocation networks, *Ind Eng Chem Res* 48:7152–7167, 2009.

Eden MR, Jørgensen SB, Gani R, El-Halwagi MM: Property integration: a new approach for simultaneous solution of process and molecular design problems, *Comp Aided Chem Eng* 10:79–84, 2002.

Eden MR, Jørgensen SB, Gani R, El-Halwagi MM: A novel framework for simultaneous separation process and product design, *Chem Eng Proc* 43(5):595–608, 2004.

El-Halwagi MM: *Process integration*, Amsterdam, 2006, Elsevier/Academic Press.

El-Halwagi MM, Glasgow IM, Eden MR, Qin X: Property integration: componentless design techniques and visualization tools, *AIChE J* 50(8):1854–1869, 2004.

Eljack FT, Abdelhady AF, Eden MR, Gabriel F, Qin X, El-Halwagi MM: Targeting optimum resource allocation using reverse problem formulations and property clustering techniques, *Comp Chem Eng* 29:2304–2317, 2005.

Eljack F, Eden M, Kazantzi V, Qin X, El-Halwagi MM: Simultaneous process and molecular design: a property-based approach, *AIChE J* 35(5):1232–1239, 2007.

Foo DCY: Automated targeting technique for batch process integration, *Ind Eng Chem* 49(2):9899–9916, 2010.

Foo DCY, Kazantzi V, El-Halwagi MM, Manan ZA: Surplus diagram and cascade analysis techniques for targeting property-based material reuse network, *Chem Eng Sci* 61:2626–2642, 2006.

Gabriel FB, Harell DA, Dozal E, El-Halwagi MM: *Pollution targeting via functionality tracking*, New Orleans, AIChE Spring Meeting, 2003a.

Gabriel F, Harell D, Kazantzi V, Qin X, El-Halwagi M: *A novel approach to the synthesis of property integration networks*, San Francisco, AIChE Annual Meeting, November 2003b.

Gani R, Pistikopoulos E: Property modeling and simulation for product and process design, *Fluid Phase Equilib* 194–197:43–59, 2002.

Glasgow IM, Eden MR, Shelley MD, Krishnagopalan G, El-Halwagi MM: *Property integration for process optimization*, Reno Nevada, AIChE Annual Meeting, 2001.

Grooms D, Kazantzi V, El-Halwagi MM: Scheduling and operation of property-interception networks for resource conservation, *Comp Chem Eng* 29:2318–2325, 2005.

Kazantzi V, El-Halwagi MM: Targeting material reuse via property integration, *Chem Eng Prog* 101(8):28–37, 2005.

Kazantzi V, Harell D, Gabriel F, Qin X, El-Halwagi MM: Property-based integration for sustainable development. In Barbosa-Povoa A, Matos H, editors: Proceedings of European Symposium on Computer-Aided Process Engineering 14 (ESCAPE 14), Elsevier, pp 1069–1074, 2004a.

Kazantzi V, Qin X, El-Halwagi M, Eljack F, Eden M: Simultaneous process and molecular design through property clustering: a visualization tool, *Ind Eng Chem Res* 46:3400–3409, 2007.

Kazantzi V, Qin X, Gabriel F, Harell D, El-Halwagi MM: Process modification through visualization and inclusion techniques for property-based integration. In Floudas CA, Agrawal R, editors: Proceedings of the Sixth Foundations of Computer Aided Design (FOCAPD), CACHE Corp., pp 279–282, 2004b.

Kheireddine H, Dadmohammadi Y, Deng C, Feng X, El-Halwagi MM: Optimization of direct recycle networks with the simultaneous consideration of property, mass, and thermal effects, *Ind Eng Chem Res* 50(7):3754–3762, 2011.

Nápoles-Rivera F, Jiménez-Gutiérrez A, Ponce-Ortega JM, El-Halwagi MM: Recycle and reuse mass exchange networks based on properties using a global optimization technique, *Comp Aided Chem Eng* 28:871–876, 2010.

Nápoles-Rivera F, Ponce-Ortega JM, El-Halwagi MM, Jiménez-Gutiérrez A: Global optimization of mass and property integration networks with in-plant property interceptors, *Chem Eng Sci* 65(15):4363–4377, 2010.

Ng DKS, Foo DCY, Rabie AH, El-Halwagi MM: Simultaneous synthesis of property based reuse/recycle and interception networks for batch processes, *AIChE J* 54(10):2624–2632, 2008.

Ng DKS, Foo DCY, Tan RR, El-Halwagi MM: Automated targeting techniques for concentration- and property-based total resource conservation networks, *Comp Chem Eng* 34(5):825–845, 2010.

Papadopoulos AI, Linke P: Efficient integration of optimal solvent and process design using molecular clustering, *Chem Eng Sci* 61:6316–6336, 2006.

Ponce-Ortega JM, El-Halwagi MM, Jiménez-Gutiérrez A: Global optimization for the synthesis of property-based recycle and reuse networks including environmental constraints, *Comp Chem Eng* 34(3):318–330, 2010.

Ponce-Ortega JM, Hortua AC, El-Halwagi MM, Jiménez-Gutiérrez A: A property-based optimization of direct-recycle networks and wastewater treatment processes, *AIChE J* 55(9):2329–2344, 2009.

Qin X, Gabriel F, Harell D, El-Halwagi M: Algebraic techniques for property integration via componentless design, *Ind Eng Chem* 43:3792–3798, 2004.

Seingheng H, Tan RR, Auresenia J, Fuchino T, Foo DCY: Synthesis of near-optimal topologically constrained property-based water network using swarm intelligence, *J Clean Techn Environ Policy* 9:27–36, 2007.

Shelley MD, El-Halwagi MM: Component-less design of recovery and allocation systems: a functionality-based clustering approach, *Comp Chem Eng* 24:2081–2091, 2000.

Solvason CC, Chemmangattuvalappil NG, Eden MR: Decomposition techniques for molecular synthesis and structured product design, *Comp Aided Chem Eng* 26:153–158, 2009.

Willets WR: Titanium pigments. In "Paper Loading Materials," TAPPI Monograph Series, No. 19, New York, New York, Technical Association of the Pulp and Paper Industry, pp 96–114, 1958.

10 DIRECT-RECYCLE NETWORKS: AN ALGEBRAIC APPROACH

Graphical approaches such as the material recycle pinch diagram presented in Chapter 4 provide insightful visualization and methodical approaches to the targeting and synthesis of direct-recycle networks. Notwithstanding the usefulness and insights of the graphical methods, it is beneficial to develop an algebraic procedure that is particularly useful in the following cases:

- Numerous sources and sinks: As the number of sources and sinks increase, it becomes more convenient to use spreadsheets or algebraic calculations to handle the targeting.
- Scaling problems: If there is a significant difference in values of flow rates and/or loads for some of the sources and/or sinks, the graphical representation becomes inaccurate because the larger flows/loads will skew the scale for the other streams.
- If the targeting is tied with a broader design task that is handled through algebraic computations, it is desirable to use consistent algebraic tools for all the tasks.

This chapter presents the algebraic analogue to the material recycle pinch diagram.

PROBLEM STATEMENT

The problem can be expressed as follows:

Given a process with a number ($N_{sources}$) of process sources (for example, process streams, wastes) that may be considered for possible recycle and replacement of the fresh material and/or reduction of waste discharge. Each source, i, has a given flow rate, W_i, and a given composition of a targeted species, y_i. Available for service is a fresh (external) resource that can be purchased to supplement the use of process sources in sinks. The sinks are N_{sinks} process units that employ a fresh resource. Each sink, j, requires a feed whose flow rate, G_j^{in}, and an inlet composition of a targeted species, z_j^{in}, must satisfy the following bounds:

[10.1] $0 \leq z_j^{in} \leq z_j^{max}$ where $j = 1, 2, ..., N_{sinks}$

Fresh (external) resource may be purchased to supplement the use of process sources in sinks. The objective is to develop a noniterative algebraic procedure aimed at minimizing the purchase of fresh resource, maximizing the usage of process sources, and minimizing waste discharge.

ALGEBRAIC TARGETING APPROACH

In this section, the algebraic procedure developed by Almutlaq and El-Halwagi (2007) and Almutlaq et al. (2007) is presented. First, it is necessary to recall the two direct-recycle optimality conditions derived in Chapter 3:

Sink composition rule: When a fresh resource is mixed with process source(s), the composition of the mixture entering the sink should be set to a value that minimizes the fresh arm. For instance, when the fresh resource is a pure substance that can be mixed with pollutant-laden process sources, the composition of the mixture should be set to the maximum admissible value.

Source prioritization rule: To minimize the usage of the fresh resource, recycle of the process sources should be prioritized in order of their fresh arms, starting with the source having the shortest fresh arm.

These rules constitute the basis for the material recycle pinch diagram, as shown in Fig. 10.1. As described in Chapter 3, the sink composite is a cumulative representation of all the sinks and corresponds to the upper bound on their feasibility region whereas the source composite curve is a cumulative representation of all process streams considered for recycle. The source composite stream may be represented anywhere and is then slid horizontally (in the case of pure fresh) on the flow rate axis until it touches the sink composite stream with the source composite below the sink composite in the overlapped region.

Now, suppose that we start plotting both composite streams from the origin point (Fig. 10.2). If the

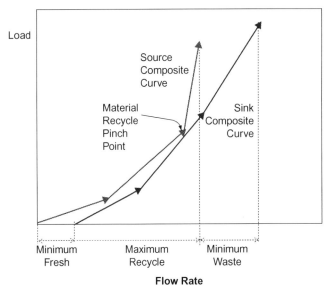

FIGURE 10-1 Material recycle pinch diagram. *Source: El-Halwagi et al. (2003).*

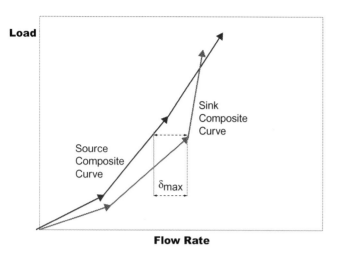

FIGURE 10-2 Sliding the source composite to the left generates infeasibility. *Source: Almutlaq and El-Halwagi (2007).*

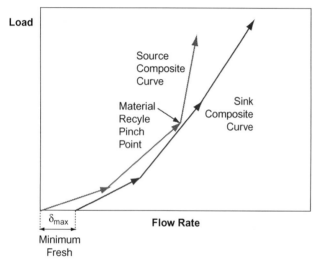

FIGURE 10-3 Minimum fresh target corresponds to maximum flow shortage. *Source: Almutlaq and El-Halwagi (2007).*

source composite is completely below the sink composite, then the process does not require a fresh resource. Nonetheless, if there is any portion of the source composite lying above the sink composite, then infeasibilities exist and must be removed by using a fresh resource. Such infeasibility may be described in a couple of ways by looking vertically and horizontally. Looking vertically at a given flow rate, if the source composite lies above the sink composite, then the source composite violates the maximum load admissible to the sink. Alternatively, by looking horizontally at a given load, if the source composite lies to the left of the sink composite, then there is a shortage of the flow rate necessary for the sink. The maximum horizontal infeasibility corresponds to the maximum shortage of flow rate, which is designated as δ_{max}. Indeed, all infeasibilities are eliminated by sliding the source composite curve to the right a distance equal to δ_{max} (Fig. 10.3). Consequently, the target for minimum fresh usage is equal to the maximum shortage; that is,

[10.2] Target for minimum fresh consumption = δ_{max}

The targeting question involves the algebraic identification of this maximum shortage without the need to resort to the graphical representation.

Let us revisit Fig. 10.2 and draw horizontal lines at corner points (kinks) of the source- and sink-composite curves (Fig. 10.4). Let us use an index k to designated those horizontal lines, starting with k = 0 at the zero load level and going up at each horizontal level. The load at each horizontal level, k, is referred to as M_k. The vertical distance between each two horizontal lines is referred to as a *load interval* and is given the index k as well. The load within interval k is calculated as follows:

[10.3] $$\Delta M_k = M_k - M_{k-1}$$

The next step is to calculate the flow rates of the source and the sink within each load interval. These flow rates

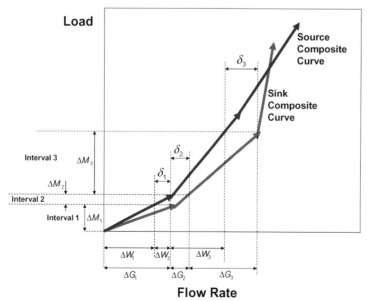

FIGURE 10-4 Load intervals, flows, and residuals. *Source: Almutlaq and El-Halwagi (2007).*

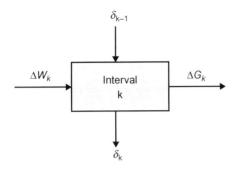

FIGURE 10-5 Flow balance around a load interval. *Source: Almutlaq and El-Halwagi (2007).*

correspond to the horizontal distances on the source- and sink-composite curves contained within the interval. Therefore, the following expressions may be used to calculate the source and sink flow rate (respectively) within the kth interval:

$$[10.4] \qquad \Delta W_k = \frac{\Delta M_k}{y_{\text{source in interval k}}}$$

and

$$[10.5] \qquad \Delta G_k = \frac{\Delta M_k}{z^{\max}_{\text{sink in interval k}}}$$

Figure 10.4 illustrates the concepts of a load interval and flow rates of sources and sinks within an interval. Additionally, Fig. 10.4 illustrates that, at any horizontal level ($\bar{k}$), the horizontal distance between the source- and the sink-composite curves is given by:

$$[10.6] \qquad \delta_{\bar{k}} = \sum_{k=1}^{\bar{k}} W_k - \sum_{k=1}^{\bar{k}} G_k$$

Equation 10.6 indicates that, at any value of load, the horizontal distance between the source- and the sink-composite curves is the difference in cumulative flow rates. As mentioned earlier, a negative value of δ implies that the source composite lies to the left of the sink composite, which is infeasible.

To illustrate Eq. 10.6, let us apply it to the first interval:

$$[10.7] \qquad \delta_1 = \Delta W_1 - \Delta G_1$$

This result can be verified by Fig. 10.4. Similarly, by applying Eq. 10.6 to the second interval, we have:

$$[10.8] \qquad \delta_2 = \Delta W_1 + \Delta W_2 - \Delta G_1 - \Delta G_2$$

Substituting from Eq. 10.7 into Eq. 10.8, we obtain

$$[10.9] \qquad \delta_2 = \delta_1 + \Delta W_2 - \Delta G_2$$

and, for the kth interval:

$$[10.10] \qquad \delta_k = \delta_{k-1} + \Delta W_k - \Delta G_k$$

with $\delta_0 = 0$. Equation 10.10 is represented by Fig. 10.5. The flow balances can be carried out for all intervals, resulting in the cascade diagram shown on Fig. 10.6. As

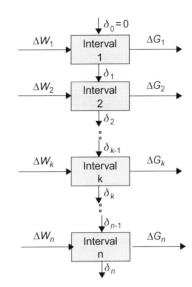

FIGURE 10-6 Cascade diagram. *Source: Almutlaq and El-Halwagi (2007).*

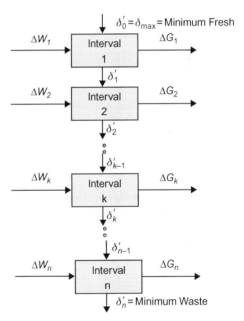

FIGURE 10-7 Revised cascade diagram. *Source: Almutlaq and El-Halwagi (2007).*

mentioned earlier, the most negative value of δ on the cascade diagram ($\delta_{\max}$) represents the target for minimum fresh consumption as indicated by Eq. 10.2. To remove the infeasibilities, a flow rate of the fresh resource equal to $\delta_{\max}$ is added to the top of the cascade (that is, $\delta'_0 = \delta_{\max}$). The residuals are accordingly increased by $\delta_{\max}$; that is,

$$[10.11] \qquad \delta'_k = \delta_k + \delta_{\max}$$

where δ'_k is the revised residual flow leaving the kth interval. Consequently, the most negative residual becomes zero, thereby designating the pinch location. Additionally, the revised residual leaving the last interval is the target for minimum wastewater discharge because it represents the unrecycled/unreused flow rates of sources. Figure 10.7 is an illustration of the revised cascade diagram. As can be seen

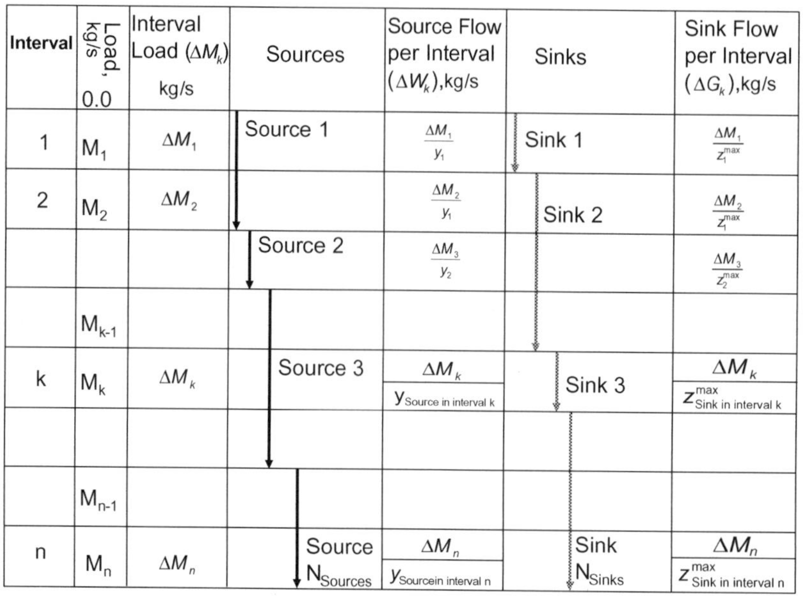

Interval	Load, kg/s 0.0	Interval Load (ΔM_k) kg/s	Sources	Source Flow per Interval (ΔW_k), kg/s	Sinks	Sink Flow per Interval (ΔG_k), kg/s
1	M_1	ΔM_1	Source 1	$\frac{\Delta M_1}{y_1}$	Sink 1	$\frac{\Delta M_1}{z_1^{max}}$
2	M_2	ΔM_2		$\frac{\Delta M_2}{y_1}$	Sink 2	$\frac{\Delta M_2}{z_1^{max}}$
			Source 2	$\frac{\Delta M_3}{y_2}$		$\frac{\Delta M_3}{z_2^{max}}$
	M_{k-1}					
k	M_k	ΔM_k	Source 3	$\frac{\Delta M_k}{y_{\text{Source in interval }k}}$	Sink 3	$\frac{\Delta M_k}{z_{\text{Sink in interval }k}^{max}}$
	M_{n-1}					
n	M_n	ΔM_n	Source N_{Sources}	$\frac{\Delta M_n}{y_{\text{Source in interval }n}}$	Sink N_{Sinks}	$\frac{\Delta M_n}{z_{\text{Sink in interval }n}^{max}}$

FIGURE 10-8 Load-interval diagram. *Source: Almutlaq and El-Halwagi (2007).*

from this illustration, the following residuals determine the targets:

[10.12] δ_{max} = Target for minimum fresh usage

[10.13] δ_n' = Target for minimum waste discharge

where n is the last load interval.

ALGEBRAIC TARGETING PROCEDURE

Based on the foregoing analysis, the algebraic procedure can be summarized as follows (Almutlaq and El-Halwagi, 2007):

1. Rank the sinks in ascending order of maximum admissible composition, $z_1^{max} \leq z_2^{max} \leq \ldots z_j^{max} \ldots \leq z_{N_{Sinks}}^{max}$
2. Rank sources in ascending order of pollutant composition; that is, $y_1 < y_2 < \ldots y_i \ldots < y_{N_{Sources}}$
3. Calculate the load of each sink ($M_j^{Sink,max} = G_j z_j^{max}$) and source ($M_i^{Source} = W_i y_i$).
4. Compute the cumulative loads for the sinks and for the sources (by summing up their individual loads).
5. Rank the cumulative loads in ascending order.
6. Develop the load-interval diagram (LID) shown in Fig. 10.8. First, the loads are represented in ascending order starting with zero load. The scale is irrelevant. Next, each source (and each sink) is represented as an arrow whose tail corresponds to its starting load and head corresponds to it ending load. Equations 10.3 through 10.5 are used to calculate the intervals load, source flow rate, and sink flow rate.
7. Based on the interval source and sink flow rates, develop the cascade diagram and carry out flow balances around the intervals to calculate the values of the flow residuals (δ_ks). The most negative δ_k is the target for minimum fresh consumption; that is,

[10.12] δ_{max} = Target for minimum fresh usage

8. Revise the cascade diagram by adding the maximum δ_k to the first interval and calculate the revised residuals. The interval with the first zero residual is the material recycle/reuse global pinch point. The residual flow leaving the last interval is the target for minimum waste discharge; that is,

[10.13] δ_n' = Target for minimum waste discharge

It is worth noting that, when the fresh source is impure, the foregoing procedure can be modified to account for the concentration of contaminants. Almutlaq et al. (2005) describes this procedure in detail. In addition to concentration constraints, property constraints can also be added and addressed in an algorithmic way (Ng et al., 2010).

CASE STUDY: TARGETING FOR WATER USAGE AND DISCHARGE IN A FORMIC ACID PLANT

Here, we revisit Example 4.1 on the direct recycle of water in a formic acid plant. The data for the problem are shown in Tables 10.1 and 10.2. The process flow sheet is illustrated by Figure 10.9 and was described in Chapter 4. The last column is calculated as the cumulative load.

SOLUTION

The LID is illustrated in Fig. 10.10. The cascade diagram is given by Fig. 10.11a. As can be seen, the most negative residual is −3667 kg/hr. Therefore, the target for minimum fresh water is 3667 kg/hr. When this value is added to the first interval (Fig. 10.11b), we can carry out the revised cascade calculations, leading to a

TABLE 10-1 Source Data for the Formic Acid Example

Source	Flow Rate, kg/hr	Mass Fraction of Impurities	Load of Impurities, kg/hr	Cumulative Load of Impurities, kg/hr
Distilled Water	4000	0.0175	70	70
Quenching Effluent	3000	0.0600	180	250

TABLE 10-2 Sink Data for the Formic Acid Example

Sink	Flow Rate, kg/hr	Maximum Inlet Mass Fraction of Impurities	Maximum Inlet Load of Impurities, kg/hr	Cumulative Maximum Inlet Load of Impurities, kg/hr
Hydrolysis Reactor	6000	0.0100	60	60
Quenching Column	3000	0.0300	90	150
Fly Ash Stabilizer	500	0.1000	50	200

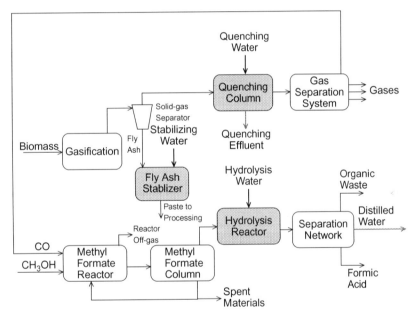

FIGURE 10-9 A block flow diagram of the formic-acid process.

Interval	Load, kg/hr	Interval Load (ΔM_k) kg/hr	Sources	Source Flow per interval (ΔW_k), kg/hr	Sinks	Sink Flow per Interval (ΔG_k), kg/hr
	0					
1	60	60	Source 1	3429	Sink 1 $z^{max}=0.01$	6000
2	70	10	$y=0.0175$	571	Sink 2 $z^{max}=0.03$	333
3	150	80	Source 2	1333		2667
4	200	50	$y=0.0600$	833	Sink 3 $z^{max}=0.10$	500
5	250	50		834		0

FIGURE 10-10 LID for the formic acid case study.

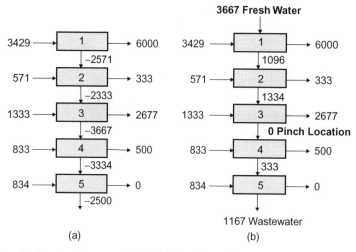

FIGURE 10-11 Cascade diagram for the formic acid case study (a) with infeasibilities (b) revised.

TABLE 10-3 Source Data for the Pulp and Paper Problem

Source	Flow Rate, tonne/day	Concentration of Impurities, ppm	Load of Impurities, kg/day
Source 1	8901	0	0
Source 2	10,995	35.8	393,621

Source: Lovelady et al. (2007).

TABLE 10-4 Sink Data for the Pulp and Paper Problem

Sink	Flow Rate, tonne/day	Maximum Inlet Concentration of Impurities, ppm	Maximum Inlet Load of Impurities, kg/day
Sink 1	1450	4.5	6525
Sink 2	13,995	6.8	95,166

Source: Lovelady et al. (2007).

TABLE 10-5 Sink Data for Hydrogen Problem

Sinks	Flow (mol/s)	Maximum Inlet Impurity Concentration (mol %)	Load (mol/s)
1	2495	19.39	483.8
2	180.2	21.15	38.1
3	554.4	22.43	124.4
4	720.7	24.86	179.2

Source: Alves and Towler (2002).

TABLE 10-6 Source Data for the Hydrogen Problem

Sources	Flow (mol/s)	Impurity Concentration (mol %)	Load (mol/s)
1	623.8	7	43.7
2	415.8	20	83.2
3	1801.9	25	450.5
4	138.6	25	34.7
5	346.5	27	93.6
6	457.4	30	137.2

Source: Alves and Towler (2002).

target of minimum waste discharge (residual leaving last interval) of 1167 kg/hr. The zero residual designates the pinch location. Hence, the material recycle pinch point is located at the horizontal lines separating intervals 3 and 4. As can be seen from the LID, this location corresponds to a cumulative load of 150 kg/hr and a source mass fraction of the impurity of the source being 0.06, and is also located at the end of sink 2 and the beginning of sink 3. These results are consistent with the ones determined graphically in Chapter 4.

HOMEWORK PROBLEMS

10.1. Solve Problem 4.1 using the algebraic method.
10.2. Solve Problem 4.2 using the algebraic method.
10.3. Solve Problem 4.3 using the algebraic method.
10.4. Solve Problem 4.4 using the algebraic method.
10.5. Solve the food processing case study (Example 4.2) using the algebraic method.
10.6. Consider the water recycle problem in a pulp and paper plant. Fresh water is available with a 3.7 ppm concentration of a chloride impurity. Tables 10.3 and 10.4 give the source and sink data.
  Hint: See the approach proposed by Almutlaq et al. (2005) for handling impure fresh resources.
10.7. Consider the problem of recycling hydrogen in a petroleum refinery based on the case study reported by Alves and Towler (2002). The refinery has four sinks that use fresh hydrogen and six sources that contain hydrogen and may be recycled to reduce the usage of fresh hydrogen. Tables 10.5 and 10.6

summarize the data for the process sinks and sources.

The available fresh hydrogen contains 0.05 mole fraction of impurities. Use an algebraic approach to determine the direct-recycle targets for minimum fresh usage of hydrogen and waste discharge.

NOMENCLATURE

G	sink (unit) flow rate, mass/time
M	load of contaminant, mass/time
$M^{sink,max}$	maximum admissible load to the sink, mass/time
$M^{Source,max}$	contaminant load in source, mass/time
$N_{sources}$	number of process streams (or sources)
N_{sinks}	number of process units (sinks)
W	sink (unit) flow, mass or volume/time
y	contaminant composition of process streams (or sources)
z	contaminant composition of process streams (or sources)
$\bar{k}$	total number of intervals

SUPERSCRIPTS

min	lower bound of allowable contaminant concentration to the sink
max	upper bound of allowable contaminant concentration to the sink

SUBSCRIPTS

i	index for sources
j	index for sinks
k	index for load intervals

GREEK LETTERS

δ	interval residual
Δ	difference between two consecutive intervals

REFERENCES

Almutlaq A, El-Halwagi MM: An algebraic targeting approach to resource conservation via material recycle/reuse, *Int J Environ Pollution (IJEP)* 29(1–3):4–18, 2007.

Almutlaq A, Kazantzi V, El-Halwagi MM: An algebraic approach to targeting waste discharge and impure-fresh usage via material recycle/reuse networks, *J Clean Technol Environ Policy* 7(4):294–305, 2005.

Alves JJ, Towler GP: Analysis of refinery hydrogen distribution systems, *Ind Eng Chem Res* 41:5759–5769, 2002.

El-Halwagi MM, Gabriel F, Harell D: Rigorous graphical targeting for resource conservation via material recycle/reuse networks, *Ind Eng Chem Res* 42:4319–4328, 2003.

Lovelady EM, El-Halwagi MM, Krishnagopalan G: An integrated approach to the optimization of water usage and discharge in pulp and paper plants, *Int J Environ Pollution (IJEP)* 29(1–3):274–307, 2007.

Ng DKS, Foo DCY, Tan RR, El-Halwagi MM: Automated targeting techniques for concentration- and property-based total resource conservation networks, *Comp Chem Eng* 34(5):825–845, 2010.

11. Synthesis of Mass-Exchange Networks: An Algebraic Approach

The graphical pinch analysis presented in Chapter 5 provides designers with a useful tool that represents the global flow of mass from the rich streams to the mass-separating agents (MSAs) and determines performance targets such as the minimum operating cost (MOC) of the MSAs. Notwithstanding the usefulness of the pinch diagram, it is subject to the accuracy problems associated with any graphical approach. This is particularly so when there is a wide range of operating compositions for the rich and the lean streams. In such cases, an algebraic method is recommended. This chapter presents an algebraic procedure that yields results equivalent to those provided by the graphical pinch analysis.

THE COMPOSITION-INTERVAL DIAGRAM

The composition-interval diagram (CID) is a useful tool for ensuring thermodynamic feasibility of mass exchange. On this diagram, $N_{sp} + 1$ corresponding composition scales are generated. First, a composition scale, y, for the rich streams is established. Then, Eq. 5.5 is employed to create N_{sp} corresponding composition scales for the process MSAs. On the CID, each process stream is represented as a vertical arrow whose tail corresponds to its supply composition and whose head represents its target composition. Next, horizontal lines are drawn at the heads and tails of the arrows. These horizontal lines define a series of composition intervals. The number of intervals is related to the number of process streams via

[11.1] $\quad N_{int} \leq 2(N_R + N_{SP}) - 1$

with the equality applying in cases where no heads or tails coincide. The composition intervals are numbered from top to bottom. The index k will be used to designate an interval with $k = 1$ being the uppermost interval and $k = N_{int}$ being the lowermost interval. Figure 11.1 provides a schematic representation of the CID. Within any interval, it is thermodynamically feasible to transfer mass from the rich streams to the MSAs. It is also feasible to transfer mass from a rich stream in an interval k to any MSA that lies in an interval $\bar{k}$ below it (that is, $\bar{k} \geq k$).

TABLE OF EXCHANGEABLE LOADS

The objective of constructing a table of exchangeable loads (TEL) is to determine the mass-exchange loads of the process streams in each composition interval. The exchangeable load of the ith rich stream that passes through the kth interval is defined as

[11.2] $\quad W^R_{i,k} = G_i(y_{k-1} - y_k),$

where y_{k-1} and y_k are the rich-scale compositions of the transferrable species that respectively correspond to the

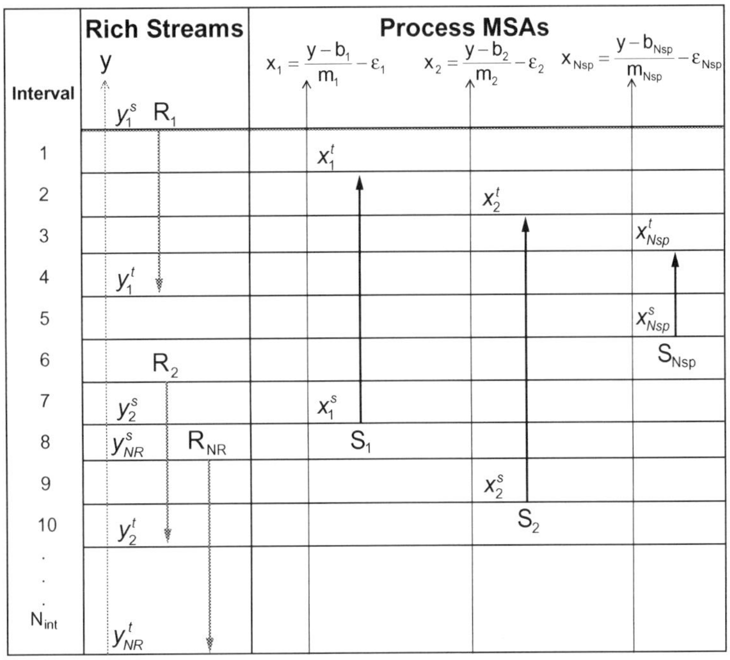

FIGURE 11-1 Composition interval diagram.

top and the bottom lines defining the kth interval. On the other hand, the exchangeable load of the jth process MSA that passes through the kth interval is computed through the following expression

$$[11.3] \qquad W_{j,k}^S = L_j^C (x_{j,k-1} - x_{j,k})$$

where $x_{j,k-1}$ and $x_{j,k}$ are the compositions on the jth lean-composition scale, which respectively correspond to the higher and lower horizontal lines bounding the kth interval. Clearly, if a stream does not pass through an interval, its load within that interval is zero.

Having determined the individual loads of all process streams for all composition intervals, one can also obtain the collective loads of the rich and the lean streams. The collective load of the rich streams within the kth interval is calculated by summing the individual loads of the rich streams that pass through that interval; that is,

$$[11.4] \qquad W_k^R = \sum_{i \text{ passes through interval } k} W_{i,k}^R.$$

Similarly, the collective load of the lean streams within the kth interval is evaluated as follows:

$$[11.5] \qquad W_k^S = \sum_{j \text{ passes through interval } k} W_{j,k}^S.$$

We are now in a position to incorporate material balance into the synthesis procedure with the objective of allocating the pinch point as well as evaluating excess capacity of process MSAs and load to be removed by external MSAs. These aspects are assessed through the mass-exchange cascade diagram.

MASS-EXCHANGE CASCADE DIAGRAM

As mentioned earlier, the CID generates a number N_{int} of composition intervals. Within each interval, it is thermodynamically as well as technically feasible to transfer a certain mass of the key targeted component from a rich stream to a lean stream. Furthermore, it is feasible to pass mass from a rich stream in an interval to any lean stream in a lower interval. Hence, for the kth composition interval, one can write the following component material balance for the key targeted component:

$$[11.6] \qquad W_k^R + \delta_{k-1} - W_k^S = \delta_k,$$

where δ_{k-1} and δ_k are the residual masses of the key targeted component entering and leaving the kth interval.

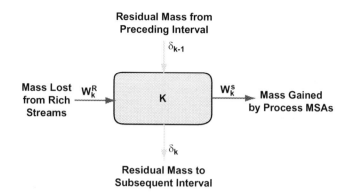

FIGURE 11-2 Component material balance around a composition interval.

Equation 11.6 indicates that the total mass input of the key component to the kth interval is due to collective load of the rich streams in that interval as well as the residual mass of the key component leaving the interval above it, δ_{k-1}. A total mass, W_k^S, of the key targeted component is transferred to the MSAs in the kth interval. Hence, a residual mass, δ_k, of the targeted component leaving the kth interval can be calculated via Eq. 11.6. This output residual also constitutes the influent residual to the subsequent interval. Figure 11.2 illustrates the component material balance for the key targeted component around the kth composition interval.

It is worth pointing out that δ_0 is zero because no rich streams exist above the first interval. In addition, thermodynamic feasibility is ensured when all the δ_ks are nonnegative. Hence, a negative δ_k indicates that the capacity of the process lean streams at that level is greater than the load of the rich streams. The most negative δ_k corresponds to the excess capacity of the process MSAs in removing the targeted component. Therefore, this excess capacity of process MSAs should be reduced by lowering the flow rate and/or the outlet composition of one or more of the MSAs. After removing the excess MSA capacity, one can construct a revised TEL in which the flow rates and/or outlet compositions of the process MSAs have been adjusted. Consequently a revised cascade diagram can be generated. On the revised cascade diagram, the location at which the residual mass is zero corresponds to the mass-exchange pinch composition. As expected, this location is the same as that with the most negative residual on the original cascade diagram. Because an overall material balance for the network must be realized, the residual mass leaving the lowest composition interval of the revised cascade diagram must be removed by external MSAs.

Example 11-1 Dephenolization of aqueous wastes

This is the same case study described by Example 5.2. Here, we will tackle the problem through the aforementioned algebraic technique. As has been previously described, the first step is to create the CID for the process streams as shown in Fig. 11.3. Then, we construct the TEL for the problem. This is shown in Table 11.1.

Next, the mass-exchange cascade diagram is generated. As can be seen in Fig. 11.4, the most negative residual mass is -0.0184 kg/s. This value corresponds to the excess capacity of process MSAs. It is worth noting that an identical result was obtained in Chapter 3, through the pinch diagram. Such excess capacity can be removed by reducing the flow rates and/or outlet compositions of the process MSAs. If we decide to eliminate this excess by decreasing the flow rate of S_2, the actual flow rate of S_2 should be 2.08 kg/s, as was calculated via Eq. 3.9. Using the adjusted flow rate of S_2, we can now construct the revised TEL for the problem as depicted by Table 11.2.

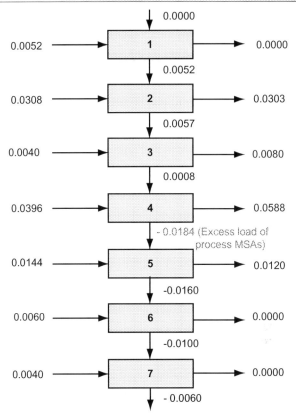

Interval	Waste Streams R_1		Process MSAs X_1	X_2
	y 0.0500		0.0240	0.0317
1		0.0474	0.0227	0.0300
2		0.0320	0.0150	0.0199
3	R_2 0.0300		0.0140	0.0186
4		0.0168	0.0074	0.0100
5		0.0120	0.0050	0.0068 S_2
6		0.0100	0.0040 S_1	0.0055
7		0.0060	0.0020	0.0029

FIGURE 11-3 CID for the dephenolization example.

FIGURE 11-4 Cascade diagram for dephenolization example.

TABLE 11-1 TEL for Dephenolization Example

	Load of Waste Streams kg phenol/s			Load of Process MSAs kg phenol/s		
Interval	R_1	R_2	$R_1 + R_2$	S_1	S_2	$S_1 + S_2$
1	0.0052	–	0.0052	–	–	–
2	0.0308	–	0.0308	–	0.0303	0.0303
3	0.0040	–	0.0040	0.0050	0.0039	0.0089
4	0.0264	0.0132	0.0396	0.0330	0.0258	0.0588
5	0.0096	0.0048	0.0144	0.0120	–	0.0120
6	0.0040	0.0020	0.0060	–	–	–
7	–	0.0040	0.0040	–	–	–

TABLE 11-2 Revised TEL for Dephenolization Example

	Load of Waste Streams kg phenol/s			Load of Process MSAs kg phenol/s		
Interval	R_1	R_2	$R_1 + R_2$	S_1	S_2	$S_1 + S_2$
1	0.0052	–	0.0052	–	–	–
2	0.0308	–	0.0308	–	0.0210	0.0210
3	0.0040	–	0.0040	0.0050	0.0027	0.0077
4	0.0264	0.0132	0.0396	0.0330	0.0179	0.0509
5	0.0096	0.0048	0.0144	0.0120	–	0.0120
6	0.0040	0.0020	0.0060	–	–	–
7	–	0.0040	0.0040	–	–	–

(Continued)

Example 11-1 Dephenolization of aqueous wastes (*Continued*)

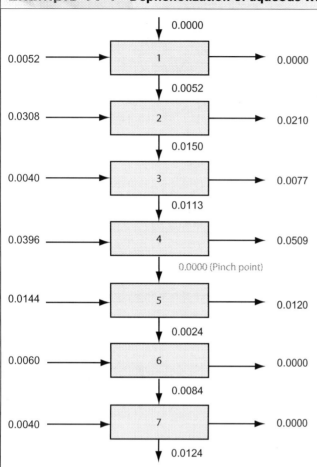

FIGURE 11-5 Revised cascade diagram for the dephenolization example.

Hence, the revised mass-exchange cascade diagram is generated as shown in Fig. 11.5. On this diagram, the residual mass leaving the fourth interval is zero. Therefore, the mass-exchange pinch is located on the line separating the fourth and the fifth intervals. As can be seen in Fig. 11.3, this location corresponds to a set of corresponding composition scales $(y, x_1, x_2) = (0.0168, 0.0074, 0.0100)$. Furthermore, Fig. 11.5 shows the residual mass leaving the bottom interval as 0.0124 kg/s. This value is the amount of targeted component to be removed by external MSAs. Again, a similar result was obtained from the pinch diagram (Fig. 5.17).

HOMEWORK PROBLEMS

11.1. Using an algebraic procedure, synthesize an optimal MEN for the benzene recovery example described by Example 5.1.

11.2. Resolve the dephenolization example presented in this chapter for the case when the two waste streams are allowed to mix.

11.3. Using the cascade diagram, solve Problem 5.3 on the desulfurization of coke-oven gas (COG).

11.4. The techniques presented in this chapter can be generalized to tackle MENs with multiple targeted components. Consider the COG-sweetening process addressed by the previous problem. Carbon dioxide often exists in COG in relatively large concentrations. Therefore, partial removal of CO_2 is sometimes desirable to improve the heating value of COG, and almost complete removal of CO_2 is required for gases undergoing low-temperature processing. The data for the CO_2-laden rich and lean streams are given in Table 11.3.

Synthesize an MOC MEN that removes hydrogen sulfide and carbon dioxide simultaneously. Hint: See El-Halwagi and Manousiouthakis (1989a, b).

TABLE 11-3 Stream Data for COG-Sweetening Problem

	Rich Streams				MSAs					
Stream	G_i (kg/s)	Supply Mass Fraction of CO_2	Target Mass Fraction of CO_2	Stream	L_j^c kg/s	Supply Mass Fraction of CO_2	Target Mass Fraction of CO_2	m_j for CO_2	b_j for CO_2	c_j \$/kg
R_1	0.90	0.0600	0.0050	S_1	2.3	0.0100	?	0.35	0.000	0.00
R_2	0.10	0.1150	0.0100	S_2	∞	0.0003	?	0.58	0.000	0.10

Source: El-Halwagi and Manousiouthakis (1989a, 1989b).

NOMENCLATURE

b_j	intercept of equilibrium line of the jth MSA
G_i	flow rate of the ith rich stream
i	index of rich streams
j	index of MSAs
k	index of composition intervals
L_j	flow rate of the jth MSA
L_j^C	upper bound on available flow rate of the jth MSA
m_j	slope of equilibrium line for the jth MSA
$N_{i,above\ pinch}$	number of independent synthesis problems above the pinch
$N_{i,below\ pinch}$	number of independent synthesis problems below the pinch
N_{int}	number of composition intervals
N_{ra}	number of rich streams immediately above the pinch
$N_{R,above\ pinch}$	number of rich streams above the pinch
N_{rb}	number of rich streams immediately below the pinch
$N_{R,below\ pinch}$	number of rich streams below the pinch
N_{sa}	number of lean streams immediately above the pinch
$N_{S,above\ pinch}$	number of lean streams above the pinch
N_{sb}	number of lean streams immediately below the pinch
$N_{S,below\ pinch}$	number of lean streams below the pinch
N_{SP}	number of process MSAs
R_i	the ith rich stream
S_j	the jth lean stream
$W_{i,k}^R$	the exchangeable lead of the ith rich stream that passes through the kth interval as defined by Eq. 11.2
$W_{j,k}^S$	the exchangeable load of the jth MSA passing through the kth interval as defined by Eq. 11.3
W_k^R	the collective exchangeable load of the rich streams in interval k as defined by Eq. 11.4
W_k^S	the collective exchangeable load of the MSAs in interval k as defined by Eq. 11.5
$x_{j,k-1}$	composition of key component in the jth MSA at the upper horizontal line defining the kth interval
$x_{j,k}$	composition of key component in the jth MSA at the lower horizontal line defining the kth interval
x_j^{pinch}	composition of key component in the jth MSA at the pinch
y_{k-1}	composition of key component in the ith rich stream at the upper horizontal line defining the kth interval
y_k	composition of key component in the ith rich stream at the lower horizontal line defining the kth interval

REFERENCES

El-Halwagi MM, Manousiouthakis V: Synthesis of mass exchange networks, *AIChE J* 35(8):1233–1244, 1989a.

El-Halwagi MM, Manousiouthakis V: Design and analysis of mass exchange networks with multicomponent targets, San Francisco, AIChE Annual Meeting, November 1989b.

CHAPTER 12

SYNTHESIS OF HEAT-INDUCED SEPARATION NETWORKS FOR CONDENSATION OF VOLATILE ORGANIC COMPOUNDS

Mass-exchange operations employ mass-separating agents to induce the transfer of targeted components from the rich phase to the lean phase. Another important class of separations involves the use of energy-separating agents to induce separations. This chapter deals with the optimal design of *heat-induced separation networks* (HISENs). Examples of HISENs include condensation, crystallization, and drying. The chapter will focus on condensation because of its importance in recovering volatile organic compounds (VOCs), which are among the most serious atmospheric pollutants. A shortcut graphical method will be presented. This method is primarily based on the work of Richburg and El-Halwagi (1995). More generalized procedures and broader applications can be found in literature (for example, Sharifzadeh et al., 2011; Cisternas et al., 2006; Hamad and Fayed, 2004; Dunn and El-Halwagi, 2003; Castier and Queiroz, 2002; Parthasarathy et al., 2001a, b; Dunn and El-Halwagi, 1994a, b; El-Halwagi et al., 1995; Dunn et al., 1995).

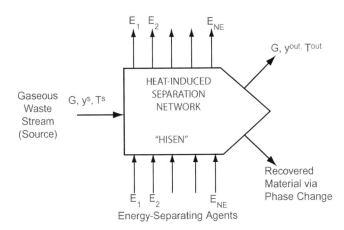

FIGURE 12-1 Schematic representation of HISEN synthesis problem.

PROBLEM STATEMENT

Given a VOC-laden gaseous stream whose flow rate is G, whose supply composition of the VOC is y^s, and whose supply temperature is T^s, it is desired to design a cost-effective condensation system that can recover a certain fraction, α, of the VOC contained in the stream.

Available for service are several refrigerants. The operating temperature for the jth refrigerant, t_j, is given. For convenience in terminology, these refrigerants are arranged in order of decreasing operating temperature; that is,

[12.1] $\quad t_1 \geq t_2 \geq \ldots t_j \ldots \geq t_{NE}$

The operating cost of the jth refrigerant (denoted by C_j, \$/kJ removed) is known. The flow rate of each refrigerant is unknown and is to be determined through optimization.

SYSTEM CONFIGURATION

Normally, the VOC-condensation system involves three units as schematically depicted in Fig. 12.1. The initial step is to cool the stream to a temperature slightly above the freezing point of water, to dehumidify the gas to prevent detrimental icing effects in subsequent stages. Next, the stream is cooled to T^* to recover the VOC. The temperature T^* is an optimization variable. To utilize the cooling capacity of the gaseous stream at T^*, it is recycled back to the system for heat integration. The remaining cooling duty is accomplished by a refrigerant.

It is worth pointing out that the bypassed portion in Fig. 12.2 is an optimization variable. The bypass is

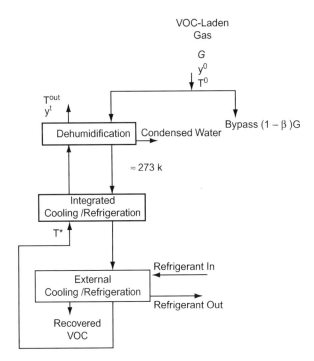

FIGURE 12-2 VOC-recovery system. *Source: Richburg and El-Halwagi (1995)*.

strongly linked to the selection of a target composition for the VOC in the outlet gas. At first glance, it may appear that the target composition is calculated via

[12.2] $\quad y^t = (1 - \alpha)y^s,$

which is not necessarily optimal. Indeed, one may pass only a fraction, β, of the gaseous stream through the condensation system such that the amount condensed from the fraction β is equal to the amount to be recovered from the whole stream; that is,

[12.3a] $$\beta G(y^s - y^t) = \alpha G y^s$$

or

[12.3b] $$y^t = \left(1 - \frac{\alpha}{\beta}\right) y^s$$

Because condensation must occur ($y^t < y^s$), the following bounds on β should be met:

[12.4] $$\alpha < \beta \le 1.00$$

The rest of the gaseous stream, $(1 - \beta)G$, is bypassed, and the net effect is that a fraction α of the VOC contained in the whole gaseous waste is recovered. Hence, the identification of the optimum value of β is part of the system optimization.

INTEGRATION OF MASS AND HEAT OBJECTIVES

The primary objective of the VOC-condensation system is to meet mass-recovery objectives. However, heat is a key element in realizing the mass objectives. Hence, the mass and heat interactions of the problem have to be identified and reconciled. This can be achieved by converting the VOC-recovery task from a mass-transfer to a heat-transfer duty, by relating the composition of the VOC to the temperature of the gaseous waste. When a VOC-laden gas is cooled, the composition of the VOC remains constant until condensation starts at a temperature T^c, defined by

[12.5] $$p^s = p^o(T^c),$$

where p^s is the supply (inlet) partial pressure of the VOC, and $p^o(T)$ is the vapor pressure of the VOC expressed as a function of the gaseous-waste temperature T. Hence, for a dilute system, the molar composition of the VOC in the gaseous emission, y, can be described as

[12.6a] $$y(T) = y^s \quad \text{if} \quad T > T^c$$

[12.6b] $$= p^o(T)/[P^{total} - p^o(T)] \quad \text{if} \quad T \le T^c,$$

where y^s is the supply (inlet) mole fraction of the VOC in the waste and P^{total} is the total pressure of the gas. Therefore, for a given target composition of the VOC, y^t, the VOC-recovery task is equivalent to cooling the stream to a separation-target temperature, T^*, which is calculated via

[12.7] $$y^t = p^o(T^*)/[P^{total} - p^o(T^*)].$$

Having converted the VOC-recovery problem into a heat-transfer task, we now proceed to develop the design procedure.

DESIGN APPROACH

The design procedure starts by identifying the minimum utility cost for a given heat-transfer driving force. Next, the fixed and operating costs are traded off by iterating over the driving forces until the minimum total annualized cost is attained.

MINIMIZATION OF EXTERNAL COOLING UTILITY

It is beneficial to develop the enthalpy expressions for the gaseous VOC-laden stream as its temperature is cooled from T^s to some arbitrary temperature T, which is below T^c. Assuming that the latent heat of the VOC remains constant over the condensation range, the enthalpy change (for example, kJ/kmol of VOC-free gaseous stream) can be evaluated through

[12.8]
$$h(T) - h(T^s) = \int_{T^s}^{T} C_{p,g}(T) dT + \int_{T^s}^{T} y(T) C_{P,V}(T) dT$$
$$+ [y^s - y(T)]\lambda + \int_{T^c}^{T} [y^s - y(T)] C_{P,L}(T) dT.$$

A similar expression (excluding the last two terms) can be derived for the enthalpy of the recycled cold gas. A convenient way of identifying the minimum utility cost of the system is the thermal pinch diagram described in Chapter 9. A temperature scale for the gaseous stream to be cooled, T, is related to a temperature scale for the recycled cold gas, t, by using a minimum driving force ($T = t + \Delta T_1^{min}$). Next, Eq. 12.8 is used to plot the enthalpy of the hot stream (VOC-laden stream) versus T. Similarly, one can plot the enthalpy of the cold gas recycled to the system against t. The cold stream is slid down until it touches the hot stream at the pinch point and, therefore, the minimum cooling requirement, Q^c, can be identified (Fig. 12.3). This is the load to be removed by the refrigerant in the bottom exchanger of Fig. 12.2.

For each value of β, Eq. 12.3b is used to determine y^t, which in turn is employed in Eq. 12.7 to calculate T^*. As can be seen from Fig. 12.3, for a given T^*, the value of Q^c is determined graphically. In addition, the outlet

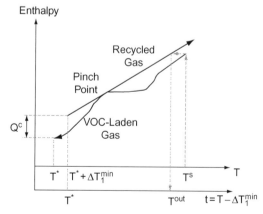

FIGURE 12-3 Pinch diagram for the VOC-condensation system. *Source: Richburg and El-Halwagi (1995).*

temperature for the gaseous stream leaving the network, T^{out}, is identified.

SELECTION OF COOLING UTILITIES

The next step is to screen the candidate refrigerants with the objective of minimizing the utility cost. This step can be accomplished by examining the cost of refrigerants that operate below T^*. Consider two refrigerants, u and v, whose costs ($/kJ removed) are C_u and C_v, where $u > v$. It is useful to recall that the refrigerants are arranged in order of decreasing operating temperatures; hence, $t_u < t_v$. If $C_u \leq C_v$, then refrigerant u is preferred over v. This rationale can be employed to compare all refrigerants below T^*, with the result of identifying the one that yields the lowest cooling cost.

TRADING OFF FIXED COST VERSUS OPERATING COST

Once the minimum utility cost has been identified, trade-offs between operating and fixed costs must be established. This step is undertaken iteratively. For given values of minimum approach temperatures, use the pinch diagram to obtain minimum cooling cost and outlet gas temperature. By conducting enthalpy balance around each unit, determine intermediate temperatures and exchanger sizing. Hence, one can evaluate the fixed cost of the system. Next, the minimum approach temperatures are altered, until the minimum total annualized cost (TAC) is identified.

SPECIAL CASE: DILUTE WASTE STREAMS

In many environmental applications, the VOC is dilute enough to render the latent heat change of the stream negligible compared to its sensible heat change. In such cases, most of the cooling utility is employed to cool the carrier gas to T^*, while only a small fraction of the cooling duty is used to condense the VOC. Therefore, the hot and cold composite lines become identical in shape (linear if specific heat is assumed constant over temperature), and the pinch diagram may be approximated as shown in

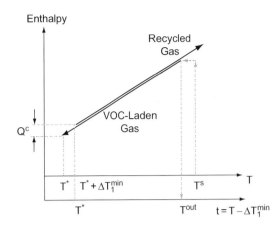

FIGURE 12-4 Pinch diagram for dilute VOC-condensation system. *Source: Richburg and El-Halwagi (1995).*

Fig. 12.4. In this diagram, the two composite lines touch over the overlapping temperature zone, so there are an infinite number of pinch points.

As can be seen from Fig. 12.4, the minimum cooling utility requirement is given by

$$Q^c = \beta G \bar{C}_{p,g} \Delta T_1^{min}, \quad [12.9]$$

while the rest of the cooling task is fully integrated with the recycled gaseous stream. By combining Equations 12.3a, 12.7, and 12.9, and noting that for dilute streams $p^o(T^*) \ll P^{total}$, we get

$$Q^c = \alpha G y^s \bar{C}_{p,g} \Delta T_1^{min} / \{y^s - [p^o(T^*)/P^{total}]\}. \quad [12.10]$$

Because $p^o(T^*)$ is a monotonically increasing function of T^*, the lower the value of T^*, the lower the Q^c. For a given refrigerant j and a minimum driving force ΔT_2^{min}, the lowest attainable gas temperature is $t_j + \Delta T_2^{min}$. Hence, the optimal value of T^* can be obtained by comparing the costs of refrigerants needed to cool the gas to $t_j + \Delta T_2^{min}$, where $j = 1, 2, ..., N$. This entails searching over a finite set of at most N temperatures to identify the minimum cooling cost. Once the optimal T^* is determined, Equations 12.3b and 12.7 are used to calculate the optimal β, and the optimum network is configured.

Example 12.1 Removal of Methyl Ethyl Ketone

A gaseous emission contains 330 ppmv of methyl ethyl ketone (MEK). It is desired to reduce the MEK content of this stream to 33 ppmv ($\alpha = 0.9$) prior to atmospheric discharge. The flow rate of the gas is 0.10 kmole/s, its pressure is 121,560 Pa (1.2 atm), and its supply temperature is 300 K. Available for service are three refrigerants: NH_3, HFC134 and liquid N_2, whose respective operating temperatures are 275, 268, and 100 K, and operating costs are 8, 19, and 31 $/$10^6$ kJ removed, respectively. The average specific heat of the gaseous waste is taken as 30 kJ/(kmol K). The latent heat of condensation for MEK is approximated to be 36,000 kJ/kmol over the condensation range. The vapor pressure of MEK is given by (Yaws, 1994)

$$\log p^o(T) = 49.8308 - 3096.5/T - 15.184 \log T + 0.007485T - 1.7084 * 10^{-13} T^2, \quad [12.11]$$

where $p^o(T)$ is in pascals and T is in kelvins. To avoid the formation of solid MEK in the system, the lowest permissible operating temperature of the gas is 190 K, which is slightly above the freezing temperature of MEK (186.5 K).

The cost of condensers/heat exchangers ($) is taken as 1500(heat transfer area in m^2)$^{0.6}$. The overall heat-transfer coefficients for the dehumidification, integrated cooling/condensation, and external cooling/condensation sections are 0.05, 0.05, and 0.10 kW/(m^{2}K), respectively. For N_2, it is assumed that no refrigeration system

(Continued)

Example 12.1 Removal of Methyl Ethyl Ketone (*Continued*)

is needed; instead, liquid nitrogen is purchased and stored in a tank whose cost is $150,000. For all units, a linear depreciation scheme is used with five years of useful life period and negligible salvage value.

SOLUTION

The supply composition of MEK is related to its partial pressure via Dalton's law:

[12.12] $$y^s = \frac{p^s}{P^{total}}.$$

Hence,

[12.13] $$p^s = 330 \times 10^{-6} \times 121{,}560 = 40\,\text{Pa}.$$

Similarly, the partial pressure corresponding to 33 ppmv is 4 Pa.

If we combine Equations 12.5, 12.11, and 12.13, the temperature at which MEK starts condensing can be calculated as follows:

$$\log 40 = 49.8308 - 3096.5/T^c - 15.184 \log T^c + 0.007485\,T^c - 1.7084 \ast 10^{-13}\,T^{c\,2};$$

that is,

[12.14] $$T^c = 214.5\,\text{K}.$$

Let us plot Eq. 12.11 over the condensation range as shown by Fig. 12.5. As can be seen from the figure, to reduce the MEK composition to 33 ppmv (partial pressure of MEK = 4 Pa), the temperature of the whole stream ($\beta = 1$) can be dropped to 194.5 K. Alternatively, only a fraction, β, of the stream may be passed through the condensation system. As indicated by Eq. 12.4, the bounds on β are

[12.15] $$0.900 < \beta \leq 1.000.$$

Because of the freezing limitations, the bounds are even more stringent than those given by Eq. 12.13. At the lowest permissible operating temperature of the gas (190 K), the vapor pressure of the MEK is about 2.3 Pa. According to Eq. 12.3b, the lower bound on β can be calculated as follows:

$$\frac{2.3}{121{,}560} = \left(1 - \frac{0.900}{\beta}\right) 330 \times 10^{-6}.$$

Hence, the smallest fraction of gas to be passed through the condensation system is

[12.16] $$\beta = 0.96$$

and Eq. 12.15 is modified to

[12.17] $$0.96 \leq \beta \leq 1.00$$

As has been mentioned before, for each value of β, there is an equivalent value of T^* that can be determined by Equations 12.3b and 12.7. Therefore, the bounds on T^* are given by

[12.18] $$190 \leq T^* \leq 194.5$$

Out of the three candidate refrigerants, only N_2 is capable of reaching this range. Therefore, N_2 is chosen as the external cooling utility. T^* is related to the operating temperature of the refrigerant ($t_3 = 100\,K$) by

[12.19] $$T^* = 100 + \Delta T_2^{min}$$

which can be combined with Eq. 12.18 to give

[12.20] $$90.0 \leq \Delta T_2^{min} \leq 94.5$$

Due to the dilute nature of the stream, we may assume that the latent heat of condensation for MEK is much smaller than the sensible heat removed from the gas. Therefore, we can apply the procedure presented in Section 12.5. Once a value is selected for ΔT_2^{min}, Eq. 12.17 can be used to determine T^*, and Equations 12.3b and 12.7 can be employed to calculate the value of β. Because the bounds on β (and consequently on ΔT_2^{min}) are tight, we will iterate over two values of β: 0.96 and 1.00 ($\Delta T_2^{min} = 90.0$ and 94.5 K). The other iterative variable is ΔT_1^{min}. Both variables are used to trade off fixed versus operating costs. As an illustration, consider the following iteration: $\Delta T_1^{min} = 5\,K$ and $\Delta T_2^{min} = 90\,K$. According to Eq. 12.17, $T^* = 190\,K$, which corresponds to $\beta = 0.96$ (Eq. 12.16). Using heat balance, the network can be configured as shown in Fig. 12.6. Let us now proceed to estimate the size and the cost of the system at

$$\Delta T_1^{min} = 5\,K \quad \text{and} \quad \Delta T_2^{min} = 90\,K.$$

The minimum cooling utility is calculated using Eq. 12.9,

[12.21] $$Q^c = 0.96 \times 0.1 \times 30 \times 5 = 14.4\,\text{kW}$$

which corresponds to the following utility cost:

[12.22] $$\text{Annual operating cost of refrigerant} = 14.4\,\frac{\text{kJ}}{\text{s}} \times 3600\,\frac{\text{s}}{\text{hr}} \times 8760\,\frac{\text{hr}}{\text{yr}} \times 31\,\frac{\$}{10^6\,\text{kJ}}$$
$$= \$14{,}000/\text{yr}.$$

The exchangers can be sized as follows:

[12.23] $$A = \frac{Q}{U \cdot \Delta T_{lm}}$$

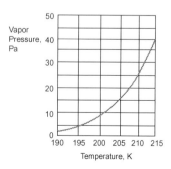

FIGURE 12-5 Vapor pressure versus temperature for MEK.

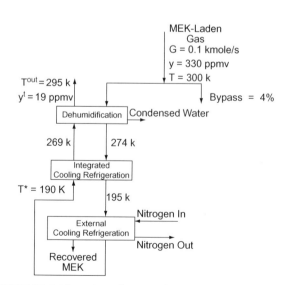

FIGURE 12-6 Network configuration for MEK removal example.

(*Continued*)

Example 12.1 Removal of Methyl Ethyl Ketone (*Continued*)

where A is the heat-transfer area, Q is the heat transferred, U is the overall heat-transfer coefficient, and ΔT_{lm} is the logarithmic mean temperature difference. For instance, the area of the first exchanger can be estimated as follows:

$$[12.24] \qquad A_1 = \frac{0.96 \times 0.1 \times 30 \times (300 - 274)}{0.05 \times 5}$$
$$= 300\,\text{m}^2$$

for a fixed cost of $1500(300^{0.6})\$46,000$. The other exchangers can be sized similarly, and the total fixed cost of the second and third exchangers is calculated to be \$89,000 and \$2000, respectively. When the cost of the nitrogen tank is added and the fixed cost is depreciated over five years, the annualized fixed cost of the system is \$57,400/yr. Combining this result with Eq. 12.22, we get a total annualized cost of \$71,500/yr. The same procedure is repeated for several values of ΔT_1^{min} and ΔT_2^{min}. The results are plotted in Fig. 12.7 for $\beta = 0.96$ (which corresponds to $\Delta T_2^{min} = 90\,\text{K}$). As can be seen from the figure, the optimal value of ΔT_1^{min} is about 5 K, and the minimum total annualized cost is approximately \$71,500/yr. Hence, the optimal system is the configuration illustrated by Fig. 12.6.

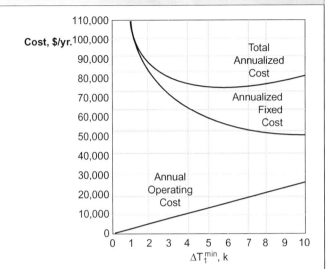

FIGURE 12-7 Cost versus ΔT_1^{min} for MEK example (at $\beta = 0.96$).

EFFECT OF PRESSURE

Pressure is an important factor in condensation. In general, the task of condensing VOCs entails the use of an energy-induced separation network (EISEN), in which pressure and cooling energies are employed to induce condensation (Dunn et al., 1995). Let us revisit the MEK example and assign a target composition of 11 ppmv MEK. With the gaseous stream operating at 1.2 atm, MEK will freeze before reaching this target. Alternatively, the stream can be pressurized to induce additional condensation. Using Eq. 12.7, at the lowest permissible gas temperature (190 K), we get

$$11 \times 10^{-6} = \frac{2.26}{P^{total}};$$

that is,

$$[12.25] \qquad P^{total} = 205{,}456\,\text{Pa (or 2.0 atm)}$$

Hence, the target composition can be achieved by compressing the gaseous waste to 2.0 atm. In some cases, following condensation, the stream may be throttled using a turbine that recovers the pressure energy and may induce additional condensation. In such cases, the costs of cooling, compression, and depressurization should be traded off (for example, see Dunn et al., 1995, and Problem 12.2).

HOMEWORK PROBLEMS

12.1 In a polyvinyl chloride plant (Richburg and El-Halwagi, 1995), a continuous air emission leaves a dryer stack. The primary pollutant in this stream is vinyl chloride (VC). The flow rate of the gas is 0.20 kmol/s, its supply temperature is 338 K, and it contains 0.5 mol/mol% of VC. It is desired to recover 90 percent of the VC in the gaseous waste. Available for service are four refrigerants: SO_2, HFC134, NH_3, and N_2, whose respective operating temperatures are 280, 265, 245, and 140 K and operating costs are 12, 10, 13, and 16 \$/10^6 kJ removed. The average specific heats of the air, VC vapor, and VC liquid are taken as 29, 50, and 85 kJ/(kmol K), respectively. The latent heat of condensation for VC is assumed to be 24,000 kJ/kmol over the operating range. The vapor pressure of VC is given by (Yaws, 1994):

$$[12.26]$$
$$\log p^\circ(T) = 52.9654 - 2.5016\,10^3/T$$
$$- 17.914 \log T + 0.0108 T - 4.531{*}10^{-14} T^2,$$

where $p^\circ(T)$ is in mm Hg and T is in K. The cost of condensers/heat exchangers (\$) is taken as 1500(heat transfer area in m^2)$^{0.6}$. The overall heat-transfer coefficients for the dehumidification, integrated cooling/condensation, and external cooling/condensation sections are 0.05, 0.05, and 0.10 kW/(m^2 K), respectively. For N_2, it is assumed that no refrigeration system is needed; instead, liquid nitrogen is purchased and stored in a tank whose cost is \$150,000. For all units, a linear depreciation scheme is used with 10 years of useful life period and negligible salvage value. Design a condensation system that features the minimum total annualized cost.

12.2 A gaseous emission has a flow rate of 0.02 kmol/s and contains 0.014 mole fraction of vinyl chloride. The supply temperature of the stream is 338 K. It is desired to recover 80 percent of the vinyl chloride using a combination of pressurization and cooling. Available for service are two refrigerants: NH_3 and N_2. Problem 12.1 and by Dunn et al. (1995) provide thermodynamic and economic data. Design a cost-effective energy-induced separation system.

12.3 The techniques presented in this chapter can be extended to streams with multicomponent VOCs. Consider a magnetic-tape manufacturing plant that disposes of a VOC-laden gaseous emission (Dunn and El-Halwagi, 1994a). This stream contains four main VOCs: tetrahydrofuran (THF), MEK, toluene, and cyclohexanone. The compositions of these compounds in the gaseous emission are 723, 865, 114, and 849 ppmv, respectively. Because of environmental regulations, the organic loading of these compounds should be reduced by 95 mol% via condensation. Six refrigerants are being considered: HCFC142b, HFC134a, HCFC22, SO_2, NH_3, and N_2. Synthesize a minimum-cost condensation system. (Hint: See Dunn and El-Halwagi, 1994a.)

NOMENCLATURE

C_j	cost of the jth refrigerant ($/kJ removed)
$C_{p,g}$	specific heat of the VOC-free gas [kJ/(kmol.K)]
$\overline{C}_{p,g}$	average specific heat of VOC-free gas [kJ/(kmol.K)]
$C_{P,L}$	specific heat of the liquid VOC [kJ/(kmol.K)]
$C_{P,V}$	specific heat of the vapor VOC [kJ/(kmol.K)]
G	flow rate of VOC-free gaseous stream (kmol/s)
h	specific enthalpy of gaseous stream (kJ/kmol of VOC-free gaseous stream)
j	index for refrigerants
N	total number of potential refrigerants
$p^o(T)$	vapor pressure of the VOC at T (mm Hg)
P^{total}	pressure of gaseous stream (mm Hg)
Q	rate of heat transfer (kW)
Q^c	minimum cooling requirement (kW)
t	temperature of cold gas to be heated (K)
t_j	operating temperature of the jth refrigerant (K)
T	temperature of gaseous stream being cooled (K)
T^c	dew temperature for the VOC (K)
T^s	supply (inlet) temperature of gaseous stream (K)
T^*	separation-target temperature to which the gaseous stream has to be cooled (K)
y^s	supply composition of VOC (kmol VOC/kmol VOC-free gaseous stream)
y^t	target composition of VOC (kmol VOC/kmol VOC-free gaseous stream)

GREEK LETTERS

α	recovery fraction of VOC in gaseous stream (kmol VOC recovered/kmol VOC in gaseous waste)
β	fraction of gaseous stream to be passed through the condensation system
ΔT_1^{min}	minimum driving force for the dehumidification and integrated cooling/condensation blocks (K)
ΔT_2^{min}	minimum driving force for the external cooling/condensation block (K)
λ	latent heat of condensation for VOC (kJ/kmole)

REFERENCES

Castier M, Queiroz EM: Energy targeting in heat exchanger network synthesis using rigorous physical property calculations, *Ind Eng Chem Res* 41(6):1511–1515, 2002.

Cisternas LA, Vasquez CM, Swaney RE: On the design of crystallization-based separation processes: review and extension, *AIChE J* 52(5):1754–1769, 2006.

Dunn RF, El-Halwagi MM: Process integration technology review: background and applications in the chemical process industry, *J Chem Tech Biotech* 78:1011–1021, 2003.

Dunn RF, El-Halwagi MM: Optimal design of multicomponent VOC Condensation Systems, *J Hazard Mater* 38:187–206, 1994a.

Dunn RF, El-Halwagi MM: Selection of optimal VOC condensation systems, *J Waste Manage* 14(2):103–113, 1994b.

Dunn RF, Zhu M, Srinivas BK, El-Halwagi MM: Optimal design of energy induced separation networks for VOC recovery, *AIChE Symp Ser* 90(303):74–85, 1995.

El-Halwagi MM, Srinivas BK, Dunn RF: Synthesis of optimal heat induced separation networks, *Chem Eng Sci* 50(1):81–97, 1995.

Hamad A, Fayed ME: Simulation-aided optimization of volatile organic compounds recovery using condensation, *Chem Eng Res Des* 82(A7):895–906, 2004.

Parthasarathy G, Dunn RF, El-Halwagi MM: Development of heat-integrated evaporation and crystallization networks for ternary wastewater systems. 2. Interception task identification for the separation and allocation network, *Ind Eng Chem Res* 40(13):2842–2856, 2001.

Parthasarathy G, Dunn RF, El-Halwagi MM: Development of heat-integrated evaporation and crystallization networks for ternary wastewater systems. 1. Design of the separation system, *Ind Eng Chem Res* 40(13):2827–2841, 2001.

Richburg A, El-Halwagi MM: A graphical approach to the optimal design of heat-induced separation networks for VOC recovery. In Biegler LT, Doherty MF, editors: *Fourth international conference on foundations of computer-aided process design*, New York, *AIChE Symp Ser* 91(304):256–259, 1995.

Sharifzadeh M, Rashtchian D, Pishvaie MR, Thornhill NF: Energy induced separation network synthesis of an olefin compression section: a case study, *Ind Eng Chem Res* 50(3):1610–1623, 2011.

Yaws CL: *Handbook of vapor pressure* (vol 1, p. 342), Houston Texas, Gulf Pub Co, 1994.

CHAPTER 13
Design of Membrane-Separation Systems

Recently, membrane processes have gained a growing level of applicability in industry. Membrane systems have several advantages. In addition to their high selectivity (which can provide concentrations as low as parts per billion), low energy consumption, and moderate cost, they are compact and modular. Therefore, membrane units can be readily added to existing plants. This chapter provides an overview of the use of membrane separation systems for environmental applications. First, the chapter will categorize pressure-driven membrane technologies. Then, modeling and design techniques will be discussed for reverse osmosis as a representative membrane technology. For more details on the subject, readers are referred to Gassner and Marechal (2010), Kim et al. (2009), Saif et al. (2008), Hamad et al. (2007), Lu et al. (2007), Kookos (2002), Marriott et al. (2003a, b), Maskan et al. (2000), El-Halwagi et al. (1996), Srinivas and El-Halwagi (1993), El-Halwagi (1993), and El-Halwagi (1992).

CLASSIFICATION OF MEMBRANE SEPARATIONS

The most common membrane systems are driven by pressure. The essence of a pressure-driven membrane process is to selectively permeate one or more species through the membrane. The stream retained at the high-pressure side is called the retentate, and the stream transported to the low-pressure side is called the permeate (Fig. 13.1). Pressure-driven membrane systems include microfiltration, ultrafiltration, reverse osmosis, pervaporation, and gas/vapor permeation. Table 13.1 summarizes the main features and applications of these systems.

Among the technologies listed in Table 13.1, reverse osmosis (RO) has gained significant commercial acceptance. Therefore, the remainder of this chapter will discuss RO systems as representative of membrane-separation technologies.

REVERSE-OSMOSIS SYSTEMS

Recently, there has been a growing industrial interest in using reverse osmosis for several objectives such as water purification and demineralization as well as environmental applications. The first step in designing the system is to understand the operating principles and modeling of RO modules.

OPERATING PRINCIPLES

When two solutions with different solute (pollutant) concentrations are separated by a semipermeable membrane,

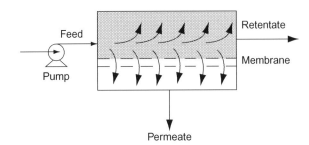

FIGURE 13-1 Schematic representation of a membrane-separation unit.

a chemical-potential difference arises across the membrane. This difference causes the carrier (typically water) to permeate through the membrane from the low-concentration (high-chemical potential) side to the high-concentration (low-chemical potential) side. This flow is referred to as the **osmotic flow** (Fig. 13.2a). Osmotic flow causes an increase in the pressure head of the high-concentration side. This flow continues until the pressure difference across the membrane balances the difference in the chemical potential across the membrane, resulting in equilibrium. The pressure head at equilibrium is termed **osmotic pressure** (Fig. 13.2b). The osmotic phenomenon can be reversed by applying a pressure greater than the osmotic pressure on the high-concentration side. Therefore, **reverse osmosis** is based on the phenomenon that takes place when an external pressure (larger than the osmotic pressure) is applied to a solution (for example, aqueous waste) causing the carrier (for example, water) to preferentially permeate, while the solute (for example, pollutant) is rejected and left in the retentate (Fig. 13.2c). Osmotic pressures can be evaluated using chemical potential calculations. However, for conceptual design purposes, approximate expressions can be employed.[1]

Typically, RO systems are preceded by pretreatment units to remove suspended solids/colloidal matter and add chemicals that control biological growth and reduce scaling and fouling. Membranes are typically made of synthetic polymers coated on a backing (skin). Examples of membrane materials include polyamides, cellulose acetate, and sulfonated polysulfone.

---

[1] For example, in the case of dilute solutions, the van't Hoff equation may be used to predict the osmotic pressure ($\pi = CRT$), where π is the osmotic pressure of the solution, C is the molar concentration of the solute, R is the universal gas constant, and T is the absolute temperature. For dissociating solutes, the concentration is that of the total ions. For example, NaCl dissociates water into two ions: Na^+ and Cl^-. Therefore, the total molar concentration of ions is *twice* the molar concentration of NaCl. A useful rule of thumb for predicting osmotic pressure of aqueous solutions is 0.01 psi/ppm of solute (Weber, 1972).

TABLE 13-1 Features of Pressure-Driven Membrane Systems for Environmental Applications

Process	Retentate	Permeate	Common Range of Feed Pressure (atm)	Membrane Type	Typical Applications/ Species
Microfiltration	Liquid	Liquid	1.5 to 7	Porous	Organic and metal suspensions, oil/water emulsions
Ultrafiltration	Liquid	Liquid	2 to 15	Porous	Oil/water emulsions, pesticides, herbicides, bivalent ions
Reverse osmosis (hyperfiltration or nanofiltration)	Liquid	Liquid	10 to 70	Porous/ nonporous	Desalination, salts, organics, ions, heavy metals
Pervaporation	Liquid	Vapor	1.5 to 60 (permeate is under vacuum)	Nonporous	Volatile organic compounds
Vapor permeation	Vapor	Vapor	Less than saturation pressure of feed (permeate is under vacuum)	Nonporous	Volatile organic compounds
Gas permeation	Gas	Gas	3.0 to 55	Nonporous	He, H_2, NO_x, CO, CO_2, hydrocarbons, chlorinated hydrocarbons

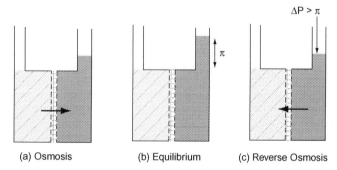

FIGURE 13-2 Direct and reverse osmosis.

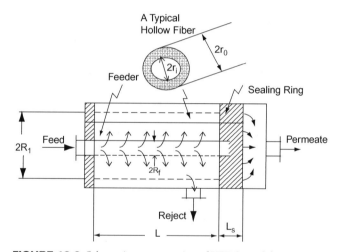

FIGURE 13-3 Schematic representation of HFRO module.

There are three main configurations of RO units: hollow fiber, tubular, and spiral wound. In particular, hollow-fiber reverse-osmosis (HFRO) systems have received considerable industrial attention. Among the different configurations of semipermeable membranes used for reverse osmosis, the hollow-fiber modules have a number of distinguishing characteristics: large surface-to-volume ratio, the self-supporting fibers, and negligible concentration polarization near the membrane surface. The next section focuses on the modeling of HFRO units.

MODELING OF HFRO UNITS

A hollow-fiber reverse-osmosis module consists of a shell that houses the hollow fibers (Fig. 13.3). The fibers are grouped together in a bundle with one end sealed and the other open to the atmosphere. The open ends of the fibers are potted into an epoxy-sealing head plate, after which the permeate is collected. The pressurized feed solution (denoted by the shell side fluid) flows radially from a central porous tubular distributor. As the feed solution flows around the outer side of the fibers toward the shell perimeter, the permeate solution penetrates through the fiber wall into the bore side by reverse osmosis. The permeate is collected at the open ends of the fibers. The reject solution is collected at the porous wall of the shell.

In modeling an RO unit, two aspects should be considered: membrane transport equations and hydrodynamic modeling of the RO module. The membrane transport equations represent the phenomena taking place at the membrane surface (for example, water permeation, solute flux, and so on). The hydrodynamic model deals with the macroscopic transport of the various species, along with their associated momentum and energy. In recent years, a number of mathematical models have been devised to simulate the hydrodynamic performance of HFRO modules. In these models, two approaches have been generally adopted to estimate the pressure variation through the shell side. In the first approach, the shell-side pressure is assumed to be constant (Gupta, 1987; Soltanieh and Gill, 1982, 1984; Ohya et al., 1977). The second approach treats the fiber bundle as a porous medium and describes the flow through this medium by Darcy's equation with an arbitrary empirical constant (Kabadi et al., 1979; Hermans, 1978; Orofino, 1977; Dandavati et al., 1975). The value of the empirical constant is strictly applicable to the geometry and operating conditions for which it has been evaluated. As a result of the pressure drop inside the fibers along the module, the pressure difference across the fiber wall may vary significantly, and thus the permeation rate may change considerably along the fiber length. Therefore, an axial component of the shell-side flow arises, in addition to the radial component. El-Halwagi et al. (1996) developed a two-dimensional model that captures the radial and axial flows within an HFRO module.

Assuming that the densities of the feed, the permeate, and the reject are the same, one can write the following overall material balance around the module:

$$[13.1] \quad q_F = q_P + q_R,$$

where q_F, q_P, and q_R are volumetric flow rates of the feed, the permeate, and the reject, respectively, per module. A component material balance on the solute is given by

$$[13.2] \quad q_F C_F = q_P C_P + q_R C_R.$$

In addition to material balance, two transport equations can be used to predict the flux of water and solute. For instance, the following simplified model can be used (Dandavati et al., 1975; Evangelista, 1986).

Water Flux:

$$[13.3] \quad N_{Water} = A\left(\Delta P - \frac{\pi_F}{C_F} C_S\right)\gamma$$

where N_{water} is water flux, ΔP is pressure difference across the membrane, π_F is osmotic pressure of feed, C_F is solute concentration in the feed, C_S is average solute concentration in the shell side, and γ is given by

$$\gamma = \frac{\eta}{1 + \dfrac{16 A \mu r_o L L_S \eta}{1.0133 \times 10^5 r_i^4}},$$

where
$$[13.4] \quad \eta = \frac{\tanh \theta}{\theta}$$
and
$$[13.4] \quad \theta = \left(\frac{16 A \mu r_o}{1.0133 \times 10^5 r_i^2}\right)^{1/2} \frac{L}{r_i}.$$

In many cases, Eq. 13.3 may be further simplified by assuming linear shell-side concentration and pressure profiles; that is,

[13.5]

$$\Delta P \approx \frac{P_F + P_R}{2} - P_P$$
$$= \frac{P_F + P_F - \textit{shell-side pressure drop per module}}{2} - P_P$$
$$= P_F - \left(\frac{\textit{shell-side pressure drop per module}}{2} + P_P\right)$$

where P_F, P_R, and P_P are pressures of feed, reject, and permeate, respectively.

Similarly,

$$[13.6] \quad C_S \approx \frac{C_F + C_R}{2}.$$

Solute Flux:

$$[13.7] \quad \begin{aligned} N_{solute} &= \left(\frac{D_{2M}}{K\delta}\right) C_S \\ &\approx \left(\frac{D_{2M}}{K\delta}\right)\left(\frac{C_F + C_R}{2}\right). \end{aligned}$$

Permeate Flow Rate:

The volumetric flow rate of the permeate per module can be obtained from the water flux as follows:

$$[13.8] \quad q_P = S_m N_{water},$$

where S_m is the surface area of the hollow fibers per module. By combining Equations 13.3, 13.6, and 13.8, we get

$$[13.9a] \quad N_{water} = A\left[\Delta P - \frac{\pi_F}{2}\left(1 + \frac{C_R}{C_F}\right)\right]\gamma$$

and

$$[13.9b] \quad q_p = S_m A\left[\Delta P - \frac{\pi_F}{2}\left(1 + \frac{C_R}{C_F}\right)\right]\gamma.$$

Permeate Concentration:

The solute concentration in the permeate may be approximated by the ratio of the solute to water fluxes; that is,

$$[13.10] \quad C_P \approx \frac{N_{solute}}{N_{water}}.$$

Equations 13.2 through 13.10 provide a complete model for describing the performance of an HFRO module. For a given set of q_F, C_F, P_F, and P_P (typically atmospheric), along with the values of the module physical properties, one can predict q_R, q_P, C_R, and C_P. A particularly useful solution scheme is in the case of highly rejecting membranes. In this case, most of the solute is retained in the reject. Therefore, Eq. 13.2 can be simplified to

$$[13.11] \quad q_F C_F \approx (q_F - q_P) C_R.$$

Combining Equations 13.9b and 13.11, we obtain

$$q_F C_F = \left\{q_F - S_m A\left[\Delta P - \frac{\pi_F}{2}\left(1 + \frac{C_R}{C_F}\right)\right]\gamma\right\} C_R.$$

Hence,

$$[13.12]$$
$$S_m A \frac{\pi_F}{2 C_F} \gamma C_R^2 + \left[q_F - S_m A\left(\Delta P - \frac{\pi_F}{2}\right)\gamma\right] C_R - q_F C_F = 0,$$

Example 13-1 Modeling the removal of an inorganic salt from water

Let us consider the following case of removing an inorganic salt from an aqueous stream. It is desired to reduce the salt content of a 26 m³/hr water stream (Q_F) whose feed concentration, C_F, is 0.035 kmol/m³ (approximately 2000 ppm). The feed osmotic pressure (π_F) is 1.57 atm. A 30 atm (P_F) booster pump is used to pressurize the feed. Sixteen hollow fiber modules are to be employed for separation. The modules are configured in parallel with the feed distributed equally among the units. The following properties are available for the HFRO modules:

Permeability (A) = $5.573 \times 10^{-8} \dfrac{m}{s \cdot atm}$

Salt flux constant $\left(\dfrac{D_{2M}}{K\delta}\right)$ = 1.82×10^{-8} m/s

Viscosity (μ) = $10^{-3} \dfrac{kg}{ms}$

Outside radius of fibers (r^o) = 42×10^{-6} m
Inside radius of fibers (r_i) = 21×10^{-6} m
Fiber length (L) = 0.750 m
Seal length (L_S) = 0.075 m
Membrane area per module (S_m) = 180 m²
Shell side pressure drop per module = 0.4 atm
Permeate pressure = 1 atm
Estimate the values of the permeate flow rate and concentration.

SOLUTION

Because the 16 modules are in parallel, they will behave the same. Therefore, we can model one module and deduce the result for the rest of the modules. The feed flow rate per module is given by

[13.13]
$$q_F = \dfrac{Q_F}{n}$$
$$= \dfrac{26.0}{(16)(3600)}$$
$$= 4.5 \times 10^{-4} \dfrac{m^3}{module.s}.$$

Pressure drop across the membrane can be calculated using Eq. 13.5:

$$\Delta P = 30.0 - \left(\dfrac{0.4}{2} + 1.0\right)$$
$$= 28.8 \, atm.$$

The module properties (Eq. 13.4) are

$$\theta = \left[\dfrac{(16)(5.573 \times 10^{-8})(0.001)(42 \times 10^{-6})}{(1.0133 \times 10^5)(21 \times 10^{-6})^2}\right]^{1/2}\left(\dfrac{0.75}{21 \times 10^{-6}}\right)$$
$$= 1.0339$$

$$\eta = \dfrac{\tanh(1.0339)}{1.0339}$$
$$= 0.75$$

$$\gamma = \dfrac{0.75}{1 + \dfrac{(16)(0.5573 \times 10^{-7})(0.001)(42 \times 10^{-6})(0.75)(0.075)(0.75)}{(1.0133 \times 10^5)(21 \times 10^{-6})^4}}$$
$$= 0.6943.$$

By substituting the values of the various parameters in Eq. 13.12, we get

$$1.562 \times 10^{-4} C_R^2 + 2.549 \times 10^{-4} C_R - 1.575 \times 10^{-5} = 0.$$

Solving for C_R, we get

$$C_R = 0.0596 \, kmol/m^3 \text{ (the other root is rejected for being negative)}$$

Using Eq. 13.6, we have

$$C_S \approx \dfrac{0.035 + 0.0596}{2}$$
$$= 0.0473 \, kmol/m^3.$$

The solute flux can be determined from Eq. 13.7:

$$N_{solute} \approx (1.82 \times 10^{-8})(0.0473)$$
$$= 8.61 \times 10^{-10} \dfrac{kmol}{m^2 s}$$

The water flux can be calculated from Eq. 13.9a as follows:

$$N_{water} = 5.573 \times 10^{-8}\left[28.8 - \dfrac{1.57}{2}\left(1 + \dfrac{0.0596}{0.035}\right)\right]0.6943$$
$$= 1.032 \times 10^{-6} \, m/s,$$

and from Eq. 13.8, the volumetric flow rate of the permeate per module is

$$q_p = 180 \times 1.032 \times 10^{-6}$$
$$= 1.86 \times 10^{-4} \dfrac{m^3}{module.s}.$$

Hence, the volumetric flow rate of the permeate from the 16 modules, Q_p, is 10.7 m³/hr (2.97 × 10⁻³ m³/s).

The solute concentration in the permeate may be obtained from Eq. 13.10 as follows

$$C_P \approx \dfrac{8.61 \times 10^{-10}}{1.032 \times 10^{-6}}$$
$$= 8.34 \times 10^{-4} \, kmol/m^3 \quad (48 \, ppm).$$

which is a quadratic equation that can be solved for C_R. After we do so, Equations 13.6, 13.7, 13.9a, 13.9b, and 13.10 are used to determine the values of C_S, N_{solute}, N_{water}, q_P, and C_P, respectively. It is worth pointing out that the validity of Eq. 13.11 can be confirmed when $q_P C_P \ll q_F C_F$ (highly rejecting membranes).

The foregoing equations assume that membrane performance is time independent. In some cases, a noticeable reduction in permeability occurs over time, primarily due to membrane fouling. In such cases, design and operational provisions are used to maintain a steady performance of the system (Zhu et al., 1997).

Example 13-2 Estimation of transport parameters

In a permeation experiment, an HFRO module with a membrane area of 200 m² is used to remove a nickel salt from an electroplating wastewater. The feed to the module has a flow rate of 5×10^{-4} m³/s, a nickel-salt composition of 4000 ppm, and an osmotic pressure of 2.5 atm. The average pressure difference across the membrane is 28 atm. The permeate is collected at atmospheric pressure. The results of the experiment indicate that the water recovery[2] is 80 percent, and the solute rejection[3] is 95 percent. Evaluate the transport parameters $A\gamma$ and $(D_{2M}/K\delta)$.

SOLUTION

Let us first evaluate the composition and flow rate of the permeate using solute rejection and permeate recovery:

$$C_p = (1 - 0.95)4000$$
$$= 200 \text{ ppm}$$

and

$$q_p = 0.8 \times 5 \times 10^{-4}$$
$$= 4 \times 10^{-4} \text{ m}^3/\text{s}$$

Therefore, according to Eq. 13.8,

$$N_{water} = \frac{4 \times 10^{-4}}{200}$$
$$= 2 \times 10^{-6} \text{ m/s}.$$

The reject flow rate can be evaluated via material balance:

$$q_R = 5 \times 10^{-4} - 4 \times 10^{-4}$$
$$= 1 \times 10^{-4} \text{ m}^3/\text{s}.$$

Using the component material balance, we get

$$C_R = \frac{5 \times 10^{-4} \times 4000 - 4 \times 10^{-4} \times 200}{1 \times 10^{-4}}$$
$$= 19{,}200 \text{ ppm}.$$

The product of the transport parameters A and γ can be calculated from Eq. 13.9a:

$$2 \times 10^{-6} = A\gamma\left[28 - \frac{2.5}{2.0}\left(1 + \frac{19200}{4000}\right)\right];$$

that is,

$$A\gamma = 9.64 \times 10^{-8} \frac{\text{m}}{\text{s} \cdot \text{atm}}.$$

Therefore, for this system the water flux Equation 13.9a is

$$N_{water} = 9.64 \times 10^{-8}\left[\Delta P - \pi_F\left(1 + \frac{C_R}{C_F}\right)\right].$$

We now turn our attention to the solute flux. Using Eq. 13.10, we get

$$N_{solute} = 200 \times 2 \times 10^{-6}$$
$$= 4 \times 10^{-4} \frac{\text{ppm} \cdot \text{m}}{\text{s}}.$$

By employing Eq. 13.7, we can evaluate the salt flux constant:

$$\left(\frac{D_{2M}}{K\delta}\right) = \frac{2 \times 4 \times 10^{-4}}{4000 + 19{,}200}$$
$$= 3.45 \times 10^{-8} \text{ m/s}.$$

---

[2] Water recovery (or conversion) is the ratio of the permeate flow rate to the feed flow rate×100%.

[3] Solute rejection is defined as $\left(1 - \frac{C_P}{C_F}\right) \times 100\%$

DESIGNING SYSTEMS OF MULTIPLE REVERSE-OSMOSIS MODULES

In most industrial applications, it is rare that a single RO module can be used to address the separation task. Instead, a **reverse-osmosis network** (RON) is employed. A RON is composed of multiple RO modules, pumps, and turbines. The following sections describe the problem of synthesizing a system of RO modules and a systematic procedure for designing an optimal RON.

SYNTHESIS OF RONs: PROBLEM STATEMENT

The task of synthesizing an optimal RON can be stated as follows: For a given feed flow rate (Q_F) and feed concentration (C_F), it is desired to synthesize a minimum cost system of reverse-osmosis modules, booster pumps, and energy-recovery turbines that can separate the feed into two streams: an environmentally acceptable permeate and a retentate (reject) stream in which the undesired species is concentrated. The permeate stream must meet the following two requirements:

1. The permeate flow rate should be no less than a given flow rate; that is,

   [13.14] $\qquad Q_P \geq Q_P^{\min}.$

2. The concentration of the undesirable species in the permeate should not exceed a certain limit (typically the environmental regulation):

   [13.15] $\qquad C_P \leq C_P^{\max}.$

The flow rate per module is typically bounded by manufacturer's constraints:

[13.16] $\qquad q_F^{\min} \leq q_F \leq q_F^{\max}.$

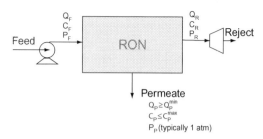

FIGURE 13-4 RON synthesis problem.

Figure 13.4 shows a schematic representation of the RON problem.

SHORTCUT METHOD FOR SYNTHESIS OF RONs

The problem of designing a RON entails determining membrane types, sizes, number, and arrangement, as well as the optimal operating conditions and the type, number, and size of any pumps and energy-recovery devices. To understand the basic principles of synthesizing an optimal RON, let us consider the class of problems for which one stage of parallel RO modules is used.[4] A booster pump is first used to raise the pressure to its optimal level. The feed is distributed among a number n of parallel modules. The reject is collected as a retentate stream, which is fed to an energy-recovery turbine (if the value of recovered energy is higher than the cost of recovering it). The permeate streams are also gathered and constitute the environmentally acceptable stream. The following design and operating variables are to be optimized:

1. Feed pressure P_F, and consequently size of booster pump
2. Number of parallel modules, n
3. Size of energy-recovery turbine

It is instructive to consider the equations specifying the system performance.

Overall material balance for the RON:

[13.17] $$Q_F = Q_P + Q_R,$$

where Q_F, Q_P, and Q_R are the volumetric flow rates of the feed, the permeate, and the reject, respectively, for the whole RON.

System and module flow rates:

[13.18] $$Q_F = n\, q_F$$

[13.19] $$Q_P = n\, q_P$$

[13.20] $$Q_R = n\, q_R$$

Solute material balance around the RON:

[13.21] $$Q_F C_F = Q_P C_P + Q_R C_R$$

Water flux:

[13.3] $$N_{Water} = A\left(\Delta P - \frac{\pi_F}{C_F}C_S\right)\gamma$$

Pressure drop across the membrane:

[13.5]
$$\Delta P \approx \frac{P_F + P_R}{2} - P_P$$
$$= \frac{P_F + P_F - \text{shell-side pressure drop per module}}{2} - P_P$$
$$= P_F - \left(\frac{\text{shell-side pressure drop per module}}{2} + P_P\right)$$

Average shell-side pressure:

[13.6] $$C_S \approx \frac{C_F + C_R}{2}$$

Solute flux:

[13.7] $$\begin{aligned}N_{solute} &= \left(\frac{D_{2M}}{K\delta}\right)C_S \\ &\approx \left(\frac{D_{2M}}{K\delta}\right)\left(\frac{C_F + C_R}{2}\right)\end{aligned}$$

Permeate flow rate per module:

[13.8] $$q_P = S_m N_{water}$$

Solute concentration in the permeate:

[13.10] $$C_P \approx \frac{N_{solute}}{N_{water}}$$

These 11 equations describe the performance of the single-stage RON. For a given Q_F, C_F, and Q_P,[5] there are 12 unknown variables (C_P, C_R, C_S, n, N_{water}, N_{solute}, P_F, q_F, q_P, q_R, Q_R, and ΔP). Therefore, fixing one variable specifies the whole system, which thus can be optimized by iteratively varying one variable and evaluating the cost at each iteration. For instance, the feed pressure can be varied and the total annualized cost of the system can be plotted versus pressure to locate the optimum feed pressure. Nonetheless, to avoid the simultaneous solution of 11 equations, it is recommended to vary C_p iteratively. The following iterative procedure can be used to optimize the system:

1. Setting the flow rate of the permeate to be Q_P^{min}, we can calculate the reject flow rate by rearranging the material balance (Eq. 13.17) as follows:

[13.17] $$Q_R = Q_F - Q_P.$$

---

[4] For the cases of multiple stages of RO modules, readers are referred to Evangelista (1986), El-Halwagi (1992, 1993) and El-Halwagi and El-Halwagi (1992).

[5] In the problem statement, it is required to meet a minimum demand on permeate flow rate. Typically, it is possible to provide an actual flow rate that is very close to the minimum required flow rate by adjusting the number of modules.

2. Select an iterative value of C_P that satisfies Eq. 13.15. Therefore, one can calculate the reject concentration by rearranging the component material balance on the solute (Eq. 13.21); that is,

$$[13.21] \quad C_R = \frac{Q_F C_F - Q_P C_P}{Q_R}$$

3. Calculate the average shell-side concentration according to Eq. 13.6:

$$[13.6] \quad C_S \approx \frac{C_F + C_R}{2}$$

4. Calculate the salt flux according to Eq. 13.7:

$$[13.7] \quad N_{solute} = \left(\frac{D_{2M}}{K\delta}\right) C_S$$

5. Evaluate the water flux using Eq. 13.10:

$$[13.10] \quad N_{water} \approx \frac{N_{solute}}{C_P}$$

6. Determine the pressure difference across the membrane, ΔP, by solving Eq. 13.3, where the only unknown is ΔP:

$$[13.3] \quad N_{Water} = A\left(\Delta P - \frac{\pi_F}{C_F}C_S\right)\gamma$$

7. Calculate the feed pressure using Eq. 13.5:

$$[13.5]$$
$$P_F \approx \Delta P + \left(\frac{shell \text{-} side\ pressure\ drop\ per\ module}{2} + P_P\right),$$

where P_P is typically one atmosphere.

8. Evaluate the permeate flow rate per module using Eq. 13.8:

$$[13.8] \quad q_P = S_m N_{water}$$

9. To determine the number of parallel modules, n, calculate Q_P/q_P and then set n equal to the next larger integer.

10. Now that the values of P_F and n have been determined, we can model the system as described in the first subsection of Section 13.2. With the proper cost functions, the total annualized cost of the system can be evaluated for this iteration. Next, a new value of C_P is selected, and steps 2 through 10 are repeated. The TAC of the system can be plotted versus C_P (or any of the 10 other variables) to determine the minimum TAC of the system.

Example 13-3 Optimization of a single-stage RON

Consider the aqueous waste described in Example 13.1 ($Q_F = 26\,m^3/hr$, $C_F = 0.035\,kmol/m^3$, and $\pi_F = 1.57\,atm$). A booster pump is used to pressurize the feed. Hollow-fiber modules are to be configured in parallel (single stage) with the feed distributed equally among the units. The following properties are available for the HFRO modules:

Permeability $(A) = 5.573 \times 10^{-8} \dfrac{m}{s \cdot atm}$

Salt flux constant $\left(\dfrac{D_{2M}}{K\delta}\right) = 1.82 \times 10^{-8}\,m/s$

Viscosity $(\mu) = 10^{-3} \dfrac{kg}{m \cdot s}$

Outside radius of fibers $(r_o) = 42 \times 10^{-6}\,m$
Inside radius of fibers $(r_i) = 21 \times 10^{-6}\,m$
Fiber length $(L) = 0.750\,m$
Seal length $(L_S) = 0.075\,m$
Membrane area per module $(S_m) = 180\,m^2$
Permeate pressure $= 1\,atm$
Shell side pressure drop per module $= 0.4\,atm$.

The flow rate per module is bounded by the following minimum and maximum flows:

$$[13.22] \quad 3.0 \times 10^{-4} \leq q_F\left[\frac{m^3}{s \cdot module}\right] \leq 2.2 \times 10^{-3}.$$

The system cost can be evaluated via the following expressions:

[13.23]   Annualized fixed cost of modules including membrane replacement ($/yr) = 2300 × Number of modules

[13.24]   Annualized fixed cost of pump ($/yr) = $6.5\,[Q_F(P_F - 1.013 \times 10^5)]^{0.65}$,

where Q_F is in m^3/s and P_F is in N/m^2.

[13.25]   Annualized fixed cost of turbine ($/yr) = $18.4\,[Q_R(P_R - 1.013 \times 10^5)]^{0.43}$,

where Q_R is in m^3/s and P_R is in N/m^2.

[13.26]   Annual operating cost of pump ($/yr) = $\dfrac{Q_F(P_F - 1.013 \times 10^5)}{\eta_{Pump}} \times \dfrac{\$0.06/kWhr \times 8760\,hr/yr}{10^3\,W/kW}$,

where Q_F is in m^3/s, P_R is in N/m^2, and $\eta_{Pump} = 0.7$.

[13.27]   Annual operating cost of pretreatment (chemicals) = $\$0.03/m^3$ of feed

[13.28]   Operating value of turbine ($/yr) = $Q_R(P_R - 1.013 \times 10^5)\,\dfrac{0.06}{\eta_{Turbine}\,10^3} \times 8760$

where Q_R is in m^3/s, P_R is in N/m^2, and $\eta_{Turbine} = 0.7$.

[13.29]
TAC = Annualized fixed cost of modules + Annualized fixed cost of pump
+ Annualized fixed cost of turbine + Annual operating cost of pump
+ Annual operating cost of chemicals − Operating value of turbine.

(Continued)

Example 13-3 Optimization of a single-stage RON (*Continued*)

Design a minimum-TAC RON that can produce at least 10.7 m³/hr of permeate at a maximum allowable permeate composition of 0.0012 kmol/m³ (70 ppm).

SOLUTION

Let us apply the procedure given earlier. We first select an iterative value of C_p. For instance, let us set $C_p = 0.0008$ kmol/m³. According to Eq. 13.17:

$$Q_R = Q_F - Q_P = 26 - 10.7 = 15.3 \text{ m}^3/\text{hr}.$$

Using the component material balance (Eq. 13.21):

$$C_R = \frac{Q_F C_F - Q_P C_P}{Q_R} = \frac{26.0 \times 0.0350 - 10.7 \times 0.0008}{15.3}$$
$$= 0.0589 \text{ kmol/m}^3.$$

Using Eq. 13.6, we have

$$C_S = \frac{0.0350 + 0.0589}{2} = 0.04695 \text{ kmol/m}^3.$$

The solute flux can be determined from Eq. 13.7:

$$N_{Solute} = (1.82 \times 10^{-8})(0.04695) = 8.545 \times 10^{-10} \frac{\text{kmol}}{\text{m}^2\text{s}}.$$

The water flux can be calculated using Eq. 13.10:

$$N_{Water} = \frac{8.545 \times 10^{-10}}{0.0008} = 1.068 \times 10^{-6} \text{ m/s}.$$

The pressure difference across the membrane can be determined from Eq. 13.9a as follows:

$$1.068 \times 10^{-6} = 5.573 \times 10^{-8} \left[\Delta P - \frac{1.57}{2}\left(1 + \frac{0.0589}{0.035}\right) \right] 0.6943,$$

i.e.,

$$\Delta P = 29.7 \text{ atm}.$$

Therefore,

$$P_F = 29.7 + \left(\frac{0.4}{2} + 1.0\right)$$
$$= 30.9 \text{ atm}$$

and from Eq. 13.8, the volumetric flow rate of permeate per module is

$$q_p = S_m N_{water} = 180 \times 1.068 \times 10^{-6} = 1.92 \times 10^{-4} \frac{\text{m}^3}{\text{module s}}.$$

Hence, the number of modules can be calculated as follows:

$$n = \frac{Q_P}{q_p} = \frac{10.7}{3600 \times 1.92 \times 10^{-4}} = 15.5 \approx 16 \text{ modules}.$$

We can now estimate the TAC of the system using Equations 13.23 through 13.29:

Annualized fixed cost of modules, \$/yr $= 2300 \times 16 = 36{,}800$ \$/yr

$$\text{Annualized fixed cost of pump, \$/yr} = 6.5 \left[\frac{26}{3600}(30.9 \times 1.013 \times 10^5 - 1.013 \times 10^5)\right]^{0.65}$$
$$= 4304 \text{ \$/yr}$$

$$\text{Annualized fixed cost of turbine, \$/yr} = 18.4 \left\{\frac{15.3}{3600}[(30.9 - 0.4) \times 1.013 \times 10^5 - 1.013 \times 10^5]\right\}^{0.43}$$
$$= 1070 \text{ \$/yr}$$

$$\text{Annual operating cost of pump, \$/yr} = \frac{26(30.9 \times 1.013 \times 10^5 - 1.013 \times 10^5)}{3600 \times 0.7}$$
$$\times \frac{0.06 \times 8760}{1000}$$
$$= 16{,}425 \text{ \$/yr}$$

Annual operating cost of pretreatment chemicals $= 0.03 \times 8760 \times 26$
$= 6833$ \$/yr

$$\text{Operating value of turbine, \$/yr} = \frac{15.3}{3600}[(30.9 - 0.4) \times 1.013 \times 10^5 -$$
$$1.013 \times 10^5]\frac{0.7 \times 0.06 \times 8760}{1000}$$
$$= 4672 \text{ \$/yr}.$$

Therefore,

$$\text{TAC} = 36{,}800 + 4304 + 1070 + 16{,}425 + 6833 - 4672 = 60{,}760 \text{ \$/yr}.$$

Next, a new value of C_p is selected and the iterative scheme is continued. The results of the iterations are shown in Figs. 13.5 and 13.6 as functions of C_p and P_F, respectively.

As can be seen from the plots, the minimum TAC is about \$55,000/yr. The optimal permeate composition is about 0.0005 kmol/m³, which corresponds to a feed pressure of 48 atm. It is interesting to note that the optimum value of C_p is significantly less than the required target composition (0.0012 kmol/m³). In other words, more separation can be obtained for less cost! It is also worth mentioning that, in some cases, environmental regulations may allow the bypass of a fraction of the feed and later mix it with the "overseparated permeate" to attain the required target composition. In such cases, lower costs than the ones shown in Figs. 13.5 and 13.6 can be attained.

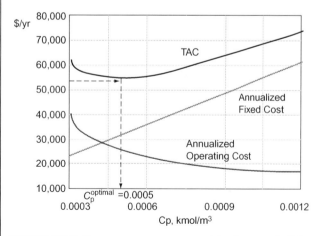

FIGURE 13-5 Cost versus permeate composition for example 13.3.

(*Continued*)

Example 13-3 Optimization of a single-stage RON (*Continued*)

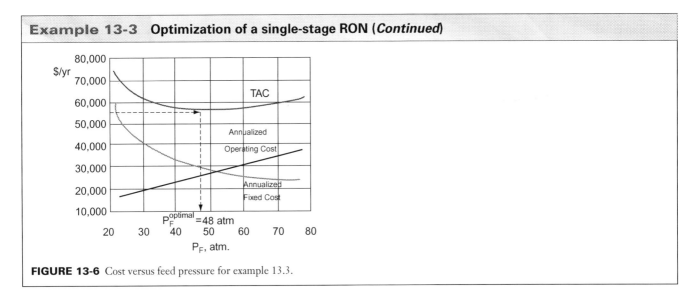

FIGURE 13-6 Cost versus feed pressure for example 13.3.

HOMEWORK PROBLEMS

13.1. An HFRO module with a membrane area of 300 m² is used to remove methyl ethyl ketone (MEK) from wastewater. The feed to the module has a flow rate of 9×10^{-4} m³/s, an MEK composition of 9500 ppm, and an osmotic pressure of 5.2 atm. The average pressure difference across the membrane is 50 atm. The permeate is collected at atmospheric pressure. The water recovery for the module is 86 percent, and the solute rejection is 98 percent. Evaluate the transport parameters $A\gamma$ and $(D_{2M}/K\delta)$.

13.2. Urea is removed from wastewater using a thin-film polysulfone reverse-osmosis membrane. When pure water was fed to the module under 68 atm, a permeate flux of 0.81 m³/m².day was obtained. For a feed pressure of 68 atm and a feed composition of 1 wt/wt% of urea, it was found that the membrane provides a permeate flux of 0.56 m³/m².day and a solute rejection of 85 percent (Matsuura, 1994). Evaluate the transport parameters $A\gamma$ and $(D_{2M}/K\delta)$ as well as the water recovery of the system.
Hint: For pure-water experiments, $A\gamma = N_{Water}/\Delta P$.

13.3. In the previous problem, it is desired to treat 12 m³/hr of wastewater containing 1 wt/wt% urea. The osmotic pressure of the feed is 7 atm. The power consumption of the system is estimated to be 5.0 kWh/m³ of permeate. Maintenance, pretreatment, and operating cost (excluding power) is $0.2/m³ of permeate. The annualized fixed cost of the system may be evaluated through the following expression:

Annualized fixed cost ($/yr) = 18,000 (Feed flow rate, m³/hr)$^{0.7}$.

Estimate the total annualized cost of the system and the cost per m³ of permeate.

13.4. In Example 13.1, the water recovery (permeate flow rate/feed flow rate) was 0.41. It is desired to increase water recovery by designing a multistage

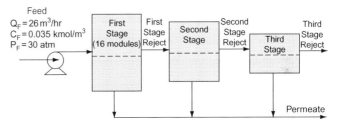

FIGURE 13-7 A three-stage tapered system for problem 13.4.

system. Two additional RO stages are to be used to process the reject from the first stage. Figure 13.7 illustrates the system configuration. The flow rate per module in the second and third stages is kept the same as in the first stage (4.5×10^{-4} m³/module.s). As a result of permeation, the flow rate entering each stage decreases successively. Hence, the number of modules also decreases successively from one stage to the next, resulting in the tapered configuration shown in Fig. 13.7. Evaluate the total water recovery of the system and the average permeate composition.
Hint: Reject flow rate, composition, and pressure for the first stage become feed flow rate, composition, and pressure (respectively) for the second stage.

13.5. A wastewater stream has a flow rate of 22 m³/hr and contains 26.0 ppm of monochlorophenol (MCP) and 3.0 ppm of trichlorophenol (TCP). Both phenolic compounds are toxic and must be reduced to acceptable levels prior to discharge. Design a cost-effective RON that satisfies the following specifications:

Minimum water recovery = 70 percent
Maximum composition of MCP in permeate = 9.5 ppm
Maximum composition of TCP in permeate = 1.2 ppm

The following data are available (El-Halwagi, 1992):

Permeability $(A) = 1.20 \times 10^{-8} \dfrac{\text{kg}}{\text{s} \cdot \text{N}}$

Solute flux constant for MCP $= 2.43 \times 10^{-4} \dfrac{\text{kg}}{\text{s} \cdot \text{m}^2}$

Solute flux constant for TCP $= 2.78 \times 10^{-4} \dfrac{\text{kg}}{\text{s} \cdot \text{m}^2}$

Outside radius of fibers $= 42 \times 10^{-6}$ m
Inside radius of fibers $= 21 \times 10^{-6}$ m
Fiber length $= 0.750$ m
Seal length $= 0.075$ m
Membrane area per module $= 180$ m^2
Pressure drop per module $= 0.405 \times 10^5$ N/m^2
Permeate pressure $= 1.013 \times 10^5$ N/m^2

The flow rate per module is bounded by the following minimum and maximum flows:

$$2.1 \times 10^{-4} \leq q_F \left(\dfrac{\text{m}^3}{\text{s} \cdot \text{module}} \right) \leq 4.6 \times 10^{-4}.$$

The system cost can be evaluated via the following expressions:

Annualized fixed cost of modules including membrane replacement ($/yr) = $1450 \times$ Number of modules

Annualized fixed cost of pump ($/yr) = $6.5 \, [Q_F (P_F - 1.013 \times 10^5)]^{0.65}$,

where Q_F is in m^3/s and P_F is in N/m^2.

Annualized fixed cost of turbine ($/yr) = $18.4 \, [Q_R (P_R - 1.013 \times 10^5)]^{0.43}$,

where Q_R is in m^3/s and P_R is in N/m^2.

Annual operating cost of pump ($/yr)

$$= \dfrac{Q_F (P_F - 1.013 \times 10^5)}{\eta_{Pump}} \times \dfrac{\$0.06/\text{kWhr} \times 8760\,\text{hr/yr}}{10^3 \, \text{W/kW}},$$

where Q_F is in m^3/s, P_R is in N/m^2, and $\eta_{Pump} = 0.7$.

Annual operating cost of pretreatment (chemicals) = $0.03/m^3 of feed.

13.6. Consider the textile process shown in Fig. 13.8 (Muralikrishnan et al., 1996). Prior to spinning the fibers, a sizing agent is deposited onto the fibers to enhance their processability during spinning. The sizing agent in this problem is polyvinyl alcohol (PVA). PVA is added as a 5.5 percent (by weight) solution in water. During spinning operation, part of the water that is added along with PVA is lost to the surroundings. After spinning, the sizing agent has to be removed from the fibers by washing the fibers with hot water in a desizing unit. The fibers are washed again in a washing unit to remove any residual PVA. Fibers leaving the washing unit pass through other finishing operations. Figure 13.8 shows the pertinent material balance data. The wastewater stream of the plant consists of water from desizing and washing units. This wastewater stream is fed to the biotreatment unit.

Because of the high value of PVA, the plant is interested in reducing PVA losses through mass integration. Using the tools described in Chapters 4, 7, and 11, devise cost-effective strategies for maximizing the value of recovered PVA − recovery cost. The devised strategies may include segregation,

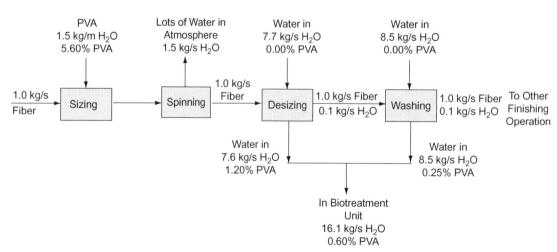

FIGURE 13-8 Textile finishing plant. *Source: Muralikrishnan et al. (1996).*

mixing, recycle, and interception using HFRO. The following information may be used in design:

Recycle Constraints

The following technical constraints should be observed in any proposed solution:
1. Desizing
   - $7.0 \leq$ Flow rate of desize feed (kg/s) ≤ 7.7
   - $0.00 \leq$ PVA concentration (wt%) ≤ 0.15
2. Washing
   - $8.5 \leq$ Flow rate of wash feed (kg/s) ≤ 9.0
   - $0.00 \leq$ PVA concentration (wt%) ≤ 0.08
3. Sizing
   - $1.5 \leq$ Flow rate of sizing feed (kgv/s) ≤ 1.6
   - $5.6 \leq$ Concentration of PVA (wt%) ≤ 5.7

Separation of PVA from Water

HFRO modules may be used for separating PVA from water. The data required for designing HFRO are given as follows:
1. Properties of HFRO modules
   Permeability $(A\gamma) = 5.573 \times 10^{-8}$ m/(s. atm)
   Solute constant $(D_{2M}/k\delta\ 12) = 3.047 \times 10^{-8}$ m/s
   Outside radius of fibers $(r_o) = 42 \times 10^{-6}$ m
   Inside radius of fibers $(r_i) = 21 \times 10^{-6}$ m
   Fiber length (L) = 0.750 m
   Seal length (L_s) = 0.075 m
   Membrane area per module $(S_m) = 180$ m^2
   Pressure drop per module $(P_d) = 0.4$ atm
   Maximum allowable flow rate per module $(q_{f,max}) = 2.2 \times 10^{-3}$ m^3/(module. s)
   Minimum allowable flow rate per module $(q_{f,min}) = 3.00 \times 10^{-4}$ m^3/(module. s)
   Maximum allowable pressure = 44.0×10^5 N/m^2
   Osmatic pressure (atm) = 5.66 [concentration (in wt%)]
   Molecular weight of PVA = 100,000 kg/kg mol
2. Cost data

Annualized fixed cost of modules including membrane replacement ($/yr) = 5000 × Number of modules

Annualized fixed cost of pump ($/yr) = $6.5[Q_F(P_F - 1.013 \times 10^5)]^{0.65}$

where Q_F is in m^3/s and P_F is in N/m^2

Annualized fixed cost of turbine ($/yr) = $18.4[Q_R(P_R - 1.013 \times 10^5)]^{0.43}$

where Q_R is in m^3/s and P_R is in N/m^2

Annualized fixed cost of piping = $\dfrac{\$50}{m.yr}$

Annual operating cost of pump ($/yr)

$$= \frac{Q_F(P_F - 1.013 \times 10^5)}{\eta_{pump}}$$
$$\times \frac{\$0.06/kWhr \times 8760\,hr/yr}{10^3\,W/kW}$$

where Q_F is in m^3/s and P_F is in N/m^2

Annual operating cost of pretreatment (chemicals) = $0.03/m^3 of feed

Operating value of turbine ($/yr) =
$Q_R(P_R - 1.013 \times 10^5)\eta_{Turbine}$
$\times \dfrac{0.06}{10^3} \times 8760$

where Q_R is in m^3/s, P_R is in N/m^2, and $\eta_{turbine} = 0.7$

PVA cost = $2.2/kg

Cost of wash water including heating = $9/m^3

Wastewater treatment cost = $2/m^3

SYMBOLS

A	Permeability (m/s.atm)
C_F	Concentration of solute in feed stream (kmol/m^3)
C_P	Concentration of solute in permeate stream (kmol/m^3)
C_R	Concentration of solute in reject stream (kmol/m^3)
C_S	Average solute concentration in shell side (kmol/m^3)
L	Fiber length (m)
LS	Seal length (m)
n	Number of modules
N_{solute}	Solute flux through the membrane (kmol/m^3)
N_{water}	Water flux through the membrane (kmol/m^3)
P_F	Pressure of feed stream entering a module (atm)
P_P	Pressure of permeate stream leaving a module (atm)
P_R	Pressure of reject stream leaving a module (atm)
ΔP	Pressure difference across the membrane (atm)
q_F	Feed flow rate per module (m^3/s.module)
q_P	Permeate flow rate per module (m^3/s.module)
q_R	Reject flow rate per module (m^3/s.module)
Q_F	Feed flow rate (m^3/s)
Q_P	Flow rate of permeate (m^3/s)
Q_R	Flow rate of reject (m^3/s)
r_i	Inside radius of fibers (m)
r^o	Outside radius of fibers (m)
S_m	Surface area of hollow fibers per module (m2)

GREEK LETTERS

γ	a constant defined by Eq. 13.4
η	a constant defined by Eq. 13.4
θ	a constant defined by Eq. 13.4

μ	Viscosity (kg/m.sec)
π_F	osmatic pressure of feed (atm)
η_{Pump}	Efficiency of pump
$\eta_{Turbine}$	Efficiency of turbine

SUBSCRIPTS

F	Feed
P	Permeate
R	Reject

SUPERSCRIPTS

min	minimum
max	maximum

REFERENCES

Dandavati MS, Doshi MR, Gill WN: Hollow fiber reverse osmosis: experiments and analysis of radial flow systems, *Chem Eng Sci.* 30:877–886, 1975.

El-Halwagi AM, El-Halwagi MM: Waste minimization via computer-aided chemical process synthesis: a new design philosophy, *Trans Egyptian Soc Chem Eng* 18(2):155–187, 1992.

El-Halwagi AM, Manousiouthakis V, El-Halwagi MM: Analysis and simulation of hollow fiber reverse osmosis modules, *Sep Sci Tech* 31(18):2505–2529, 1996.

El-Halwagi MM: Synthesis of reverse osmosis networks for waste reduction, *AIChE J* 38(8):1185–1198, 1992.

El-Halwagi MM: Optimal design of membrane hybrid systems for waste reductionm, *Sep Sci Technol* 28(1–3):283–307, 1993.

Evangelista F: Improved graphical analytical method for the design of reverse osmosis desalination plants, *Ind Eng Chem Process Des Dev* 25(2):366–375, 1986.

Gassner M, Marechal F: Combined mass and energy integration in process design at the example of membrane-based gas separation systems, *Comp Chem Eng* 34:2033–2042, 2010.

Gupta SK: Design and analysis of a radial-flow hollow-fiber reverse-osmosis system, *Ind Eng Chem Res* 26:2319–2323, 1987.

Hamad A, Al-Fadala H, Warsame A: Optimum waste interception in liquefied natural gas processes, *Int J Environ & Poll* 29:47–69, 2007.

Hermans JJ: Physical aspects governing the design of hollow fiber modules, *Desalination* 26:45–62, 1978.

Kabadi VN, Doshi MR, Gill WG: Radial flow hollow fiber reverse osmosis: experiments and theory, *Chem Eng Commun* 3:339–365, 1979.

Kim YM, Kim SJ, Kim YS, Lee S, Kim IS, Kim JH: Overview of systems engineering approaches for a large-scale seawater desalination plant with a reverse osmosis network, *Desalination* 238:312–332, 2009.

Kookos IK: A targeting approach to the synthesis of membrane networks for gas separations, *J Memb Sci* 208:193–202, 2002.

Lu YY, Hu YD, Zhang XL, Wu LY, Liu QZ: Optimum design of reverse osmosis system under different feed concentration and product specification, *J Memb Sci* 287:219–229, 2007.

Marriott J, Sorensen E: A general approach to modelling membrane modules, *Chem Eng Sci* 58:4975–4990, 2003.

Marriott J, Sorensen E: The optimal design of membrane systems, *Chem Eng Sci* 58:4991–5004, 2003.

Maskan F, Wiley DE, Johnston LPM, Clements DJ: Optimal design of reverse osmosis module networks, *AIChE J* 46:946–954, 2000.

Muralikrishnan G, Crabtree EW, El-Halwagi MM: Design of membrane hybrid systems for pollution prevention, New Orleans, AIChE Spring Meeting, February 1996.

Ohya H, Nakajima H, Takagi K, Kagawa S, Negishi Y: An analysis of reverse osmotic characteristics of B-9 hollow module, *Desalination* 21:257–274, 1977.

Orofino TA: Technology of hollow fiber reverse osmosis systems. In Sourirajan S, editor: *Reverse osmosis and synthetic membranes*, Ottawa, Canada, 1977, National Research Council, pp 313–341.

Saif Y, Elkamel A, Pritzker M: Optimal design of reverse-osmosis networks for wastewater treatment, *Chem Eng & Proc* 47:2163–2174, 2008.

Soltanieh M, Gill WN: Analysis and design of hollow fiber reverse osmosis systems, *Chem Eng Commun* 18:311–330, 1982.

Soltanieh M, Gill WN: An experimental study of the complete mixing model for radial flow hollow fiber reverse osmosis systems, *Desalination* 49:57–88, 1984.

Srinivas BK, El-Halwagi MM: Optimal design of pervaporation systems for waste reduction, *Comp Chem Eng* 17(10):957–970, 1993.

Weber WJ: *Physicochemical processes for water quality control*, New York, Wiley Interscience, 1972, p. 312

Zhu M, El-Halwagi MM, Al-Ahmad M: Optimal design and scheduling of flexible reverse osmosis networks, *J Memb Sci* 129:161–174, 1997.

CHAPTER 14

OVERVIEW OF OPTIMIZATION

Optimization refers to the identification of the "best" solution from among the set of candidate solutions. An optimization formulation typically involves the minimization or maximization of an objective function subject to a number of constraints. Optimization is a very effective and powerful tool that aids in the systematic solution of process integration problems. The principles of optimization theory and algorithms are covered by various books (for example, Horst and Tuy, 2010; Eiselt and Sandblom, 2010; Hendrix and Tóth, 2010; Ravindran et al., 2006; Diwekar, 2003; Tawarmalani, and Sahinidis, 2003; Floudas and Pardalos, 2001; Edgar and Himmelblau, 2001). Additionally, for a retrospective and future prospective on optimization, readers are referred to the following publications (Floudas and Gounaris, 2009; Biegler and Grossmann, 2004; Grossmann and Biegler, 2004). This chapter presents an overview of formulating and solving optimization problems.

WHAT IS MATHEMATICAL PROGRAMMING?

Mathematical programming deals with the formulation, solution, and analysis of optimization problems or mathematical programs. A mathematical program (or a mathematical programming model, an optimization problem, an optimization model, or an optimization formulation) can be represented by:

[P14.1] $\quad \min (\text{or max}): f(x_1, x_2, \ldots, x_N)$,

where $f(x_1, x_2, \ldots, x_N)$ is the objective function and $x_1, x_2, \ldots, x_N$ are the optimization (or decision) variables.

The minimization or maximization of the objective function is subject to:

Inequality Constraints:

$$g_1(x_1, x_2, \ldots, x_N) \leq 0$$
$$g_2(x_1, x_2, \ldots, x_N) \leq 0$$
$$\vdots$$
$$g_m(x_1, x_2, \ldots, x_N) \leq 0$$

Of course, the inequality constraints may be written as greater than or equal to zero (by multiplying any of the above constraints by -1). Additionally, the right-hand side does not have to be zero.

Equality Constraints:

$$h_1(x_1, x_2, \ldots, x_N) = 0$$
$$h_2(x_1, x_2, \ldots, x_N) = 0$$
$$\vdots$$
$$h_E(x_1, x_2, \ldots, x_N) = 0 \quad E \leq N$$

In a vector notation, a mathematical program can be expressed as:

[P14.2a] $\quad \min (\text{or max}) \, f(x)$,

where $x^T = [x_1, x_2, \ldots, x_N]$,
subject to:

$$g(x) \leq 0$$

where $g^T = [g_1, g_2, \ldots, g_m]$

$$h(x) = 0$$

where $h^T = [h_1, h_2, \ldots, h_E]$

The vector x is referred to as the vector of optimization variables. A vector x satisfying all the constraints is called a **feasible point**. The combination of all feasible points is referred to as the **feasibility region**. An optimum point must be feasible. Therefore, the optimization search is conducted over the feasibility region.

An optimization problem in which the objective function as well as all the constraints are linear is called a linear program (LP); otherwise, it is termed a nonlinear program (NLP). If all the variables in the mathematical program are integers (for example, 0, 1, 2, 3), the optimization program is referred to as an integer program (IP). The most commonly used integer variables are the zero/one binary integer variables. An optimization formulation that contains continuous (real) variables (for example, pressure, temperature, or flow rate) as well as integer variables is called a mixed-integer program (MIP). Depending on the linearity or nonlinearity of MIPs, they are designated as mixed-integer linear programs (MILPs) and mixed-integer nonlinear programs (MINLPs).

HOW TO FORMULATE AN OPTIMIZATION MODEL?

The mathematical formulation of an optimization model entails the following steps (El-Halwagi, 2006):

1. Determine the objective function
   * Determine the quantity to be optimized (the objective function).

- Identify the decision (optimization) variables that are needed to describe the objective function.
- Express the objective function as a mathematical function in terms of the optimization variables.
2. Develop your game plan to tackle the problem
   - Determine an approach to address the problem.
   - Determine your rationale.
   - Determine the key concepts necessary to transform your thoughts and approach into a working formulation.
3. Develop the constraints
   - Transform your approach to a mathematical framework.
   - Develop a search space that is rich enough to embed the solution alternatives.
   - Identify all explicit relations, restrictions, and limitations that are needed to describe the approach. Express them mathematically as equality and inequality constraints.
   - Incorporate subtle constraints (for example, nonnegativity or integer requirements on the input variables).
4. Improve formulation
   - Avoid highly nonlinear constraints or terms that lead to difficulties in convergence or solution.
- Use insights to simplify formulation or to provide useful bounds on variables.
- Enhance the clarity of the model to become easy to understand, to debug, and to reveal important information.

Formulating process integration and design tasks as optimization problems or mathematical-programming models has several advantages:

- It provides an effective framework for solving large-scale problems and highly interactive tasks that cannot be readily addressed by graphical and algebraic techniques.
- An optimization formulation may be parametrically solved to establish optimality conditions that can be used in developing visualization and algebraic solutions.
- In properly formulating an optimization problem, it is necessary to establish a model that accurately describes the task using mathematical relationships that capture the essence of the problem. In so doing, many important interactions are explicitly described with the result of providing better understanding of the system.
- When the solution of a mathematical program is implemented using computer-aided tools, it is possible to effectively examine what-if scenarios and conduct sensitivity analysis.

Example 14-1 Optimization of fuel blending

A company intends to mix biodiesel and petro-diesel to produce a cost-effective, technically viable, and environmentally friendly fuel. Table 14.1 gives data of the available biodiesel and petro-diesel to be mixed. These data include the cost of each type of fuel, the heating value, and the extent of greenhouse gas (GHG) emission upon combustion. The densities of both fuels are assumed to be roughly the same.

For technical considerations, the heating value of the blend should not be less than 119,500 Btu/gal. For environmental reasons, the GHG emission associated with burning the blend should not exceed 6540 gm CO_2/gal. What is the optimal mixing ratio of the two fuels that will minimize the cost of fuel blend ($/gal) while satisfying the technical and environmental constraints?

SOLUTION

Consider a basis of 1.0 gal of the blended fuel. What is our objective function? Here, we intend to minimize the cost of the blended fuel. What is it a function of? It is a function of how much should be purchased from each of the two fuels.

Let:

$X1$ be amount of biodiesel (gal)
$X2$ be amount of petro-diesel (gal)

Now, we can express the objective function in terms of these two optimization variables. For a basis of 1 gal of the blended fuel, let us define our objective function as:

Z = cost of 1 gal of the blended fuel ($)

TABLE 14-1 Data for the Fuel-Blending Problem

Characteristic	Biodiesel	Petro-Diesel
Cost ($/gal)	3.80	3.20
Heating value (Btu/gal)	117,000	129,000
GHG emission (gm CO_2/gal)	2100	9500

The objective function can be mathematically described as follows:

[14.1a] $$\text{minimize } Z = 3.80*X1 + 3.20*X2$$

Now, what is our game plan to tackle the problem? The cost has been addressed by defining the objective function in terms of the optimization variables. Petro-diesel is cheaper than biodiesel but, upon combustion, petro-diesel generates more than the maximum allowable limit on GHG emission. On the other hand, selecting biodiesel only will violate the constraint on minimum heating value. Therefore, our approach should entail tracking the GHG emissions and the heating value of the blended fuel as a function of the quantities of the two mixed fuels. This can be accomplished by representing a blending operation where the two fuels are mixed and using a mixing rule to predict the resulting heating value and GHG emission. Fig. 14.1 represents the blending process.

Assuming linear mixing rules for the heating value and GHG emission, the constraints are written as follows:

[14.2a] $$117{,}000*X1 + 129{,}000*X2 \geq 119{,}500$$

[14.3a] $$2100*X1 + 9500*X2 \leq 6540$$

For a basis of 1 gal of the blended fuel, the mass balance (which can be written as a volume balance because of the equal densities of the mixed fuels) is given by:

[14.4a] $$X1 + X2 = 1$$

To ensure a sound solution, the following nonnegativity constraints are added:

[14.5a] $$X1 \geq 0$$

[14.6a] $$X2 \geq 0$$

Expressions 14.1 through 14.6 represent the optimization formulation for the blending problem. This mathematical program can be solved graphically, algebraically, and using a computer-aided software. Here, a graphical approach is used. Because this is a two-variable problem, we create an X2 versus X1 representation of the constraints and the objective function.

(Continued)

Example 14-1 Optimization of fuel blending (*Continued*)

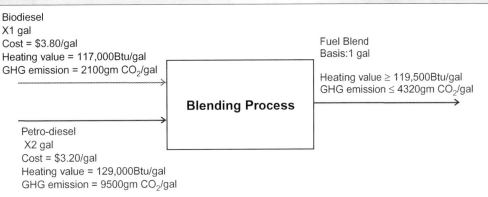

FIGURE 14-1 The fuel-blending process for example 14.1.

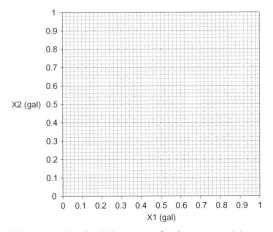

FIGURE 14-2a The feasibility region for the nonnegativity constraints.

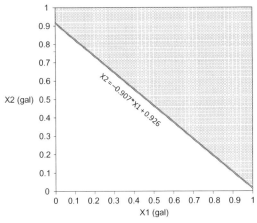

FIGURE 14-2b The feasibility region for the nonnegativity and the heating-value constraints.

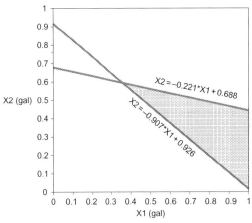

FIGURE 14-2c The feasibility region satisfying the nonnegativity, heating-value, and GHG constraints.

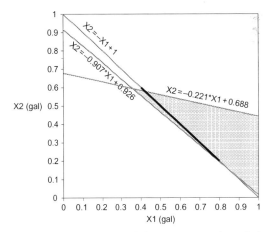

FIGURE 14-2d Passing the mass-balance constraint through the feasibility region.

The feasibility region for the nonnegativity Constraints 14.5 and 14.6 is shown by the first Cartesian quadrant (Fig. 14.2a).

The heating-value constraint (14.2a) can be rewritten as:

[14.2b] $\qquad X2 \geq -0.907*X1 + 0.926$

To satisfy this constraint along with the two nonnegativity constraints, the feasibility region is shown by the shaded region on Fig. 14.2b, which lies in the first quadrant and above the straight line given by:

[14.2c] $\qquad X2 = -0.907*X1 + 0.926$

The GHG constraint (14.3a) can also be rearranged to give:

[14.3b] $\qquad X2 \geq -0.221*X1 + 0.688$,

which is represented on the X2 versus X1 graph by the region below the following straight line:

[14.3c] $\qquad X2 = -0.221*X1 + 0.688$

Adding this constraint to the nonnegativity and the heating-value constraints, the feasibility region satisfying these four constraints is shown by the shaded region on Fig. 14.2c.

(*Continued*)

Example 14-1 Optimization of fuel blending (Continued)

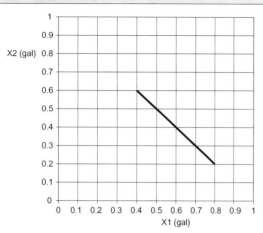

FIGURE 14-2e The feasibility region for the fuel-blending problem.

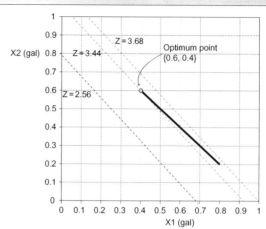

FIGURE 14-2f Identification of the optimal solution of the fuel-blending problem.

Finally, the mass-balance constraint given by Eq. 14.4a is rewritten as:

[14.4b] $\qquad X2 = -X1 + 1,$

which is passed through the feasibility region of the rest of the constraints as shown by Fig. 14.2d. The intersection of all the constraints results in the linear segment shown by Fig. 14.2e, which represents the feasibility region of the problem.

Next, we turn our attention to the objective function, which can be rewritten as:

[14.1b] $\qquad X2 = -1.188*X1 + 0.313*Z$

The objective function can be represented as a family of parallel straight lines having a slope of -1.188 and an intercept of $0.313*Z$. For each value of the objective function, Z, an intercept is calculated, and the straight line described by Eq. 14.1b is plotted. For instance, when Z is 2.56, the intercept is 0.8, and Eq. 14.1b is represented by the line shown (with the legend Z = 2.56) on Fig. 14.2f. Nonetheless, this line does not intersect with the feasibility region. Therefore, the intercept is increased to move the objective function up until it first intersects with the feasibility region at the point (0.4, 0.6) when Z = 3.44. Higher values of Z will lead to objective-function lines intersecting with the feasibility region until Z is increased to 3.68. Beyond this value, the objective function does not intersect with the feasibility region. As can be seen from Fig. 14.2f, the lowest feasible value of the objective function corresponding to the minimum cost of the blend is Z = \$3.44/gal, with the optimal blending ratio being 0.4 gal of biodiesel to 0.6 gal of petro-diesel.

USING THE SOFTWARE LINGO TO SOLVE OPTIMIZATION PROBLEMS

LINGO is an optimization software that solves linear, nonlinear, and mixed integer linear and nonlinear programs. It is developed by LINDO Systems Inc. A free trial version can be downloaded from www.LINDO.com. The following is a brief description of how to get started using LINGO. Additional information can be obtained through the help menu on the LINGO software or from LINGO User's Guide published by Schrage (2006).

The following is some basic information on writing a LINGO optimization model. The general form of the model is as follows:

Min (or Max) = Objective function;
Constraints, each followed by;

The semicolon at the end of the objective function and each constraint is the delimiter that designates the end of the line. If the line for minimizing or maximizing an objective function is not included, LINGO will solve the model as a set of equations provided that the degrees of freedom are appropriate. In writing constraints, the equalities and inequalities can be described as follows:

=   The expression to the left must equal the one on the right.
<=   The expression to the left must be less than or equal to the expression on the right.
>=   The expression to the left must be greater than or equal to the expression on the right.
<   The expression to the left must be strictly less than the expression on the right.
>   The expression to the left must be strictly greater than the expression on the right.

In writing constraints, the following symbols are used for arithmetic operations:

+   Addition
−   Subtraction
*   Multiplication
/   Division
^   Power

The following are some of the mathematical functions used by LINGO:

@ABS(X)	Returns the absolute value of X
@EXP(X)	Returns the constant e (2.718281...) to the power X
@LOG(X)	Returns the natural logarithm of X
@SIGN(X)	Returns −1 if X is less than 0, returns +1 if X is greater than or equal to 0
@BND(L, X, U)	Limits the variable X to greater or equal to L and less than or equal to U
@BIN(X)	Limits the variable X to a binary integer value (0 or 1)
@GIN(X)	Limits the variable X to only integer values

INTERPRETING DUAL PRICES IN THE RESULTS OF A LINGO SOLUTION

In the results report, LINGO displays the values of "dual prices" of constraints. The dual prices have an interesting economic interpretation, and their numerical values are often of interest. To indicate this interpretation, consider the problem

[P14.2b] $$\min_x f(x)$$

Subject to a set of equality constraints:

$$h_i(x) = b_i \quad i = 1, 2, \ldots, m,$$

where the numerical value b_i is regarded as the amount of some scarce resources that is limited according to the constraint $h_i(x)$. The dual prices correspond to the Lagrange multipliers commonly used in optimization theory and applications. As the amount of the resource b_i varies, the optimal solution also changes. Based on characteristics of the Lagrange multipliers, one can show that at the optimum solution and for a small perturbation Δb in the value of the right-hand side of the constraint, the following observation can be made:

[14.8] $$\frac{\Delta f}{\Delta b_i} \approx -\text{Dual price of constraint i}$$

If we think of the objective function as dollars of return, then it may be interpreted as dollars per unit of the ith source. Another interpretation that follows directly from the previous equation is that the dual prices represent sensitivity coefficients. This means that the marginal value of the objective function for a change in the value of the ith resource is given by negative the value of the dual price of the ith constraint. This observation is only valid near the optimum solution and is useful for conducting sensitivity analyses.

Example 14-2 Using LINGO to solve the fuel-blending problem

Let us revisit Example 14.1 on fuel blending. The objective function is expressed as:

[14.1] $$\text{minimize } Z = 3.80*X1 + 3.20*X2$$

and the constraints are given by:

[14.2a] $$117{,}000*X1 + 129{,}000*X2 \geq 119{,}500$$

[14.3a] $$2100*X1 + 9500*X2 \leq 6540$$

[14.4a] $$X1 + X2 = 1$$

[14.5] $$X1 \geq 0$$

[14.6] $$X2 \geq 0$$

This optimization formulation can be written as the following LINGO program:

[P14.3]
```
Min = 3.80*X1 + 3.20*X2;
117000*X1 + 129000*X2 >= 119500;
2100*X1 + 9500*X2 <= 6540;
X1 + X2 =1;
X1 >= 0;
X2 >= 0;
```

The solution is obtained by clicking on the "Solve" option under the LINGO drop-down menu. The following is a summary of the solution results as reported by LINGO.

```
Global optimal solution found.
Objective value: 3.440000
Variable  Value
X1        0.4000000
X2        0.6000000
```

Example 14-3 Sensitivity analysis and interpretation of dual prices

Consider the following optimization model:

[P14.4]
```
min = (x1 - 2)^2 + (x2 - 2)^2;
x1 + x2 = 6.00;
Optimal solution found at step: 4
Objective value: 2.000000
Variable  Value       Reduced Cost
X1        3.000000    0.0000000E+00
X2        3.000000    0.6029251E-07

Row  Slack or Surplus   Dual Price
1    2.000000           1.000000
2    0.0000000E100     -2.000000
```

It is worth noting that the second row in the program is the constraint $x1 + x2 = 6.0$ (the objective function is the first row). Therefore, the dual price of the constraint is -2.0. This means that at the optimum solution and for small perturbation of the right-hand side of the constraint, the ratio of change in value of the objective function to the change in the value of the right-hand side of the constraint is $-(-2) = 2$.

For instance, if we increase the right-hand side of the constraint by 0.01 (that is, b = 6.01), we should get an increase in the objective function by $2*0.01 = 0.02$ (that is, $f^* = 2.02$). Similarly, if we decrease the right-hand side of the constraint by 0.01 (that is, b = 5.99), we should get a decrease in the objective function by 0.02 (that is, $f^* = 1.98$). Let us test the model using LINGO:

[P14.5]
```
min = (x1 - 2)^2 + (x2 - 2)^2;
x1 + x2 = 6.01;
Optimal solution found at step: 4
Objective value: 2.020050
Variable  Value       Reduced Cost
X1        3.005000    0.0000000E100
X2        3.005000    0.6155309E-07

Row  Slack or Surplus   Dual Price
1    2.020050           1.000000
2   -0.2288818E-06     -2.010000
```

[P14.6]
```
min = (x1 - 2)^2 + (x2 - 2)^2;
x1 + x2 = 5.99;

Optimal solution found at step: 4
Objective value: 1.980050
```

(Continued)

Example 14-3 Sensitivity analysis and interpretation of dual prices (Continued)

Variable	Value	Reduced Cost
X1	2.995000	0.0000000E100
X2	2.995000	-0.5999458E-07

Row	Slack or Surplus	Dual Price
1	1.980050	1.000000
2	-0.2288818E-06	-1.990000

These results confirm the meaning of dual prices. It is worth noting that such observations are valid for small perturbations in the value of b_i. For instance, if we increase b by 1.0, we get the following model:

[P14.7]
```
min = (x1 - 2)^2 + (x2 - 2)^2;
x1 + x2 = 7;
```

Optimal solution found at step: 4
Objective value: 4.500000

Variable	Value	Reduced Cost
X1	3.500000	0.0000000E100
X2	3.500000	0.1111099E-06

Row	Slack or Surplus	Dual Price
1	4.500000	1.000000
2	0.0000000E100	-3.000000

Notice that when we varied b by 1.0, f* changed by 2.5 (not 2.0). Consequently, the sensitivity analysis should be carried out for small perturbations in b_i.

Example 14-4 Minimizing the cost of producing hydrogen from a biorefinery

Let us revisit Example 2.14 (accounting for plant size in determining the minimum product price via the break-even analysis). In this example, hydrogen is to be produced from biomass. The larger the plant size, the lower the fixed-cost per-unit production (because of the economy of scale such as the sixth-tenths-factor rule) but the higher the operating cost (because of the increased transportation cost needed to get the mass from farther locations). The following data and expressions were discussed in Example 2.14:

- Hydrogen yield = 70 kg H_2 produced from the process/tonne dry biomass fed to the process
- The minimum selling price of hydrogen is determined through the break-even expression:

[14.9a]   Annual sales of hydrogen = Total annualized cost of the plant

- [14.10]

Annual sales of hydrogen (ASH) = $C^{hydrogen\ min}$ ($/kg hydrogen)*Flow rate of produced hydrogen (kg/day)*365(day/yr),

where $C^{hydrogen\ min}$ is the minimum selling price of hydrogen at the selected production rate:

- [14.11]

Total annualized cost of the plant (TAC) = Annual fixed charges + Annualized fixed cost + Annual operating cost + Annual transportation cost

- [14.12]   Annual fixed charges = $10.3*10^6$/yr

- [14.13]

Annualized fixed cost = $0.2698*10^6$*(Flow rate of biomass in tonne/day)$^{0.6}$

- [14.14]   Annual operating cost = 26,061*Flow rate of biomass (tonne/day)

- [14.15]

Annual transportation cost = 0.4*(Flow rate of biomass in tonne/day − 2000)* (Flow rate of biomass in tonne/day)

The size of the biorefinery is to be determined but it should be limited to a processing capacity between 2000 and 20,000 tonne biomass/day. What is the optimal size of the biorefinery (tonne biomass/day) that will minimize the selling price of hydrogen?

SOLUTION

The objective function of the program is:

[14.16]   Minimize $C^{hydrogen\ min}$

Subject to the following constraints:

Break-even expression for pricing the hydrogen:

[14.9b]   ASH = TAC

where

[14.17]

ASH = 365*$C^{hydrogen\ min}$ ($/kg hydrogen)*Flow rate of produced hydrogen (kg/day)
TAC = $10.3*10^6$/yr + $0.2698*10^6$*(Flow rate of biomass in tonne/day)$^{0.6}$
+ 26,061*Flow rate of biomass (tonne/day) + 0.4*(Flow rate of biomass in tonne/day − 2000)*(Flow rate of biomass in tonne/day)

Hydrogen yield:

[14.18]

Flow rate of produced hydrogen (kg/day) = 70*Flow rate of biomass (tonne/day)

Capacity limits:

[14.19]   2000 ≤ Flow rate of biomass (tonne/day) ≤ 20,000

Let CHydrogen be the price of hydrogen ($/kg), FHydrogen be the flow rate of produced hydrogen (kg/day), and FBiomass be the flow rate of processed biomass (tonne/day). Therefore, the optimization formulation given by Equations 14.16 through 14.19 can be written as the following LINGO program:

[P14.8]
```
Min = CHydrogen;
ASH = TAC;
ASH = 365*CHydrogen*FHydrogen;
TAC = 10.3*10^6 + 0.2698*10^6*FBiomass^0.6 +
26061*FBiomass + 0.4*(FBiomass - 2000)*FBiomass;
FHydrogen = 70*FBiomass;
FBiomass >= 2000;
FBiomass <= 20000;
```

Solving using LINGO, we get:

```
Global optimal solution found.
Objective value: 1.450589
```

Variable	Value
CHYDROGEN	1.450589
ASH	0.3551130E+09
TAC	0.3551130E+09
FHYDROGEN	670701.6
FBIOMASS	9581.451

The solution indicates that the minimum selling price of hydrogen is $1.45/kg at an optimum capacity of the biorefinery being 9581 tonne/day of processed biomass. These results are consistent with the graphical solution obtained in Example 2.14.

Example 14-5 Reduction of discharge in a tire-to-fuel plant

This case study was described in details in Example 3.6, and is adapted from El-Halwagi (1997) and Noureldin and El-Halwagi (2000). The optimization formulation and solution are taken from El-Halwagi (2006). The problem involves a processing facility that converts scrap tires into fuel via pyrolysis. Figure 14.3 is a simplified flow sheet of the process. The amount of water generated by chemical reaction is a function of the reaction temperature, T_{rxn}, through the following correlation:

[14.20] $\quad W_{rxn} = 0.152 + (5.37 - 7.84 \times 10^{-3} T_{rxn})e^{(27.4 - 0.04 T_{rxn})}$,

where W_{rxn} is in kg/s and T_{rxn} is in K. At present, the reactor is operated at 690 K, which leads to the generation of 0.12 kg water/s. To maintain acceptable product quality, the reaction temperature should be maintained within the following range:

[14.21] $\quad 690 \leq T_{rxn}(K) \leq 740$.

Wastewater leaves the plant from three outlets: W_1 from the decanter, W_2 from the seal pot, and W_3 with the wet cake. Fresh water is used in two locations: shredding and the seal pot. The flow rate of water-jet makeup depends on the applied pressure coming out of the compression stage "P_{comp}" via the following expression:

[14.22] $\quad G_1 = 0.47\, e^{-0.009 P_{comp}}$,

where G_1 is in kg/s and P_{comp} is in atm. To achieve acceptable shredding, the jet pressure may be varied within the following range:

[14.23] $\quad 70 \leq P_{comp}(atm) \leq 90$.

At present, P_{comp} is 70 atm, which requires a water-jet makeup flow rate of 0.25 kg/s.

The flow rate of the water stream passing through the seal pot is referred to as G_2 and it has a flow rate of 0.15 kg/s. An equivalent flow rate of wastewater stream is withdrawn from the seal pot; that is,

[14.24] $\quad W_2 = G_2$

The water lost in the cake is related to the mass flow rate of the water-jet makeup through:

[14.25] $\quad W_3 = 0.4\, G_1$.

At present, the values of W_1, W_2, and W_3 are 0.27, 0.15, and 0.10 kg/s, respectively. The cost of wastewater transportation and treatment is $0.10/kg, leading to a wastewater treatment cost of approximately $1.33 million/yr. If minor process modifications (changes in reaction temperature and pressure of the jet shredder) are allowed (without recycle or rerouting), what is the target for reduction in flow rate of terminal wastewater discharges?

SOLUTION

The objective of this problem is to minimize wastewater discharge; that is,

[14.26] $\quad$ Minimize $W1 + W2 + W3$

To minimize these three wastewater streams, it is necessary to relate them to the flow rates of fresh water and to the possible variations in the allowed changes: reaction temperature and jet pressure. Although these changes are not directly related to the flow rate of the waste streams, they can be related to the flow rates of fresh water used. Consequently, water balances can be employed to relate the flow rates of the fresh, wastewater, and generation by chemical reaction.

Let us now proceed and transform this approach to mathematical constraints. An overall water balance is given by:

[14.27] $\quad G_1 + G_2 + W_{rxn} = W_1 + W_2 + W_3$

where

[14.20] $\quad W_{rxn} = 0.152 + (5.37 - 7.84 \times 10^{-3} T_{rxn})e^{(27.4 - 0.04 T_{rxn})}$

[14.22] $\quad G_1 = 0.47\, e^{-0.009 P_{comp}}$

and

[14.28] $\quad G_2 = 0.15$

Next, we relate W_2 and W_3 to G_2 and G_3:

[14.24] $\quad W_2 = G_2$

[14.25] $\quad W_3 = 0.4\, G_1$

Additionally, there are bounds on reaction temperature and jet pressure given by:

[14.21] $\quad 690 \leq T_{rxn}(K) \leq 740$

[14.23] $\quad 70 \leq P_{comp}(atm) \leq 90$

Therefore, we can formulate the optimization program as the following LINGO input file:

```
[P14.9]
min = W1 + W2 + W3;
G1 + G2 + Wrxn = W1 + W2 + W3;
```

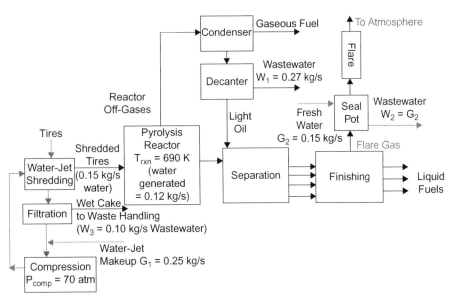

FIGURE 14-3 Simplified flow sheet of tire-to-fuel plant. *Source: El-Halwagi (2006).*

(Continued)

Example 14-5 Reduction of discharge in a tire-to-fuel plant (*Continued*)

```
Wrxn = 0.152 + (5.37 - 0.00784*Trxn)*
@exp(27.4 - 0.04*Trxn);
G1 = 0.47*@exp(-0.009*Pcomp);
G2 = 0.15;
W2 = G2;
W3 = 0.4*G1;
Trxn > 690;
Trxn < 740;
Pcomp > 70;
Pcomp < 90;
```

Using LINGO, we get the following solution:

```
Objective value: 0.4388316
```

Variable	Value
W1	0.2051983
W2	0.1500000
W3	0.8363332E-01
G2	0.1500000
G1	0.2090833
PCOMP	90.00000
WRXN	0.7974833E-01
TRXN	709.9490

These results are consistent with the ones obtained in Chapter 3 (Fig. 3.24). The value of G_1 is reduced to 0.205 kg/s and the minimum value of W_3 is 0.084 kg/s. Furthermore, the optimum reaction temperature is 710 K.

Example 14-6 Yield targeting in acetaldehyde production through ethanol oxidation

Consider the process of producing acetaldehyde via ethanol oxidation described in Example 3.7 (Chapter 3). Fig. 14.4 shows the process flow sheet. The process description and data are based on the case study reported by Al-Otaibi and El-Halwagi (2006). In this case study, the following symbols are used: "S," "A," and "E," to respectively refer to a process stream, the flow of acetaldehyde (in tonne per year or tpy), and the flow of ethanol (in tpy). Ethanol feedstock (50 percent ethanol) is vaporized in a flash column. Some ethanol is lost in the bottoms stream of the flash drum. The ratio of ethanol in the bottoms stream (E2) to ethanol in the feed (E1) of the flash is given as a function of the flash temperature:

[14.29a]
$$\alpha = \frac{E2}{E1} = 10.5122 - 0.0274 * T_{flash}$$

where T_{flash} is the temperature of the flash drum in K.

Equivalently, using the component mass balance for ethanol around the flash column (E1 = E2 + E3), Eq. 14.29a may be written in terms of the ratio of the ethanol in the top stream (E3) and the feed (E1) to the flash drum:

[14.29b]
$$(1 - \alpha) = \frac{E3}{E1}$$

The acceptable range for T_{flash} is given by the following bounds:

[14.30]
$$380.0 \leq T_{flash}(K) \leq 383.5$$

At present, the flash temperature is 380 K.

The vapor stream leaving the flash is mixed with air, and fed to an oxidation reactor where acetaldehyde and water are produced according to the following chemical reaction:

$$CH_3CH_2OH + \tfrac{1}{2}O_2 \rightarrow CH_3CHO + H_2O$$

Assuming that the aforementioned chemical reaction is only reaction taking place in the reactor, the ethanol consumed in the reactor can be related to acetaldehyde formed in the reactor through stoichiometry and molecular weights. Therefore,

[14.31] Ethanol consumed in the reactor = (46/44) *acetaldehyde formed in the reactor

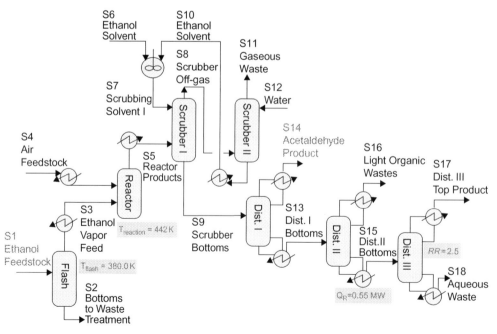

FIGURE 14-4 Schematic representation of base case for the acetaldehyde process. *Source: Al-Otaibi and El-Halwagi (2006).*

(*Continued*)

Example 14-6 Yield targeting in acetaldehyde production through ethanol oxidation (*Continued*)

The reactor yield is defined as:

[14.32a] $$Y_{reactor} = \frac{\text{Mass rate of acetaldehyde formed in the reactor}}{\text{Mass flow rate of ethanol fed to the reactor}}$$

Because there is no acetaldehyde entering the reactor, therefore, the mass rate of acetaldehyde formed in the reactor is equal to A5. Hence, Eq. 14.32a may be rewritten as:

[14.32b] $$Y_{reactor} = \frac{A5}{E3}$$

The dependence of the reactor yield on the reaction temperature is given by:

[14.33] $$Y_{reactor} = 0.33 - 4.2*10^{-6}*(T_{rxn} - 580)^2,$$

where T_{rxn} is the reactor temperature (K), which is constrained by the following range of acceptable operating temperature:

[14.34] $$300 \leq T_{rxn}(K) \leq 860$$

At present the reactor is operated at 442 K.

Scrubbing in two absorption columns is carried out using water and ethanol to recover the product and unreacted materials. The liquid leaving the bottom of the first absorption column is separated in three distillation columns. Acetaldehyde is recovered from the top of the first column. The acetaldehyde recovered in the first distillation column is a function of the heat duty of the reboiler of that column. The relationship is given by:

[14.35] $$A14 = \beta*A9$$

where

[14.36] $$\beta = 0.14*Q_R + 0.89$$

where Q_R is the reboiler heat duty (heat flow rate) in MW. The range of the reboiler duty is:

[14.37] $$0.55 \leq Q_R(MW) \leq 0.76$$

For the base case, the reboiler duty (heat flow rate) is 0.55 MW.

Organic wastes are collected from the top of the second column. The third column produces a distillate that contains ethanol, water, and minor components. The bottom product is primarily an aqueous waste and is fed to the biotreatment facility. To reduce ethanol losses (with the aqueous waste going to biotreatment), the reflux ratio of the third distillation column may be manipulated. The following relations may be used:

[14.38] $$E17 = \gamma*E15$$

where

[14.39] $$\gamma = 0.653*e^{(0.085*RR)}$$

where RR is the reflux ratio in the third distillation column. Currently, the reflux ratio for the column is 2.5 and the working range for the reflux ratio is:

[14.40] $$2.5 \leq RR \leq 5.0$$

The objective of this case study is to maximize the overall process yield without adding new process equipment. The overall process yield is defined as:

[14.41a] $$\text{Process yield} = \frac{\text{Mass flow rate of acetaldehyde in final product stream}}{\text{Mass flow rate of fresh ethanol fed to process as feedstock}}$$

The use of the phrase "ethanol fed to the process as a feedstock" excludes the use of ethanol for nonreactive purposes such as solvents (in stream S6). Therefore, the process yield is given by:

[14.41b] $$\text{Process yield} = \frac{A14}{E_{Fresh}}$$

where A14 is the mass flow rate of acetaldehyde in the final acetaldehyde product (in stream S14) and E_{Fresh} is the mass flow rate of ethanol in the fresh feedstock to the process (in stream S1 in the current process configuration).

The following flows assumptions are used throughout the case study (even after process changes):
- E6 = 400 tpy of ethanol
- Acetaldehyde leaves the process in two streams: S14 and S16
- Ethanol leaves the process in two streams: S17 and S18
- Direct recycle of ethanol is allowed only from the top of the third distillation column to the flash column

At present, the plant produces 100,000 tpy of acetaldehyde (that is, A14 = 100,000). It is desired to explore increasing the process yield using existing units and without adding new pieces of equipment. As can be seen from Eq. 14.41b, enhancing the process yield involves increasing the A14 and/or decreasing E_{Fresh}. In this case study, the desired production level of acetaldehyde is to be maintained at 100,000 tpy. What is the target for maximum overall process yield and what are the necessary process changes to reach the yield target?

SOLUTION

The first step is to calculate the current "base-case" process yield. Here is the LINGO program for the base case. Lines preceded by "!" are comment/explanatory statements that are ignored by LINGO.

```
[P14.10]
!Base case program;
Max = A14/Efresh;
! Base case data;
TFlash = 380;
TR = 442;
QR = 0.55;
RR = 2.5;
ERecycle = 0;
A14 = 100000;
! Modeling equations;
E1 + E6 = EReactor + E2 + E17 + E18;
AReactor = A14 + A16;
AReactor = A5;
E1 = Efresh + ERecycle;
E2 = Alfa*E1;
Alfa = 10.5122 - 0.0274*TFlash;
TFlash <= 383.5;
TFlash >= 380.0;
E3 = E1 - E2;
E6 = 400;
YReactor = 0.33 - 0.0000042*(TR - 580)*(TR - 580);
TR >= 300;
TR <= 860;
AReactor = YReactor*E3;
EReactor = (46/44)*AReactor;
E5 = E3 - EReactor;
E9 = E5 + E6;
A9 = A5;
A14 = Beta*A9;
Beta = 0.14*QR + 0.89;
QR >= 0.55;
QR <= 0.76;
E15 = E9;
E17 = Gamma*E15;
Gamma = 0.653*@exp(0.085*RR);
RR>=2.5;
RR<=5;
```

The following solution is obtained using LINGO:

```
Global optimal solution found.

Objective value: 0.2175399
```

(*Continued*)

264 Sustainable Design Through Process Integration

Example 14-6 Yield targeting in acetaldehyde production through ethanol oxidation (*Continued*)

Variable	Value
A14	100000.0
EFRESH	459685.8
TFLASH	380.0000
TR	442.0000
QR	0.5500000
RR	2.500000
E6	400.0000
EREACTOR	108113.2
E2	46060.52
E17	120956.7
E18	184955.4
AREACTOR	103412.6
A16	3412.616
A5	103412.6
E1	459685.8
ERECYCLE	0.000000
ALFA	0.1002000
E3	413625.3
YREACTOR	0.2500152
E5	149371.5
E4	257484.7
E9	149771.5
A9	103412.6
BETA	0.9670000
E15	149771.5
GAMMA	0.8076083

Therefore, the base-case yield of the process is 0.218 kg acetaldehyde/kg ethanol. Next, we develop the optimization program for the yield-maximization problem.

```
[P14.11]
! Yield Optimization;
Max = A14/Efresh;
A14 = 100000;
E1 + E6 = EReactor + E2 + E17 + E18;
AReactor = A14 + A16;
AReactor = A5;
! The following constraint is the ethanol balance
around the mixing point of the recycle and fresh feed;
Efresh = E1 - ERecycle;
! Direct recycle is only allowed from E17;
ERecycle = E17;
E2 = Alfa*E1;
Alfa = 10.5122 - 0.0274*TFlash;
TFlash <= 383.5;
TFlash >= 380.0;
E3 = E1 - E2;
E6 = 400;
YReactor = 0.33 - 0.0000042*(TR - 580)*(TR - 580);
TR >= 300;
TR <= 860;
AReactor = YReactor*E3;
EReactor = (46/44)*AReactor;
E5 = E3 - EReactor;
E9 = E5 + E6;
A9 = A5;
A14 = Beta*A9;
Beta = 0.14*QR + 0.89;
QR >= 0.55;
QR <= 0.76;
E15 = E9;
E17 = Gamma*E15;
Gamma = 0.653*@exp(0.085*RR);
RR>=2.5;
RR<=5;
```

The solution of this program using LINGO gives the following results:

```
Global optimal solution found.
Objective value: 0.9427595
```

Variable	Value
A14	100000.0
EFRESH	106071.6

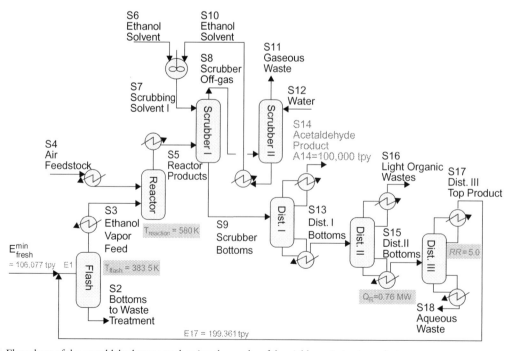

FIGURE 14-5 Flow sheet of the acetaldehyde process showing the results of the yield-maximization solution.

(*Continued*)

Example 14-6 Yield targeting in acetaldehyde production through ethanol oxidation (*Continued*)

E1	305438.5	YREACTOR	0.3300000
E6	400.0000	TR	580.0000
EREACTOR	104923.2	E5	199202.0
E2	1313.386	E9	199602.0
E17	199367.0	A9	100361.3
E18	235.0225	BETA	0.9964000
AREACTOR	100361.3	QR	0.7600000
A16	361.3007	E15	199602.0
A5	100361.3	GAMMA	0.9988225
ERECYCLE	199367.0	RR	5.000000
ALFA	0.4300000E-02		
TFLASH	383.5000		
E3	304125.2		

The results are consistent with those obtained in Chapter 3 and presented by Fig. 14.5.

Example 14-7 The transportation problem

The transportation problem deals with the allocation of certain commodities from sources to destinations. For instance, consider the case of a company that owns three biorefineries that use hexane as a solvent in the process. The three biorefineries are located in towns A, B, and C. Hexane is available from four suppliers with hexane-manufacturing facilities in cities 1, 2, 3, and 4. Table 14.2a lists the daily requirements for hexane in the three biorefineries, and Table 14.2b gives the production capacities of the four suppliers.

The cost of shipping one tonne of hexane from each supplying plant to each consuming biorefinery is given in Table 14.3. The number listed at the intersection of a supplier and consumer is the transportation cost per tonne.

Develop and solve an optimization formulation that determines the optimal allocation of hexane from the suppliers to the biorefineries to minimize the total transportation cost while meeting the requirements of the biorefineries.

SOLUTION

The optimization variables are the daily tonnes of hexane transported from each supplier to each consuming biorefinery. Table 14.4 shows the terms used to denote these optimization variables (tonne/day). There are 12 decision variables.

[P14.12]
Objective function:

$$\text{Minimize transportation cost} = 20*X1A + 10*X1B + 50*X1C + \\ 30*X2A + 0*X2B + 80*X2C + \\ 110*X3A + 60*X3B + 150*X3C + \\ 70*X4A + 10*X4B + 90*X4C$$

These are subject to the following constraints:
Hexane-demand constraints for the biorefineries:

$$X1A + X2A + X3A + X4A = 6$$
$$X1B + X2B + X3B + X4B = 1$$
$$X1C + X2C + X3C + X4C = 10$$

Availability of hexane from the supplier:

$$X1A + X1B + X1C \leq 7$$
$$X2A + X2B + X2C \leq 5$$
$$X3A + X3B + X3C \leq 3$$
$$X4A + X4B + X4C \leq 6$$

TABLE 14-2a Daily Hexane Requirements for the Biorefineries

Biorefinery	Daily Requirement (tonne)
A	6
B	1
C	10

Nonnegativity constraints

$$X1A \geq 0, X1B \geq 0, \ldots\ldots, X4C \geq 0$$

The LINGO formulation is given by the following program:

[P14.12]

```
Min = 20*X1A + 10*X1B + 50*X1C + 30*X2A + 0*X2B
+ 80*X2C + 110*X3A + 60*X3B + 150*X3C + 70*X4A +
10*X4B + 90*X4C;
X1A + X2A + X3A + X4A = 6;
X1B + X2B + X3B + X4B = 1;
```

TABLE 14-2b Daily Production Capacities of the Hexane Suppliers

Production Facilities	Daily Production (tonne)
1	7
2	5
3	3
4	6

TABLE 14-3 Transportation Costs, $/tonne

Supplier	Consuming Biorefinery A	Consuming Biorefinery B	Consuming Biorefinery C
1	20	10	50
2	30	0	80
3	110	60	150
4	70	10	90

TABLE 14-4 Terms Used for Transported Quantities (tonne/day)

Supplier	Consumer A	Consumer B	Consumer C
1	X1A	X1B	X1C
2	X2A	X2B	X2C
3	X3A	X3B	X3C
4	X4A	X4B	X4C

(*Continued*)

Example 14-7 The transportation problem (*Continued*)

```
X1C + X2C + X3C + X4C = 10;
X1A + X1B + X1C <= 7;
X2A + X2B + X2C <= 5;
X3A + X3B + X3C <= 3;
X4A + X4B + X4C <= 6;
X1A >=0;
X1B >=0;
X1C >=0;
X2A >=0;
X2B >=0;
X2C >=0;
X3A >=0;
X3B >=0;
X3C >=0;
X4A >=0;
X4B >=0;
X4C >=0;
```

Solving using LINGO, we get the following optimum solution:

```
Objective value: 840.0000
Variable      Value
X1A           1.000000
X1B           0.000000
X1C           6.000000
X2A           5.000000
X2B           0.000000
X2C           0.000000
X3A           0.000000
X3B           0.000000
X3C           0.000000
X4A           0.000000
X4B           1.000000
X4C           4.000000
```

It is interesting to note that the transportation cost from supplier 2 to biorefinery B is zero (indicating that they lie in close proximity). At first glance, one may have thought that the daily requirement of biorefinery B (1 tonne/day) should come from supplier 2. The optimal solution indicates otherwise. The daily requirement of biorefinery B is transported from supplier 4, which charges $10/tonne. If we add the constraint X2B = 1 to the program to force the allocation of 1 tonne/day from supplier 2 to biorefinery B, the following solution is obtained.

```
Objective value: 860.0000
Variable      Value
X1A           2.000000
X1B           0.000000
X1C           5.000000
X2A           4.000000
X2B           1.000000
X2C           0.000000
X3A           0.000000
X3B           0.000000
X3C           0.000000
X4A           0.000000
X4B           0.000000
X4C           5.000000
```

Therefore, forcing an allocation that is seemingly (and misleadingly) optimal for supplier 2 and biorefinery B increases the overall transportation cost for the company that owns the three biorefineries. This observation underscores the importance of a holistic view of the problem, which is a hallmark of process integration and systems optimization. Some of the individual decisions may be suboptimal in their own right but serve as a part of an overall optimization solution for the whole system.

A BRIEF INTRODUCTION TO SETS, CONVEX ANALYSIS, AND SYMBOLS USED IN OPTIMIZATION

This section presents common concepts and terms used in formulating optimization problems. These include sets, convex analysis, and symbols used in optimization formulations.

SETS

A **set** is a collection of objects and is typically denoted by upper-case letters: A, B, X, Y, and so on. The objects comprising the set are called members or elements and are usually denoted by lowercase letters: a, b, x, y, and so on. The statement "p is an element or a member of A, or equivalently p belongs to A" is written as: Conversely, p ∉ A indicates that p is not a member of A. There are essentially two ways to specify a particular set:

- Listing all the members within braces (curly brackets). For instance, the set of even numbers that are positive and less than 10: A = {2,4,6,8} denotes the set A whose elements are the numbers 2, 4, 6, 8.
- Stating the properties that characterize the elements in the set. For instance, A = {x:x is an even positive number, x < 10} or A = {x|x is an even positive number, x < 10} is read as follows: A is the set of x where (or such that) x is an even positive number and x is less than 10.

It is also common to describe indexed families of sets. Suppose that the index n belongs to a set N; then we can define the family of sets A_n as $\{A_n: n \in N\}$. For example, consider sets {1,3}, {2,4}, {3,5} and {4,6}. They are a family of four sets that can be indexed as follows: $A_n = \{n, n + 2\}$ for each $n \in N = \{1,2,3,4\}$. Hence, the family of sets is: {{n, n + 2} : n ∈ N = {1,2,3,4}}.

CONVEX ANALYSIS

Convex analysis deals with the assessment of the attributes of convex sets and convex functions. So, what are convex sets and convex functions?

A set of m-dimensional vectors (for example, the feasibility region of an optimization formulation) is called a **convex set**; whenever two vectors belong to the set then so too does the line segment between the vectors (for example, line segment between p and q). Figure 14.6 illustrates examples of a convex set and a nonconvex set.

On the other hand, a function f(x) is called a **convex function** over a certain region R if the following condition holds: for any two vectors x_a and x_b lying in the region R:

$$[14.42a] \quad \lambda f(x_a) + (1 - \lambda)f(x_b) \geq f[\lambda x_a + (1 - \lambda)x_b]$$

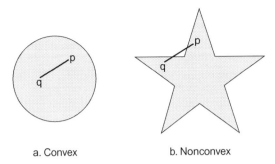

FIGURE 14-6 Examples of convex and nonconvex sets (or feasibility regions).

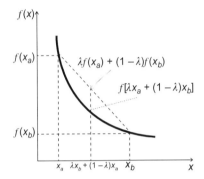

FIGURE 14-7 Example of a convex function.

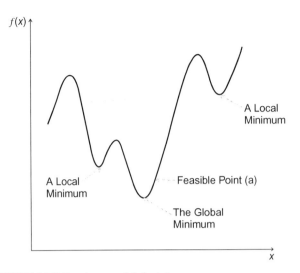

FIGURE 14-8 Local versus global minimum.

where λ is a scalar between 0 and 1. Therefore, for a convex function, the line segment connecting any two points on the graph of the function never lies below the graph. Otherwise, the function is nonconvex. Figure 14.7 shows an example of a convex function.

A function f(x) is called **strictly convex** over the region R if any two vectors x_a and x_b lie in the region R:

[14.42b] $\quad \lambda f(x_a) + (1 - \lambda)f(x_b) > f[\lambda x_a + (1 - \lambda)x_b]$

where λ is a scalar between 0 and 1.

The primary reason for interest in convex analysis in optimization is the desire to obtain global solutions. Figure 14.8 shows a nonconvex unconstrained nonlinear minimization problem with an objective function that exhibits multiple optimum points (two local minima and one global minimum). The relevant solution is the global optimum point. In fact, a feasible point (such as point a) may outperform the two local minima. Clearly, the task becomes far more complex for problems involving multiple optimization variables and constraints. If the objective function is a convex function and the feasibility region is a convex set, the optimization program is called a **convex program** and the local minimum is also the **global minimum**. In formulating optimization problems, effort should be made to develop convex programs. A special class of convex programs is the linear program. Therefore, the solution of a linear program is always global. It is worth noting that LINGO has a "Global Solver" that can be activated by going to the menu bar, clicking on "LINGO," selecting "Options" from the drop-down menu, and then selecting the "Global Solver" tab and checking the "Use Global Solver" box.

There are several useful properties of convex functions and sets:

- The intersection of any number of convex sets is a convex set. Therefore, if all the constraints are convex, then the feasibility region is convex. If one of the constraints is nonconvex, then the whole feasibility region is rendered nonconvex.
- The summation of convex functions yields a convex function.
- If a function g(x) is convex, then the feasibility region defined by the set: $R = \{x : g(x) \leq \delta\}$ is convex for all scalars δ. Therefore, if a constraint is convex in the equality form, it is also convex in the inequality form.
- A convex function multiplied times a positive scalar gives a convex function.

These properties are useful in formulating and assessing convex programs.

SYMBOLS USED IN OPTIMIZATION FORMULATIONS

Table 14.5 provides a listing of some of the commonly used symbols in set and logic expressions and their meanings.

THE USE OF 0–1 BINARY-INTEGER VARIABLES

An important and very common use of 0–1 binary integer variables is to represent discrete decisions and what-if scenarios. Consider an event that may or may not occur, and suppose that it is part of the problem to decide between these two alternatives. To model such a dichotomy, we use a binary variable $I \in \{0, 1\}$ and let:

[14.43a] $\quad I = 1$ if the event occurs

TABLE 14-5 Commonly Used Symbols

Symbol	Meaning
$\forall$	For every, for all, for any
$\exists$	There exists or such that
$\neg$	Negation
$\wedge$	The AND operator (a logical conjunction)
$\vee$	The OR operator (a logical disjunction). At least one of the two choices must be made.
$\underline{\vee}$	The exclusive OR operator (an exclusive disjunction). Only one of the two choices must be made.
$\Rightarrow$	Implication: if … then …
$\Leftrightarrow$	Equivalence: if and only if (iff)

and

[14.43b] $I = 0$ if the event does not occur

Additionally, binary integers can be instrumental in modeling logical statements, what-if scenarios, and Boolean logic. Boolean algebra (named after nineteenth-century mathematician Boole) is a form of algebra for which the values are reduced to one of two statements: *true* or *false*. Binary integer variables are naturally suited to represent Boolean algebra because of their discrete nature where 1 may be assigned to *true* and 0 assigned to *false* (or the other way around). In the case of scenarios involving multiple choices, binary integer variables are useful in modeling the selection process. Suppose that there is a set $C = \{i \mid i =$ an event or an option, $i = 1$, 2, …, $N_{Choices}\}$. For each event or option (with an index i), a binary integer variable, I_i, is defined to take the value of 1 when the option is selected and zero when it is not. Consider the following cases:

- Selection of exactly one option from set C: This is the case of using the exclusive OR operator: $I_1 \underline{\vee} I_2 \underline{\vee} \cdots I_{N_{Choices}}$. Therefore, the constraint is written as:

[14.44] $$\sum_{i=1}^{N_{Choices}} I_i = 1$$

- Selection of at least one option from set C: This is the case of using the OR operator: $I_1 \vee I_2 \vee \cdots I_{N_{Choices}}$. Therefore, the constraint is given by:

[14.45] $$\sum_{i=1}^{N_{Choices}} I_i \geq 1$$

- Selection of option v if option u is chosen: This is a case of implication: u$\Rightarrow$v. Hence, the constraint is expressed as:

[14.46] $$I_v \geq I_u$$

- Selection of option u if and only if option v is selected. This is the case of equivalence: u$\Leftrightarrow$v, which can be modeled as:

[14.47] $$I_v = I_u$$

There are numerous applications for the use of binary integers. The following are some examples.

Example 14-8 The assignment problem

A classical optimization problem is the assignment problem. It involves the assignment of n people (or machines or streams, and so on) to m jobs (or spaces or units, and so on). Each job must be done by exactly one person; also, each person can do at most one job. The cost of person i doing job j is $C_{i,j}$. The problem is to assign the people to the jobs so as to minimize the total cost of completing all the jobs.

To formulate this problem, which is known as the assignment problem, we introduce 0–1 variables $X_{i,j}$ where $i = 1, 2, \ldots, n$ and $j = 1, 2, \ldots, m$ corresponding to the event of assigning person j to do job i. Therefore, the assignment problem can be posed through the following MILP formulation:

[P14.13] $$\text{Minimize assignment cost} = \sum_{i=1}^{n}\sum_{j=1}^{m} C_{i,j} X_{i,j}$$

Because exactly one person must do job j, we have the following constraint

[14.48] $$\sum_{j=1}^{m} X_{i,j} = 1 \quad \forall i$$

Because each person can do no more than one job, we have the following constraint:

[14.49] $$\sum_{i=1}^{n} X_{i,j} \leq 1 \quad \forall j$$

As an illustration, consider the following assignment problem on plant layout. Four new reactors (R1, R2, R3, and R4) are to be installed in a chemical plant.

TABLE 14-6 Assignment Costs ($MM)

Reactor	Space 1	Space 2	Space 3	Space 4
R1	15	11	13	15
R2	13	9	12	17
R3	14	15	10	14
R4	17	13	11	16

There are four vacant spaces (1, 2, 3, and 4 available). Table 14.6 gives the cost of assigning reactor i to space j. Assign the reactors to the spaces so as to minimize the total cost.

SOLUTION

Let Xij be a binary integer variable to denote existence (or absence) of reactor i in space j (when Xij = 1, reactor i exists in space j; when Xij = 0, reactor i does not exist in space j).

Objective Function:

[P14.14] Minimize assignment cost = 15*X11 + 11*X12 + 13*X13 + 15*X14 + 13*X21 + 9*X22 + 12*X23 + 17*X24 + 14*X31 + 15*X32 + 10*X33 + 140000*X34 + 17*X41 + 13*X42 + 11*X43 + 16*X44

(Continued)

Example 14-8 The assignment problem (*Continued*)

Subject to the following constraints:
On reactors, each space must be assigned to one and only one reactor:

X11 + X12 + X13 + X14 = 1 for reactor R1
X21 + X22 + X23 + X24 = 1 for reactor R2
X31 + X32 + X33 + X34 = 1 for reactor R3
X41 + X42 + X43 + X44 = 1 for reactor R4

On spaces, each reactor must be assigned to one and only one space (because there are four spaces and four reactors, there is no need to use less than or equal constraint. Instead, we use an equality constraint):

X11 + X21 + X31 + X41 = 1 for space 1
X12 + X22 + X32 + X42 = 1 for space 2
X13 + X23 + X33 + X43 = 1 for space 3
X14 + X24 + X34 + X44 = 1 for space 4

The input file for LINGO can be expressed as:

```
[P14.15]
min = 15*X11 + 11*X12 + 13*X13 + 15*X14 + 13*X21 +
9*X22 + 12*X23 + 17*X24 + 14*X31 + 15*X32 + 10*X33
+ 140000*X34 + 17*X41 + 13*X42 + 11*X43 + 16*X44;
X11 + X12 + X13 + X14 = 1;
X21 + X22 + X23 + X24 = 1;
X31 + X32 + X33 + X34 = 1;
X41 + X42 + X43 + X44 = 1;
X11 + X21 + X31 + X41 = 1;
X12 + X22 + X32 + X42 = 1;
X13 + X23 + X33 + X43 = 1;
X14 + X24 + X34 + X44 = 1;
@bin(X11);
@bin(X12);
@bin(X13);
@bin(X14);
@bin(X21);
@bin(X22);
@bin(X23);
@bin(X24);
@bin(X31);
@bin(X32);
@bin(X33);
@bin(X34);
@bin(X41);
@bin(X42);
@bin(X43);
@bin(X44);
End
```

This program can be solved using LINGO. The following are the results of the optimal solution.

```
Objective value: 49.00
Variable         Value
X11              0.000000
X12              1.000000
X13              0.000000
X14              0.000000
X21              1.000000
X22              0.000000
X23              0.000000
X24              0.000000
X31              0.000000
X32              0.000000
X33              0.000000
X34              1.000000
X41              0.000000
X42              0.000000
X43              1.000000
X44              0.000000
```

Therefore, the minimum total assignment cost associated with this assignment policy is $49.00 MM. An optimal assignment policy is:

[14.50a]   Reactor R1 to space 2

[14.50b]   Reactor R2 to space 1

[14.50c]   Reactor R3 to space 4

[14.50d]   Reactor R4 to space 3

ENUMERATING MULTIPLE SOLUTIONS USING INTEGER CUTS

It is worth mentioning that there may be more than one solution attaining the optimal value of the objective function. In such cases, it is necessary to enumerate all of these solutions. This can be achieved by using *integer-cut* constraints. The basic idea is to add a constraint that forbids the formation of an optimal solution that has already been identified. The optimization program with the additional constraint is run, and if the identified solution is equal to the solution without the integer cut, then both solutions are equivalent in terms of the objective function but provide different implementations. The procedure is repeated by adding integer cuts until an inferior solution is obtained or no solution is found.

As an illustration, let us reconsider the previous distillation-column assignment example, in which the optimal solution involves the following matches:

[14.51a]   X12 = 1

[14.51b]   X21 = 1

[14.51c]   X34 = 1

[14.51d]   X43 = 1

Therefore, the following integer-cut constraint may be added to the optimization formulation:

[14.52]   X12 + X21 + X34 + X43 ≤ 3

Consequently, for this constraint to be satisfied, the combination of these four assignments must be excluded from any subsequent optimization. By adding this constraint and rerunning LINGO, we get the following optimal solution:

```
Objective value: 49.00
Variable         Value
X11              0.000000
X12              0.000000
```

```
X13    0.000000
X14    1.000000
X21    0.000000
X22    1.000000
X23    0.000000
X24    0.000000
X31    1.000000
X32    0.000000
X33    0.000000
X34    0.000000
X41    0.000000
X42    0.000000
X43    1.000000
X44    0.000000
```

This solution provides the same value of the objective function but yields another integer combination:

[14.53a] $X14 = 1$

[14.53b] $X22 = 1$

[14.53c] $X31 = 1$

[14.53d] $X43 = 1$

Hence, we add the following integer cut:

[14.54] $X14 + X22 + X31 + X43 \leq 3$

By rerunning LINGO with the two integer cuts, we get:

```
Objective value: 49.00
Variable    Value
X11    1.000000
X12    0.000000
X13    0.000000
X14    0.000000
X21    0.000000
X22    1.000000
X23    0.000000
X24    0.000000
X31    0.000000
X32    0.000000
X33    0.000000
X34    1.000000
X41    0.000000
X42    0.000000
X43    1.000000
X44    0.000000
```

This solution provides the same value of the objective function but yields another integer combination:

[14.55a] $X11 = 1$

[14.55b] $X22 = 1$

[14.55c] $X34 = 1$

[14.55d] $X43 = 1$

Hence, we add the following integer cut:

[14.56] $X11 + X22 + X34 + X43 \leq 3$

By rerunning LINGO, we get

```
Objective value: 50.00
Variable    Value
X11    1.000000
X12    0.000000
X13    0.000000
X14    0.000000
X21    0.000000
X22    1.000000
X23    0.000000
X24    0.000000
X31    0.000000
X32    0.000000
X33    1.000000
X34    0.000000
X41    0.000000
X42    0.000000
X43    0.000000
X44    1.000000
```

Because the identified optimal solution of the last program is $50.00 MM, which is higher than the already found solution of $49.00 MM, then we stop, and no additional alternatives yield the same value of the global optimum solution.

MODELING DISJUNCTIONS AND DISCONTINUOUS FUNCTIONS WITH BINARY INTEGER VARIABLES

Binary (0–1) integer variables are very useful in developing models in cases that involve disjunctions for constraints and discontinuous function. Effective techniques such as *convex-hull relaxation*, *Big-M constraints*, and *generalized disjunctive programming*, (Balas, 1985; Türkay and Grossmann, 1996; Grossmann and Lee, 2003) have been developed to systematize the modeling of such cases. Some of these techniques are described in this section.

DISCONTINUOUS FUNCTIONS

Discontinuous functions can be used to describe abrupt changes over a certain decision variable. For instance, consider a cost model that favors the selection of technology A up to a certain production capacity (F_{Switch}), and then moves to technology B for production capacities immediately above F_{Switch}; that is,

[14.57a] $Cost = Cost\_A$ for $F < F_{Switch}$

[14.57b] $Cost = Cost\_B$ for $F \geq F_{Switch}$

These cost functions may be constant numbers (for example, Fig. 14.9a) or variables depending on the production capacity (for example, Fig. 14.9b). Binary integer variables can be used to model these functions. For instance, the cost function may be transformed to the following mixed-integer formulation using a binary integer variable (I):

[14.58] $Cost = Cost\_A*I + Cost\_B*(1 - I)$

[14.59a]   where $I = 1$ for $F < F_{Switch}$

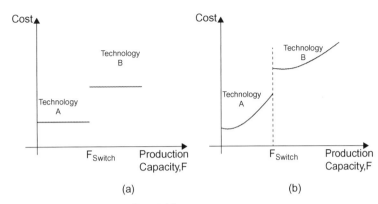

FIGURE 14-9 Discontinuous functions: (a) constant costs, (b) variable costs.

and

[14.59b] $I = 0$ for $F \geq F_{Switch}$

Let us examine this formulation:

When $I = 1$,

[14.60a] $Cost = Cost\_A*1 + Cost\_B*(1-1) = Cost\_A$

and when $I = 0$,

[14.60b] $Cost = Cost\_A*0 + Cost\_B*(1-0) = Cost\_B$

But how do we model the conditions that assign the values of I to be 0 or 1 based on the production capacity? Consider the following constraints:

[14.61] $(L - F_{Switch})*I < F - F_{Switch} \leq (U - F_{Swtich})*(1 - I)$ $I \in \{0,1\}$

where L and U are lower and upper bounds, respectively, to any feasible value of F. Therefore, $L - F_{Switch}$ is always negative and $U - F_{Switch}$ is always positive. When $F < F_{Switch}$, the term $F - F_{Switch}$ becomes negative, and I is forced to be 1, so that $(L - F_{Switch})*I$ becomes negative to satisfy $(L - F_{Switch}) * I < F - F_{Switch}$. Otherwise, if I is 0, the value of $(L - F_{Switch})*$ becomes 0, which renders $(L - F_{Switch}) * I < F - F_{Switch}$ infeasible. On the other hand, when $F \geq F_{Switch}$, the term $F - F_{Switch}$ becomes nonnegative and I is forced to be 0 to satisfy $F - F_{Switch} \leq (U - F_{Switch})*(1 - I)$.

Of particular importance to design problems is the use of discontinuous functions to model the fixed and total capital investments. For instance, consider a piece of equipment or a plant that will only be selected if its size, x, exceeds a certain lower bound, L, and falls below a certain upper bound, U. The fixed capital investment (FCI) function is given by:

[14.62a] $FCI = c + ax^b$ for $L \leq x \leq U$

and

[14.62b] $FCI = 0$ for $x = 0$

where c is a constant representing the fixed charges (for example, civil work, electric connections) that will have to be spent regardless of the size of the equipment (or the plant). The constants a and b respectively represent the cost coefficient and the sizing exponent (for example, b = 0.6 in the case of the sixth-tenths-factor rule). This cost expression can be modeled using an integer variable, I, as follows:

[14.63a] $FCI = c*I + ax^b$

[14.63b] $L*I \leq x \leq U*I$

[14.63c] $I \in \{0,1\}$

When x is zero, the left-hand side of Constraint 14.63b forces I to become 0 and the FCI calculated from Eq. 14.63a becomes 0. When x is between the lower and upper bounds, the right-hand side of Constraint 14.63b forces I to become 1 and the FCI calculated from Eq. 14.63a becomes identical to Eq. 14.62a. It is also worth noting that Constraint 14.63b is very useful in modeling a **disjoint domain** for x. No x values will be generated for the range higher than 0 and lower than L. If such values were generated, the left-hand side of Constraint 14.63b would require I to be 0 while the right-hand side of 14.63b will require I to be 1 and an infeasibility is generated. Hence, if x does not lie in the [L U] domain, its only feasible value is 0, which will be associated with I being 0 and which causes the left- and right-hand sides of Constraint 14.63b to be satisfied.

BIG-M REFORMULATION

Let us consider the case of a **disjunction** of two constraints whereby only one of two constraints may be used (that is, $g_1(x) \leq 0 \ \vee \ g_2(x) \leq 0$). An example of such disjunction is when a constraint, $g_1(x) \leq 0$, applies over a certain range of x while the other constraint, $g_2(x) \leq 0$, applies over another range of x. One way of modeling this case is through the **Big-M formulation** (or reformulation because it changes the original formulation of the optimization program), which is expressed as follows:

[14.64a] $I_1 + I_2 = 1$

where the binary variable I_1 is assigned a value of 1 constraint when g_1) is used, while the binary variable I_2 is assigned a value of 1 when constraint $g_2(x)$ is used:

[14.64b] $$g_1(x) \leq M_1(1 - I_1)$$

[14.64c] $$g_2(x) \leq M_2(1 - I_2)$$

where M_1 and M_2 are sufficiently large numbers that constitute upper bounds on any possible values of $g_1(x)$ and $g_2(x)$, respectively. Therefore, when $I_1 = 1$, Constraint 14.64a forces I_2 to become 0 and causes Constraint 14.64b to revert to its original form ($g_1(x) \leq 0$), whereas Constraint 14.64b becomes:

[14.65] $$g_2(x) \leq M_2$$

which will be satisfied (essentially neglecting Constraint 14.64c) because the value of M_2 was chosen to be a sufficiently large number that constitutes an upper bound on any possible value of $g_2(x)$. The same rationale applies to ignoring Constraint 14.64b when $I_2 = 1$.

If the constraints are in the form of inequalities less than a constant, that is, $g_1(x) \leq a_1$ and $g_2(x) \leq a_2$, the Big-M formulation becomes:

[14.66a] $$g_1(x) \leq a_1 + M_1(1 - I_1)$$

[14.66b] $$g_2(x) \leq a_2 + M_2(1 - I_2)$$

Furthermore, when the constraints are in the form of $a_1 \leq g_1(x)$ and $a_2 \leq g_2(x)$, the Big-M formulation becomes:

[14.67a] $$a_1 - M_1(1 - I_1) \leq g_1(x)$$

[14.67b] $$a_2 - M_2(1 - I_2) \leq g_2(x)$$

CONVEX-HULL REFORMULATION

Let us revisit the case of a disjunction of two constraints previously addressed in the Big-M formulation. Only one of two constraints may be used (that is, $g_1(x) \leq 0 \ \lor \ g_2(x) \leq 0$). The convex-hull formulation uses a binary integer variable for each of the disjunction constraints. It is similar to the Big-M formulation:

[14.64a] $$I_1 + I_2 = 1$$

where the binary variable I_1 is assigned a value of 1 constraint when $g_1(x)$ is used, while the binary variable I_2 is assigned a value of 1 when constraint $g_2(x)$ is used. The convex-hull formulation also uses a *disaggregation* of variables. Each variable, x, is disaggregated into a number of variables equal to the number of disjoint constraints. Therefore, in the case of two disjoint constraints,

[14.68] $$x = v_1 + v_2$$

where v_1 and v_2 are the disaggregated variables for the g_1 and g_2 constraints, respectively. Therefore, the constraints are written as:

[14.69a] $$g_1(v_1) \leq 0$$

[14.69b] $$g_2(v_2) \leq 0$$

with the selection of the disaggregated variables determined using the integer variables; that is,

[14.70a] $$L_1 * I_1 \leq v_1 \leq U_1 * I_1$$

and

[14.70b] $$L_2 * I_2 \leq v_2 \leq U_2 * I_2$$

Consider the case when the objective function is in the form of Eq. 14.62a, the convex-hull reformulation is given by:

[14.71a] $$FCI = FCI_1 + FCI_2$$

[14.71b] $$FCI_1 = c_1 * I_1 + a_1 v_1^{b_1}$$

[14.71c] $$FCI_2 = c_2 * I_2 + a_2 v_2^{b_2}$$

[14.71d] $$L_1 * I_1 \leq v_1 \leq U_1 * I_1$$

[14.71e] $$L_2 * I_2 \leq v_2 \leq U_2 * I_2$$

[14.71f] $$I_1 + I_2 = 1$$

[14.71g] $$I_1 \in \{0,1\} \quad \text{and} \quad I_2 \in \{0,1\}$$

Such convex-hull formulations provide tighter feasibility region than the Big-M formulation but introduce more variables (Türkay and Grossmann, 1996).

Another particularly useful application of convex-hull reformulation is the piecewise linearization of nonlinear functions over discretized domains. Suppose that we have a nonlinear function $f(x)$, such as the one shown by Fig. 14.10. It is possible to approximate the nonlinear function with discretized line segments. There are various ways of linearizing the function in a piecewise manger. Figure 14.10 shows an underestimation piecewise linearization of the function. Underestimation means that the line segments lie under the original function and, therefore, underestimate its value. If an infinite number of underestimators are drawn, then these infinite piecewise linear underestimators will fit the function perfectly. Of course, this is unrealistic. Hence, a reasonable number of line segments is selected to provide a proper balance between accuracy in representing the function versus the number of linear segments (which will impact the size of the optimization problem). Suppose that we choose to represent the function by the two line segments shown by Fig. 14.10, and define the variables x_1 and x_2 to represent the variable x over the two domains. The following are the expressions of the two linear functions:

[14.72a] $$f_1(x_1) = m_1 * x_1 + b_1 \quad \text{for } L_1 \leq x_1 \leq U_2$$

[14.72b] $$f_2(x_2) = m_2 * x_2 + b_2 \quad \text{for } L_2 \leq x_2 \leq U_2$$

where $L_1 = x_a$, $U_1 = x_b = L_2$, and $U_2 = x_c$, and m and b are the slope and the intercept of a linear segment. Constraints 14.72a and 14.72b can be modeled via the convex-hull reformulation as follows:

[14.73a] $$x = x_1 + x_2$$

[14.73b] $$f = f_1 + f_2$$

[14.73c] $$f_1 = m_1 * x_1 + b_1 * I_1$$

[14.73d] $$f_2 = m_2 * x_2 + b_2 * I_2$$

[14.73e] $$L_1 * I_1 \leq x_1 \leq U_1 * I_1$$

[14.73f] $$L_2 * I_2 \leq x_2 \leq U_2 * I_2$$

[14.73g] $$I_1 + I_2 = 1$$

[14.73h] $$I_1 \in \{0,1\} \quad \text{and} \quad I_2 \in \{0,1\}$$

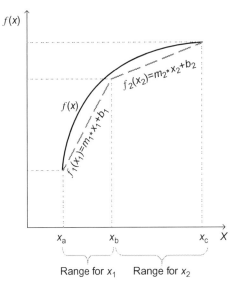

FIGURE 14-10 Piecewise linearization with convex-hull reformulation.

Example 14-9 Maximizing revenue of gas-processing facility (El-Halwagi, 2006)

A processing facility utilizes an industrial gas in making several products. The value added by the industrial gas is described by the following function:

[14.74] $\qquad$ Value added (\$/hr) $= 14.0 * F^{0.8}$

where F is the flow rate of purchased gas in 1000 SCF/hr (thousand standard cubic feet per hour).

The processing facility purchases the gas from an adjacent supplier. According to the contractual arrangement between the facility and the supplier, the supplier provides the gas at \$4.0/1000 SCF if the purchase is less than 300,000 SCF/hr and at \$3.5/1000 SCF if the purchase exceeds 300,000 SCF/hr. The maximum processing capacity of the plant is 500,000 SCF/hr.

What should be the flow rate of purchased gas necessary to maximize the gas-processing revenue (described as value added minus purchased gas cost)?

SOLUTION

The objective function is given by:

[14.75] $\qquad$ Revenue = Value Added − Purchased Gas Cost

The purchased gas cost can be described using a binary integer variable (I), which is defined depending on the value of the gas flow rate (F, expressed in 1000 SCF/hr) as:

[14.76a] $\qquad$ I = 0 if F ≥ 300

and

[14.76b] $\qquad$ I = 1 if F < 300

Choosing the lower and upper bounds on the values of F to be 0 and 500, and using Constraint 14.61, we get:

[14.77] $\qquad$ $(0 - 300)*I < F - 300 \leq (500 - 300)*(1 - I)$

According to Eq. 14.58, the cost of the purchased gas is given by:

[14.78] $\qquad$ Purchased gas cost $= 4.0 * F * I + 3.5 * F * (1 - I)$

We can now formulate the problem as an MINLP. The following LINGO input file is given:

```
(P14.16)
max = value_added - Gas_cost;
Value_added = 14*F^0.8;
Gas_Cost = 4.0*F*I + 3.5*(1-I)*F;
-300*I < F - 300;
F - 300 <= 200*(1-I);
@bin(I);
```

The solution to this program gives the following results:

```
Objective value: 293.6013

Variable      Value
VALUE_ADDED   1468.006
GAS_COST      1174.405
F             335.5443
I             0.000000
```

Example 14-10 Inclusion of disjoint constraints in the gas-processing case study

Let us expand the scope of Example 14.9 by incorporating the effect of GHG emissions. The net GHG emission of the gas-processing facility depends on the flow rate of the processed gas (F in 1000 SCF/hr) as follows:

[14.79a] GHG emission (in tonne CO_2 equivalent/hr) = $0.060*F^{0.90}$
if $F \geq 300$

[14.79b] $= 0.045*F^{0.95}$ if $F < 300$

The maximum permissible GHG emission of the process is given by:

[14.80] Maximum permissible GHG emission (in tonne CO_2 equivalent/hr) = $0.033*F$

a. Determine the flow rate of purchased gas necessary to maximize the gas-processing revenue (described as value added − purchased gas cost) while satisfying the GHG-emission requirement.
b. Reformulate the GHG constraints using the Big-M formulation.
c. Reformulate the GHG constraints using the convex-hull formulation.
d. Reformulate the gas-cost constraint using the convex-hull formulation.

SOLUTION

a. Determining Flow Rate

One way is to create a combined expression for GHG emission using Equations Equations 14.79a and 14.77b and the integer variable, I, defined by Equations 14.74a and 14.74b:

[14.81] GHG emission (in tonne CO_2 equivalent/hr) = $0.045*I*F^{0.95} + 0.060*(1-I)*F^{0.90}$

The resulting program is given below:

[P14.17]
```
max = value_added - Gas_cost;
Value_added = 14*F^0.8;
Gas_Cost = 4.0*F*I + 3.5*(1-I)*F;
-300*I < F - 300;
F - 300 <= 200*(1 - I);
GHG = 0.045*F^0.95*I + 0.06*F^0.9*(1-I);
GHG <= 0.033*F;
@bin(I);
```

The solution shows the following results:

```
Objective value: 290.1753
Variable        Value
VALUE_ADDED     1671.962
GAS_COST        1381.787
F               394.7962
I               0.000000
GHG1            13.02828
GHG2            0.000000
GHG             13.02828
```

Therefore, the addition of GHG constraints has reduced the revenue from $293.6 MM/yr to $290.2 MM/yr and has shifted the optimal flow rate from 335.5 to 394.8 1000 SCF/hr.

b. Big-M Reformulation

The added Constraint 14.81 involves bilinear terms that are nonconvex and can pose numerical challenges to identifying the global solution. Therefore, it is beneficial to explore the possibility of replacing this constraint with a linear formulation. Constraints 14.79a, 14.79b, and 14.78 may be rewritten as the following disjoint constraints:

[14.82a] $0.060*F^{0.90} - 0.033*F \leq 0$ if $F \geq 300$

[14.82b] $0.045*F^{0.95} - 0.033*F \leq 0$ if $F < 300$

Let us use the Big-M formulation. Although M can be any sufficiently large number to serve as an upper bound to the constraint, the selection of tight upper bounds reduces the size of the feasibility region and consequently reduces the computational time. Therefore, let us identify the maximum feasible values for Constraints 14.82a and 14.82b by solving maximization problems for the constraints. Starting with Constraint 14.82a, the maximum feasible value of the constrain is determined as:

[P14.18]
```
max=0.06*F^0.9 - 0.033*F;
F >=300;
F <= 500;
```

which yields the following solution:

```
Objective value: 0.2756083
Variable   Value
F          300.0000
```

Hence, we choose M_1 to be 0.28. Next, we solve the upper-bound problem for Constraint 14.80b:

[P14.19]
```
max = 0.045*F^0.95 - 0.033*F;
F <=300;
```

which gives the following solution:

```
Objective value: 0.3077542
Variable   Value
F          177.1918
```

Therefore, let us select M_2 to be 0.31. Next, the binary integer variables for the two disjoint constraints are related through Constraint 14.64a:

[14.64a] $I_1 + I_2 = 1$ $I_1 \in \{0,1\}$ and $I_2 \in \{0,1\}$

where the binary variable I_1 is assigned a value of 1 constraint when $g_1(x)$ is used while the binary variable I_2 is assigned a value of 1 when constraint $g_2(x)$ is used.
Revising Constraints 14.77 to be written in terms of I_2 (because it assumes a value of 1 when Constraint 14.80b is used):

[14.77] $(0 - 300)*I_2 < F - 300 \leq (500 - 300)*(1 - I_2)$

Following the Big-M Constraints 14.64b and 14.64c, we get:

[14.83a] $0.060*F^{0.90} - 0.033*F \leq 0.28*(1 - I_1)$

[14.83b] $0.045*F^{0.95} - 0.033*F \leq 0.31*(1 - I_2)$

The following is the LINGO program of the Big-M formulation:

[P14.20]
```
max = value_added - Gas_cost;
Value_added = 14*F^0.8;
Gas_Cost = 4.0*F*I2 1 3.5*(1-I2)*F;
-300*I2 < F - 300;
F - 300 <= 200*(1 - I2);
I1 + I2 = 1;
0.06*F^0.9 - 0.033*F <= 0.28*(1 - I1);
0.045*F^0.95 - 0.033*F <= 0.31*(1 - I2);
@bin(I1);
@bin(I2);
```

(Continued)

Example 14-10 Inclusion of disjoint constraints in the gas-processing case study (*Continued*)

which is solved to get:

```
Objective value: 290.1753
Variable        Value
VALUE_ADDED     1671.962
GAS_COST        1381.787
F               394.7963
I2              0.000000
I1              1.000000
```

These are the same results obtained from Program P14.17.

c. Convex-Hull Reformulation of the GHG Constraints:

First, the flow rate variable, F, is disaggregated into two variables: F1 and F2 to be used in Constraints 14.82a and 14.82b, respectively (that is, F_1 will represent the flows between 300 and 500, and F_2 will represent the flows between 0 and 300). Hence,

[14.84a] $$F = F_1 + F_2$$

The definitions of the integer variables I_1 and I_2 are the same as in the Big-M formulation. Applying Constraints 14.69a, 14.69b, 14.70a, and 14.70b to the two disjoint Constraints 14.82a and 14.82b, we obtain:

[14.84b] $$0.060 * F_1^{0.90} - 0.033 * F_1 \leq 0$$

[14.84c] $$0.045 * F_2^{0.95} - 0.033 * F \leq 0$$

with the selection of the disaggregated variables determined using the integer variables; that is,

[14.84d] $$300 * I_1 \leq F_1 \leq 500 * I_1$$

[14.84e] $$0 \leq F_2 \leq 300 * I_2$$

and as before,

[14.64a] $$I_1 + I_2 = 1 \quad I_1 \in \{0,1\} \ \text{and} \ I_2 \in \{0,1\}$$

The following is the LINGO program for the convex-hull formulation:

```
(P14.21a)
max = value_added - Gas_cost;
Value_added = 14*F^0.8;
Gas_Cost = 4.0*F*I2 + 3.5*(1-I2)*F;
-300*I2 < F - 300;
F - 300 <= 200*(1 - I2);
I1 + I2 = 1;
F = F1 + F2;
0.06*F1^0.9 - 0.033*F1 <= 0;
0.045*F2^0.95 - 0.033*F2 <= 0;
300*I1 <=F1;
F1 <= 500*I1;
0<=F2;
F2 <=300*I2;
@bin(I1);
@bin(I2);
```

The solution gives the following results:

```
Objective value: 290.1753
Variable        Value
VALUE_ADDED     1671.962
GAS_COST        1381.787
F               394.7963
I2              0.000000
I1              1.000000
F1              394.7963
F2              0.000000
```

Again, these are the same results obtained from Programs P14.17 and P14.20.

d. Convex-Hull Reformulation of the Gas-Cost Constraint

The convex hull reformulation can also be applied to the gas-cost function to avoid the bilinear terms in Eq. 14.78. It is important to note that Constraint 14.84d limits the values of F_1 to either the range of 300 to 500 or otherwise to 0 (disjoint domains). Similarly, Constraint 14.84e limits the values of F_2 to the range of 0 to 300. Because Constraint 14.64a will allow either F_1 or F_2 to exist, Constraint 14.82 will calculate the proper value of the flow rate. Consequently, the gas-cost function can be expressed as:

[14.85] $$\text{Purchased gas cost} = 3.5 * F_1 + 4.0 * F_2$$

This linear expression is preferred to the nonconvex bilinear cost function of Eq. 14.76. Program P14.21b shows the LINGO formulation:

```
[P14.21b]
max = value_added - Gas_cost;
Value_added = 14*F^0.8;
Gas_Cost = 3.5*F1 + 4.0*F2;
I1 + I2 = 1;
F = F1 + F2;
0.06*F1^0.9 - 0.033*F1 <= 0;
0.045*F2^0.95 - 0.033*F2 <= 0;
300*I1 <=F1;
F1 <= 500*I1;
0<=F2;
F2 <=300*I2;
@bin(I1);
@bin(I2);
```

This program is solved to get the following solution, which is identical to the results obtained earlier through the bilinear and the Big-M formulations:

```
Objective value: 290.1753
Variable        Value
VALUE_ADDED     1671.962
GAS_COST        1381.787
F               394.7963
F2              0.000000
F1              394.7963
I1              1.000000
I2              0.000000
```

Example 14-11 Water desalination

A seawater desalination plant is needed to provide water for drinking and industrial use. Two technologies are considered: reverse osmosis (RO) and multi-effect distillation (MED). Only one technology is to be selected.

Table 14.7 gives the techno-economic data for the two technologies. Because of infrastructure limitations, there are practical constraints on the size of each technology given by:

[14.86a] $$100{,}000 \leq W_{In}^{RO} \leq 320{,}000$$

where W_{In}^{RO} is the flow rate (m³/day) of seawater entering the RO plant. Additionally,

[14.86b] $$50{,}000 \leq W_{In}^{MED} \leq 400{,}000$$

where W_{In}^{MED} is the flow rate (m³/day) of seawater entering the MED plant.

a. Determine the optimal size of the desalination plant that maximizes the annual revenue defined as:

[14.87] Annual revenue = Annual value of desalinated water − Total annualized cost of desalination

(*Continued*)

Example 14-11 Water desalination (Continued)

TABLE 14-7 Techno-Economic Data for the Two Desalination Technologies

Technology	Annualized Fixed Cost (AFC, $/yr)	Operating Cost ($/m³ seawater)	Water Recovery (m³ desalinated water/m³ feed seawater)	Value of Desalinated Water ($/m³ desalinated water)
RO	$2.0*10^6 + 1166*$(Flow rate of seawater, m³/day)$^{0.8}$	0.18	0.55	0.88
MED	$13.0*10^6 + 2227*$(Flow rate of seawater, m³/day)$^{0.7}$	0.24	0.65	0.82

Source: Adapted from Atilhan et al. (2011)

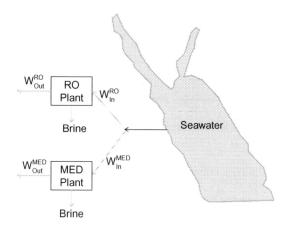

FIGURE 14-11 Representation of the desalination problem.

b. The thermal energy for the MED plant is supplied by a coupled power plant. As a result of changes in the power plant, it will provide less thermal energy to MED such that the maximum permissible flow rate of seawater entering the MED plant will be reduced to 300,000 m³/day. Re-solve the problem with this new constraint.

SOLUTION

a. Figure 14.11 illustrates a schematic representation of the problem. The terms W_{Out}^{RO} and W_{Out}^{MED} represent the flow rates (m³/day) of desalinated water from the RO and the MED plants. Although only one technology is to be used, seawater is fed to the RO and the MED plants. A no-flow solution to a plant corresponds to that plant not selected.

Let us define two binary integer variables:

[14.88] I_{RO} = 1 if RO is selected
  = 0 if RO is not selected

To account for the acceptable flow rate range for RO and to select the proper value of the integer variable, the following constraints are used based on Constraint 14.63b:

[14.89] $100{,}000*I_{RO} \leq W_{In}^{RO} \leq 320{,}000*I_{RO}$

Because the AFC contains a fixed-charge term, it is necessary to use the integer variable as described by Constraint 14.63a to set the AFC equal to 0 when there is no flow going through the RO plant; that is,

[14.90] $AFC_{RO} = 2{,}000{,}000*I_{RO} + 1{,}166*(W_{In}^{RO})^{0.8}$

Adding the operating cost and assuming an annual operation of 365 days, we get the total annualized cost (TAC_{RO}) to be:

[14.91] $TAC_{RO} = 2{,}000{,}000*I_{RO} + 1166*(W_{In}^{RO})^{0.8} + 0.18*365*W_{In}^{RO}$

Similarly, for the MED plant:

[14.92] I_{MED} = 1 if MED is selected
  = 0 if MED is not selected

[14.93] $50{,}000*I_{MED} \leq W_{In}^{MED} \leq 400{,}000*I_{MED}$

[14.94] $TAC_{MED} = 13{,}000{,}000*I_{MED} + 2227*(W_{In}^{MED})^{0.8} + 0.24*265*W_{In}^{MED}$

Because only desalination technology is to be selected,

[14.87] $I_{RO} + I_{MED} = 1$

Given the recovery ratio of RO and MED, the flow rates of desalinated water can be related to the flow rates of incoming seawater as follows:

[14.95] $W_{Out}^{RO} = 0.55*W_{In}^{RO}$

[14.96] $W_{Out}^{MED} = 0.65*W_{In}^{MED}$

[14.97] The annual sales of desalinated water from RO = $0.88*365*W_{Out}^{RO}$

and

[14.98] The annual sales of desalinated water from MED = $0.82*365*W_{Out}^{MED}$

The objective function is defined as the sum of the annual revenues for RO and MED (realizing that one of the two terms will be 0 in the solution). The following is the LINGO program for the problem:

```
[P14.22]
MAX = REV;
REV = RO_REV*I_RO + MED_REV*I_MED;
I_RO 1 I_MED = 1;
100000*I_RO <= W_IN_RO;
W_IN_RO <= 320000*I_RO;
50000*I_MED <= W_IN_MED;
W_IN_MED <= 400000*I_MED;
TAC_RO = 2000000*I_RO + 0.18*365*W_IN_RO + 1166*W_
IN_RO^0.8;
TAC_MED = 13000000*I_MED + 0.24*365*W_IN_MED +
2227*W_IN_MED^0.7;
W_OUT_RO = 0.55*W_IN_RO;
W_OUT_MED = 0.65*W_IN_MED;
W_SALES_RO = 0.88*365*W_OUT_RO;
W_SALES_MED = 0.82*365*W_OUT_MED;
RO_REV = W_SALES_RO - TAC_RO;
MED_REV = W_SALES_MED - TAC_MED;
@bin(I_RO);
@bin(I_MED);
```

(Continued)

Example 14-11 Water desalination (Continued)

Solving using LINGO, we get:

```
Global optimal solution found.
Objective value: 0.1119302E+08

Variable        Value
REV             0.1119302E+08
RO_REV          0.000000
I_RO            0.000000
MED_REV         0.1119302E+08
I_MED           1.000000
W_IN_RO         0.000000
W_IN_MED        400000.0
TAC_RO          0.000000
TAC_MED         0.6662498E+08
W_OUT_RO        0.000000
W_OUT_MED       260000.0
W_SALES_RO      0.000000
W_SALES_MED     0.7781800E+08
```

The solution shows that MED is selected to process 400,000 m³/day of seawater and to generate $11.19 MM of annual revenue.

b. The upper bound constraint for MED is changed to (W_IN_MED <= 300000*I_MED), and the program is re-solved to get the following output:

```
Objective value: 3939433.

Variable        Value
REV             3939433.
RO_REV          3939433.
I_RO            1.000000
MED_REV         0.000000
I_MED           0.000000
W_IN_RO         320000.0
W_IN_MED        0.000000
TAC_RO          0.5259177E+08
TAC_MED         0.000000
W_OUT_RO        176000.0
W_OUT_MED       0.000000
W_SALES_RO      0.5653120E+08
W_SALES_MED     0.000000
```

With the new upper bound on the size of the MED plant, optimization now prefers the selection of the RO plant to process 320,000 m³/day and to yield a revenue of $3.94 MM/yr.

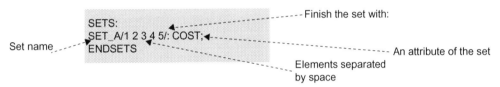

FIGURE 14-12a LINGO format for a set (Set_A) with an attribute (COST).

```
SETS:
SET_A/1..5/: COST;
ENDSETS
```

FIGURE 14-12b Short LINGO format for a set (Set_A) with an attribute (COST).

USING SET FORMULATIONS IN LINGO

As has been seen in some problems such as the transportation and the assignment problems, generic formulations can be expressed in compact forms. The use of set formulations is attractive in dealing with large problems with repetitive constraints. The generic set formulation is developed once different scenarios can be run by altering the data. The following section explains how some of the key commands in set formulation can be coded using LINGO.

SUMMATION

Consider the summation over I, which belongs to a set called SET_A:

[14.99a] $$\sum_{i \in SET\_A}$$

In LINGO syntax, it is expressed as:

[14.99b] @SUM (SET_A (i) :

Now, let us take the summation of a function of i such as COST(i):

[14.100a] $$\sum_{i \in SET\_A} COST(i)$$

which is written in LINGO syntax as:

[14.100b] @SUM (SET_A (i) : COST(i));

DEFINING SETS

To define the A set, we always start with SETS, and finish with ENDSETS. One way of representing sets is

[14.101] **SET_Name / List of members / : List of attributes separated by commas ;**

For instance, consider SET_A, which has five elements given by SET_A = {i | i = 1,2, . . . , 5}. Also, consider COST(i), which is an attribute of SET_A. The values of the COST(i) are = {2, 5, 6, 8, 11}. Figure 14.12a illustrates how the set and its attribute can be expressed.

Instead of listing all the elements of Set_A, we can use the first number and the last number separated by two period (that is, 1..5) as shown by Fig. 14.12b.

Example 14-12 Expressing summation and entering data for a set with one attribute

Using the set formulation of LINGO, describe the following problem:

[P14.23a] $$Min = \sum_{i \in SET\_A} COST(i)$$

with

$$SET\_A = \{i | i = 1,2,\ldots,5\}$$
$$COST(i) = \{2,5,6,8,11\}$$

SOLUTION

The program can be expressed as shown by Program (P14.23b):

```
[P14.23b]
SETS:
SET_A/ 1..5 / : COST;
ENDSETS
DATA:
COST = 2 5 6 8 11;
ENDDATA
min = @SUM (SET_A (i): COST(i));
```

The solution is obtained to be:

```
Objective value: 32.00000
Variable    Value
COST( 1)    2.000000
COST( 2)    5.000000
COST( 3)    6.000000
COST( 4)    8.000000
COST( 5)   11.00000
```

Example 14-13 Expressing summation and entering data for a set with two attributes

Using the set formulation of LINGO, describe the following problem:

[P14.24a] $$Min = \sum_{i \in SET\_A} FACTOR(i) * COST(i)$$

with $SET\_A = \{i | i = 1,2,\ldots,5\}$ having two attributes: $FACTOR(i) = \{3,2,1,0,10\}$ and $COST(i) = \{2,5,6,8,11\}$.

SOLUTION

The program can be written as illustrated by Program (P14.24b).

```
[P14.24b]
SETS:
SET_A/ 1..5 / : FACTOR, COST;
ENDSETS
DATA:
FACTOR = 3 2 1 0 10;
COST = 2 5 6 8 11;
ENDDATA
min = @SUM (SET_A (i): FACTOR(i)*COST(i) );
```

The following is the solution:

```
Objective value: 132.0000
Variable    Value
FACTOR(1)   3.000000
FACTOR(2)   2.000000
FACTOR(3)   1.000000
FACTOR(4)   0.000000
FACTOR(5)  10.00000
COST(1)     2.000000
COST(2)     5.000000
COST(3)     6.000000
COST(4)     8.000000
COST(5)    11.00000
```

```
DATA:
Attribute = data1 data2   datan;
ENDDATA
```

FIGURE 14-13a Format for entering one-dimensional data of a single attribute.

```
DATA:
COST = 2 5 6 8 11;
ENDDATA
```

FIGURE 14-13b Entering data.

```
DATA:
COST = 2,5, 6,8,11;
ENDDATA
```

FIGURE 14-13c Inserting commas between data for readability.

ENTERING DATA

- To enter data for the attributes, we always start with DATA: and finish with ENDDATA
- For one-dimensional data of a single attribute, the LINGO syntax is shown by Fig. 14.13a.

For instance, for $SET\_A = \{i \mid i = 1,2, \ldots, 5\}$ with attribute $COST(i)$ and the values of $COST(i)$ are $= \{2, 5, 6, 8, 11\}$, Fig. 14.13b shows how these data are entered in LINGO.

For readability, the values of the data may be optionally separated by commas (,) as shown by Fig. 14.13c.

THE @FOR COMMAND

The @FOR command is used to repeat a constraint for all members of a set. Consider $SET\_A = \{i | i = 1,2,3,4\}$ and the following constraint:

[P14.25a] $Cost(i) \leq Budget(i) \; i \in SET\_A$

```
@FOR (Set_A(i):
        Cost(i) <= Budget(i));
```

FIGURE 14-14 Using the @FOR Command.

This translates to:

```
Cost(1) <= Budget(1);
Cost(2) <= Budget(2);
Cost(3) <= Budget(3);
Cost(4) <= Budget(4);
```

The use of the @FOR command in this case is shown by Fig. 14.14.

The program is written as illustrated by Program 14.25b.

```
[P14.25b]
SETS:
Set_A/1..4/: Cost, Budget;
ENDSETS

@FOR (Set_A(i):
        Cost(i) <= Budget(i));
```

DEALING WITH DOUBLE SUMMATIONS

Suppose that there are double summations over i (i ∈ SOURCES) and j (j ∈ DESTINATIONS) as given by Eq. 14.101:

$$[14.101a] \qquad \sum_{i \in SOURCES} \sum_{j \in DESTINATIONS}$$

The double summations create pairs of (i, j). Let us call the set of all these pairs CONNECTIONS(i,j). For example, suppose that the two sets have the following elements: SOURCES = {1, 2, 3, 4} and DESTINATIONS = {A, B, C}. Therefore, the set CONNECTIONS has the following elements:

CONNECTIONS = {(1,A), (1,B), (1,C),(2,A), (2,B), (2,C), (3,A), (3,B), (3,C), (4,A), (4,B), (4,C)}

Suppose that there is one attribute for the set SOURCES (called Capacity), one attribute of set DESTINATIONS (called Demand), and two attributes for set CONNECTIONS (called Cost and Flow Rate). Program 14.26a shows the LINGO formulation for modeling the three sets:

```
[P14.26a]
SETS:
SOURCES/1..4/: Capacity;
DESTINATIONS/A B C/:Demand;
CONNECTIONS (SOURCES, DESTINATIONS): Cost, Flow
rate;
ENDSETS
```

Notice that the expression CONNECTIONS (SOURCES, DESTINATIONS) creates the new set CONNECTIONS whose elements are all the combinations of elements from the two sets: SOURCES and DESTINATIONS. If there is a need to include only some elements of the combination, then they should be explicitly listed. For example, if it is required to include only the combinations (1,A), (1,C), and (2,B), then the formulation should be described as that given by Program P14.26b.

```
DATA:
Cost=   20   10    50
        30    0    80
       110   60   150
        70   10    90;
ENDDATA
```

FIGURE 14-15 Entering two-dimensional data.

```
[P14.26b]
SETS:
SOURCES/1..4/: Capacity;
DESTINATIONS/A B C/:Demand;
CONNECTIONS (SOURCES, DESTINATIONS)/ 1 A, 1 C, 2 B/:
Cost, Flow rate;
ENDSETS
```

Now, the double summations described by Eq. 14.101a can be expressed in LINGO syntax as:

[14.101b] @SUM (CONNECTIONS(i, j):

Next, let us include an objective function. For instance, consider the objective function to be the product of the cost and the flow rate for all the pairs included in the double summations; that is,

$$[14.102a] \qquad \min \sum_{i \in SOURCES} \sum_{j \in DESTINATIONS} Cost(i,j)*Flowrate(i,j)$$

Which can be described in LINGO as:

```
[14.102b]
Min = @SUM (CONNECTIONS (i,j): Cost(i,j)*Flow
rate(i,j));
```

ENTERING TWO-DIMENSIONAL DATA

Consider the cost data given by Table 14.3 of Example 14.7. The two-dimensional data are for the attribute Cost(i,j) over sets: SUPPLIERS = {1,2,3,4} and CONSUMERS = {A,B,C}.

To enter these data for the Cost(I,j) attributes over all the pairs of elements from the two sets SUPPLIERS and CONSUMERS, the LINGO syntax shown by Fig. 14.15 is used. Notice that the last data entry is followed by a semicolon.

USING @FOR IN THE CASE OF REPEATING CONSTRAINTS WITH TWO-DIMENSIONAL VARIABLES

A common constraint is the nonnegativity requirement given by Constraint 14.103a, which can be expressed in LINGO through Constraint 14.103b:

[14.103a]
$$x(i,j) \geq 0 \qquad i \in Set\_A \quad j \in Set\_B$$

```
[14.103b]
@FOR( Set_A(i): @FOR( Set_B(j): (x(i,j)>= 0) ) );
```

Alternatively, we can define the set CONNECTIONS of all the (i,j) combinations and then carry out the @FOR command over the set CONNECTIONS as shown by Program P14.27:

```
[P14.27]
SETS:
Set_A/1..4/:;
Set_B/A B C/:;
CONNECTIONS (Set_A, Set_B):X;
ENDSETS
@FOR (CONNECTIONS(i,j): (x(i,j))>=0);
```

Another common constraint is the definition of binary integer variables given by Constraint 14.104a, which can be expressed in LINGO through Constraint 14.104b:

[14.104a]
$$x(i,j) = \{0,1\} \quad i \in Set\_A \; j \in Set\_B$$

[14.104b]
```
@FOR( Set_A(i): @FOR( Set_B(j): @BIN(x(i,j)) ) );
```

Alternatively, we can define the set CONNECTIONS of all the (i,j) combinations and then carry out the @FOR command over the set CONNECTIONS as shown by Program P14.28:

```
[P14.28]
SETS:
Set_A/1..4/:;
Set_B/A B C/:;
CONNECTIONS (Set_A, Set_B):X;
ENDSETS
@FOR (CONNECTIONS(i,j): @BIN(x(i,j)));
```

Example 14-14 Developing a set formulation for the transportation problem of Example 14.7

Example 14.7 dealt with the transportation of hexane from suppliers in cities 1, 2, 3, and 4 to consuming biorefineries that are located in towns A, B, and C. Table 14.2a listed the daily requirements for hexane in the three biorefineries, and Table 14.2b gave the production capacities of the four suppliers.

Table 14.3 gives the costs ($/tonne) of shipping hexane from each supplying plant to each consuming biorefinery.

Develop and solve an optimization program using set formulations to determine the optimal allocation of hexane from the suppliers to the biorefineries to minimize the total transportation cost while meeting the requirements of the biorefineries.

SOLUTION

Program P14.29a gives the optimization formulation. To view the program in details, go to the LINGO drop-down menu, select "Generate," and then select "Display model" as shown by Fig. 14.16. The displayed model is shown by Program 14.29b.

```
[P14.29a]
SETS:
SOURCES/1..4/: Capacity;
DESTINATIONS/A  B  C/  :Demand;
CONNECTIONS (SOURCES, DESTINATIONS): Cost,
Flowrate;
ENDSETS
Min = @SUM (CONNECTIONS (i,j):
Cost(i,j)*Flowrate(i,j));
DATA:
Capacity = 7 5 3 6;
Demand = 6 1 10;
Cost =    20        10          50
          30         0          80
         110        60         150
          70        10          90;
ENDDATA
! Supply Availability Constraints;
@FOR (SOURCES (i):
    @SUM (DESTINATIONS (j):
        Flowrate (i,j) )<= Capacity(i));
! Demand Constraints;
@FOR (DESTINATIONS (j):
    @SUM (SOURCES (i):
        Flowrate(i,j) )= Demand (j));
@FOR( SOURCES(i): @FOR( DESTINATIONS(j):
Flowrate(i,j)>= 0) ) ;
```

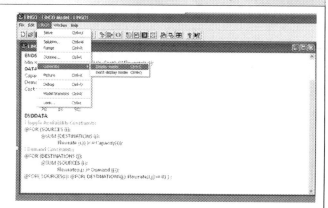

FIGURE 14-16 Displaying the model.

```
[P14.29b]
MODEL:
[_1] MIN= 20 * FLOWRATE_1_A + 10 * FLOWRATE_1_B
+ 50 * FLOWRATE_1_C + 30
* FLOWRATE_2_A + 80 * FLOWRATE_2_C + 110 *
FLOWRATE_3_A + 60 *
FLOWRATE_3_B + 150 * FLOWRATE_3_C + 70 *
FLOWRATE_4_A + 10 *
FLOWRATE_4_B + 90 * FLOWRATE_4_C ;
[_2] FLOWRATE_1_A + FLOWRATE_1_B + FLOWRATE_1_C <=
7 ;
[_3] FLOWRATE_2_A + FLOWRATE_2_B + FLOWRATE_2_C <=
5 ;
[_4] FLOWRATE_3_A + FLOWRATE_3_B + FLOWRATE_3_C <=
3 ;
[_5] FLOWRATE_4_A + FLOWRATE_4_B + FLOWRATE_4_C <=
6 ;
[_6] FLOWRATE_1_A + FLOWRATE_2_A + FLOWRATE_3_A +
FLOWRATE_4_A = 6 ;
[_7] FLOWRATE_1_B + FLOWRATE_2_B + FLOWRATE_3_B +
FLOWRATE_4_B = 1 ;
[_8] FLOWRATE_1_C + FLOWRATE_2_C + FLOWRATE_3_C +
FLOWRATE_4_C = 10 ;
[_9] FLOWRATE_1_A >= 0 ;
[_10] FLOWRATE_1_B >= 0 ;
[_11] FLOWRATE_1_C >= 0 ;
[_12] FLOWRATE_2_A >= 0 ;
[_13] FLOWRATE_2_B >= 0 ;
```

(Continued)

Example 14-14 Developing a set formulation for the transportation problem of Example 14.7 (*Continued*)

```
[_14] FLOWRATE_2_C >= 0 ;
[_15] FLOWRATE_3_A >= 0 ;
[_16] FLOWRATE_3_B >= 0 ;
[_17] FLOWRATE_3_C >= 0 ;
[_18] FLOWRATE_4_A >= 0 ;
[_19] FLOWRATE_4_B >= 0 ;
[_20] FLOWRATE_4_C >= 0 ;
END
```

The following is the solution using LINGO. It is consistent with the solution obtained for Example 14.7.

```
  Objective value: 840.0000
  Variable         Value
CAPACITY(1)        7.000000
CAPACITY(2)        5.000000
CAPACITY(3)        3.000000
CAPACITY(4)        6.000000
DEMAND(A)          6.000000
DEMAND(B)          1.000000
DEMAND(C)          10.00000
COST(1, A)         20.00000
COST(1, B)         10.00000
COST(1, C)         50.00000
COST(2, A)         30.00000
COST(2, B)         0.000000
COST(2, C)         80.00000
COST(3, A)         110.0000
COST(3, B)         60.00000
COST(3, C)         150.0000
COST(4, A)         70.00000
COST(4, B)         10.00000
COST(4, C)         90.00000
FLOWRATE(1, A)     1.000000
FLOWRATE(1, B)     0.000000
FLOWRATE(1, C)     6.000000
FLOWRATE(2, A)     5.000000
FLOWRATE(2, B)     0.000000
FLOWRATE(2, C)     0.000000
FLOWRATE(3, A)     0.000000
FLOWRATE(3, B)     0.000000
FLOWRATE(3, C)     0.000000
FLOWRATE(4, A)     0.000000
FLOWRATE(4, B)     1.000000
FLOWRATE(4, C)     4.000000
```

Alternatively, the program can be formulated using the @FOR command over a set CONNECTIONS that contains all the elements in the sets SOURCES and DESTINATIONS. Program P14.29c shows the formulation.

```
[P14.29c]
SETS:
SOURCES/1..4/: Capacity;
DESTINATIONS/A B C/ :Demand;
CONNECTIONS (SOURCES, DESTINATIONS): Cost,
Flowrate;
ENDSETS
Min = @SUM (SOURCES (i):
  @SUM(DESTINATIONS(j): Cost(i,j)*Flowrate(i,j)));
DATA:
Capacity = 7 5 3 6;
Demand = 6 1 10;
Cost =  20         10         50
        30          0         80
       110          6        150
        70         10         90;
ENDDATA
! Supply Availability Constraints;
@FOR (SOURCES (i):
  @SUM (DESTINATIONS (j):
    Flowrate (i,j) )<= Capacity(i));
! Demand Constraints;
@FOR (DESTINATIONS (j):
  @SUM (SOURCES (i):
    Flowrate(i,j) )= Demand (j));
@FOR( CONNECTIONS (i,j): Flowrate(i,j)>= 0) ;
```

Example 14-15 Developing a set formulation for the assignment problem of Example 14.8

Example 14.8 dealt the assignment of four new reactors (R1, R2, R3, and R4) to four vacant spaces (1, 2, 3, and 4). Table 14.6 gives the cost of assigning reactor i to space j. Let us develop a set formulation to assign the reactors to the spaces while minimizing the total assignment cost.

SOLUTION

Program P14.30 shows the set formulation for the assignment problem. The last constraint with double summations can be replaced with an @FOR command over the set CONNECTIONS; that is, @FOR(CONNECTIONS(i,j): @BIN(x(i,j))).

```
[P14.30]
min=cost;
SETS:
REACTOR /1..4/;
SPACE /A B C D/;
CONNECTIONS (REACTOR,SPACE): X, Cost_Space;
ENDSETS
DATA:
Cost_Space = 15    11         13         15
             13     9         12         17
             14    15         10         14
             17    13         11         16;
ENDDATA
COST = @SUM (CONNECTIONS (i,j): Cost Space(i,j)*X(i,j));
@FOR (REACTOR(i):
  @SUM( SPACE(j): x(i,j)) = 1);
@FOR (SPACE(j):
  @SUM( REACTOR(i): x(i,j) ) = 1);
@FOR( Reactor(i): @FOR( Space(j): @BIN(x(i,j)) ) );
```

The following is the solution to this program:

```
Objective value: 49.00000
Variable   Value   Reduced Cost
COST       49.00000
X( 1, A)   0.000000
X(1, B)    1.000000
```

(*Continued*)

Example 14-15 Developing a set formulation for the assignment problem of Example 14.8 (Continued)

X(1, C)	0.000000		COST_SPACE(1, C)	13.00000
X(1, D)	0.000000		COST_SPACE(1, D)	15.00000
X(2, A)	1.000000		COST_SPACE(2, A)	13.00000
X(2, B)	0.000000		COST_SPACE(2, B)	9.000000
X(2, C)	0.000000		COST_SPACE(2, C)	12.00000
X(2, D)	0.000000		COST_SPACE(2, D)	17.00000
X(3, A)	0.000000		COST_SPACE(3, A)	14.00000
X(3, B)	0.000000		COST_SPACE(3, B)	15.00000
X(3, C)	0.000000		COST_SPACE(3, C)	10.00000
X(3, D)	1.000000		COST_SPACE(3, D)	14.00000
X(4, A)	0.000000		COST_SPACE(4, A)	17.00000
X(4, B)	0.000000		COST_SPACE(4, B)	13.00000
X(4, C)	1.000000		COST_SPACE(4, C)	11.00000
X(4, D)	0.000000		COST_SPACE(4, D)	16.00000
COST_SPACE(1, A)	15.00000			
COST_SPACE(1, B)	11.00000			

As mentioned in Section 14.7, multiple solutions can be generated using integer cuts.

ADDING LOGICAL OPERATORS

Table 14.8 describes some of the key logical operators recognized by LINGO.

TABLE 14-8 Some of the Logical Operators Recognized by LINGO

Symbol	Meaning
#EQ#	Equal
#NE#	Not equal
#GE#	Greater than or equal
#GT#	Greater than
#LT#	Less than
#LE#	Less than or equal to

Example 14-16 Adding logical constraints to the assignment problem

Suppose that we would like to limit the assignment of reactors to spaces to reactors 2, 3, and 4 and that we would also like to limit the assignment to spaces 2, 3, and 4. How should Program P14.30 be revised?

SOLUTION

Two logical constraints should be added to the assignment constraints: $i \geq 2$ and $j \geq 2$. These can be included in the @FOR statements as shown by Program P14.31:

[P14.31]
```
min=cost;
SETS:
REACTOR /1..4/;
SPACE /A B C D/;
CONNECTIONS (REACTOR,SPACE): X, Cost_Space;
ENDSETS
DATA:
Cost_Space = 15   11   13   15
             13    9   12   17
             14   15   10   14
             17   13   11   16;
ENDDATA
COST = @SUM (CONNECTIONS (i,j): Cost_
  Space(i,j))*X(i,j));
!Reactor 1, can go into no more than one space;
@SUM(SPACE(j): x(1,j) ) = 1;
!For Reactors 2 & 3;
@FOR(REACTOR(i) | i #GE# 2: @SUM(SPACE(j):
  x(i,j) )=1);  ← To limit the constraint to i>=2
!Space 1 holds no more than one reactor;
@SUM( REACTOR(i): x(i,1) ) = 1 ;
!For Spaces 2 & 3;
@FOR (SPACE(j) | j #GE# 2:
@SUM(REACTOR(): x(i,j) ) = 1 );  ← To limit the
  constraint to j>=2
@FOR( Reactor(i): @FOR( Space(j): @BIN(x(i,j)) ) );
```

HOMEWORK PROBLEMS

14.1. Consider the blending of a fresh solvent and a recyclable hydrocarbon waste so that the blend can be used as a cleaning agent in a degreasing unit. The data for the problem (Shelley and El-Halwagi, 2000) are as follows:

The fresh organic solvent has the following properties (S is wt% of sulfur, ρ is density, and RVP is Reid vapor pressure):

[14.105a] $\quad S_{fresh} = 0.1$ (wt%)

[14.105b] $\quad \rho_{fresh} = 610$ kg/m^3

[14.105c] $\quad RVP_{fresh} = 2.1$ atm

The properties of the recyclable waste to be mixed with the fresh solvent are:

[14.106a] $\quad S_{Waste} = 0.5$ (wt%)

[14.106b] $\quad \rho_{Waste} = 580$ kg/m^3

[14.106c] $\quad RVP_{Waste} = 5.2$ atm

The maximum available flow rate of the recyclable waste stream is 30.0 kg/min.

The degreaser has the following constraints on the properties and flow rates:

[14.107a] $\quad 0.0 \leq S_{degreaser}$(wt%) ≤ 1.0

[14.107b] $\quad 555 \leq \rho_{degreaser}$(kg/m^3) ≤ 615

[14.107c] $\quad 2.1 \leq RVP_{degreaser}$(atm) ≤ 4.0

[14.107d] $\quad 36.6 \leq F_{degreaser}$(kg/min) ≤ 36.8

The following mixing rules can be used to evaluate the properties resulting from mixing the fresh solvent with the recyclable waste to form the blend:

[14.108a]
Total flow rate of the blend*Sulfur wt% in the blend = Flow rate of the fresh solvent*Sulfur wt% in the fresh solvent + Flow rate of the recyclable waste*Sulfur wt% in the recyclable waste

[14.108b]
Total flow rate of the blend/Density of the blend = Flow rate of the fresh solvent/Density of the fresh solvent + Flow rate of the recyclablewaste/Density of the recyclable waste

TABLE 14-9 Utility Requirements for the Two Polymers (MW/tonne per hour of produced polymer)

Utility	Polymer A	Polymer B
Refrigeration	4	2
Heating	2	5

TABLE 14-10 Utility Requirements for the Two Polymers (MW/tonne per hour of produced polymer)

Utility	Polymer A	Polymer B
Refrigeration	5	3
Heating	5	6

[14.108c]
Total flow rate of the blend*(RVP of the blend)$^{1.44}$ = Flow rate of the fresh solvent*(RVP of the fresh solvent)$^{1.44}$ + Flow rate of the recyclable waste*(RVP of the recyclable waste)$^{1.44}$

Develop an optimization formulation and solve it to determine the minimum required flow rate of the fresh solvent to be used in the degreaser.

14.2. A polymer plant consists of two production lines that yield two polymer products: A and B. Both lines use the same monomer (M) as a feedstock. Due to the formation of by-products and process losses, each tonne produced of either polymer requires four tonnes of M. The maximum supply of monomer M to the plant is 75 tonnes/hr. The process requires refrigeration and heating. Table 14.9 lists the consumption of these utilities required for both products.

Available for service are 80 MW of refrigeration and 60 MW of heating. The net profits for producing A and B are 1000 and 1500 $/tonne produced, respectively. What is the optimal production rate of each polymer that will maximize the net profit of the plant?

14.3. A polymer plant consists of two production lines that yield two polymer products: A and B. Both lines use the same monomer (M) as a feedstock. Due to the formation of by-products and process losses, each tonne produced of either polymer requires four tonnes of M. The maximum supply of monomer M to the plant is 45 tonnes/hr. The process requires refrigeration and heating. Table 14.10 lists the consumption of these utilities required for both products.

Available for service are 60 MW of refrigeration and 75 MW of heating. The net profits for producing A and B are 1000 and 1500 $/tonne produced, respectively. What is the optimal production rate of each polymer that will maximize the net profit of the plant?

TABLE 14-11 Unit Transportation Costs ($/gal) for Problem 14.6

Supplier/Customer	T_1	T_2	T_3	T_4
S_1	0.13	–	0.97	0.10
S_2	0.85	0.91	–	–
S_3	0.10	0.89	1.00	0.98

TABLE 14-12 Hours Needed by Each Student for Each Task

Task/Student	Lama	Ramo	Ila
Literature Search	20	25	30
Simulation	10	10	10
Report Writing	15	20	20

14.4. The two plants described in Problems 14.2 and 14.3 are now owned by the same parent company that is trying to maximize the combined net profit of both plants. The parent company has relaxed the restrictions on monomer supply to each plant (75 tonnes/hr for the first plant and 45 tonnes/hr for the second plant). Instead, the parent company can provide a combined supply of the monomer to both plants of 120 tonnes/hr at most. Therefore, it is necessary to optimally distribute the available monomer between the two plants. What are the optimal production capacities of polymers A and B in each plant that will maximize the combined net profit of both plants? Comment on the results of Problems 14.2 and 14.3.

14.5. A major oil company wants to build a refinery that will be supplied from three port cities. Port B is located 186 miles (300 km) east and 249 miles (400 km) north of Port A, while Port C is 249 miles (400 km) east and 62 miles (100 km) south of Port B. Determine the location of the refinery so that the total amount of pipe required to connect the refinery to the ports is minimized.

14.6. Three biodiesel-manufacturing plants (S_1, S_2, and S_3) are used to provide four customers (T_1, T_2, T_3, and T_4) with their biodiesel requirements. The yearly capacities of suppliers S_1, S_2, and S_3 are 135, 56, and 93 MM gal/year, respectively. The yearly requirements of the customers T_1, T_2, T_3, and T_4 are 62, 83, 39, and 91 MM gal/year, respectively. Table 14.11 gives the unit costs ($/gal) for supplying each customer from each supplier (a dash in the table indicates the impossibility of certain supplies for certain customers).

What are the optimal transportation quantities from suppliers to customers that will minimize the total transportation cost?

14.7. Three students (Lama, Ramo, and Ila) are working in the same group on a design project. The design project involves three tasks: literature search, simulation, and report writing. Because the three students did not get along well, they decided that each person would do one of three assigned tasks without any cooperation with the other two. In an attempt to identify the best way to distribute tasks among the three students, they all indicated the number of hours that they would need to finish each task. Table 14.12 gives these estimates.

To minimize the total person-hours involved in the project, who should do what? Can there be more than one solution? If yes, what are they?

14.8. Sudoku (which is an abbreviation of a Japanese phrase that means "the digits should remain single") is a logic-based number-placement puzzle. It usually takes the form of a 9 × 9 matrix with some of the cells given. The objective of the puzzle is to fill out all the matrix cells with numbers from 1 to 9 such that the following three rules are satisfied:
 – Every row contains each of the numbers 1 to 9 exactly once.
 – Every column contains each of the numbers 1 to 9 exactly once.
 – Each of the nine 3 × 3 submatrices contains each of the numbers 1 to 9 exactly once.

1. Develop a generic optimization formulation to solve the Sudoku puzzle.
2. Write a LINGO set formulation to code your formulation.
3. Pick a published Sudoku puzzle and apply your code to solve it. Use integer cuts to check for the possibility of multiple solutions.

14.9. A chemical plant produces polyamide synthetic membranes that can be used in pervaporation modules. To ensure the quality of the product, at least 1500 m² of the polymeric membrane must be checked in an eight-hour shift. The plant has two grade inspectors (I and II) to undertake quality control inspection. Grade I inspector can check 20 m² in an hour with an accuracy of 96 percent. Grade II inspector checks 14 m² an hour with an accuracy of 93 percent. The maximum available number of grade I and II inspectors are 10 and 15 per shift, respectively. The hourly wages of grade I inspectors are $20 per hour whereas those grade II inspectors earn $16 per hour. Any error an inspector makes costs the plant $12 per m². Find the optimal assignment of inspectors that minimizes the daily inspection cost.

14.10. A chemical plant has budgeted $25 MM for the development of new waste-treatment systems. Seven waste-treatment systems are being considered. Each system can handle a certain capacity (indicated in the following table) and accomplish the required waste-treatment task for that particular capacity. Table 14.13 gives the projected capital costs for these systems. Multiple units of the same technology may be purchased. The operating costs for all the systems are almost identical. The following additional constraints are imposed:
 – If stripping is used, a biological treatment must be purchased to handle the occasional shutdowns on the stripping process.

TABLE 14-13 Capacity and Cost of Each Waste Treatment Alternative

System	A	B	C	D	E	F	G
Description	Biological Treatment	Membrane Separation	Adsorption	Oxidation	Precipitation	Stripping	Solvent Extraction
Capacity, tons/hr	20	17	15	15	10	8	5
Cost, $MM	14.5	9.2	7.0	7.0	8.4	1.4	4.7

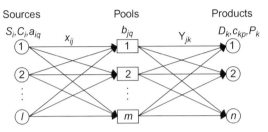

FIGURE 14-17 The pooling problem. *Source: Pham et al. (2009).*

TABLE 14-14 Production Capacities and Demand (tonne per month) for Problem 14.13

	Month 1	Month 2	Month 3	Month 4
Production during Regular Shift	110	150	190	160
Production during Overtime Shift	60	75	105	80
Demand	80	240	290	210

- If solvent extraction is used, the biological treatment cannot be used.
- For each additional unit of the same type, the plant gets a 20 percent discount on the price. For example, the first precipitation unit costs $10 MM. If a second precipitation unit is purchased, it will cost $8 MM, and so on. Which systems should the plant select?

14.11. Use piecewise linearization and convex-hull reformulation to transform the optimization program of the gas-processing facility of Example 14.9 into a globally solvable MILP. Explore different numbers of linearization segments and comment on your results.

14.12. Consider the pooling problem shown by Fig. 14.17 (for example, Haverly, 1978; Pham et al., 2009), which may be stated as follows: "Given is a set SOURCES = $\{i | i = 1,2,\ldots,l\}$ of intermediate streams. Each source has a given available capacity, S_i, one unit cost, C_i, and known values of N_q characterizing qualities, $a_{i,q}$, where q is an index for properties. The amount (or flow rate) from source i to pool j is denoted by a variable X_{ij}. The sources must be blended because there are not enough pools ($m < n$) or there are insufficient pool capacities to store them separately. Sources can be sent to some or all pools.

"As a result of mixing the sources, each pool, j, has unknown values for the qualities $b_{j,q}$. The flow from pool j to sale product k is y_{jk} and is to be determined. The n products with price P_k have constraints on known demand D_k and minimum values of desired quality specifications c_{kq}." For the pooling problem, it is desired to carry out the following tables.

1. Develop an optimization formulation to minimize the cost of the sources.
2. Develop a convex-hull global optimization procedure for the pooling problem and assess its performance by applying it to published test problems for pooling. Hint: See Pham et al. (2009).

14.13. A plant produces a pharmaceutical ingredient in two shifts (a regular working shift and an overtime shift) to meet specific demands of other plant units for the present and the future. Over the next four months, the production capacities during regular working and overtime as well as monthly demands (tonne product per month) are given in Table 14.14.

The cost of production is $2.0 per kg of the ingredient if done in regular working or $2.6 per kg of the ingredient if done in overtime. The produced ingredient can be stored before delivery at a cost of $0.40 per month per kg of the ingredient. What is the optimal production schedule for the pharmaceutical ingredient of the product so as to meet present and future demands at minimum cost?

REFERENCES

Al-Otaibi M, El-Halwagi M: Targeting techniques for enhancing process yield, *Trans. IChemE, Part A: Chemical Engineering Research and Design* 84(A10):943–951, 2006.

Atilhan S, Linke P, Abdel-Wahab A, El-Halwagi MM: A systems integration approach to the design of regional water desalination and supply networks, *Int J Proc Sys Eng (IJPSE)* 1(2):125–135, 2011.

Balas E: Disjunctive programming and a hierarchy of relaxations for discrete optimization, *SIAM J Algebra Discr* 6(3):466–486, 1985.

Biegler LT, Grossmann IE: Retrospective on optimization, *Comp Chem Eng* 28:1169–1192, 2004.

Diwekar UM: *Introduction to applied optimization*, Dordrecht, the Netherlands, 2003, Kluwer Academic Publishers.

Edgar TF, Himmelblau DM: *Optimization of chemical processes*, ed 2, New York, 2001, McGraw Hill.

Eiselt HA, Sandblom CL: *Operations research: a model-based approach*, Berlin Heidelberg, 2010, Springer-Verlag.

El-Halwagi MM: *Pollution prevention through process integration: systematic design tools*, San Diego, 1997, Academic Press.

El-Halwagi MM: *Process integration*, Amsterdam, 2006, Elsevier.

Floudas CA, Gounaris CEA: A review of recent advances in global optimization, *J Global Optim* 45:3–38, 2009.

Floudas CA, Pardalos PM: *Encyclopedia of global optimization*, Dordrecht, the Netherlands, 2001, Kluwer Academic Publishers.

Grossmann IE, Biegler LT: Future perspective on optimization, Part II, *Comp Chem Eng* 28:1193–1218, 2004.

Grossmann IE, Lee S: Generalized convex disjunctive programming: nonlinear convex hull relaxation, *Comput Optim Appl* 26(1):83–100, 2003.

Haverly CA: Studies of the behaviour of recursion for the pooling problem, *ACM SIGMAP Bull* 25:29–32, 1978.

Hendrix EMT, Tóth BG: *Introduction to nonlinear global optimization*, New York, 2010, Springer.

Horst R, Tuy H: *Global optimization: deterministic approaches*, ed 3, Berlin, 2010, Springer-Verlag.

Noureldin MB, El-Halwagi MM: Pollution-prevention targets through integrated design and operation, *Comp Chem Eng* 24:1445–1453, 2000.

Pham V, Laird C, El-Halwagi MM: Convex hull discretization approach to the global optimization of pooling problems, *Ind Eng Chem Res* 48(4):1973–1979, 2009.

Ravindran A, Ragsdell KM, Reklaitis GV: *Engineering optimization: methods and application*, ed 2, New York, 2006, Wiley.

Schrage L: *Optimization modeling with LINGO*, ed 6, Chicago, 2006, LINDO Systems.

Shelley MD, El-Halwagi MM: Component-less design of recovery and allocation systems: a functionality-based clustering approach, *Comput Chem Eng* 24:2081–2091, 2000.

Tawarmalani M, Sahinidis NV: *Convexification and global optimization in continuous and mixed integer nonlinear programming*, Dordrecht, the Netherlands, 2003, Kluwer Academic Publishers.

Türkay M, Grossmann IE: Disjunctive programming techniques for the optimization of process systems with discontinuous investment costs-multiple size regions, *Ind Eng Chem Res* 35:2611–2623, 1996.

CHAPTER 15
AN OPTIMIZATION APPROACH TO DIRECT RECYCLE

Chapters 4 and 10 provided graphical and algebraic approaches to the targeting and synthesis of direct-recycle networks. Graphical tools may become cumbersome for problems with numerous sources or sinks and for problems with a wide range of flow rate or load scale. Additionally, algebraic techniques cannot easily handle problems with multiple fresh resources. Therefore, it is beneficial to develop a mathematical-programming approach to the targeting of material recycle. Such an approach can also be integrated with other optimization techniques. This chapter provides an optimization-based formulation for the mathematical solution of the direct-recycle problem. First, the problem addressed in Chapters 4 and 10 will be restated. Then, a structural representation of the solution alternatives will be presented. Next, the mathematical-programming formulation will be discussed along with its solution applied to a case study.

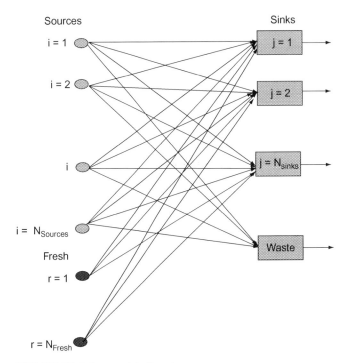

FIGURE 15-1 Source-sink allocation.

PROBLEM STATEMENT

The direct-recycle problem can be expressed as follows:
Given a process with:

- A set of process sinks (units): SINKS = {j = 1, 2,..., N_{Sinks}}. Each sink requires a feed with a given flow rate, G_j, and a composition, z_j^{in}, that satisfies the following constraint:

  [15.1] $\quad z_j^{min} \leq z_j^{in} \leq z_j^{max} \quad$ where $j = 1, 2, ..., N_{Sinks}$

  where z_j^{min} and z_j^{max} are given lower and upper bounds on admissible compositions to unit j.

- A set of process sources: SOURCES = {i = 1, 2,..., $N_{Sources}$} which can be recycled/reused in process sinks. Each source has a given flow rate, W_i, and a given composition, y_i.
- Available for service is a set of fresh (external) resources: FRESH = {r = 1, 2,..., N_{Fresh}}, which can be purchased to supplement the use of process sources in sinks. The cost of the rth fresh resource is C_r ($/kg), and its composition is x_r. The flow rate of each fresh resource (F_r) is to be determined so as to minimize the total cost of the fresh resources.

PROBLEM REPRESENTATION

The first step in the analysis is to represent the problem through a source-sink representation, shown in Fig. 15.1. Each source is split into fractions (of unknown flow rate) that are allocated to the various sink. An additional sink is placed to account for unrecycled/unreused material. This sink is referred to as the "waste" sink. The fresh sources are also allowed to split and are allocated to the sinks. Clearly, there is no need to allocate a portion of a fresh resource to the waste sink.

OPTIMIZATION FORMULATION

The objective is to minimize the cost of the fresh resources; that is,

[15.2a] $\quad\quad$ Minimize $\sum_{r=1}^{N_{Fresh}} C_r * F_r$

If the intent is to minimize the flow rate of the fresh, then the objective function can be expressed as:

[15.2b] $\quad\quad$ Minimize $\sum_{r=1}^{N_{Fresh}} F_r$

Each source, i, is split into N_{Sinks} fractions that can be assigned to the various sinks (Fig. 15.2). The flow rate of each split is denoted by $w_{i,j}$. Additionally, one split is forwarded to the waste sink. The stream accounts for the unrecycled flow rate and is denoted by $w_{i,waste}$. Similarly, each fresh source is split into N_{Sinks} fractions that can be

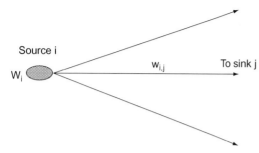

FIGURE 15-2 Splitting of sources.

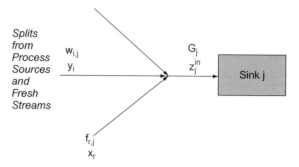

FIGURE 15-3 Mixing of split fractions and assignment to sinks.

assigned to the various sinks. The flow rate of each fresh split is denoted by $f_{r,j}$.

Therefore, the splitting constraint can be written as:

$$[15.3] \quad W_i = \sum_{j=1}^{N_{Sinks}} w_{i,j} + w_{i,waste} \quad i = 1, 2, \ldots, N_{Sources}$$

A similar constraint can be written for the splitting of the rth fresh resource:

$$[15.4] \quad F_r = \sum_{j=1}^{N_{Sinks}} f_{r,j} \quad r = 1, 2, \ldots, N_{Fresh}$$

$$[15.5] \quad Waste = \sum_{i=1}^{N_{Sources}} w_{i,waste}$$

Next, we examine the opportunities for mixing these splits and assigning them to sinks. Figure 15.3 shows the mixing of the split fractions into a feed to the jth sink. The split fractions come from the process sources and the fresh streams.

Mixing for the jth sink:

$$[15.6] \quad G_j = \sum_{i=1}^{N_{Sources}} w_{i,j} + \sum_{r=1}^{N_{Fresh}} f_{r,j} \quad \text{where } j = 1, 2, \ldots, N_{Sinks}$$

$$[15.7] \quad G_j * z_j^{in} = \sum_{i=1}^{N_{Sources}} w_{i,j} * y_i + \sum_{r=1}^{N_{Fresh}} f_{r,j} * x_r$$
where $j = 1, 2, \ldots, N_{Sinks}$

where

$$[15.1] \quad z_j^{min} \leq z_j^{in} \leq z_j^{max} \quad \text{where } j = 1, 2, \ldots, N_{Sinks}$$

It is worth pointing out that, in cases of multiple components, the same formulation can be used by writing Constraints 15.1 and 15.7 for all components.

To ensure the nonnegativity of the fresh flows and the fractions of source allocated to a sink and for flow of fresh resources, the following constraints are written:

$$[15.8] \quad f_{r,j} \geq 0 \quad \text{where } r = 1, 2, \ldots, N_{Fresh} \text{ and } j = 1, 2, \ldots, N_{Sinks}$$

$$[15.9] \quad w_{i,j} \geq 0 \quad \text{where } i = 1, 2, \ldots, N_{Fresh} \text{ and } j = 1, 2, \ldots, N_{Sinks}$$

The foregoing formulation is a linear program that can be solved globally to identify the minimum cost (or flow rate) of the fresh streams, an optimum assignment of process sources to sinks, and the discharged waste.

The results are consistent with those obtained graphically using the material recycle pinch diagram (shown by

---

Example 15-1 Water recycle in a formic-acid plant

Consider the process of manufacturing formic acid (H–C(=O)–OH) described by Example 15.1 and shown by Fig. 15.4. Biomass is gasified to produce a syngas mixture. A solid-gas separator is used to remove the particulates referred to as the fly ash. To avoid the problem of solid-waste disposal of fly ash while adding value, water is mixed with the fly ash to yield a paste that can be used in the manufacture of construction materials (for example, in making bricks or providing a filler for asphalt). The hot gases leaving the solid-gas separator are to be separated to different streams, including a carbon monoxide stream that is used as a reactant in making formic acid. Water quenching is used to cool the hot gases, leaving the solid-gas separator. The carbon monoxide stream leaving the gas-separation system is reacted with methanol to give methyl formate

(H–C(=O)–O–CH$_3$) through the following carbonylation reaction over a strong-base catalyst:

$$[15.10a] \quad CH_3OH + CO \rightarrow HCOOCH_3$$

Methyl formate is recovered using a distillation column. Next, water is used to hydrolyze methyl formate to formic acid and methanol as follows:

$$[15.10b] \quad HCOOCH_3 + H_2O \rightarrow HCOOH + CH_3OH$$

(Continued)

Example 15-1 Water recycle in a formic-acid plant (Continued)

A separation network is used to recover formic acid and to produce a stream of distilled water and another stream of organics.

There are three process sinks that use fresh water: the fly ash stabilizer, the quenching column, and the hydrolysis reactor. At present, a total of 9500 kg/hr of fresh water is used in quenching and hydrolysis. Two process streams (sources) are considered for recycle: the quenching effluent and distilled water. A total of 7000 kg/hr of wastewater is discharged as quenching effluent and distilled water. It is desired to use direct recycle to reduce fresh-water usage and wastewater discharge. The following constraints define the acceptable ranges for flow rates and concentrations fed to the two process sinks.

Fly Ash Stabilizer:

[15.11] $\quad 500 \leq$ Flow rate of stabilizing water fed to the fly ash stabilizer (kg/hr) ≤ 600

[15.12] $\quad 0.0 \leq$ Concentration of the impurities (wt%) ≤ 10.0

Quenching Column:

[15.13] $\quad 3000 \leq$ Flow rate of the quenching water fed to the column (kg/hr) ≤ 3500

[15.14] $\quad 0.0 \leq$ Concentration of the impurities (wt%) ≤ 3.0

Hydrolysis Reactor:

[15.15] $\quad 6000 \leq$ Flow rate of hydrolysis water fed to the reactor (kg/hr) ≤ 6500

[15.16] $\quad 0.0 \leq$ Concentration of the impurities (wt%) ≤ 1.0

Table 15.1 summarizes the data for the process sinks. As mentioned earlier in the procedure for constructing the material-recycle pinch diagram, the sinks are listed in ascending order of the maximum allowable inlet mass fraction of impurities. Furthermore, because the objective is to minimize the fresh usage, the lower bound of the acceptable flow rate of each sink is selected as the required flow rate. Table 15.2 provides the data for the process sources arranged in ascending order of concentration of impurities.

SOLUTION

The objective here is to minimize the usage of fresh water. Let us designate the flow rate of acetic acid as Fresh. Therefore the objective function can be expressed as:

TABLE 15.1 Sink Data for the Formic Acid Example

Sink	Flow Rate, kg/hr	Maximum Inlet Mass Fraction of Impurities	Maximum Inlet Load of Impurities, kg/hr
Hydrolysis Reactor	6000	0.0100	60
Quenching Column	3000	0.0300	90
Fly Ash Stabilizer	500	0.1000	50

TABLE 15.2 Source Data for the Formic Acid Example

Source	Flow Rate, kg/hr	Mass Fraction of Impurities	Load of Impurities, kg/hr
Distilled Water	4000	0.0175	70
Quenching Effluent	3000	0.0600	180

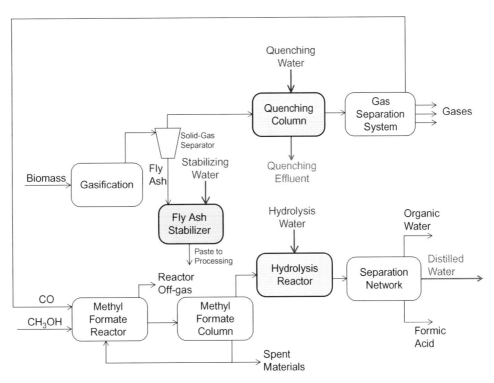

FIGURE 15.4 A block flow diagram of the formic-acid process.

(*Continued*)

Example 15-1 Water recycle in a formic-acid plant (*Continued*)

[15.17] Objective: Minimize Fresh

The flow rate of the fresh is the sum of the flow rates of the fresh used in the three process sinks (hydrolysis reactor, quenching column, and fly ash stabilizer), which are referred to as Fresh_Hydrolysis, Fresh_Quenching, Fresh_Stabilizer, respectively. Therefore,

[15.18] Fresh = Fresh_Hydrolysis + Fresh_Quenching + Fresh_Stabilizer

When considering direct recycle, the problem has two process sources (distilled water and quenching effluent). Each of the process sources is split and fed to the three process sinks. Additionally, unrecycled process sources are discharged to the waste treatment system. Referring to the flow rate of the distilled water as F_Distilled and to the distilled water flow rates fed to the three process sinks and wastewater treatment as: F_Distilled_to_Hydrolysis, F_Distilled_to_Quenching, F_Distilled_to_Stabilizer, and F_Distilled_Discharged. Therefore, the source-splitting constraint for distilled water can be written as:

[15.19]
F_Distilled = F_Distilled_to_Hydrolysis + F_Distilled_to_Quenching + F_Distilled_to_Stabilizer + F_Distilled_Discharged

Similarly, the splitting constraint for the effluent discharge source is given by:

[15.20]
F_Effluent = F_Effluent_to_Quenching + F_Effluent_to_Hydrolysis + F_Effluent_to_Stabilizer + F_Effluent_Discharged

Next, we represent the mixing of the split sources and assign them to sinks. The following are the flow and component balance constraints for the water and the pollutant entering each sink:

Hydrolysis Reactor:

[15.21] F_Distilled_to_Hydrolysis + F_Effluent_to_Hydrolysis + Fresh_Hydrolysis = 6000

[15.22] F_Distilled_to_Hydrolysis*0.0175 + F_Effluent_to_Hydrolysis*0.06 <= 6000*0.01

Quenching Column:

[15.23] F_Distilled_to_Quenching + F_Effluent_to_Quenching + Fresh_Quenching = 3000

[15.24] F_Distilled_to_Quenching*0.0175 + F_Effluent_to_Quenching*0.06 <= 3000*0.03

Fly Ash Stabilizer:

[15.25] F_Distilled_to_Stabilizer + F_Effluent_to_Stabilizer + Fresh_Stabilizer = 500

[15.26] F_Distilled_to_Stabilizer*0.0175 + F_Effluent_to_Stabilizer*0.06 <= 500*0.1

Similarly, the total flow rate going to waste treatment is given by:

[15.27] Waste = F_Distilled_Discharged + F_Effluent_Discharged

Additional constraints are also needed to ensure the nonnegativity of the variables and the proper bounds on the feed to the sinks. The following is a LINGO program of this formulation:

[P15.1]
```
Min=Fresh;
Fresh=Fresh_Hydrolysis+Fresh_Quenching+Fresh_
    Stabilizer;
F_Distilled=4000;
F_Distilled=F_Distilled_to_Hydrolysis+F_Distilled_
    to_Quenching+F_Distilled_to_Stabilizer+F_
    Distilled_Discharged;
F_Effluent=3000;
F_Effluent=F_Effluent_to_Quenching+F_Effluent_to_
    Hydrolysis+F_Effluent_to_Stabilizer+F_Effluent_
    Discharged;
F_Distilled_to_Hydrolysis+F_Effluent_to_
    Hydrolysis+Fresh_Hydrolysis=6000;
F_Distilled_to_Hydrolysis*0.0175+F_Effluent_to_
    Hydrolysis*0.06<=6000*0.01;
F_Distilled_to_Quenching+F_Effluent_to_
    Quenching+Fresh_Quenching=3000;
F_Distilled_to_Quenching*0.0175+F_Effluent_to_
    Quenching*0.06<=3000*0.03;
F_Distilled_to_Stabilizer+F_Effluent_to_
    Stabilizer+Fresh_Stabilizer=500;
F_Distilled_to_Stabilizer*0.0175+F_Effluent_to_
    Stabilizer*0.06<=500*0.1;
Waste=F_Distilled_Discharged+F_Effluent_Discharged;
```

Solving using LINGO, we get the following solution:

```
Objective value: 3666.667
Variable                        Value
FRESH                           3666.667
FRESH_HYDROLYSIS                2571.429
FRESH_QUENCHING                 1095.238
FRESH_STABILIZER                0.000000
F_DISTILLED                     4000.000
F_DISTILLED_TO_HYDROLYSIS       3428.571
F_DISTILLED_TO_QUENCHING        571.4286
F_DISTILLED_TO_STABILIZER       0.000000
F_DISTILLED_DISCHARGED          0.000000
F_EFFLUENT                      3000.000
F_EFFLUENT_TO_QUENCHING         1333.333
F_EFFLUENT_TO_HYDROLYSIS        0.000000
F_EFFLUENT_TO_STABILIZER        500.0000
F_EFFLUENT_DISCHARGED           1166.667
WASTE                           1166.667
```

Therefore, the targets are found to be:

[15.28a] Minimum fresh-water usage = 3667 kg/hr

[15.28b] Minimum wastewater discharge = 1167 kg/hr

These results are consistent with the ones determined graphically in Chapter 4 and algebraically in Chapter 10. Figure 15.5a shows the implementation of this solution.

As mentioned in Chapter 4, this is not the only possible implementation of the targets. Let us examine another arrangement. Suppose that we assign all of the fresh targets (3667 kg/hr) to the hydrolysis reactor. Therefore, we add the constraint:

[15.29] Fresh_Hydrolysis = 3666.667;

Then, we solve the program using LINGO to get:

```
Objective value: 3666.667
Variable          Value
FRESH             3666.667
FRESH_HYDROLYSIS  3666.667
```

(*Continued*)

Example 15-1 Water recycle in a formic-acid plant (*Continued*)

```
FRESH_QUENCHING            0.000000
FRESH_STABILIZER           0.000000
F_DISTILLED             4000.000
F_DISTILLED_TO_HYDROLYSIS  1882.352
F_DISTILLED_TO_QUENCHING   2117.647
F_DISTILLED_TO_STABILIZER     0.000000
F_DISTILLED_DISCHARGED        0.000000
F_EFFLUENT              3000.000
F_EFFLUENT_TO_QUENCHING    882.3529

F_EFFLUENT_TO_HYDROLYSIS   450.9805
F_EFFLUENT_TO_STABILIZER   499.9995
F_EFFLUENT_DISCHARGED     1166.667
WASTE                     1166.667
```

Figure 15.5b shows this solution.

Other solutions can be generated by assigning various values of a split assigned to a sink and resolving the revised optimization program.

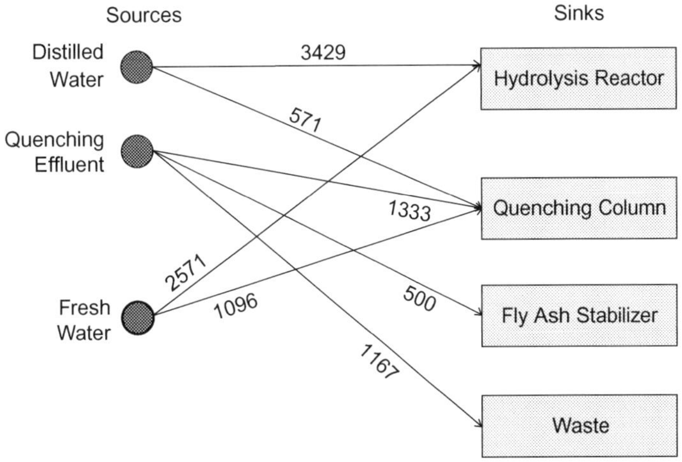

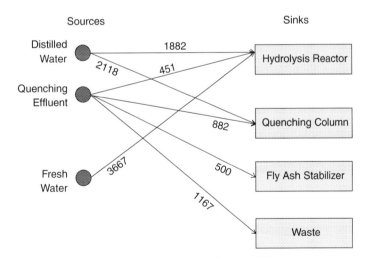

FIGURE 15.5a An implementation of the targets for the formic-acid process (all numbers represent assigned flow rate from source to sink in kg/hr).

FIGURE 15-5b An alternate implementation of the targets for the formic-acid process (all numbers represent assigned flow rate from source to sink in kg/hr).

Example 15-2 Using set formulation to solve the water–recycle problem in the formic-acid plant

The objective of this example is to solve the problem of Example 15.1 using the LINGO set formulation described in Chapter 14. The following sets are first defined:

FLOW_CONNECTIONS (SOURCE, SINKS): Set of all connections from sources to sinks
SINKS: Set of process sinks
SOURCE: Set of process sources

with the following indices for the sources and the sink:
i: Index for sources
j: Index for sinks

Next, we define the following symbols:
$F(j)$: Flow rate of fresh water entering sink j
Fresh: Flow rate of the fresh water used in all the sinks
$G(j)$: Flow rate entering sink j

(*Continued*)

Example 15-2 Using set formulation to solve the water–recycle problem in the formic-acid plant (*Continued*)

Split(I,j): Flow rate split from source I to sink j
W(i): Flow rate of a source i
Waste(i): Flow rate of waste from source i
Waste_Total: Total flow rate of the waste from all sources
Xf: Concentration of the impurity in the fresh feed
Y(i): Concentration of the impurity in source i
Z(j): Concentration of the impurity entering sink j
Zmax(j): Maximum allowable concentration of impurity entering sink j

The following is the LINGO set formulation:

```
[P15.2]
min=Fresh;
SETS:
SOURCES /1..2/: W, Y, Waste;
SINKS /1..3/: G, Zmax, Z, F;
FLOW_CONNECTIONS (SOURCES, SINKS): Split;
ENDSETS

DATA:
W=4000 3000;
Y=0.0175 0.0600;
G=6000 3000 500;
ZMAX=0.01 0.03 0.10;
ENDDATA

Fresh=@SUM(SINKS(j): F(j));
XF=0;
@FOR (SOURCES(i):
  @SUM(SINKS(j): Split(i,j))1Waste(i)=W(i));
@FOR (SINKS(j):
  @SUM(SOURCES(i): Split(i,j))1F(j)=G(j));
@FOR (SINKS(j):
  @SUM( SOURCES(i): Split(i,j)*Y(i))1F(j)*Xf=G(j)*
  Z(j));
@FOR (SINKS(j): Z(j)<=Zmax(j));
Waste_Total=@SUM(SOURCES(i): Waste(i));
```

To see the model in detail, one can use the LINGO menu bar and go to "LINGO," then "Generate," and then "Display Model" to get:

```
[P15.3]
MODEL:
[_1] MIN= FRESH;
[_2] FRESH-F_1-F_2-F_3=0;
[_4] SPLIT_1_1+SPLIT_1_2+SPLIT_1_3+WASTE_1=4000;
[_5] SPLIT_2_1+SPLIT_2_2+SPLIT_2_3+WASTE_2=3000;
[_6] SPLIT_1_1+SPLIT_2_1+F_1=6000;
[_7] SPLIT_1_2+SPLIT_2_2+F_2=3000;
[_8] SPLIT_1_3+SPLIT_2_3+F_3=500;
[_9] 0.0175*SPLIT_1_110.06*SPLIT_2_1-6000*Z_1=0;
[_10] 0.0175*SPLIT_1_210.06*SPLIT_2_2-3000*Z_2=0;
[_11] 0.0175*SPLIT_1_310.06*SPLIT_2_3-500*Z_3=0;
[_12] Z_1<=0.01;
[_13] Z_2<= 0.03;
[_14] Z_3<= 0.1;
[_15] WASTE_TOTAL - WASTE_1 - WASTE_2=0;
END
```

The solution is obtained as follows:

```
Objective value: 3666.667
Variable            Value
FRESH               3666.667
XF                  0.000000
WASTE_TOTAL         1166.667
W(1)                4000.000
W(2)                3000.000
Y(1)                0.1750000E-01
Y(2)                0.6000000E-01
WASTE(1)            0.000000
WASTE(2)            1166.667
G(1)                6000.000
G(2)                3000.000
G(3)                500.0000
ZMAX(1)             0.1000000E-01
ZMAX(2)             0.3000000E-01
ZMAX(3)             0.1000000
Z(1)                0.1000000E-01
Z(2)                0.3000000E-01
Z(3)                0.6000000E-01
F(1)                3666.667
F(2)                0.000000
F(3)                0.000000
SPLIT(1, 1)         1882.353
SPLIT(1, 2)         2117.647
SPLIT(1, 3)         0.000000
SPLIT(2, 1)         450.9804
SPLIT(2, 2)         882.3529
SPLIT(2, 3)         500.0000
```

This is the same solution shown by Fig. 15.5b.

Example 15-3 Vinyl acetate case study

The block flow diagram of a vinyl acetate monomer (VAM) plant is shown by Fig. 15.6 (El-Halwagi, 2006). The primary chemical reaction for VAM production is:

C_2H_4 + $0.5 O_2$ + $C_2H_4O_2$ = $C_4H_6O_2$ + H_2O
(ethylene) (oxygen) (acetic acid) (VAM) (water)

Consider the process shown in the next figure. In an acid tower, 10,000 kg/hr of acetic acid (AA) along with 200 kg/hr of water are evaporated. The vapor is fed with oxygen and ethylene to the reactor where 7000 kg/hr of acetic acid are reacted and 10,000 kg/hr of VAM are formed. The reactor off-gas is cooled and fed to the first absorber, where AA (5100 kg/hr) is used as a solvent. Almost all the gases leave from the top of the first absorption column together with 1200 kg/hr of AA. This stream is fed to the second absorption column where water (200 kg/hr) is used to scrub acetic acid. The bottom product of the first absorption column is fed to the primary distillation tower where VAM is recovered as a top product (10,000 kg/hr) along with small amount of AA that is not worth recovering (100 kg/hr) and water (200 kg/hr). This stream is sent to final finishing. The bottom product of the primary tower (6800 kg/hr of AA and 2300 kg/hr of water) is mixed with the bottom product of the second absorption column (1200 kg/hr of AA and 200 kg/hr of water). The mixed waste is fed to a neutralization system followed by biotreatment.

The following technical constraints should be observed in any proposed solution:

Neutralization:

[15.30] $0 \leq$ Flow rate of feed to neutralization (kg/hr) $\leq 11{,}000$

(*Continued*)

Example 15-3 Vinyl acetate case study (*Continued*)

[15.31] $0 \leq$ AA in feed to neutralization (wt%) $\leq 85\%$

Acid Tower:

[15.32] $10{,}200 \leq$ Flow rate of feed to acid tower (kg/hr) $\leq 11{,}200$

[15.33] $0.0 \leq$ Water in feed to acid tower (wt%) ≤ 10.0

First Absorber:

[15.34] $5100 \leq$ Flow rate of feed to absorber I (kg/hr) ≤ 6000

[15.35] $0.0 \leq$ Water in feed to absorber I (wt%) ≤ 5.0

It can be assumed that the process performance will not significantly change as a result of direct-recycle activities.

What is the target for minimum usage (kg/hr) of fresh acetic acid in the process if segregation, mixing, and direct recycle are used? Can you find more than one strategy to reach the target?

SOLUTION

The objective here is to minimize the usage of fresh acetic acid. Let us designate the flow rate of acetic acid as FreshAA. Therefore the objective function can be expressed as:

[15.36] Objective: Minimize FreshAA

When considering direct recycle, the problem has two process sources (bottoms of absorber II and bottoms of primary tower) and two process sinks (absorber I and acid tower). Additionally, unrecycled process sources are fed to the waste treatment system. Referring to the flow rate of absorber II as WAbs2, the flow rate of the primary tower as WPrimary, and the flow rate from source i to sink j as wij, the source-splitting constraints can be written as:

[15.37] $WAbs2 = w11 + w12 + Waste1$

[15.38] $WPrimary = w21 + w22 + Waste2$

where Waste1 and Waste2 represent the unrecycled flow rates of sources 1 and 2. Similarly, the fresh flow rate is split between sinks 1 and 2. Hence,

[15.39] $FreshAA = FreshAA1 + FreshAA2$

where FreshAA1 and FreshAA2 are the flow rates of the fresh acetic acid used in sinks 1 and 2, respectively.

Next, we represent the mixing of the split sources and assign them to sinks. Let the flow rate entering absorber I and the acid tower be referred to as GAbs1 and GAcid, and the inlet composition of acetic acid be referred to as zAbs1 and zAcid. Therefore, the overall and component material balances at the inlets of the two process sinks are written as follows:

[15.40] $GAbs1 = w11 + w21 + FreshAA1$

[15.41] $GAcid = w12 + w22 + FreshAA2$

[15.42] $GAbs1*zAbs1 = w11*0.14 + w21*0.25$

[15.43] $GAcid*zAcid = w12*0.14 + w22*0.25$

Similarly, the total flow rate going to waste treatment is given by:

[15.44] $Waste = Waste1 + Waste2$

Additional constraints are also needed to ensure the nonnegativity of the variables and the proper bounds on the feed to the sinks. The following is a LINGO program of the aforementioned formulation:

```
min=FreshAA;
! Splitting of process sources and fresh;
WAbs2=w11+w12+Waste1;
WPrimary=w21+w22+Waste2;
FreshAA=FreshAA1+FreshAA2;
! Mixing at the inlet of the sinks;
GAbs1=w11+w2+1FreshAA1;
GAcid=w12+w22+FreshAA2;
GAbs1*zAbs1=w11*0.141w21*0.25;
GAcid*zAcid=w12*0.141w22*0.25;
Waste=Waste11Waste2;
! Problem data and nonnegativity constraints;
WAbs2=1400;
```

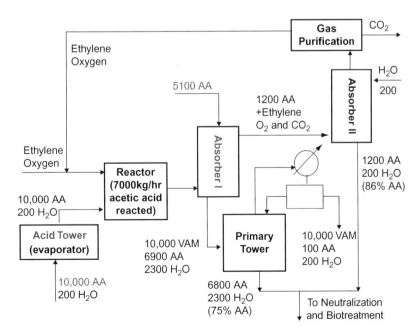

FIGURE 15-6 Schematic flow sheet of VAM process (all numbers are in kg/hr). *Source: El-Halwagi (2006).*

(*Continued*)

Example 15-3 Vinyl acetate case study (*Continued*)

```
WPrimary=9100;
GAbs1=5100;
GAcid=10200;
zAbs1>=0.00;
zAbs1<=0.05;
zAcid>=0.00;
zAcid<=0.10;
w11>= 0;
W21>=0;
W12>=0;
W22>=0;
FreshAA1>=0;
FreshAA2>=0;
End
```

Solving using LINGO, we get the following solution:

```
Objective value: 9584.000
Variable    Value
FRESHAA     9584.000
W11         0.000000
W12         1400.000
WASTE1      0.000000
WABS2       1400.000
W21         1020.000
W22         3296.000
WASTE2      4784.000
WPRIMARY    9100.000
FRESHAA1    4080.000
FRESHAA2    5504.000
GABS1       5100.000
GACID       10200.00
ZABS1       0.0500000
ZACID       0.1000000
WASTE       4784.000
```

Therefore, the target for minimum fresh AA is 9584 kg/hr, and the target of minimum waste discharge is 4784 kg/hr. The implementation of this solution is shown in Fig. 15.7a.

As mentioned earlier, it is possible in some cases to have multiple (sometimes infinite) direct-recycle strategies that can be implemented to attain the same target. Mathematical programming can be used to generate such solutions. For instance, let us examine the case when all of the source from the second absorber tower is assigned to the first absorber. This can be accomplished by adding the following constraint to the previous optimization formulation:

[15.45] $W11 = 1400;$

Solving the revised program, we get:

```
Objective value: 9584.000
Variable    Value
FRESHAA     9584.000
W11         1400.000
W12         0.000000
WASTE1      0.000000
WABS2       1400.000
W21         236.0000
W22         4080.000
WASTE2      4784.000
WPRIMARY    9100.000
FRESHAA1    3464.000
FRESHAA2    6120.000
GABS1       5100.000
GACID       10200.00
ZABS1       0.0500000
ZACID       0.1000000
WASTE       4784.000
```

Figure 15.7b shows this solution.

An additional solution can be determined by assigning some percentage (for example, 50 percent) of the first source to the first sink. This can be realized by adding the following constraint to the original optimization formulation:

[15.46] $W11 = 0.5*1400;$

Solving the revised program, we get:

```
Objective value: 9584.000
Variable    Value
FRESHAA     9584.000
W11         700.0000
W12         700.0000
WASTE1      0.000000
WABS2       1400.000
W21         628.0000
W22         3688.000
WASTE2      4784.000
WPRIMARY    9100.000
FRESHAA1    3772.000
FRESHAA2    5812.000
```

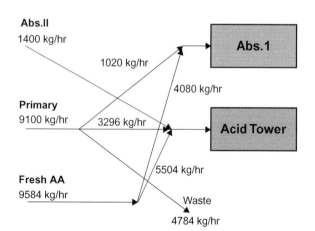

FIGURE 15-7a Direct-recycle configuration when all of source from the second absorber tower is fed to acid tower. *Source: El-Halwagi (2006).*

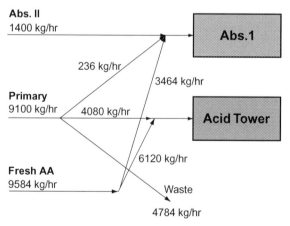

FIGURE 15-7b Direct-recycle configuration when all of source from the second absorber tower is fed to the first absorber. *Source: El-Halwagi (2006).*

(*Continued*)

Example 15-3 Vinyl acetate case study (Continued)

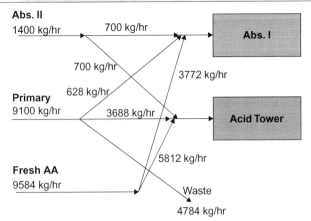

```
GABS1      5100.000
GACID     10200.00
ZABS1      0.5000000E-01
ZACID      0.1000000
WASTE      4784.000
```

This solution is shown by Fig. 15.7c.
Other solutions can be generated by assigning various values of a split assigned to a sink and resolving the revised optimization program.

FIGURE 15-7c Direct-recycle configuration when half of source second absorber tower is fed to the first absorber.

Example 15-4 Dealing with impurities in the fresh feed: hydrogen recycle in a refinery

Let us revisit Example 4.3, which deals with the recycle of hydrogen in a petroleum refinery based on the case study reported by Alves and Towler (2002). The refinery has four sinks that use fresh hydrogen and six sources that contain hydrogen and may be recycled to reduce the usage of fresh hydrogen. Tables 15.3 and 15.4 summarize the data for the process sinks and sources.

The available fresh hydrogen contains 0.05 mole fraction of impurities. Determine the direct-recycle targets for minimum fresh usage of hydrogen and waste discharge.

SOLUTION

A key advantage of the set formulation is its reusability. By revising the data in Program P15.2, based on Tables 15.3 and 15.4, we get:

```
[P15.5]
min=Fresh;

SETS:
SOURCES /1..6/: W, Y, Waste;
SINKS /1..4/: G, Zmax, Z, F;
FLOW_CONNECTIONS (SOURCES, SINKS): Split;
ENDSETS

DATA:
W=623.8 415.8 1801.9 138.6 346.5 457.4;
Y=0.07 0.20 0.25 0.25 0.27 0.30;
G=2495 180.2 554 720.7;
ZMAX=0.1939 0.2115 0.2243 0.2486;
ENDDATA
```

```
Fresh=@SUM(SINKS(j): F(j));
XF=0.05;
@FOR (SOURCES(i):
  @SUM(SINKS(j): Split(I,j))+Waste(i)=W(i));
@FOR (SINKS(j):
  @SUM(SOURCES(i): Split(I,j))+F(j)=G(j));
@FOR (SINKS(j):
  @SUM(SOURCES(i): Split(I,j)*Y(i))+F(j)*Xf=G(j)*Z
  (j));
@FOR (SINKS(j): Z(j)<=Zmax(j));
Waste_Total=@SUM(SOURCES(i): Waste(i));
```

Using LINGO, we get the following solution:

```
Objective value: 268.6999
Variable      Value
FRESH         268.6999
XF            0.5000000E-01
WASTE_TOTAL   102.7999
W(1)          623.8000
W(2)          415.8000
W(3)          1801.900
W(4)          138.6000
W(5)          346.5000
W(6)          457.4000
```

TABLE 15.3 Sink Data for Hydrogen Problem

Sinks	Flow (mol/s)	Maximum Inlet Impurity Concentration (mol %)	Load (mol/s)
1	2495	19.39	483.8
2	180.2	21.15	38.1
3	554.4	22.43	124.4
4	720.7	24.86	179.2

Source: Alves and Towler (2002).

TABLE 15.4 Source Data for the Hydrogen Problem

Sources	Flow (mol/s)	Impurity Concentration (mol %)	Load (mol/s)
1	623.8	7	43.7
2	415.8	20	83.2
3	1801.9	25	450.5
4	138.6	25	34.7
5	346.5	27	93.6
6	457.4	30	137.2

Source: Alves and Towler (2002).

(Continued)

Example 15-4 Dealing with impurities in the fresh feed: hydrogen recycle in a refinery (*Continued*)

Y(1)	0.7000000E-01	F(3)	0.000000
Y(2)	0.2000000	F(4)	0.000000
Y(3)	0.2500000	SPLIT(1, 1)	402.0956
Y(4)	0.2500000	SPLIT(1, 2)	0.000000
Y(5)	0.2700000	SPLIT(1, 3)	137.1426
Y(6)	0.3000000	SPLIT(1, 4)	84.56175
WASTE(1)	0.000000	SPLIT(2, 1)	415.8000
WASTE(2)	0.000000	SPLIT(2, 2)	0.000000
WASTE(3)	0.000000	SPLIT(2, 3)	0.000000
WASTE(4)	0.000000	SPLIT(2, 4)	0.000000
WASTE(5)	0.000000	SPLIT(3, 1)	1304.493
WASTE(6)	102.7999	SPLIT(3, 2)	145.5115
G(1)	2495.000	SPLIT(3, 3)	0.000000
G(2)	180.2000	SPLIT(3, 4)	351.8956
G(3)	554.0000	SPLIT(4, 1)	138.6000
G(4)	720.7000	SPLIT(4, 2)	0.000000
ZMAX(1)	0.1939000	SPLIT(4, 3)	0.000000
ZMAX(2)	0.2115000	SPLIT(4, 4)	0.000000
ZMAX(3)	0.2243000	SPLIT(5, 1)	0.000000
ZMAX(4)	0.2486000	SPLIT(5, 2)	0.000000
Z(1)	0.1939000	SPLIT(5, 3)	346.5000
Z(2)	0.2115000	SPLIT(5, 4)	0.000000
Z(3)	0.2243000	SPLIT(6, 1)	0.000000
Z(4)	0.2486000	SPLIT(6, 2)	0.000000
F(1)	234.0114	SPLIT(6, 3)	70.35739
F(2)	34.68850	SPLIT(6, 4)	284.2427

Fig. 4.32) and identify the targets of 269 and 103 mol/s for usage and discharge of hydrogen, respectively.

ADDITIONAL READINGS

Much research has been carried out in the area of optimizing direct-recycle networks for purposes of mass conservation, especially water management. The following is a brief discussion of general categories of research and sample references. There are several review articles such as the ones by Jezowski (2010), Foo (2009), Dunn and El-Halwagi (2003), and Bagajewicz (2000). A number of algorithmic approaches and superstructure-based optimization have been developed to systematize and automate the targeting and synthesis of material recycle networks (for example, Foo, 2010, 2007; Faria and Bagajewicz, 2009; Bandyopadhyay, 2009; Liu et al., 2009; Liu et al., 2004; Bagajewicz and Savelski, 2001; Yang et al., 2000; Alva-Argaez et al., 1999; and Sorin and Bedard, 1999). Multiple contaminants have been handled systematically (for example, Rubio-Castro et al., 2011; Liu et al., 2009; Majozi and Gouws, 2009; and Doyle and Smith, 1997). In addition to direct-recycle strategies, interception devices have also been considered (for example, Mohammadnejad et al., 2011; Karthick et al., 2010; Nápoles-Rivera et al., 2010b; Gabriel and El-Halwagi, 2005; Noureldin and El-Halwagi, 1999, and El-Halwagi et al., 1996). Systems approaches to cooling-water systems have been developed for fresh water (Kim and Smith, 2001) and seawater (Bin Mahfouz et al., 2006). In addition to concentration-based constraints, thermal effects have been considered (Polley et al., 2010; Friedler, 2010; Kim et al., 2009; Bogataj and Bagajewicz, 2008; Savulescu et al., 2005), property-based constraints (Ng et al., 2010, 2008; Nápoles-Rivera et al., 2010a, b; Ponce-Ortega et al., 2010, 2009; Kazantzi and El-Halwagi, 2005; El-Halwagi et al., 2004; Shelley and El-Halwagi, 2000). Furthermore, material recycle networks have been designed with simultaneous consideration of mass, heat, and properties (Kheireddine et al., 2011). Techniques have been developed for batch systems (Majozi and Gouws, 2009; Gouws and Majozi, 2008; Ng et al., 2008; and Rabie and El-Halwagi, 2008). Global optimization approaches have been developed for the solution of the nonconvex material-recycle problems (Ahmetovic and Grossmann, 2011; Rubio-Castro et al., 2011, 2010; Nápoles-Rivera et al., 2010b; Ponce-Ortega et al., 2010; Karuppiah and Grossmann, 2006; Gabriel and El-Halwagi, 2005; Noureldin and El-Halwagi, 1999).

HOMEWORK PROBLEMS

15.1. Use an optimization formulation to solve Problem 4.1.
15.2. Use a linear-programming formulation to solve Problem 4.3.
15.3. Use mathematical programming to solve Problem 4.4.
15.4. Use LINGO's set formulation to solve Problem 4.5.
15.5. Figure 15.8 shows a schematic representation of the phenol production process from cumene hydroperoxide (CHP). More details on the process can be found in literature (Ponce-Ortega et al., 2009; Kheireddine et al., 2011). Phenol is chosen as the key pollutant due to its health and environmental hazards. In addition to constraints on the

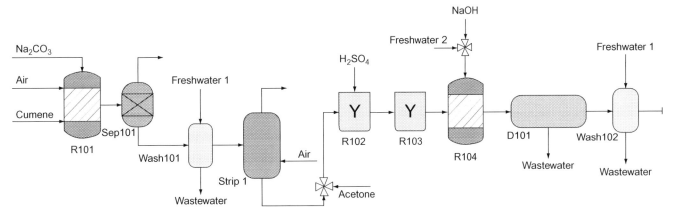

FIGURE 15-8 Cumene-to-phenol process. *Source: Ponce-Ortega et al. (2009); Kheireddine et al. (2011).*

concentration of phenol, there are also constraints on temperature, vapor pressure, and the pH of any stream to be recycled. The following mixing rules are used for the pH and the vapor pressure:

[15.47] $$10^{pH} = \sum_i x_i 10^{pH_i}$$

and

[15.48] $$p = \sum_i x_i p_i$$

where x_i is the fractional contribution of stream i.

Following is a list of sources, sinks, and available fresh water sources:

- *Process sinks*
  1. Waterwash cumene peroxidation section (Wash101)
  2. Neutralizer (R104)
  3. Waterwash cleavage section (Wash102)
- *Process sources*
  1. Stream 8 from Wash101
  2. Stream 22 from Decanter (D101)
  3. Stream 25 from Wash102
- *Fresh water sources*
  1. Freshwater1: 0.000 impurity concentration
  2. Freshwater2: 0.012 impurity concentration (mass fraction)

Tables 15.5 and 15.6 summarize the data for the sources and the sinks.

TABLE 15.5 Data for the Sources and Fresh Water of the Phenol Plant

Source	Flow Rate (kg/hr)	Impurity Concentration, z_j (Mass Fraction)	Temperature, T(°C)	Vapor Pressure (kPa)	Cost ($/tonne)
Washer101	3661	0.016	75	38	
Decanter101	1766	0.024	65	25	
Washer102	1485	0.220	40	7	
Freshwater1		0.000	25	3	1.32
Freshwater2		0.012	35	6	0.88

Source: Kheireddine et al. (2011).

TABLE 15.6 Data for the Sinks of the Phenol Plant

Sinks	Water Flow Rate (kg/hr)	Maximum Inlet Impurity Concentration (Mass Fraction) z_j^{max}	Minimum Temperature, T(°C)	Maximum Temperature, T(°C)	Minimum Vapor Pressure (kPa)	Maximum Vapor Pressure (kPa)
Wash101	2718	0.013	60	80	20	47
Wash102	1993	0.013	30	75	4	38
Neutralizer R104	1127	0.1	25	65	3	25

Source: Kheireddine et al. (2011).

Develop an optimization formulation and solve it to identify the optimal water recycle strategies for the phenol plant.

REFERENCES

Alves JJ, Towler GP: Analysis of refinery hydrogen distribution systems, *Ind Eng Chem Res* 41:5759–5769, 2002.

Ahmetovic E, Grossmann IE: Global superstructure optimization for the design of integrated process water networks, *AIChE J* 57:434–457, 2011.

Alva-Argaez A, Vallianatos A, Kokossis A: A multi-contaminant transhipment model for mass exchange networks and wastewater minimisation problems, *Comput Chem Eng* 23:1439–1453, 1999.

Bagajewicz M: A review of recent design procedures for water networks in refineries and process plants, *Comput Chem Eng* 24:2093–2113, 2000.

Bagajewicz M, Savelski M: On the use of linear models for the design of water utilization systems in process plants with a single contaminant, *Chem Eng Res Des* 79:600–610, 2001.

Bandyopadhyay S: Targeting minimum waste treatment flow rate, *Biochem Eng J* 152:367–375, 2009.

Bin Mahfouz AS, El-Halwagi MM, Abdel-Wahab A: Process integration techniques for optimizing seawater cooling systems and biocide discharge, *J Clean Tech & Env Policy* 8:203–215, 2006.

Bogataj M, Bagajewicz MJ: Synthesis of non-isothermal heat integrated water networks in chemical processes, *Comput Chem Eng* 32:3130–3142, 2008.

Doyle SJ, Smith R: Targeting water reuse with multiple contaminants, *Process Saf Environ Protect* 75:181–189, 1997.

Dunn RF, El-Halwagi MM: Process integration technology review: background and applications in the chemical process industry, *J Chem Technol Biotechnol* 78:1011–1021, 2003.

El-Halwagi MM: *Process integration*, Amsterdam, 2006, Academic Press/Elsevier.

El-Halwagi MM, Glasgow IM, Eden MR, Qin X: Property integration: componentless design techniques and visualization tools, *AIChE J* 50(8):1854–1869, 2004.

El-Halwagi MM, Hamad AA, Garrison GW: Synthesis of waste interception and allocation networks, *AIChE J* 42:3087–3101, 1996.

Faria DC, Bagajewicz MJ: Profit-based grassroots design and retrofit of water networks in process plants, *Comput Chem Eng* 33:436–453, 2009.

Foo DCY: Water cascade analysis for single and multiple impure fresh water feed, *Chem Eng Res Des* 85:1169–1177, 2007.

Foo DCY: State-of-the-art review of pinch analysis techniques for water network synthesis, *Ind Eng Chem Res* 48:5125–5159, 2009.

Foo DCY: Automated targeting technique for batch process integration, *Ind Eng Chem Res* 49:9899–9916, 2010.

Friedler F: Process integration, modelling and optimisation for energy saving and pollution reduction, *Appl Therm Eng* 30:2270–2280, 2010.

Gabriel FB, El-Halwagi MM: Simultaneous synthesis of waste interception and material reuse networks: Problem reformulation for global optimization, *Environ Prog* 24:171–180, 2005.

Gouws JF, Majozi T: A mathematical technique for the design of near-zero-effluent batch processes, *Water Sa* 34:291–295, 2008.

Jezowski J: Review of water network design methods with literature annotations, *Ind Eng Chem Res* 49:4475–4516, 2010.

Karthick R, Kumaraprasad G, Sruti B: Hybrid optimization approach for water allocation and mass exchange network, *Resour Conserv Recycl* 54:783–792, 2010.

Karuppiah R, Grossmann IE: Global optimization for the synthesis of integrated water systems in chemical processes, *Comput Chem Eng* 30:650–673, 2006.

Kazantzi V, El-Halwagi MM: Targeting material reuse via property integration, *Chem Eng Prog* 101:28–37, 2005.

Kheireddine H, Dadmohammadi Y, Deng C, Feng X, El-Halwagi MM: Optimization of direct recycle networks with the simultaneous consideration of property, mass, and thermal effects, *Ind Eng Chem Res* 50(7):3754–3762, 2011.

Kim J, Yoo C, Moon I: A simultaneous optimization approach for the design of wastewater and heat exchange networks based on cost estimation, *J Clean Prod* 17:162–171, 2009.

Kim JK, Smith R: Cooling water system design, *Chem Eng Sci* 56:3641–3658, 2001.

Liu YA, Lucas B, Mann J: Up-to-date tools for water-system optimization, *Chem Eng* 111:30–41, 2004.

Liu ZY, Yang Y, Wan LZ, Wang X, Hou KH: A heuristic design procedure for water-using networks with multiple contaminants, *AIChE J* 55:374–382, 2009.

Majozi T, Gouws JF: A mathematical optimisation approach for wastewater minimisation in multipurpose batch plants: multiple contaminants, *Comput Chem Eng* 33:1826–1840, 2009.

Mohammadnejad S, Bidhendi GRN, Mehrdadi N: Water pinch analysis in oil refinery using regeneration reuse and recycling consideration, *Desalination* 265:255–265, 2011.

Nápoles-Rivera F, Jiménez-Gutiérrez A, Ponce-Ortega JM, El-Halwagi MM: Recycle and reuse mass exchange networks based on properties using a global optimization technique, *Comput-Aided Chem Eng* 28:871–876, 2010.

Nápoles-Rivera F, Ponce-Ortega JM, El-Halwagi MM, Jiménez-Gutiérrez A: Global optimization of mass and property integration networks with in-plant property interceptors, *Chem Eng Sci* 65(15):4363–4377, 2010.

Ng DKS, Foo DCY, Tan RR, El-Halwagi M: Automated targeting technique for concentration- and property-based total resource conservation network, *Comput Chem Eng* 34:825–845, 2010.

Ng DKS, Foo DCY, Rabie AH, El-Halwagi MM: Simultaneous synthesis of property-based reuse/recycle and interception networks for batch processes, *AIChE J* 54(10):2624–2632, 2008.

Noureldin MB, El-Halwagi MM: Interval-based targeting for pollution prevention via mass integration, *Comput Chem Eng* 23:1527–1543, 1999.

Polley GT, Picon-Nunez M, Lopez-Maciel JD: Design of water and heat recovery networks for the simultaneous minimisation of water and energy consumption, *Appl Therm Eng* 30:2290–2299, 2010.

Ponce-Ortega JM, El-Halwagi MM, Jimenez-Gutierrez A: Global optimization for the synthesis of property-based recycle and reuse networks including environmental constraints, *Comput Chem Eng* 34:318–330, 2010.

Ponce-Ortega JM, Hortua AC, El-Halwagi M, Jimenez-Gutierrez A: A property-based optimization of direct recycle networks and wastewater treatment processes, *AIChE J* 55:2329–2344, 2009.

Rabie A, El-Halwagi MM: Synthesis and scheduling of optimal batch water-recycle networks, *Chin J Chem Eng* 16(3):474–479, 2008.

Rubio-Castro E, Ponce Ortega JM, Serna-González M, Jiménez-Gutiérrez A, El-Halwagi MM: A global optimal formulation for water integration in eco-industrial parks considering multiple pollutants, *Comp Chem Eng* 2011, in press, doi 10:1016/j.compchemeng 2011.03.010.

Rubio-Castro E, Ponce-Ortega JM, Napoles-Rivera F, El-Halwagi MM, Serna-Gonzalez M, Jimenez-Gutierrez A: Water integration of eco-industrial parks using a global optimization approach, *Ind Eng Chem Res* 49:9945–9960, 2010.

Savulescu L, Kim JK, Smith R: Studies on simultaneous energy and water minimization: Part II: Systems with maximum re-use of water, *Chem Eng Sci* 60:3291–3308, 2005.

Shelley MD, El-Halwagi MM: Componentless design of recovery and allocation systems: a functionality-based clustering approach, *Comp Chem Eng* 24:2081–2091, 2000.

Sorin M, Bedard S: The global pinch point in water reuse networks, *Process Saf Environ Prot* 77:305–308, 1999.

Yang YH, Lou HH, Huang YL: Synthesis of an optimal wastewater reuse network, *Waste Manage* 20:311–319, 2000.

CHAPTER 16
SYNTHESIS OF MASS-EXCHANGE NETWORKS: A MATHEMATICAL PROGRAMMING APPROACH

Chapters 5 and 11 presented graphical and algebraic approaches to the synthesis of mass-exchange networks (MENs). This chapter will show how optimization techniques enable designers to accomplish the following:

- Simultaneously screen all mass-separating agents (MSAs), even when there are no process MSAs.
- Determine the minimum operating cost (MOC) solution and locate the mass-exchange pinch point.
- Determine the best outlet composition for each MSA.
- Construct a network of mass exchangers with the least number of units that realize the MOC solution.

GENERALIZATION OF THE COMPOSITION INTERVAL DIAGRAM

The construction of a composition interval diagram (CID) has been discussed in the section entitled "The Composition-Interval Diagram" in Chapter 11. This representation tool will now be generalized by incorporating external MSAs. In the generalized CID, $N_S + 1$ composition scales are created. First, a single composition scale, y, is established for the waste streams. Next, Eq. 5.5 is utilized to generate N_S corresponding composition scales (N_{SP} for process MSAs and N_{SE} for external MSAs). The locations corresponding to the supply and target compositions of the streams determine a sequence of composition intervals. The number of these intervals depends on the number of streams through the following inequality:

[16.1] $$N_{int} \leq 2(N_R + N_S) - 1$$

The construction of the CID allows the evaluation of exchangeable loads for each stream in each composition interval. Hence, one can create a TEL for the waste streams in which the exchangeable load of the ith waste stream within the kth interval is defined as

[16.2] $$W_{i,k}^R = G_i(y_{k-1} - y_k)$$

when stream i passes through interval k, and

[16.3] $$W_{i,k}^R = 0$$

when stream i does not pass through interval k. The collective load of the waste streams within interval k, W_k^R, can be computed by summing up the individual loads of the waste streams that pass through that interval; that is,

[16.4] $$W_k^R = \sum_{i \text{ passes through interval } k} W_{i,k}^R$$

On the other hand, because the flow rate of each MSA is unknown, exact capacities of MSAs cannot be evaluated. Instead, one can create a *TEL per unit mass of the MSAs* for the lean streams. In this table, the exchangeable load *per unit mass of the MSA* is determined as follows:

[16.5] $$w_{j,k}^S = x_{j,k-1} - x_{j,k}$$

for the jth MSA passing through interval k, and

[16.6] $$w_{j,k}^S = 0$$

when the jth MSA does not pass through the kth interval.

PROBLEM FORMULATION

The section entitled "Mass-Exchange Cascade Diagram" in Chapter 11 presented a technique for evaluating thermodynamically feasible material balances among process streams by using the mass-exchange cascade diagram. This technique will now be generalized to include external MSAs. Once again, the objective is to minimize the cost of MSAs that can remove pollutant from waste streams in a thermodynamically feasible manner. Because the flow rates of the MSAs are not known, the objective function as well as the material balances around composition intervals have to be written in terms of these flow rates. The solution of the optimization program determines the optimal flow rate of each MSA. Hence, the task of identifying the MOC of the problem can be formulated through the following optimization program (El-Halwagi and Manousiouthakis, 1990a):

[P16.1] $$\min \sum_{j=1}^{N_S} C_j L_j$$

subject to
Mass balance around each composition interval

$$\delta_k - \delta_{k-1} + \sum_{j \text{ passes through interval } k} L_j w_{j,k}^S = W_k^R$$

$$k = 1, 2, \ldots, N_{int}$$

Non-negativity and limits on flow rates

$$L_j \geq 0, \quad j = 1, 2, \ldots, N_S$$
$$L_j \leq L_j^c, \quad j = 1, 2, \ldots, N_S$$

Non-negativity and values of residuals

$$\delta_0 = 0,$$
$$\delta_{N_{int}} = 0,$$
$$\delta_k \geq 0 \quad k = 1, 2, \ldots, N_{int} - 1.$$

The preceding program (P16.1) is a linear program that seeks to minimize the objective function of the operating cost of MSAs where C_j is the cost of the jth ($/kg of recirculating MSA, including regeneration and makeup costs) and L_j is the flow rate of the jth MSA. The first set of constraints represents successive material balances around each composition interval where δ_{k-1} and δ_k are the residual masses of the key pollutant entering and leaving the kth interval. The second and third sets of constraints guarantee that the optimal flow rate of each MSA is nonnegative and is less than the total available quantity of that lean stream. The fourth and fifth constraints ensure that the overall material balance for the problem is realized. Finally, the last set of constraints enables the waste streams to pass the mass of the pollutant downward if it does not fully exchange it with the MSAs in a given interval. This transfer of residual loads is thermodynamically feasible due to the way in which the CID has been constructed.

The solution of program (P16.1) yields the optimal values of all the L_js ($j = 1, 2, \ldots, N_S$) and the residual mass-exchange loads δ_ks ($k = 1, 2, \ldots, N_{int} - 1$). The location of any pinch point between two consecutive intervals, k and $k + 1$, is indicated when the residual mass-exchange load δ_k vanishes. This is a *generalization of the concept of a mass-exchange pinch point* discussed in Chapters Five and Eleven. Because the plant may not involve the use of any process MSAs, external MSAs can indeed be used above the pinch to obtain an MOC solution. However, the pinch point still maintains its significance as the most thermodynamically constrained region of the network at which all mass-transfer duties take place with driving forces equal to the minimum allowable composition differences.

Example 16-1 The dephenolization case study revisited

The dephenolization problem was described by Example 5.2. Tables 16.1 and 16.2 summarize the data for the rich (waste) and the lean streams.

The first step in determining the MOC is to construct the CID for the problem to represent the waste streams along with the process and external MSAs. The CID is shown in Fig. 16.1, for the case when the minimum allowable composition differences are 0.001. Hence, one can evaluate the exchangeable loads for the two waste streams over each composition interval. These loads are calculated through Equations 16.2 and 16.3. Table 16.3 illustrates the results.

Next, using Equations 16.5 and 16.6, the TEL for the lean streams per unit mass of the MSA is created. Table 16.4 depicts these loads.

We are now in a position to formulate the problem of minimizing the cost of MSAs. By adopting the linear-programming formulation (P16.1), one can write the following optimization program:

[P16.2]
$$\min 0.081 L_3 + 0.214 L_4 + 0.060 L_5$$

subject to

$\delta_1 = 0.0052$
$\delta_2 - \delta_1 + 0.0101 L_2 = 0.0308$
$\delta_3 - \delta_2 + 0.0010 L_1 + 0.0013 L_2 = 0.0040$
$\delta_4 - \delta_3 + 0.0066 L_1 + 0.0086 L_2 = 0.0396$
$\delta_5 - \delta_4 + 0.0024 L_1 + 0.0537 L_4 = 0.0144$
$\delta_6 - \delta_5 + 0.0222 L_4 = 0.0060$
$\delta_7 - \delta_6 + 0.0444 L_4 = 0.0040$
$\delta_8 - \delta_7 + 0.0420 L_4 = 0.0000$
$\delta_9 - \delta_8 + 0.0510 L_3 + 0.0114 L_4 = 0.0000$
$\delta_{10} - \delta_9 + 0.0555 L_3 + 0.0123 L_4 + 0.0277 L_5 = 0.000$
$\delta_{11} - \delta_{10} + 0.0025 L_3 + 0.0013 L_5 = 0.0000$
$-\delta_{11} + 0.0010 L_3 = 0.0000$
$\delta_k \geq 0, \quad k = 1, 2, \ldots, 11$
$L_j \geq 0, \quad j = 1, 2, \ldots, 5$
$L_1 \leq 5,$
$L_2 \leq 3.$

TABLE 16-1 Data for Waste Streams in Dephenolization Example

Stream	Description	Flow Rate, G_i (kg/s)	Supply Composition, y_i^s	Target Composition, y_i^t
R_1	Condensate from first stripper	2	0.050	0.010
R_2	Condensate from second stripper	1	0.030	0.006

TABLE 16-2 Data for MSAs in Dephenolization Example

Stream	Description	Upper Bound on Flow Rate L_j^C (kg/s)	Supply Composition, x_j^s	Target Composition, x_j^t	Equilibrium Distribution Coefficient, $m_j = y/x_j$	Cost C_j ($/kg of Recirculation MSA)
S_1	Gas oil	5	0.005	0.015	2.00	0.000
S_2	Lube oil	3	0.010	0.030	1.53	0.000
S_3	Activated carbon	∞	0.000	0.110	0.02	0.081
S_4	Ion-exchange resin	∞	0.000	0.186	0.09	0.214
S_5	Air	∞	0.000	0.029	0.04	0.060

(Continued)

Example 16-1 The dephenolization case study revisited (Continued)

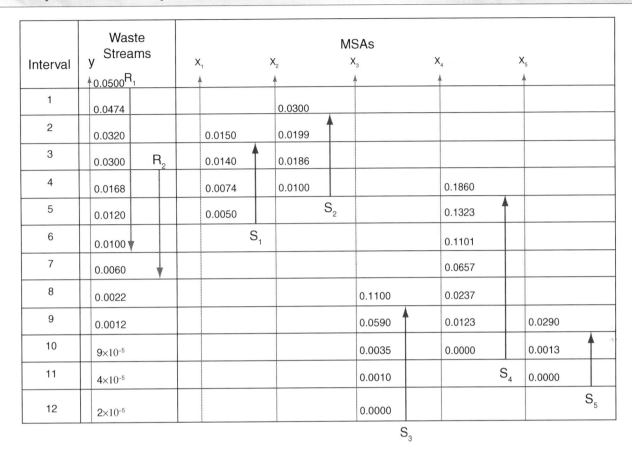

FIGURE 16-1 CID for dephenolization example.

TABLE 16-3 TEL for Waste Streams

	Load of Waste Streams (kg phenol/s)		
Interval	R_1	R_2	$R_1 + R_2$
1	0.0052	–	0.0052
2	0.0308	–	0.0308
3	0.0040	–	0.0040
4	0.0264	0.0132	0.0396
5	0.0096	0.0048	0.0144
6	0.0040	0.0020	0.0060
7	–	0.0040	0.0040
8	–	–	–
9	–	–	–
10	–	–	–
11	–	–	–
12	–	–	–

In terms of LINGO input, program P16.2 can be written as:

```
min=0.081*L3+0.214*L4+0.060*L5;
delta1=0.0052;
delta2-delta1+0.0101*L2=0.0308;
delta3-delta2+0.001*L1+0.0013*L2=0.0040;
delta4-delta3+0.0066*L1+0.0086*L2=0.0396;
delta5-delta4+0.0024*L1+0.0537*L4=0.0144;
delta6-delta5+0.0222*L4=0.0060;
delta7-delta6+0.0444*L4=0.0040;
delta8-delta7+0.0420*L4=0.0000;
delta9-delta8+0.051*L3+0.0114*L4=0.000;
delta10-delta9+0.0555*L3+0.0123*L4+0.0277*L5=0.000;
delta11-delta10+0.0025*L310.0013*L5=0.0000;
-delta11+0.0010*L3=0.000;
delta1>=0.0;
delta2>=0.0;
delta3>=0.0;
delta4>=0.0;
delta5>=0.0;
delta6>=0.0;
delta7>=0.0;
delta8>=0.0;
delta9>=0.0;
delta10>=0.0;
delta11>=0.0;
L1>=0.0;
L2>=0.0;
L3>=0.0;
L4>=0.0;
L5>=0.0;
L1<=5.0;
L2<=3.0;
```

(Continued)

Example 16-1 The dephenolization case study revisited (Continued)

TABLE 16-4 TEL (kg phenol/kg MSA) for MSAs

Interval	Capacity of Lean Streams per Unit Mass of MSA (kg phenol/kg MSA)				
	S_1	S_2	S_3	S_4	S_5
1	–	–	–	–	–
2	–	0.0101	–	–	–
3	0.0010	0.0013	–	–	–
4	0.0066	0.0086	–	–	–
5	0.0024	–	–	0.0537	–
6	–	–	–	0.0222	–
7	–	–	–	0.0444	–
8	–	–	–	0.0420	–
9	–	–	0.0510	0.0114	–
10	–	–	0.0555	0.0123	0.0277
11	–	–	0.0025	–	0.0013
12	–	–	0.0010	–	–

The following is the solution report generated by LINGO:

```
Objective value: 0.9130909E-02

Variable        Value
L3              0.1127273
L4              0.0000000E+00
L5              0.0000000E+00
DELTA1          0.5200000E-02
DELTA2          0.1499200E-01
L2              2.080000
DELTA3          0.1128800E-01
L1              5.000000
DELTA4          0.0000000E+00
DELTA5          0.2399999E-02
DELTA6          0.8399999E-02
DELTA7          0.1240000E-01
DELTA8          0.1240000E-01
DELTA9          0.6650909E-02
DELTA10         0.3945454E-03
DELTA11         0.1127273E-03
```

As can be seen from the results, the solution to the linear program yields the following values for L_1, L_2, L_3, L_4, and L_5: 5.0000, 2.0800, 0.1127, 0.0000, and 0.0000, respectively. The optimum value of the objective function is 9.13×10^{-3}/s (approximately 288×10^3/yr). It is worth pointing out that the same optimum value of the objective function can also be achieved by other combinations of L_1 and L_2 along with the same value of L_3 (because both L_1 and L_2 are virtually free). The solution of P16.2 also yields a vanishing δ_4, indicating that the mass-exchange pinch is located at the line separating intervals 4 and 5. All these findings are consistent with the solutions obtained in Chapters 5 and 11.

It is also useful to use LINGO's set formulation to solve the problem. Let us define the following sets: LEAN to represent the lean stream, INTERVAL to represent the composition intervals, and LEAN_INTERVAL with members representing the lean streams over the composition intervals and with the attribute $w_{k,j}^S$ representing the load of the jth lean stream per unit flow in interval k. Because the mass residual (designated by Delta(k)) entering the first interval and leaving the last interval must be set to zero, we write two separate equations for the first and last intervals to enforce these zero residuals. The rest of the component mass balances around the intervals are, therefore, written for intervals greater than or equal to interval 2 (given by the command k #GE# 2 as described in Chapter 14). The following is the set formulation of the program:

[P16.3]

```
SETS:
LEAN /1..5/: C, L, Lmax;
INTERVAL /1..12/: Delta, WR;
LEAN_INTERVAL (INTERVAL,LEAN): WS;
ENDSETS
DATA:
C = 0 0 0.081 0.214 0.060;
Lmax = 5 3 100 100 100;
WS=0         0        0       0       0
   0         0.0101   0       0       0
   0.0010    0.0013   0       0       0
   0.0066    0.0086   0       0       0
   0.0024    0        0       0.0537  0
   0         0        0       0.0222  0
   0         0        0       0.0444  0
   0         0        0       0.0420  0
   0         0        0.0510  0.0114  0
   0         0        0.0555  0.0123  0.0277
   0         0        0.0025  0       0.0013
   0         0        0.0010  0       0;
WR=0.0052 0.0308 0.0040 0.0396 0.0144 0.0060 0.0040
   0 0 0 0 0;
ENDDATA
Min=Cost;
Cost=@SUM(LEAN(j): C(j)*L(j));
@FOR(INTERVAL(k)|k #GE# 2:
    @SUM(LEAN(j): -WS(k,j)*L(j))+WR(k)+Delta
    (k-1)=Delta(k));
@SUM(LEAN(j): -WS(1,j)*L(j))+WR(1)=Delta(1);
@SUM(LEAN(j): -WS(12,j)*L(j))+WR(12)+Delta(11)=0;
@FOR(LEAN(j): L(j)<=Lmax(j));
@FOR(INTERVAL(k): Delta(k)>=0);
```

(Continued)

Example 16-1 The dephenolization case study revisited (*Continued*)

The following solution is given:

Objective value: 0.9130909E-02

Variable	Value
COST	0.9130909E-02
C(1)	0.000000
C(2)	0.000000
C(3)	0.8100000E-01
C(4)	0.2140000
C(5)	0.6000000E-01
L(1)	5.000000
L(2)	2.080000
L(3)	0.1127273
L(4)	0.000000
L(5)	0.000000
LMAX(1)	5.000000
LMAX(2)	3.000000
LMAX(3)	100.0000
LMAX(4)	100.0000
LMAX(5)	100.0000
DELTA(1)	0.5200000E-02
DELTA(2)	0.1499200E-01
DELTA(3)	0.1128800E-01
DELTA(4)	0.000000
DELTA(5)	0.2400000E-02
DELTA(6)	0.8400000E-02
DELTA(7)	0.1240000E-01
DELTA(8)	0.1240000E-01
DELTA(9)	0.6650909E-02
DELTA(10)	0.3945455E-03
DELTA(11)	0.1127273E-03
DELTA(12)	0.000000
WR(1)	0.5200000E-02
WR(2)	0.3080000E-01
WR(3)	0.4000000E-02
WR(4)	0.3960000E-01
WR(5)	0.1440000E-01
WR(6)	0.6000000E-02
WR(7)	0.4000000E-02
WR(8)	0.000000
WR(9)	0.000000
WR(10)	0.000000
WR(11)	0.000000
WR(12)	0.000000
WS(1, 1)	0.000000
WS(1, 2)	0.000000
WS(1, 3)	0.000000
WS(1, 4)	0.000000
WS(1, 5)	0.000000
WS(2, 1)	0.000000
WS(2, 2)	0.1010000E-01
WS(2, 3)	0.000000
WS(2, 4)	0.000000
WS(2, 5)	0.000000
WS(3, 1)	0.1000000E-02
WS(3, 2)	0.1300000E-02
WS(3, 3)	0.000000
WS(3, 4)	0.000000
WS(3, 5)	0.000000
WS(4, 1)	0.6600000E-02
WS(4, 2)	0.8600000E-02
WS(4, 3)	0.000000
WS(4, 4)	0.000000
WS(4, 5)	0.000000
WS(5, 1)	0.2400000E-02
WS(5, 2)	0.000000
WS(5, 3)	0.000000
WS(5, 4)	0.5370000E-01
WS(5, 5)	0.000000
WS(6, 1)	0.000000
WS(6, 2)	0.000000
WS(6, 3)	0.000000
WS(6, 4)	0.2220000E-01
WS(6, 5)	0.000000
WS(7, 1)	0.000000
WS(7, 2)	0.000000
WS(7, 3)	0.000000
WS(7, 4)	0.4440000E-01
WS(7, 5)	0.000000
WS(8, 1)	0.000000
WS(8, 2)	0.000000
WS(8, 3)	0.000000
WS(8, 4)	0.4200000E-01
WS(8, 5)	0.000000
WS(9, 1)	0.000000
WS(9, 2)	0.000000
WS(9, 3)	0.5100000E-01
WS(9, 4)	0.1140000E-01
WS(9, 5)	0.000000
WS(10, 1)	0.000000
WS(10, 2)	0.000000
WS(10, 3)	0.5550000E-01
WS(10, 4)	0.1230000E-01
WS(10, 5)	0.2770000E-01
WS(11, 1)	0.000000
WS(11, 2)	0.000000
WS(11, 3)	0.2500000E-02
WS(11, 4)	0.000000
WS(11, 5)	0.1300000E-02
WS(12, 1)	0.000000
WS(12, 2)	0.000000
WS(12, 3)	0.1000000E-02
WS(12, 4)	0.000000
WS(12, 5)	0.000000

These results are consistent with the solution obtained earlier using the detailed linear-programming formulation as well as the graphical and algebraic solutions presented in Chapters 5 and 11.

OPTIMIZATION OF OUTLET COMPOSITIONS

As discussed in Chapter 5, the target compositions are only upper bounds on the outlet compositions. Therefore, it may be necessary to optimize the outlet compositions.[1] A short cut method of optimizing the outlet composition is the use of "lean substreams" (El-Halwagi, 1993; Garrison et al., 1995). Consider an MSA j, whose target composition is given by x_j^t. To determine the optimal outlet composition, x_j^{out}, a number ND_j of substreams are assumed. Each substream, d_j, where $d_j = 1, 2, \ldots, ND_j$, is a decomposed portion of the MSA that extends from the given x_j^s to a selected value of outlet composition, $x_{j,d}^{out}$, which lies between x_j^s and x_j^t. The flow rate of each substream, L_{j,d_j}, is unknown and

---
[1] More rigorous techniques for optimizing outlet composition are described by El-Halwagi and Manousiouthakis (1990b) and Garrison et al. (1995), and are beyond the scope of this book.

is to be determined as part of the optimization problem. The number of substreams is dependent on the level of accuracy needed for the MEN analysis. Theoretically, an infinite number of substreams should be used to cover the whole composition span of each MSA. However, in practice, few (typically less than five) substreams are needed. On the CID, the various substreams are represented against their composition scale. The formulation (P16.1) can therefore be revised to

[P16.4] $$\min \sum_{j=1}^{N_S} C_j \sum_{d_j=1}^{ND_j} L_{j,d_j}$$

subject to
Mass balances around composition intervals

$$\delta_k - \delta_{k-1} + \sum_{\substack{j \text{ passes through} \\ \text{interval } k}} \sum_{d_j=1}^{ND_j} L_{j,d_j} w_{j,k}^S = W_k^R$$

$$k = 1, 2, \ldots, N_{int} \quad L_{j,d_j} \geq 0, j = 1, 2, \ldots, N_S$$

Splitting of substreams

$$\sum_{d_j=1}^{ND_j} L_{j,dj} \leq L_j^C, \quad j = 1, 2, \ldots, N_S$$

Residual constraints

$$\delta_0 = 0$$
$$\delta_{N_{int}} = 0$$
$$\delta_k \geq 0, \quad k = 1, 2, \ldots, N_{int} - 1.$$

The preceding program (P16.3) is a linear program that minimizes the operating cost of MSAs. The solution of this program determines the optimal flow rate of each substream and, consequently, the optimal outlet compositions. If more than one of the substreams are selected, the total flow rate can be obtained by summing up the individual flow rates of the substreams, and the outlet composition may be determined by averaging the outlet compositions as follows:

[16.7] $$L_j = \sum_{d_j=1}^{ND_j} L_{j,d_j},$$

[16.8] $$x_j^{out} = \frac{\sum_{d_j=1}^{ND_j} L_{j,d_j} x_{j,d_j}^{out}}{L_j}$$

Example 16-2 Optimizing the outlet compositions for the benzene removal problem

To demonstrate this procedure, let us revisit Example 5.1, on the recovery of benzene from a gaseous emission of a polymer facility. Instead of determining the outlet composition of S_1 graphically, we will determine it mathematically. Tables 16.5 and 16.6 give the stream data for the waste stream and for the lean streams.

To determine the optimal outlet composition of S_1, several substreams are created to span an outlet composition between x_1^s and x_1^t. Let us select six substreams with outlet compositions of 0.0060, 0.0055, 0.0050, 0.0045, 0.0040, and 0.0035. Figure 16.2 shows the CID for the problem.

In terms of the LINGO input, the problem can be formulated via the following linear program:

```
MODEL:
MIN=0.05*L3;

D1+0.0005*L2=5.0E-05;
D2-D110.0005*L11+0.00025*L2=2.5E-05;
D3-D210.0005*L11+0.00025*L2+0.0005*L12=2.5E-05;
D4-D310.0005*L11+0.0005*L12+0.0005*L13=2.5E-05;
D5-D410.0005*L11+0.0005*L12+0.0005*L13+0.0005*L14=2.5E-05;
D6-D5+0.0005*L11+0.0005*L12+0.0005*L13+0.0005*L14+0.0005*L15=2.5E05;
D7-D6+0.0005*L11+0.0005*L12+0.0005*L13+0.0005*L14+0.0005*L15+0.0005*L16=2.5E-05;
D8-D7=2.6E-05;
-D8+0.0077*L3=0.000154;
D1>0.0;
D2>0.0;
D3>0.0;
D4>0.0;
D5>0.0;
D6>0.0;
D7>0.0;
D8>0.0;
L11>0.0;
L12>0.0;
L13>0.0;
L14>0.0;
L15>0.0;
L16>0.0;
L2>0.0;
L2<0.05;
```

TABLE 16-5 Data for Waste Stream in Benzene Removal Example

Stream	Flow Rate G_i (kg mol/s)	Supply Composition (mole fraction) y_i^s	Target Composition (mole fraction) y_i^t
R_1	0.2	0.0020	0.0001

TABLE 16-6 Data for Lean Streams in Benzene Removal Example

Stream	Upper Bound on Flow Rate L_j^C (kg mol/s)	Supply Composition of Benzene (mole fraction) x_j^s	Target Composition of Benzene (mole fraction) x_j^t	m_j	C_j (\$/kmol)	ε_j
S_1	0.08	0.0030	0.0060	0.25	0.00	0.0010
S_2	0.05	0.0020	0.0040[a]	0.50	0.00	0.0010
S_3	∞	0.0008	0.0085	0.10	0.05	0.0002

[a]This value is located above the inlet of R_1 on the CID and, therefore, must be reduced. Because R_1 has a supply composition of 0.002, the maximum practically feasible value of x_2^t is $(0.002/0.5) - 0.001 = 0.003$.

(Continued)

Example 16-2 Optimizing the outlet compositions for the benzene removal problem (Continued)

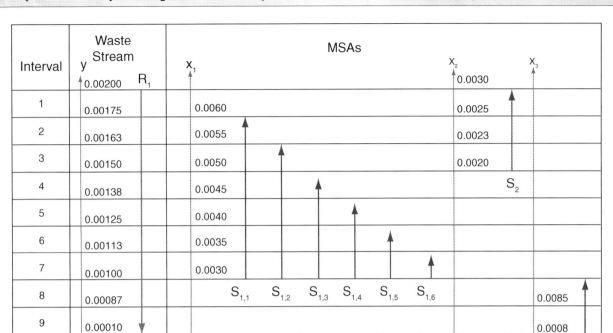

FIGURE 16-2 CID for benzene recovery example with lean substreams.

```
L3>0.0;
L11+L12+L13+L14+L15+L16<0.08;
END
```

The solution to this program yields the following results:

```
Objective value: 0.1168831E-02
Variable       Value
L3             0.2337662E-01
D1             0.5000000E-04
L2             0.0000000E+00
D2             0.4300000E-04
L11            0.6400000E-01
D3             0.3600000E-04
L12            0.0000000E+00
D4             0.2900000E-04
```

Variable	Value
L13	0.0000000E+00
D5	0.2200000E-04
L14	0.0000000E+00
D6	0.1500000E-04
L15	0.0000000E+00
D7	0.0000000E+00
L16	0.1600000E-01
D8	0.2600000E-04

This solution indicates that the minimum operating cost is $33,700/yr ($0.00117/s), which corresponds to an optimal flow rate of S_3 of 0.0234 kg mol/s. By applying Equations 16.7 and 16.8, we can determine the flow rate of S_1 to be 0.08 kg mol/s and the outlet composition to be 0.0055. The pinch location corresponds to the vanishing residual mass at the line separating intervals 7 and 8 (y = 0.001). All these results are consistent with those obtained graphically in Chapter 3.

STREAM MATCHING AND NETWORK SYNTHESIS

Having identified the values of all the flow rates of lean streams as well as the pinch location, we can now minimize the number of mass exchangers for an MOC solution. As mentioned earlier, when a pinch point exists, the synthesis problem can be decomposed into two subnetworks, one above the pinch and one below the pinch. The subnetworks will be denoted by SN_m, where $m = 1,2$. It is therefore useful to define the following subsets:

[16.9] $R_m = \{i \mid i \in R, \text{stream } i \text{ exists in } SN_m\}$

[16.10] $S_m = \{j \mid j \in S, \text{stream } j \text{ exists in } SN_m\}$

[16.11] $R_{m,k} = \{i \mid i \in R_m, \text{stream } i \text{ exists in interval } \bar{k} \leq k; \bar{k}, k \in SN_m\}$

[16.12]
$S_{m,k} = \{j \mid j \in S_m, \text{stream } j \text{ exists in interval } k \in SN_m\}$.

For a rich stream, i,

$$\delta_{i,k} - \delta_{i,k-1} + \sum_{j \in S_{m,k}} W_{i,j,k} = W_{i,k}^R.$$

Within any subnetwork, the mass exchanged between any two streams is bounded by the smaller of the two loads. Therefore, the upper bound on the exchangeable mass between streams i and j in SN_m is given by

$$[16.13] \quad U_{i,j,m} = \min\left\{\sum_{k \in SN_m} W_{i,k}^R, \sum_{k \in SN_m} W_{j,k}^S\right\}.$$

Now, we define the binary variable $E_{i,j,m}$, which takes the values of 0 when there is no match between streams i and j in SN_m and takes the value of 1 when there exists a match between streams i and j (and hence an exchanger) in SN_m. Based on Eq. 16.13, one can write

$$[16.14] \quad \sum_{k \in SN_m} W_{i,j,k} - U_{i,j,m} E_{i,j,m} \leq 0$$
$$i \in R_m, \quad j \in S_m, \quad m = 1, 2,$$

where $W_{i,j,k}$ denotes the mass exchanged between the ith rich stream and the jth lean stream in the kth interval. Therefore, the problem of minimizing the number of mass exchangers can be formulated as a mixed-integer linear program (MILP) (El-Halwagi and Manousiouthakis, 1990a):

$$[P16.15] \quad \text{minimize} \sum_{m=1,2} \sum_{i \in R_m} \sum_{j \in S_m} E_{i,j,m},$$

subject to the following:

Material balance for each rich stream around composition intervals:

$$\delta_{i,k} - \delta_{i,k-1} + \sum_{j \in S_{m,k}} W_{i,j,k} = W_{i,k}^R \quad i \in R_{m,k},$$
$$k \in SN_m, \quad m = 1, 2$$

Material balance for each lean stream around composition intervals:

$$\sum_{i \in R_{m,k}} W_{i,j,k} = W_{j,k}^S \quad j \in S_{m,k}, k \in SN_m, \quad m = 1, 2$$

Matching of loads:

$$\sum_{k \in SN_m} W_{i,j,k} - U_{i,j,m} E_{i,j,m} \leq 0 \quad i \in R_m, j \in S_m, m = 1, 2$$

Nonnegative residuals:

$$\delta_{i,k} \geq 0 \quad i \in R_{m,k}, k \in SN_m, m = 1, 2$$

Nonnegative loads:

$$W_{i,j,k} \geq 0 \quad i \in R_{m,k}, j \in S_{m,k}, k \in SN_m, m = 1, 2$$

Binary integer variables for matching streams:

$$E_{i,j,m} = 0/1 \quad i \in R_m, j \in S_m, m = 1, 2$$

The preceding program is an MILP that can be solved (for example, using the computer code LINGO) to provide information on the stream matches and exchangeable loads. It is interesting to note that the solution of Program P16.4 may not be unique. It is possible to generate all integer solutions to P16.4 by adding constraints that exclude previously obtained solutions from further consideration. For example, any previous solution can be eliminated by requiring that the sum of $E_{i,j,m}$ that is nonzero in that solution be less than the minimum number of exchangers. Also, if the costs of the various exchangers are significantly different, the objective function can be modified by multiplying each integer variable by a weighting factor that reflects the relative cost of each unit.

Example 16-3 Network synthesis for dephenolization problem

Let us revisit the dephenolization problem described in Example 16.1. The objective is to synthesize a MOC-MEN with the least number of units. First, CID (Fig. 16.3) and the tables of exchangeable loads (Tables 16.7 and 16.8) are developed based on the MOC solution identified in Example 16.1. Because neither S_4 nor S_5 were selected as part of the MOC solution, there is no need to include them. Furthermore, because the optimal flow rates of S_1, S_2, and S_3 have been determined, the TEL for the MSAs can now be developed with the total loads of MSAs and not per kilogram of each MSA.

Because the pinch decomposes the problem into two subnetworks, it is useful to calculate the exchangeable load of each stream above and below the pinch. Tables 16.9 and 16.10 present these values.

We can now formulate the synthesis task as an MILP whose objective is to minimize the number of exchangers. Above the pinch (subnetwork $m = 1$), there are four possible matches: $R_1 - S_1$, $R_1 - S_2$, $R_2 - S_1$, and $R_2 - S_2$. Hence, we need to define four binary variables ($E_{1,1,1}$, $E_{1,2,1}$, $E_{2,1,1}$, and $E_{2,2,1}$). Similarly, below the pinch (subnetwork $m = 2$), we have to define four binary variables ($E_{1,1,2}$, $E_{1,3,2}$, $E_{2,1,2}$, and $E_{2,3,2}$) to represent the potential matches between $R_1 - S_1$, $R_1 - S_3$, $R_2 - S_1$, and $R_2 - S_3$. Therefore, the objective function is described by:

Minimize $E_{1,1,1} + E_{1,2,1} + E_{2,1,1} + E_{2,2,1} + E_{1,1,2} + E_{1,3,2} + E_{2,1,2} + E_{2,3,2}$

subject to the following constraints:

Material Balances for R_1 Around Composition Intervals

$$\delta_{1,1} = 0.0052$$
$$\delta_{1,2} - \delta_{1,1} + W_{1,2,2} = 0.0308$$
$$\delta_{1,3} - \delta_{1,2} + W_{1,1,3} + W_{1,2,3} = 0.0040$$
$$\delta_{1,4} - \delta_{1,3} + W_{1,1,4} + W_{1,2,4} = 0.0264$$
$$\delta_{1,5} - \delta_{1,4} + W_{1,1,5} = 0.0096$$
$$\delta_{1,6} - \delta_{1,5} = 0.0040$$
$$\delta_{1,7} - \delta_{1,6} = 0.0000$$
$$\delta_{1,8} - \delta_{1,7} = 0.0000$$
$$-\delta_{1,8} + W_{1,3,9} = 0.0000$$

Material Balances for R_2 Around Composition Intervals

$$\delta_{2,4} + W_{2,1,4} + W_{2,2,4} = 0.0132$$
$$\delta_{2,5} - \delta_{2,4} + W_{2,1,5} = 0.0048$$
$$\delta_{2,6} - \delta_{2,5} = 0.0020$$
$$\delta_{2,7} - \delta_{2,6} = 0.0040$$
$$\delta_{2,8} - \delta_{2,7} = 0.0000$$
$$-\delta_{2,8} + W_{2,3,9} = 0.0000$$

(Continued)

Example 16-3 Network synthesis for dephenolization problem (*Continued*)

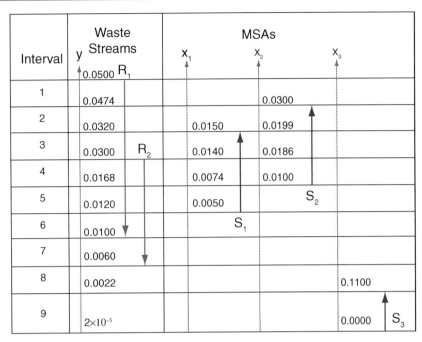

FIGURE 16-3 CID for dephenolization problem.

TABLE 16-7 TEL for Waste Streams in Dephenolization Example

	Load of Waste Streams (kg phenol/s)	
Interval	R_1	R_2
1	0.0052	–
2	0.0308	–
3	0.0040	–
4	0.0264	0.0132
5	0.0096	0.0048
6	0.0040	0.0020
7	–	0.0040
8	–	–
9	–	–

Pinch (between intervals 4 and 5)

TABLE 16-8 TEL for MSAs in Dephenolization Example

	Load of MSAs (kg phenol/s)		
Interval	S_1	S_2	S_3
1	–	–	–
2	–	0.0210	–
3	0.0050	0.0027	–
4	0.0330	0.0179	–
5	0.0120	–	–
6	–	–	–
7	–	–	–
8	–	–	–
9	–	–	0.0124

Pinch (between intervals 4 and 5)

TABLE 16-9 Exchangeable Loads above the Pinch

Stream	Load (kg phenol/s)
R_1	0.0664
R_2	0.0132
S_1	0.0380
S_2	0.0416
S_3	0.0000

TABLE 16-10 The Exchangeable Loads Below the Pinch

Stream	Load (kg phenol/s)
R_1	0.0136
R_2	0.0108
S_1	0.0120
S_2	0.0000
S_3	0.0124

Material Balances for S_1 Around Composition Intervals

$$W_{1,1,3} = 0.0050$$
$$W_{1,1,4} + W_{2,1,4} = 0.0330$$
$$W_{1,1,5} + W_{2,1,5} = 0.0120$$

Material Balances for S_2 Around Composition Intervals

$$W_{1,2,2} = 0.0210$$
$$W_{1,2,3} = 0.0027$$
$$W_{1,2,4} + W_{2,2,4} = 0.0179$$

(*Continued*)

Example 16-3 Network synthesis for dephenolization problem (*Continued*)

Material Balances for S_2 Around the Ninth Interval

$$W_{1,3,9} + W_{2,3,9} = 0.0124$$

Matching of Loads

$$W_{1,1,3} + W_{1,1,4} \leq 0.0380 E_{1,1,1}$$
$$W_{1,2,2} + W_{1,2,3} + W_{1,2,4} \leq 0.0416 E_{1,2,1}$$
$$W_{2,1,4} \leq 0.0132 E_{2,1,1}$$
$$W_{2,2,4} \leq 0.0132 E_{2,2,1}$$
$$W_{1,1,5} \leq 0.0120 E_{1,1,2}$$
$$W_{2,1,5} \leq 0.0108 E_{2,1,2}$$
$$W_{1,3,9} \leq 0.0124 E_{1,3,2}$$
$$W_{2,3,9} \leq 0.0108 E_{2,3,2}$$

with the nonnegativity and integer constraints.
In terms of LINGO input, this program can be written as follows:

[P16.6]
```
MIN=E111+E121+E211+E221+E112+E132+E212+E232;

D11=0.0052;
D12-D11+W122=0.0308;
D13-D12+W113+W123=0.0040;
D14-D13+W114+W124=0.0264;
D15-D14+W115=0.0096;
D16-D15=0.0040;
D17-D16=0.0000;
D18-D17=0.0000;
-D18+W139=0.0000;
D24-W214+W224=0.0132;
D25-D24+W215=0.0048;
D26-D25=0.0020;
D27-D26=0.0040;
D28-D27=0.0000;
-D28+W239=0.0000;
W113=0.0050;
W114+W214=0.0330;
W115+W215=0.0120;
W122=0.0210;
W123=0.0027;
W124+W224=0.0179;
W139+W239=0.0124;
W113+W114<=0.0136*E111;
W122+W123+W124<=0.0416*E121;
W214<=0.0132*E211;
W224<=0.0132*E221;
W115<=0.012*E112;
W215<=0.0108*E212;
W139<0.0124*E132;
W239<0.0108*E232;
D11>0.0;
D12>0.0;
D13>0.0;
D14>0.0;
D15>0.0;
D16>0.0;
D17>0.0;
D18>0.0;
D24>0.0;
D25>0.0;
D26>0.0;
D27>0.0;
D28>0.0;
W122>0.0;
W113>0.0;
W123>0.0;
W124>0.0;
W139>0.0;
W214>0.0;
W224>0.0;
W215>0.0;
W239>0.0;
@BIN(E111);
@BIN(E121);
@BIN(E211);
@BIN(E221);
@BIN(E112);
@BIN(E132);
@BIN(E212);
@BIN(E232);
END
```

This MILP can be solved using LINGO to yield the following results:

Objective value: 7.000000

Variable	Value
E111	1.000000
E121	1.000000
E211	0.000000
E221	1.000000
E112	1.000000
E132	1.000000
E212	1.000000
E232	1.000000
D11	0.5200000E-02
D12	0.1500000E-01
W122	0.2100000E-01
D13	0.1130000E-01
W113	0.5000000E-02
W123	0.2700000E-02
D14	0.0000000E+00
W114	0.3300000E-01
W124	0.4699999E-02
D15	0.2400000E-02
W115	0.7200001E-02
D16	0.6400000E-02
D17	0.6400000E-02
D18	0.6400000E-02
W139	0.6400000E-02
D24	0.0000000E+00
W214	0.00000000E+00
W224	0.1320000E-01
D25	0.0000000E+00
W215	0.4800000E-02
D26	0.2000000E-02
D27	0.6000000E-02
D28	0.6000000E-02
239	0.6000000E-02

(*Continued*)

Example 16-3 Network synthesis for dephenolization problem (*Continued*)

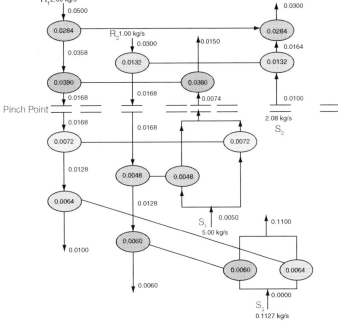

FIGURE 16-4 MOC network for dephenolization example.

These results indicate that the solution features seven units that represent the matches $R_1 - S_1$, $R_1 - S_2$, and $R_2 - S_2$ above the pinch and $R_1 - S_1$, $R_1 - S_3$, $R_2 - S_1$, and $R_2 - S_3$ below the pinch. The load for each exchanger can be evaluated by simply adding up the exchangeable loads within the same subnetwork. For instance, the transferable load from R_1 to S_2 above can be calculated as follows:

$$\text{Exchangeable Load for } E_{1,2,1} = W_{1,2,2} + W_{1,2,3} + W_{1,2,4}$$
$$= 0.0210 + 0.0027 + 0.0047$$
$$= 0.0284 \text{ kg phenol/s}$$

[16.15]

Hence, one can use these results to construct the network shown in Fig. 16.4. This is the same configuration obtained using the algebraic method illustrated by Fig. 5.13. However, the loads below the pinch are distributed differently. This is consistent with the earlier observation that the problem can have multiple solutions. The final design should be based on considerations of total annualized cost, safety, flexibility, operability, and controllability. As has been discussed in Chapter 5, the minimum TAC can be attained by trading off fixed versus operating costs by optimizing driving forces, stream mixing, and mass-load paths.

HOMEWORK PROBLEMS

16.1. Using linear programming, resolve the dephenolization example presented in this chapter for the case when the two waste streams are allowed to mix.
16.2. Using mixed-integer programming, find the minimum number of mass exchangers in the benzene recovery example described by Example 5.1.
16.3. Use optimization to solve Problem 5.1.
16.4. Apply the optimization-based approach presented in this chapter to solve Problem 5.3.
16.5. Employ linear programming to solve Problem 5.4.
16.6. Solve Problem 5.5 using optimization.
16.7. Consider the metal pickling plant shown in Fig. 16.5 (El-Halwagi and Manousiouthakis, 1990a). The objective of this process is to use a pickle solution (for example, HCl) to remove corrosion products, oxides, and scales from the metal surface. The spent pickle solution contains zinc chloride and ferrous chloride as the two primary contaminants. Mass exchange can be used to selectively recover zinc chloride from the spent liquor. The zinc-free liquor is then forwarded to a spray furnace in which ferrous chloride is converted to hydrogen chloride and iron oxides. The hydrogen chloride is absorbed and recycled to the pickling path. The metal leaving the pickling path is rinsed off by water to remove the clinging film of drag-out chemicals that adheres to the metal workpiece surface. The rinse wastewater contains zinc chloride as the primary pollutant that must be recovered for environmental and economic purposes.

The purpose of the problem is to systematically synthesize a cost-effective MEN that can recover zinc chloride from the spent pickle liquor, R_1, and the rinse wastewater, R_2. Two mass-exchange processes are proposed for recovering zinc: solvent

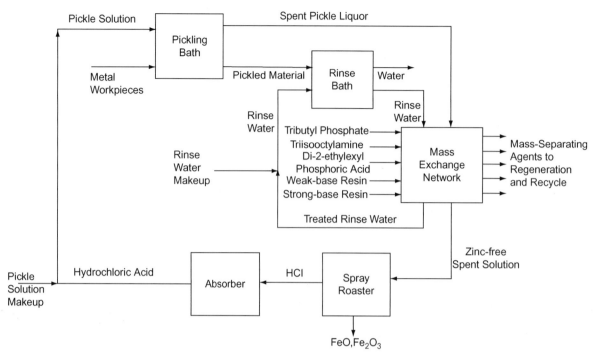

FIGURE 16-5 Zinc recovery from metal pickling plant. *Source: El-Halwagi and Manousiouthakis (1990a).*

TABLE 16-11 Stream Data for Zinc Recovery Problem

	Rich Stream				MSAs					
Stream	G_i (kg/s)	y_i^s	y_i^t	Stream	L_j^c (kg/s)	x_j^s	x_j^t	m_j	b_j	c_j ($/kg)
R_1	0.2	0.08	0.02	S_1	∞	0.0060	0.0600	0.845	0.000	0.02
R_2	0.1	0.03	0.001	S_2	∞	0.0100	0.0200	1.134	0.010	0.11
				S_3	∞	0.0090	0.0500	0.632	0.020	0.04
				S_4	∞	0.0001	0.0100	0.376	0.0001	0.05
				S_5	∞	0.0040	0.0150	0.362	0.002	0.13

extraction and ion exchange. For solvent extraction, three candidate MSAs are suggested: tributyl phosphate, S1; triisooctyl amine, S2; and di-2-ethylhexylphosphoric acid, S3. For ion exchange, two resins are proposed: a strong-base resin, S4, and a weak-base resin, S5. Table 16.11 summarizes the data for all the streams. All compositions are given in mass fractions. Assume a value of 0.0001 for the minimum allowable composition difference for all lean streams.

16.8. Etching of copper, using an ammoniacal solution, is an important operation in the manufacture of printed circuit boards for the microelectronics industry (El-Halwagi and Manousiouthakis, 1990a). During etching, the concentration of copper in the ammoniacal solution increases. Etching is most efficiently carried out for copper concentrations between 10 and 13 w/w% in the solution, whereas etching efficiency almost vanishes at higher concentrations (15 to 17 w/w%). To maintain the etching efficiency, copper must be continuously removed from the spent ammoniacal solution through solvent extraction. The regenerated ammoniacal etchant can then be recycled to the etching line.

The etched printed circuit boards are washed out with water to dilute the concentration of the contaminants on the board surface to an acceptable level. The extraction of copper from the effluent rinse water is essential for both environmental and economic reasons because decontaminated water is returned to the rinse vessel.

A schematic of the etching process is shown in Fig. 16.6. The proposed copper recovery scheme is to feed both the spent etchant and the effluent rinse water to a MEN in which copper is transferred to some selective solvents. Two extractants are recommended for this separation task: LIX63 (an aliphatic hydroxyoxime), S_1, and P1 (an aromatic hydroxyoxime), S_2. The former solvent appears to work most efficiently at moderate copper concentrations, whereas the latter extractant offers remarkable extraction efficiencies at low copper concentrations. Table 16.12 summarizes the stream data for the problem.

Two types of contractors will be utilized: a perforated-plate column for S_1 and a packed column for S_2. The basic design and cost data that

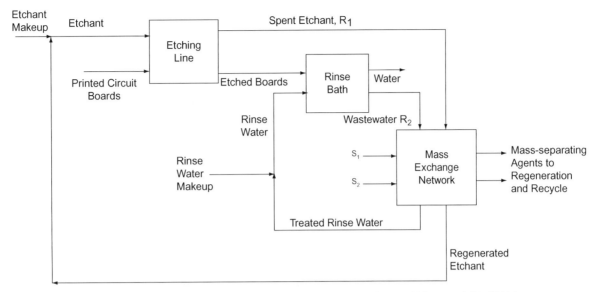

FIGURE 16-6 Recovery of copper from liquid effluents of an etching plant. *Source: El-Halwagi and Manousiouthakis (1990a).*

TABLE 16-12 Stream Data for Copper Etching Problem

Rich Streams				MSAs						
Stream	G_i (kg/s)	y_i^s	y_i^t	Stream	L_j^C (kg/s)	x_j^s	x_j^t	m_j	b_j	C_j ($/kg)
R_1	0.25	0.13	0.10	S_1	∞	0.030	0.070	0.734	0.001	0.01
R_2	0.10	0.06	0.02	S_2	∞	0.001	0.020	0.111	0.013	0.12

Source: El-Halwagi and Manousiouthakis (1990a).

should be employed in this problem are given by El-Halwagi and Manousiouthakis (1990a).

It is desired to synthesizes an optimum MEN that features minimum total annualized cost (TAC), where

TAC = annualized fixed cost + operating cost.

(Hint: Vary the minimum allowable composition differences to trade off fixed versus operating costs iteratively.)

16.9. Styrene can be produced by the dehydrogenation of ethylbenzene using live steam over an oxide catalyst at temperatures of 600°C to 650°C (Stanley and El-Halwagi, 1995). Figure 16.7 shows the process flow sheet. The first step in the process is to convert ethylene and benzene into ethylbenzene. The reaction products are cooled and separated. One of the separated streams is an aqueous waste (R_1). The main pollutant in this stream is benzene. Ethylbenzene leaving the separation section is fed to the styrene reactor whereby styrene and hydrogen are formed. Furthermore, by-products (benzene, ethane, toluene, and methane) are generated. The reactor product is then cooled and decanted. The aqueous layer leaving the decanter is a wastewater stream (R_2) that consists of steam condensate saturated with benzene. The organic layer, consisting of styrene, benzene, toluene, and unreacted ethylbenzene, is sent to a separation section.

There are two primary sources for aqueous pollution in this process: the condensate streams R1 (1000 kg/hr) and R2 (69,500 kg/hr). Both streams have the same supply composition, which corresponds to the solubility of benzene in water, 1770 ppm (1.77 × 10^{-3} kg benzene/kg water). Consequently, they may be combined as a single stream. The target composition is 57 ppb as dictated by the VOC environmental regulations called National Pollutant Discharge Elimination System (NPDES).

Three mass-exchange operations are considered: steam stripping, air stripping, and adsorption using granular activated carbon. Table 16.13 gives the stream data.

Using linear programming, determine the MOC solution of the system.

16.10. In the previous problem, it is desired to compare the total annualized cost of the benzene-recovery system to the value of recovered benzene. The total annualized cost for the network is defined as:

TAC = Annual operating cost + 0.2 × Fixed capital cost.

The fixed cost ($) of a moving-bed adsorption or regeneration column is given by 30,000$V^{0.57}$, where V is the volume of the column (m^3) based

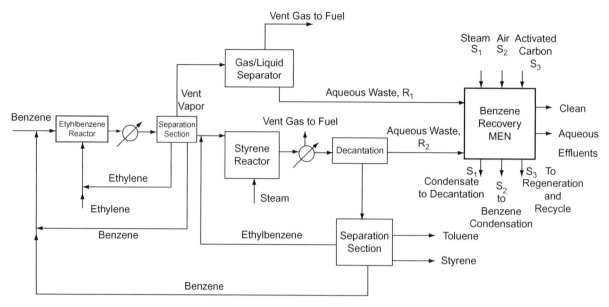

FIGURE 16-7 Schematic flow sheet of a styrene plant. *Source: Stanley and El-Halwagi (1995).*

TABLE 16-13 Data for MSAs in Styrene Plant Problem

Stream	L_j^c (kg/s)	Supply Composition x_j^s	Target Composition x_j^t	m_j	ε_j	C_j ($/kg MSA)
Steam (S_1)	∞	0	1.62	0.5	0.15	0.004
Air (S_2)	∞	0	0.02	1.0	0.01	0.003
Carbon(S_3)	∞	3×10^{-5}	0.20	0.8	5×10^{-5}	0.026

(All compositions and equilibrium data are in mass ratios, kg benzene/kg benzene-free MSA.)

on a 15-minute residence time for the combined flow rate of carbon and wastewater (or steam). A steam stripper already exists on site with its piping, ancillary equipment, and instrumentation. The column will be salvaged for benzene recovery. The only changes needed involve replacing the plates inside the column with new sieve trays. The fixed cost of the sieve trays is $1750/plate. The overall column efficiency is assumed to be 65 percent.

If the value of recovered benzene is taken as $0.20/kg, compare the annual revenue from recovering benzene to the TAC of the MEN.

16.11. In many situations, there is a trade-off between reducing the amount of generated waste at the source and recovering it via a separation system. For instance, in Problem 16.9, live steam is used in the styrene reactor to enhance the product yield. However, the steam eventually constitutes the aqueous waste, R_2. Hence, the higher the flow rate of steam, the larger the cost of the benzene recycle/reuse MEN. These opposing effects call for simultaneous consideration of source reduction of R_2 and its recycle/reuse. One way of approaching this problem is by invoking economic criteria. Let us define the economic potential of the process ($/yr) as follows (Stanley and El-Halwagi, 1995):

Economic potential = Value of produced styrene
+ Value of recovered ethylbenzene
− Cost of ethylbenzene − cost of steam
− TAC of the recycle/reuse network

Determine the optimal steam ratio (kg steam/kg ethylbenzene) that should be used in the styrene reactor to maximize the economic potential of the process.

NOMENCLATURE

C_j unit cost of the jth MSA including regeneration and makeup ($/kg of recirculating MSA)
d_j index for substreams of the jth MSA
$E_{i,j,m}$ a binary integer variable designating the existence or absence of an exchanger between rich stream i and lean stream j in subnetwork m
G_i flow rate of the ith waste stream
i index of waste streams
j index of MSAs
k index of composition intervals
L_j flow rate of the jth MSA (kg/s)
L_{j,d_j} flow rate of substream of d_j the jth MSA (kg/s)
L_j^c upper bound on available flow rate of the jth MSA (kg/s)

m	subnetwork index (given the index value one above the pinch and two below the pinch)
m_j	slope of equilibrium line for the jth MSA
N_{int}	number of composition intervals
N_R	number of waste streams
N_S	number of MSAs
ND_j	number of substreams for the jth MSA
R_i	the ith waste stream
R_m	a set defined by Eq. 16.9
$R_{m,k}$	a set defined by Eq. 16.11
S_j	the jth MSA
S_m	a set defined by Eq. 16.10
$S_{m,k}$	a set defined by Eq. 16.12
SN_m	subnetwork m
$U_{i,j,m}$	upper bound on exchangeable mass between i and j in subnetwork m as defined by Eq. 16.13 (kg/s)
$W_{i,j,k}$	exchangeable load between the ith waste stream and the jth MSA in the kth interval (kg/s)
$W_{i,k}^R$	exchangeable load of the ith waste stream, which passes through the kth interval as defined by Equations 16.2 and 16.3 (kg/s)
$w_{j,k}^S$	exchangeable load of the jth MSA, which passes through the kth interval as defined by Equations 16.5 and 16.6 (kg/s)
W_k^R	the collective exchangeable load of the waste streams in interval k as defined by Eq. 16.4 (kg/s)
$x_{j,k-1}$	composition of key component in the jth MSA at the upper horizontal line defining the kth interval
$x_{j,k}$	composition of key component in the jth MSA at the lower horizontal line defining the kth interval
x_j^{out}	outlet composition of the jth MSA (defined by Eq. 16.8)
x_j^s	supply composition of the jth MSA
x_j^t	target composition of the jth MSA
y_{k-1}	composition of key component in the ith waste stream at the upper horizontal line defining the kth interval
y_k	composition of key component in the ith waste stream at the lower horizontal line defining the kth interval

REFERENCES

El-Halwagi MM: A process synthesis approach to the dilemma of simultaneous heat recovery, waste reduction and cost effectiveness. In El-Sharkawy AI, Kummler RH, editors: *Proceedings of the Third Cairo International Conference on Renewable Energy Sources*, vol 2, pp. 579–594, 1993.

El-Halwagi MM, Manousiouthakis V: Automatic synthesis of mass-exchange networks with single-component targets, *Chem Eng Sci* 45(9):2813–2831, 1990a.

El-Halwagi MM, Manousiouthakis V: Simultaneous synthesis of mass exchange and regeneration networks, *AIChE J* 36(8):1209–1219, 1990b.

Garrison GW, Cooley BL, El-Halwagi MM: Synthesis of mass exchange networks with multiple target mass separating agents, *Dev Chem Eng Miner Proc* 3(1):31–49, 1995.

Stanley C, El-Halwagi MM: Synthesis of mass-exchange networks using linear programming techniques. In Rossiter AP, editor: *Waste minimization through process design*, New York, McGraw Hill, pp 209–225, 1995.

CHAPTER 17

SYNTHESIS OF REACTIVE MASS-EXCHANGE NETWORKS

Chapters 5, 11, and 16 covered the synthesis of physical mass exchange networks. In these systems, the targeted species was transferred from the rich phase to the lean phase in an intact molecular form. In some cases, it may be advantageous to convert the transferred species into other compounds using reactive mass-separating agents (MSAs). Typically, reactive MSAs have a greater capacity and selectivity to remove an undesirable component than physical MSAs. Furthermore, because they react with the undesirable species, it may be possible to convert pollutants into other species that may either be reused within the plant itself or sold.

The synthesis of a network of reactive mass exchangers involves the same challenges described in synthesizing physical MENs. The problem is further compounded by virtue of the reactivity of the MSAs. Driven by the need to address this important problem, Srinivas and El-Halwagi introduced the problem of synthesizing *reactive mass-exchange networks* (REAMENs) and developed systematic techniques for its solution (Srinivas and El-Halwagi, 1994; El-Halwagi and Srinivas, 1992). This chapter provides the basic principles of synthesizing REAMENs. The necessary thermodynamic concepts are covered. Chemical equilibrium is then tackled in a manner that renders the REAMEN synthesis task close to the MEN problem. Finally, an optimization-based approach is presented and illustrated by a case study.

OBJECTIVES OF REAMEN SYNTHESIS

The problem of synthesizing REAMENs can be stated as follows (El-Halwagi and Srinivas, 1992):

> Given a number N_R of waste (rich) streams and a number N_S of lean streams (physical and reactive MSAs), it is desired to synthesize a cost-effective network of physical and/or reactive mass exchangers that can preferentially transfer a certain undesirable species, A, from the waste streams to the MSAs whereby it may be reacted into other species. Given also are the flow rate of each waste stream (G_i), its supply (inlet) composition (y_i^s), and its target (outlet) composition (y_i^t), where $i = 1, 2, \ldots, N_R$. In addition, the supply and target compositions $(x_j^s$ and $x_j^t)$ are given for each MSA, where $j = 1, 2, \ldots, N_S$. The flow rate of any lean stream (L_j) is unknown but is bounded by a given maximum available flow rate of that stream; that is,

$$[17.1] \qquad L_j \leq L_j^C.$$

Figure 17.1 is a schematic illustration of the REAMEN synthesis problem.

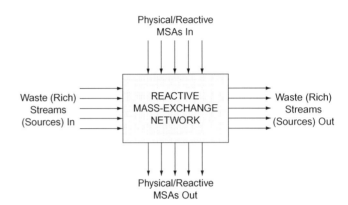

FIGURE 17-1 Schematic representation of the REAMEN synthesis problem.

As previously discussed in the synthesis of physical MENs, several design decisions are to be made:

- Which mass-exchange operations should be used (absorption, adsorption, and so on)?
- Which MSAs should be selected (for example, physical/reactive transfer, which solvents, adsorbents)?
- What is the optimal flow rate of each MSA?
- How should these MSAs be matched with the waste streams (that is, stream pairings)?
- What is the optimal system configuration (for example, how should these mass exchangers be arranged? Is there any stream splitting and mixing?)?

The first step in synthesizing a REAMEN is to establish the conditions for which the reactive mass exchange is thermodynamically feasible. The next section covers this issue.

CORRESPONDING COMPOSITION SCALES FOR REACTIVE MASS EXCHANGE

The fundamentals of reactive mass exchange, design of individual units, chemical equilibrium, and kinetics are covered in the literature (for example, Friedly, 1991; El-Halwagi, 1971, 1990; Kohl and Reisenfeld, 1997; Astarita et al., 1983). This section presents the salient basics of these systems.

To establish the conditions for thermodynamic feasibility of reactive mass exchange, it is necessary to invoke the basic principles of mass transfer with chemical reactions. Consider a lean phase j that contains a set $B_j = \{B_{z,j} \mid z = 1, \ldots, NZ_j\}$ of reactive species (that is, the set B_j contains NZ_j reactive species, each denoted by $B_{z,j}$, where the index z assumes values from 1 to NZ_j). These species react with the transferrable key solute, A, or among themselves via Q_j independent chemical reactions that may be represented by

$$[17.2] \qquad A + \sum_{z=1}^{NZ_j} \nu_{1,z,j} B_{z,j} = 0$$

and

$$[17.3] \qquad \sum_{z=1}^{NZ_j} \nu_{q_j,z,j} B_{z,j} = 0, \qquad q_j = 2, 3, \ldots, Q_j,$$

where the stoichiometric coefficients $\nu_{q,z,j}$ are positive for products and negative for the reactants. It is worth noting that stoichiometric equations can be mathematically handled as algebraic equations. Therefore, although component A may be involved in more than the first reaction, one can always manipulate the stoichiometric equations algebraically to keep A in the first reaction and eliminate it from the other stoichiometric equations.

Compositions of the different species can be tracked by relating them to the extents of the reactions through the following expression:

$$[17.4] \qquad b_{z,j} = b_{z,j}^o - \sum_{q_j=2}^{Q_j} \nu_{q_j,z,j} \xi_{q_j} \qquad (z = 1, 2, \ldots, NZ_j),$$

where $b_{z,j}$ is the composition of species $B_{z,j}$ in the jth lean phase, $b_{z,j}^o$ is the admissible composition of species $B_{z,j}$ in the jth lean phase, and ξ_{q_j} is the extent of the q_jth reaction (or the q_jth reaction coordinate). The extent of the reaction is defined for reactions 2 through Q_j. The reason for not defining it for the first reaction is that the variable u_j plays indirectly the role of the extent of reaction for the first reaction.[1] The admissible compositions may be selected as the lean-phase composition at some particular instant of time, or any other situation that is compatible with stoichiometry and mass-balance bounds.

The equilibrium constant of a reaction is the product of compositions of reactants and products each raised to its stoichiometric coefficient. Hence, for the reaction described by Eq. 17.2, one may write

$$K_{1,j} = \frac{1}{a_j} \prod_{z=1}^{NZ_j} \left(b_{z,j}^o - \sum_{q_j=2}^{Q_j} \nu_{q_j,z,j} \xi_{q_j} \right)^{\nu_{1,z,j}},$$

that is,

$$[17.5] \qquad a_j = \frac{1}{K_{1,j}} \prod_{z=1}^{NZ_j} \left(b_{z,j}^o - \sum_{q_j=2}^{Q_j} \nu_{q_j,z,j} \xi_{q_j} \right)^{\nu_{1,z,j}}$$

and for the reactions given by Eq. 17.3:

---

[1] For this reason, whenever there is a single reaction taking place in the jth MSA, no extent of reaction is defined. Instead, the fractional saturation, u_j, is employed.

$$[17.6] \qquad K_{q_j,j} = \prod_{z=1}^{NZ_j} \left(b_{z,j}^o - \sum_{q_j=2}^{Q_j} \nu_{q_j,z,j} \xi_{q_j} \right)^{\nu_{1,z,j}}, \qquad q_j = 2, 3, \ldots, Q_j,$$

where $K_{q_j,j}$ is the equilibrium constant for the q_jth reaction and a_j is the composition of the physically dissolved A in lean phase j.

It is now useful to recall the concepts of *molarity* and *fractional saturation* (Astarita et al., 1983). The molarity m_j of a reactive MSA is the total equivalent concentration of species that may react with component A. On the other hand, the fractional saturation u_j is a variable that represents the degree of saturation of chemically combined A in the jth lean phase. Therefore, $u_j m_j$ is the total concentration of chemically combined A in the jth MSA. Hence, the total concentration of A in MSA j can be expressed as

$$[17.7] \qquad x_j = a_j + u_j m_j, \qquad j \in S,$$

where the physically dissolved concentration of A, a_j, equilibrates with the rich-phase composition through a distribution function F_j; that is,

$$[17.8] \qquad y_i^* = F_j(a_j), \qquad j \in S.$$

Equations 17.4 through 17.8 represent a complete mathematical description of the chemical equilibrium between a rich phase and the jth MSA. Simultaneous solution of these nonlinear equations (for instance, by using the software LINGO) yields the equilibrium compositions of both phases in the following form:

$$[17.9] \qquad y_i^* = f_j(x_j^*) \qquad j \in S$$

For any mass-exchange operation to be thermodynamically feasible, the following conditions must be satisfied:

$$[17.10a] \qquad x_j < x_j^*, \qquad j \in S,$$

and/or

$$[17.10b] \qquad y_i > y_i^*, \qquad i \in R,$$

that is,

$$[17.11a] \qquad y = f(x_j + \varepsilon_j^S), \qquad j \in S,$$

where

$$[17.11b] \qquad y = y_i - \varepsilon_i^R, \qquad i \in R,$$

where ε_i^R and ε_j^S are positive quantities called, respectively, the rich and the lean minimum allowable composition differences. The parameters ε_i^R and ε_j^S are optimizable quantities that can be used for trading off capital versus operating costs (see Chapters 5 and 1). Equations 17.11a and 17.11b provide a correspondence among the rich and the lean composition scales for which mass exchange is practically feasible. This is the reactive equivalent to Eq. 5.5, used for establishing the corresponding composition scales for physical MENs.

Example 17-1 Absorption of H_2S in aqueous sodium hydroxide

Consider an aqueous caustic soda solution whose molarity $m_1 = 5.0 \, \text{kmol/m}^3$ (20 wt% NaOH). This solution is to be used in absorbing H_2S from a gaseous waste. The operating range of interest is $0.0 \leq x_1(\text{kmol/m}^3) \leq 5.0$. Derive an equilibrium relation for this chemical absorption over the operating range of interest.

SOLUTION

The absorption of H_2S in this solution is accompanied by the following chemical reactions (Astarita and Gioia, 1964):

[17.12] $$H_2S + NaOH = NaHS + H_2O$$

[17.13] $$H_2S + 2NaOH = 2NaHS + H_2O$$

As has been described earlier, the stoichiometric reactions should be manipulated algebraically to retain the transferable species (H_2S) only in the first equation. Therefore, H_2S can be eliminated from Eq. 17.13 by subtracting Eq. 17.12 from Eq. 17.13 to get

[17.14] $$NaHS + NaOH = Na_2S + H_2O.$$

Equations 17.12 and 17.14 can be written in ionic terms as follows:

[17.15] $$H_2S + OH^- = HS^- + H_2O$$

[17.16] $$HS^- + OH^- = S^{2-} + H_2O.$$

Let us denote the three ionic species as follows:

$$OH^- \equiv B_{1,1}$$
$$HS^- \equiv B_{2,1}$$
$$S^{2-} \equiv B_{3,1},$$

with aqueous-phase concentrations referred to as $b_{1,1}$, $b_{2,1}$, and $b_{3,1}$, respectively. Also, let us denote the composition of the physically dissolved H_2S in the aqueous solution as a_1.

For cases when the concentration of water remains almost constant with respect to the other species, one can define the following reaction equilibrium constants for Equations 17.15 and 17.16, respectively:

[17.17] $$K_{1,1} = \frac{b_{2,1}}{a_1 \cdot b_{1,1}}$$

and

[17.18] $$K_{2,1} = \frac{b_{3,1}}{b_{2,1} \cdot b_{1,1}},$$

where $K_{1,1} = 9.0 \times 10^6 \, \text{m}^3/\text{kmol}$ and $K_{2,1} = 0.12 \, \text{m}^3/\text{kmol}$.

The distribution coefficient for the physically dissolved H_2S is given by

[17.19] $$\frac{y_1}{a_1} = 0.368,$$

where y_1 is the composition of H_2S in the gaseous stream.

It is useful to relate the molarity of the aqueous caustic soda ($m_1 = 5.0 \, \text{kmol NaOH/m}^3$) to that of the other reactive species. Once the reactions start, the composition of NaOH will decrease. However, it is possible to relate the molarity of the solution to the concentration of the reactive species at any reaction coordinate. Suppose that after a certain extent of reaction (see Equations 17.15 and 17.16), an analyzer is placed in the solution to measure the compositions of OH^-, HS^-, and S^{2-}, with the measured concentrations denoted by $b_{1,1}$, $b_{2,1}$, and $b_{3,1}$, respectively. These measured concentrations are related to the molarity as follows. According to Eq. 17.15, $b_{2,1}$ kmol OH^- must have reacted to yield $b_{2,1}$ kmol HS^-. Similarly, according to Eq. 17.16, $b_{3,1}$ kmoles of OH^- must have reacted with $b_{3,1}$ kmol HS^- to yield $b_{3,1}$ kmoles of S^{2-}. But the $b_{3,1}$ kmoles of HS^- must have resulted from the reaction of $b_{3,1}$ kmol OH^- (according to Eq. 17.15). Hence, $2b_{3,1}$ kmol OH^- are consumed in producing $b_{3,1}$ kmol S^{2-}. Therefore, the molarity of the aqueous caustic soda can be related to the concentrations of the reactive ions as follows:

[17.20] $$5 = b_{1,1} + b_{2,1} + 2b_{3,1}.$$

There are two forms of reacted H_2S in the solution: HS^- and S^{2-}. By recalling that $u_1 m_1$ is the total concentration of chemically combined H_2S in the aqueous caustic soda and conducting an atomic balance on S over Equations 17.15 and 17.16, we get

[17.21] $$5u_1 = b_{2,1} + b_{3,1}.$$

As discussed earlier, the admissible compositions may be selected as the lean-phase composition at some particular instant of time, or any other situation that is compatible with stoichiometry and mass–balance bounds, such as Equations 17.20 and 17.21. Let us arbitrarily select the admissible composition of S^{2-} to be 0; that is,

[17.22] $$b_{3,1}^0 = 0.$$

This selection automatically fixes the corresponding values of $b_{1,1}^0$ and $b_{2,1}^0$. According to Equations 17.20 and 17.22, we get

[17.23] $$b_{2,1}^0 = 5u_1.$$

Similarly, according to Equations 17.21 through 17.23, we obtain:

[17.24] $$b_{1,1}^0 = 5(1 - u_1).$$

As mentioned earlier, the extent of reaction is defined for all reactions except the first one. Hence, we define ξ_2 as the extent of reaction for Eq. 17.16. Equation 17.4 can now be used to describe the compositions of the reactive species as a function of ξ_2 and the admissible compositions; that is,

[17.25] $$b_{1,1} = 5(1 - u_1) - \xi_2$$

[17.26] $$b_{2,1} = 5u_1 - \xi_2$$

[17.27] $$b_{3,1} = \xi_2.$$

Substituting from Equations 17.19, 17.25, 17.26, and 17.27 into Eq. 17.17, we get

$$9.0 \times 10^6 = \frac{5u_1 - \xi_2}{\frac{y_1}{0.368}[5(1 - u_1) - \xi_2]}$$

or

[17.28] $$y_1 = 4.09 \times 10^{-8} \frac{5u_1 - \xi_2}{[5(1 - u_1) - \xi_2]}.$$

Substituting from Equations 17.25 through 17.27 into Eq. 17.18, we obtain

$$0.12 = \frac{\xi_2}{[5u_1 - \xi_2][5(1 - u_1) - \xi_2]},$$

which can be rearranged to

[17.29] $$\xi_2^2 - 13.33\xi_2 + 25u_1(1 - u_1) = 0.$$

According to Eq. 17.7, the total concentration of H_2S in the aqueous caustic soda (physically dissolved and chemically reacted) can be expressed as follows:

[17.30a] $$x_1 = a_1 + 5u_1.$$

(Continued)

Example 17-1 Absorption of H_2S in aqueous sodium hydroxide (*Continued*)

In this system, the physically dissolved H_2S is much less than the chemically combined[2]; that is,

[17.30b] $$a_1 \ll 5u_1,$$

which simplifies Eq. 17.30a to

[17.30c] $$x_1 = 5_1.$$

Equations 17.28, 17.29, and 17.30c can be used to develop an expression for reactive mass exchange of H_2S. First, a value of fractional saturation is selected (where $\frac{x_1^s}{5} \le u_1 \le \frac{x_1^t}{5}$). Then, Eq. 17.30c is used to calculate the corresponding x_1. Next, Eq. 17.29 is solved to determine the value of ξ_2. Finally, Eq. 17.28 is solved to evaluate y_1. The pair (y_1, x_1) is in equilibrium. The same procedure is repeated for several values of u_1 (between $\frac{x_1^s}{5}$ and $\frac{x_1^t}{5}$) to yield pairs of (y_1, x_1) that are in equilibrium. Nonlinear regression can be employed to derive an equilibrium expression for these pairs.

To illustrate this procedure, let us start with a fractional saturation of $u_1 = 0.1$. According to Eq. 17.30c, $x_1 = 0.5 \, kmol/m^3$. By substituting for $u_1 = 0.1$ in Eq. 17.29, we obtain

$$\xi_2^2 - 13.33\xi_2 + 2.25 = 0,$$

which can be solved to get

[17.31] $$\xi_2 = \frac{13.33 - \sqrt{13.33 \times 13.33 - 4 \times 2.25}}{2}$$
$$= 0.171 \, kmol/m^3 \text{ (the other root is rejected)}$$

[2] This assumption can be numerically verified by comparing values of a_1 with $u_1 m_1$ after the equilibrium equation is generated.

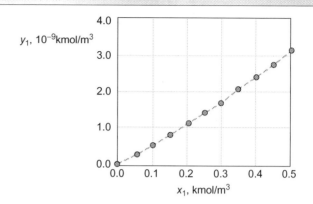

FIGURE 17-2 Equilibrium data for example of absorbing H_2S in caustic soda.

The values of u_1 and ξ_2 are plugged into Eq. 17.28 to get $y_1 = 3.1 \times 10^{-9} \, kmol/m^3$. Hence, the pair $(3.1 \times 10^{-9}, 0.5)$ is in equilibrium.[3] The same procedure is repeated for values of u_1 between 0.0 and 0.1. The results are plotted as shown in Fig. 17.2. The plotted data are slightly convex and can be fitted to the following quadratic function (all compositions in $kmol/m^3$):

[17.32] $$y_1 = 2.364 \times 10^{-9} x_1^2 + 5.022 \times 10^{-9} x_1$$

with a correlation coefficient r^2 of 0.9999.

[3] We can now test the validity of Eq. 17.30b. According to Eq. 17.19, $a_1 = 8.45 \times 10^{-9} \, kmol/m^3$, which is indeed much less than $u_1 m_1 = 0.5$.

SYNTHESIS APPROACH

Now that a procedure for establishing the corresponding composition scales for the rich-lean pairs of stream has been outlined, it is possible to develop the CID. The CID is constructed in a manner similar to that described in Chapter 11. However, it should be noted that the conversion among the corresponding composition scales may be more laborious because of the nonlinearity of equilibrium relations. Furthermore, a lean scale, x_j, represents all forms (physically dissolved and chemically combined) of the pollutant. First, a composition scale, y, for component A in any rich stream is created. This scale is in one-to-one correspondence with any composition scale of component A in the ith rich stream, y_i, via Eq. 17.11b. Then, Eq. 17.11a is used to generate N_S composition scales for component A in the lean streams. Next, each stream is represented by an arrow whose tail and head correspond to the supply and target compositions, respectively, of the stream. The partitions corresponding to these heads and tails establish the composition intervals. As with the CID for physical MENs, within any composition interval it is thermodynamically feasible to transfer component A from the rich streams to the lean streams. Also, according to the second law of thermodynamics, it is spontaneously possible to transfer component A from the rich streams in a given composition interval to any lean stream within a lower composition interval.

Note that the foregoing composition partitioning procedure ensures thermodynamic feasibility only when all the equilibrium relations described by Eq. 17.9 are convex. In this case, by merely satisfying Eq. 17.11 at both ends of a composition interval, Eq. 17.10 is automatically satisfied throughout that interval (Fig. 17.3a). On the other hand, when at least one of the equilibrium relations expressed by Eq. 17.9 is nonconvex, the satisfaction of Eq. 17.11 at both ends of an interval does not necessarily imply the realization of inequalities (Eq. 17.10) throughout that interval (Fig. 17.3b). In such a case, additional composition partitioning is needed. This can be achieved by discretizing the nonconvex portions of the equilibrium curves through linear overestimators.[4] Clearly, the number of these linear segments depends upon the desired degree of accuracy. For instance, Fig. 17.3b may be convexified by introducing two linear parts at the concavity inflection point (Fig. 17.3c). Consequently, additional intervals will be created within the CID by partitioning

[4] Srinivas and El-Halwagi (1994) and Chen and Hung (2005) have developed more rigorous methods of tackling nonconvex equilibrium.

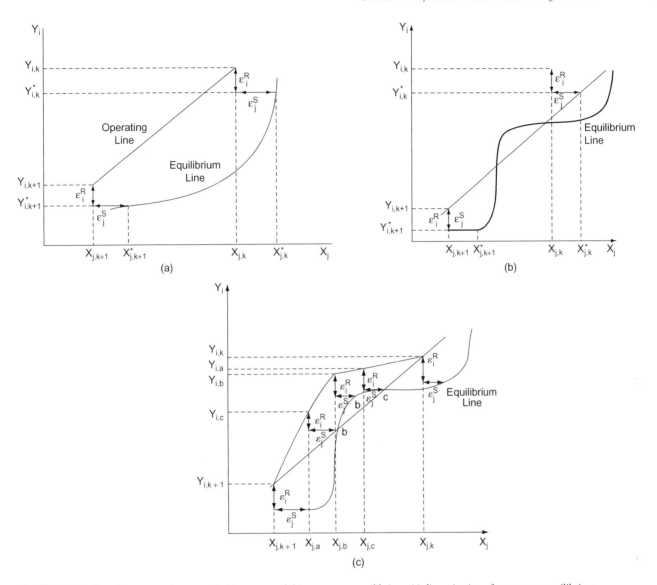

FIGURE 17-3 Reactive mass exchanger with (a) convex and (b) nonconvex equilibrium; (c) discretization of nonconvex equilibrium.
Source: El-Halwagi and Srinivas (1992).

the composition scales at the locations corresponding to points a, b, and c. The end result of this partitioning procedure is that the entire composition range is divided into n_{int} composition intervals, with $k = 1$ being the highest and $k = n_{int}$ being the lowest.

Having established a one-to-one thermodynamically feasible correspondence among all the composition scales, we can now solve the REAMEN problem via a transshipment formulation similar to that described in Chapter 16 for the synthesis of physical MENs.

Example 17-2 Removal of H$_2$S from kraft pulping process

Kraft pulping is a common process in the paper industry. Figure 17.4 shows a simplified flow sheet of the process. In this process, wood chips are reacted (cooked) with white liquor in a digester. White liquor (which contains primarily NaOH, Na$_2$S, Na$_2$CO$_3$, and water) is employed to dissolve lignin from the wood chips. The cooked pulp and liquor are passed to a blow tank where the pulp is separated from the spent liquor, "weak black liquor," which is fed to a recovery system for conversion to white liquor. The first step in recovery is concentration of the weak liquor via multiple effect evaporators. The concentrated solution is sprayed in a furnace. The smelt from the furnace is dissolved in water to form green liquor, which is reacted with lime (CaO) to produce white liquor and calcium carbonate "mud." The recovered white liquor is mixed with makeup materials and recycled to the digester. The calcium carbonate mud is thermally decomposed in a kiln to produce lime, which is used in the causticizing reaction.

There are several gaseous wastes emitted from the process (see Dunn and El-Halwagi, 1993, and Problem 17.5). In this example, we focus on the gaseous waste leaving the multiple effect evaporators, R$_1$, whose primary pollutant is H$_2$S. Table 17.1 gives stream data for this waste stream. A rich-phase minimum allowable composition difference, ε_i^R, of 1.5×10^{-10} kmol/m^3 is used.

A process lean stream and an external MSA are considered for removing H$_2$S. The process lean stream, S$_1$, is a caustic soda solution that can be used as

(Continued)

Example 17-2 Removal of H₂S from kraft pulping process (*Continued*)

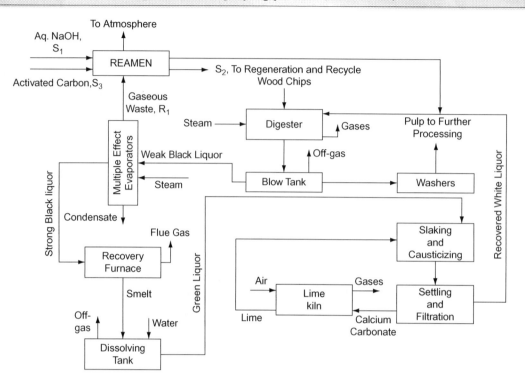

FIGURE 17-4 Kraft pulping process.

TABLE 17-1 Data for Gaseous Emission of Kraft Pulping Process

Stream	Description	Flow Rate G_i, m³/s	Supply Composition (10^{-10} kmol/m³) y_i^s	Target Composition (10^{-10} kmol/m³) y_i^t
R_1	Gaseous waste from evaporators	16.2	1600	3.0

TABLE 17-2 Data for MSAs of Kraft Pulping Example

Stream	Upper Bound on Flow Rate L_j^c m³/s	Supply Composition (kmol/m³) x_j^s	Target Composition (kmol/m³) x_j^t	ε_j^s (kmol/m³)	C_j $/m³ MSA
S_1	0.01	0.000	0.500	0.10	—
S_2	∞	0.010	0.025	0.01	2900

a solvent for the reactive separation of H_2S. A bonus for using the process MSA is the conversion of a portion of the absorbed H_2S into Na_2S, which is needed for white-liquor makeup. In other words, H_2S ("the pollutant") is converted into a valuable chemical that is needed in the process. The external MSA, S_2, is a polymeric adsorbent. Table 17.2 gives the data for the candidate MSAs. The equilibrium data for the transfer of H_2S from the waste stream to the adsorbent is given by

[17.33] $$y_1 = 2 \times 10^{-9} x_2,$$

where y_1 and x_2 are given in kmol/m³.

For the given data, determine the minimum operating cost of the REAMEN, and construct a network with the minimum number of exchangers.

SOLUTION

The following expression for the equilibrium data of H_2S in caustic soda has been derived in Eq. 17.32 of Example 17.1 (all compositions are in kmol/m³):

[17.32] $$y_1 = 2.364 \times 10^{-9} x_1^2 + 5.022 \times 10^{-9} x_1.$$

By invoking Eq. 17.11, we get the following equation for the practical feasibility curve:

[17.34] $$y_1 - 1.5 \times 10^{-10} = 2.364 \times 10^{-9}(x_1 + 0.1)^2 + 5.022 \times 10^{-9}(x_1 + 0.1).$$

Similarly, for activated carbon:

[17.35] $$y_1 - 1.5 \times 10^{-10} = 2 \times 10^{-9}(x_2 + 0.01).$$

(*Continued*)

Example 17-2 Removal of H₂S from kraft pulping process (*Continued*)

The CID for the problem is shown in Fig. 17.5. Tables 17.3 and 17.4 show the exchangeable loads for the waste streams and the MSAs, respectively.

According to the two-stage targeting procedure, we first minimize the annual operating cost of the MSAs (given by $3600 \times 8760 \times 2100 \times L_2 = 6.62256 \times 10^{10} L_2$ when we assume 8760 operating hours per year). By applying the linear programming formulation described in Chapter 16, one can write the following optimization program:

[P17.1] $\qquad \min 6.62256 \times 10^{10} L_2,$

subject to

$$\delta_1 = 25{,}270 \times 10^{-10}$$
$$\delta_2 - \delta_1 + 0.5\, L_1 = 559 \times 10^{-10}$$
$$\delta_3 - \delta_2 = 62 \times 10^{-10}$$
$$\delta_4 - \delta_3 = 0.0$$
$$-\delta_4 + 0.015\, L_2 = 0.0$$
$$\delta_k \geq 0 \quad k = 1,2,3,4$$
$$L_j \geq 0 \quad j = 1,2$$
$$L_1 \leq 0.01 \times 10^{-10}.$$

It is worth pointing out that the wide range of coefficients may cause computational problems for the optimization software. This is commonly referred to as the "scaling" problem. One way of circumventing this problem is to define scaled flow rates of MSAs in units of $10^{-10}\,m^3/s$ and scaled residual loads in units of $10^{-10}\,kmol/s$; that is, let

[17.36] $\qquad L_j^{scaled} = 10^{10} L_j, \quad j = 1,2$

[17.37] $\qquad \delta_k^{scaled} = 10^{10} \delta_k, \quad k = 1,2,3,4.$

With the new units, the scaled program becomes

[P17.2] $\qquad \min 6.62256\, L_2^{scaled}$

subject to

$$\delta_1^{scaled} = 25{,}270$$
$$\delta_2^{scaled} - \delta_1^{scaled} + 0.5\, L_1^{scaled} = 559$$
$$\delta_3^{scaled} - \delta_2^{scaled} = 62$$
$$\delta_4^{scaled} - \delta_3^{scaled} = 0.0$$
$$-\delta_4^{scaled} + 0.015\, L_2^{scaled} = 0.0$$
$$\delta_k^{scaled} \geq 0 \quad k = 1,2,3,4$$
$$L_j^{scaled} \geq 0 \quad j = 1,2$$
$$L_1^{scaled} \leq 0.01$$

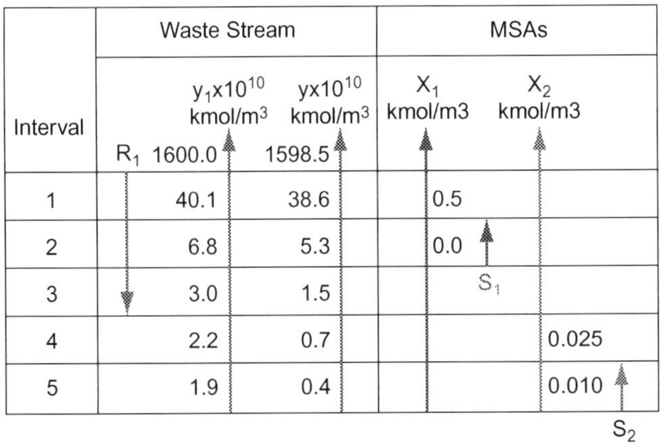

Interval	Waste Stream $y_1 \times 10^{10}$ kmol/m³ R₁ 1600.0	$y \times 10^{10}$ kmol/m³ 1598.5	MSAs X₁ kmol/m³	X₂ kmol/m³
1	40.1	38.6	0.5	
2	6.8	5.3	0.0	
3	3.0	1.5	S₁	
4	2.2	0.7		0.025
5	1.9	0.4		0.010

FIGURE 17-5 CID for kraft pulping example.

In terms of LINGO input, the scaled program can be written as follows (with the S in each variable indicating that it is a scaled variable):

```
model:
min = 6.62256*LS2;
deltaS1 = 25270;
deltaS2 - deltaS1 + 0.5*LS1 = 559;
deltaS3 - deltaS2 = 62;
deltaS4 - deltaS3 = 0.0;
deltaS4 + 0.015*LS2 = 0.0;
deltaS1 >= 0.0;
deltaS2 >= 0.0;
deltaS3 >= 0.0;
deltaS4 >= 0.0;
LS1 >= 0.0;
LS2 >= 0.0;
end
```

The solution report from LINGO gives the following results:

```
Objective value:        27373
Variable                Value
LS2                     4133.333
DELTAS1                 25270.00
DELTAS2                 0.0000000E+00
LS1                     51658.00
DELTAS3                 62.00000
DELTAS4                 62.00000
```

Therefore, the minimum operating cost is approximately \$27,000/yr, and the pinch location is between the second and third composition intervals. The REAMEN involves two exchangers: one above the pinch matching R₁ with S₁, and one below the pinch matching R₁ with S₂.

TABLE 17-3 The TEL for the Gaseous Waste

Interval	Load of R₁ (10^{-10} kmol H₂S/s)
1	25,270
2	559
3	62
4	0
5	0

TABLE 17-4 TEL for Lean Streams

	Capacity of Lean Streams per m³ of MSA (kmol H₂S/m³ MSA)	
Interval	S₁	S₂
1	—	—
2	0.5	—
3	—	—
4	—	—
5	—	0.015

HOMEWORK PROBLEMS

17.1. Derive an equilibrium expression for the reactive absorption of H_2S in diethanolamine (DEA). The molarity of DEA is $2 \, kmol/m^3$. The following reaction takes place:

$$H_2S + (C_2H_5)_2NH = HS^- + (C_2H_5)_2NH_2^+,$$

whose equilibrium constant is given by (Lal et al., 1985)

[17.38]
$$K = 234.48 = \frac{[HS^-][(C_2H_5)_2NH_2^+]}{[H_2S][(C_2H_5)_2NH]}$$

with all concentrations in $kmol/m^3$. The physical distribution coefficient is

[17.39]
$$0.363 = \frac{\text{Composition of } H_2S \text{ in gas } (kmol/m^3)}{\text{Composition of physically dissolved } H_2S \text{ in DEA } (kmol/m^3)}$$

17.2. Develop equilibrium equations for the reactive absorption of CO_2 into the following:
 1. Aqueous potassium carbonate
 2. Monoethanolamine

(Hint: See Astarita et al., 1983, pp. 68–79.)

17.3. Coal may be catalytically hydrogenated to yield liquid transportation fuels. A simplified process flow diagram of a coal-liquefaction process is shown in Fig. 17.6. Coal is mixed with organic solvents to form a slurry that is reacted with hydrogen. The reaction products are fractionated into several transportation fuels. Hydrogen sulfide is among the primary gaseous pollutants of the process (Warren et al., 1995). Hence, it is desired to design a cost-effective H_2S recovery system.

Two major sources of H_2S emissions from the process are the acid gas stream evolving from hydrogen manufacture, R_1, and the gaseous waste emitted from the separation section of the process, R_2, as shown in Fig. 17.6. Table 17.5 summarizes stream data for these acid gas streams.

Six potential MSAs should be simultaneously screened. These include absorption in water, S_1; adsorption onto activated carbon, S_2; absorption in chilled methanol, S_3; and the use of the following reactive solvents: diethanolamine (DEA), S_4; hot

TABLE 17-5 Data for the Waste Streams of the Coal-Liquefaction Problem

Stream	Flow Rate, L_i^C, m^3/s	Supply Composition, y_j^s, $kmol/m^3$	Target Composition, y_j^t, $kmol/m^3$
R_1	121.1	3.98×10^{-4}	2.1×10^{-7}
R_2	28.9	71.6×10^{-4}	2.1×10^{-7}

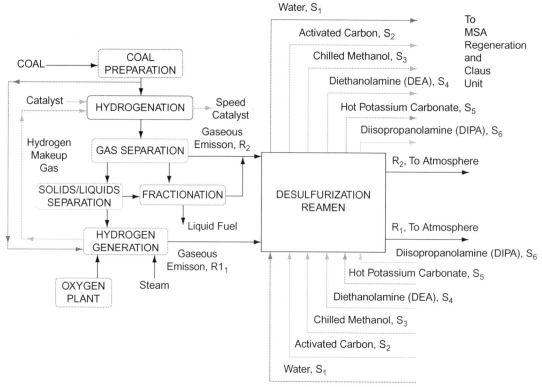

FIGURE 17-6 Coal-liquefaction process. *Source: Warren et al. (1995).*

potassium carbonate, S_5; and diisopropanolamine (DIPA), S_6.

Equilibrium relations governing the transfer of hydrogen sulfide from the gaseous waste streams to the various separating agents can be approximated over the range of operating compositions (kmol/m³) as follows:

[17.40] $\quad y = 0.398\, x_1$

[17.41] $\quad y = 0.015\, x_2$

[17.42] $\quad y = 0.027\, x_3$

[17.43] $\quad y = 0.079\, x_4^{2.6}$

[17.44] $\quad y = 0.013\, x_5$

[17.45] $\quad y = 0.010\, x_6$

TABLE 17-6 Data for MSAs of Coal-Liquefaction Problem

Stream	L_j^c m³/s	Supply Composition, x_j^s, 10^{-6} kmol/m³	Target Composition, x_j^t, kmol/m³
S_1	∞	1000	0.0150
S_2	∞	2.00	0.4773
S_3	∞	2.15	0.2652
S_4	∞	3.40	0.1310
S_5	∞	3.20	0.0412
S_6	∞	1.30	0.7160

The unit operating costs of water, S_1; activated carbon, S_2; chilled methanol, S_3; diethanolamine, S_4; hot potassium carbonate, S_5; and diisopropanolamine, S_6, are 0.001, 8.34, 2.46, 5.94, 3.97, and 4.82 in \$/m³, respectively. These costs include the cost of regeneration and makeup. Table 17.6 gives stream data for the MSAs. The values of ε_i^R and ε_j^S are taken to be 0.0 and 2×10^{-7} kmol/m³, respectively.

Synthesize a REAMEN that features the minimum number of units that realize the minimum operating cost.

17.4. Most of the world's rayon is produced through the viscose process (El-Halwagi and Srinivas, 1992). Figure 17.7 is a schematic representation of the process, in which cellulose pulp is treated with caustic soda, then reacted with carbon disulfide to produce cellulose xanthate. This compound is dissolved in dilute caustic soda to give a viscose syrup, which is fed to vacuum-flash boiling deaerator to remove air. The gaseous stream leaving the deaerator, R_1, should be treated for H_2S removal prior to its atmospheric discharge. In spinning, a viscose solution is extruded through fine holes submerged in an acid bath to produce the rayon fibers. The acid-bath solution contains sulfuric acid, which neutralizes caustic soda and decomposes xanthate and various sulfur-containing species, thus producing H_2S as the major hazardous compound in the exhaust gas stream, R_2.

It is desired to synthesize a REAMEN for treating the gaseous wastes (R_1 and R_2) of a viscose rayon plant. Three MSAs are available to select from: caustic soda, S_1 (a process stream already existing in the plant with $m_1 = 5.0$ kmol/m³),

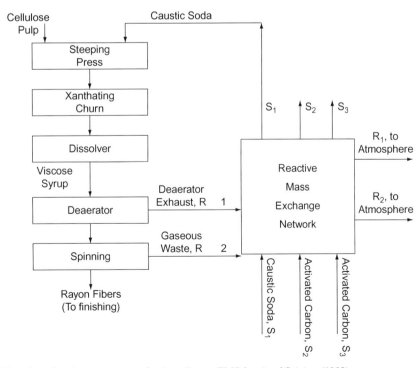

FIGURE 17-7 Simplified flow sheet for viscose-rayon production. *Source: El-Halwagi and Srinivas (1992).*

TABLE 17-7 Stream Data for the Viscose-Rayon Example

	Rich Streams				Lean Streams		
Stream	G_i	y_i^S (kmol/m³)	y_i^t (kmol/m³)	Stream	L_j^C (m³/s)	x_j^S (kmol/m³)	x_j^t (kmol/m³)
R_1	0.87	1.3×10^{-5}	2.2×10^{-7}	S_1	2.0×10^{-4}	0.0	0.1
R_2	0.10	0.9×10^{-5}	2.2×10^{-7}	S_2	∞	2.0×10^{-6}	1.0×10^{-3}
				S_3	∞	1.0×10^{-6}	3.0×10^{-6}

diethanolamine, S_2 (with $m_2 = 2.0$ kmol/m³), and activated carbon, S_3. The unit costs for S_2 and S_3, including stream makeup and subsequent regeneration, are 64.9 \$/m³ and 169.4 \$/m³, respectively. Table 17.7 gives stream data.

The chemical absorption of H_2S into caustic-soda solution (Astarita and Gioia, 1964) involves the following two reactions:

[17.46] $\quad H_2S + OH^- \rightarrow HS^- + H_2O$

[17.47] $\quad HS^- + OH^- \rightarrow S^{2-} + H_2O.$

Because the concentration of water remains approximately constant, one can define the following two reaction equilibrium constants:

[17.48] $\quad K_{1,1} = \dfrac{[HS^-]}{[H_2S][OH^-]}$

and

[17.49] $\quad K_{2,1} = \dfrac{[S^{2-}]}{[HS^-][OH^-]}$

where $K_{1,1} = 9.0 \times 10^6$ m³/kmol and $K_{2,1} = 0.12$ m³/kmol. The distribution coefficient for the physically dissolved portion of H_2S between rich-phase and caustic-soda solution is given by

[17.50] $\quad y_i/a_1 = 0.368.$

The overall reaction of hydrogen sulfide with diethanol amine (Lal et al., 1985) is given by

[17.51]
$$H_2S + (C_2H_4OH)_2NH \rightleftharpoons HS^- + (C_2H_4OH)_2NH_2^+,$$

for which the equilibrium constant is given by

[17.52] $\quad K_{1,2} = 234.48 = \dfrac{[HS^-][R_2NH_2^+]}{[H_2S][R_2NH]}$

and the physical distribution coefficient is given by (Kent and Eisenberg, 1976)

[17.53] $\quad y_i/a_2 = 0.363.$

The adsorption isotherm for H_2S on activated carbon is represented by (Valenzuela and Myers, 1989):

[17.54] $\quad y_i/a_3 = 0.015.$

The following values for the minimum allowable composition differences are selected:

$$\varepsilon_1^S = 3.50, \quad \varepsilon_2^S = 0.014, \quad \varepsilon_3^S = 10^{-6},$$
$$\varepsilon_1^R = 10^{-7}, \quad \varepsilon_2^R = 1.5 \times 10^{-7} \text{ kmol/m}^3$$

17.5. Consider the kraft pulping process shown in Fig. 17.8 (Dunn and El-Halwagi, 1993). The first step in the process is digestion in which wood chips, containing primarily lignin, cellulose, and hemicellulose, are "cooked" in white liquor (NaOH, Na_2S, Na_2CO_3, and water) to solubilize the lignin. The off-gases leaving the digester contain substantial quantities of H_2S. The dissolved lignin leaves the digester in a spent solution referred to as the "weak black liquor." This liquor is processed through a set of multiple-effect evaporators designed to increase the solid content of this stream from approximately 15 percent to approximately 65 percent. At the higher concentration, this stream is referred to as strong black liquor. The contaminated condensate removed through the evaporation process can be processed through an air stripper to transfer sulfur compounds (primarily H_2S) to an air stream prior to further treatment and discharge of the condensate stream. The strong black liquor is burned in a furnace to supply energy for the pulping processes and to allow the recovery of chemicals needed for subsequent pulp production. The burning of black liquor yields an inorganic smelt (Na_2CO_3 and Na_2S) that is dissolved in water to produce green liquor (NaOH, Na_2S, Na_2CO_3, and water), which is reacted with quick lime (CaO) to convert the Na_2CO_3 into NaOH. The conversion of the Na_2CO_3 into NaOH is referred to as the causticizing reaction and involves two reactions. The first reaction is the conversion of calcium oxide to calcium hydroxide in the presence of water in an agitated slaker. The calcium hydroxide subsequently reacts with Na_2CO_3 to form NaOH and a calcium carbonate precipitate. The calcium carbonate is then heated in the lime kiln to regenerate the calcium oxide and release carbon dioxide. These reactions result in the formation of the original white liquor for reuse in the digesting process.

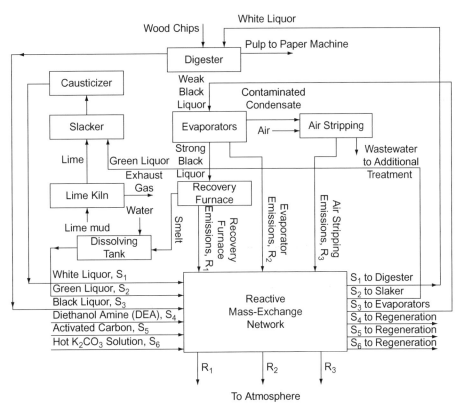

FIGURE 17-8 Simplified flow sheet of the kraft pulping process. *Source: Dunn and El-Halwagi (1993).*

TABLE 17-8 Data for Gaseous Emissions of Kraft Pulping Process

Stream	Description	Flow Rate, G_i, m³/s	Supply Composition (10^{-5} kmol/m³) y_i^s	Target Composition (10^{-7} kmol/m³) y_i^t
R_1	Gaseous waste from recovery furnace	117.00	3.08	2.1
R_2	Gaseous waste from evaporator	0.43	8.2	2.1
R_3	Gaseous waste from air stripper	465.80	1.19	2.1

TABLE 17-9 Data for MSAs of Kraft Pulping Example

Stream	Description	Upper Bound on Flow Rate, L_j^c, m³/s	Supply Composition (kmol/m³) x_j^s	Target Composition (kmol/m³) x_j^t
S_1	White liquor	0.040	0.320	3.100
S_2	Green liquor	0.049	0.290	1.290
S_3	Black liquor	0.100	0.020	0.100
S_4	DEA	∞	2×10^{-6}	0.020
S_5	Activated carbon	∞	1×10^{-6}	0.002
S_6	Hot potassium carbonate	∞	0.003	0.280

Three major sources in the kraft process are responsible for the majority of the H_2S emissions. These involve the gaseous waste streams leaving the recovery furnace, the evaporator, and the air stripper, respectively denoted by R_1, R_2, and R_3. Table 17.8 summarizes stream data for the gaseous wastes. Several candidate MSAs are screened. These include three process MSAs and three external MSAs. The process MSAs are the white, green, and black liquors (referred to as S_1, S_2, and S_3, respectively). The external MSAs include DEA, S_4; activated carbon, S_5; and 30 wt% hot potassium carbonate solution, S_6. Table 17.9 summarizes stream data for the MSAs. Synthesize an MOC REAMEN that can accomplish the desulfurization task for the three waste streams.

NOMENCLATURE

a_j	physically dissolved concentration of A in lean stream j (kmol/m^3)
A	key transferable component
$b_{z,j}$	composition of reactive species $B_{z,j}$ in lean phase j (kmol/m^3)
$b^o_{z,j}$	admissible composition of reactive species $B_{z,j}$ in lean phase j (kmol/m^3)
B_j	set of reactive species in lean stream j
$B_{z,j}$	index of reactive species in lean stream j
c_j	unit cost of the jth MSA j ($/kg)
f_j	equilibrium distribution function between rich-phase composition and total content of A in lean phase j, defined in Eq. 17.9
F_j	equilibrium distribution function between rich-phase composition and physically dissolved A in lean phase j, defined in Eq. 17.8
G_i	flow rate of rich stream i (kmol/s)
i	index for rich streams
j	index for lean streams
k	index for composition intervals
$K_{q_j,j}$	equilibrium constant for the q_jth reaction in lean phase j
L_j	flow rate of lean stream j (kmol/s)
L^c_j	upper bound on flow rate of lean stream j, kmol/s index for subnetworks
m_j	molarity of lean stream j (kmol/m^3)
n_p	number of mass-exchange pinch points in the problem
N_R	number of rich streams
N_S	number of lean streams
NZ_j	number of reactive species in lean stream j
q_j	index for the independent reactions in lean stream j
Q_j	number of independent reactions in lean stream j
r^2	correlation coefficient
R	set of rich streams
R_i	the ith rich stream
S	set of lean streams
S_j	the jth lean stream
u_j	fractional saturation of chemically combined A in the jth MSA
x_j	composition of key component in lean stream j (kmol/m^3)
x^s_j	supply composition of key component in lean stream j (kmol/m^3)
x^t_j	target composition of key component in lean stream j (kmol/m^3)
$x_{j,k}$	the upper bound composition for interval k on the scale S_j (kmol/m^3)
x^*_j	equilibrium composition of key component in lean stream j (kmol/m^3)
y	composition of key component in any rich stream (kmol/m^3)
y_i	composition of key component in rich stream i (kmol/m^3)
y^s_i	supply composition of key component in rich stream i (kmol/m^3)
y^t_i	target composition of key component in rich stream i (kmol/m^3)
y^*	equilibrium composition of key component in any rich stream (kmol/m^3)
z	index for the reactive species

GREEK LETTERS

δ_k	residual load leaving interval k (kmol/s)
$\delta_{i,k}$	residual load leaving interval k for rich stream i (kmol/s)
ε^R_i	rich-phase minimum allowable composition difference (kmol/m^3)
ε^S_j	lean-phase minimum allowable composition difference (kmol/m^3)
$\nu_{q_j,z,j}$	stoichiometric coefficient of reactive species z in reaction q_j in lean phase j
ξ_{q_j}	extent of reaction q_j in the jth MSA

SPECIAL SYMBOL

[ ]	concentration (kmol/m^3)

REFERENCES

Astarita G, Gioia F: Hydrogen sulfide chemical absorption, *Chem Eng Sci* 19:963–971, 1964.

Astarita G, Savage DW, Bisio A: *Gas treating with chemical solvents*, New York, Wiley, 1983.

Chen CL, Hung PS: Simultaneous synthesis of mass exchange networks for waste minimization, *Comp Chem Eng* 29:1561–1576, 2005.

Dunn RF, El-Halwagi MM: Optimal recycle/reuse policies for minimizing the wastes of pulp and paper plants, *J Environ Sci Health* A28(1):217–234, 1993.

El-Halwagi MM: An engineering concept of reaction rate, *Chem Eng* May, pp. 75–78, 1971.

El-Halwagi MM: Optimization of bubble column slurry reactors via natural delayed feed addition, *Chem Eng Commun* 92:103–119, 1990.

El-Halwagi MM, Srinivas BK: Synthesis of reactive mass-exchange networks, *Chem Eng Sci* 47(8):2113–2119, 1992.

Friedly JC: Extent of reaction in open systems with multiple heterogeneous reactions, *AIChE J* 37(5):687–693, 1991.

Kent RL, Eisenberg B: Better data for amine treating, *Hydrocarbon Process* February, pp. 87–90, 1976.

Kohl A, Reisenfeld F: Gas purification editors: (ed 5), Houston, Texas, Gulf Publ. Co., 1997.

Lal D, Otto FD, Mather AE: The solubility of H_2S and CO_2 in a diethanolamine solution at low partial pressures, *Can J Chem Eng* 63:681–685, 1985.

Srinivas BK, El-Halwagi MM: Synthesis of reactive mass-exchange networks with general nonlinear equilibrium functions, *AIChE J* 40(3):463–472, 1994.

Valenzuela DP, Myers A: *Gas Purification Adsorption equilibrium data handbook*, Englewood Cliffs, New Jersey, Prentice Hall, 1989.

Warren A, Srinivas BK, El-Halwagi MM: Design of cost-effective waste-reduction systems for synthetic fuel plants, *J Environ Eng* 121(10):742–747, 1995.

CHAPTER 18
MATHEMATICAL OPTIMIZATION TECHNIQUES FOR MASS INTEGRATION

Chapter 6 presented a graphical approach to the development of combined mass-integration strategies. This chapter presents mathematical optimization techniques for mass integration including mass-exchange interception and rerouting of targeted species throughout the process. Because the interception devices in this chapter are limited to mass-exchange operations, the waste-interception network becomes a mass-exchange network (MEN). Other separation networks (for example, heat-induced separation and membrane systems described in Chapters 12 and 13) can be similarly included in the interception network.

The emphasis in this chapter is on targeting pollution at the heart of the plant rather than dealing with pollutants in terminal waste streams. The chapter also provides a mathematical framework for tackling pollution as a multimedia (multiphase) problem.

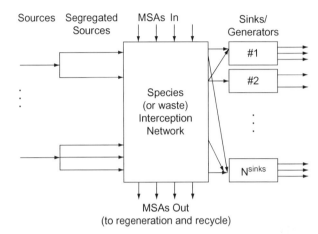

FIGURE 18-1 Schematic representation of mass-integration framework with MSA-induced interception. *Source: El-Halwagi et al. (1996).*

PROBLEM STATEMENT AND CHALLENGES

The problem addressed in this chapter can be briefly stated as follows:

Given a process with terminal gaseous and liquid wastes that contain a certain pollutant, it is desired to identify minimum-cost strategies for segregation, mixing, *in-plant* pollutant interception using mass-exchange operations, recycle, and sink/generator manipulation that can reduce the pollutant load and concentration in the terminal waste streams to a specified level (Fig. 18.1).

To identify the optimum solution for the aforementioned problem, one should be able to answer the following challenging questions:

- Which phase(s) (gaseous, liquid) should be intercepted?
- Which process streams should be intercepted?
- To what extent should the pollutant be removed from each process stream?
- Which separation operations should be used for interception?
- Which mass-separating agents (MSAs) should be selected for interception?
- What is the optimal flow rate of each MSA?
- How should these separating agents be matched with the sources (that is, stream pairings)?
- Which units (sinks/generators) should be manipulated for source reduction? By what means?
- Should any streams be segregated or mixed? Which ones?
- Which streams should be recycled/reused? To what units?

Because of the complexity of these questions, the solution approach will be presented in stages. The next section presents a systematic method for in-plant interception using MSAs. Later, interception will be integrated with the other mass-integration strategies.

SYNTHESIS OF MSA-INDUCED INTERCEPTION NETWORKS

In Chapters 5, 11, and 16, the MEN-synthesis techniques dealt with cases where the separation task was defined as part of the design task. Streams to be separated were given along with their supply and target compositions. Furthermore, separation of one source (rich stream) had no effect on the other sources. As has been discussed in the section entitled "Problem Statement and Challenges," when mass integration involving in-plant interception is considered, designers have to determine which streams are to be intercepted and the extent of separation for each stream. Moreover, interception of one source may affect the other sources. Hence, designers should understand the global flow of pollutants and mass interactions throughout the process and employ this understanding in setting optimal interception strategies. Toward this end, a particularly useful tool is the path diagram.

THE PATH DIAGRAM

The **path diagram** (for example, El-Halwagi et al., 1996; El-Halwagi and Spriggs, 1996; El-Halwagi and Spriggs, 1998a, 1998b; Garrison et al., 1996; Hamad et al., 1996) is a representation tool that captures the overall flow of the pollutant throughout the plant. It also relates the flow

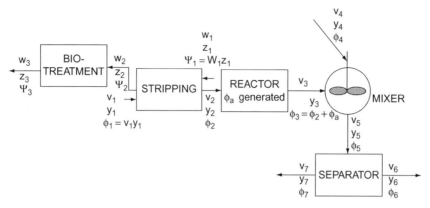

FIGURE 18-2 Loads and compositions of pollutant throughout a process. *Source: El-Halwagi et al. (1996).*

of the pollutant to the performance of the different processing units. A path diagram represents the load of the targeted species throughout the process as a function of its composition in carrying streams. Instead of considering the whole flow sheet, one should keeps track of the units involving the targeted species. Hence, the pollutant is tracked throughout the process via material balances and unit modeling equations. A path diagram is created for each targeted species in each phase (gaseous, liquid, solid). Each pollutant-laden stream (source) is represented by a node on a load-composition diagram. These nodes are connected with composition profiles of the streams within the units. The exact shape of the composition profile within a unit is typically not needed unless modifications within the units are considered. Therefore, these profiles can be approximated by arrows of any shape. The directionality of these arrows reflects the orientation of mass flow: arrow tails emanate from inputs to units, and arrow heads point toward outputs of units. A unit may have multiple inputs and outputs, so each node may be associated with multiple arrow heads and tails.

To demonstrate the construction of the path diagram, consider Fig. 18.2 (El-Halwagi et al., 1996), which illustrates a section of a process involving the pollutant-laden streams. Figure 18.2 also shows the various loads and compositions of the pollutant throughout the process. For a given pollutant-laden stream v, the term V_v is the flow rate of the stream, y_v is the composition of the pollutant, and Φ_v is the load of the pollutant in the stream, defined as

[18.1] $$\Phi_v = V_v y_v.$$

Similarly, on the liquid path, the load of the pollutant in the wth liquid source is given by

[18.2] $$\Psi_w = W_w z_w,$$

where Ψ_w, W_w, and z_w are the load, flow rate, and composition of the wth liquid source (node).

The path diagram for this process is given by Fig. 18.3. The first step in Fig. 18.3 is a stripping process in which the pollutant is transferred from a liquid stream (source $w = 1$) to a gaseous stream (source $v = 1$). Because of the countercurrent contact of the two streams, the path profile for each stream (Ψ_w, W_w, z_w and Φ_v versus y_v) is given by a linear arrow extending between inlet and outlet compositions and having a slope equal to the flow rate of the stream. The gaseous stream leaving the stripper is then processed in a continuously stirred tank reactor where some additional mass of the pollutant, Φ_a, is generated by chemical reaction. Owing to the complete mixing in the reactor, the concentration of the pollutant instantaneously changes from the inlet concentration, y_2, to the outlet concentration, y_3. Furthermore, the pollutant loading increases to $\Phi_3 = \Phi_2 + \Phi_a$. The effluent from the reactor is mixed with another process stream ($v = 4$) to give a composition, y_5, and mass loading, Φ_5. The composition y_5 can be graphically determined using the lever-arm principle. The final operation involves the separation of the mixed stream into two terminal streams ($v = 6$ and 7).

Similarly, one can develop the liquid path diagram as shown in Fig. 18.4. It involves three nodes, $w = 1–3$, which are related by the stripper and the biotreatment units.

The path diagram provides the big picture for mass flow. This is a fundamentally different vision from the equipment-oriented description of a process (the flow sheet), in which the big picture is lost. The path diagram can also be used to determine the effect on the rest of the diagram of manipulating any node. In addition, as will be shown later, it provides a systematic way for identifying where to remove the pollutants and to what extent they should be removed.

INTEGRATION OF THE PATH AND THE PINCH DIAGRAMS

As mentioned earlier, the path diagram can be used to predict the effect of intercepting one stream on the rest of the streams. To quantify this relationship, let us consider that all gaseous and liquid nodes on the path diagrams are intercepted. Upon interception, the composition and load of the vth gaseous node are altered from y_v and Φ_v to y_v^{int} and Φ_v^{int}. Similarly, the composition and load of the wth liquid node are altered from z_w and Ψ_w to z_w^{int} and Ψ_w^{int}. The pinch diagram is used to determine optimal interceptions, whereas the path diagram is responsible for tracking the consequences of interceptions. The relationship between the path and the pinch diagrams can be envisioned as shown by Fig. 18.5, which illustrates the back-and-forth passage of streams between the two diagrams. Consider the gaseous nodes ($v = 1, 2,\ldots,$ NV) that

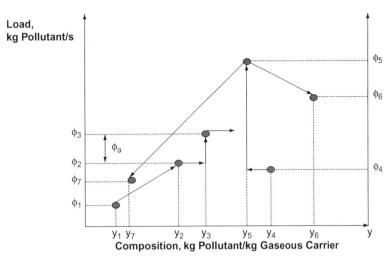

FIGURE 18-3 Gaseous path diagram. *Source: El-Halwagi et al. (1996).*

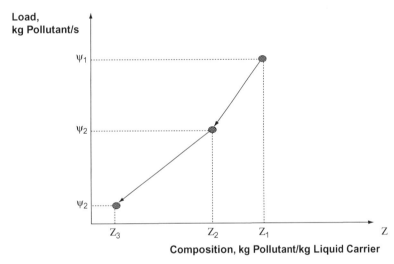

FIGURE 18-4 The liquid path diagram. *Source: El-Halwagi et al. (1996).*

are passed from the path diagram to the pinch diagram. Upon interception, these streams change composition from y_v to the intercepted compositions y_v^{int} where $v = 1, 2,..., NV$. Each intercepted stream is returned to the path diagram. According to the path diagram equations, this interception propagates throughout the whole path diagram, affecting the other nodes. In turn, this propagation in the path diagram affects the pinch diagram by changing the waste composite stream. Consequently, the lean composite stream is adjusted to respond to the changes in the waste composite stream. The path and the pinch diagrams can be integrated using mixed-integer nonlinear programming (El-Halwagi et al., 1996). Alternatively, one can use the shortcut method described in the following section.

SCREENING OF CANDIDATE MSAs USING A HYBRID OF PATH AND PINCH DIAGRAMS

The selection of interception technologies can be aided by useful concepts stemming from the path and the

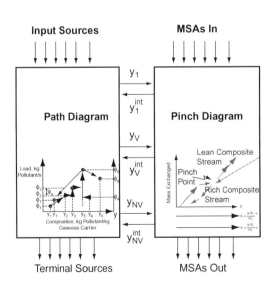

FIGURE 18-5 Integration of the path and the pinch diagrams. *Source: El-Halwagi et al. (1996).*

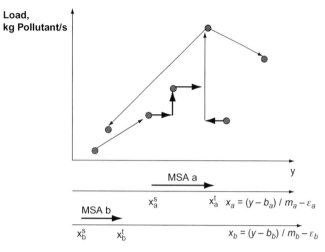

FIGURE 18-6 Screening MSAs using the path diagram.

pinch diagrams. After the path diagram is plotted for the sources, the pinch diagram is superimposed through a horizontal projection. Composition scales for the MSAs are plotted in one-to-one correspondence with the composition scale for the sources using Eq. 3.5. Each MSA is then represented versus its composition scale as a horizontal arrow extending between its supply and target compositions (Fig. 18.6).

Several useful insights can be gained from this hybrid diagram. Let us consider two MSAs a and b, whose costs (\$/kg of recirculating MSA) are c_a and c_b. These costs can be converted into \$/kg of removed pollutant, c_a^r and c_b^r, as follows:

$$[18.3a] \qquad c_a^r = \frac{c_a}{x_a^t - x_a^s}$$

and

$$[18.3b] \qquad c_b^r = \frac{c_b}{x_b^t - x_b^s}$$

If arrow b lies completely to the left of arrow a, and c_b^r is less than c_a^r, MSA b is chosen, because it is thermodynamically and economically superior to MSA a. Another useful observation can be inferred from the relative location of the sources and the MSAs. If an MSA lies to the right of a source node, the MSA is not a candidate for intercepting this node, because mass exchange is infeasible.

Once optimal interception strategies are prescreened for each node, one can formulate the synthesis task as an optimization program whose objective is to minimize interception cost subject to the path diagram equations and the prescreened interception strategies. The following example illustrates this approach.

Example 18-1 Interception of chloroethanol in an ethyl chloride process

Ethyl chloride (C_2H_5Cl) can be manufactured by catalytically reacting ethanol and hydrochloric acid (El-Halwagi et al., 1996). Figure 18.7 is a simplified flow sheet of the process. The facility involves two integrated processes: the ethanol plant and the ethyl chloride plant. First, ethanol is manufactured by the catalytic hydration of ethylene. Compressed ethylene is heated with water and reacted to from ethanol. Ethanol is separated using distillation followed by membrane separation (pervaporation). The aqueous waste from distillation is fed to a biotreatment unit in which the organic content of the wastewater is used as a bionutrient. The separated ethanol is reacted with hydrochloric acid in a multiphase reactor to form ethyl chloride. The reaction takes place primarily in the liquid phase. A by-product of the reaction is chloroethanol (CE, also referred to as ethylene chlorohydrin, C_2H_5OCl). This side reaction will be referred to as the *oxychlorination reaction*. The rate of chloroethanol generation via oxychlorination (approximated by a pseudo zero-order reaction) is given:

$$[18.4] \qquad r_{oxychlorination} = 6.03 \times 10^{-6} \text{ kg chloroethanol/s}.$$

Whereas ethyl chloride is one of the least toxic of all chlorinated hydrocarbons, CE is a toxic pollutant. The off-gas from the reactor is scrubbed with water in two absorption columns. The first column is intended to recover the majority of unreacted ethanol, hydrogen chloride, and CE. The second scrubber purifies the product from traces of unreacted materials and acts as a backup column in case the first scrubber is out of operation. Each scrubber contains two sieve plates and has an overall column efficiency of 65 percent (that is, NTP = 1.3). Following the scrubber, ethyl chloride is finished and sold. The aqueous streams leaving the scrubbers are mixed and recycled to the reactor. A fraction of the CE recycled to the reactor is reduced to ethyl chloride. This side reaction will be called the **reduction reaction**. The rate of CE depletion in the reactor due to this reaction can be approximated by the following pseudo first-order expression:

$$[18.5] \qquad r_{reduction} = 0.090 z_5 \text{ kg chloroethanol/s},$$

where z_5 is in units of mass fraction.

The compositions of CE in the gaseous and liquid effluents of the ethyl chloride reactor are related through an equilibrium distribution coefficient as follows:

$$[18.6] \qquad \frac{y_1}{z_6} = 5.$$

Because of the toxicity of CE, the aqueous effluent from the ethyl chloride reactor, R_1, causes significant problems for the biotreatment facility. The objective of this case study is to optimally intercept CE-laden streams so as to reduce the CE content of R_1 to meet the following regulations:

Target composition:

$$[18.7] \qquad z_6^{int} = z_1^{Terminal\ target} \leq 7 \text{ ppmw}$$

Target load:

$$[18.8] \qquad \text{Load of CE in } R_1 \leq 1.05 \times 10^{-6} \text{ kgCE/s}$$

Six MSAs are available for removing CE: three for gaseous streams and three for liquid streams (see Tables 18.1 and 18.2).

SOLUTION

This case study involves four units, three gaseous sources (nodes), and six liquid sources. The flow rates of all gaseous and liquid sources are shown in Fig. 18.7. Also, the compositions of the two liquid nodes corresponding to entering fresh water are given ($z_1 = z_3 = 0$). The first step in the analysis is the development of the path diagram equations to quantify the relationship of the seven unknown compositions. As shown by Fig. 18.7, overall material balances around the four

(Continued)

Example 18-1 Interception of chloroethanol in an ethyl chloride process (Continued)

units ($u = 1$–4, assuming constant flow rate of carriers) provide the following flow rates of the carrier gas and liquid:

[18.9] $V_1 = V_2 = V_3 = 0.150$ kg/s

[18.10] $W_1 = W_2 = W_3 = W_4 = 0.075$ kg/s

[18.11] $W_5 = W_6 = 0.150$ kg/s

Component material balance for chloroethanol around the reactor ($u = 1$):

[18.12a]
Chloroethanol in recycled reactants
 + Chloroethanol generated due to oxychlorination reaction
= Chloroethanol in off-gas + Chloroethanol in wastewater W_j
 + Chloroethanol depleted by reduction reaction.

TABLE 18-1 Data for MSAs That Can Remove CE from Gaseous Streams

Stream	Description	Supply Composition x_j^s (ppmw)	Target Composition x_j^t (ppmw)	m_j	ε_j ppmw	C_j $/kg MSA	C_j^r $/ kg CE Removed
SV_1	Polymeric resin	2	10	0.03	5	0.08	10,000
SV_2	Activated carbon	5	30	0.06	10	0.10	4000
SV_3	Oil	200	300	0.80	20	0.05	500

TABLE 18-2 Data for MSAs That Can Remove CE from Liquid Streams

Stream	Description	Supply Composition X_j^s (ppmw)	Target Composition X_j^t (ppmw)	M_j	ε_j ppmw	C_j $/kg MSA	C_j^r $/ kg CE Removed
SW_1	Zeolite	3	15	0.09	15	0.70	58,333
SW_2	Air	0	10	0.10	100	0.05	5000
SW_3	Steam	0	15	0.80	50	0.12	8000

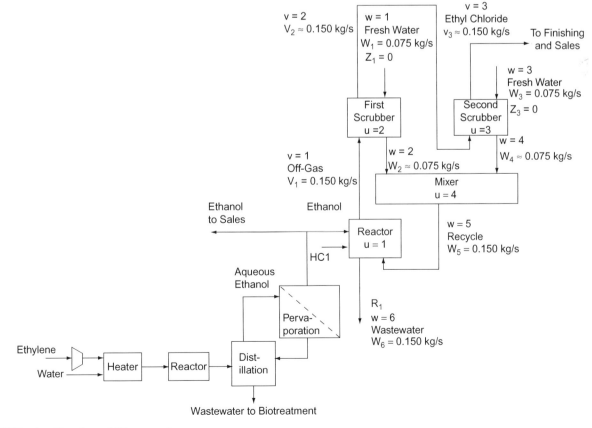

FIGURE 18-7 Flowsheet of CE case study.

(Continued)

Example 18-1 Interception of chloroethanol in an ethyl chloride process (Continued)

Hence,

[18.12b] $$W_5 z_5 + 6.03 \times 10^{-6} = V_1 y_1 + W_6 z_6 + 0.09 z_5.$$

But, as discussed in the problem statement, the compositions of CE in the gaseous and liquid effluents of the ethyl chloride reactor are related through an equilibrium distribution coefficient as follows:

[18.13] $$\frac{y_1}{z_6} = 5.$$

Therefore, Eq. 18.12b can be further simplified by invoking Eq. 18.13 and substituting for the numerical values of V_1, W_5, and W_6 to get

[18.14a] $$0.180 y_1 - 0.060 z_5 = 6.03 \times 10^{-6}$$

with y_1 and z_5 in mass fraction units. Therefore, the previous equation can be rewritten as:

[18.14b] $$0.180 y_1 - 0.060 z_5 = 6.030$$

with y_1 and z_5 in units of ppmw.

Component material balance for chloroethanol around the first scrubber ($u = 2$):

[18.15a] $$V_1 y_1 + W_1 z_1 = V_2 y_2 + W_2 z_2$$

Substituting for the values of V_1, V_2, W_1, and W_2, and noting that z_1 is 0 (fresh water), we get

[18.15b] $$2(y_1 - y_2) - z_2 = 0.$$

The Kremser equation can be used to model the scrubber:

[18.16a] $$NTP = \frac{\ln\left[\left(1 - \frac{HV_1}{W_1}\right)\left(\frac{y_1 - Hz_1}{y_2 - Hz_1}\right) + \frac{HV_1}{W_1}\right]}{\ln\left(\frac{W_1}{HV_1}\right)}$$

where Henry's coefficient $H = 0.1$ and $NTP = 1.3$. Therefore,

$$1.3 = \frac{\ln\left(0.8 \frac{y_1}{y_2} + 0.2\right)}{\ln(5)}$$

that is,

[18.16b] $$y_2 = 0.10 y_1.$$

Similar to Equations 18.15b and 18.16b, one can derive the following two equations for the second scrubber ($u = 3$):

[18.17] $$2(y_2 - y_3) - z_4 = 0$$

and

[18.18] $$y_3 = 0.10 y_2.$$

Component material balance for chloroethanol around the mixer ($u = 4$)

[18.19a] $$W_2 z_2 + W_4 z_4 = W_5 z_5$$

By plugging the values of W_1, W_2, and W_3 into Eq. 18.19a, we get

[18.19b] $$z_2 + z_4 - 2 z_5 = 0$$

Hence, the gaseous and liquid path diagram (Equations 18.13 through 18.19) can be summarized by the following model (with all compositions in ppmw):

For $u = 1$,

$$0.180 y_1 - 6.030 = 0.060 z_5$$

$$y_1 - 5 z_6 = 0.$$

For $u = 2$,

[P18.1] $$2 y_2 + z_2 = 2 y_1$$

$$y_2 = 0.10 y_1.$$

For $u = 3$,

$$2 y_3 + z_4 = 2 y_2$$

$$y_3 = 0.10 y_2.$$

For $u = 4$,

$$2 z_5 = z_2 + z_4.$$

As mentioned earlier, the gaseous path diagram for CE involves three unknown compositions (y_1, y_2, and y_3), whereas the liquid path diagram for CE has four unknown compositions (z_2, z_4, z_5, and z_6). The foregoing seven equations can be solved simultaneously to get these compositions. In terms of LINGO input, Model P18.1 can be written as follows:

```
0.180*y1-0.060*z5=6.030;
y1-5*z6=0.0;
2*y2+z2-2*y1=0.0;
y2-0.10*y1=0.0;
2*y31z4-2*y2=0.0;
y3-0.10*y2=0.0;
2*z5-z2-z4=0.0;
```

The solution to this model gives the following compositions (in ppmw CE) prior to interception:

$y_1 = 50.0$
$y_2 = 5.0$
$y_3 = 0.5$
$z_2 = 90.0$
$z_4 = 9.0$
$z_5 = 49.5$
$z_6 = 10.0$

Figure 18.8 is a schematic representation of the path diagram for the liquid sources.

As can be seen from Fig. 18.9, the objective of this problem is to move the terminal wastewater node ($w = 6$) to the targeted location designated by the X. The challenge is how to move the other nodes to meet this target at minimum cost.

To incorporate interception, Fig. 18.10 is developed in a manner similar to that described by Fig. 18.6. Next, we use the criteria of thermodynamic feasibility and cost to prescreen interception strategies for each node. For instance, because air lies to the left of steam and has a lower cost (per kg CE removed), it is chosen in favor of steam. Nodes lying to the right of air ($w = 2, 5$) can be intercepted down to 10 ppmw CE using air stripping. Any interception of these nodes below 10 ppmw CE should be handled by air (up to 10 ppmw CE) followed by zeolite. Similarly, nodes $w = 4$ and 6 can be intercepted by zeolite down to 1.6 ppmw CE.

Similarly, Fig. 18.11 represents the gaseous path diagram and MSAs. Because of thermodynamic infeasibility, oil cannot be used to intercept any node. Activated carbon should be used to intercept nodes $v = 1$ and 2 down to 0.9 ppmw CE. Any interceptions below 0.9 ppmw CE may be handled by the polymeric resin, which can reduce the CE content to 0.21 ppmw.

Having prescreened the interception strategies, we should integrate these strategies with the path diagram as illustrated by Fig. 18.5. First, the path diagram

(Continued)

Example 18-1 Interception of chloroethanol in an ethyl chloride process (*Continued*)

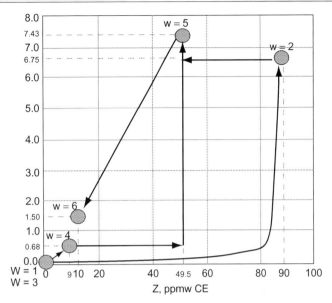

FIGURE 18-8 Liquid path diagram for CE case study.

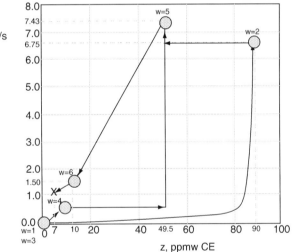

FIGURE 18-9 Liquid path diagram for CE case study with "X" marking waste-reduction target.

Equation P18.1 should be revised to include potential interception of all nodes as follows:

(P18.2)
$$0.180 y_1 - 6.030 = 0.060 z_5^{int}$$

$$y_1 - 5 z_6 = 0$$

$$2 y_2 + z_2 = 2 y_1^{int}$$

$$y_2 = 0.10 y_1^{int}$$

$$2 y_3 + z_4 = 2 y_2^{int}$$

$$y_3 = 0.10 y_2^{int}$$

$$2 z_5 = z_2^{int} + z_4^{int}.$$

As part of the optimization procedure, we will have to determine which nodes should be intercepted and the optimal values of the intercepted nodes. Next, let us consider the case when only one[1] node is intercepted at a time.

Interception of w = 2

As determined earlier, air stripping is the optimal interception strategy for $w = 2$. Therefore, the following LINGO model can be written to minimize the stripping cost[2]:

```
min=0.05*3600*8760*Lair;
0.180*y1-0.060*z5=6.030;
```

---
[1] Simultaneous interception of multiple sources may be considered by allowing intercepted compositions of multiple nodes to be treated as optimization variables in the same formulation. See El-Halwagi et al. (1996) for more details.
[2] Here, the objective is to minimize operating cost followed by trading off with fixed cost as discussed in Appendix II. If needed, one can directly minimize total annualized cost by adding a fixed-cost term to the objective function. This fixed-cost term is expressed in terms of z_7^{int} as described in Appendix II.

(*Continued*)

Example 18-1 Interception of chloroethanol in an ethyl chloride process (Continued)

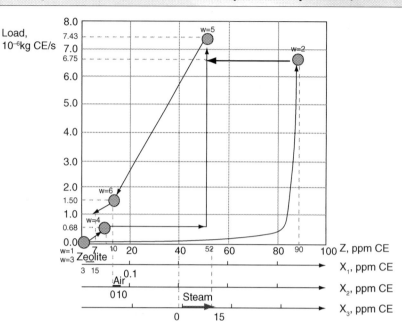

FIGURE 18-10 Prescreening MSAs for intercepting liquid path diagram for CE case study.

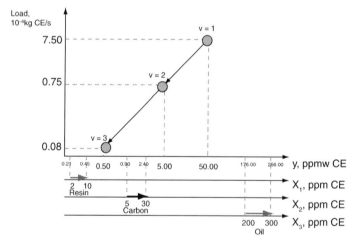

FIGURE 18-11 MSAs for intercepting gaseous path diagram for CE case study (schematic representation, not to scale).

```
y1-5*z6=0.0;
2*y2+z2-2*y1=0.0;
y2-0.10*y1=0.0;
2*y3+z4-2*y2=0.0;
y3-0.10*y2=0.0;
2*z5-zINT2-z4=0.0;
z6<=7.0;
10*Lair-0.075*(90.0-zINT2)=0.0;
```

The following is a LINGO output summarizing the solution:

```
Objective value: 1032410

Variable    Value
LAIR        0.6547501
Y1          35.00000
Z5          4.499997
Z6          7.000000
```

```
Y2          3.500000
Z2          63.00000
Y3          0.3500000
Z4          6.300000
ZINT2       2.699993
```

The solution indicates that, to reduce the CE content of the terminal wastewater stream to 7 ppmw, air stripping can be used to intercept the bottom product of the first scrubber ($w = 2$) from 90 ppmw CE to 2.7 ppmw CE, at a cost of $1,032,410/yr. The same procedure can be repeated for the other nodes as shown below.

Interception of $w = 4$:

```
min=0.70*3600*8760*LZeolite;
0.180*y1-0.060*z5=6.030;
y1-5*z6=0.0;
2*y2+z2-2*y1=0.0;
y2-0.10*y1=0.0;
```

(Continued)

Example 18-1 Interception of chloroethanol in an ethyl chloride process (*Continued*)

```
2*y3+z4-2*y2=0.0;
y3-0.10*y2=0.0;
2*z5-z2-zINT4=0.0;
z6<=7.0;
12*LZeolite-0.075*(9.0-zINT4)=0.0;
```

Solution: INFEASIBLE
Because no feasible solution was found, there is no way that $w = 4$ can be intercepted to reduce the CE content of the terminal wastewater stream to 7 ppmw. Indeed, even if CE in $w = 4$ is completely eliminated, the terminal wastewater stream will drop its content of CE to only 9.6 ppmw, as shown by the following model:

```
0.180*y1-0.060*z5=6.030;
y1-5*z6=0.0;
2*y2+z2-2*y1=0.0;
y2-0.10*y1=0.0;
2*y3+z4-2*y2=0.0;
y3-0.10*y2=0.0;
2*z5-z2-zINT4=0.0;
zINT4=0.0;
```

Variable	Value
Y1	47.85714
Z5	43.07143
Z6	9.571429
Y2	4.785714
Z2	86.14286
Y3	0.4785714
Z4	8.614286
ZINT4	0.0000000E+00

Interception of *w* = 5

```
min=0.05*3600*8760*Lair;
0.180*y1-0.060*zINT5=6.030;
y1-5*z6=0.0;
2*y2+z2-2*y1=0.0;
y2-0.10*y1=0.0;
2*y3+z4-2*y2=0.0;
y3-0.10*y2=0.0;
2*z5-z2-z4=0.0;
Z6<=7.0;
10*Lair-0.15*(49.5-zINT5)=0.0;
```

Objective value: 1064340.

Variable	Value
LAIR	0.6750001
Y1	35.00000
ZINT5	4.499997
Z6	7.000000
Y2	3.500000
Z2	63.00000
Y3	0.3500000
Z4	6.300000
Z5	34.65000

Because z_5 was intercepted to 4.5 ppmw, the separation task cannot be done by air alone (air can reduce the composition to 10 ppmw). Hence, the program is adjusted to include zeolite.

```
min=0.05*3600*8760*Lair10.70*3600*7860*LZeolite;
0.180*y1-0.060*zINT5=6.030;
y1-5*z6=0.0;
2*y2+z2-2*y1=0.0;
y2-0.10*y1=0.0;
2*y3+z4-2*y2=0.0;
y3-0.10*y2=0.0;
2*z5-z2-z4=0.0;
Z6<=7.0;
10*Lair-0.15*(49.5-10.0)=0.0;
12*LZeolite-0.15*(10.0-zINT5)=0.0;
```

Objective value: 2296000.

Variable	Value
LAIR	0.5925000
LZEOLITE	0.6875004E-01
Y1	35.00000
ZINT5	4.499997
Z6	7.000000
Y2	3.500000
Z2	63.00000
Y3	0.3500000
Z4	6.300000
Z5	34.65000

Interception of *w* = 6 (terminal wastewater)

```
min=0.70*3600*8760*LZeolite;
0.180*y1-0.060*z5=6.030;
y1-5*z6=0.0;
2*y2+z2-2*y1=0.0;
y2-0.10*y1=0.0;
2*y3+z4-2*y2=0.0;
y3-0.10*y2=0.0;
2*z5-z2-z4=0.0;
zINT6<=7.0;
12*LZeolite-0.150*(10.0-zINT6)=0.0;
```

Objective value: 827820.0

Variable	Value
LZEOLITE	0.3750000E-01
Y1	50.00000
Z5	49.50000
Z6	10.00000
Y2	5.000000
Z2	90.00000
Y3	0.5000000
Z4	9.000000
ZINT6	7.000000

Alternatively,

```
min=0.70*3600*8760*LZeolite;
zINT6<=7.0;
12*LZeolite-0.150*(10.0-zINT6)=0.0;
```

Objective value: 827820.0

Variable	Value
LZEOLITE	0.3750000E-01
ZINT6	7.000000

Interception of *v* = 1

```
min=0.10*3600*8760*LCarbon;
0.180*y1-0.060*z5=6.030;
y1-5*z6=0.0;
2*y2+z2-2*yINT1=0.0;
y2-0.10*yINT1=0.0;
2*y3+z4-2*y2=0.0;
y3-0.10*y2=0.0;
2*z5-z2-z4=0.0;
z6 <= 7.0;
25*LCarbon-0.150*(35.0-yINT1)=0.0;
```

(*Continued*)

Example 18-1 Interception of chloroethanol in an ethyl chloride process (Continued)

```
Objective value: 576248.8
Variable     Value
LCARBON      0.1827273
Y1           35.00000
Z5           4.499997
Z6           7.000000
Y2           0.4545451
Z2           8.181812
YINT1        4.545451
Y3           0.4545451E-01
Z4           0.8181812
```

Interception of $v = 2$

```
min=0.10*3600*8760*LCarbon;
0.180*y1-0.060*z5=6.030;
y1-5*z6=0.0;
2*y2+z2-2*y1=0.0;
y2-0.10*y1=0.0;
2*y3+z4-2*yINT2=0.0;
y3-0.10*yINT2=0.0;
2*z5-z2-z4=0.0;
z6 <= 7.0;
25*LCarbon-0.150*(5.0-yINT2)=0.0;
```

INFEASIBLE
[This is a situation similar to the interception of $w = 4$].

Based on these results, the optimal single interception for the problem is to use activated-carbon adsorption to separate CE from the gaseous stream leaving the reactor ($v = 1$) and reduce its composition to $y_1^{int} = 4.55$ ppmw CE (which corresponds to removing 4.57×10^{-6} kg CE/s from $v = 1$). The optimal solution has a minimum operating cost of approximately \$576,250/yr. Several important observations can be drawn from the list of generated solutions:

- Waste reduction is a multimedia (multiphase) problem. The solution to a wastewater problem may lie in the gas phase.
- In-plant interception may be superior to terminal waste separation. For instance, separating CE from the terminal wastewater stream incurs an annual operating cost of \$827,820/year, which is 44 percent more expensive than the optimal solution.
- More mass may be removed for less cost. For instance, intercepting $v = 1$ involves the removal of 4.57×10^{-6} kg CE/s, whereas the more expensive interception of $w = 6$ involves the removal of 0.45×10^{-6} kg CE/s. This result can be explained based on thermodynamic and mass-transfer arguments. Removal of the pollutant from more concentrated streams is more thermodynamically favorable than removal from less concentrated streams. Therefore, it is typically less expensive to use low-cost MSAs to eliminate the pollutant from in-plant streams than it is to employ high-cost MSAs to remove traces of pollutant from the dilute terminal streams. It is also worth noting the nonlinear propagation of mass of the pollutant throughout the process. For a given effect on the pollutant content of a terminal stream, different masses may be removed from different in-plant streams. Another reason for this mass–cost observation is that removal of the pollutant in one medium can be less expensive than separation from another medium. For instance, a certain amount of the pollutant may be removed from air at a lower cost than that of recovering a smaller quantity from wastewater.

DEVELOPING STRATEGIES FOR SEGREGATION, MIXING, AND DIRECT RECYCLE

The previous sections have focused on in-plant interception for waste reduction. In this section, we target other mass-integration strategies, including stream segregation, mixing, and direct recycle (without interception). Later in this chapter, interception strategies will be incorporated. Without interception, Fig. 18.1 is revised to the structural representation shown in Fig. 18.12. This representation embeds potential configurations of interest by allowing each source to be segregated, mixed, allocated to a unit, and recycled back to the process. This is an extended version of the direct-recycle problem described in Chapter 15. Here, we also account for the effect of the process model on the recycle strategies and vice versa. The optimization task is to determine the flow rate and composition

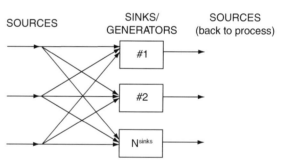

FIGURE 18-12 Structural representation of segregation, mixing, and direct-recycle options.

of each stream in Fig. 18.12, while considering the path diagram expressions to incorporate the interactions among units and streams. This task can be mathematically formulated as shown in the following section.

Example 18-2 Case study revisited: Segregation, mixing, and recycle for chloroethanol case study

The scope of the previously addressed CE case study is now altered to allow for stream segregation, mixing, and recycle within the ethyl chloride plant. There are five sinks: the reactor ($u = 1$), the first scrubber ($u = 2$), the second scrubber ($u = 3$), the mixing tank ($u = 4$), and the biotreatment facility for effluent treatment ($u = 5$). There are six sources of CE-laden aqueous streams ($w = 1$ to 6). There is the potential for segregating two liquid sources ($w = 2, 4$). The following process constraints should be considered:

[18.20] Composition of aqueous feed to reactor ≤ 65 ppmw CE

[18.21] Composition of aqueous feed to first scrubber ≤ 8 ppmw CE

[18.22] Composition of aqueous feed to second scrubber $= 0$ ppmw CE

[18.23] $0.090 \leq$ Flow rate of aqueous feed to reactor (kg/s) ≤ 0.150

[18.24] $0.075 \leq$ Flow rate of aqueous feed to first scrubber (kg/s) ≤ 0.090

(Continued)

Example 18-2 Case study revisited: Segregation, mixing, and recycle for chloroethanol case study (*Continued*)

[18.25] $0.075 \le$ Flow rate of aqueous feed to second scrubber
(kg/s) ≤ 0.085

The objective of this case study is to determine the target for minimizing the total load (flow rate × composition) of CE discharged in terminal wastewater of the plant using segregation, mixing, and recycle strategies.

SOLUTION

Figure 18.13 is structural representation of segregation, mixing, and direct-recycle candidate strategies for the problem. Each source is split into several fractions that can be fed to a sink. The flow rate of the streams passed from source w to sink u is referred to as f_{wu}. The terms F_u^{in}, Z_u^{in}, and F_u^{out} represent the inlet flow rate, inlet composition, and outlet flow rate of the streams associated with unit u. Because mixing is embedded, there is no need to include the mixing tank ($u = 4$) or the source that it generates ($w = 5$) in the analysis. Unless recycle of biotreatment effluent is considered, there is no need to represent the biotreatment sink in Fig. 18.13. However, streams allocated to biotreatment should be represented, and their flow rates are referred to as f_{w5} ($u = 5$ is the biotreatment sink). Finally, fresh water may be used in any unit u at a flow rate of Fresh$_u$.

Because the flow rates of the streams fed to the sinks are to be determined as part of the optimization problem, the path equations represented by Model P18.1 should be revised as follows to allow for variable flow rate of the streams:

$$(0.15 + 0.2 * F_1^{in})y_1 - (F_1^{in} - 0.09)Z_1^{in} = 6.030$$

$$0.15(y_1 - y_2) - F_2^{in}(z_2 - Z_2^{in}) = 0.0$$

[P18.3]
$$\left(\frac{F_2^{in}}{0.015}\right)^{1.3} - \frac{\left(1 - \frac{0.015}{F_2^{in}}\right)(y_1 - 0.1 Z_2^{in})}{y_2 - 0.1 Z_2^{in}} - \frac{0.015}{F_2^{in}} = 0.0$$

$$0.15(y_2 - y_3) - F_3^{in}(z_4 - Z_3^{in}) = 0.0$$

$$\left(\frac{F_3^{in}}{0.015}\right)^{1.3} - \frac{\left(1 - \frac{0.015}{F_3^{in}}\right)(y_2 - 0.1 Z_3^{in})}{y_3 - 0.1 Z_3^{in}} - \frac{0.015}{F_3^{in}} = 0.0.$$

The revised path diagram is integrated with material allocation equations to form the constraints for the mathematical formulation. The following model presents the optimization program as a LINGO file. The commented-out lines (preceded by !) are explanatory statements that are not part of the formulation.

! Objective is to minimize load of CE in terminal wastewater streams

```
min=F65*z6+F25*z2+F45*z4;
```

! Water balances around inlets of sinks

```
F1In - f21 - f41 - f61 - Fresh1=0.0000;
F2In - f22 - f42 - f62 - Fresh2=0.0000;
F3In - f23 - f43 - f63 - Fresh3=0.0000;
```

! CE balances around inlets of sinks

```
F1In*Z1In - f21*z2 - f41*z4 - f61*z6=0.0000;
F2In*Z2In - f22*z2 - f42*z4 - f62*z6=0.0000;
F3In*Z3In - f23*z2 - f43*z4 - f63*z6=0.0000;
```

! Water balances around sinks

```
F1In - F1Out=0.0000;
F2In - F2Out=0.0000;
F3In - F3Out=0.0000;
```

! Water balances for sources to be split

```
F1Out - f65 - f61 - f62 - f63=0.0000;
F2Out - f25 - f21 - f22 - f23=0.0000;
F3Out - f45 - f41 - f42 - f43=0.0000;
```

! Path-diagram equations

```
(0.15+0.2*F1In)*y1-(F1In  - 0.09)*Z1In=6.030;
0.15*(y1-y2)-F2In*(z2-Z2In)=0.0000;
(F2In/0.015)^1.3-(1-0.015/F2In)*(y1-0.1*Z2In)/(y2-
    0.1*Z2In)  - 0.015/F2In=0.0000;
0.15*(y2-y3)-F3In*(z4-Z3In)=0.0000;
(F3In/0.015)^1.3-(1-0.015/F3In)*(y2-0.1*Z3In)/(y3-
    0.1*Z3In)  - 0.015/F3In=0.0;
y1-5*z6=0.0000;
```

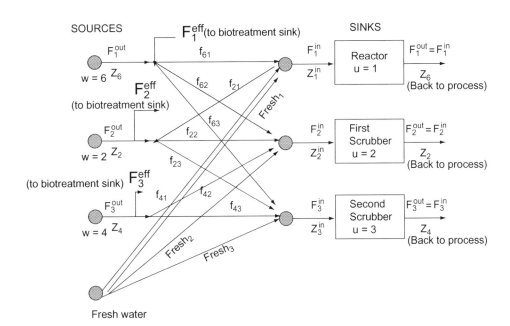

FIGURE 18-13 Structural representation of segregation, mixing, and direct-recycle options for CE case study.

(*Continued*)

Example 18-2 Case study revisited: Segregation, mixing, and recycle for chloroethanol case study (*Continued*)

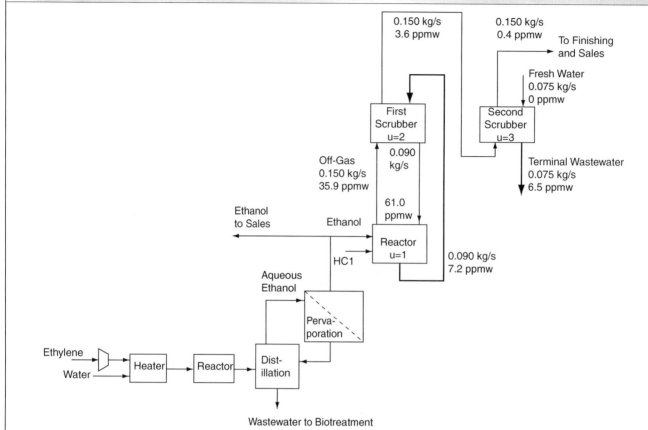

FIGURE 18-14 Solution to CE case study using segregation, mixing, and direct recycle.

```
! Restrictions on what can be recycled

Z1In <= 65.0000;
Z2In <8.0000;
Z3In =0.0000;
F1In <= 0.1500;
F1In >= 0.0900;
F2In <= 0.0900;
F2In >= 0.07500;
F3In <= 0.0850;
F3In >= 0.0750;
```

! If solution gives negative flows or compositions, the nonnegativity constraints can then be added

This is a nonlinear program that can be solved using LINGO to get the following answer:

```
Objective value: 0.4881247

Variable   Value
F65        0.0000000E+00
Z6         7.178572
F25        0.0000000E+00
Z2         60.96842
F45        0.7500000E-01
Z4         6.508329
F1IN       0.9000000E-01
F21        0.9000000E-01
F41        0.0000000E+00
F61        0.0000000E+00
FRESH1     0.0000000E+00
F2IN       0.9000000E-01
F22        0.0000000E+00
F42        0.0000000E+00
F62        0.9000000E-01
FRESH2     0.0000000E+00
F3IN       0.7500000E-01
F23        0.0000000E+00
F43        0.0000000E+00
F63        0.0000000E+00
FRESH3     0.7500000E-01
Z1IN       60.96842
Z2IN       7.178572
Z3IN       0.0000000E+00
F1OUT      0.9000000E-01
F2OUT      0.9000000E-01
F3OUT      0.7500000E-01
Y1         35.89286
Y2         3.618950
Y3         0.3663237
```

These results can be used to construct the solution as shown in Fig. 18.14. The target for minimum CE discharge through segregation, mixing, and direct recycle is 0.488 kg/s (about 15,000 tons/yr). The solution indicates that the optimal policy is to segregate the effluents of the two scrubbers, pass the effluent of the first scrubber to the reactor, recycle the aqueous effluent of the reactor to the first scrubber, and dispose of the second scrubber effluent as the terminal wastewater stream.

Note that this case study has presented mass-integration opportunities via intraprocess integration (within the ethyl chloride plant). The scope of the case study can be broadened to incorporate interprocess integration (for example, between the ethanol plant and the ethyl chloride plant). For instance, the

(*Continued*)

Example 18-2 Case study revisited: Segregation, mixing, and recycle for chloroethanol case study (*Continued*)

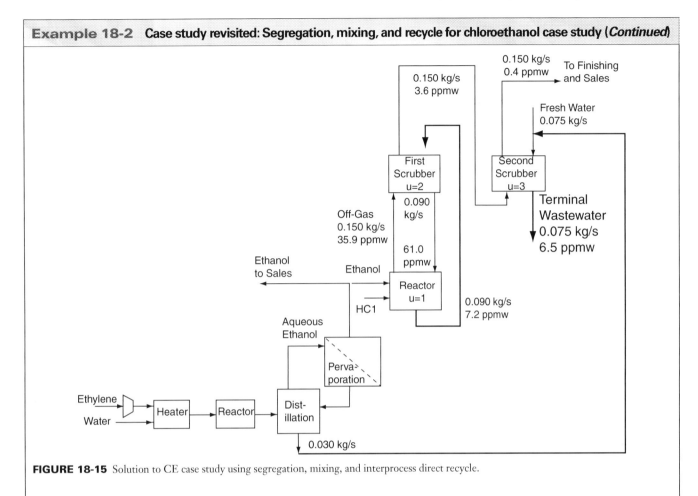

FIGURE 18-15 Solution to CE case study using segregation, mixing, and interprocess direct recycle.

wastewater leaving the distillation column (0.030 kg/s) does not contain any CE. If the composition of the various species contained in this wastewater is also within tolerated recycle limits for the second scrubber, it can be used to replace some of the fresh water. Figure 18.15 shows this solution. Once these integration opportunities have been identified, a cost-benefit analysis should be undertaken to determine the applicability of these strategies.

INTEGRATION OF INTERCEPTION WITH SEGREGATION, MIXING, AND RECYCLE

The procedures presented heretofore can be combined to allow the integration of interception with other mass-integration strategies. For instance, suppose that we wish to use interception to cut the discharged load of CE in Fig. 18.15 by 50 percent, by seeking a target composition of 3.3 ppmw CE in the aqueous effluent of the second scrubber (discharged load = 0.25×10^{-6} kg CE/s). Which nodes should be intercepted with which separating agents to meet the target at minimum cost? The procedure presented in the sections on "Synthesis of MSA-Induced Wins" and "Developing Strategies for Segregation, Mixing, and Direct Recycle" can be directly applied to yield an optimal solution (Fig. 18.15) that uses activated carbon to intercept the gaseous stream leaving the first scrubber ($v = 2$) and reduce its composition to $y_2^{int} = 1.83$ ppmw. The annual operating cost of the system is approximately $32,000/yr. This interception propagates through the process, resulting in a terminal wastewater composition of 3.3 ppmw CE.

It is interesting to compare the optimal configuration shown in Fig. 18.16 to end-of-pipe solutions. Suppose we retain the identified segregation, mixing, and direct-recycle strategies shown in Fig. 18.14, but intercept the wrong stream. For instance, if the CE content of the terminal wastewater stream is to be reduced from 6.5 ppmw to 3.3 ppmw, the cheapest practically feasible MSA is zeolite (this choice can be made based on Fig. 18.10). This end-of-pipe separation (after segregation, mixing, and recycle) requires an annual operating cost of $441,500/yr, which is 14 times more expensive than the optimal solution. If end-of-pipe separation is employed without segregation, mixing, and recycle, the targeted disposal limit of 0.25×10^{-6} kg CE/s can be met by reducing the CE content of the reactor effluent (the original wastewater stream of Fig. 18.7) from 10 to 1.8 ppmw CE. Again, zeolite is used for this separation based on Fig. 18.10. The operating cost associated with this end-of-pipe solution is about $2.3 million/yr. These nonoptimal options meet the CE-loading target and provide the same waste-reduction effect as the optimal solution. However, as can be seen from this case study, the costs of these solutions can be

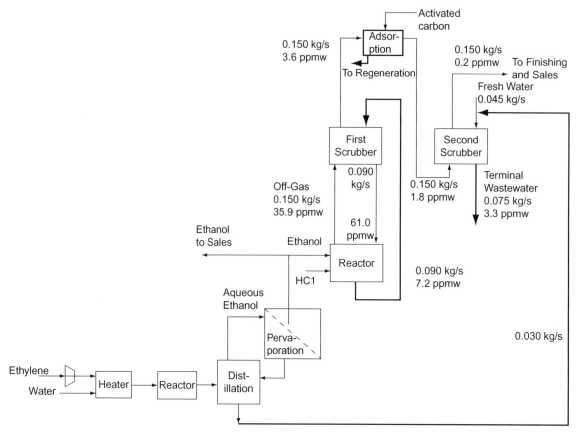

FIGURE 18-16 Solution to CE case study using segregation, mixing, interception, and recycle.

significantly higher than those of solutions obtained using mass integration.

HOMEWORK PROBLEMS

18.1. Solve Problem 6.1 using mathematical optimization.
18.2. Employing mathematical programming, solve Problem 6.2.
18.3. The kraft pulping process is a common technology for converting wood chips to pulp that is subsequently made into paper. Figure 18.17 is a schematic representation of a kraft process (Lovelady et al., 2007). In this process, 6000 tonne per day (tpd) of wood chips are cooked in a caustic solution referred to as the white liquor (composed primarily of NaOH and Na_2S). The reaction takes place in a digester. The produced pulp is washed in brown-stock washers and then fed or sold to a papermaking plant. After the brown-stock washing, the spent cooking solution is referred to as weak black liquor. This liquor is concentrated by a number of multiple-effect evaporators (MEE) and a concentrator to provide a strong black liquor. The vapor streams from the MEE and the concentrator are discharged as wastewater. The strong black liquor is burned in a recovery boiler to generate heat and chemicals (referred to as smelt). The recovery furnace is connected to an electrostatic precipitator (ESP) to remove the ashes from the off-gas prior to discharge. The smelt from the recovery furnace is composed primarily of Na_2S, Na_2CO_3, and some Na_2SO_4. The smelt is dissolved to form green liquor that is clarified to remove any undissolved materials (known collectively as dregs). The clarified green liquor is fed to a slaker where lime and water react to form slaked lime:

[18.26]
$$CaO + H_2O \rightarrow Ca(OH)_2$$
Molecular weights 56   18   74

Slaking is the primary reaction involving water in the process. The slaking reaction consumes 256 tpd of calcium oxide along with the stoichiometrically equivalent quantity of water. Therefore, the amount of water depleted in the pulping process due to the slaking reaction is as follows:

[18.27]
$$= 256 \text{ tpd CaO} * \frac{18 \text{ tonne } H_2O}{56 \text{ tonne CaO}}$$
$$= 168 \text{ tpd}$$

Slaking is followed by a causticizing reaction between the slaked lime and sodium carbonate to yield white liquor:

[18.28] $Ca(OH)_2 + Na_2CO_3 \rightarrow 2NaOH + CaCO_3$

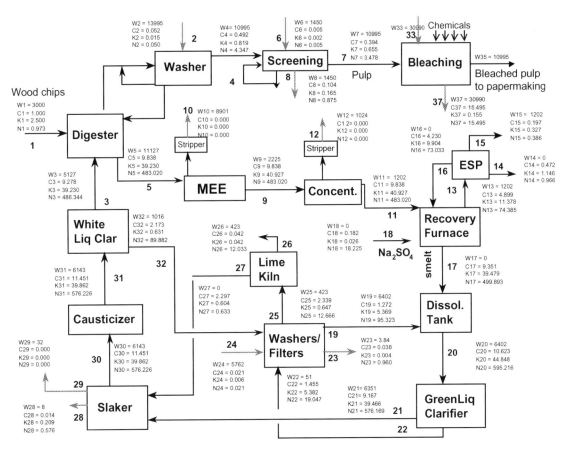

FIGURE 18-17 The kraft pulping process. *Note: W, C, K, and N designate the flow rate in tonne/day of water and the ionic species of chlorine, potassium, and sodium, respectively. Source: Lovelady et al. (2007).*

The white liquor from the white-liquor clarifier (stream S3) is fed back to the digester for cooking the wood chips, and the chemical cycle for the white liquor is closed. Meanwhile, calcium carbonate from the causticizing reaction is processed through the wasters/filters and fed to the lime kiln where the following thermal decomposition reaction takes place:

[18.29] $\quad CaCO_3 \rightarrow CaO + CO_2$

The produced calcium oxide is recycled back to the slaker for the reaction given by Eq. 18.26, and the calcium-oxide cycle is closed.

Water usage and discharge is one of the largest environmental problems associated with the kraft pulping process. Several aqueous effluents are disposed of from the plant, including wastewater from the bleaching plant, brown-stock washers, and the evaporator. Figure 18.17 gives the relevant material balance data. To partially close the water loop of the process and reduce wastewater, several aqueous streams must be recycled. The difficulty involved in water recycle is that certain ionic species such as chlorine, potassium, silica, and sodium tend to accumulate in the process as partial closure is carried out. This leads to problems such as corrosion and plugging. The primary constraint on the buildup of non-process elements (NPEs) is associated with the "stick temperature" for the recovery furnace. It can be related to the Cl, K, and Na through the following constraints:

[18.30]
$$\frac{K_{11} + K_{16} + K_{18}}{39.1} \leq 0.1 \frac{N_{11} + N_{16} + N_{18}}{23}$$

[18.31]
$$\frac{C_{11} + C_{16} + C_{18}}{35} \leq 0.02 \left(\frac{N_{11} + N_{16} + N_{18}}{23} + \frac{K_{11} + K_{16} + K_{18}}{39} \right)$$

where C_i, N_i, and K_i are the ionic loads of Cl, Na, and K (respectively) in the ith source.

The objective of this problem is to employ mass integration techniques to reduce wastewater discharge while alleviating any buildup of ionic species.

18.4. Consider the tricresyl phosphate process shown in Fig. 18.18. Tricresyl phosphate, $(CH_3C_6H_4O)_3PO$, can be produced by reacting cresol with phosphorus oxychloride. First, phosphorus oxychloride and creosol are mixed and heated, and then fed

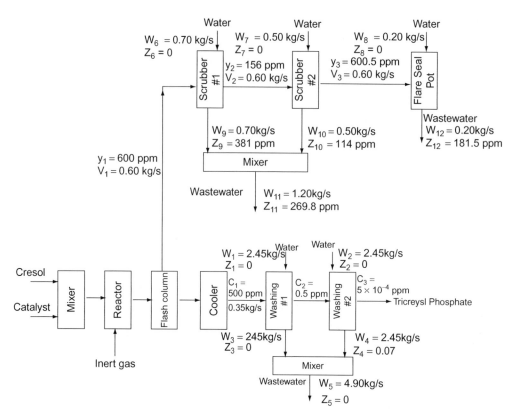

FIGURE 18-18 Tricresyl phosphate process. *Source: Hamad et al. (1996).*

continuously and slowly to a reactor. Inert gas is added to the reactor as a diluent. The exit product stream from the reactor is cooled and washed twice to improve the purity of the product. Washing the product results in reducing the concentration of cresol from 500 ppm to .007 ppm in the final product. To ensure proper wetting, the mass flow rate of washing water in the first and second washing units (W_1 and W_2) should be at least seven times the mass flow rate of the product stream. The water streams used in the first and second washing units result in wastewater streams W_3 and W_4, which are mixed to result in a terminal wastewater stream (W_5).

The following constraints are imposed on the quality and quantity of washing water:

Constraints on flow rate of washing water:

[18.31]   First washing unit: $W_1 = 2.45\,kg/s$

[18.32]   Second washing unit: $W_2 = 2.45\,kg/s$

Constraints on concentration of cresol in washing water:

[18.33]   First washing unit: $0 \leq z_1 \leq 5\,ppm$

[18.34]
Second washing unit: $z_2 = 0\,ppm$ (fresh water must be used for washing)

The gaseous stream exiting the reactor has a flow rate of 0.6 kg/s and contains 600 ppm cresol. It passes through two water scrubbers before it goes finally to a flare system for burning. To prevent backward propagation of fire in the flare, water is utilized in a water valve to seal the flare. Water utilized in scrubber #1 (W_6), scrubber #2 (W_7), and flare seal pot (W_8) possess the following constraints:

Constraints on feed water flow rate:

[18.35]   First scrubber: $0.7 \leq W_6 \leq 0.84, kg/s$

[18.36]   Second scrubber: $0.5 \leq W_7 \leq 0.6, kg/s$

[18.37]   Flare seal pot: $0.2 \leq W_8 \leq 0.25, kg/s$

Constraints on the concentration of cresol in feed water

[18.38]   First scrubber: $0 \leq z_6 \leq 30\,ppm$

[18.39]   Second scrubber: $0 \leq z_7 \leq 30\,ppm$

[18.40]   Flare seal pot: $0 \leq z_8 \leq 100\,ppm$

Wastewater streams W_9 and W_{10} are mixed to form a terminal wastewater stream W_{11}. Terminal wastewater stream W_{12} is generated in the flare system.

The concentration of cresol in the product stream is denoted by c in Fig. 18.18. The following modeling correlations describe the performance of the involved units:

TABLE 18-3 Data for MSAs That Can Remove Cresol from Liquid Streams

Stream	Description	Supply Composition (ppmw)	Target Composition (ppmw)	m_j	ε_j ppmw	cost $/kg MSA
SW_1	Oil	0	343	0.0584	685	0.04
SW_2	Light gas	15	61	0.108	13	0.02

TABLE 18-4 Data for MSAs That Can Remove Cresol from Gaseous Streams

Stream	Description	Supply Composition (ppmw)	Target Composition (ppmw)	m_j	ε_j ppmw	cost $/kg MSA
SV_1	Polymeric resin	0	1023	0.185	1135	0.03
SV_2	Adsorbent	4	560	0.027	1292	0.02

For first washing unit (u = 1):

[18.41] $\quad c_2 = .001*c_1 + .8*z_1$

[18.42] $\quad 0.34*(c_1 - c_2) - W_1*(z_3 - z_1) = 0$

For second washing unit (u = 2):

[18.43] $\quad c_3 = .001*c_2 + .8*z_2$

[18.44] $\quad 0.34*(c_2 - c_3) - W_1*(z_4 - z_2) = 0$

For first scrubber (u = 3):

[18.45]
$$0.45 = \frac{\ln\left[\left(1 - \frac{HV_1}{W_6}\right)\left(\frac{y_1^{int} - Hz_6}{y_2 - Hz_6}\right) + \frac{HV_1}{W_6}\right]}{\ln\left(\frac{W_6}{HV_1}\right)}$$

For second scrubber (u = 4):

[18.46]
$$0.35 = \frac{\ln\left[\left(1 - \frac{HV_2}{W_7}\right)\left(\frac{y_2^{int} - Hz_7}{y_3 - Hz_7}\right) + \frac{HV_2}{W_7}\right]}{\ln\left(\frac{W_7}{HV_2}\right)}$$

For flare seal pot (u = 5):

[18.47]
$$5.4 = \frac{\ln\left[\left(1 - \frac{HV_3}{W_8}\right)\left(\frac{y_3^{int} - Hz_8}{y_4 - Hz_8}\right) + \frac{HV_3}{W_8}\right]}{\ln\left(\frac{W_8}{HV_3}\right)},$$

where y_4 is the concentration of the gaseous stream after passing through the water seal pot in the flare system, and Henry's coefficient (H) is 0.064 (in units of weight fraction by weight fraction). Cresol is a toxic chemical that can be absorbed via the skin and may cause damage to the kidneys, liver, and nervous system. The objective of this problem is to reduce cresol concentration in any discharged wastewater stream to 5 ppmw or less.

The following data (Hamad et al., 1996) may be used to solve the problem:

Interception techniques:
To reduce cresol content in discharged wastewater streams, four MSAs are available: two MSAs to remove cresol from gaseous streams and two MSAs to remove cresol from liquid streams. Tables 18.3 and 18.4 provide data for the candidate MSAs.

Piping and pumping cost:
The cost for piping and pumping is assumed to be linear with the mass flow rate of the stream. The cost of piping (c_{pipe}) is approximated by

[18.48] $c_{pipe} = \$ 3.3 \times 10^{-4}/(m.kg/s)$.

Table 18.5 provides the distances among units needed for piping.

Equipment cost data:
The mass-exchange columns are assumed to be packed and have a depreciation period of three years. One-meter diameter columns are used. The following cost may be used:

[18.49] cost of packing ($) = $5250/m^3$.year

[18.50] cost of column ($) = 12,600*height of column shell

where

[18.51] $\quad$ height of column shell = 1.3*tray spacing* number of trays

A tray spacing of 0.41 m may be employed. All columns are assumed to have an overall efficiency of 25

TABLE 18-5 Distances among Units

Source	Sink	Distance, m
Washing unit # 2 (W_4)	Washing unit # 1	30
Washing unit # 2 (W_4)	Scrubber # 1	100
Washing unit # 2 (W_4)	Scrubber # 2	110
Washing unit # 2 (W_4)	Flare seal pot	90
Washing unit # 1 (W_3)	Scrubber # 1	80
Washing unit # 1 (W_3)	Scrubber # 2	90
Washing unit # 1 (W_3)	Flare seal pot	70
Washing unit # 1 (W_3)	Mass exchanger*	20
Scrubber # 1 (W_9)	Scrubber # 2	20
Scrubber # 1 (W_9)	Flare seal pot	95
Scrubber # 1 (W_9)	Mass exchanger*	20
Scrubber # 2 (W_{10})	Flare seal pot	75
Scrubber # 2 (W_{10})	Mass exchanger*	20
Flare seal pot (W_{12})	Mass exchanger*	80

*In case additional mass exchangers are added to the plant to tackle the streams leaving these units.

percent. The fixed cost of the column is obtained by multiplying the equipment cost by a factor of 5 to account for installation, instrumentation, and other ancillary devices.

NOMENCLATURE

b_j	intercept of equilibrium line for mass-separating agent j
f_{wu}	flow rate of stream allocated from source w to unit u (kg/s)
H	Henry's coefficient
l_j	mass flow rate of mass-separating agent j for pollutant-laden gaseous streams (kg/s)
L_j	mass flow rate of mass-separating agent j for pollutant-laden liquid streams (kg/s)
m_j	slope of equilibrium line for mass-separating agent j for the gas phase
M_j	slope of equilibrium line for mass-separating agent j for the liquid phase
NTP	number of theoretical stages
u	index for pollutant-processing unit (sink/generator)
v	index for pollutant-laden gaseous stream (source)
V_v	flow rate of gaseous stream (source) v (kg/s)
w	index for pollutant-laden liquid stream (source)
W_w	flow rate of liquid source w (kg/s)
x_j	mass fraction of pollutant in mass-separating agent j
x_j^s	supply composition of pollutant in mass-separating agent j for pollutant-laden gaseous streams
x_j^t	target composition of mass-separating agent j for pollutant-laden gaseous streams
X_j^s	supply composition of pollutant in mass-separating agent j for pollutant-laden liquid streams
X_j^t	target composition of pollutant in mass-separating agent j for pollutant-laden liquid streams
y_v	composition of pollutant in gaseous stream v
y_v^{int}	composition of pollutant in gaseous stream v after interception
$z_i^{Terminal}$	composition of pollutant in terminal liquid waste stream I
z_w	composition of pollutant in liquid stream w
z_w^{int}	composition of pollutant in liquid stream w after interception

SUBSCRIPTS

i	waste stream i
j	mass separating agent j
u	pollutant-processing (sink/generator) unit u
v	pollutant-laden gaseous stream v
w	pollutant-laden liquid stream w

SUPERSCRIPTS

in	inlet
int	intercepted
out	outlet
s	supply
t	target
Terminal	denotes terminal streams

GREEK LETTERS

ε_j	minimum allowable composition for mass-separating agent j between equilibrium and operating lines
φ_a	mass generated of pollutant in the reactor
φ_v	mass load of pollutant in gaseous stream v
ψ_w	mass load of pollutant in liquid stream w

REFERENCES

El-Halwagi MM, Hamad AA, Garrison GW: Synthesis of waste interception and allocation networks, *AIChE J* 42(11):3087–3101, 1996.

El-Halwagi MM, Spriggs HD: Solve design puzzles with mass integration, *Chem Eng Prog* 94(August):25–44, 1998.

El-Halwagi MM, Spriggs HD: Educational tools for pollution prevention through process integration, *Chem Eng Educ* 32(4):246–253, 1998.

El-Halwagi M.M., Spriggs H.D.: An integrated approach to cost and energy efficient pollution prevention. *Proceedings of Fifth World Congr of Chem Eng*, San Diego, vol III, pp 344–349, 1996.

Garrison G.W., Spriggs H.D., El-Halwagi M.M.: A global approach to integrating environmental, energy, economic and technological objectives. *Proceedings of Fifth World Congr of Chem Eng*, San Diego, vol I, pp 675–680, 1996.

Hamad A.A., Garrison G.W., Crabtree E.W., El-Halwagi M.M.: Optimal design of hybrid separation systems for waste reduction. *Proceedings of Fifth World Congr of Chem Eng*, San Diego, vol III, pp 453–458, 1996.

Lovelady EM, El-Halwagi MM, Krishnagopalan G: An integrated approach to the optimization of water usage and discharge in pulp and paper plants, *Int J of Environ and Pollution (IJEP)* 29(1–3):274–307, 2007.

CHAPTER 19
MATHEMATICAL TECHNIQUES FOR THE SYNTHESIS OF HEAT-EXCHANGE NETWORKS

Chapter 7 provided graphical and algebraic procedures for the synthesis of heat-exchange networks (HENs). This chapter presents a mathematical programming approach to the synthesis of HENs. First, the HEN formulation will be presented and solved to attain minimum heating and cooling utility cost and for selecting the optimum utilities. Next, a formulation is presented to synthesize a network of heat exchangers that can meet the minimum utility targets. The optimization formulations in this chapter are based on the transshipment formulation of Papoulias and Grossmann (1983). Numerous other methods have been developed for the synthesis of HENs. These methods have been reviewed by Rossiter (2010), Majozi (2010), Kemp (2009), Ponce-Ortega et al. (2008), Al-Thubaiti et al. (2008), Smith (2005), Furman and Sahinidis (2002), Shenoy (1995), Linnhoff (1993), Gundersen and Naess (1988).

TARGETING FOR MINIMUM HEATING AND COOLING UTILITIES

This section is based on the transshipment formulation of Papoulias and Grossmann (1983). According to this formulation, heat is to be transshipped from sources (hot streams) to destination (cold streams). To ensure thermodynamic feasibility of heat exchange, heat must go through "warehouses." These are represented by the temperature intervals through which heat transfer is feasible. Chapter 9 discussed the construction of these temperature intervals as part of developing the temperature interval diagram (TID). Once the TID is constructed, an optimization-based cascade analysis can be developed as a generalization to the algebraic procedure described in Chapter 7.

Consider a process with N_H process hot streams, N_C process cold streams, N_{HU} heating utilities, and N_{CU} cooling utilities. The index u is used for process hot streams and heating utilities while the index v is used for process cold streams and cooling utilities. The index z is used for temperature intervals. As mentioned in Chapter 9, on the TID, two corresponding temperature scales are generated: hot and cold separated by a minimum temperature driving force. Streams are represented by vertical arrows extending between supply and target temperatures. Next, horizontal lines are drawn at the heads and tails of the arrows, thereby defining a series of temperature intervals $z = 1, 2,..., n_{int}$. Heat can be transferred within the same interval or passed downward. Next, the table of exchangeable heat loads (TEHL) is constructed (as described in Chapter 9) to determine the heat-exchange loads of the process streams in each temperature interval. The exchangeable load of the uth hot stream (losing sensible heat), which passes through the zth interval, is defined as

[19.1] $$HH_{u,z} = F_u C_{p,u}(T_{z-1} - T_z),$$

where T_{z-1} and T_z are the hot-scale temperatures at the top and the bottom lines defining the zth interval. On the other hand, the exchangeable capacity of the vth cold stream (gaining sensible heat), which passes through the zth interval, is computed through

[19.2] $$HC_{v,z} = f_v c_{p,v}(t_{z-1} - t_z),$$

where t_{z-1} and t_z are the cold-scale temperatures at the top and the bottom lines defining the zth interval.

By summing up the heating loads and cooling capacities, we get

[19.3] $$HH_z^{Total} = \sum_{\substack{u \text{ passes through interval } z \\ \text{where } u=1, 2, \ldots, N_H}} HH_{u,z}.$$

and

[19.4] $$HC_z^{Total} = \sum_{\substack{v \text{ passes through interval } z \\ \text{and } v=1, 2, \ldots, N_C}} HC_{v,z}$$

Next, we move to incorporating heating and cooling utilities. For temperature interval z, the heat load of the uth heating utility is given by

[19.5]
$$HHU_{u,z} = FU_u C_{p,u}(T_{z-1} - T_z)$$
$$\text{where } u = N_H + 1, N_H + 2, \ldots, N_H + N_{HU}$$

where FU_u is the flow rate of the uth heating utility.

The sum of all heating loads of the heating utilities in interval is expressed as

[19.6] $$HHU_z^{Total} = \sum_{\substack{u \text{ passes through interval } z \\ \text{where } u=N_H+1, N_H+2, \ldots, N_H+N_{HU}}} HUU_{u,z}$$

The total heating load of the uth utility in the HEN may be evaluated by summing up the individual heat loads over intervals:

[19.7] $$QH_u = \sum_z HHU_{u,z}$$

Sustainable Design Through Process Integration.
© 2012 Elsevier Inc. All rights reserved.

Similarly, the cooling capacities of the vth cooling utility in the zth interval is calculated as follows:

$$\text{HCU}_{v,z} = fU_v c_{p,v}(t_{z-1} - t_z),$$

where

[19.8] $v = N_C + 1, N_C + 2, \ldots, N_C + N_{CU}$

and where fU_v is the flow rate of the vth cooling utility.

The sum of all cooling capacities of the cooling utilities is expressed as

[19.9]
$$\text{HCU}_z^{Total} = \sum_{\substack{v \text{ passes through interval } z \\ \text{where } v = N_C+1, N_C+2, \ldots, N_C + N_{CU}}} \text{HCU}_{v,z}$$

The total cooling capacity of the uth utility in the HEN may be evaluated by summing up the individual cooling loads over intervals:

[19.10] $QC_v = \sum_z \text{HCU}_{v,z}$

For the zth temperature interval, one can write the following heat-balance equation (see Fig. 19.1):

[19.11]
$$HH_z^{Total} - HC_z^{Total} = HHU_z^{Total} - HHC_z^{Total} + r_{z-1} - r_z, \quad z = 1, 2, \ldots, n_{int}.$$

where

[19.12] $r_0 = r_{n_{int}} = 0$

[19.13] $r_z \geq 0, \quad z = 1, 2, \ldots, n_{int} - 1$

[19.14]
$$FU_u \geq 0, \quad u = N_H + 1, N_H + 2, \ldots, N_H + N_{HU}$$

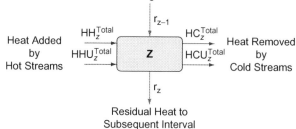

FIGURE 19-1 Heat balance around temperature interval including utilities.

[19.15]
$$fU_v \geq 0 \quad v = N_C + 1, N_C + 2, \ldots, N_C + N_{CU}.$$

The objective here is to minimize the cost of heating and cooling utilities. Let CH_u designate the cost of the uth heating utility and CC_v designate the cost of the vth cooling utility. If this cost is given in terms of $/unit flow of utility, then the objective function can be expressed as

[19.16a]
$$\text{Minimize} \sum_{u=N_H+1}^{N_H+N_{HU}} CH_u * FU_u + \sum_{v=N_C+1}^{N_C+N_{CU}} CC_v * fU_v$$

On the other hand, if this cost is given in terms of $/unit heat added or removed by the utility, then the objective function can be expressed as

[19.16b]
$$\text{Minimize} \sum_{u=N_H+1}^{N_H+N_{HU}} CH_u * QH_u + \sum_{v=N_C+1}^{N_C+N_{CU}} CC_v * QC_v.$$

This objective function can be minimized subject to the set of aforementioned constraints. This formulation is a linear program that can be solved using commercially available software (for example, LINGO).

Example 19-1 Pharmaceutical facility

Consider the pharmaceutical processing facility described by Example 7.1 and illustrated by Fig. 19.2. The feed mixture (C_1) is first heated to 550 K, and then fed to an adiabatic reactor where an endothermic reaction takes place. The off-gases leaving the reactor (H_1) at 520 K are cooled to 330 K prior to being forwarded to the recovery unit. The mixture leaving the bottom of the reactor is separated into a vapor fraction and a slurry fraction. The vapor fraction (H_2) exits the separation unit at 380 K and is to be cooled to 300 K prior to storage. The slurry fraction is washed with a hot immiscible liquid at 380 K. The wash liquid is purified and recycled to the washing unit. During purification, the temperature drops to 320 K. Therefore, the recycled liquid (C_2) is heated to 380 K. Two utilities—HU_1 and CU_1—are available for service. The cost of the heating and cooling utilities ($/10^6$ kJ) are 6 and 8, respectively. Table 19.1 gives stream data.

In the current operation, the heat exchange duties of H_1, H_2, C_1, and C_2 are fulfilled using the cooling and heating utilities. Therefore, the current annual operating cost of utilities is

$$\left[(4750 + 120)\text{kW} \times 6 \times 10^{-6} \frac{\$}{\text{kJ}} + (1900 + 400)\text{kW} \times 8 \times 10^{-6} \frac{\$}{\text{kJ}} \right]$$
$$\times 3600 \times 8760 \frac{s}{\text{yr}} = \$1,501,744/\text{yr}.$$

The objective of this case study is to use linear programming to determine the minimum operating cost. A value of $\Delta T^{min} = 10$ K is used.

(Continued)

Example 19-1 Pharmaceutical facility (Continued)

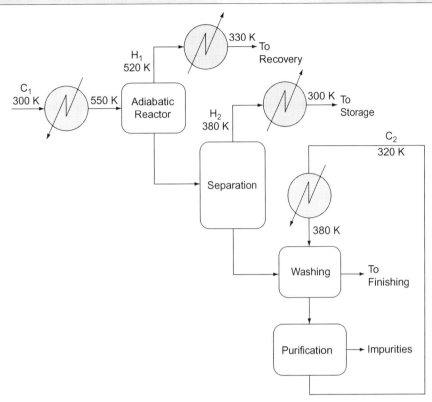

FIGURE 19-2 Simplified flow sheet of the pharmaceutical process.

TABLE 19-1 Stream Data for Pharmaceutical Process

Stream	Flow Rate × Specific Heat, kW/K	Supply Temperature, K	Target Temperature, K	Enthalpy Change, kW
H_1	10	520	330	−1900
H_2	5	380	300	−400
HU_1	?	560	520	?
C_1	19	300	550	4750
C_2	2	320	380	120
CU_1	?	290	300	?

SOLUTION

The first step is to create the TID, including process streams and utilities (Fig. 19.3). Next, the tables of exchangeable loads are developed for the process hot and cold streams, as shown by Tables 19.2 and 19.3.

Next, the problem is formulated as an optimization program as follows:

[P19.1]

$$\text{minimize}(6 \times 10^{-6} Q^{min}_{Heating} + 8 \times 10^{-6} Q^{min}_{Cooling})3600 \times 8760 = $$
$$189.216 Q^{min}_{Heating} + 252.288 Q^{min}_{Cooling}$$

subject to

$$r_1 - Q^{min}_{Heating} = -760$$
$$r_2 - r_1 = -1170$$
$$r_3 - r_2 = -110$$
$$r_4 - r_3 = -300$$
$$r_5 - r_4 = -280$$
$$-r_5 + Q^{min}_{Cooling} = 50$$

with all the nonnegativity constraints. In terms of LINGO input, the program can be written as follows:

```
min=189.216*Qhmin+252.288*Qcmin;
r1-Qhmin=-760;
r2-r1=-1170;
r3-r2=-110;
r4-r3=-300;
r5-r4=-280;
-r5+Qcmin=50;
Qhmin>=0;
Qcmin>=0;
r1>=0;
r2>=0;
r3>=0;
r4>=0;
r5>=0;
```

(Continued)

Example 19-1 Pharmaceutical facility (Continued)

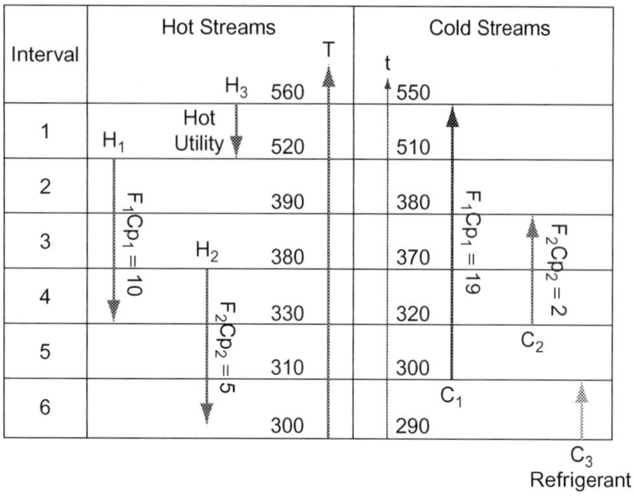

FIGURE 19-3 The temperature interval diagram for the pharmaceutical example.

The solution obtained using LINGO is as follows:

```
Objective value: 508360.3
Variable    Value
QHMIN       2620.000
QCMIN       50.00000
R1          1860.000
R2          690.0000
R3          580.0000
R4          280.0000
R5          0.0000000E+00
```

TABLE 19-2 TEHL for Process Hot Streams

Interval	Load of H_1 (kW)	Load of H_2 (kW)	Total Load (kW)
1	–	–	–
2	1300	–	1300
3	100	–	100
4	500	250	750
5	–	100	100
6	–	50	50

TABLE 19-3 TEHL for Process Cold Streams

Interval	Capacity of C_1 (kW)	Capacity of C_2 (kW)	Total Capacity (kW)
1	760	–	760
2	2470	–	2470
3	190	20	210
4	950	100	1050
5	380	–	380
6	–	–	–

These results are identical to those obtained using the graphical and the algebraic methods in Chapter 7.

Example 19-2 Using LINGO's set formulation to solve Example 19.1

In this example, LINGO's set formulation is used to solve the previous example. A set INTERVALS of temperature intervals is defined. Attributes of this set include hot and cold loads of the process streams and utilities in each interval. The data from Tables 19.2 and 19.3 are entered for the hot and cold loads of the process streams. Also, hot and cold utility factors are characterized as 1 or 0, depending on whether or not the utility passes through the interval. Also, the input heat residual to the first interval and from the last interval are set to 0. The following is the set formulation for the minimum operating cost problem.

```
[P19.2]
min= CH*QHU+CC*QCU;
CH=189.216; CC=252.288;

SETS:
INTERVAL /1..6/: Hot_Load, Cold_Load, HotU_Load,
ColdU_Load, Factor_HotU, Factor_ColdU, residual;
ENDSETS

DATA:
Hot_Load=0 1300 100 750 100 50;
Cold_Load=760 2470 210 1050 380 0;
! Factor_HotU designates presence or absence of
  heating utility in interval;
Factor_HotU=1 0 0 0 0 0;
Factor_ColdU=0 0 0 0 0 1;

ENDDATA
```

```
QHU=@SUM(INTERVAL(z): HotU_Load(z));
QCU=@SUM(INTERVAL(z): ColdU_Load(z));
@FOR (INTERVAL(z)|z #GE# 2:
    Hot_Load(z)+HotU_Load(z)+Residual(z-1)
= Cold_Load(z)+ColdU_Load(z)+Residual(z));
Hot_Load(1)+HotU_Load(1)=Cold_Load(1)+ColdU_
  Load(1)+Residual(1);
Hot_Load(6)+HotU_Load(6)+Residual(5)
= Cold_Load(6)+ColdU_Load(6);
@FOR (INTERVAL(z): Residual(z) >= 0);
```

Using the "Display Model" option (selecting "LINGO" from the menu bar, and then "Generate," then "Display Model"), we get

```
MODEL:
[_1] MIN= 189.216 * QHU+252.288 * QCU;
[_4] QHU-HOTU_LOAD_1-HOTU_LOAD_2-HOTU_LOAD_3-HOTU_
     LOAD_4-HOTU_LOAD_5-HOTU_LOAD_6=0;
[_5] QCU-COLDU_LOAD_1-COLDU_LOAD_2-COLDU_LOAD_
     3-COLDU_LOAD_4-COLDU_LOAD_5-COLDU_LOAD_6=0;
[_6] RESIDUAL_1+HOTU_LOAD_2-COLDU_LOAD_2-RESIDUAL_
     2=1170;
[_7] RESIDUAL_2+HOTU_LOAD_3-COLDU_LOAD_3-RESIDUAL_
     3=110;
[_8] RESIDUAL_3+HOTU_LOAD_4-COLDU_LOAD_4-RESIDUAL_
     4=300;
[_9] RESIDUAL_4+HOTU_LOAD_5-COLDU_LOAD_5-RESIDUAL_
     5=280;
```

(Continued)

Example 19-2 Using LINGO's set formulation to solve Example 19.1 (*Continued*)

```
[_10] RESIDUAL_5+HOTU_LOAD_6-COLDU_LOAD_6-RESIDUAL_
       6 =-50;
[_11] HOTU_LOAD_1-COLDU_LOAD_1-RESIDUAL_1=760;
[_12] RESIDUAL_5+HOTU_LOAD_6-COLDU_LOAD_6 =-50;
[_13] RESIDUAL_1>=0;
[_14] RESIDUAL_2>=0;
[_15] RESIDUAL_3>=0;
[_16] RESIDUAL_4>=0;
[_17] RESIDUAL_5>=0;
[_18] RESIDUAL_6>=0;
END
```

The following is the solution:

```
Objective value: 508360.3
Variable           Value
CH                189.2160
QHU              2620.000
CC                252.2880
QCU                50.00000
HOT_LOAD(1)        0.000000
HOT_LOAD(2)     1300.000
HOT_LOAD(3)      100.0000
HOT_LOAD(4)      750.0000
HOT_LOAD(5)      100.0000
HOT_LOAD(6)       50.00000
COLD_LOAD(1)     760.0000
COLD_LOAD(2)    2470.000
COLD_LOAD(3)     210.0000
COLD_LOAD(4)    1050.000
COLD_LOAD(5)     380.0000
COLD_LOAD(6)       0.000000
HOTU_LOAD(1)     760.0000
HOTU_LOAD(2)    1170.000
HOTU_LOAD(3)     110.0000
HOTU_LOAD(4)     300.0000
HOTU_LOAD(5)     280.0000
HOTU_LOAD(6)       0.000000
COLDU_LOAD(1)      0.000000
COLDU_LOAD(2)      0.000000
COLDU_LOAD(3)      0.000000
COLDU_LOAD(4)      0.000000
COLDU_LOAD(5)      0.000000
COLDU_LOAD(6)     50.00000
FACTOR_HOTU(1)     1.000000
FACTOR_HOTU(2)     0.000000
FACTOR_HOTU(3)     0.000000
FACTOR_HOTU(4)     0.000000
FACTOR_HOTU(5)     0.000000
FACTOR_HOTU(6)     0.000000
FACTOR_COLDU(1)    0.000000
FACTOR_COLDU(2)    0.000000
FACTOR_COLDU(3)    0.000000
FACTOR_COLDU(4)    0.000000
FACTOR_COLDU(5)    0.000000
FACTOR_COLDU(6)    1.000000
RESIDUAL(1)        0.000000
RESIDUAL(2)        0.000000
RESIDUAL(3)        0.000000
RESIDUAL(4)        0.000000
RESIDUAL(5)        0.000000
RESIDUAL(6)        0.000000
```

STREAM MATCHING AND HEN SYNTHESIS

Now that the values of the minimum heating and cooling utilities have been identified, it is desired to synthesize a network of heat exchangers that satisfies the minimum utility targets with the minimum number of heat exchangers. The objective of a minimum number of units is intended to indirectly minimize the fixed cost by assuming that the lower number of units entails lower fixed cost (for example, six-tenths-factor rule). As mentioned earlier, when a pinch point exists, the synthesis problem can be decomposed into two subnetworks, one above the pinch and one below the pinch. The subnetworks will be denoted by SN_m, where $m = 1, 2$. It is therefore useful to define the following subsets:

[19.17] $H_m = \{u | u \in H, \text{hot stream } u \text{ exists in } SN_m\}$,

where H_u is the set of hot streams existing in SN_m;

[19.18] $C_m = \{v | v \in S, \text{cold stream } u \text{ exists in } SN_m\}$,

where C_u is the set of cold streams existing in SN_m;

[19.19] $H_{m,z} = \{u | u \in H_m, \text{stream } u \text{ exists in interval } \bar{z} \leq z; \bar{z}, z \in SN_m\}$,

where $H_{m,z}$ is the set of hot streams existing in interval z in SN_m; and

[19.20] $C_{m,z} = \{v | v \in S_m, \text{cold stream } v \text{ exists in interval } z \in SN_m\}$,

where $C_{m,z}$ is the set of cold streams existing in interval z in SN_m.

For hot stream, u, the heat balance around temperature interval z is given by

[19.21] $r_{u,z} - r_{u,z-1} + \sum_{v \in S_{m,z}} Q_{u,v,z} = Q^H_{u,z}$,

where $r_{u,z}$ is the residual heat leaving interval z, $r_{u,z-1}$ is the residual heat entering interval z, $Q_{u,v,z}$ is the heat exchanged from hot stream u to cold stream v in interval z, and $Q^H_{u,z}$ is the heat exchanged from hot stream u in interval z.

Within any subnetwork, the heat exchanged between any two streams is bounded by the smaller of the two heat loads. Therefore, the upper bound on the exchangeable heat between streams u and v in SN_m (designated by $U_{u,v,m}$) is given by

[19.22] $U_{u,v,m} = \min \left\{ \sum_{z \in SN_m} Q^H_{u,z}, \sum_{z \in SN_m} Q^C_{v,z} \right\}$, where $Q_{v,z}^C$ is the heat exchanged from cold stream v in interval z.

Next, we define the binary variable $E_{u,v,m}$, which takes the values of 0 when there is no match between streams u and v in SN_m, and takes the value of 1 when there exists a match between streams u and v (and hence an exchanger) in SN_m. Based on Eq. 19.22, one can write

[19.23]
$$\sum_{z \in SN_m} Q_{u,v,z} \leq U_{u,v,m} E_{u,v,m}$$
$$u \in H_m, v \in C_m, m = 1, 2$$

where $Q_{u,v,z}$ denotes the mass exchanged between the uth hot stream and the vth cold stream in the zth interval. Therefore, the problem of minimizing the number of heat exchangers can be formulated as a mixed-integer linear program (MILP) (Papoulias and Grossmann, 1983):

[P19.3]
$$\text{minimize} \sum_{m=1,2} \sum_{u \in H_m} \sum_{v \in C_m} E_{u,v,m}$$

subject to the following:

Heat balance for each hot stream around each temperature interval:

$$r_{u,z} - r_{u,z-1} + \sum_{v \in S_{m,z}} Q_{u,v,z} = Q_{u,z}^H$$
$$u \in H_{m,z}, z \in SN_m, m = 1, 2$$

Heat balance for each cold stream around each temperature interval:

$$\sum_{u \in S_{m,z}} Q_{u,v,z} = Q_{v,z}^C \quad v \in C_{m,z}, z \in SN_m, m = 1, 2$$

Matching of loads:

$$\sum_{z \in SN_m} Q_{u,v,z} \leq U_{u,v,m} E_{u,v,m}$$
$$u \in H_m, v \in C_m, m = 1, 2$$

Nonnegativity constraints for the residuals and the loads:

$$r_{u,z} \geq 0 \quad \forall u, \forall z$$

$$Q_{u,v,z} \geq 0 \quad \forall u, \forall v, \forall z$$

Binary integer variables for matching the streams:

$$E_{u,v,m} \in \{0,1\} \quad \forall u, \forall v, \forall m$$

The preceding program is a mixed-integer linear program that can be solved using optimization tools (for example, LINGO) to provide information on the stream matches and exchangeable loads. It is interesting to note that the solution of Program P19.3 may not be unique. It is possible to generate all integer solutions to P19.4 by adding integer-cut constraints that exclude previously obtained solutions from further consideration. For example, any previous solution can be eliminated by requiring that the sum of $E_{u,v,m}$ that were nonzero in that solution be less than the minimum number of exchangers. Also, if the costs of the various exchangers are significantly different, the objective function can be modified by multiplying each integer variable by a weighting factor that reflects the relative cost of each unit.

Example 19-3 Network synthesis for Example 19.1

The objective of this example is to synthesize a network with a minimum number of heat exchangers satisfying the minimum utility targets. First, let us construct the table of exchangeable loads for the hot and the cold streams over all the temperature intervals. Table 19.4 shows the results. Next, the exchangeable loads above and below the pinch are calculated by summing up the loads over the intervals above and below the pinch, respectively. Tables 19.5 and 19.6 give the results.

Above the pinch, there are six possible matches between the hot and the cold streams. Using the symbol $E_{u,v,1}$ to designate a binary-integer variable for a match between a hot stream u and a cold stream v above the pinch (subnetwork $m = 1$), then we have the following possible matches above the pinch: E_{111}, E_{121}, E_{211}, E_{221}, E_{311}, and E_{321}. On the other hand, below the pinch, there is only one possible match between H_2 ($u = 2$) and the cooling utility ($v = 3$) in subnetwork $m = 2$; that is, E_{232}. The following is the LINGO formulation:

[P19.4]
```
min=E111+E211+E121+E221+E311+E321+E232;

!Heat Balances for H1;
r12+Q112=1300.0;
r13-r12+Q113+Q123=100.0;
r14-r13+Q114+Q124=500.0;
r15-r14+Q115=0.0;
r15+Q116+Q136 =0.0;

!Heat Balances for H2;
r24+Q214+Q224=250.0;
r25-r24+Q215 =100.0;
r25+Q216+Q236=50.0;

!Heat Balances for HU1(H3);
r31+Q311=2620.0;
r32-r31+Q312=0.0;
r33-r32+Q313+Q323=0.0;
r34-r33+Q314+Q324=0.0;
r35-r34+Q315=0.0;
r35=0.0;

!Heat Balances for C1;
Q311=760;
Q112+Q312=2470.0;
Q113+Q313=190.0;
Q114+Q214+Q314=950.0;
Q115+Q215+Q315=380.0;

!Heat Balances for C2;
```

(Continued)

Example 19-3 Network synthesis for Example 19.1 (*Continued*)

TABLE 19-4 Table of Exchangeable Loads

	Load of Hot Streams (kW)			Load of Cold Streams (kW)		
	H_1	H_2	HU (H_3)	C_1	C_2	CU (C_3)
Interval						
1	0	0	2620	760	0	0
2	1300	0	0	2470	0	0
3	100	0	0	190	20	0
4	500	250	0	950	100	0
5	0	100	0	380	0	0
Pinch						
6	0	50	0	0	0	50

TABLE 19-5 Exchangeable Loads above the Pinch

Stream	Load (kW)
H_1	1900
H_2	350
HU (H_3)	2620
C_1	4750
C_2	120
CU (C_3)	0

TABLE 19-6 Exchangeable Loads below the Pinch

Stream	Load (kW)
H_1	0
H_2	50
HU_1 (H_3)	0
C_1	0
C_2	0
CU (C_3)	50

```
Q123+Q323=20.0;
Q124+Q224+Q324=100.0;

!Heat Balance for CU1(C3);
Q236=50.0;

!Matching of Loads;
Q112+Q113+Q114+Q115<=1900.0*E111;
Q214+Q215<=350.0*E211;
Q123+Q124<=120.0*E121;
Q224<=120.0*E221;
Q311+Q312+Q313+Q314+Q315<=2620.0*E311;
Q323+Q324<=120.0*E321;
Q236=50.0*E232;

! Nonnegativity constraints;

r12>=0.0;
r13>=0.0;
r14>=0.0;
r15>=0.0;

r24>=0.0;
r25>=0.0;

r31>=0.0;
r32>=0.0;
r33>=0.0;
r34>=0.0;
r35>=0.0;

Q112>=0.0;
Q113>=0.0;
Q114>=0.0;
Q115>=0.0;
Q116>=0.0;
```

```
Q123>=0.0;
Q124>=0.0;
Q136>=0.0;

Q214>=0.0;
Q215>=0.0;
Q216>=0.0;

Q224>=0.0;
Q236>=0.0;
Q312>=0.0;
Q313>=0.0;
Q314>=0.0;
Q315>=0.0;

Q323>=0.0;
Q324>=0.0;

! Binary integers;

@BIN(E111);
@BIN(E211);
@BIN(E121);
@BIN(E221);
@BIN(E311);
@BIN(E321);
@BIN(E232);
```

The following solution of this MILP is obtained using LINGO:

```
  Objective value: 5.000000

  Variable     Value
  E111         1.000000
  E211         1.000000
  E121         0.000000
  E221         0.000000
```

(*Continued*)

Example 19-3 Network synthesis for Example 19.1 (Continued)

TABLE 19-7 Heat-Exchange Loads between Pairs of Streams above the Pinch

Hot Stream	Cold Stream	Heat-Exchange Load (kW)
H_1	C_1	$1300 + 100 + 220 + 280 = 1900$
H_1	C_2	0
H_2	C_1	$250 + 100 = 350$
H_2	C_2	0
HU (H_3)	C_1	$760 + 1170 + 90 + 480 = 2500$
HU (H_3)	C_2	$20 + 100 = 120$

```
E311      1.000000
E321      1.000000
E232      1.000000
R12       0.000000
Q112      1300.000
R13       0.000000
Q113      100.0000
Q123      0.000000
R14       280.0000
Q114      220.0000
Q124      0.000000
R15       0.000000
Q115      280.0000
Q116      0.000000
Q136      0.000000
R24       0.000000
Q214      250.0000
Q224      0.000000
R25       0.000000
Q215      100.0000
Q216      0.000000
Q236      50.00000
R31       1860.000
Q311      760.0000
R32       690.0000
Q312      1170.000
R33       580.0000
Q313      90.00000
Q323      20.00000
R34       0.000000
Q314      480.0000
Q324      100.0000
R35       0.000000
Q315      0.000000
```

The results indicate that five exchangers exist: E111, E211, E311, E321 (above the pinch), and E232 (below the pinch). The heat duty of each exchanger is calculated by summing up the exchangeable loads between each pair of streams over the temperature intervals. Table 19.7 presents these summations above the pinch. Below the pinch, there is only one match transferring 50 kW from H_2 to the cooling utility (C_3).

The network configuration with the exchanged heat duties is shown by Fig. 19.4. Each heat exchanger is represented by an ellipse with the heat-transfer rate noted inside it. Temperatures are placed next to each stream. The intermediate temperatures can be calculated using heat balances. For example, consider the exchanger matching H_1 and C_1 and transferring 1900 kW. The heat balance for C_1 (with $fc_p = 19$ kW/K) gives:

[19.24] Outlet temperature of C_1 leaving the exchanger with $H_1 = 300 + \dfrac{350}{19}$
$= 318.4\,K$

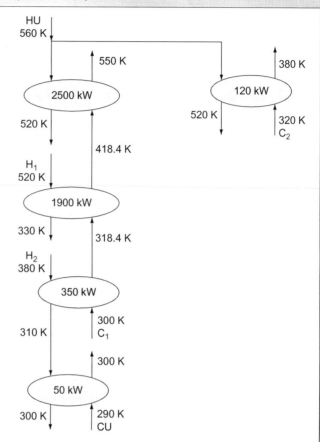

FIGURE 19-4 HEN configuration for the pharmaceutical process.

To generate an alternate solution, an integer-cut constraint is added by requiring that the sum of the fixed matches in the previous solution be less than or equal to four; that is,

[19.25] $E111 + E211 + E311 + E321 + E232 \Leftarrow 4;$

Solving the MILP with the additional integer-cut constraint, we get

```
Global optimal solution found.
Objective value: 5.000000
Variable Value
E111      1.000000
E211      1.000000
E121      1.000000
E221      0.000000
E311      1.000000
E321      0.000000
E232      1.000000
R12       0.000000
Q112      1300.000
R13       0.000000
Q113      80.00000
Q123      20.00000
R14       280.0000
Q114      120.0000
Q124      100.0000
R15       0.000000
Q115      280.0000
Q116      0.000000
Q136      0.000000
```

(Continued)

Example 19-3 Network synthesis for Example 19.1 (*Continued*)

```
R24     0.000000
Q214    250.0000
Q224    0.000000
R25     0.000000
Q215    100.0000
Q216    0.000000
Q236    50.00000
R31     1860.000
Q311    760.0000
R32     690.0000
Q312    1170.000
R33     580.0000
Q313    110.0000
Q323    0.000000
R34     0.000000
Q314    580.0000
Q324    0.000000
R35     0.000000
Q315    0.000000
```

This solution also has five matches: E111, E211, E121, E311 and E232. Figure 19.5 shows the network configuration.

Again, we add another integer-cut constraint:

[19.26]   E111 + E211 + E121 + E311 + E232 $\Leftarrow$ 4;

No feasible solution is found, indicating that the next configuration alternative will have more than five exchangers. It is worth noting that, within the same set of matches, there are multiple solutions for distributing the exchanged loads. These can be generated by adding constraints limiting the extent of heat transferred between a pair of streams. Furthermore, some implementations may be constructed by having parallel versus series configurations. These different alternatives should eventually be compared based on economic criteria.

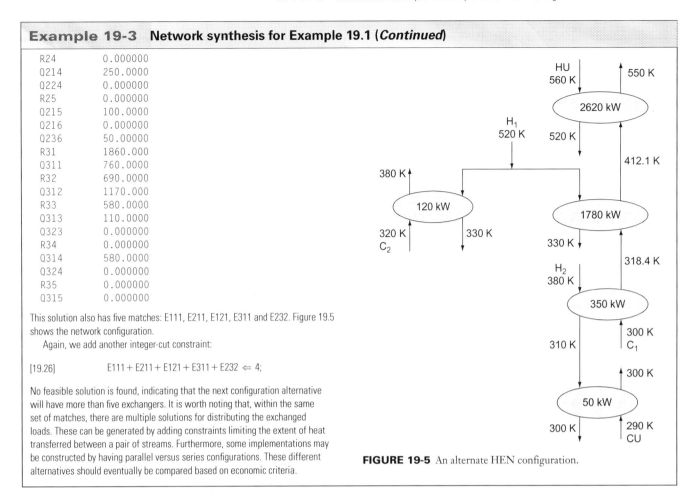

FIGURE 19-5 An alternate HEN configuration.

HANDLING SCHEDULING AND FLEXIBILITY ISSUES IN HEN SYNTHESIS

The nominal designs of HENs and other process subsystems address base-case inputs, outputs, and performance that are set to certain steady-state values (Fig. 19.6a). Designing for the base case is a crucial step as it addresses a likely scenario around which the operation will be carried out. Nonetheless, fluctuations will occur. Regardless of such fluctuations, the process is expected to perform within bounds of acceptable operation. Given that process inputs are likely to change, it is important to account for such variations. Lack of appropriate process flexibility can lead to unacceptable performance and outputs violating the required specifications (Fig. 19.6b). When the process is designed with proper accounting for expected fluctuations, the outputs will meet the specifications with the use of the necessary control systems (Fig. 19.6c). The nominal design can also be changed through retrofitting when a new set of base-case input data and desired outputs are defined (Fig. 19.6d). Such retrofitting may include adding new units, increasing the capacities of existing units, rerouting streams, and modifying design and/or operating variables. It is also possible to intentionally change the operating mode over time periods to account for certain changes in feedstocks, product demand, and product type and to manage preplanned transitions and production schemes. This is achieved via scheduling (Fig. 19.6e) that allows the HEN and other process components to respond to the expected variations in meaningful and effective manners. Numerous mathematical techniques have been developed for the systematic optimization of process scheduling (for example, Al-Mutairi and El-Halwagi, 2010, 2009; Majozi, 2010; Li and Ierapetritou, 2010; Shaik and Floudas, 2008; Ryu and Pistikopoulos, 2007; Liu and Karimi, 2007; Maravelias and Grossmann, 2003; Shang et al., 1999; Pekny and Reklaitis, 1998; Kondili et al., 1993).

One approach for synthesizing HENs while accounting for scheduling is described by Al-Mutairi and El-Halwagi (2009). Scheduling is considered over a certain time horizon during which the operation is discretized into a number of periods. For each period, the HEN synthesis optimization formulation is developed. Then, all these models are combined into a multi-period optimization formulation that accounts for the HEN synthesis over the multiple periods and requires that one configuration be flexible enough to address the scheduling changes. Figure 19.7 summarizes the procedure, and Fig. 19.8 shows the

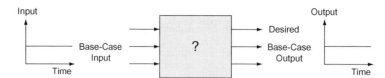

FIGURE 19-6a Nominal (base-case) grassroot design. *Source: Al-Mutairi and El-Halwagi (2010).*

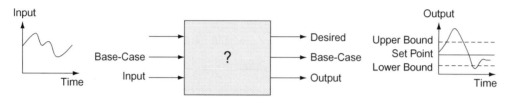

FIGURE 19-6b Inflexible design. *Source: Al-Mutairi and El-Halwagi (2010).*

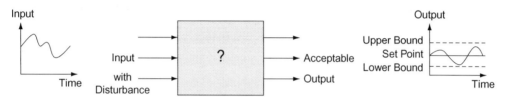

FIGURE 19-6c Including flexibility in design with proper process control and operation. *Source: Al-Mutairi and El-Halwagi (2010).*

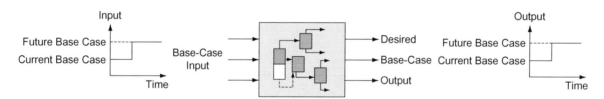

FIGURE 19-6d Base-case design retrofitting. *Source: Al-Mutairi and El-Halwagi (2010).*

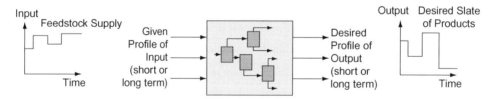

FIGURE 19-6e Process scheduling. *Source: Al-Mutairi and El-Halwagi (2010).*

simultaneous consideration of the multiple periods using a single temperature interval diagram. The result is a cost-effective HEN that performs well under the planned schedules over the multiperiod time horizon.

HOMEWORK PROBLEMS

19.1 Resolve Problem 7.1 using linear programming.
19.2 Resolve Problem 7.2 using linear programming.
19.3 Resolve Problem 7.3 using linear programming.

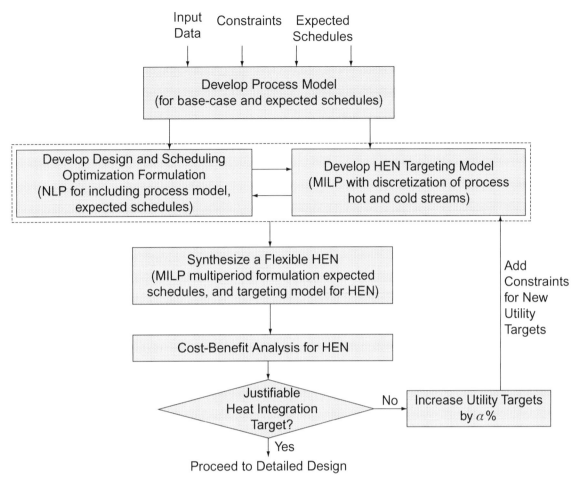

FIGURE 19-7 Design procedure for simultaneous heat integration and scheduling. *Source: Al-Mutairi and El-Halwagi (2009).*

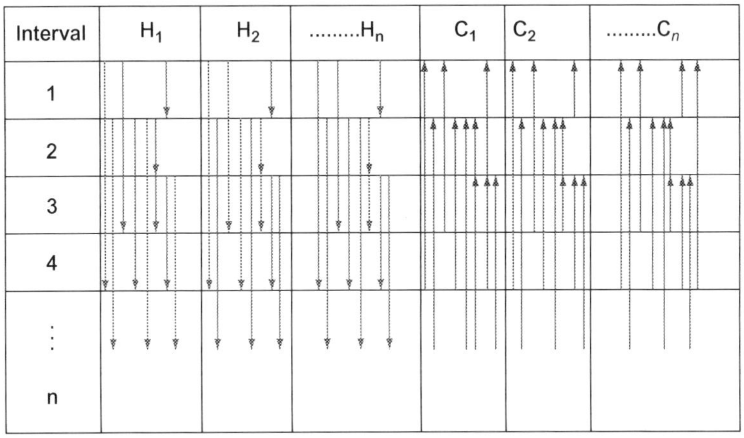

FIGURE 19-8 The temperature-interval diagram for the discretized streams over multiple periods. *Source: Al-Mutairi and El-Halwagi (2009).*

19.4 Resolve Problem 7.4 using an optimization formulation.

19.5 Resolve Problem 7.5 using an optimization formulation.

19.6 Consider the fluid catalytic cracking (FCC) system of a refinery with a base case of receiving 35,000 barrels per day (BPD) of feed and producing 31,859 BPD of gasoline (Al-Mutairi and El-Halwagi, 2009). Table 19.8 gives the base-case problem data for the hot and cold streams. It is desired to schedule half of the year at the base-case data and the other half for a feed of 50,000 BPD. Determine the

TABLE 19-8 Base-Case Data for the FCC

Stream	FCp (MW/°C)	T^s (°C)	T^t (°C)	ΔH (MW)
H1	0.184839	343	281	11.46
H2	0.1819355	343	281	11.28
H3	0.3445	281	232	16.88
H4	0.0082	343	121	1.82
H5	0.0057	263	49	1.23
H6	0.0195238	202	97	2.05
H7	0.0256	97	49	1.23
H8	0.0662295	254	193	4.04
H9	0.0571	193	179	0.8
H10	0.056569	179	77	5.77
H11	0.125	163	157	0.75
H12	0.13	157	49	14.04
H13	0.3563	111	60	18.17
H14	0.2283	60	37	5.25
C1	0.124565	182	274	11.46
C2	0.0757047	125	274	11.28
C3	0.2365	274	360	20.34
C4	0.082	51	76	2.05
C5	0.0824	76	125	4.04

Source: CANMET (2004).

minimum annual heating and cooling utility targets for the schedule of the two cases.

NOMENCLATURE

$C_{m,z}$ the set of cold streams existing in interval z in SN_m
$C_{P,u}$ specific heat of hot stream u [kJ/(kg K)]
$c_{P,v}$ specific heat of cold stream v [kJ/(kg K)]
C_u set of cold streams existing in SN_m
$E_{u,v,m}$ binary integer variable that takes the value of 0 when there is no match between streams u and v in SN_m and takes the value of 1 when there is a match
f flow rate of cold stream (kg/s)
F flow rate of hot stream (kg/s)
$HC_{v,z}$ cold load in interval z
$HH_{u,z}$ hot load in interval z
$H_{m,z}$ set of hot streams existing in interval z in SN_m
H_u set of hot streams existing in SN_m
N_C number of process cold streams
N_{CU} number of cooling utilities
N_H number of process hot streams
N_{HU} number of process cold streams
$Q_{u,v,z}$ heat exchanged from hot stream u to cold stream v in interval z
$Q_{v,z}^C$ heat exchanged from cold stream v in interval z
$Q_{u,z}^H$ heat exchanged from hot stream u in interval z
r_z residual heat leaving interval z
t temperature of cold stream (K)
t_v^s supply temperature of cold stream v (K)
t_v^t target temperature of cold stream v (K)
T_u^s supply temperature of hot stream u (K)
T_u^t target temperature of hot stream u (K)
T temperature of hot stream (K)
u index for hot streams
$U_{u,v,m}$ upper bound on the exchangeable heat between streams u and v in SN_m
v index for cold streams
z temperature interval

GREEK LETTER

ΔT^{min} minimum approach temperature (K)

REFERENCES

Al-Mutairi E, El-Halwagi MM: Integration method for considering scheduling in design of heat exchange networks, *Appl Therm Eng* 29:3482–3490, 2009.

Al-Mutairi E, El-Halwagi MM: Environmental-impact reduction through simultaneous design, scheduling, and operation, *Clean Technol Environ Policy* 12(5):537–545, 2010.

Al-Thubaiti M, Al-Azri N, El-Halwagi M: Optimize heat transfer networks: an innovative method applies heat integration to cost effectively retrofit bottlenecks in utility systems, *Hydrocarbon Process* March:109–115, 2008.

CANMET Pinch analysis: for the efficient use of energy, water & hydrogen, oil refining industry, energy recovery at a fluid catalytic cracking unit, Varennes, Quebec, 2004, Canada Centre for Mineral and Energy Technology. (The example described in the CANMET publication was initially developed by Rossiter and Associates.)

Furman KC, Sahinidis NV: A critical review and annotated bibliography for heat exchanger network synthesis in the 20th century, *Ind Eng Chem Res* 41(10):2335–2370, 2002.

Gundersen T, Naess L: The synthesis of cost optimal heat exchanger networks: an industrial review of the state of the art, *Comp Chem Eng* 12(6):503–530, 1988.

Kemp I: *Pinch analysis and process integration: a user guide on process integration for the efficient use of energy*, ed 2, Butterworth-Heinemann, 2009.

Kondili E, Pantelides CC, Sargent RWH: A general algorithm for short-term scheduling of batch-operation: I. MILP formulation, *Comp Chem Eng* 17:211–227, 1993.

Li ZK, Ierapetritou MG: Rolling horizon based planning and scheduling integration with production capacity consideration, *Chem Eng Sci* 65:5887–5900, 2010.

Linnhoff B: Pinch analysis: a state of the art overview. *Trans Inst Chem Eng Chem Eng Res Des* 71(A5): 503–522, 1993.

Liu Y, Karimi IA: Scheduling multistage, multiproduct batch plants with nonidentical parallel units and unlimited intermediate storage, *Chem Eng Sci* 62:1549–1566, 2007.

Majozi T: *Batch chemical process integration: analysis, synthesis, and optimization*, Heidelberg, Springer, 2010.

Maravelias CT, Grossmann IE: New general continuous-time state-task network formulation for short-term scheduling of multipurpose batch plants, *Ind Eng Chem Res* 42:3056–3074, 2003.

Papoulias SA, Grossmann IE: A structural optimization approach in process synthesis II. Heat recovery networks, *Comp Chem Eng* 7(6):707–721, 1983.

Pekny JF, Reklaitis GV: Towards the convergence of theory and practice: A technology guide for scheduling/planning methodology. In Pekny JF, Blau GE, editors: *Third International Conference on Foundations of Computer-Aided Process Operations*, vol 94, New York, Amer Inst Chemical Engineers, pp 91–111, 1998.

Ponce-Ortega JM, Jimenez-Gutierrez A, Grossmann IE: Optimal synthesis of heat exchanger networks involving isothermal process streams, *Comp Chem Eng* 32:1918–1942, 2008.

Rossiter AP: Improve energy efficiency via heat integration, *Chem Eng Prog* 106(12):33–42, 2010.

Ryu JH, Pistikopoulos EN: A novel approach to scheduling of zero-wait batch processes under processing time variations, *Comp Chem Eng* 31:101–106, 2007.

Shaik MA, Floudas CA: Unit-specific event-based continuous-time approach for short-term scheduling of batch plants using RTN framework, *Comp Chem Eng* 32:260–274, 2008.

Shang ZG, Kokossis AC, Hui CW: Synthesis, optimal planning and scheduling for process plant utility systems, *Comp Chem Eng* 23:S137–S142, 1999.

Shenoy UV: *Heat exchange network synthesis: process optimization by energy and resource analysis*, Houston, Texas, Gulf Publ. Co., 1995.

Smith R: *Chemical process design and integration*, New York, Wiley, 2005.

CHAPTER 20
SYNTHESIS OF COMBINED HEAT AND REACTIVE MASS-EXCHANGE NETWORKS

Heretofore, the presented mass-exchange (physical and reactive) synthesis techniques were applicable to the cases where mass-exchange temperatures are known ahead of the synthesis task. In a typical mass-exchange network (MEN), there is a strong interaction between mass and heat. For instance, the mass-exchange equilibrium relation of a mass-separating agent (MSA) is affected by its temperature. Therefore, heating or cooling may be beneficial to the performance of the MEN. However, heating/cooling may incur an additional cost. Indeed, there is a trade-off between the mass and heat objectives. Ideally, it is desired to minimize the total cost of heat and mass exchange. Because the mass-exchange equilibrium relations are dependent upon temperature and the selection of optimal mass-exchange temperatures is an important element of design, the selection of these temperatures should involve a trade-off between the cost of the MSAs and the cost of the heating/cooling utilities. This leads to the problem of synthesizing combined heat and reactive mass-exchange networks (CHARMEN) introduced by Srinivas and El-Halwagi (1994). The problem statement and solution approach are presented next.

SYNTHESIS OF COMBINED HEAT AND REACTIVE MASS-EXCHANGE NETWORKS

The CHARMEN synthesis problem can be stated as follows (Srinivas and El-Halwagi, 1994): Given a number N_R of rich (for example, waste) streams and a number N_S of lean streams (physical and reactive MSAs), it is desired to synthesize a cost-effective network of physical and/or reactive mass exchangers that can preferentially transfer certain undesirable species from the rich streams to the MSAs. Given also are the flow rate of each rich stream, G_i; its supply (inlet) composition, y_i^s; its target (outlet) composition, y_i^t; and the supply and target compositions, x_j^s and x_j^t, for each MSA. In addition, available for service are hot and cold streams (process streams as well as utilities) that can be used to optimize the mass-exchange temperatures.

Because of the interaction of mass and heat in the CHARMEN, one should synthesize the MEN and the heat-exchange network (HEN) simultaneously. This section presents an optimization-based method for the synthesis of CHARMENs.[1] Other approaches are available in literature (for example, Isafiade and Fraser, 2009; Hamad et al., 2007; Grossmann et al., 1999; and Srinivas and El-Halwagi, 1994).

Two key assumptions are invoked:

1. Each mass exchanger operates isothermally. Although each lean or rich stream can assume several values of temperature that differ from one mass exchanger to another, it is assumed that each stream passes isothermally within each mass exchanger.
2. In the range of operating temperatures and compositions, the equilibrium relations are monotonic functions of temperature of the MSA. This is typically true. For instance, normally in gas absorption Henry's coefficient monotonically decreases as the temperature of the MSA is lowered, whereas for stripping, the gas–liquid distribution coefficient monotonically increases as the temperature of the stripping agent is increased.

The concept of "lean substreams" (El-Halwagi, 1992; Srinivas and El-Halwagi, 1994) is useful in trading off mass versus heat objectives. Each MSA, j, is assumed to split into ND_j lean substreams. A lean substream S_{j,d_j} is defined as the d_jth (where $d_j = 1, 2, \ldots, ND_j$) portion of MSA j whose composition and temperature vary between a supply value of (x_j^s, T_j^s) and a target value of (x_j^t, T_j^t), and has a flow rate of L_{j,d_j} that does not split or mix with other substreams. It can be shown that a minimum operation cost (MOC) solution of a CHARMEN is realized when each lean substream exchanges mass at a single temperature T_{j,d_j}^* (for the proof, readers are referred to Srinivas and El-Halwagi 1994). The identification of temperature T_{j,d_j}^* and flow rate L_{j,d_j} for each lean substream is part of the optimization problem.

In setting up the CHARMEN-synthesis formulation, we first choose a number ND_j of lean substreams for MSA j, each operating at a selected temperature T_{j,d_j}^* that lies within the admissible temperature range for the MSA. The number of substreams is dependent on the level of accuracy needed for equilibrium dependence on temperature. Theoretically, an infinite number of substreams should be used to cover the whole temperature span of each MSA. However, in practice, few (typically less than five) substreams are needed. Because each temperature corresponds to an equilibrium relation, each lean substream should have its own composition scale on the composition interval diagram (CID). On the CID, the various substreams are represented against their composition scale. For each lean substream, two heat exchange tasks take place: one before mass exchange and one after. First, the temperature of the lean substream is changed from T_j^s to T_{j,d_j}^*. Mass exchange occurs at T_{j,d_j}^*. Then, the temperature of the substream has to be altered from T_{j,d_j}^* to T_j^t. These are heat-exchange

---
[1] More rigorous techniques for optimizing outlet composition are described by Srinivas and El-Halwagi (1994).

tasks that can be handled through the synthesis of HENs, as discussed in Chapters 7 and 19.

Therefore, the CHARMEN synthesis problem can be formulated by combining the HEN formulation of Chapter 19 and the REAMEN synthesis equations developed in Chapter 17 after adjustment to incorporate the notion of substreams. For instance, the cost of MSAs can be expressed as

$$[20.1] \quad \text{Cost of MSAs} = \sum_{j=1}^{N_S} C_j \sum_{d_j=1}^{ND_j} L_{j,d_j}$$

and the material balances around the composition intervals become

$$[20.2] \quad \delta_k - \delta_{k-1} + \sum_{j}\sum_{\substack{d_j \text{ passes through} \\ \text{interval } k}} L_{j,d_j} w^S_{j,k} = W^R_k$$

$$k = 1, 2, \ldots, N_{int}$$

where

$$[20.3] \quad L_{j,d_j} \geq 0 \quad j = 1, 2, \ldots, N_S$$

and

$$[20.4] \quad \sum_{d_j=1}^{ND_j} L_{j,d_j} \leq L^C_j \quad j = 1, 2, \ldots, N_S$$

with the mass-residual constraints

$$[20.5] \quad \delta_0 = \delta_{N_{int}} = 0$$

$$[20.6] \quad \delta_k \geq 0, \quad k = 1, 2, \ldots, N_{int} - 1.$$

These equations, coupled with the HEN-targeting Equations 19.1 through 19.16, represent the constraints of the CHARMEN-synthesis formulation. The objective is to minimize the cost of MSAs and heating/cooling utilities. This is a linear-programming formulation whose solution determines the optimal flow rate and temperature of each substream and heating/cooling utilities. To demonstrate this formulation, let us consider the following example.

Example 20-1 CHARMEN synthesis for ammonia removal from a gaseous emission

A gaseous emission is to be treated for the removal of ammonia. Table 20.1 provides the stream data. Two scrubbing agents are considered for the removal of ammonia: water, S_1, and an inorganic solvent, S_2. The absorption of ammonia in water is coupled with the following chemical reaction:

$$[20.7] \quad NH_3 + H_2O = NH_4^+ + OH^-$$

Within the considered range of operation, the equilibrium relation for ammonia scrubbing in water is dependent on the temperature as follows:

$$[20.8] \quad y = (0.053T_1 - 14.5)x_1,$$

where y is the mass fraction of ammonia in the gas, T_1 is the water temperature in K, and x_1 is the mass fraction of ammonia in the water. Table 20.2 summarizes the data for the lean streams.

Two cooling utilities are available: CU_1 and CU_2. The specific heat for both coolants is assumed to be that of water. A value of $\Delta T^{min} = 5\,K$ is used. Table 20.3 gives the data for the two cold streams.

The objective of this case study is to synthesize a CHARMEN that has a minimum operating cost (cost of MSAs + cost of cooling utilities).

To cover the temperature span for water, let us choose four substreams operating at 283, 288, 293, and 298 K. At each temperature, the equilibrium constant is evaluated. The CID for the problem is shown in Fig. 20.1, and the temperature interval diagram (TID) in Fig. 20.2.

The objective function is to minimize the operating cost of the MSAs and the cooling utilities; that is,

$$\min\,[0.005(L_{1,1} + L_{1,2} + L_{1,3} + L_{1,4}) + 0.160L_2 + 0.002fU_1 + 0.006fU_2]3600 \times 8760$$
$$= 157{,}680(L_{1,1} + L_{1,2} + L_{1,3} + L_{1,4}) + 5{,}045{,}760L_2 + 63{,}072fU_1 + 189{,}216fU_2,$$

subject to the following constraints:

Material balances around composition intervals

$$\delta_1 + 0.0320L_2 = 0.0032$$

$$\delta_2 - \delta_1 + 0.0012L_{1,4} + 0.0160L_2 = 0.0016$$

$$\delta_3 - \delta_2 + 0.0015L_{1,3} + 0.0013L_{1,4} + 0.0160L_2 = 0.0016$$

$$\delta_4 - \delta_3 + 0.0021L_{1,2} + 0.0016L_{1,3} + 0.0012L_{1,4} + 0.0160L_2 = 0.0016$$

TABLE 20-1 Data for Waste Stream in Ammonia Removal Example

Stream	Flow Rate, G_i (kg/s)	Supply Composition (mass fraction) y_i^s	Target Composition (mass fraction) y_i^t
R_1	1.0	0.011	0.001

TABLE 20-2 Data for Lean Streams in Ammonia Removal Example

Stream	Supply Composition of Ammonia (mass fraction) x_j^s	Target Composition of Ammonia (mass fraction) x_j^t	m_j	ε_j	C_j $/kg	T_j^s (K)	T_j^t (K)	T_j^{LB} (K)	T_j^{UB} (K)
S_1	0.0000	0.0050	0.053T−14.5 (where T is in K)	0.0010	0.005	298	≤ 298	283	298
S_2	0.0040	0.1090	0.100	0.0010	0.160	295	295	295	295

(Continued)

Example 20-1 CHARMEN synthesis for ammonia removal from a gaseous emission (*Continued*)

TABLE 20-3 Stream Data for Coolants

Stream	Supply Temperature, K	Target Temperature, K	Cost, $/kg Coolant
C_1	288	293	0.002
C_2	278	283	0.006

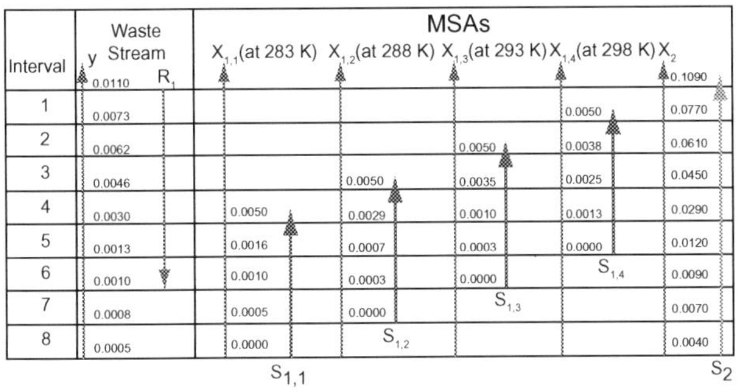

FIGURE 20-1 Composition interval diagram for ammonia removal case study.

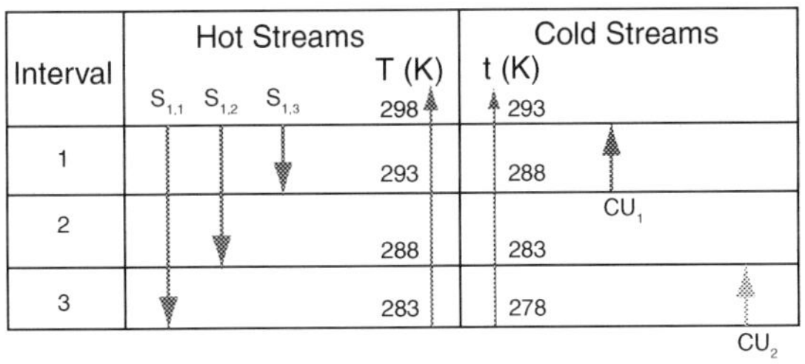

FIGURE 20-2 Temperature interval diagram for ammonia removal case study.

$$\delta_5 - \delta_4 + 0.0034L_{1,1} + 0.0022L_{1,2} + 0.0016L_{1,3} + 0.0013L_{1,4} + 0.0170L_2 = 0.0017$$

$$\delta_6 - \delta_5 + 0.0005L_{1,1} + 0.0004L_{1,2} + 0.0003L_{1,3} + 0.0030L_2 = 0.0003$$

$$\delta_7 - \delta_6 + 0.0005L_{1,1} + 0.0003L_{1,2} + 0.0020L_2 = 0.0000$$

$$-\delta_7 + 0.0005L_{1,1} + 0.0030L_2 = 0.0000.$$

Heat balances around composition intervals

$$r_1 + 5.0 \times 4.18 f U_1 - 5.0 \times 4.18(L_{1,1} + L_{1,2} + L_{1,3}) = 0.0$$

$$-r_2 + 5.0 \times 4.18 f U_2 - 5.0 \times 4.18 L_{1,1} = 0.0$$

$$r_2 - r_1 - 5.0 \times 4.18(L_{1,1} + L_{1,2}) = 0.0.$$

These equations coupled with the nonnegativity constraints form a linear program that can be modeled on LINGO as follows:

[P20.1]
```
MIN=157680*L11+157680*L12+157680*L13+157680*L14+
  5045760*L2+63072*FU1+189216*FU2;
D1+0.0320*L2=0.00320;
D2-D1+0.0012*L14+0.0160*L2=0.00160;
D3-D2+0.0015*L13+0.0013*L14+0.0160*L2=0.00160;
D4-D3+0.0021*L12+0.0016*L13+0.0012*L14+0.0160
  *L2=0.00160;
D5-D4+0.0034*L11+0.0022*L12+0.0016*L13+0.0013*L14+
  0.0170*L2=0.00170;
D6-D5+0.0005*L11+0.0004*L12+0.0003*L13+0.0030
  *L2=0.00030;
D7-D6+0.0005*L11+0.0003*L12+0.0020*L2=0.00000;
-D7+0.0005*L11+0.0030*L2=0.00000;

r1+20.90*FU1-20.90*(L11+L12+L13)=0.0;
r2-r1-20.90*(L11+L12)=0.0;
-r2+20.90*FU2-20.90*L11=0.0;

D1>=0.0;
D2>=0.0;
D3>=0.0;
D4>=0.0;
D5>=0.0;
D6>=0.0;
D7>=0.0;
D8>=0.0;
D9>=0.0;
r1>0.0;
r2>0.0;
r3>0.0;
L11>=0.0;
L12>=0.0;
L13>=0.0;
L14>=0.0;
L2>=0.0;
FU1>=0.0;
FU2>=0.0;
```

The following solution is obtained using LINGO:

```
Objective value: 378432.0
Variable    Value
L11         0.0000000E+00
L12         0.0000000E+00
L13         1.000000
L14         0.9999999
L2          0.0000000E+00
FU1         1.000000
FU2         0.0000000E+00
D1          0.3200000E-02
D2          0.3600000E-02
D3          0.2400000E-02
D4          0.1200000E-02
D5          0.0000000E+00
D6          0.0000000E+00
D7          0.0000000E+00
R1          0.0000000E+00
R2          0.0000000E+00
D8          0.0000000E+00
D9          0.0000000E+00
R3          0.0000000E+00
```

The solution involves the use of two water substreams for scrubbing, at 293 K and 298 K. The first coolant is employed to undertake the cooling duty. Several conceptual flow sheets can be developed to implement this solution. Figures 20.3a and b illustrate such configurations. The developed mass exchanger is composed of two sections. The top section employs $S_{1,3}$ to reduce the ammonia composition in the gas from 0.13 percent to 0.10 percent and operates at an average water temperature of 293 K. In Fig. 20.3a, temperature is reduced by indirect contact cooling of the column internals. On the other hand, Fig. 20.3b uses a heat exchanger to cool $S_{1,3}$ to 293 K. The rest of the column achieves the remaining separation (1.10 to 0.13 percent ammonia in gas) and involves the use of $S_{1,4}$ and $S_{1,3}$. Following the synthesis task, an analysis study is needed to size the column, predict the actual temperature profile in the column, and validate or refine the water flow and system performance.

(*Continued*)

Example 20-1 CHARMEN synthesis for ammonia removal from a gaseous emission (*Continued*)

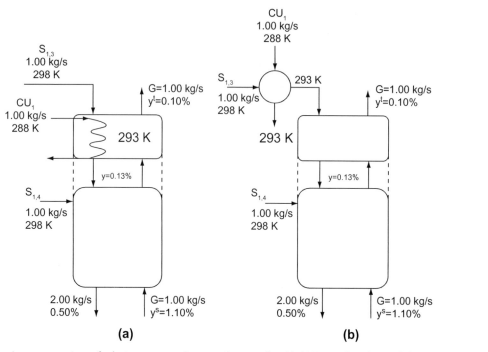

FIGURE 20-3 Schematic representations of solution to ammonia removal case study with (a) Internal cooling and (b) External cooling.

Example 20-2 Incorporation of CHARMEN synthesis into mass integration for an ammonium nitrate plant

Ammonium nitrate is manufactured by reacting ammonia with nitric acid. Consider the process shown by Fig. 20.4. First, natural gas is reformed and converted into hydrogen, nitrogen, and carbon dioxide. Hydrogen and nitrogen are separated and fed to the ammonia synthesis plant. A fraction of the produced ammonia is employed in nitric acid formation. Ammonia is first oxidized with compressed air, and then absorbed in water to form nitric acid. Finally, nitric acid is reacted with ammonia to produce ammonium nitrate.

The plant disposes of two waste streams: gaseous and aqueous. The gaseous emission results from the ammonia and the ammonium nitrate plants. It is fed to an incinerator prior to atmospheric disposal. In the incinerator, ammonia is converted into NO_x. Because of more stringent NO_x regulations, the composition of ammonia in the feed to the incinerator has to be reduced from 0.57 to 0.07 wt%. The lean streams presented in Table 20.2 may be employed to remove ammonia. The main aqueous waste of the process results from the nitric acid plant. The nitric acid waste is neutralized with an aqueous ammonia solution before biotreatment.

The objective of this case study is to identify a cost-effective solution that can reduce the ammonia content of the feed to the incinerator to 0.07 wt%.

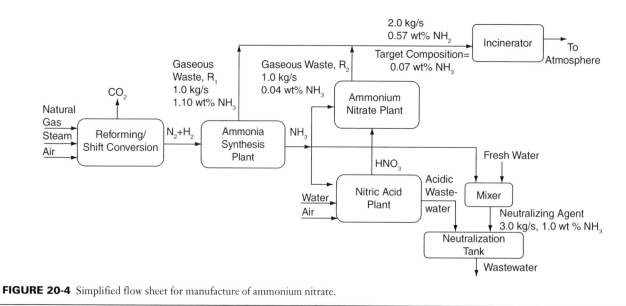

FIGURE 20-4 Simplified flow sheet for manufacture of ammonium nitrate.

(*Continued*)

Example 20-2 Incorporation of CHARMEN synthesis into mass integration for an ammonium nitrate plant (*Continued*)

SOLUTION

Let us first segregate the two sources forming the feed to the incinerator. As can be seen from the source—sink mapping diagram (Fig. 20.5)—the gaseous emission from the ammonium nitrate process (R_2) is within the acceptable zone for the incinerator. Therefore, it should not be mixed with R_1 (the gaseous emission from ammonia synthesis plant), and then separated. Instead, the ammonia content of R_1 should be reduced to 0.10 wt%, and then mixed with R_1 to provide an acceptable feed to the incinerator as shown by Fig. 20.5. The task of removing ammonia from R_1 from 1.10 to 0.10 wt% is identical to the case study solved in Example 20.1. Hence, the solution presented in Fig. 20.3 can be used.

Instead of disposing the water leaving the scrubber for R_1 into the biotreatment facility, it is beneficial to use it in a process sink. An appropriate sink is the mixer preceding the neutralizer (used in neutralizing the wastewater from the nitric acid). Hence, the water leaving the R_1 scrubber can be used to replace the fresh water consumption in the mixer by 2.0 kg/s (173 tons/day). Because of its ammonia content (0.50 wt%), the amount of ammonia needed for neutralization is reduced by 0.01 kg/s. For an ammonia cost of $100/ton, the ammonia savings are about $32,000/year. Figure 20.6 illustrates the identified solution.

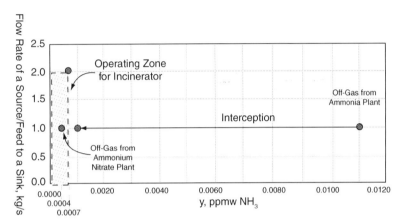

FIGURE 20-5 Gaseous source-sink mapping diagram for ammonium nitrate example.

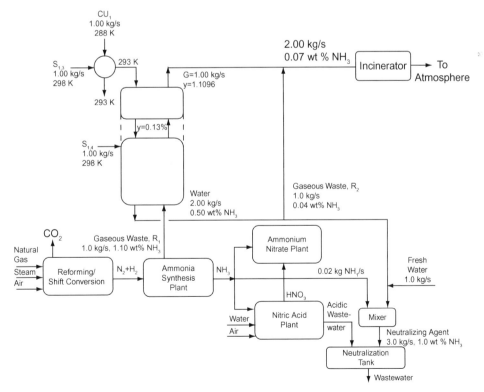

FIGURE 20-6 Solution of ammonium nitrate case study.

HOMEWORK PROBLEM

20.1. Consider a pulping process that results in two gaseous emissions that are rich in H_2S (R_1 and R_2) and must be desulfurized. The task of desulfurizing the two gaseous streams within the plant may be accomplished by the use of one or more of the following reactive mass-exchange processes: reactive absorption of H_2S in the white liquor (S_1) to produce Na_2S (a necessary ingredient for kraft pulping); absorption in N-methyl-2-pyrrolidone (S_2); and absorption in methyldiethanolamine (MDEA) (S_3). The operating cost of MSAs (primarily regeneration and makeup) is $0.000, 0.007, and 0.001/kg for S_1, S_2, and S_3, respectively. Also, heating/cooling utilities are available in the form of heating oil, cooling water, and chilled nitrogen with costs of 363×10^{-3}, 1.3×10^{-5}, and 3.10×10^{-3} $/kg, respectively. The equilibrium data for H_2S in the MSAs as a function of temperature can be found in literature (Srinivas and El-Halwagi, 1994). The minimum allowable composition differences, ε_j, for MSAs S_1, S_2, and S_3 are 1.0×10^{-3}, 1.0×10^{-6}, and 1.0×10^{-3}, respectively. Thermodynamic feasibility for heat exchange was ensured by minimum allowable temperature difference, ΔT^{min}, of 10 K. Tables 20.4 through 20.6 give stream data.

Synthesize an optimum CHARMEN that can transfer the H_2S from the waste streams (R_1 and R_2) to one or more of the MSAs (S_1 to S_3).

NOMENCLATURE

d_j	index for substreams of the jth MSA
G_i	flow rate of the ith rich stream
i	index of rich streams
j	index of MSAs
k	index of composition intervals
L_j	flow rate of the jth MSA (kg/s)
$L_{j,dj}$	flow rate of substream dj of the jth MSA (kg/s)
L_j^c	upper bound on available flow rate of the jth MSA (kg/s)
m_j	slope of equilibrium line for the jth MSA
N_{int}	number of composition intervals
N_R	number of rich streams
N_S	number of MSAs
ND_j	number of substreams for the jth MSA
R_i	the ith rich stream
S_j	the jth MSA
$S_{j,dj}$	the d_jth substream of the jth MSA
t	temperature of cold stream (K)
$W_{i,k}^R$	exchangeable load of the ith rich stream that passes through the kth interval (kg/s)
$w_{j,k}^S$	exchangeable load of the jth MSA that passes through the kth interval (kg/s)
W_k^R	the collective exchangeable load of the rich streams in interval k (kg/s)
$x_{j,k}$	composition of key component in the jth MSA at the lower horizontal line defining the kth interval
x_j^s	supply composition of the jth MSA
x_j^t	target composition of the jth MSA
z	temperature interval

GREEK LETTERS

δ_k	mass residual from interval k (kg/s)
ΔT^{min}	minimum approach temperature (K)
ε_j	minimum allowable composition difference for the jth MSA

TABLE 20-4 Data for Waste Streams in Pulping Case Study

Stream	Flow Rate (kg/s)	y_i^s (mass fraction)	y_i^t (mass fraction)
R_1	104	8.83×10^{-4}	5.00×10^{-6}
R_2	442	7.00×10^{-4}	5.00×10^{-6}

TABLE 20-5 Supply and Target Compositions for MSAs in Pulping Case Study

MSA	Description	L_j^c (kg/s)	x_j^s (mass fraction)	x_j^t (mass fraction)
S_1	White liquor	40	0.07557	0.115
S_2	15 wt% N-methyl-2-pyrrolidone	∞	1.0×10^{-6}	1.0×10^{-5}
S_3	15 wt% MDEA	∞	0.001	0.01

TABLE 20-6 Thermal Data for Streams in Pulping Case Study

Stream	Supply Temperature (K)	Target Temperature (K)	Lower-Bound Temperature (K)	Upper-Bound Temperature (K)	Average Specific Heat (kJ/kg K)
R_1	298	298	288	313	1.00
R_2	298	298	288	313	1.00
S_1	368	368	368	368	2.50
S_2	313	313	298	313	2.40
S_3	310	310	280	330	2.40
Heating oil	373	343	343	373	2.2
Cooling water	288	298	288	298	4.2
Chilled nitrogen	100	100	100	100	

REFERENCES

El-Halwagi MM: A process synthesis approach to the dilemma of simultaneous heat recovery, waste reduction and cost effectiveness. In El-Sharkawy AI, Kummler RH, editors: *Proceedings of the Third Cairo International Conference on Renewable Energy Sources*, vol 2, pp. 579–594, 1992.

Grossmann IE, Caballero JA, Yeomans H: Mathematical programming approaches to the synthesis of chemical process systems, *Korean J Chem Eng* 16:407–426, 1999.

Hamad A, Alfadala H, Warsame A: Optimum waste interception in liquefied natural gas processes, *Int J Environ Pollut* 29:47–69, 2007.

Isafiade AJ, Fraser DM: Interval based MINLP superstructure synthesis of combined heat and mass exchanger networks, *Chem Eng Res Des* 87:1536–1542, 2009.

Srinivas BK, El-Halwagi MM: Synthesis of combined heat and reactive mass-exchange networks, *Chem Eng Sci* 49:2059–2074, 1994.

CHAPTER 21
DESIGN OF INTEGRATED BIOREFINERIES

The escalating consumption of nonrenewable energy sources, the dwindling fossil-fuel reserves, and the increasing levels of greenhouse gas (GHG) emissions are strong motivators for the use of sustainable energy sources. Biofuels are among the most attractive options for a renewable energy source. In addition to being renewable, biofuels also provide an advantage toward the reduction of GHG emissions because of CO_2 sequestration by the growing biomass through photosynthesis. A biorefinery is a processing facility that converts biomass feedstocks into value-added products such as fuels and chemicals (Ng et al., 2009). Reviews of biorefining technologies are available in literature (for example, Cardona et al., 2010; Basu, 2010; Clark and Deswarte, 2008; Kamm et al., 2006, Klass, 1996; El-Halwagi, 1986). This chapter provides a brief discussion of sample approaches for conceptual design of a biorefinery and techno-economic assessment of a biorefinery.

CONCEPTUAL DESIGN OF A BIOREFINERY

An effective design approach of biorefineries should involve the synthesis, integration, and screening of plausible alternatives. Given the enormous number of alternatives, it is crucial to have systematic methods for the quick synthesis and evaluation of candidate technologies and flow sheets. Research efforts included the computer-aided processing of routes prior to establishing the optimal product for optimal energy savings in the process (for example, Alvarado-Morales et al., 2009; Gosling, 2005). Tay et al. (2011) presented a graphical approach to the design of biorefineries using ternary carbon-hydrogen-oxygen (C-H-O) diagrams and lever-arm rules for the selection and quantification of pathways. Optimization approaches have been used to develop and evaluate biorefinery configurations (for example, Pham and El-Halwagi, 2011; Bao et al., 2011 and 2009; Ojeda et al., 2011; Martin and Grossmann, 2010; Kokossis et al., 2010; Ng et al., 2009; Tan et al., 2009; Sammons et al., 2008). The following is an example of an optimization-based approach for the conceptual design of a biorefinery. It is based on the method developed by Pham and El-Halwagi (2011) and is referred to as the *"forward-backward"* approach. The problem (represented by Fig. 21.1) may be stated as follows:

"Given a set of biomass feedstocks with known flow rates and characteristics and a desired final product with specifications, it is desired to develop a systematic methodology for the generation of optimal configurations from feedstocks to products. Available for service is a set of conversion technologies with known performance. Various objectives may be considered such as the highest yield, the highest energy efficiency, the shortest route (the least number of processing steps), the minimum-cost route, or the most sustainable route (as characterized by sustainability metrics)."

The following is a brief description of the procedure:

1. Forward and backward branching
   The design procedure starts with the synthesis of configurations of interest using forward synthesis from the feedstocks and backward synthesis from the products and identifications of the chemical species associated with the forward and backward synthesis. (As an illustration, consider the conversion of biomass to ethanol shown by Fig. 21.2). The feedstock may be converted to methane (via anaerobic digestion), sugar (by enzymatic hydrolysis), syngas (through gasification), and so on. Next, another layer of nodes for chemical species is created with the chemicals that may be manufactured from the chemicals in the previous layer. For instance, acetylene may be produced from methane (through cracking). These two layers have been created with forward synthesis from feedstocks to intermediates. Conversely, the backward synthesis starts from the final product (ethanol) and identifies the species and technologies that can lead to ethanol. As an example, ethanol may be produced from bromoethane (C_2H_5Br) through hydrolysis, and bromoethane in turn may be produced from ethylene via by hydrobromination.

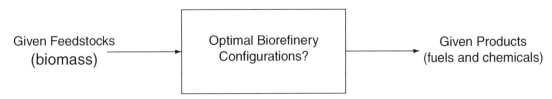

FIGURE 21-1 Schematic representation of the problem of conceptual design of a biorefinery. *Source: Pham and El-Halwagi (2011).*

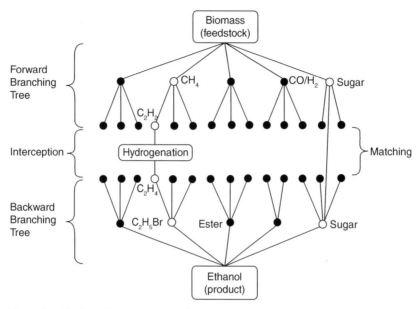

FIGURE 21-2 Examples of forward and backward branching trees. *Source: Pham and El-Halwagi (2011).*

2. Matching and interception

   When there are two nodes with identical chemical species, they are connected in what is referred to as **matching**, and a continuous pathway is created. An example is sugar created from the forward branching and also from the backward branching, as shown by Fig. 21.2. On the other hand, when a chemical species of a certain node can be converted to another node, the conversion technology is referred to as **interception**. An interception branch connecting the two nodes is created, and a continuous pathway is created. For instance, acetylene (C_2H_2) from the forward branching can be intercepted using hydrogenation to produce ethylene (C_2H_4), which comes from backward branching. The interception using hydrogenation leads to a continuous path (biomass→methane →acetylene→ethylene→bromoethane→ethanol). The stoichiometric equations are listed to identify all involved species, and performance models are obtained for the individual steps. The number of steps may be limited (for example, two forward layers, two backward layers, and one interception layer) to avoid the generation of too complex configurations.

   The result of the matching and interception is the generation of various pathways connecting the feedstock and the products. Figure 21.3 shows an example, where the letters represent the nodes and the arcs with numbers represent the conversion technologies.

3. Prescreening

   Two nodes may be directly or indirectly connected. In the prescreening step, the different routes connecting two specific nodes are compared based on key performance indicators (for example, cost, energy consumption, GHG emissions, number of processing steps, yield, and so on) and using models with appropriate level of details. Hence, one connection is selected as the optimum between the two pairs. Such prescreening is used to reduce the size of the optimization problem by making judicious decisions between each two nodes. Table 21.1 describes the conversion step eliminated in prescreening due to low yield. The processing technologies 6, 14, 15, 19, and 24 are eliminated due to their low yields. Because Arc 24 has been removed, then Arc 37 will also have to be removed because it no longer has a continuous path from the feedstock to the product. An economic analysis leads to the elimination of Arc 17 because Arc 7 is cheaper than Arc 17 (coupled with Arc 3 or 4). Similar prescreening steps are undertaken for the rest of the superstructure. Table 21.2 summarizes the outcome of prescreening, and Fig. 21.4 shows the resulting superstructure following prescreening.

4. Pathway selection

   Next, an optimization problem is formulated and solved to select the optimal pathway based on performance criteria. Integer-cut constraints can be used to generate the next-to-best solution and so forth. For instance, if the cost data shown by Table 21.3 are used in the objective function, the selected pathways are: mixed alcohols production via acid fermentation and esterification (A→B→J→O; $15.14/GJ of products); mixed alcohol production via acid fermentation and ketonization (A→C→H→O; $15.28/GJ); and mixed butanol and ethanol production via ABE fermentation (A→O; $15.47/GJ). The values of the objective functions for the generated solutions are relatively close. Therefore, a more detailed analysis should be carried out. The value of the procedure is in generating and screening the initial conceptual design. Subsequently, detailed

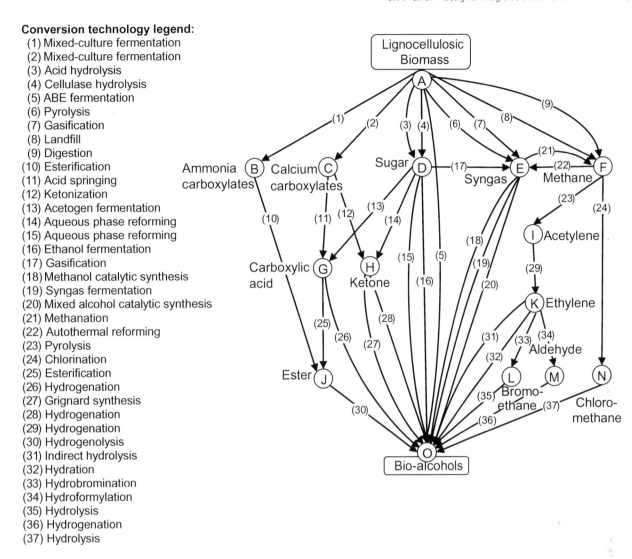

Conversion technology legend:
(1) Mixed-culture fermentation
(2) Mixed-culture fermentation
(3) Acid hydrolysis
(4) Cellulase hydrolysis
(5) ABE fermentation
(6) Pyrolysis
(7) Gasification
(8) Landfill
(9) Digestion
(10) Esterification
(11) Acid springing
(12) Ketonization
(13) Acetogen fermentation
(14) Aqueous phase reforming
(15) Aqueous phase reforming
(16) Ethanol fermentation
(17) Gasification
(18) Methanol catalytic synthesis
(19) Syngas fermentation
(20) Mixed alcohol catalytic synthesis
(21) Methanation
(22) Autothermal reforming
(23) Pyrolysis
(24) Chlorination
(25) Esterification
(26) Hydrogenation
(27) Grignard synthesis
(28) Hydrogenation
(29) Hydrogenation
(30) Hydrogenolysis
(31) Indirect hydrolysis
(32) Hydration
(33) Hydrobromination
(34) Hydroformylation
(35) Hydrolysis
(36) Hydrogenation
(37) Hydrolysis

FIGURE 21-3 The generated superstructure. *Source: Pham and El-Halwagi (2011).*

TABLE 21-1 Conversion Steps Eliminated in the Prescreening Based on Low Yield

Conversion Steps	Feed	Product	Yield	Yield Base
(6) Pyrolysis	Biomass	Syngas	29.2%	Biomass weight
(14) Aqueous phase reforming	Sugar	Ketones	23.7%	Fed carbon weight
(15) Aqueous phase reforming	Sugar	Alcohols	8.7%	Fed carbon weight
(19) Syngas fermentation	Syngas	Alcohols	53.1%	Carbon monoxide weight
(24) Chlorination	Methane	Chloro-methane	12%	Methane weight

Source: Pham and El-Halwagi (2011).

techno-economic analysis is undertaken for the attractive alternatives.

Because the calculated costs of these three configurations are relatively close, additional analyses must be carried out. This is consistent with the nature of a short-cut conceptual design approach that is intended to generate promising alternatives which can then be screened using a detailed analysis.

TECHNO-ECONOMIC ASSESSMENT OF A BIOREFINERY

As mentioned in the previous section, once a conceptual configuration of a biorefinery is synthesized, focus can be given to a more detailed techno-economic analysis. Modeling and sizing of the units, as well as an economic analysis along the lines of the techniques mentioned in Chapter 2, can be used. Figure 21.5 illustrates one

TABLE 21-2 Data for the Pathway Nodes and Prescreening Results

Nodes	Feed	Product	Route[a]
A→F	Biomass	Methane	(7) Gasification & (21) Methanation
			(8) Landfill
			(9) Digestion
A→E	Biomass	Syngas	**(7) Gasification**
			(9) Digestion & (22) Autothermal reforming
H→O	Ketones	Alcohols	(27) Grignard synthesis
			(28) Hydrogenation
K→O	Ethylene	Ethanol	(31) Indirect hydrolysis
			(32) Hydration
			(33) Hydrobromination & (35) Hydrolysis
			(34) Hydroformylation & (36) Hydrogenation
G→O	Acid carboxylic	Alcohols	**(25) Esterification & (30) Hydrogenolysis**
			(26) Hydrogenation
E→O	Syngas	Methanol	**(18) Methanol synthesis**
	Syngas	Alcohols	(20) Mixed alcohol synthesis

Source: Pham and El-Halwagi (2011).
Note: [a] The numbers in parentheses correspond to the conversion technologies with the numbers listed in Fig. 5. Bold routes are optimal ones connecting the two nodes based on prescreening.

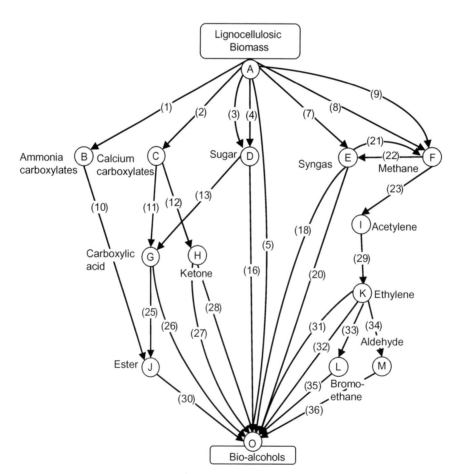

FIGURE 21-4 The superstructure of the synthesized pathways after the prescreening step. *Source: Pham and El-Halwagi (2011).*

TABLE 21-3 Cost Data for the Pathways

Pathways	Description	Product	Base Year					Year 2010	
			Capacity MMGPY[a]	Product cost[b] $/gal	Energy density MJ/gal	Date	PPI[c]	Alcohols cost $/gal	$/GJ
A→B→J→O	Mixed alcohols production via acid fermentation and esterification	Mixed alcohols	45	1.21	92	2007	213.7	1.39	15.14
A→C→H→O	Mixed alcohols production via acid fermentation and ketonization	Mixed alcohols	35	1.44	101	2009	229.4	1.54	15.28
A→D→G→J→O	Ethanol production via acetogen fermentation and ester synthesis	Ethanol	n/a	n/a	n/a	n/a	n/a	n/a	n/a
A→D→O	Ethanol production via hydrolysis and yeast fermentation	Ethanol	50	1.03	79	2003	161.8	1.56	19.69
A→O	Mixed butanol and ethanol production via ABE fermentation	Mixed alcohols	n/a	1.50	110	2007	214.8	1.71	15.47
A→E→O	Methanol production via biomass gasification	Methanol	24	0.61	59	2001	151.8	1.18	19.98
A→F→I→K→O	Ethanol production via syntheses of methane, acetylene, and ethylene	Ethanol	13	2.74	79	2008	245.5	2.73	34.43

Source: Pham and El-Halwagi (2011).
Note: [a]MMGPY: Million gallons per year of products. Online operation is 8000 hours per year.
[b]All the product costs exclude feedstock costs.
[c]PPI: Producer price index for chemicals and allied products. PPI in 2010 (245.1) is the average from January to July.

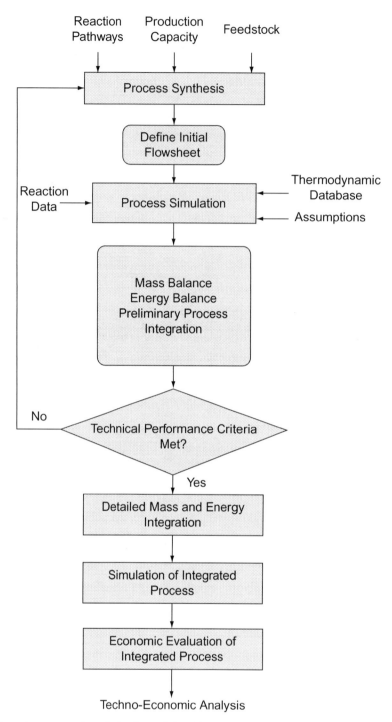

FIGURE 21-5 A procedure for techno-economic analysis. *Source: Pokoo-Aikins et al. (2010a).*

approach to conducting the techno-economic analysis, which includes process synthesis, simulation, integration, and economic evaluation. Facililty location can also be incorporated in the procedure (e.g., Bowling et al., 2011).

As an example of techo-economic evaluation, consider the case of producing hydrocarbon fuels from biomass through the carboxylate platform described in the previous section followed by dehydration and oligomerization of the produced mixed alcohols. Figure 21.6 is an illustration of a flow sheet describing the pathway. For the base-case capacity of a biomass feedstock of 40 tonne biomass (dry basis)/hr, the fixed capital investment (FCI) is $131 MM. Figure 21.7 shows the percentage distribution of the key processing systems to the FCI. More detailed analysis can then be dedicated to the large-cost items, and integration opportunities should be sought to reduce the fixed and operating costs. Figure 21.8 shows the dependence of the FCI and the minimum selling price on the plant size expressed as the biomass flow rate (tonne/hr).

Let us consider another example on the production of biodiesel. Several techno-economic studies have been carried out for several feedstocks that can be converted

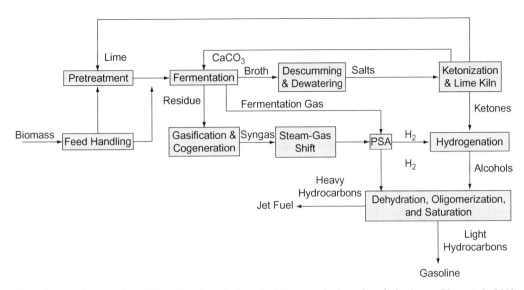

FIGURE 21-6 A flow sheet implementation of the carboxylate platform for biomass to hydrocarbon fuels. *Source: Pham et al. (2010).*

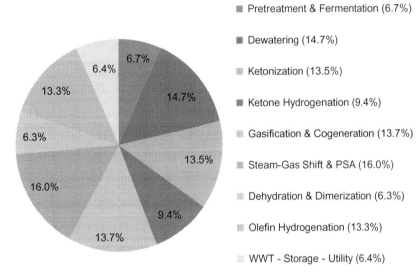

FIGURE 21-7 Distribution of the FCI for the carboxylate-platform process. *Source: Pham et al. (2010).*

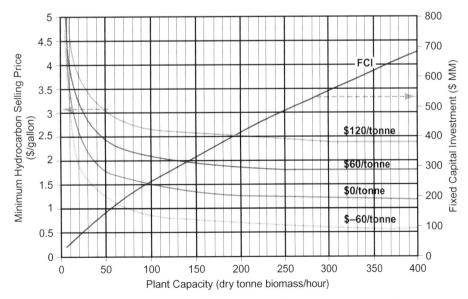

FIGURE 21-8 Economic analysis of FCI and minimum selling price as a function of biomass-to-hydrocarbon fuel plant capacity at different costs of biomass feedstock. *Source: Pham et al. (2010).*

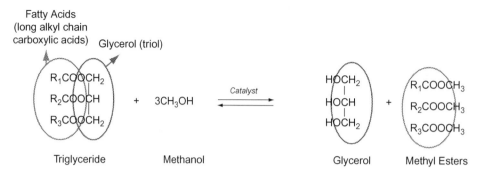

FIGURE 21-9 The overall transesterification route for producing biodiesel from triglycerides. *Source: Elms and El-Halwagi (2009).*

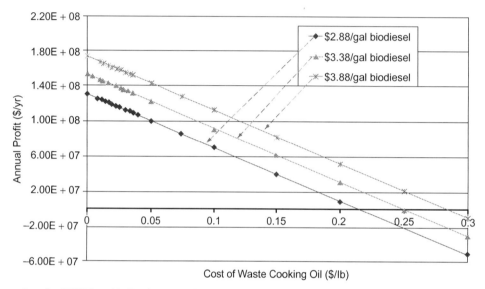

FIGURE 21-10 Annual profit of WCO-to-biodiesel process. *Source: Elms and El-Halwagi (2009).*

to biodiesel. These feedstocks include soybean oil (for example, Myint and El-Halwagi, 2009), sewage sludge (for example, Pokoo-Aikins et al., 2010b), waste cooking oil (for example, Elms and El-Halwagi, 2010 and 2009), and algae (for example, Pokoo-Aikins et al., 2010a). The predominant reaction pathway for producing biodiesel is the transesterification route shown by Fig. 21.9.

Figure 21.10 is an example of the obtained techno-economic analysis. It shows the annual profit of a 40 MM gal/yr waste cooking oil (WCO) plant producing biodiesel. The annual profit is evaluated at different selling prices of biodiesel and different costs of WCO. A decrease in the selling price of biodiesel or an increase in the cost of WCO (for instance, due to an increased demand to use in biodiesel production) will lower the profit and can render the process economically unviable. These sensitivity issues are extremely important in the biofuel market.

REFERENCES

Alvarado-Morales M, Terra J, Gernaey KV, Woodley JM, Gani R: Biorefining: computer-aided tools for sustainable design and analysis of bioethanol production, *Chem Eng Res Des* 87:1171–1183, 2009.

Bao B, Ng DKS, Tay DHS, Jiménez-Gutiérrez A, El-Halwagi MM: A shortcut method for the preliminary synthesis of process-technology pathways: an optimization approach and application for the conceptual design of integrated biorefineries, *Comp Chem Eng* 35(8):1374–1383, 2011.

Bao B, Ng DKS, Tay DHS, El-Halwagi MM: *Synthesis of technology pathways for an integrated biorefinery*, Nashville, Tennessee, AIChE Annual Meeting, November 2009.

Basu P: *Biomass gasification and pyrolysis: practical design and theory*, Burlington, USA, Academic Press/Elsevier, 2010.

Bowling IM, Ponce-Ortega JM, El-Halwagi MM: Facility location and supply chain optimization for a biorefinery, *Ind Eng Chem Res* 50(10):6276–6286, 2011.

Cardona CA, Sánchez OJ, Gutiérrez LF: *Process synthesis for fuel ethanol production*, Boca Raton, CRC Press, 2010.

Clark JH, Deswarte FI, editors: *Introduction to chemicals from biomass*, John Wiley and Sons, Ltd, 2008.

El-Halwagi MM, editor: *Biogas technology, transfer, and diffusion*, Essex, UK, Elsevier Applied Science Publishers Ltd, 1986.

Elms RD, El-Halwagi MM: The effect of greenhouse gas policy on the design and scheduling of biodiesel plants with multiple feedstocks, *Clean Technol Environ Policy* 12(5):547–560, 2010.

Elms RD, El-Halwagi MM: Optimal scheduling and operation of biodiesel plants with multiple feedstocks, *Int J Process Syst Eng* 1(1):1–28, 2009.

Gosling I: Process simulation and modeling for industrial bioprocessing: tools and techniques, *Ind Biotechnol* 1(2):106–109, 2005.

Kamm B Gruber PR, Kamm M, editors: *Biorefineries: industrial processes and production*, vol 2, Status quo and future directions, Weinheim, Germany, Wiley-VCH, 2006.

Klass DL: *Biomass for renewable energy, fuels, and chemicals*, San Diego, Academic Press/Elsevier, 1998.

Kokossis AC, Yang AD, Tsakalova M, Lin TC: A systems platform for the optimal synthesis of biomass based manufacturing systems. In Pierucci S, Ferraris BG, editors: *20th European Symposium on Computer Aided Process Engineering*, vol 28, 2010, pp 1105–1110.

Martin M, Grossmann IE: Superstructure optimization of lignocellulosic bioethanol plants. In Pierucci S, Ferraris BG, editors: *20th European symposium on computer aided process engineering*, Amsterdam, Elsevier Science BV, vol 28, pp 943–948, 2010.

Myint LL, El-Halwagi MM: Process analysis and optimization of biodiesel production from soybean oil, *J Clean Tech Env Policies* 11(3):263–276, 2009.

Ng DK, Pham V, El-Halwagi MM, Jiménez-Gutiérrez A, Spriggs HD: A hierarchical approach to the synthesis and analysis of integrated biorefineries. In El-Halwagi MM, Linninger AA, editors: *Design for energy and the environment: Proceedings of the 7th International conference on the foundations of computer-aided process design (FOCAPD)*, CRC Press, Taylor & Francis, pp 425–432, 2009.

Ojeda KA, Sanchez EL, Suarez J, Avila O, Quintero V, El-Halwagi MM, Kafarov V: Application of computer-aided process engineering and exergy analysis to evaluate different routes of biofuels production from lignocellulosic biomass, *Ind Eng Chem Res* 50:2768–2772, 2011.

Pham V, El-Halwagi MM: Process synthesis and optimization of biorefinery configurations, *AIChE J.* (2011, in press DOI: 10.1002/aic.12640).

Pham V, Holtzapple MT, El-Halwagi MM: Techno-economic analysis of biomass to fuel via the MixAlco process, *J Ind Microbiol Biotechnol* 37(11):1157–1168, 2010.

Pokoo-Aikins G, Nadim A, Mahalec V, El-Halwagi MM: Design and analysis of biodiesel production from algae grown through carbon sequestration, *Clean Technol Environ Policy* 12:239–254, 2010.

Pokoo-Aikins G, Heath A, Mentzer RA, Mannan MS, Rogers WJ, El-Halwagi MM: A multi-criteria approach to screening alternatives for converting sewage sludge to biodiesel, *J Loss Prev Process Ind* 23(3):412–420, 2010.

Sammons NE, Jr., Yuan W, Eden MR, Aksoy B, Cullinan HT: Optimal biorefinery resource utilization by combining process and economic modeling, *Chem Eng Res Des* 86:800–808, 2008.

Tan RR, Ballacillo JAB, Aviso KB, Culaba AB: A fuzzy multiple-objective approach to the optimization of bioenergy system footprints, *Chem Eng Res Des* 87:1162–1170, 2009.

Tay DHS, Ng DKS, Kheireddine H, El-Halwagi MM: Synthesis of an integrated biorefinery via the C–H–O ternary diagram, *Clean Techn Environ Policy* (2011 in press), DOI 10.1007/s10098-011-0354-4.

CHAPTER 22
MACROSCOPIC APPROACHES OF PROCESS INTEGRATION

The hallmark of process integration is the holistic perspective. The system boundaries can be drawn around a direct-recycle network, a mass-exchange network, a heat-exchange network, a cogeneration network, or the whole plant. In this chapter, the boundaries are expanded to go beyond the plant and to include the surrounding environmental systems. Because of the general applicability of process integration, there is a consistent approach in treating macroscopic systems in way similar to how individual process plants are treated. The chapter will focus on the following areas:

- Eco-industrial parks
- Material flow analysis and reverse problem formulation for watersheds
- Process integration as an enabling tool in environmental impact assessment
- Process integration in life cycle analysis

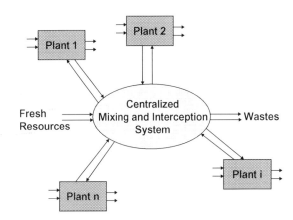

FIGURE 22-1 A mass-integration representation of the EIP problem. *Source: Spriggs et al. (2004).*

ECO-INDUSTRIAL PARKS

The resources and wastes of multiple plants may be integrated in different ways. It is possible to exchange by-products and wastes. It is also possible to use common utility and treatment systems. Such "industrial symbiosis" opportunities call for the consideration of **eco-industrial parks** (EIPs). An EIP may be defined "as a cluster of processing and other facilities together with supporting infrastructure which create the circular economy and promote sustainable development. Such a park will be developed and managed as a real estate development, seeking high environmental, economic, and social benefits as well as business excellence. It will implement conditions for participation and will recruit members who in total achieve the desired objective. This park will be bottom line driven and the tenants selected in such a way that economic performance increases as the level of integration and, therefore, efficiency increases. Participation will provide clear economic and other advantages over the current stand-alone processing model" (Spriggs et al., 2004). The creation of an EIP entails technical and economic challenges that are best addressed through a process integration framework. There are also organization challenges in coordinating the operations of multiple companies. Spriggs et al. (2004) proposed a mass-integration representation of the EIP problem, which allows for the exchange of wastes, by-products, fresh resources among several plants using a centralized facility that allows mixing of different streams, and interception using separation and treatment units (Fig. 22.1).

The problem of designing an EIP may be stated as follows (Lovelady and El-Halwagi, 2009):

Given a set PROCESSES = $\{p|p = 1, 2,..., N_{Process}\}$ of industrial processes. For each process, the following are given:

- A set of process sinks (units): $SINKS_p = \{j_p|j_p = 1, 2, ..., N_{sinks,\,p}\}$. Each sink requires a certain flow rate, G_{j_p}, and a given pollutant composition, $z^{in}_{j_p}$, which must satisfy the following constraint:

[22.1] $\quad z^{min}_{j_p} \leq z^{in}_{j_p} \leq z^{max}_{j_p} \quad \forall j_p \in \{1 ... N_{sinks,p}\}$

where $z^{min}_{j_p}$ and $z^{max}_{j_p}$ are given lower and upper bounds on permissible compositions to unit j_p.

- A set of process effluent wastewater streams referred to as sources: SOURCES = $\{i_p|i_p = 1, 2,..., N_{sources,p}\}$. Each source has a given flow rate, F_{i_p}, and a given pollutant composition, $y^{in}_{i_p}$.

It is desired to design an EIP that treats various sources from the $N_{processes}$ and assigns them to different sinks. The EIP will require the installation of a set of interception units: INTERCEPTORS = $\{k|k = 1, 2,..., N_{Int}\}$ to be used for treating the effluents by reducing the composition of the targeted species to allow them to be assigned to various process sinks.

Available for service is a fresh (external) resource (for example, fresh water) that can be purchased to supplement the use of process sources.

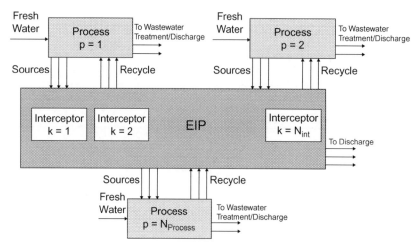

FIGURE 22-2 Schematic representation of the EIP design problem. *Source: Lovelady and El-Halwagi (2009).*

The design challenges include answering the following questions:

- Which streams should be recycled within the same process and which streams should be sent to the EIP? What changes will they undergo in the EIP (for example, mixing, separation, extent of separation)?
- What interception technologies should be used? What are their tasks? Which streams should be assigned to these interception units?
- How much fresh should be used? Where?
- How much waste should be discharged? Where?

Several process integration techniques have been developed to answer the aforementioned questions and to synthesize cost-effective EIPs. Spriggs et al. (2004) proposed a graphical technique based on the material recycle pinch diagram developed by El-Halwagi et al. (2003). Algorithmic and optimization-based approaches have also been introduced for the synthesis of several categories of the EIP-synthesis problem using (for example, Rubio-Castro et al., 2010 and 2011; Aviso et al., 2011; Bandyopadhyay et al., 2010; Lim and Park, 2010; Chew and Foo, 2009; Lovelady et al., 2009a; Lovelady and El-Halwagi, 2009; Chew et al., 2008).

The following is an illustration of an optimization approach Lovelady and El-Halwagi (2009) developed based on the source-interception-sink framework for mass integration (introduced by El-Halwagi et al., 1996, and described in Chapter 18). The representation shown by Fig. 22.2 includes several processes, sources, sinks, and interceptors as well as the fresh resources and environmental discharges involved in synthesizing the EIP.

To simplify the terminology, a single index, i, is given to all the sources. Therefore, $i_1 = 1$ is designated to be $i = 1$, $i_1 = 2$ is designated to be $i = 2$, and so on until $i_1 = N_{Sources,1}$, which is referred to as $i = N_{Sources,1}$, and then $i_2 = 1$, which is given the index $i = N_{Sources,1} + 1$, and so on. Similarly, a single index, j, is given to all the sinks. With this terminology, a source-interception-sink representation (Fig. 22.3) is used as a superstructure that includes the various configurations of interest. Sources are split and allocated to the EIP to be intercepted by candidate interception technologies and then assigned to process sinks or discharged as waste. Fresh resources are used in the sinks as needed.

Several objectives may be sought (for example, minimum fresh water, minimum waste discharged, minimum cost, and so on). Let us consider the following optimization formulation with the objective of minimizing the total annualized cost of the interception devices in the EIP, the cost of fresh water, and waste treatment (Lovelady and El-Halwagi, 2009):

Minimize total annualized cost =

[22.2]
$$\sum_{k=1}^{N_{int}} Interception\_Cost_k + C_{Fresh} \sum_{j=1}^{N_{sinks}} Fresh_j + C_{waste} \cdot waste$$

where $Interception\_Cost_k$ is the total annualized cost associated with interception device k, C_{Fresh} is the cost of the fresh water ($/unit mass of water), $Fresh_j$ is the amount of fresh water fed to the jth sink (mass per year), C_{waste} is the annual waste treatment cost, and $waste$ is the total amount of flow going to waste (mass/yr). The following show model constraints.

Splitting of sources leaving each process to the interception units and the wastewater treatment system:

[22.3] $\quad F_i = \sum_{k=1}^{N_{int}} w_{i,k} + w_i^{waste} \quad \forall i \in \{1 \ldots N_{Sources}\}$

where F_i is the flow rate of the ith source.

Flow balance around the mixing point entering the kth interceptor:

[22.4] $\quad W_k = \sum_{i=1}^{N_{sources}} w_{i,k} \quad \forall k \in \{1 \ldots N_{int}\}$

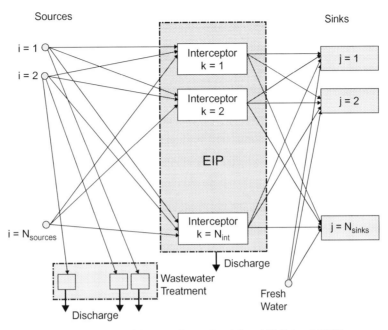

FIGURE 22-3 A source-interception-sink representation of the EIP. *Source: Lovelady and El-Halwagi (2009).*

Component material balance for the mixed sources before entering the kth interceptor:

$$[22.5] \quad W_k \cdot Y_k^{in} = \sum_{i=1}^{N_{sources}} w_{i,k} \cdot y_i^{in} \quad \forall k \in \{1 \ldots N_{int}\}$$

Design model for the kth interceptor:

$$[22.6] \quad y_k^{out} = f_k(W_k, Y_k^{in}, D_k, P_k) \quad \forall k \in \{1 \ldots N_{int}\}$$

where D_k and P_k are the design and operating variables of the kth interceptor.

Splitting of the f intercepted streams from the EIP to the process sinks:

$$[22.7] \quad W_k = \sum_{j=1}^{N_{sink}} g_{k,j} \quad \forall k \in \{1 \ldots N_{int}\}$$

Flow balances around the mixing point before the jth sink:

$$[22.8] \quad G_j = F_j + \sum_{k=1}^{N_{int}} g_{k,j} \quad \forall j \in \{1 \ldots N_{sinks}\}$$

Component material balance for the pollutant(s) entering the process sinks:

$$[22.9] \quad G_j \cdot z_j^{in} = F_j \cdot y_{fresh} + \sum_{k=1}^{N_{int}} g_{k,j} \cdot y_k^{out}$$
$$\forall j \in \{1 \ldots N_{sinks}\}$$

Composition constraints for the feeds to the process sinks:

$$[22.1] \quad z_j^{min} \leq z_j^{in} \leq z_j^{max} \quad \forall j \in \{1 \ldots N_{sinks}\}$$

It is worth noting that the unused streams from the EIP are discharged; that is,

$$[22.10] \quad EIP_{discharge} = \sum_{k=1}^{N_{int}} g_{k, j=Waste}$$

Nonnegativity constraints:

$$[22.11] \quad g_{k,j} \geq 0 \quad \forall k \in \{1 \ldots N_{int}\} \text{ and } \forall j \in \{1 \ldots N_{sinks}\}$$

$$[22.12] \quad w_{i,k} \geq 0 \quad \forall i \in \{1 \ldots N_{sources}\} \text{ and } \forall k \in \{1 \ldots N_{int}\}$$

$$[22.13] \quad F_j \geq 0 \quad \forall j \in \{1 \ldots N_{sinks}\}$$

$$[22.14] \quad EIP_{discharge} \geq 0$$

The solution of this optimization program determines the distribution of the streams and the types and sizes of the interception technologies so as to minimize the total annualized cost of the EIP.

Example 22-1 Design of an EIP

Consider the industrial complex Lovelady and El-Halwagi (2009) describe, with five plants as shown by Fig. 22.4. Tables 22.1 to 22.3 give the data for the sources, sinks, and interceptors. The cost of fresh water is $0.6/tonne, and the EIP is operated 8760 hours/year.

By formulating the problem as an optimization program using the aforementioned formulation and solving using LINGO, the solution is obtained to be $4.04 MM/yr. One interception technology is selected; Fig. 22.5 shows the allocation of streams. The integrated solution should be compared with the original cost of the fresh before the EIP was used and without integration to be $47.30 MM/yr. These significant results are attributed to the use of a centralized EIP and the conservation of fresh water via recycle/reuse, interception, and interplant integration.

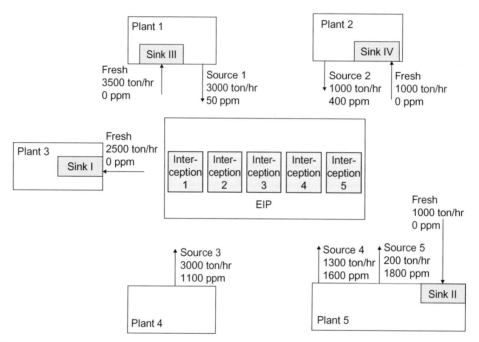

FIGURE 22-4 Schematic representation of the EIP case study. *Source: Lovelady and El-Halwagi (2009).*

TABLE 22-1 Sink Data for the EIP Case Study

Sink	Flow Rate, ton/hr	Maximum Inlet Composition of Pollutant (ppm)
I	2500	40
II	2000	225
III	3500	500
IV	1000	760

Source: Lovelady and El-Halwagi (2009).

TABLE 22-2 Source Data for the EIP Case Study

Source	Flow Rate, ton/hr	Inlet Composition of Pollutant (ppm)
I	3000	50
II	1000	400
III	3000	1100
IV	1300	1600
V	200	1800

Source: Lovelady and El-Halwagi (2009).

TABLE 22-3 Interception Data for the EIP Case Study

Interception Technology	Minimum Inlet Composition of Intercepted Stream (ppm)	Minimum Outlet Composition of Intercepted Streams (ppm)	Interception Cost ($/kg removed)
I	1000	450	0.05
II	800	300	0.06
III	370	250	0.08
IV	300	100	0.09
V	280	30	0.16

Source: Lovelady and El-Halwagi (2009).

(Continued)

Example 22-1 Design of an EIP (Continued)

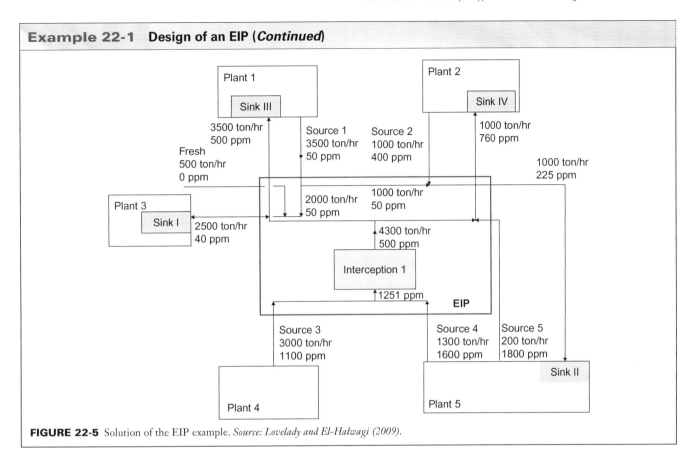

FIGURE 22-5 Solution of the EIP example. *Source: Lovelady and El-Halwagi (2009).*

MATERIAL FLOW ANALYSIS AND REVERSE PROBLEM FORMULATION FOR WATERSHEDS

Industrial facilities interact strongly with the surroundings. An example of such interaction is the influence of industrial water usage and discharge on the surrounding watersheds. A watershed covers a geographical region that includes one or more sources of water flowing through various surfaces and/or underground drainage pathways. Typically, a watershed involves a number of tributaries that feed into reaches, streams, or rivers and finally lead to a catchment area (for example, lake, sea, ocean). Figure 22.6 shows a typical tributary with inputs from rural areas, population centers, industrial processes, precipitation (rain), and agricultural drainage. All these inputs interact within the same tributary, and the inputs to all the tributaries interact within the same watershed. To understand the characteristics of watersheds, the impact of various inputs and outputs on the watersheds, and the impact of the watersheds on the surroundings, it is important to develop quantitative models that track the flows and compositions of various pollutants throughout the watershed. In this context, a useful tool is the material flow analysis (MFA), which is developed to track flows and concentrations of targeted components (for example, pollutants, nutrients). An MFA model accounts for all relevant activities, water sources, water users, pollution sources, as well as models for physical, chemical, and biological activities taking place within the watershed. Figure 22.7 shows an example of a nitrogen cycle. Detailed MFA watershed models have been developed for regional application (for example, Lira-Barragán et al., 2011a, b; Lovelady et al., 2009b; El-Baz, 2005a, b).

Models such as the MFA interact well with a mass-integration framework. The MFA model provides the data needed for mass integration (for example, flows, concentrations of sources and sinks, impact of a change in input on the output characteristics, and so on). In return, mass integration is used to devise solution strategies for certain objectives of the watershed. Figure 22.8 shows the interaction.

Predictive water-quality models such as the MFA are carried out in the **forward mode**; that is, a given scenario is posed and its consequences are tracked via the MFA. As an illustration, let us consider the scenario of a new industrial facility to be installed on a watershed (Fig. 22.9a). The exact plant site is to be determined. There are several environmental regulations for the process discharges. Furthermore, there are overall environmental regulations for the watershed. The process has certain process discharges with a given composition of a certain pollutant, $y^{Process}$. How will the installation of the new process impact the watershed, and will the environmental regulations for the watershed be satisfied? If the typical MFA forward mode is used, a construction location is chosen and the MFA model is used to track the process discharge throughout the watershed. If the discharge to the lake meets the environmental regulations for the lake, the environmental discharge of the process is deemed

380 Sustainable Design Through Process Integration

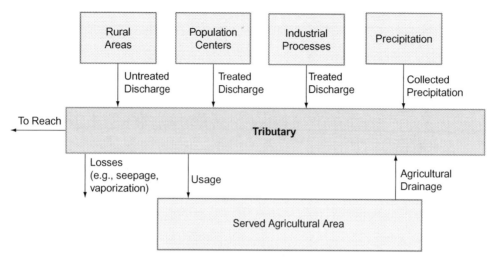

FIGURE 22-6 Inputs and outputs of a typical tributary. *Source: El-Baz et al. (2005a).*

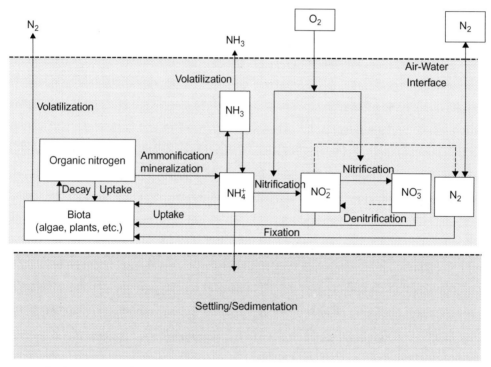

FIGURE 22-7 Accounting for the nitrogen cycle in a watershed. *Source: El-Baz et al. (2005a).*

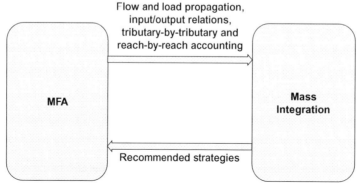

FIGURE 22-8 Interaction between MFA and mass integration. *Source: El-Baz et al. (2005b).*

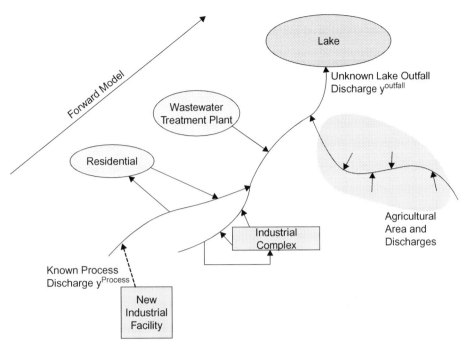

FIGURE 22-9A Forward model for tracking discharges of a new process. *Source: Lovelady et al. (2009b).*

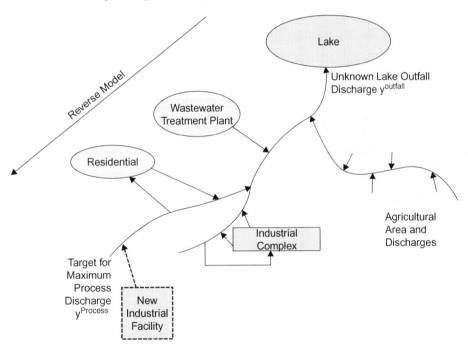

FIGURE 22-9B Inverse approach to integrating the process and the watershed. *Source: Lovelady et al. (2009b).*

satisfactory. Otherwise, a tedious trial-and-error procedure is carried out to alter the plant design, wastewater treatment systems, and location until a satisfactory discharge to the lake is realized.

A more effective approach to the forward problem is to "start with the end in mind." Such an MFA approach is referred to as **reverse problem formulation** (Lovelady et al., 2009b). According to this approach, the desired characteristics of the discharge to the lake are set to the satisfactory limits, and the MFA model is included in an optimization formulation that seeks to determine the maximum acceptable discharge from the process that will achieve the satisfactory limits for the lake while accounting for all interactions throughout the watershed. Figure 22.9b is a schematic representation of the reverse problem formulation. The same concept can be extended to the scenario of having multiple new plants to be installed while having multiple constraints at different locations of the watershed as shown by Fig. 22.9c. Examples of optimization formulations for the reverse MFA models combined with mass-integration models can be found in literature (Lira-Barragán et al., 2010, 2011; Lovelady et al., 2009b).

Let us consider the MFA reverse problem formulation, including process integration developed by Lovelady et al.

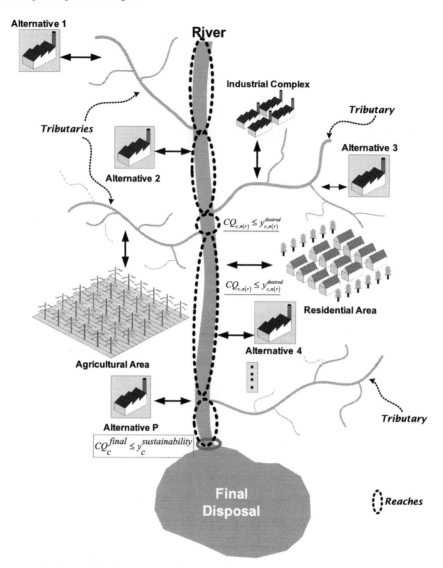

FIGURE 22-9C Inverse approach to integrating the process and the watershed with multiple plants and constraints throughout the watershed. *Source: Lira-Barragán et al. (2011a).*

(2009b) and described by the following model. The watershed is discretized into a number of reaches. Each reach is given the index i. The MFA model is developed over multiple periods. Each period is given the index t. The following constraints are constructed:

Flow Balance for the Reaches

$$[22.15] \quad Q_{i,t} = Q_{i-1,t} + P_{i,t} + D_{i,t} + H_{i,t} + \sum_{j=1}^{N_{Trib,t}} T_{j,i,t} - L_{i,t} - U_{i,t}$$

where

$Q_{i,t}$ = Flow rate leaving the ith reach, m^3/s
$Q_{i-1,t}$ = Flow rate entering the ith reach, m^3/s
$P_{i,t}$ = Precipitation flow to the ith reach, m^3/s
$L_{i,t}$ = Net losses from the ith reach (for example, seepage, vaporization, use, and so on), m^3/s
$D_{i,t}$ = Nontributary direct discharge to the ith reach, m^3/s

$H_{i,t}$ = Total discharge (for example, industrial discharge + sanitary discharge, and so on) to the reach m^3/sec
$T_{j,i,t}$ = Tributary discharge from the jth tributary to the ith reach, m^3/s
$U_{i,t}$ = Usage discharge from the reach ith, m^3/sec

Pollutant Balance for the Reaches

$$[22.16] \quad Q_{i,t} * CQ_{i,t} = Q_{i-1,t} * CQ_{i-1,t} + H_{i,t} * CH_{i,t} + P_{i,t} * CP_{i,t} + D_{i,t} * CD_{i,t} + \sum_{j=1}^{N_{Trib,t}} T_{j,it} * CT_{j,i,t} - L_{i,t} * CL_{i,t} - U_{i,t} * CU_{i,t} - \int_{V=0}^{V_{i,t}} r_{i,t}\, dV_{i,t}$$

where

$Q_{i,t} * CQ_{i,t}$ = Load rate (of the targeted species) leaving reach i with the output flow (where $CQ_{i,t}$ is the concentration of the targeted species in the output flow from reach i)

$Q_{i-1,t} * CQ_{i-1,t}$ = Load rate entering the reach via convective flow

$H_{i,t} * CH_{i,t}$ = Total load rate of discharges over reach i

$P_{i,t} * CP_{i,t}$ = Total load rate of precipitation (rain) over reach i

$D_{i,t} * CD_{i,t}$ = Total load rate of drainage over reach i

$L_{i,t} * CL_{i,t}$ = Total load of losses (for example, seepage and vaporization) over reach i

$U_{i,t} * CU_{i,t}$ = Total load of usage over reach i

The reaction rate is calculated from the kinetics of the targeted species. For instance, if the reaction is modeled through a first-order kinetics, we get

$$[22.17] \quad \int_{V=0}^{V_{i,t}} r_{i,t}\, dV_{i,t} = k_{i,t} * CQ_{i,t} * V_{i,t},$$

where $k_{i,t}$ is the Arrhenius reaction rate constant. Hence,

[22.18]

$$Q_{i,t} * CQ_{i,t} = Q_{i-1,t} * CQ_{i-1,t} + H_{i,t} * CH_{i,t} + P_{i,t} * CP_{i,t} + D_{i,t} * CD_{i,t} + \sum_{j=1}^{N_{Trib,i}} T_{j,it} * CT_{j,i,t} - L_{i,t} * CL_{i,t} - U_{i,t} * CU_{i,t} - k_{i,t} * CQ_{i,t} * V\, dV_{i,t}$$

Pollutant Balance for the Tributaries

[22.19]

$$T_{j,i,t} * CT_{j,i,t} = S_{j,i,t}^{untreated} * CS_{j,i,t}^{untreated} + S_{j,i,t}^{treated} * CS_{j,i,t}^{treated} + I_{j,i,t} * CI_{j,i,t} + P_{j,i,t} * CP_{j,i,t} + D_{j,i,t} * CD_{j,i,t} - L_{j,i,t} * CL_{j,i,t} - U_{j,i,t} * CU_{j,i,t} - r_{j,i,t} * V_{j,i,t}$$

where

$S_{j,i,t}^{untreated}$ = Untreated sewage (sanitary waste) discharged to the jth tributary, m^3/s

$S_{j,i,t}^{treated}$ = Treated sewage (sanitary waste) discharged to the jth tributary, m^3/s

$I_{j,i,t}$ = Industrial flow of wastewater discharged to the jth tributary, m^3/s

$P_{j,i,t}$ = Precipitation flow discharged to the jth tributary, m^3/sec

$L_{j,i,t}$ = Net losses from the jth tributary (for example, seepage, vaporization, use, and so on), m^3/s

$D_{j,i,t}$ = Agricultural drainage discharged to the jth tributary, m^3/s

To simplify the terminology for modeling the reaches and tributaries over a given averaged time period, these equations are rewritten as

$$[22.20] \quad (Q_u, y_u) = \psi_u(Q_u^{in}, y_u^{in}, I_u^{in}, C_u^{in}),$$

where u is an index used for all tributaries and reaches. Also, Q_u and y_u are the flow rate and pollutant concentration leaving the uth reach or tributary. Both Q_u and y_u are modeled through a function of vectors of input flow and pollutant concentrations (Q_u^{in} and y_u^{in}), and all inputs from the surroundings (for example, residential, industrial, agricultural) in the form of flow rate and pollutant concentration vectors (I_u^{in} and C_u^{in}).

Suppose that a new plant is to be installed at location $\bar{u}$ or that the effluent flow rate or concentration will be changed. The flow rate and pollutant concentration of the process effluent are designated as $P_{\bar{u}}$ and $y_{\bar{u}}^{Process}$. Hence, the MFA model at location $\bar{u}$ is expressed as

$$[22.21] \quad (Q_{\bar{u}}, y_{\bar{u}}) = \psi_{\bar{u}}(Q_{\bar{u}}^{in}, y_{\bar{u}}^{in}, I_{\bar{u}}^{in}, C_{\bar{u}}^{in}, P_{\bar{u}}, y_{\bar{u}}^{Process}).$$

This expression can be inverted to have $y_{\bar{u}}^{Process}$ on the left-hand side; that is,

$$[22.22] \quad y_{\bar{u}}^{Process} = \Omega_{\bar{u}}(Q_{\bar{u}}, y_{\bar{u}}, Q_{\bar{u}}^{in}, y_{\bar{u}}^{in}, I_{\bar{u}}^{in}, C_{\bar{u}}^{in}, P_{\bar{u}})$$

The desired environmental target for the pollutant concentration at the outfall (or catchment area of the watershed) is given by $y^{Outfall, desired}$.

The objective of the reverse problem formulation is to identify the maximum concentration or load of the pollutant(s) that will not lead to the violation of the environmental constraint for the pollutant discharge at the outfall or catchment area of the watershed or other constraints throughout the watershed. For instance, if the flow rate of the process discharge is given, then the model can be written as (Lovelady et al., 2009b)

[P22.1]

$$\text{Maximize } y_u^{Process}$$

subject to

$$(Q_u, y_u) = \psi_u(Q_u^{in}, y_u^{in}, I_u^{in}, C_u^{in}) \quad \forall u \neq \bar{u}$$

$$y_{\bar{u}}^{Process} = \Omega_{\bar{u}}(Q_{\bar{u}}, y_{\bar{u}}, Q_{\bar{u}}^{in}, y_{\bar{u}}^{in}, I_{\bar{u}}^{in}, C_{\bar{u}}^{in}, P_{\bar{u}})$$

$$y^{outfall} \leq y^{Outfall, desired}$$

The solution identifies the maximum allowable pollutant concentration in the process effluent and tracks the consequences of a new plant (or a different discharge) at location $\bar{u}$ on the watershed.

As an illustration of the application of the reverse problem formulation, consider the problem of identifying the location of a new fertilizer plant with four candidate locations over the Bahr El-Baqar drainage system leading to Lake Manzala in Egypt (Fig. 22.10). The effluent from the fertilizer plant will have a flow rate of 2.0 m^3/s and a concentration of 12.5 ppm of phosphorus concentration of 12.5 ppm. To prevent lake eutrophication resulting from the excessive growth of biomass based on a high load of nutrients like phosphorus, the phosphorus input to Lake Manzala is limited to a maximum of 1.3 ppm. Two costs are to be considered for comparison: phosphorus treatment cost (using interception devices to lower phosphorus composition to an acceptable level) and cost of land. The

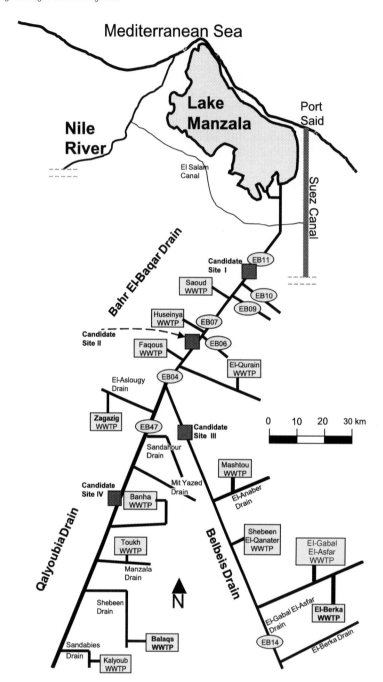

FIGURE 22-10 Candidate site locations for the fertilizer plant. *Source: Lovelady et al. (2009b).*

problem is posed as a reverse problem optimization formulation and solved to give the results shown by Table 22.4.

Another example is the case study given Lira-Barragán et al. (2011a,b) for the Balsas watershed system located in Mexico. This watershed is highly impacted by several industrial zones and their effluent discharges. Figure 22.11 shows the main streams of the river along with 20 possible sites for a new plant—which discharges 5.0 m³/s of an industrial effluent containing 20 ppm of a hazardous pollutant—to be installed. Environmental constraints are imposed throughout the watershed.

An optimization program was developed using the reverse problem formulation of Lira-Barragán et al. (2011b) to account for various levels of pollutant concentration at the final discharge location. A trade-off can be established (as shown by the Pareto curve of Fig. 22.12) to trade off the final discharge concentration and the total annualized cost of the system. This trade-off gives the decision makers a valuable tool for deciding to what extent the pollution should be reduced and the associated costs. It is worth noting that, by changing the desired concentration of the final discharge, the plant location changes.

TABLE 22-4 Results of Reverse Problem Formulation and Associated Economics

Location	Maximum Target for Phosphorus in Effluent, mg/L (calculated through reverse problem formulation)	Minimum Phosphorus Load (kg/yr) to Be Removed to Satisfy the Desired Target for Lake Manzala	Interception Cost, $MM/yr	Total Annualized Cost, $MM/yr
Site I	5.6	435,384	9.14	19.14
Site II	10.21	144,387	3.03	20.03
Site III	13.93	0	0	35
Site IV	18.73	0	0	18

Source: Lovelady et al. (2009b).

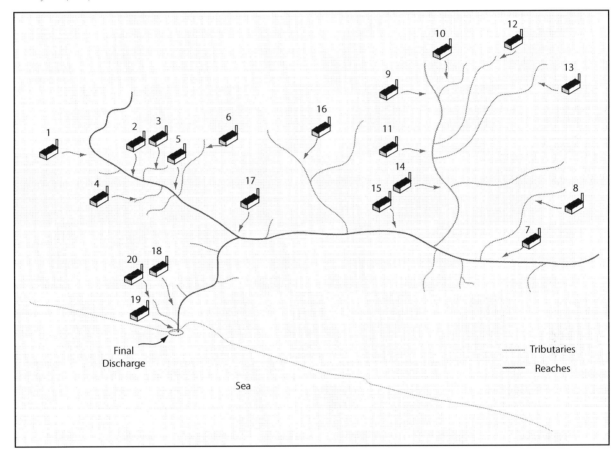

FIGURE 22-11 The Balsas watershed and possible plant locations. *Source: Lira-Barragán et al. (2011b).*

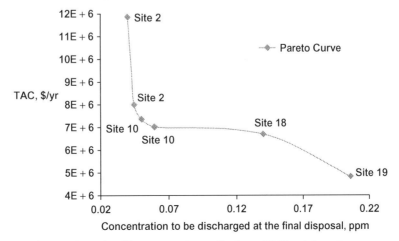

FIGURE 22-12 The Pareto curve showing the trade-off between total annualized cost (TAC) and the pollutant concentration at the final discharge location. *Source: Lira-Barragán et al. (2011b).*

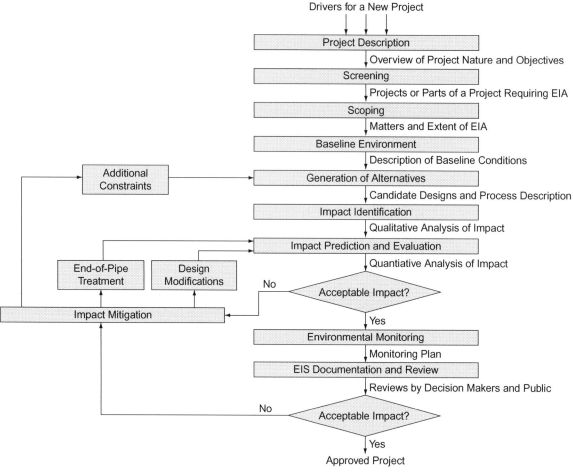

FIGURE 22-13 Steps in a typical EIA procedure. *Source: El-Halwagi et al. (2009).*

PROCESS INTEGRATION AS AN ENABLING TOOL IN ENVIRONMENTAL IMPACT ASSESSMENT

Environmental impact assessment (EIA) is a procedure for evaluating and mitigating the biological, physical, chemical, economic, and social consequences of a proposed project on the surrounding environment. Typically, an EIA involves a variation of the following steps (El-Halwagi et al., 2009):

1. Project description
2. Screening to determine which projects or parts of a project require an EIA
3. Scoping which involves a top-level analysis for the identification of the primary matters to be included in the EIA and the extent to which these matters should be addressed
4. Description of the baseline environment before the project is implemented
5. Generation of alternatives and process description
6. Identification of impact on the environment carried out at top level to determine which priority-area parts of the project should be studied in more details
7. Prediction and evaluation of impact on the environment to study the priority areas and determine (quantitatively whenever possible) the consequences of the planned activities on the environment
8. Impact mitigation
9. Environmental monitoring plan
10. Documentation and review of the environmental impact statement (EIS), which includes a description of the expected impact on the environment and the possible alternative scenarios or actions

Figure 22.13 shows a schematic representation of the key steps in a typical EIA procedure.

There are extensive process engineering activities involved in preparing an EIS. The activities include the generation of design alternatives, the techno-economic analysis, safety assessment (for example, hazard and operability analysis [HAZOP] or hazard identification [HAZID]), and the evaluation of the environmental consequences of each alternative. Consequently, a design alternative may be accepted, altered, or rejected. This tedious procedure (shown by Fig. 22.14) is costly and time consuming.

The EIA procedure can be significantly improved using process integration in guiding the process-engineering activities associated with the EIA through the following strategies (El-Halwagi et al., 2009):

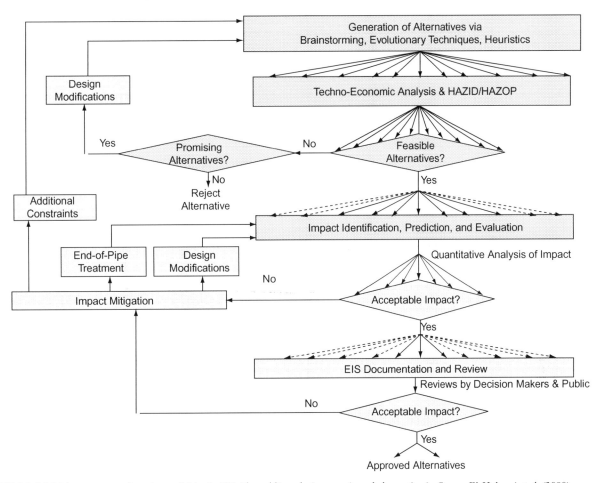

FIGURE 22-14 Main process engineering activities in EIA (dotted lines designate rejected alternatives). *Source: El-Halwagi et al. (2009).*

1. Systematic generation of alternatives and targeting for benchmarking environmental impact ahead of detailed design (for example, overall mass targeting, direct-recycle targets, mass-exchange targets, minimum heating and cooling utility targets, cogeneration targets, and so on)
2. Reverse problem formulation (as described in the section on "Material Flow Analysis and Reverse Problem Formulation for Watersheds")
3. Integration of alternatives with the rest of the process (for example, the tools given in Chapters 6 and 18)

These strategies are shown by the procedure illustrated by Fig. 22.15.

PROCESS INTEGRATION IN LIFE CYCLE ANALYSIS

Life cycle analysis (LCA) is an approach to the assessment of environmental impacts throughout the cradle-to-grace ecological cycle of a product, which involves extraction of the raw materials, processing, manufacturing, distribution, use, and discharge or recycle. Figure 22.16a shows examples of cradle-to-grave ecological cycles for switchgrass to power, and Fig. 22.16b shows for soybean to biodiesel. There are standard techniques for performing an LCA (for example, Horne et al., 2009). This section provides a brief discussion on the inclusion of process integration in LCA.

Process integration results should be coupled with LCA to evaluate the environmental benefits beyond the process itself using a cradle-to-grave assessment. Chouinard-Dussault et al. (2011) proposed a procedure for coupling process integration with LCA for tracking GHG emissions. The procedure shown by Fig. 22.17 starts by obtaining the process data on base-case mass and energy balances with focus on feedstocks, products, by-products, and wastes, utilities, and associated GHG emissions. For a given scenario (that is, a level of energy and mass integration), process integrations tools are used to determine the minimum utility requirements and materials usage and discharge. Then, the total energy consumption and emissions are determined though an LCA tool such as the GREET (Greenhouse gases, Regulated Emissions, and Energy use in Transportation) model (GREET, 2007). The GREET model can be used to generate "well-to-wheels" estimates of GHG emissions for different energy sources. The evolved process integration scenarios are compared. Facility location can also be incorporated (e.g., Bowling et al., 2011).

This approach can be used to compare various alternatives. For instance, it can be used to compare different

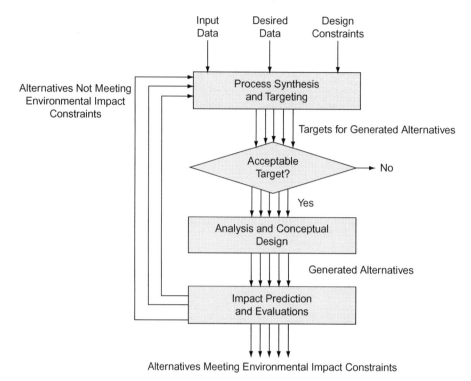

FIGURE 22-15 Using process synthesis and targeting in EIA. *Source: El-Halwagi et al. (2009).*

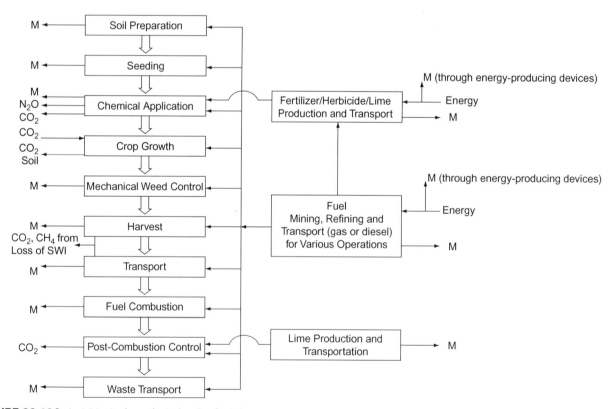

FIGURE 22-16A Activities in the ecological cycle of switchgrass to power (M represents a mixture of GHGs). *Source: Qin et al. (2006).*

biofuel production routes including processes with the scenarios before and after process integration. The results shown by Fig. 22.18 cover a comparison between petroleum gasoline, integrated and unintegrated corn ethanol, ethanol from corn stover, petroleum diesel, and biodiesel from soybeans using integrated and unintegrated processes.

Another example is the assessment of different extents of heat integration. Although achieving minimum heating and

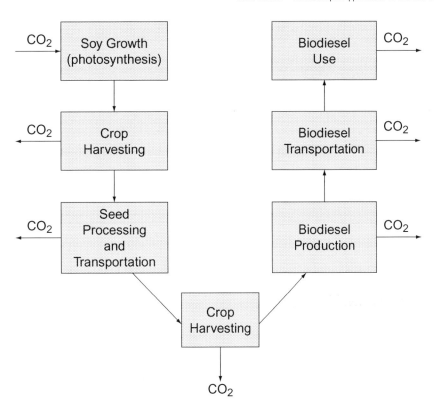

FIGURE 22-16B Ecological cycle for soy to biodiesel showing CO_2 sequestration and emissions. *Source: Elms and El-Halwagi (2010).*

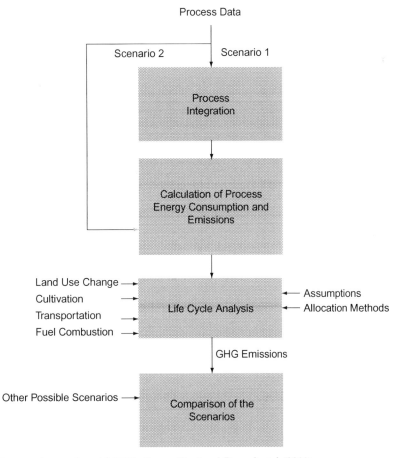

FIGURE 22-17 Coupling of process integration with LCA. *Source: Chouinard-Dussault et al. (2011).*

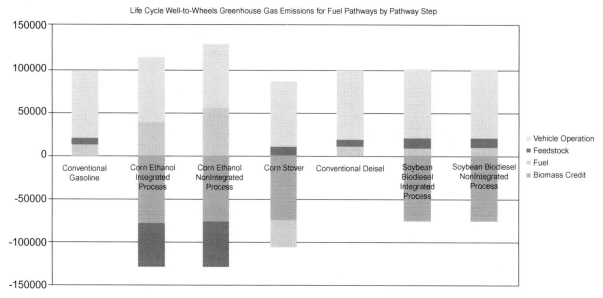

FIGURE 22-18 Comparison of the well-to-wheels GHG emissions for alternate pathways. *Source: Chouinard-Dussault et al. (2011).*

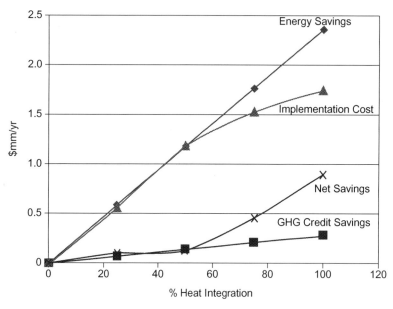

FIGURE 22-19 Analysis for the energy integration of the corn ethanol synthesis process. *Source: Chouinard-Dussault et al. (2011).*

cooling utilities provides minimum energy usage, there are other factors (for example, implementation cost such as total annualized cost, GHG credits resulting from the reduction in energy usage, and so on). Figure 22.19 shows this analysis for a corn ethanol plant. The minimum heating and cooling utilities are determined, and then different scenarios of implementation are examined with each scenario corresponding to a certain percentage of the ultimate heat integration (leading to minimum heating and cooling utilities). This analysis is useful if characterizing the trade-offs between savings, associated costs, and GHG emissions on a life cycle basis.

NOMENCLATURE

C_{Fresh} fresh cost ($/tonne)
$CD_{i,t}$ concentration of the targeted species in the drainage to reach i
$CH_{i,t}$ concentration of the targeted species in the total discharge to reach i
$CL_{i,t}$ concentration of the targeted species in the losses from reach i
$CP_{i,t}$ concentration of the targeted species in the precipitation (rain) to reach i

$CQ_{i,t}$	concentration of the targeted species in the output flow from reach i
$CU_{i,t}$	concentration of the targeted species in the water used from reach i
Cwaster	waste cost ($/tom)
$D_{i,t}$	nontributary direct discharge to the ith reach, m³/s
$D_{j,i,t}$	agricultural drainage discharged to the jth tributary, m³/s
D_k	design variable of interceptor k
F_j	flow rate of the fresh to sink j (tonne/hr)
$Fresh_j$	amount of fresh resource fed to the sink j (tonne/yr)
G_j	flow rate demand for sink j (tons/hr)
$g_{k,j}$	amount of flow from interceptor k to sink j (tonne/hr)
$H_{i,t}$	total discharge (for example, industrial discharge + sanitary discharge, and so on) to the reach m³/sec
$I_{j,i,t}$	industrial flow of wastewater discharged to the jth tributary, m³/s
Interception_Cost$_k$	total annualized fixed equipment cost associated with interception device k
$L_{i,t}$	net losses from the ith reach (for example, seepage, vaporization, use, and so on), m³/s
$L_{j,i,t}$	net losses from the jth tributary (for example, seepage, vaporization, use, and so on), m³/s
N_{sinks}	number of sinks
$N_{sources}$	number of sources
N_{int}	number of interceptors
$P_{i,t}$	precipitation flow to the ith reach, m³/s
$P_{j,i,t}$	precipitation flow discharged to the jth tributary, m³/sec
P_k	operating variable of interceptor k
$Q_{i,t}$	flow rate leaving the ith reach, m³/s
$Q_{i-1,t}$	flow rate entering the ith reach, m³/s
$S_{j,i,t}^{untreated}$	untreated sewage (sanitary waste) discharged to the jth tributary, m³/s
$S_{j,i,t}^{treated}$	treated sewage (sanitary waste) discharged to the jth tributary, m³/s
$T_{j,i,t}$	tributary discharge from the jth tributary to the ith reach, m³/s
$U_{i,t}$	usage discharge from the reach ith, m³/sec
y_k^{out}	outlet composition of interceptor k (ppm)
$w_{i,k}\nabla$	amount of flow from source i to interceptor k (tonne/hr)
W_k	total amount of flow going to interceptor k (tonne/hr)
w_{i,k_i}	flow entering from source i to interceptor k (tonne/hr)
waste	total amount of flow going to waste (tonne/hr)
y_i^s	inlet (supply) composition of source i (ppm)
y_{fresh}	impurity composition in the fresh resource (ppm)
y_k^{in}	inlet composition of source interceptor k (ppm)
y_k^{out}	outlet composition of interceptor k (ppm)
$y^{Outfall, desired}$	desired environmental target for the pollutant concentration at the outfall (or catchment area of the watershed), ppm
z_j^{min}	lower bound composition to sink j (ppm)
z_j^{in}	inlet composition to sink j (ppm)
z_j^{max}	upper bound composition to sink j (ppm)

INDICES

i	source for EIP or reach for watersheds
j	sink for EIP or tributary for watersheds
k	interceptors
t	time
u	location on the watershed

REFERENCES

Aviso KB, Tan R/R, Culaba AB, Foo DCY, Hallale N: Fuzzy optimization of topologically constrained eco-industrial resource conservation networks with incomplete information, *Eng Optim* 43:257–279, 2011.

Bandyopadhyay S, Sahu GC, Foo DCY, Tan RR: Segregated targeting for multiple resource networks using decomposition algorithm, *AIChE J* 56(5):1235–1248, 2010.

Bowling IM, Ponce-Ortega JM, El-Halwagi MM: Facility location and supply chain optimization for a biorefinery, *Ind Eng Chem Res* 50(10):6276–6286, 2011.

Chew IML, Foo DCY: Automated targeting for inter-plant water integration, *Chem Eng J* 153(1–3):23–36, 2009.

Chew IML, Tan R, Ng DKS, Foo DCY, Majozi T, Gouws J: Synthesis of direct and indirect interplant water network, *Ind Eng Chem Res* 47(23):9485–9496, 2008.

Chouinard-Dussault P, Bradt L, Ponce-Ortega JM, El-Halwagi MM: Incorporation of process integration into life cycle analysis for the production of biofuels, *Clean Techn Environ Policy* (2011 in press), DOI: 10.1007/s10098-010-0339-8.

El-Baz AA, Ewida KT, Shouman MA, El-Halwagi MM: Material flow analysis and integration of watersheds and drainage systems: I. Simulation and application to ammonium management in Bahr El-Baqar drainage system, *Clean Techn Environ Policy* 7:51–61, 2005a.

El-Baz AA, Ewida KT, Shouman MA, El-Halwagi MM: Material flow analysis and integration of watersheds and drainage systems: II. Integration and solution strategy with application to ammonium management in Bahr El-Baqar drainage system, *Clean Techn Environ Policy*, 7:78–86, 2005b.

El-Halwagi MM, Lovelady EM, Abdel-Wahab A, Linke P, Alfadala HE: Apply process integration to environmental impact assessment, *Chem Eng Prog*:36–42, February 2009.

El-Halwagi MM, Gabriel F, Harell D: Rigorous graphical targeting for resource conservation via material recycle/reuse networks, *Ind Eng Chem Res* 42:4319–4328, 2003.

El-Halwagi MM, Hamad AA, Garrison GW: Synthesis of waste interception and allocation networks, *AIChE J* 42(11):3087–3101, 1996.

GREET: Greenhouse gases, regulated emissions, and energy use in transportation, Software Version 2.7, Chicago, University of Chicago/Argonne National Lab, 2007.

Horne R, Grant R, Verghese K: *Life cycle assessment: principles, practice, and prospects*, Collingwood, Australia, CSIRO Publishing, 2009.

Lim SR, Park JM: Interfactory and intrafactory water network system to remodel a conventional industrial park to a green eco-industrial park, *Ind Eng Chem Res* 49(3):1351–1358, 2011b.

Lira-Barragán LF, Ponce-Ortega JM, Serna-González M, El-Halwagi MM: Synthesis of water networks considering the sustainability of the surrounding watershed, *Comp Chem Eng* (2011a, in press), DOI: 10.1016/j.compchemeng.2011.03.021.

Lira-Barragán LF, Ponce-Ortega JM, Serna-González M, El-Halwagi MM: An MINLP model for the optimal location of a new industrial plant with simultaneous consideration of economic and environmental criteria, *Ind Eng Chem Res* 50(2):953–964, 2011b.

Lovelady EM, El-Halwagi MM: Design and integration of eco-industrial parks, *Environ Prog Sustainable Energy* 28(2):265–272, 2009b.

Lovelady EM, El-Halwagi MM, Chew IMI, Ng DKS, Foo DCY, Tan RR: A property-integration approach to the design and integration of eco-industrial parks. In El-Halwagi MM, Linninger AA, editors: *Design for energy and the environment: Proceedings of the 7th international conference on the foundations of computer-aided process design (FOCAPD)*, CRC Press, Taylor & Francis, pp 559–568, 2009a.

Lovelady EM, El-Baz A, El-Monayeri D, El-Halwagi MM: Reverse problem formulation for integration process discharges with watershed and drainage systems: optimization framework and application to managing phosphorus in Lake Manzala, *J Ind Ecol* 13(6):914–927, 2009.

Qin X, Mohan T, El-Halwagi MM, Cornforth F, McCarl BA: Switchgrass as an alternate feedstock for power generation: integrated environmental, energy, and economic life cycle analysis, *J Clean Techn Env Polic* 8(4):233–249, 2006.

Rubio-Castro E, Ponce-Ortega JM, Nápoles-Rivera F, El-Halwagi MM, Serna-González M, Jiménez-Gutiérrez A: Water integration of eco-industrial parks using a global optimization approach, *Ind Eng Chem Res* 49(20):9945–9960, 2010.

Rubio-Castro E, Ponce Ortega JM, Serna-González M, Jiménez-Gutiérrez A, El-Halwagi MM: A global optimal formulation for water integration in eco-industrial parks considering multiple pollutants, *Comp Chem Eng* (2011, in press).

Spriggs HD, Lowe EA, Watz J, Lovelady EM, El-Halwagi MM: Design and development of eco-industrial parks, New Orleans, AIChE Spring Meeting, Paper #109a, April 2004.

CHAPTER 23

CONCLUDING THOUGHTS: LAUNCHING SUCCESSFUL PROCESS-INTEGRATION INITIATIVES AND APPLICATIONS

By now (unless you have skipped too many chapters!), you have seen a wide variety of process integration (PI) tools and techniques that systematically provide unique insights, performance benchmarks, and best-in-class implementations. Transforming process integration into value-added applications requires more than just the tools. It requires an effective work process that considers and integrates various technical and business matters. This chapter discusses key items associated with the successful initiation, development, and implementation of successful PI applications.

COMMERCIAL APPLICABILITY

Numerous case studies have been presented throughout this book. The results are typically insightful, intuitively nonobvious, and present a step change from the "business-as-usual" approach that relies mostly on brainstorming, "cherry-picking" for ideas, evolutionary techniques, and heuristics. But do these PI tools work that well in "real life"? The answer is a *resounding yes!* Many successful PI applications around the world attest to the power, uniqueness, and effectiveness of PI. Table 23.1 summarizes a sample of PI applications.

As can be seen from Table 23.1, for several objectives and various industries, PI consistently performs well. Regardless of the type and size of the process, there are typically significant opportunities for improvement. Why? El-Halwagi and Spriggs (1998) attribute these opportunities to three main factors:

> **Organization.** Because of the complexity of a processing facility, there is typically the tendency to subdivide the process into smaller, tractable portions. Sometimes, these subdivisions are at the level of sections of the plant and sometimes they are even at the level of individual units. The net result is that the focus shifts to specific, localized, and stratified portions of the process so as to manage the complex in a streamlined manner. This subdivision works against a holistic perspective of the process and quite often leads to missing major opportunities that are only identified when the whole process is understood from an integrated viewpoint and when all the necessary interactions are properly understood and utilized. Additionally, these subdivisions normally lead to solutions that address the symptoms of the problems while failing to address the root causes.
>
> **Changing business requirements.** The process industries are constantly evolving and changing. There are several reasons for such changes. These include changing market conditions, supply, and demand, varying availability, qualities and costs of raw materials and products, growing stringency in environmental regulations, needs for expansion, incorporation of new production lines, replacement of existing units and technologies, and need for safer operations. Even for a process that has been recently "optimized," there are numerous opportunities for improvement because of the changing business requirements.
>
> **Integrated tools.** A processing facility is a complex system of technologies, chemistries, units, and streams. Without an integrated approach to optimizing these systems, it becomes extremely difficult to understand the global insights and interactions of the process, identify broad opportunities, and develop sitewide implementation. The process integration tools described in this book are just becoming available for widespread use. They should result in a step change in identifying and realizing opportunities.

PITFALLS IN IMPLEMENTING PROCESS INTEGRATION

In the commercial implementation of PI, there are some serious pitfalls that must be avoided. The following are some of these pitfalls:

- **Improper data extraction.** An important concept in PI is invoking "the right level of details." Insufficient data may cause an unrealistic targeting for the problem. Too much data collection may considerably slow down the implementation. Proper data types and extent have been discussed throughout the book.
- **Improper simulation.** Analysis using actual process performance, models, and computer-aided simulation is important for a good design. Nonetheless, simulation should be carried out judiciously to extract data that are truly needed or to assess the performance of a PI implementation. All-inclusive simulation of the whole plant and of any emerging idea is unwarranted and is counterproductive as it diverts important human resources from the necessary work.
- **Inappropriate accounting for uncertainties and dynamic changes.** Process data and performance are uncertain. Also, dynamic changes are inevitable in the process. The base-case tools presented in this book should be used to carry out a sensitivity analysis to examine the effect of which uncertainties and dynamic changes should be accounted for. As an example, let us consider a direct-recycle application for which the material recycle pinch diagram is shown by Fig. 23.1a (as described in Chapter 4). This

TABLE 23-1 Examples of Industrial Applications

Type of Process	Project Objectives	Motivation	Approach	Key Results
Chemical and Polymer Complex	• Reduction of losses of volatile organic compounds (VOCs) • Reduction of thermal load of wastewater effluent	Meet environmental regulations, reduce losses, and allow future expansions	Heat and mass integration	• VOC losses reduced by 50% • Thermal load reduced by 90% • Payback period: 1.5 years
Specialty Chemicals Process	• Water-usage reduction • Solvent-usage reduction • Yield enhancement	Profitability enhancement and debottlenecking	Mass integration with direct recycle and interception	• Water-usage reduction: 33% • Solvent-usage reduction: 25% • Yield enhancement: 8%
Specialty Chemicals Process	Debottlenecking of the process and hydrogen management	Sold out product with no additional capacity and significant cost for hydrogen consumption	Systematic elimination of two primary bottlenecks and sitewide integration of hydrogen generation, usage, and discharge	12% additional capacity and 25% reduction in hydrogen cost with a payback period of less than one year
Kraft Pulping Process	Water management and conservation	High usage of water and buildup of nonprocess elements upon recycle	Sitewide tracking of water and nonprocess elements followed by a mass integration study for water minimization	Key results: 55% reduction in water usage with a payback period of less than two years
Resin Production Facility	Production debottlenecking	Sold out product with more market demands but a capped production capacity (bottleneck)	Mass integration techniques to determine subtle causes of process bottlenecks and eliminate them at minimum cost	Increase in capacity by process debottlenecking: 4% (>$1million/yr additional revenue)
Organic Chemicals Production Process	Identification of sitewide water stream recycle opportunities to reduce river water discharges	Pressure from local environmentalists and the need to meet more stringent environmental permit requirements	Sitewide tracking of water followed by a mass integration study for water recycle opportunities and potential land treatment and reverse osmosis treatment of select wastewater streams	Nine process designs selected for implementation, including one separation system resulting in 670 gpm wastewater reduction with a payback period of one year
Polymer and Monomer Production Processes	Identification of sitewide energy conservation opportunities to reduce energy costs	Reduction in operating costs for manufacturing processes and the need for additional steam generation for production capacity expansion	Sitewide tracking of energy usage followed by a heat integration study to identify energy conservation opportunities	A heat-exchange network and utility optimization process design implemented resulting in a 10% reduction in site utility costs, a 10% reduction in site wastewater hydraulic load, and a 5% production capacity increase. Annual savings are in excess of $2.5 million/year
Specialty Chemicals Production Process	Identification of sitewide energy conservation opportunities to reduce energy costs	High operating costs for utilities	Sitewide tracking of energy usage followed by a heat integration study to identify energy conservation opportunities	Five process designs implemented leading to a 25% reduction in energy usage with a payback period of less than one year
Metal Finishing Process	Reduction of cost of industrial solvent	Major solvent losses leading to a large operating cost and environmental problems	Synthesis of an energy-efficient heat-induced separation network	Recovery of 80% of lost solvent with a payback period of three years
Papermaking Process	Recovery of lost fibers and management of water system	7% losses of purchased fibers during processing and high usage of water	Integrated matching of properties of broken fibers with demands of paper machines (property integration)	Recovery and reuse of 60% of lost fibers and reduction in water usage by 30% with a payback period of less than one year
Polymer Production Processes	Identification of sitewide wastewater stream recycle opportunities	Future expansion (wastewater discharge system expected to exceed its maximum capacity during the production process expansion)	Sitewide tracking of water followed by a mass integration study for water recycle opportunities and reverse osmosis treatment of select wastewater streams	24 process designs implemented, resulting in a 30% reduction in site wastewater discharge and with a payback period of less than one year
Petrochemical Facility	Develop power cogeneration strategies and optimize utility systems	Significant usage of steam for process uses and high cost of power usage	Energy integration with emphasis on combined heat and power optimization	25% reduction in steam cost and cogeneration of 20% of power requirement for the process. Payback period is four years

Source: Taken from El-Halwagi and Spriggs (1998) and Dunn and El-Halwagi (2003).

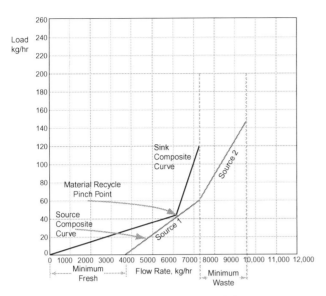

FIGURE 23-1A Base-case material recycle pinch diagram.

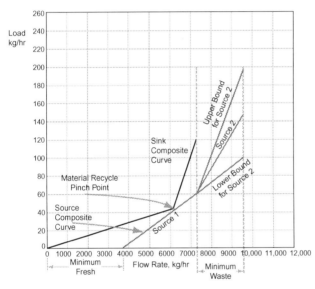

FIGURE 23-1B The material recycle pinch diagram under uncertainty.

diagram is constructed using the nominal or base-case data. Suppose that there is an uncertainty or a potential dynamic change in the concentration of Source 2 between a lower and an upper bound. The diagram can be drawn to account for the potential variation in concentration as shown by Fig. 23.1b. As can be seen, these variations have no impact on the targets for minimum fresh and minimum waste. As such, trying to firm up the data for Source 2 is unwarranted. On the other hand, any change in the concentration of Source 1 will shift the pinch. Therefore, effort should be focused on the data for Source 1 not Source 2.

- **Bypassing the targeting step.** Benchmarking the process performance ahead of detailed design is a hallmark of successful PI projects. Not only does targeting provide the benchmarks but it also aids designers in deciding which objectives to prioritize based on the target values (for example, which targeted species, heat integration versus mass integration, and so on).
- **Unreconciled short- and long-term actions.** In implementing PI projects, it is necessary to develop a consistent plan of action whereby the short-term implementations conform to the overall long-term plan.
- **Lack of appropriate in-depth knowledge.** A mere belief in PI without appropriate grounding in the fundamentals and tools can be dangerous. Well-intentioned "integration" projects that violate cardinal rules of PI can be counterproductive. Example 4.2 gives an example of such situations where a seemingly good project for water reduction turned out to have negative consequences that prevent the plant from reaching its true potential.
- **Insufficient consideration of constraints and objectives.** In extracting the data and setting up the PI framework, it is crucial to include the necessary business, safety, health, environmental, and technical constraints and objectives. Enough communications must be established with the people in charge of these areas to ensure that the evolving solutions are implementable.

STARTING AND SUSTAINING PI INITIATIVES AND PROJECTS

Tools, procedures, and insights are very important in applying PI but are not sufficient in starting and sustaining PI initiatives and projects. In this context, personal initiatives, human resources, work processes, and working cultures are all necessary. The following are key building blocks that are needed to start and sustain a successful PI initiative (Dunn and El-Halwagi, 2003):

- Understand the company's short- and long-term goals and align your PI initiative with them.
- Before selecting a specific set of problems and projects, perform a preliminary targeting analysis to determine where opportunities exist, their potential magnitude, and priority areas of work. At this stage, use approximate data that are appropriate for a targeting exercise. Remember that targets are upper bounds on performance. Although targets are excellent metrics to identify and prioritize opportunities, actual implementation may not reach the exact targets.
- Review any relevant work undertaken in the company to be aware of previous efforts and to avoid redundancy.
- Define the tasks needed and the required human, technical, and financial resources.
- Describe anticipated constraints, unique aspects of the process, and potential challenges.
- Get enthusiastic support from senior management.
- Get buy-in and enthusiastic support from managers, process engineers, and operators.
- Recruit local champions from among the process experts and the stakeholders.
- Form task-driven teams that involve the right members.
- Encourage an open environment that fosters creativity and out-of-the-box integrated thinking where the dominating culture is "how do we make it happen?"

TABLE 23-2 Examples of Discouraging Attitudes about Process Integration and Suggested Responses

Discouraging Attitudes	Response
We don't have the resources to support this process integration initiative.	Let us create resources that match the anticipated results or let us do the best we can within the available resources.
We have tried something similar before and it did not work.	Let us study the previous effort and see indeed if no more progress can be made.
These concepts will not work in my plant. We have a very unique operation.	There is now a track record of many successful process integration projects that have been applied to a wide variety of industrial processes, each of which is unique in its own right.
Has anyone else applied it before?	See the previous response.
Our process is too big/too small for this approach.	See the previous response.
I am the process expert; there is no way that someone else can do better.	Let us incorporate your experience in a process integration framework. Time and again, track record has indicated that when proper process experience is incorporated into a process integration framework, significant and intuitively nonobvious benefits have accrued.
You really don't understand the issues and problems that we face.	See the two previous responses.
Sounds great but you need to speak to someone else.	Get suggestions on "the someone else" but also see if there is a legitimate role for the individual.
I don't wish to participate in an initiative where I don't feel comfortable with the tools and techniques.	Provide appropriate training to develop the proper comfort level and understanding.
Not now! We will include it in our long-term strategic planning.	Each day without process integration implies missed opportunities.

Source: Dunn and El-Halwagi (2003).

instead of "why it won't work." It is critical in the early stages of a PI project to "grow the tree" of ideas. Later, this tree will be trimmed and the appropriate projects will be defined with all the necessary details.
- Measure, analyze, use process integration tools to develop, improve, synthesize, offer feedback, refine, and sustain projects and strategies. Do not collect more data than you should. The tools covered in this book are associated with certain data that enable the analysis to proceed forward. Typically, little data are needed in the beginning and, as the PI analysis proceeds, specific data are needed as guided by PI.
- Develop aggregate and composite representations of the process and the solutions (as described throughout this book). Present the proposed changes in a way that focuses on gained insights, is easy to follow, and highlights the key characteristics of the findings.
- Focus on the key findings and the broad issues that are emerging.
- Consult with relevant individuals all along to capture process know-how, and ensure that appropriate details are included and hurdles are overcome. This is particularly important in detailing the solutions and in getting buy-in and approval of the proposed changes.

Another important aspect is to take initiative and not be discouraged by resisting responses. Table 23.2 presents a list of the "top ten" discouraging statements and attitudes described by Dunn and El-Halwagi (2003).

Finally, the gap between the status quo and the true potential of the process is typically large, and there are tremendous opportunities for continuous process improvement on the road to sustainability. Indeed, you can create win-win situations that enhance the economic metrics of the process while improving its technical and environmental performance. A combination of the right framework, tools, approach, work process, human resources, mind-set, and attitude is a recipe for success. It is hoped that you use the PI platform to launch initiatives that will effect transformative changes in the process industries. Such initiatives are good for the business and will ultimately lead to a more sustainable world. You can make a difference, so please do!

REFERENCES

Dunn RF, El-Halwagi MM: Process integration technology review: background and applications in the chemical process industry, *J Chem Tech Biotech* 78:1011–1021, 2003.

El-Halwagi MM, Spriggs HD: Solve design puzzles with mass integration, *Chem Eng Prog* 94(August):25–44, 1998.

APPENDIX I

CONVERSION RELATIONSHIPS FOR CONCENTRATIONS AND CONVERSION FACTORS FOR UNITS

This appendix provides:

- Basic relationships for converting concentrations.
- Key conversion factors for different sets of units.

BASIC RELATIONSHIPS FOR CONVERTING CONCENTRATIONS

Pollutant concentration may be reported in various ways. It is very important to be able to relate the different composition units.

MASS VERSUS MOLAR COMPOSITIONS

Mass can be related to moles via molecular weight; that is,

$$[I.1] \qquad n_i = \frac{m_i}{M_i}$$

where

m_i = mass of species i
M_i = molecular weight of species i
n_i = number of moles of species i

Therefore, molar ratios of two species, i and j, can be related to their mass ratios as follows:

$$[I.2] \qquad \frac{n_i}{n_j} = \left(\frac{M_j}{M_i}\right)\left(\frac{m_i}{m_j}\right)$$

and mole fractions can be related to mass fractions as follows:

$$[I.3] \qquad y_i = \left(\frac{\bar{M}}{M_i}\right) x_i$$

where

y_i = mole fraction of species i
$\bar{M}$ = average molecular weight of mixture ($\sum_{all\ species} y_i M_i$)
x_i = mass fraction of species i

GAS COMPOSITION VERSUS PARTIAL PRESSURE

To relate partial pressures to compositions, one may use the ideal gas law:

$$[I.4a] \qquad p_i V = n_i RT$$

$$[I.4b] \qquad = \frac{m_i}{M_i} RT$$

where V is volume and p_i is partial pressure of species i, which can be related to total pressure of the gaseous mixture, P, via Dalton's law:

$$[I.5] \qquad y_i = \frac{p_i}{P}$$

The term R is the gas constant, which has the following values in different units:

8314.3 m³ Pa/kg-mole K
8.3143 m³ kPa/kg-mole K
8.3143 J/gm-mole K
0.082057 m³ atm/kg-mole K
0.082057 lit atm/gm-mole K
82.057 cm³ atm/gm-mole K
1.9872 cal/gm-mole K
1.9872 Btu/lb-mole °R
10.731 psia ft³/lb-mole °R
21.9 ft³ (inch Hg)/lb-mole °R
0.7302 atm ft³/lb-mole °R
1545.3 ft lb$_f$/lb-mole °R

Equation I.4b can be rearranged to give mass density of an ideal gas, ρ, as

$$[I.6] \qquad \rho_i = \frac{M_i}{V_i} = \frac{p_i M_i}{RT}$$

For instance, density of air at 1 atm and 298 K is

$$\rho_{air} = \frac{1 \times 29}{0.082057 \times 298} \approx 1.2 \text{ kg/m}^3$$

PARTS PER MILLION

In describing concentrations of pollutants in gaseous and aqueous wastes, a commonly confusing issue is the definition of parts per million (ppm). Sometimes, it is specified whether these are compositions based on weight, moles, or volume (ppmw, ppmm, or ppmv, respectively). However, in most cases, it is reported as ppm. Depending on the phase of the stream, ppm implies the basis of weight for liquids and mole or volume for gases as explained in the following sections.

For gases:

[I.7a]

$$\text{ppm of species } i = \frac{\text{mole of gaseous species } i}{\text{moles of gaseous mixture}} \times 10^6$$

[I.7b] $\quad = $ mole fraction of species $i \times 10^6$

where the mole fraction of species i in a gaseous mixture may be evaluated via Eq. I.5. Hence, one can combine Equations I.4 to I.7 to develop the following expression for converting from ppm in the gas phase to mass concentration of species i:

[I.8] $\quad c_i = \dfrac{m_i}{V} = \dfrac{(\text{ppm of species } i)\, PM_i \times 10^{-6}}{RT}$

For instance, 1 ppm of ozone (O_3, molecular weight = 48) in air at 1 atm and 298 K can be converted into kg/m^3 as follows:

$$\text{mass concentration (kg/m}^3) = \dfrac{1 \times 1 \times 48 \times 10^{-6}}{0.082057 \times 298}$$
$$= 1.96 \times 10^{-6} \text{ kg/m}^3$$
$$(\text{or } 1960\, \mu\text{g/m}^3)$$

For liquids:

[I.9a] $\quad \text{ppm} = \dfrac{\text{mass of pollutant}}{\text{mass of liquid mixture}} \times 10^6$

[I.9b] $\quad = $ mass fraction of pollutant in liquid $\times 10^6$

Of particular importance is the case of wastewater with dilute pollutants at ambient temperature. In this case, one can use the following conversion from ppm to mass concentration:

[I.10a] $\quad\quad 1\,\text{ppm} = 1\,\text{gm/m}^3$

[I.10b] $\quad\quad\quad\quad = 1\,\text{mg/lit}$

KEY CONVERSION FACTORS FOR DIFFERENT SETS OF UNITS

Length:
- 1 km = 1000 m
- 1 m = 100 cm
- 1 cm = 10^{-2} m
- 1 mm = 10^{-3} m
- 1 micron (μ) = 10^{-6} m
- 1 angstrom (A) = 10^{-10} m
- $\quad\quad\quad\quad\quad$ = 0.1 NM
- 1 in. = 2.540 cm
- 1 ft = 30.48 cm
- 1 mile = 1.609 km
- 1 centimeter = 0.3937 in.
- 1 meter = 39.37 in.

Volume:
- 1 L (Liter) = 1000 cm^3
- 1 m^3 = 1000 L
- $\quad\quad$ = 264.17 U.S. gal
- 1 U.S. gal = 4 qt
- $\quad\quad\quad\quad$ = 3.7854 L
- 1 Imperial (UK) gal = 1.201 U.S. gal
- $\quad\quad\quad\quad\quad\quad\quad$ = 4.54609 L
- $\quad\quad\quad\quad\quad\quad\quad$ = 277.4 in^3
- 1 ft^3 = 28.317 L
- $\quad\quad$ = 0.028317 m^3
- $\quad\quad$ = 7.481 U.S. gal
- 1 in.^{3} = 16.387 cm^3
- 1 bbl (barrel) = 5.6146 ft^3
- $\quad\quad\quad\quad\quad$ = 42 U.S. gal
- $\quad\quad\quad\quad\quad$ = 0.15899 m^3

Mass:
- 1 lb$_m$ = 453.59 gm
- $\quad\quad$ = 0.45359 kg
- $\quad\quad$ = 16 oz
- $\quad\quad$ = 7000 grains
- 1 kg = 1000 gm
- $\quad\quad$ = 2.2046 lb$_m$
- 1 tonne = 1000 kg
- 1 ton (short) = 2000 lb$_m$
- 1 ton (long) = 2240 lb$_m$

Density:
- 1 lb$_m$/ft^3 = 16.018 kg/m^3
- 1 kg/m^3 = 0.062428 lb$_m$/ft^3
- 1 gm/cm^3 = 62.427961 lb$_m$/ft^3
- 1 Grain/ft^3 = 2.288 gm/m^3
- lb/gal(US) = 0.11983 gm/cm^3
- $\quad\quad\quad\quad$ = 119.83 kg/m^3

Pressure:
- 1 atm = 760 mm Hg at 0°C (millimeters mercury pressure)
- $\quad\quad$ = 29.921 in. Hg at 0°C
- $\quad\quad$ = 14.696 psia
- $\quad\quad$ = 1.01325 $\times$ 10^5 N/m^2
- $\quad\quad$ = 1.01325 $\times$ 10^5 Pa (pascal)
- $\quad\quad$ = 101.325 kPa
- 1 bar = 1 $\times$ 10^5 N/m^2
- $\quad\quad$ = 100 kPa
- 1 psia = 6.89476 $\times$ 10^3 N/m^2

Temperature:

To convert from	To	Use this formula
Celsius	Fahrenheit	°F = °C $\times$ 1.8 + 32
Celsius	Kelvin	K = °C + 273.15
Celsius	Rankine	R = °C $\times$ 1.8 + 32 + 459.67
Fahrenheit	Celsius	°C = (°F − 32)/1.8
Fahrenheit	Kelvin	K = (°F + 459.67)/1.8
Fahrenheit	Rankine	R = °F + 459.67
Kelvin	Celsius	°C = K − 273.15
Kelvin	Fahrenheit	°F = K $\times$ 1.8 − 459.67
Kelvin	Rankine	R = K $\times$ 1.8
Rankine	Celsius	°C = (R − 32 − 459.67)/1.8
Rankine	Fahrenheit	°F = R − 459.67
Rankine	Kelvin	K = R/1.8

Energy:

1 J (Joule)	= 1 N.m
	= 1 kg.m^2/s^2
	= 10^7 gm.cm^2/s^2 (erg)
	= 9.48 × 10^{-4} Btu
	= 0.73756 lb$_f$.ft
1 cal (thermochemical)	= 4.1840 J
1 kWh	= 3.6 MJ
1 Btu	= 252.16 cal (thermochemical)
	= 778.17 lb$_f$.ft
	= 1.05506 kJ
1 quadrillion Btu	= 10^{15} Btu
	= 2.93 3 1011 kWh
	= 172 × 10^6 barrels of oil equivalent
	= 36 × 10^6 metric tons of coal equivalent
	= 0.93 × 10^{12} cubic feet of natural gas equivalent

Power:

1 W (watt)	= 1 J/s
1 kW	= 1000 W
1 MW	= 10^6 W
1 hp (horsepower)	= 0.74570 kW
1 quadrillion Btu/year	= 0.471 × 10^6 barrels of oil equivalent/day
1 Btu/hr	= 0.293071 W
1 ton of refrigeration	= 12,000 Btu/hr
	= 2.51685 kW

APPENDIX II: MODELING OF MASS-EXCHANGE UNITS FOR ENVIRONMENTAL APPLICATIONS

Mass-exchange operations are indispensable for pollution prevention. Within a mass-integration framework, mass-exchange operations are employed in intercepting sources by selectively transferring certain undesirable species from a number of waste streams (sources) to a number of mass-separating agents (MSAs). The objective of this appendix is to provide an overview of the basic modeling principles of mass-exchange units. For a more comprehensive treatment of the subject, readers are referred to Seader et al. (2010), Benitez (2009), Wankat (2006), Geankoplis (2003), McCabe et al. (2000), King (1980), and Treybal (1980).

WHAT IS A MASS EXCHANGER?

A mass exchanger is any direct-contact mass-transfer unit that employs an MSA (or a lean phase) to selectively remove certain components (for example, pollutants) from a rich phase (for example, a waste stream). The MSA should be partially or completely immiscible in the rich phase. When the two phases are in intimate contact, the solutes are redistributed between the two phases, leading to a depletion of the rich phase and an enrichment of the lean phase. Although various flow configurations may be adopted, emphasis will be given to countercurrent systems because of their industrial importance and efficiency. The realm of mass exchange includes the following operations:

Absorption. A process in which a liquid solvent is used to remove certain compounds from a gas by virtue of their preferential solubility. Examples of absorption involve desulfurization of flue gases using alkaline solutions or ethanolamines, recovery of volatile organic compounds using light oils, and removal of ammonia from air using water.

Adsorption. A process that utilizes the ability of a solid adsorbent to adsorb specific components from a gaseous or a liquid solution onto its surface. Examples of adsorption include the use of granular activated carbon for the removal of benzene/toluene/xylene mixtures from underground water, the separation of ketones from aqueous wastes of an oil refinery, and the recovery of organic solvents from the exhaust gases of polymer manufacturing facilities. Other examples include the use of activated alumina to adsorb fluorides and arsenic from metal-finishing emissions.

Extraction. A process that employs a liquid solvent to remove certain compounds from another liquid using the preferential solubility of these solutes in the MSA. For instance, wash oils can be used to remove phenols and polychlorinated biphenyls (PCBs) from the aqueous wastes of synthetic-fuel plants and chlorinated hydrocarbons from organic wastewater.

Ion exchange. A process in which cation and/or anion resins are used to replace undesirable anionic species in liquid solutions with nonhazardous ions. For example, cation-exchange resins may contain nonhazardous, mobile, positive ions (for example, sodium, hydrogen) that are attached to immobile acid groups (for example, sulfonic or carboxylic). Similarly, anion-exchange resins may include nonhazardous, mobile, negative ions (for example, hydroxyl or chloride) attached to immobile basic ions (for example, amine). These resins can be used to eliminate various species from wastewater, such as dissolved metals, sulfides, cyanides, amines, phenols, and halides.

Leaching. A process that is the selective solution of specific constituents of a solid mixture when brought in contact with a liquid solvent. It is particularly useful in separating metals from solid matrices and sludge.

Stripping. A process that corresponds to the desorption of relatively volatile compounds from liquid or solid streams using a gaseous MSA. Examples include the recovery of volatile organic compounds from aqueous wastes using air, the removal of ammonia from the wastewater of fertilizer plants using steam, and the regeneration of spent activated carbon using steam or nitrogen.

EQUILIBRIUM

Consider a lean phase, j, that is in intimate contact with a rich phase, i, in a closed vessel in order to transfer a certain solute. The solute diffuses from the rich phase to the lean phase. Meanwhile, a fraction of the diffused solute back-transfers to the rich phase. Initially, the rate of rich-to-lean solute transfer surpasses that of lean to rich leading to a net transfer of the solute from the rich phase to the lean phase. However, as the concentration of the solute in the rich phase increases, the back-transfer rate also increases. Eventually, the rate of rich-to-lean solute transfer becomes equal to that of lean to rich, resulting in a dynamic equilibrium with zero net interphase transfer. Physically, this situation corresponds to the state at which both phases have the same value of chemical potential for the solute. In the case of ideal systems, the transfer of one component is indifferent to the transfer of other species. Hence, the composition of the solute in the rich phase, y_i, can be related to its composition in the lean phase, x_j, via an equilibrium distribution function, f_j^*, which is a function of the system characteristics including temperature and pressure. Hence, for a given rich-stream composition, y_i, the maximum attainable composition of the solute in the lean phase, x_j^*, is given by

$$[\text{II.1}] \qquad y_i = f_j^*(x_j^*).$$

Many environmental applications involve dilute systems whose equilibrium functions can be linearized over the operating range to yield

$$[\text{II.2}] \qquad y_i = m_j x_j^* + b_j$$

Special cases of Eq. II.2 include Raoult's law for absorption:

$$[\text{II.3}] \qquad y_i = \frac{p_{solute}^o(T)}{P_{total}} x_j^*,$$

where y_i and x_j^* are the mole fractions of the solute in the gas and the liquid phases, respectively, $p_{solute}^o(T)$ is the vapor pressure of the solute at a temperature T, and P_{total} is the total pressure of the gas.

Another example of Eq. II.2 is Henry's law for stripping:

$$[\text{II.4}] \qquad y_i = H_j x_j^*,$$

where y_i and x_j^* are the mole fractions of the solute in the stripping gas and the liquid waste, respectively,[1] and H_j is Henry's coefficient, which may be theoretically approximated by the following expression:

$$[\text{II.5}] \qquad H_j = \frac{P_{total} \cdot y_i^{solubility}}{p_{solute}^o(T)},$$

in which $p_{solute}^o(T)$ is the vapor pressure of the solute at a temperature T, P_{total} is the total pressure of the stripping gas, and $y_i^{solubility}$ is the liquid-phase solubility of the pollutant at temperature T (expressed as mole fraction of the pollutant in the liquid waste).

An additional example of Eq. II.2 is the distribution function commonly used in solvent extraction:

$$[\text{II.6}] \qquad y_i = K_j x_j^*,$$

where y_i and x_j^* are the compositions of the pollutant in the liquid waste and the solvent, respectively, and K_j is the distribution coefficient.

Accurate experimental results provide the most reliable source for equilibrium data. If not available, empirical correlations for predicting equilibrium data may be invoked. These correlations are particularly useful at the conceptual-design stage. Several literature sources provide compilations of equilibrium data and correlations: for example, Lo et al. (1983) as well as Francis (1963) for solvent extraction, Reid et al. (1987) for vapor-liquid and liquid-liquid systems, Hwang et al. (1992) for steam stripping, Mackay and Shiu (1981), Fleming (1989), Clark (1990), and Yaws (1992) for air stripping, U.S. Environmental Protection Agency (1980), Cheremisinoff and Ellerbusch (1980), Perrich (1981), Yang (1987), Valenzuela and Myers (1989), Stenzel and Merz (1989), Stenzel (1993), and Yaws et al. (1995) for adsorption, Kohl and Riesenfeld (1985) for gas adsorption and absorption, and Astarita et al. (1983) for reactive absorption.

INTERPHASE MASS TRANSFER

Whenever the rich and the lean phases are not in equilibrium, an interphase concentration gradient and a mass-transfer driving force develop, leading to a net transfer of the solute from the rich phase to the lean phase. A common method of describing the rates of interphase mass transfer involves the use of overall mass-transfer coefficients that are based on the difference between the bulk concentration of the solute in one phase and its equilibrium concentration in the other phase. Suppose that the bulk concentrations of a pollutant in the rich and the lean phases are y_i and x_j, respectively. For the case of linear equilibrium, the pollutant concentration in the lean phase, which is in equilibrium with y_i, is given by

$$[\text{II.7}] \qquad x_j^* = (y_i - b_j)/m_j$$

and the pollutant concentration in the rich phase, which is in equilibrium with x_j, can be represented by

$$[\text{II.8}] \qquad y_i^* = m_j x_j + b_j.$$

Let us define two overall mass-transfer coefficients: one for the rich phase, K_y, and one for the lean phase, K_x. Hence, the rate of interphase mass transfer for the pollutant $N_{pollutant}$ can be defined as

$$[\text{II.9a}] \qquad N_{pollutant} = K_y(y_i - y_i^*)$$

$$[\text{II.9b}] \qquad = K_x(x_j^* - x_j).$$

Correlations for estimating overall mass-transfer coefficients can be found in Seader et al. (2010), Benitez (2009), Green and Perry (2007), Wankat (2006), Geankoplis (2003), McCabe et al. (2000), King (1980), and Treybal (1980).

TYPES AND SIZES OF MASS EXCHANGERS

The main objective of a mass exchanger is to provide appropriate contact surface for the rich and the lean phases. Such contact can be accomplished by using various types of mass-exchange units and internals. In particular, there are two primary categories of mass-exchange

---

[1] Throughout this book, several mass-exchange operations are considered simultaneously. It is therefore necessary to use a unified terminology such that y is always the composition in the rich phase and x is the composition in the lean phase. Readers are cautioned here that this terminology may be different from other literature, in which y is used for gas-phase composition and x is used for liquid-phase composition.

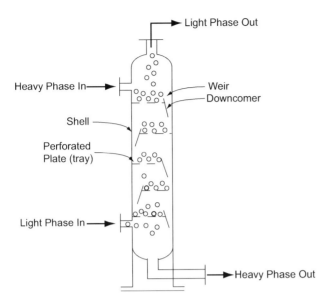

FIGURE II-1 A multistage tray column.

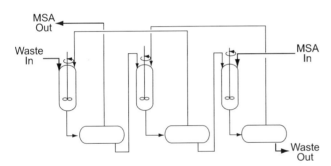

FIGURE II-2 A three-stage mixer-settler system.

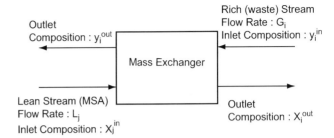

FIGURE II-3 A mass exchanger.

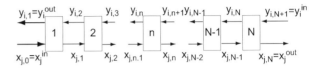

FIGURE II-4 A schematic diagram of a multistage mass exchanger.

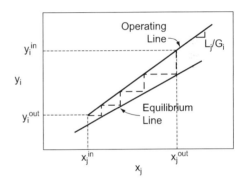

FIGURE II-5 The McCabe-Thiele diagram.

devices: multistage and differential contactors. In a **multistage mass exchanger,** each stage provides intimate contact between the rich and the lean phases followed by phase separation. Because of the thorough mixing, the pollutants are redistributed between the two phases. With sufficient mixing time, the two phases leaving the stage are essentially in equilibrium, hence the name *equilibrium stage*. Examples of a multiple-stage mass exchanger include tray columns (Fig. II.1) and mixer settlers (Fig. II.2).

To determine the size of a multiple-stage mass exchanger, let us consider the isothermal mass exchanger shown in Fig. II.3. The rich (waste) stream, i, has a flow rate G_i and its content of the pollutant must be reduced from an inlet composition, y_i^{in}, to an outlet composition, y_i^{out}. An MSA (lean stream), j, (whose flow rate is L_j, inlet composition is x_j^{in}, and outlet composition is x_j^{out}) flows countercurrently to selectively remove the pollutant.[2] Figure II.4 is a schematic representation of the multiple stages of this mass exchanger. If the nth block is an equilibrium stage, then the compositions $y_{i,n}$ and $x_{j,n}$ are in equilibrium. On the other hand, the two compositions on the same end of the stage (for example, $y_{i,n-1}$ and $x_{j,n}$) are said to be operating with each other.

One way of calculating the number of equilibrium stages (or number of theoretical plates, NTP) for a mass exchanger is the graphical McCabe-Thiele method. To illustrate this procedure, let us assume that over the operating range of compositions, the equilibrium relation governing the transfer of the pollutant from the waste stream to the MSA can be represented by the linear expression described by Eq. II.2. A material balance on the pollutant that is transferred from the waste stream to the MSA may be expressed as

$$[II.10] \quad G_i(y_i^{in} - y_i^{out}) = L_j(x_j^{out} - x_j^{in}).$$

On a y-x (McCabe-Thiele) diagram, this equation represents the operating line that extends between the points (y_i^{in}, x_j^{out}) and (y_i^{out}, x_j^{in}) and has a slope of L_j/G_i, as shown in Fig. II.5. Furthermore, each theoretical stage can be represented by a step between the operating line and the equilibrium line. Hence, NTP can be determined by "stepping off" stages between the two ends of the exchanger, as illustrated by Fig. II.5.

---

[2] Once again, because of the unified approach of this text to all mass-exchange operations, it is important to emphasize that the symbols G and L will be used to designate the flow rates of the rich stream and the MSA, respectively, as compared to flow rates of gas and liquid.

Alternatively, for the case of isothermal, dilute mass exchange with linear equilibrium, NTP can be determined through the Kremser (1930) equation:

[II.11]
$$NTP = \frac{\ln\left[\left(1 - \frac{m_j G_i}{L_j}\right)\left(\frac{y_i^{in} - m_j x_j^{in} - b_j}{y_i^{out} - m_j x_j^{in} - b_j}\right) + \frac{m_j G_i}{L_j}\right]}{\ln\left(\frac{L_j}{m_j G_i}\right)}$$

Other forms of the Kremser equation include

[II.12]
$$NTP = \frac{\ln\left[\left(1 - \frac{L_j}{m_j G_i}\right)\left(\frac{x_i^{in} - x_j^{out,*}}{x_i^{out} - x_j^{out,*}}\right) + \frac{L_i}{m_j G_i}\right]}{\ln\left(\frac{m_j G_i}{L_j}\right)},$$

where

[II.13]
$$x_j^{out,*} = \frac{y_i^{in} - b_j}{m_j}.$$

Also,

[II.14]
$$\frac{y_i^{in} - m_j x_j^{out} - b_j}{y_i^{out} - m_j x_j^{in} - b_j} = \left(\frac{L_j}{m_j G_i}\right)^{NTP}.$$

If contact time is not enough for each stage to reach equilibrium, one may calculate the number of actual plates (NAP) by incorporating mass-transfer efficiency. Two principal types of efficiency may be employed: overall and stage. The overall exchanger efficiency, η_o, can be used to relate NAP and NTP as follows:

[II.15]
$$NAP = NTP/\eta_o.$$

The stage efficiency may be defined based on the rich phase or the lean phase. For instance, when the stage efficiency is defined for the rich phase, η_y, Eq. II.11 becomes

[II.16]
$$NTP = \frac{\ln\left[\left(1 - \frac{m_j G_i}{L_j}\right)\left(\frac{y_i^{in} - m_j x_j^{in} - b_j}{y_i^{out} - m_j x_j^{in} - b_j}\right) + \frac{m_j G_i}{L_j}\right]}{-\ln\left\{1 + \eta_y\left[\left(\frac{m_j G_i}{L_j}\right) - 1\right]\right\}}.$$

The second type of mass-exchange unit is the **differential (or continuous) contactor**. In this category, the two phases flow through the exchanger in continuous contact throughout, without intermediate phase separation and recontacting. Examples of differential contactors include packed columns (Fig. II.6), spray towers (Fig. II.7), and mechanically agitated units (Fig. II.8).

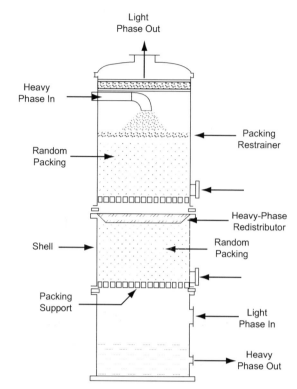

FIGURE II-6 Schematic diagram of a countercurrent packed column.

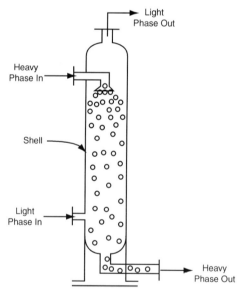

FIGURE II-7 A spray column.

The height of a differential contactor, H, may be estimated using

[II.17a]
$$H = HTU_y NTU_y$$

[II.17b]
$$= HTU_x NTU_x,$$

where HTU_y and HTU_x are the overall height of transfer units based on the rich and the lean phases, respectively, while NTU_y and NTU_x are the overall number of transfer units based on the rich and the lean phases, respectively. The overall height of a transfer unit may be provided by

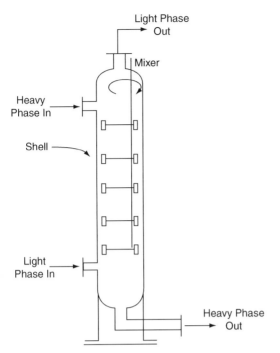

FIGURE II-8 A mechanically agitated mass exchanger.

the packing (or unit) manufacturer or estimated using empirical correlations (typically by dividing superficial velocity of one phase by its overall mass transfer coefficient). On the other hand, the number of transfer units can be theoretically estimated for the case of isothermal, dilute mass exchangers with linear equilibrium as follows:

[II.18a]
$$NTU_y = \frac{y_i^{in} - y_i^{out}}{(y_i - y_i^*)_{\log mean}},$$

where

[II.18b]
$$(y_i - y_i^*)_{\log mean} = \frac{\left(y_i^{in} - m_j x_j^{out} - b_j\right) - \left(y_i^{out} - m_j x_j^{in} - b_j\right)}{\ln\left[\dfrac{y_i^{in} - m_j x_j^{out} - b_j}{y_i^{out} - m_j x_j^{in} - b_j}\right]}$$

and

[II.19a]
$$NTU_x = \frac{x_j^{in} - x_j^{out}}{(x_j - x_j^*)_{\log mean}},$$

where

[II.19b]
$$(x_j - x_j^*)_{\log mean} = \frac{\left[x_j^{out} - \left(\dfrac{y_i^{in} - b_j}{m_j}\right)\right] - \left[x_j^{in} - \left(\dfrac{y_i^{out} - b_j}{m_j}\right)\right]}{\ln\left\{\dfrac{\left[x_j^{out} - \left(\dfrac{y_i^{in} - b_j}{m_j}\right)\right]}{\left[x_j^{in} - \left(\dfrac{y_i^{out} - b_j}{m_j}\right)\right]}\right\}}.$$

If the terminal compositions or L_j/G_i are unknown, it is convenient to use the following form:

[II.20]
$$NTU_y = \frac{\ln\left[\left(1 - \dfrac{m_j G_i}{L_j}\right)\left(\dfrac{y_i^{in} - m_j x_j^{in} - b_j}{y_i^{out} - m_j x_j^{in} - b_j}\right) + \dfrac{m_j G_i}{L_j}\right]}{1 - \left(\dfrac{m_j G_i}{L_j}\right)}.$$

The column diameter is normally determined by selecting a superficial velocity for one (or both) of the phases. This velocity is intended to ensure proper mixing while avoiding hydrodynamic problems such as flooding, weeping, or entrainment. Once a superficial velocity is determined, the cross-sectional area of the column is obtained by dividing the volumetric flow rate by the velocity.

MINIMIZING COST OF MASS-EXCHANGE SYSTEMS

Chapter Two provides an overview of process economics. Two principal categories of expenditure are particularly important: fixed and operating costs. The fixed cost ($) can be distributed over the service life of the equipment as an annualized fixed cost ($/yr). The total annualized cost (TAC) of the system is given by

[II.21]

TAC = annual operating cost + annualized fixed cost.

To minimize TAC, one can iteratively vary mass-transfer driving force to trade off annualized fixed cost versus annual operating cost. To demonstrate this concept, let us consider the isothermal mass exchanger shown in Fig. II.3. The rich (waste) stream, i, has a given flow rate G_i and a known inlet composition of pollutant y_i^{in}. The outlet composition of the pollutant in the waste stream is denoted by y_i^{out}. An MSA (lean stream) j that has a given inlet composition x_j^{in} but whose flow rate L_j is not fixed flows countercurrently to selectively remove the pollutant. Consider that over the operating range of compositions, the equilibrium relation governing the transfer of the pollutant from the waste stream to the MSA can be represented by the linear expression in Eq. II.2.

A material balance on the pollutant that is transferred from the waste stream to the MSA may be expressed as in Eq. II.10:

$$G_i(y_i^{in} - y_i^{out}) = L_j(x_j^{out} - x_j^{in})$$

On a y–x (McCabe-Thiele) diagram, this equation represents the operating line that extends between the points y_i^{in} and x_j^{out}, and y_i^{out} and x_j^{in}, and has a slope of L_j/G_i. Equation II.10 has three unknowns: y_i^{out}, L_j, and x_j^{out}. Let us first fix the outlet composition of the waste stream,

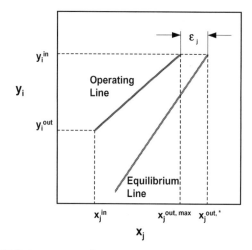

FIGURE II-9 Minimum allowable composition difference at the rich end of a mass exchanger.

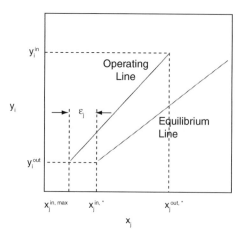

FIGURE II-10 Minimum allowable composition difference at the lean end of a mass exchanger.

y_i^{out}. As depicted by Fig. II.9, the maximum theoretically attainable outlet composition of the MSA, $x_j^{out,*}$, is in equilibrium with y_i^{in}. An infinitely large mass exchanger, however, will be needed to undertake this mass-transfer duty because of the vanishing mass-exchange driving force at the rich end of the unit. Hence, it is necessary to assign a minimum driving force between the operating and the equilibrium lines at the rich end of the exchanger (Fig. II.9) such that

[II.22] $$x_j^{out,max} = x_j^{out,*} - \varepsilon_j$$

but

[II.23] $$y_i^{in} = m_j x_j^{out,*} + b_j$$

Combining Equations II.21 and II.22, one obtains

[II.24] $$x_j^{out,max} = \frac{y_i^{in} - b_j}{m_j} - \varepsilon_j$$

where ε_j is referred to as the "minimum allowable composition difference," and $x_j^{out,max}$ is the maximum practically feasible outlet composition of the MSA that satisfies the assigned driving force, ε_j.

On the other hand, when x_j^{out} is fixed and x_j^{out} is left as an unknown, the minimum theoretically attainable inlet composition of the waste stream, $y_i^{out,*}$, is in equilibrium with x_j^{in} (Fig. II.10); that is,

[II.25] $$y_i^{out,*} = m x_j^{in} + b_j$$

By employing a minimum allowable composition difference of ε_j, at the lean end of the exchanger, one can identify the minimum practically feasible outlet composition of the waste stream to be $y_j^{out,min}$, which is given by

[II.26] $$y_i^{out,min} = m_j(x_j^{in} + \varepsilon_j) + b_j$$

The selection of the mass-transfer driving forces throughout the exchanger determines the trade-off between the fixed and the operating costs of the system. In many cases (particularly when the equilibrium line is either linear or convex, which is typical in environmental applications), the mass-transfer driving forces at both ends of the exchanger can be used to characterize the driving forces throughout the column. Therefore, the parameter ε_j provides a convenient way for trading off fixed versus operating costs of a mass exchanger. For instance, for given values of the inlet and outlet compositions of the rich stream and inlet composition of the lean stream, the value of ε_j at the rich end of the exchanger can be used to trade off fixed versus operating costs. As ε_j at the rich end of the exchanger is increased, the slope of the operating line (L_j/G_i) becomes larger and, consequently, the required flow rate of the MSA increases for a given value of the flow rate of the waste stream. Typically, the flow rate of the MSA is the most important element of the operating cost of a mass exchanger. As the flow rate of the MSA increases, additional makeup, regeneration, and materials handling (pumping, compression, and so on) are needed, leading to an increase in the operating cost of the system. On the other hand, when the value of ε_j at the rich end is increased, the column size (number of stages, height, and so on) decreases, which typically results in a reduction in the fixed cost of the mass exchanger. Hence, designers can iteratively vary ε_j at the rich end until the minimum TAC is identified (see Example II.2).

Example II-1 Removal of toluene from an aqueous waste

Air stripping is used to remove 90 percent of the toluene (molecular weight = 92) dissolved in a 10 kg/s (159 gpm) wastewater stream. The inlet composition of toluene in the wastewater is 500 ppm. Air (essentially free of toluene) is compressed to 202.6 kPa (2 atm) and bubbled through a stripper that contains sieve trays. To avoid fire hazards, the concentration of toluene in the air leaving the stripper is taken as 50 percent of the lower flammability limit (LFL) of toluene in air. The toluene-laden air exiting the stripper is fed to a condenser that recovers almost all the toluene. Figure II.11 shows a schematic representation of the process. Calculate the annual operating cost and the fixed capital investment for the system.

The following physical and economic data are available:

Physical data

- The stripping operation takes place isothermally at 298 K and follows Henry's law.
- The vapor pressure of toluene at 298 K is 3.8 kPa.
- The solubility of toluene in water at 298 K is 9.8×10^{-3} mol/mol%.
- Lower flammability limit for toluene in air is 12,000 ppm (Crowl and Louvar, 1990).

Stripper sizing criteria

- The overall efficiency of the column, η_o, is 0.25.
- Sieve plates are used with 0.41-m (16-inch) tray spacing.
- Height of column shell:

[II.27] $H = 1.3 \times \text{number of trays} \times \text{tray spacing}$

(that is, 30 percent additional height is used to allow for gas distribution, wastewater feed, gas–liquid disengagement, additional surge volume for wastewater, and so on).

- The maximum allowable superficial velocity of air in the stripper can be estimated using the following expression (based on the data provided by Peters and Timmerhaus [1991], for 0.41-m tray spacing):

[II.28]

$$\text{Maximum allowable superficial velocity of air (m/s)} = 0.068 \sqrt{\frac{\rho_{water} - \rho_{air}}{\rho_{air}}}.$$

Operating cost

- Operating cost is primarily attributed to air compression and toluene condensation.
- The operating cost for air compression is basically the electric utility needed for the isentropic compression. Electric energy needed to compress air may be calculated using the following expression:

[II.29] $\text{Compression Energy (kJ/kg)} = \left(\frac{\gamma}{\gamma - 1}\right)\left(\frac{RT_{in}}{M_{air}\eta_{isentropic}}\right)\left[\left(\frac{P_{out}}{P_{in}}\right)^{\left(\frac{\gamma-1}{\gamma}\right)} - 1\right],$

where $\gamma = 1.4$ and $\eta_{isentropic} = 0.60$. The electric energy cost is \$0.04/kWhr.

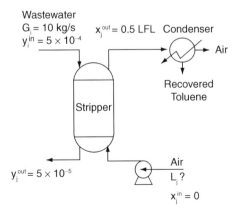

FIGURE II-11 Stripping of toluene from wastewater.

- The operating cost for the refrigerant needed for toluene condensation is 6.0×10^{-3}/kg of toluene-laden air.
- The system is to be operated for 8000 hr/annum.

Equipment cost

- The cost, \$, of the stripper (including installation and auxiliaries, but excluding the sieve trays) is given by

[II.30] $\text{Cost of column shell} = 1800H^{0.85}D^{0.95}$

where H is the column height (m) and D is the column diameter (m).

- The installed cost of a sieve tray, \$, is

[II.31] $\text{Cost} = 800D^{0.86}$

where D is the column diameter (m).

- The installed cost of the air blower, \$, is

[II.32] $\text{Cost} = 12,000L_j^{0.6}$

where L_j is the flow rate of air (kg/s).

- The installed cost of the condenser, \$, is

[II.33] $\text{Cost} = 62,000L_j^{0.6}$

where L_j is the flow rate of air (kg/s).

SOLUTION

Let us first determine the values of some of the physical properties needed in solving the problem.

The air density can be determined as follows (see Appendix I):

[II.34] $\rho_{air} = \frac{PM_{air}}{RT}.$

Therefore, density of air at 2 atm and 298 K is

$$\rho_{air} = \frac{2 \times 29}{0.082057 \times 298} \approx 2.4 \text{ kg/m}^3.$$

Because the problem will be worked out in mass units, it is necessary to evaluate compositions and Henry's constant on a mass basis.

The LFL of toluene is 12,000 ppm, which corresponds to a mole fraction of 0.012 (see Appendix I) and can be converted into mass fraction as follows:

$$\text{LFL (in mass fraction units)} = 0.012 \times \frac{92}{29}$$
$$= 0.038 \text{ kg toluene/kg air.}$$

Because the outlet composition of toluene in air is taken as 50 percent of the LFL, then

$$x_j^{out} = 0.5 \times 0.038$$
$$= 0.019 \text{ kg toluene/kg air.}$$

Inlet composition of wastewater is 500 ppm, which corresponds to a mass fraction of $y_i^{in} = 5 \times 10^{-4}$ (see Appendix I). Because 90 percent of the toluene is removed from the feed, $y_i^{out} = 5 \times 10^{-5}$.

The component material balance on toluene is

$$L_j(0.019 - 0.000) = 10 (5 \times 10^{-4} - 5 \times 10^{-5})$$

that is, $L_j = 0.237$ kg air/s

Using Eq. II.5, one may estimate Henry's constant in molar units (mole fraction of toluene in water/mole fraction of toluene in air) to be

$$\frac{9.8 \times 10^{-5} \times 202.6}{3.8}$$

$= 5.22 \times 10^{-3}$ mole fraction of toluene in water/mole fraction of toluene in air

(*Continued*)

Example II-1 Removal of toluene from an aqueous waste (*Continued*)

which can be converted into mass units (mass fraction of toluene in water/mass fraction of toluene in air) as follows:

$$m_j = 5.22 \times 10^{-3} \left(\frac{29}{18}\right)$$

$$= 0.0084 \frac{\text{mass fraction of toluene in water}}{\text{mass fraction of toluene in air}} \quad {}^3$$

Column sizing

We are now in a position to size the column. When Eq. II.11 is employed, we get

$$NTP = \frac{\ln\left[\left(1 - \frac{0.0084 \times 10}{0.237}\right)\left(\frac{5 \times 10^{-4} - 0}{5 \times 10^{-5} - 0}\right) + \frac{0.0084 \times 10}{0.237}\right]}{\ln\left(\frac{0.237}{0.0084 \times 10}\right)}$$

$$= 1.85 \text{ stages.}$$

Therefore,

$$NAP = 1.85/0.25 = 7.4.$$

That is, we need eight stages and the column height can now be evaluated using Eq. II.27

$$H = 1.3 \times 8 \times 0.41 = 4.3\text{m}.$$

Using Eq. II.28, we get

$$\text{Maximum allowable superficial velocity of air (m/s)} = 0.068 \sqrt{\frac{1000 - 2.4}{2.4}}$$

$$= 1.386 \text{m/s}.$$

---

[3] This value is in good agreement with the experimental constant reported by Mackay and Shiu (1981) to be 0.673 kPa.m³/gm mol. It is instructive to demonstrate the conversion between different ways of reporting Henry's coefficient. First, the reported value is inverted to be in the units of composition in the rich phase divided by composition in the lean phase, that is, 1.486 gm mol/(kPa.m³), which can be converted into units of mole fraction as follows:

$$1.486 \left(\frac{\text{gm mole toluene}}{\text{kPa toluene.m}^3 \text{water}}\right)(202.6\,\text{kPa air}) \left(\frac{\text{m}^3 \text{water}}{10^6\,\text{gm water}}\right)\left(\frac{18\,\text{gm water}}{\text{gm mole water}}\right) =$$

$$5.419 \times 10^{-3} \frac{\text{mole fraction of toluene in water}}{\text{mole fraction of toluene in air}}.$$

Hence, in terms of mass fraction units

$$5.419 \times 10^{-3} \frac{\text{mole fraction of toluene in water}}{\text{mole fraction of toluene in air}} \times \frac{29\,\text{kg air}}{\text{kg mol air}} \times \frac{\text{kg mol water}}{18\,\text{kg water}}$$

$$= 0.0087 \frac{\text{mass fraction of toluene in water}}{\text{mass fraction of toluene in air}}.$$

But

$$\text{Minimum column diameter} = \sqrt{\frac{4(\text{Volumetric Flow Rate of Air})}{\pi(\text{Maximum Allowable Superficial Velocity of Air})}}$$

$$= \sqrt{\frac{4 \times 0.237/2.4}{3.14 \times 1.386}}$$

$$= 0.3\text{m}.$$

Now that the size of the column has been determined, it is possible to evaluate the equipment cost.

Estimation of equipment cost

Using Equations II.30 to II.33, we get

$$\text{Installed equipment cost} = (1800)(4.3)^{0.85}(0.3)^{0.95} + (8)(800)(0.3)^{0.86}$$

$$+ (12{,}000 + 62{,}000)(0.237)^{0.6}$$

$$= \$35{,}450.$$

As for operating cost, according to Eq. II.29,

$$\text{Compression energy} = \left(\frac{1.4}{1.4 - 1}\right)\left(\frac{8.314 \times 300}{29 \times 0.6}\right)\left[\left(\frac{2}{1}\right)^{\left(\frac{1.4-1}{1.4}\right)} - 1\right]$$

$$= 110 \text{ kJ/kg of air.}$$

Because the electric energy cost is $0.04/kWhr, then

$$\text{Compression cost} = 110 \frac{\text{kJ}}{\text{kg}} \times \frac{\text{kWhr}}{3600\,\text{kJ}} \times \frac{\$0.04}{\text{kWhr}}$$

$$= 1.2 \times 10^{-3} \,\$/\text{kg of air}.$$

But, operating cost is primarily attributed to compression and condensation. Hence,

$$\text{Annual operating cost} = (1.2 \times 10^{-3} + 6.0 \times 10^{-3})\frac{\$}{\text{kg air}} \times 0.237 \frac{\text{kg air}}{s}$$

$$\times 3600 \times 8000 \frac{s}{\text{yr}}$$

$$= \$49{,}100/\text{yr}.$$

It is worth pointing out that the value of recovered toluene (at 0.2/kg) is $25,900/yr.

Example II-2 Recovery of benzene from a gaseous emission

Benzene is to be removed from a gaseous emission by contacting it with an absorbent (wash oil, molecular weight 300). The gas flow rate is 0.2 kg-mole/s (about 7700 ft³/min) and it contains 0.1 mol/mol% (1000 ppm) of benzene. The molecular weight of the gas is 29, its temperature is 300 K, and it has a pressure of 141 kPa (approximately 1.4 atm). It is desired to reduce the benzene content in the gas to 0.01 mol/mol% using the system shown in Fig. II.12. Benzene is first absorbed into oil. The oil is then fed to a regeneration system in which oil is heated and passed to a flash column that recovers benzene as a top product. The bottom product is the regenerated oil, which contains 0.08 mol/mol% benzene. The regenerated oil is cooled and pumped back to the absorber.

What is the optimal flow rate of recirculating oil that minimizes the TAC of the system?

The following data may be used.

Equilibrium data

- The absorption operation is assumed to be isothermal (at 300 K) and to follow Raoult's law.
- The vapor pressure of benzene, p^o, is given by

[II.35] $$\ln p^o = 20.8 - \frac{2789}{T - 52},$$

where p^o is in Pascal and T is in Kelvin.

(*Continued*)

Example II-2 Recovery of benzene from a gaseous emission (*Continued*)

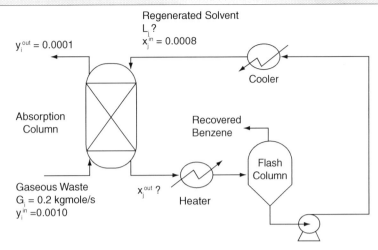

FIGURE II-12 Recovery of benzene from a gaseous emission.

Absorber sizing criteria
- The overall-gas height of transfer unit for the packing is 0.6 m.
- The superficial velocity of the gas in the absorber is taken as 1.5 m/s to avoid flooding.
- The mass velocity of oil in the absorber should be kept above 2.7 kg/m².s to ensure proper wetting.

Cost information
- The operating cost (including pumping, oil makeup, heating, and cooling) is $0.05/kg-mole of recirculating oil.
- The system is to be operated for 8000 hrs/annum.
- The installed cost, $, of the absorption column (including auxiliaries, but excluding packing) is given by

[II.36] $\quad$ Installed cost of column $= 2300\, H^{0.85} D^{0.95}$,

where H is the packing height (m) and D is the column diameter (m).
- The packing cost is $800/m³.
- The oil-regeneration system is to be salvaged from a closing unit in the plant. Hence, its fixed cost will not be accounted for in the optimization calculations.
- The absorber and packing are assumed to depreciate linearly over five years with negligible salvage values.

SOLUTION
At 300 K, the vapor pressure of benzene can be calculated from Eq. II.35:

$$\ln p^o = 20.8 - \frac{2789}{300 - 52}$$

that is,

$$p^o = 14{,}101\, \text{Pa}.$$

Because the system is assumed to follow Raoult's law, then Eq. II.3 can be used to give

$$m = \frac{14{,}101}{141{,}000}$$

$$\approx 0.1 \frac{\text{mole fraction of benzene in air}}{\text{mole fraction of benzene in oil}}.$$

As has been previously mentioned, the minimum TAC can be identified by iteratively varying ε. Because the inlet and outlet compositions of the rich stream as well as the inlet composition of the MSA are fixed, one can vary ε at the rich end of the exchanger (and consequently the outlet composition of the lean stream)

to minimize the TAC of the system. To demonstrate this optimization procedure, let us first select a value of ε at the rich end of the exchanger equal to 1.5×10^{-3} and evaluate the system size and cost for this value.

Outlet composition of benzene in oil
Let us set the outlet mole fraction of benzene in oil equal to its maximum practically feasible value given by Eq. II.23; that is,

$$x^{out} = \frac{10^{-3}}{0.1} - 1.5 \times 10^{-3}$$
$$= 8.5 \times 10^{-3}.$$

Flow rate of oil
A component material balance on benzene gives

$$L(8.5 \times 10^{-3} - 8 \times 10^{-4}) = 0.2\,(10^{-3} - 10^{-4})$$

or

$$L = 0.0234\, \text{kgmol/s}.$$

Operating cost

$$\text{Annual operating cost} = 0.05 \frac{\$}{\text{kg mol oil}} \times 0.0234 \frac{\text{kg mol oil}}{\text{s}} \times 3600 \times 8000 \frac{\text{s}}{\text{yr}}$$
$$= \$33{,}700/\text{yr}.$$

Column height
According to Eq. II.18b,

$$(y - y^*)_{\log\, mean} = \frac{(10^{-3} - 0.1 \times 8.5 \times 10^{-3})\,(10^{-4} - 0.1 \times 8 \times 10^{-4})}{\ln\left(\dfrac{10^{-3} - 0.1 \times 8.5 \times 10^{-3}}{10^{-4} - 0.1 \times 8 \times 10^{-4}}\right)}$$
$$= 6.45 \times 10^{-5}$$

Therefore, NTU_y can be calculated using Eq. II.18a as

$$NTU_y = \frac{10^{-3} - 10^{-4}}{6.45 \times 10^{-5}}$$
$$= 13.95,$$

(*Continued*)

Example II-2 Recovery of benzene from a gaseous emission (Continued)

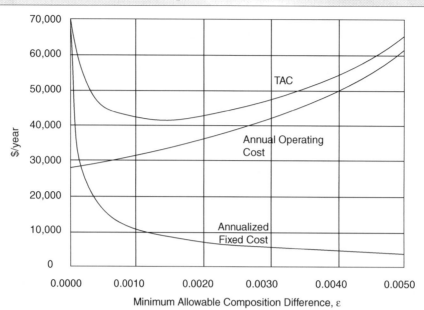

FIGURE II-13 Using mass transfer driving force to trade off fixed cost versus operating cost.

and the height is obtained from Eq. II.17a:

$$H = 0.6 \times 13.95$$
$$= 8.37 \, m.$$

Column diameter

$$D = \sqrt{\frac{4 \,(\text{volumetric flow rate of gas})}{\pi \,(\text{gas superficial velocity})}}$$

But

$$\text{molar density of gas} = \frac{P}{RT} \quad (\text{see Appendix I})$$
$$= \frac{141}{8.3143 \times 300}$$
$$= 0.057 \, kg\,mole/m^3.$$

Therefore,

$$\text{Volumetric flow rate of gas} = \frac{0.2 \frac{kg\,mole}{s}}{0.057 \frac{kg\,mole}{m^3}}$$
$$= 3.51 \, m^3/s$$

and

$$D = \sqrt{\frac{4 \times 3.51}{3.14 \times 1.5}}$$
$$= 1.73 \, m.$$

It is worth pointing out that the mass velocity of oil is

$$\frac{0.0234 \frac{kg\,mol}{s} \times \frac{300 \, kg}{kg\,mole}}{\frac{\pi}{4}(1.73)^2} \approx 3 \, kg/s,$$

which is acceptable because it is greater than the minimum wetting velocity (II.7 $\frac{kg}{m^2\,s}$).

Fixed cost

$$\text{Fixed cost of installed shell and auxiliaries} = 2300(8.37)^{0.85}(1.73)^{0.95}$$
$$= \$23,600.$$
$$\text{Cost of packing} = (800)(1.73)^2(8.37)$$
$$= \$15,700.$$

Total annualized cost

$$TAC = \text{annual operating cost} + \text{annualized fixed cost}$$
$$= 33,700 + \frac{(23,600 + 15,700)}{5}$$
$$= \$41,560/yr.$$

This procedure is carried out for various values of ε until the minimum TAC is identified. The results shown in Fig. II.13 indicate that the value of $\varepsilon = 1.5 \times 10^{-3}$ used in the preceding calculations is the optimum one leading to a minimum TAC of \$41,560 /yr.

HOMEWORK PROBLEMS

II.1. A flue gas is to be desulfurized using water scrubbing in a stagewise column. It is desired to remove 95 percent of SO_2 from the flue gas containing 100 ppm SO_2. The flow rate of the gas is 0.03 kg-mole/s (approximately 2 million standard cubic foot per day (MMscfd)). The water entering the absorber is free of SO_2. The equilibrium data for transferring SO_2 from the gas to the water may be described by

[II.37] $$y_i = 24 x_j,$$

where y_i and x_j are the mole fractions of SO_2 in gas and water, respectively.

1. What is the minimum flow rate of water needed to perform the desulfurization task?
   (Hint: Set the minimum allowable composition difference at the rich end of the absorber equal to 0.)
2. If the flow rate of water is twice the minimum, what is the value of minimum allowable composition difference at the rich end of the absorber? How many theoretical stages are required?

II.2. A packed column is used in the desulfurization operation described in the previous problem. The overall height of transfer unit based on the gas phase is 0.7 m. When the flow rate of water is twice the minimum, what height of packing is needed?

II.3. A countercurrent moving-bed adsorption column is used to remove benzene from a gaseous emission. Activated carbon is employed as the adsorbent. The flow rate of the gas is 1.2 kg/s and it contains 0.027 wt/wt% of benzene. It is desired to recover 99 percent of this pollutant. The activated carbon entering the column has 2×10^{-4} wt/wt% of benzene. Over the operating range, the adsorption isotherm (Yaws et al., 1995) is linearized to

$$[II.38] \qquad y_i = 0.0014\, x_j,$$

where y_i and x_j are the mass fractions of benzene in gas and activated carbon, respectively. The column height is 4 m and the value of HTU_y is 0.8 m. What should be the flow rate of activated carbon?

II.4. A multistage extraction column uses gas oil for the preliminary removal of phenol from wastewater. The flow rate of wastewater is 2.0 kg/s and its inlet mass fraction of phenol is 0.0358. The mass fraction of phenol in the wastewater exiting the column is 0.0168. Five kg/s of gas oil are used for extraction. The inlet mass fraction of phenol in gas oil is 0.0074. The equilibrium relation for the transfer of phenol from wastewater to gas oil is given by

$$[II.39] \qquad y_i = 2.0\, x_j,$$

where y_i and x_j are the mass fractions of phenol in wastewater and gas oil, respectively. The overall column efficiency is 55 percent. How many actual stages are needed?

II.5. Air stripping is used to remove 95 percent of the trichloroethylene (TCE, molecular weight = 131.4) dissolved in a 200 kg/s (3180 gpm) wastewater stream. The inlet composition of TCE in the wastewater is 100 ppm. Air (essentially free of TCE) is compressed to 202.6 kPa (2 atm) and is diffused through a packed stripper. The TCE-laden air exiting the stripper is fed to the plant boiler, which burns almost all TCE. A schematic representation of the process is shown in Fig. II.14. Calculate the annual operating cost and the fixed capital investment for the system.
   The following physical and economic data are available:

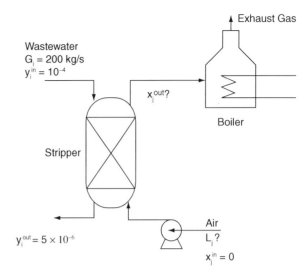

FIGURE II-14 Stripping of TCE from wastewater.

Physical data

- The stripping operation takes place isothermally at 293 K and follows Henry's law.
- The equilibrium relation for stripping TCE from water is theoretically predicted using Eq. II.5 to be

$$y_i = 0.0063\, x_j,$$

where y_i is the mass fraction of TCE in wastewater and x_i is the mass fraction of TCE in air (the value of Henry's coefficient corresponds to that reported in Mackay and Shiu [1981], as $1.065 \frac{gm\,mole}{kPa.m^3}$).

- The air-to-water ratio is recommended by the packing manufacturer to be $25 \frac{m^3\,air}{m^3\,water}$.

Stripper sizing criteria

- The maximum allowable superficial velocity of wastewater in the stripper is taken as 0.02 m/s (approximately 30 gpm/ft^2; see Cummins and Westrick, 1990).
- Overall height of transfer unit based on the liquid phase is given by

$$HTU_y = \text{Superficial velocity of wastewater}/K_y a$$

where K_y is the water-phase overall mass-transfer coefficient, and a is the surface area per unit volume of packing. The value of $K_y a$ is provided by the manufacturer to be $0.02\,s^{-1}$.

Cost information

- The operating cost for air compression is basically the electric utility needed for the isentropic compression. Electric energy needed to compress air may be calculated using Eq. II.29. The isentropic efficiency of

the compressor is taken as 60 percent, and the electric energy cost is $0.06/kWhr.
- The system is to be operated for 8000 hr/annum.
- The fixed cost, $, of the stripper (including installation and auxiliaries, but excluding packing) is given by

[II.41]   Fixed cost of column = $4700 HD^{0.9}$,

where H is the column height (m) and D is the column diameter (m).

[II.42]   The cost of packing is $700/m^3.

[II.43]   The fixed cost of the blower, $, is $12,000 L_j^{0.6}$,

where L_j is the flow rate of air (kg/s).
- Assume negligible salvage value and a five-year linear depreciation.

1. Estimate the column size, fixed cost, and annual operating cost.
2. Due to the potential error in the theoretically predicted value of Henry's coefficient, it is necessary to assess the sensitivity of your results to variations in the value of Henry's coefficient. Plot the column height, annualized fixed cost, and annual operating cost versus α, the relative deviation from the nominal value of Henry's coefficient, for $0.5 \leq \alpha \leq 2.0$. The parameter α is defined by

[II.44] α = Value of Henry's coefficient/0.0063.

3. Your company is planning to undertake extensive experimentation to obtain accurate values of Henry's coefficient that can be used in designing and evaluating the cost of this stripper. Based on your results, what would you recommend regarding the undertaking of these experiments?

II.6. Ammonia is to be absorbed from an air stream by contacting it with water in a packed column. The gas flow rate is 500 kg mol/hr and it contains 0.2 mol/mol% (2000 ppm) of ammonia. The molecular weight of the gas is 29, its temperature is 300 K, and it has a pressure of 160 kPa. It is desired to reduce the ammonia content in the gas to 0.01 mol/mol% using the system shown in Fig. II.15. Ammonia is first absorbed in water. The water is then fed to a wastewater treatment facility in which ammonia is used as a bionutrient.

What is the optimal flow rate of water that minimizes the TAC of the system?
The following data may be used.

Equilibrium data

- The absorption operation is assumed to be isothermal with the following equilibrium relation (King, 1980):

[II.45] $y_i = 1.41 x_j$,

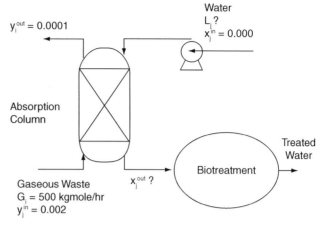

FIGURE II-15 Ammonia absorption using water.

where y_i and x_j are the mole fractions of ammonia in gas and water, respectively.

Absorber sizing criteria

- The overall-gas height of transfer unit for the packing is 0.8 m.
- The superficial velocity of the gas in the absorber is taken as 1.5 m/s to avoid flooding.
- The manufacturer of the packing has recommended a range for the mass velocity of water between 2000 and 10,000 $\frac{kg}{hr.m^2}$.

Cost information

- The operating cost of the water (including pumping and wastewater treatment) is $0.005/kgmole water.
- The system is to be operated for 8000 hr/annum.
- The purchased cost, $, of the absorption column (fiberglass reinforced plastic shell and auxiliaries excluding packing) is given by (Vatavuk, 1995)

[II.46]   Purchased cost of column ($) = $1313 S$,

where S is the absorber surface area (m^2), which can be calculated from

[II.47] $S = \pi D H + \frac{\pi}{2} D^2$,

where D is the column diameter (m) and H is the column height (m) taken as 1.2 times the packing height.

The packing cost is $600/m^3.

- The ratio of the fixed capital investment to the purchased equipment cost (Lang factor) is taken as 4.83.
- The absorber and packing are assumed to depreciate linearly over three years with negligible salvage values.

REFERENCES

Astarita G, Savage DW, Bisio A: *Gas treating with chemical solvents*, New York, Wiley, 1983.

Benitez J: *Principles and modern applications of mass transfer operations*, New York, Wiley Intersciences, 2009.

Cheremisinoff PN, Ellerbusch F: *Carbon adsorption handbook*, Ann Arbor, MI, Ann Arbor Sci. Publ., 1980.

Clark RM: Unit process research for removing volatile organic chemicals from drinking water: an overview. In Ram NM, Christman RF, Cantor KP, editors: *Significance and treatment of volatile organic compounds in water supplies*, Chelsea, MI, 1990, Lewis Publishers.

Crowl DA, Louvar JF: *Chemical process safety*, Engelwood Cliffs, NJ, Prentice Hall. p 161, 1990.

Cummins MD, Westrick JW: Treatment technologies and costs for removing VOCs from water. In Ram NM, Christman RF, Cantor KP, editors: *Significance and treatment of volatile organic compounds in water supplies*, Chelsea, MI, Lewis Publishers, pp 277, 1990.

Fleming JL: *Volatilization technologies for removing organics from water*, Park Ridge, NJ, Noyes Data Corp., 1989.

Francis AW: *Liquid-liquid equilibrium*, New York, Interscience, 1963.

Geankoplis CJ: *Transport processes and separation processes* (includes unit operations), ed 4, Prentice Hall, 2003.

Green D, Perry RH: *Perry's chemical engineers' handbook*, ed 8, New York, McGraw Hill, 2007.

Hwang Y, Olson JD, Keller GE II: Steam stripping for removal of organic pollutants from water: Parts 1 and 2, *Ind Eng Chem Res* 31(7):1753–1768, 1992.

King CJ, editor: *Separation processes* (ed 2), New York, 1980, McGraw Hill.

Kohl A, Riesenfeld F, editors: *Gas purification* (ed 5), Houston, TX, Gulf Publ. Co., 1997.

Kremser A: Theoretical analysis of absorption process, *Nat Pet News* 22(21):42–43, 1930.

Lo TC, Baird MHI, Hanson C: *Handbook of solvent extraction*, New York, Wiley-Interscience, 1983.

Mackay D, Shiu WY: A critical review of Henry's law constants for chemicals of environmental interest, *J Phys Chem Ref Data* 10(4):1175–1199, 1981.

McCabe WL, Smith JC, Harriot P: *Unit operations of chemical engineering*, ed 5, New York, McGraw Hill, 2000.

Perrich JR: *Activated carbon adsorption for wastewater treatment*, Boca Raton, FL, CRC Press, 1981.

Peters MS, Timmerhaus KD: *Plant design and economics for chemical engineers*, ed 4, New York, McGraw Hill, 1991.

Reid RC, Prausnitz JM, Poling BE: *The properties of gases and liquids*, ed 4, New York, McGraw-Hill, 1987.

Seader JD, Henley EJ, Roper DK: *Separation process principles*, ed 3, New York, Wiley, 2010.

Stenzel MH: Remove organics by activated carbon adsorption, *Chem Eng Prog* 89(4):36–43, 1993.

Stenzel MH, Merz WJ: Use of carbon adsorption processes in groundwater treatment, *Environ Prog* 8(4):257–264, 1989.

Treybal RE, editor: *Mass transfer operations* (ed 2), New York, McGraw Hill, 1980.

U. S. Environmental Protection Agency Carbon adsorption isotherms for toxic organics, EPA-600/8-80-023, Cincinnati, OH, USEPA, 1980.

Valenzuela DP, Myers A: *Adsorption equilibrium data handbook*, Englewood Cliffs, NJ, Prentice Hall, 1989.

Vatavuk WM: A potpourri of equipment prices, *Chem Eng* August:68–73, 1995.

Wankat PC: *Separation process engineering*, ed 2, Prentice Hall, 2006.

Yang RT: *Gas separation by adsorption processes*, Boston, Butterworth, 1987.

Yaws CL: *Thermodynamic and physical property data*, Houston, TX, Gulf Publ. Co., 1992.

Yaws CL, Bu L, Nijhawan S: Adsorption capacity data for 283 organic compounds, *Environ. Eng. World* May–June:16–19, 1995.

INDEX

Note: Page numbers followed by "f" indicate figures, "t" indicate tables, and "b" indicate boxes.

A

Absorption 401
Absorption coefficient 220
Absorption refrigeration (AR) cycles 179, 190–194
Absorption-cycle heat pump 191f
Acceptability 133
Acetaldehyde case study 83f
Acetaldehyde process 80f
Acetaldehyde production, yield targeting through ethanol oxidation 79–82, 262–265
Acrylonitrile (AN)
  benchmarking water usage and discharge for 10f
  case study optimal solution 11f
  direct-recycle strategies 140f
  manufactured via the vapor-phase ammoxidation of propylene 134
  manufacturing 5f
  process 5–12
  production 135f
  sink composite diagram 138f
  source-sink mapping diagram 139f
Actionable task 3
Actions, unreconciled short- and long-term 395
Actual stoichiometric targets 63–64
Adiabatic reactivity 11
Adjusted temperature, of a heating utility 154–156
Adsorption 403f
AFC (annualized fixed cost) 27, 31, 40
Algebraic analogue, to material recycle pinch diagram 223
Algebraic cascade diagram 153
Algebraic targeting approach 223–226
Algebraic targeting procedure, summarized 226
Allocation of sources 209
Alternatives
  analysis of selected 4
  comparison of 55–58
  generation of 3–4
  infinite numbers of 5–8
  selection of 4
American National Standards Institute, shorthand notations 39
Ammonia removal from a gaseous emission, CHARMEN synthesis for 358–359

Ammonium nitrate plant, incorporation of CHARMEN synthesis into mass integration for 360–361
AN *see* Acrylonitrile (AN)
Annual after-tax (nondiscounted) cash flows 50t
Annual capital charge ratio 40
Annual cost 55
Annual cost/revenue, using to compare alternatives 57
Annual revenue 55
Annualized fixed cost (AFC) 27, 31, 40
Annuity 38–40
  future sum of 40
  present sum of 40
Annuity installments, conversion to future worths 39f
Approximate correlations, modeling performance of a steam turbine in a Rankine cycle 173–174
Aqueous sodium hydroxide, absorption of H_2S in 317–318
Aqueous wastes, dephenolization of 118–120, 233–234
Arithmetic operations, in LINGO 258
Arrhenius reaction rate constant 383
Asset turnover ratio 20–21
Assignment problem 268–269
  adding logical constraints to 282
  set formulation for 281–282
Association for the Advancement of Cost Engineering (AACE) International 15–16
Augmented property index (AUP) 207, 210, 212

B

Bahr El-Baqar drainage system 383–384
Balsas watershed system 384, 385f
Base case, designing for 353
Base-case material recycle pinch diagram 395f
Benchmarking, performance of an industrial facility 63
Benzene, recovery 115–118, 403f
Benzene removal problem 304–305
BFR (boundaries of the feasibility region) 209–210, 209f, 210f

Big-M formulation 271–272, 274
Binary integer variables
  definition of 280
  for matching streams 350
  modeling disjunctions and discontinuous functions with 270–277
  representing discrete decisions and what-if scenarios 267–268
Biodiesel
  producing from triglycerides 372f
  production of 370–372
Biomass feedstocks, converting into value-added products 365
Biorefineries
  conceptual design of 365–367, 365f
  design of integrated 365
  minimizing cost of producing hydrogen from 260
  techno-economic assessment of 367–372
Biotreatment facility, debottlenecking 5–8
Boiler
  rate of heat supplied by 171
  recycling to 7f
Boiler feed water, recycling to substitute 7f
Book value, of a property 27
Boolean algebra 268
Boundaries of the feasibility region (BFR) 209–210, 209f, 210f
Bounding feed composition 90f
Bounding feed flow rate 90f
Brainstorming 5–8, 10–11
Brayton cycle 180
Brayton cycle heat pump 180f
Break-even analysis 31–33
Break-even point (BEP) 31
Break-even production rate 32
Business requirements, process integration changing 393

C

Candidate mass-separating agents (MSAs), screening using a hybrid of path and pinch diagrams 329–336
Capacity ratio with exponent 16, 17, 23
Capital cost 15
  estimation 15–21
Capital ratio 20–21

Capital recovery factor 40
Carnot cycle 165–166
  as an idealistic system 169, 180–181
  pressure-volume representation of 166f
  for a steam power plant on the T-S mollier diagram 169f
  steps constituting 165–166
  temperature-entropy representation of 166f
Carnot heat engine
  efficiency and work of 167
  expression for the efficiency of 167
Carnot heat pump, offering highest value of COP 181
Cascade diagram 152, 155f, 225f
  for dephenolization example 233f
  for direct recycle 225f
  for the HEN 174, 178f
  for the MEN 232
  for pharmaceutical case study 153f
  for specialty chemical processing facility 159f
Cash flow diagrams 38
  for annuity schemes 39f
  timeline of 38
CCHP (combined cooling, heating, and power) 198
Centrifugal pump (carbon steel), purchased cost of 25f
CHARMEN *see* Combined heat and reactive mass-exchange networks (CHARMEN)
Chemical and polymer complex process 394t
Chemical Engineering Plant Cost Index 17, 18t
Chemical processes, cost estimation and economics of 15
Chloroethanol (CE) 330
  gaseous path diagram for 332
  liquid path diagram for 332, 333f
Chloroethanol case study, segregation, mixing, and recycle for 336–339
CHP (combined heat and power) 165
CID *see* Composition interval diagram (CID)
Class life, for each type of equipment/industry 28

415

416 Index

Closed-cycle vapor-compression heat pumps 179–185
Cluster-based source-sink mapping diagram 209–210
Clustering techniques, for multiple properties 207–209
Clusters 207, 218–219
Cluster-to-property mapping 211–218
CO_2 methanation 67
Coefficient of performance (COP) 178
Cogeneration 194
Cogeneration target 194, 196
Cogeneration targeting 194–198
Cold composite stream, for pharmaceutical process 151f
Cold streams
 matching with hot streams 156
 representing 149f
Collective cooling capacity, of cold process streams 152
Collective load, of hot process streams 152
Combined cooling, heating, and power (CCHP) 198
Combined depreciation method 28
Combined heat and power (CHP) 165
Combined heat and power systems, integration of 165
Combined heat and reactive mass-exchange networks (CHARMEN)
 ammonia removal from a gaseous emission 358–359
 incorporation into mass integration for ammonium nitrate plant 360–361
 synthesis problem, formulating 358
 synthesizing 357–362
Commercial applicability 393
Component material balance around a composition interval 232f
 equation 218
Componentless design, notion of 207
Composite streams, plotting on the same diagram 114
Composition constraints 106
Composition interval diagram (CID) 231
 for ammonia removal case study 359f
 for dephenolization example 233f, 301f
 for dephenolization problem 307f
 developing 318
 generalization of 299
 for kraft pulping example 321f
 schematic representation of 231
Composition intervals, defining a series of 231
Compound interest
 defined 37

with a monthly interest rate 37
 of a single payment 36–38
Compression power, calculating 188
Computer-aided tools
 estimating FOB cost of equipment 21–23
 evaluating FCI 16
Concentrations, relationships for converting 397–398
Condensation, recovering volatile organic compounds (VOCs) 237
Condenser, cooling duty of 171
Constraint propagation, for sinks 90, 90f
Constraints
 dual prices of 259
 insufficient consideration of 395
 repeating 278, 279
Construction rules, for BFR 209–210
Consumer Price Index (CPI) 33–35
Conversion factors, for different sets of units 398–399
Convex analysis 266–267
Convex equilibrium relations 318–319
Convex function 266–267, 267f
Convex program 267
Convex sets 266
Convex-hull formulation 272
Convex-hull reformulation 272
 of GHG constraints 275
 piecewise linearization with 273f
Cooling utilities
 benchmarking 12f
 minimization of external 238–239
 minimum requirement 152–154, 239
 selection of 239
 sum of all cooling capacities of 346
 targeting for minimum 345–349
COP (coefficient of performance) 178
Copolymerization plant 115–118, 115f
Corn, composition of 68t
Corn ethanol synthesis process 390f
Corn price 69f
Corn stover 20
Corn-to-ethanol plant 68–69
Corresponding composition scales 112–113, 112f
Cost charts, for equipment 24–25
Cost estimation studies 15–16, 16t
Cost indices
 in adjusted cost estimates 17
 for equipment 23
 updating FCI 18
Countercurrent packed column 404f
Cresol 343
 in feed water 342
 in washing water 342

D

Dalton's law 239–240
Darcy's equation 244–245
Data
 entering 278f
 entering for attributes 278
 entering two-dimensional 279f
 inserting commas between 278f
Data extraction, improper 393
DCF ROI see Discounted cash flow return on investment (DCF ROI)
Decanter wastewater, hybrid separation technologies for 9f
Declining-balance (fixed-percentage) method 28
Deficit composite line 195–196
Deflation 33–35
Delivered equipment cost 21
Density, conversion factors 398
Density-mixing operators 202
Dephenolization, of aqueous wastes 118–120, 233–234
Dephenolization case study, revisited 300–303
Dephenolization problem, network synthesis for 306–309
Depreciation 26–31
Desalination problem, representation of 276f
Design procedure, for simultaneous heat integration and scheduling 355f
Design rules, describing 210
Designs, adopting/evolving earlier 11
Differential (continuous) contactor 404
Dilute waste streams, as a special case 239
Direct recycle 89
 optimality conditions 223
 optimization approach to 287
 strategies for AN example 140f
Direct-recycle networks
 algebraic approach 223
 design of 89–90
 graphical approach 89
 optimizing for mass conservation 296
Direct-recycle problem
 expressing 287
 schematic representation of 89f
Disaggregation, of variables 272
Discontinuous functions 270–271
Discount factor 45
Discount rate 45
Discounted cash flow diagram 47f, 48f
Discounted cash flow return on investment (DCF ROI) 48–52
 calculation of 49
 determination of 55f
 determining minimum selling price 50–51
Discounted cash flows 45

Discounted payback period (DPBP) 52
Discouraging attitudes, about process integration and suggested responses 396t
Disjoint constraints, in gas-processing case study 274–275
Disjoint domain 271
Disjunction, of two constraints 271–272
Distillation column, recycling 6f
Distillation configuration, with MVR 187f
Double declining-balance (DDB) method 28, 29
Double summations, dealing with 279
DPBP (discounted payback period) 52
Dual prices, sensitivity analysis and interpretation of 259–260
Dual-purpose heat pump, for residential applications 181–185
Dual-purpose vapor-compression heat pump 183–184
Dynamic changes, inappropriate accounting for 393–395

E

Eco-efficiency 2–3
Eco-industrial parks (EIPs) 375–379
 case study 378f, 379f
 design problem 376f
 mass-integration representation of problem 375f
 source-interception-sink representation 376, 377f
Economic aspect, of any sustainable design 15
Economic gross potential (EGP) 64–65
Effluent wastewater streams, set of process 375
EIA see Environmental impact assessment (EIA)
EIPs see Eco-industrial parks (EIPs)
EIS (environmental impact study) 386
EISEN (energy-induced separation network) 241
Empirical correlations
 estimating FCI 20
 for rough and quick estimates of FCI 19
End-of-pipe separation 339–340
Energy
 conversion factors 399
 use of different forms of 147
Energy integration 4–5, 165
Energy intensity index 2
Energy utilities 26
Energy-induced separation network (EISEN) 241

Energy-separating agents, use of 237
Engineering News Record Construction Index 17
Enthalpy expressions, for gaseous VOC-laden stream 238
Enthalpy-entropy Mollier diagram 170, 170f
Environmental discharges 2
Environmental impact assessment (EIA) 386
 improving using process integration 386–387
 main process engineering activities 387f
 process integration as an enabling tool in 386–387
 process synthesis and targeting in 388f
 steps in a typical 386f
Environmental impact study (EIS) 386
Environmental subsidies, ROI determining 43
Equality constraints 255
Equilibrium 403f
Equilibrium data, compilations of 402
Equilibrium line, operating line versus 123
Equilibrium relations, as monotonic functions of temperature of MSA 357
Equilibrium stages 402–403
Equipment, scaling exponents for selected pieces of 23t
Equipment cost
 estimation of 21–25
 methods for estimating 21
 types of 21
Equivalence, case of 268
Estimation studies, uncertainty in most 15–16
Ethanol production from glucose 65
Ethyl chloride process, interception of chloroethanol in 330–336
Excess process mass-separating agents (MSAs) 114
Exclusive OR operator 268
Exergy 198
Expenses, for operating cost 25–26
External heat, needed to drive AR 194
External mass-separating agents (MSAs)
 above the pinch 114–115
 cost estimation and screening of 119–120
 screening 121f
Extractable energy 195
Extractable power, versus flow rate plot 197f
Extractable power surplus and deficit composite curve 196f
Extractable power surplus composite curve 196f
Extraction 401

F

FCI see Fixed capital investment (FCI)
Feasibility criteria 124f
Feasibility region 255, 258f
Feasible point 255
Feed water flow rate, constraints on 342
Feeding sources, into a sink 212
Feedstocks, converting to biodiesel 370–372
Fertilizer plant, identifying location of a new 383–384
Fixed and total capital investment, use of discontinuous functions to model 271
Fixed capital investment (FCI) 370–372
 computer-aided tools evaluating 16
 cost indices updating 18
 defined 15
 economic analysis of 371f
 empirical correlations estimating 20
 estimating 19
 estimation based on factors of delivered equipment cost 20
 function 271
 as a function of time 17
 Lang factor, values of 18t
 updating using a cost index 18
Fixed charges 27, 31
Fixed cost, versus operating cost 125–127, 239
Flexibility, including in design 354f
Floating-head shell-and-tube heat exchanger 24f
Flow balance
 around a load interval 225f
 for reaches 382
Flow rate * specific heat, rules for matching 157–160
Flow rates
 within each load interval 224–225
 passing through pinch point 93f
 of washing water 342
@FOR command 278
 using 279–282, 279f
Formic acid example, matching sources and sinks for 103–105
Formic acid plant
 solving water-recycle problem in 291–292
 targeting for water usage and discharge 226
 water recycle in 95–98, 288–291
Forward and backward branching step 365
Forward and backward branching trees 366f
Forward mode 379–381
Forward model, for tracking discharge 381f
Forward-backward approach 365

Fractional saturation, of a reactive MSA 316
Free on board (FOB) cost 21
Fresh feed
 dealing with impurities in 295–296
 impurities in 100
 minimizing usage through recycle 74
Fresh line, locus of 202–203
Fresh loads, recycles reducing 91
Fresh material utilities 71–72
Fresh resources
 available for service 287
 minimizing cost of 287
Fuel blending, optimization of 256–258
Fuel blending problem, using LINGO to solve 259
Future sum of the annuity 39
Future value
 general formula for 37, 37t
 present worth of 38
 of a single payment 37

G

Gas composition, versus partial pressure 397
Gas constant, values of 402
Gas processing facility, maximizing revenue of 273
Gaseous path diagram 329f
 for CE 332
Gas-processing case study, inclusion of disjoining constraints in 274–275
GCC see Grand composite curve (GCC)
GHGs see Greenhouse gases (GHGs)
Global minimum 267
Global optimum point 267
Global solutions, desire to obtain 267
"Global Solver," in LINGO 267
Global warming index (GWI) 1–2
Global warming potential (GWP) 1–2
Grand composite curve (GCC) 185–186
 analysis of 193, 194f
 construction of 155f
 with integrated pockets and optimal placement of utilities 155f
 screening multiple utilities 154
 using for utility selection 156
 using to determine excess heat to be recovered for usage in AR 190–194
 using to target AR integration 193–194
Grand composite representation, screening of multiple utilities using 154–156
Graphical approach, to design of biorefineries 365

Greenhouse gases (GHGs)
 for alternate pathways 390f
 impact of emission of 1–2, 274
GREET (Greenhouse gases, Regulated Emissions, and Energy use in Transportation) model 387
GWI (global warming index) 1–2
GWP
 examples of 2t
 global warming potential 1–2

H

Hand method 18–19, 19t
Header balance 195f
Header efficiency 195
Header enthalpies 195f
Heat
 discharging 165
 extracting from a low-temperature heat source 178
 transshipping from sources to destination 345
Heat balance
 around temperature interval 152f, 346f
 for each cold stream 350
 for each hot stream 350
Heat deficit of the HEN 174–175
Heat engines 165–167
 approximating performance of a Carnot heat engine 176
 components of 165
 efficiency of 165
 industrial application of 165
 integration with the HEN 177
 placement of 174–178, 175f
 schematic representation of 165f
Heat exchange, integrating 11–12
Heat exchangers
 attaining targets for minimum heating and cooling utilities 13f
 estimating cost of 24
 including materials factor in estimating cost of 25
 minimum number of 349
 placed inside the home and outdoors 181–185
 placing above the pinch 157f
Heat integration 147
 assessment of different extents of 388–390
 determining heating steam requirements 194–195
 examples of 12
 in industrial facilities 165
 infeasible 150f
 partial 150f
Heat pumps 178–179
 in cooling mode 182f
 elements of 178
 in heating mode 182f
 placing above the pinch 185
 placing across the pinch 185, 186f

Heat pumps (*Continued*)
  placing below the pinch 185, 186f
  schematic representation of 178, 179f
  types of 179
Heat sink 165
Heat source 165
Heat supply, to AR via a heat-recovery network and a hot-water loop 192f
Heat-exchange driving force 148
Heat-exchange networks (HENs)
  cascade diagram for 174, 178f
  configuration for pharmaceutical process 352f
  handling scheduling and flexibility issues in 353–354
  heat deficit of 174–175
  heat engines integration with 177
  integrated heat engine and 178f
  mathematical programming approaches to synthesis of 345
  mathematical techniques for synthesis of 345
  stream matching and synthesis of 156–160
  synthesis of 147
  synthesis problem 147–148
  synthesizing with MEN simultaneously 357
  thermal energy integration through synthesis 165
Heat-exchange operation, transferring heat from a hot stream to a cold stream 148
Heat-exchanger network, network synthesis of 158–159
Heat-induced separation networks (HISENs) 237
Heating and cooling utilities benchmarking 12f
  targeting for minimum 345–349
Heating utilities
  excessive usage of external 147
  sum of all heating loads of 345
Heat-transfer driving force, increasing 148
Henry's coefficient 343, 402
Henry's law for stripping 402
HENs *see* Heat-exchange networks (HENs)
Heuristics 11
HFRO module *see* Hollow-fiber reverse osmosis (HFRO) module
Hierarchical yield-targeting procedure, flowchart for 74f
High pressure (HP) 194–195
Holistic perspective, of process integration 375
Hollow-fiber modules, distinguishing characteristics 244
Hollow-fiber reverse osmosis (HFRO) module 244
  modeling of units 244–247

  schematic representation of 244f
  transport parameters 247
Hot and cold streams, matching 156
Hot composite stream, for pharmaceutical process 151f
Hurdle rate 42
Hydrocarbon fuels
  implementation of carboxylate platform for biomass to 371f
  producing from biomass 367–370
Hydrodynamic model 244–245
Hydrogen recycle, in a refinery 295–296

I

ICARUS 21–23
ICIS Chemical Business 26
Ideal gas law 402
Impact diagrams 143f
Impure fresh, extension to case of 94, 94f
Incremental return on investment (IROI) 57, 58
Incremental return on sustainability (IROS) 2–3
In-depth knowledge, lack of appropriate 395
Indicators, assessing sustainability 1–3
Industrial applications, of process integration 394t
Industrial facilities, interacting with surroundings 379
Industrial processes, impacts on the ecosystem 1
Industrial symbiosis opportunities 375
Inequality constraints 255
Infeasibilities, removing using a fresh resource 223–224
Inflation 33–35
Inflation rate 33–35
In-plant pollutant interception 327
Installed equipment cost 21
Integer cuts, enumerating multiple solutions 269–270
Integer program 255
Integrated heat engine 178f
Integrated mass exchange 114
Integration
  of mass and heat objectives 238
  of path and pinch diagrams 328–329, 329f
Interception 134–143
  integration with segregation, mixing, and recycle 339–340
Interception branch 366
Interception devices 203–204, 209
Interception system, task identification for 211f
Interception techniques, reducing cresol content 343
Interception technologies, selection of 329–330

Internal rate of return (IRR) 48
Interphase concentration gradient 402
Interphase mass transfer 403f
Inter-stream conservation 208
Intra-stream conservation 208, 208f
Inverse approach, to integrating process and watershed 382f
Ion exchange 401
IROI (incremental return on investment) 57, 58
IROS (incremental return on sustainability) 2–3
IRR (internal rate of return) 48
Isentropic compression, formulas for 188
Isentropic efficiency 170
Isentropic pump 171
Isentropic-enthalpy change, in the turbine 170
Isothermal compression 166
Isothermal expansion 165
Isothermal mass exchanger 405

K

Kraft pulping process 75–76, 341f
  as application of PI 394t
  removal of H_2S from 319–321
  simplified flow sheet of 325f
  water usage and discharge 341
Kremser equation 404

L

Labor cost 26
Lagrange multipliers 259
Lake eutrophication, preventing 383–384
Lake Manzala, phosphorous input to 383–384
Land, used for an industrial purpose 2–3
Lang, method for estimating FCI of a chemical plant 17–18
Lang factors, revised 18, 19t
Leaching 401
Lean composite stream 114f, 117f
Lean end 114–115
Lean phase 111
Lean streams, collective load of 232
Lean substreams 357
Learning outcomes, of this book 12–13
Length, conversion factors 398
Level arm rule
  for clusters 208
  for a ternary source-sink diagram 106, 106f
Lever arm rule 98–101
  application of 101
  for mixing 101f
Lever arm source-prioritization rule 210
Life cycle analysis (LCA)
  coupling of process integration with 389f
  process integration in 387–390
Light absorption coefficient 220
LINDO Systems Inc. 258
Linear depreciation 27
Linear program (LP) 255, 267
LINGO
  dual prices in results of 259–266
  format for a set 277f
  logical operators recognized by 282–283
  optimization software 258
  set formulation 302, 348–349
  set formulations in 277–283
LINGO optimization model, writing 258
LINGO programs
  Big-M formulation 274–275
  convex-hull formulation 275
Liquid path diagram 329f
  for CE 332, 333f
Load interval 224
Load-interval diagram (LID)
  developing 226
  for formic acid case study 227f
Local versus global minimum 267f
Logical constraints, adding to assignment problem 282
Logical operators, adding 282–283
Low pressure (LP) 194–195

M

M&S recent values of equipment cost index 24t
Macroscopic systems, consistent approach in treating 375
MACRS method *see* Modified Accelerated Cost Recovery System (MACRS) depreciation method
Maintenance, types of 26
Manufacturer's quotation, of capital costs 16
Manufacturer's quotation method 21
Manufacturing FCI 15
Marshall and Swift (M&S) equipment cost index 23
Mass
  conversion factors 398
  exchanged by two rich streams 113f
  versus molar compositions 403f
  representation of exchanged 114f
Mass exchange, operations included 403f
Mass exchangers 111, 403f
  immediately above the pinch 123f
  minimizing number of for an MOC 305
  minimum allowable composition difference 406f
  operating isothermally 357

Index   **419**

types and sizes of 402–405
Mass flow, path diagram providing big picture for 328
Mass integration 4–5, 63, 66–67
 application of to debottleneck an acrylonitrile process 134–142
 described 133–134
 design philosophy of 142
 determining nonheating steam demands 194–195
 examples of 12
 mathematical optimization techniques for 327
 as a special case of property integration 218
Mass integration project
 payback period for 44
 ROI for 42
Mass intensity index 2
Mass targeting 63
Mass-exchange cascade diagram 232, 299
Mass-exchange driving forces, varying 126
Mass-exchange networks (MENs)
 algebraic approach to synthesis of 231
 analogy with HENs 148t
 construction of with minimum number of exchangers 122–125
 graphical approach to synthesis of 111
 mathematical programming approaches to synthesis of 299
 strong interaction between mass and heat 357
 synthesis problem 111, 111f
 synthesizing 111
 targeting approach 112
 waste-interception networks (WINs) becoming 327
Mass-exchange operations, indispensable for pollution prevention 401
Mass-exchange pinch diagram 113–120, 114f
Mass-exchange pinch point 114
Mass-exchange systems, minimizing costs of 405–410
Mass-exchange units, modeling for environmental applications 401
Mass-exchanger units, minimum number of 112
Mass-integration strategies
 for attaining targets 83
 combining 133
 hierarchy of 83, 84f, 133f
Mass-integration targeting 66–83
Mass-load paths
 defined 126–127
 elimination of process MSAs using 127f
 reduction of number of exchangers using 127f
 using 126–127

Mass-separating agents (MSAs) 111
 of draft pulping example 320t
 incorporating external in generalized CID 299
 induced wins 327–336
 intercepting liquid path diagram for CE case study 334f
 minimizing cost of 113
 minimizing operating cost of 304–305
 screening using path diagram 330f
 selectively removing certain components from a rich phase 403f
Mass-transfer driving force 402
Matching and interception step 366
Matching branch 366
Material balance
 for each lean stream 306
 for each rich stream 306
Material consumption metrics 2
Material flow analysis (MFA) 379
 interaction with mass integration 380f
 model including in an optimization formulation 381
 model providing data needed for mass integration 379
 reverse problem formulation 381–383
Material recycle pinch diagram 91–93, 93f, 223, 223f
 design rules from 93–94
 for formic acid example 97f
 before interception 94–98, 95f
 under uncertainty 395f
Material utilities 26, 72f
Materials factor, typical values of 24–25, 25t
Mathematical functions, used by LINGO 258–259
Mathematical optimization techniques, for mass integration 327
Mathematical programming 255
 targeting of material recycle 287
Maximum allowable pollutant concentration, in process effluent 383
Maximum overall process yield 83
Maximum product yields, targeting for 72–83
Maximum property load, for a sink 202
Maximum recyclable load
 of limiting reactant 74
 of targeted species 70
McCabe-Thiele diagram 403f
McCabe-Thiele method 403
Mechanical vapor recompression 186–190
Mechanically agitated mass exchanger 405f
Mechanically agitated units 404
Medium pressure (MP) 194–195

MEK
 network configuration for removal 240f
 supply composition of 239–240
 vapor pressure versus temperature for 240f
Membrane separations, classification of 243
Membrane systems, advantages of 243
Membrane transport equations 244–245
Membrane-separation systems, design of 243
Membrane-separation unit, schematic representation of 243f
MENs see Mass-exchange networks (MENs)
Metal degreasing, solvent recycle in 205–207
Metal finishing process, as application of PI 394t
Methanol plant, stoichiometric targeting for 66
Methyl ethyl ketone, removal of 240–241
Metric for Inspecting Sales and Reactants (MISR) 65
Metrics, assessing sustainability 1–3
MFA see Material flow analysis (MFA)
Minimum allowable composition difference 112–113
Minimum approach temperature, role of 148f
Minimum cost, of MSAs 112
Minimum discharge, of targeted species 71f
Minimum interception, locus for 211f
Minimum net generation, of targeted species 70
Minimum operating cost (MOC) 122
 minimizing number of mass exchangers for 305
 network for dephenolization example 309f
 satisfying 156
 solution 112, 122
Minimum waste discharge
 procedure for identifying 71f
 targeting 67–71
Minimum-utility targets 147–148
Mixed-integer linear program (MILP) 255, 350
Mixed-integer nonlinear program (MINLP) 255
Mixed-integer program (MIP) 255
Mixer-settler system, three-stage 403f
Modeling principles, of mass-exchange units 401
Modified Accelerated Cost Recovery System (MACRS) depreciation method 28–31

calculation of annual depreciation 50t
calculations of 30
Molarity, of a reactive MSA 316
Moles, relating to mass 403f
Mollier diagram, forms of 167
Money, time-value of 33–40
MSAs see Mass-separating agents (MSAs)
Multiple reverse osmosis modules, designing systems of 247–251
Multistage mass exchanger 402–403, 403f
Multistage tray column 403f
MVR
 distillation configuration with 187f
 reducing heating and cooling utilities 188–189

N

Negative NPV 55
Nelson-Farrar Refinery Construction Index 17
Net generation, of a targeted species 67–70
Net present value (NPV) 43–48
 calculation of 54t
 comparing alternatives 55
 comparison of alternative based on 56
 of a project 46
 with uniform annual after-tax cash flows 47
 worksheet for 51t
Net present worth (NPW) 43–48
Network of heat exchangers, synthesizing 349
Network structure, illustrating graphically 158–159
Network synthesis 123–125, 350–353
Nitrogen, in a watershed 380f
No-/low-cost strategy 89
Nonconvex equilibrium relations 318–319
Nondiscounted annual after tax cash flows 47t
Nondiscounted cash flow diagram 47f
Nonlinear program (NLP) 255
Nonnegativity constraints, for residuals and loads 350
Nonnegativity requirement 279
NPV see Net present value (NPV)
Number of actual plates (NAP), calculating 404

O

Objectionable material (OM) 220
Objectives, insufficient consideration of 395
Off-gas condensation, intercepting 140f

Oil recycling plant, schematic representation of 118f
150 percent declining-balance method 28
One-dimensional data of a single attribute, format for entering 278f
One-dimensional metrics 1–2
On-stream efficiency 31–32
Open-cycle mechanical vapor recompression 179, 186–190
Operating costs 15
　estimation of 25–26
　versus fixed cost 125–127, 239
Operating line, versus equilibrium line 123
Operating principles 243–244
Opportunities, using process integration 393
Optimal solution, of fuel-blending problem 258f
Optimality criteria 92
Optimization
　approach 376
　described 255
　formulation 287–296
　model, formulating 255–258
　overview of 255
　problems using LINGO to solve 258–259
　program, solution of 299
　symbols used in 267, 268t
Optimization-based approach, for conceptual design of a biorefinery 365
Optimum solutions 10–11
OR operator 268
Organic chemicals production process 394t
Organization, created by process integration 393
Osmotic flow 243
Osmotic pressure 243
Outlet compositions, optimization of 303–305
Overall mass balance, for targeted species 70
Overall mass targeting, benchmarking process performance through 63
Overall mass targets, examples of 63
Overall material balance equation 218
Oxychlorination reaction 330

P

Packed columns 404
Papermaking process 220–221
Partial pressures, relating to compositions 402
Parts per million (ppm) 397–398, 404
Path diagram 327–328
　construction of 328
　tracking consequences of interceptions 328–329

Pathway selection step 366–367
Payback period, defined 43
Payback period (PBP) 44
Payout period 43
PBP (payback period) 44
Permeate 243
Permeate concentration 245
Permeate flow rate 245
Permeate stream, requirements of 247
Petrochemical facility, as application of PI 394t
Pharmaceutical facility
　example 150–151, 346–348
　revisiting case study on using algebraic cascade diagram 153–154
Pharmaceutical process, simplified flow sheet of 12f, 347f
Pharmaceutical processing facility 11
PI see Process integration (PI)
Pinch diagram
　constructing without process MSAs 120–122
　design rules for direct recycle 93
　design rules for HENs 150
　design rules for MENs 115
　determining optimal interception 328–329
　for cogeneration 196f
　for dilute VOC-condensation system 239f
　for EIPs 376
　for HENs 150f
　for material recycle 93f
　for material recycle with interception 95
　for MENs 114f
　for property-based direct recycle 203f
　for VOC-condensation system 238f
Pinch exchangers 122–123, 157
Pinch matches 122–123, 157
Pinch point
　distinguishing two zones for direct recycle 93
　distinguishing two zones for HENs 156
　distinguishing two zones for MENs 122, 305
　feasibility criteria for MENs 122, 232
　for cogeneration 196
　for direct recycle 93, 226, 395
　for HENs 150
　for MENs 114, 232, 300
　for property-based direct recycle 203
　for REAMENs 321
Piping and pumping, cost for 343
Plant scaling exponents, for selected processes 17t
Plant size, determining minimum product price via break-even analysis 34–35
Pollutant balance

　for reaches 382
　for tributaries 383
Pollutant concentration 397
Pollutants
　global flow of 327
　loads and compositions throughout a process 328f
Polymer and monomer production processes, as application of PI 394t
Polymer production facility 115–118
Polymer production processes 394t
Pooling problem 285f
Power
　conversion factors 399
　produced by the turbine 171
Power plant, with steam turbine 167–174
Practical feasibility line 112–113
Prescreening step 366
Present sum of the annuity 39
Present time 43–45, 45f
Pressure
　conversion factors 398
　as important factor in condensation 241
Pressure factor, values of 24–25
Pressure-driven membrane process 243
Pressure-driven membrane systems 244t
Pressure-temperature expression 188
Pressure-volume diagram 165–166
Preventive maintenance 26
Principal, of a single payment 37
Problem-table algorithm 152
Process
　with reaction and separation systems 63f
　schematic representation of 72, 73f
Process analysis, problems 4, 4f
Process economics, overview of 15, 405
Process equipment, categories of 18–19
Process industries, range of commodities 1
Process integration (PI) 3
　activities 3–4
　applications 393
　described 13
　in LCA 387
　macroscopic approaches of 375–390
　pitfalls in implementing 393–395
　starting and sustaining initiatives and projects 395–396
　techniques 376
　tools 393
　transforming into value-added applications 393
Process mass-separating agents (MSAs), excess 114

　making maximum use of 113
　maximizing use of 120
Process modification
　based on property-based pinch diagram 203–204
　moving a source across the pinch 204f
　reducing load of a source 204f
Process modifications, insights for 94–98
Process representation, from a mass-integration species perspective 133–143
Process scheduling 353, 354f
Process sinks 133–134
　pure fresh resources used in 89
　set of 287, 375
Process sources, set of 287
Process streams, utilization of 89
Process synthesis
　defined 3–4
　problems 4f
Process yield 72
Producer Prices Index for Chemical and Allied Products 36
Product
　rerouting from undesirable outlets to 73–74
　routing from undesirable to desirable outlets 74f
Production cost, estimation of 26
Profitability analysis 40–58
Profitability criteria
　comparison of different 53
　with time-value of money 42–55
　without time-value of money 41–42
Property clusters, characteristics of 208
Property integration 4–5, 201
Property load, of a sink 202
Property-based design rules, for recycle and interception 210–211
Property-based direct recycle 201
Property-based material pinch diagram 203–204, 203f
Property-based material recycle pinch diagram 201–203
Property-based material reuse problem 201f
Property-based pinch diagram, for degreasing case study 206f
Property-based problems, examples of 201
Property-mixing operators 202
Pulping mill, minimizing fresh water usage in 75–76
Pure fresh resource, replaced by process sources 91–92
Pyrolysis reactions 77–78

R

Rankine cycle
　differences with Carnot cycle 169

performance of a steam power plant in 172–173
for a steam power plant 170f
steps associated with 169
Raoult's law for absorption 402
Ratio factors, based on delivered equipment cost 17–18
Raw materials 26, 73f
Reaches, watersheds discretized into 381–383
Reaction system 63
Reaction yield, maximization of 73f
Reactive mass exchange, composition scales for 315–316
Reactive mass exchanger, with convex and nonconvex equilibrium 319f
Reactive mass-exchange networks (REAMENs)
problem of synthesizing 315
synthesis of 315, 315f
Reactive mass-separating agents (MSAs), converting transferred species into other compounds using 315
Reactor, evaluating feed to 73f
Reactor yield, 72
maximizing 73
one factor in maximizing overall process yield 83
REAMENs see Reactive mass-exchange networks (REAMENs)
Recovery period 27
Recycle 89
Recycle routes, selection of 90–91
Reduction reaction 330
Refinery, hydrogen recycle in 100, 295–296
Reflectivity 220
Refrigerants
annual operating cost of 240
examples of 179
Refrigeration duty 183
Reid vapor pressure (RVP)
of the condensate 215f
examining 205
Resin production facility, as application of PI 394t
Responsive maintenance 26
Retentate stream 248
Retente 243
Return on investment (ROI)
calculation of 42
defined 41
minimum acceptable value of 42
Reverse osmosis 243
Reverse problem formulation 381, 383
Reverse-osmosis network (RON) 247
Reverse-osmosis systems 243–247
Reversible adiabatic (isentropic) compression 166
Reversible adiabatic (isentropic) expansion 166

Reversible isothermal process 166
Rich composite stream
for benzene recovery example 116f
constructing 113, 114f
Rich end 114–115
Rich streams
collective load of 232
mixing 126, 126t
Rich-to-lean solute transfer 401–402
Right level of details, invoking 393
RO units, configurations of 244
ROI see Return on investment (ROI)
RONs
designing 248
equations describing performance of 248–251
optimization of 249–250
reverse-osmosis network 247
shortcut method for synthesis of 248–251
synthesis of 247–248

S

Salvage value 27
Scaling problem, for optimization software 321
Scenarios, solution through 10–11
Scheduling 285, 353
Scrubber, recycling to 7f
Scrubber water, recycling to replace 6f
Seawater desalination plant 275
Sensitivity analysis, via break-even analysis 33
Separation and recycle, combined 8f
Separation systems 134–143
Separation technologies
defining 9f
switching order of 10f
Service life 27
Set formulations, in LINGO 277–283
Set with one attribute 278
Set with two attributes 278
Sets 266, 277
Sieve tray, purchased cost of 25f
Simple payback period 43
Simulation, improper 393
Single-stage RON
equations describing performance of 248–251
optimization of 249–250
Sink and source composite curves 93f
Sink composite curve 92
developing 92f, 202f
for formic acid example 96f
Sink composite diagram, for the AN example 138f
Sink composition rule 102, 223
Sink-load rule 92

Sinks
selection of 90–91
with ternary constraints 106f
Sixth-tenths-factor rule 23
Sizing criteria, equipment cost related to 24–25
Slaking reaction 75
Solar-assisted trigeneration system 198f
Solid MEK, avoiding formation of 241
Solute flux, predicting 245
Solution through scenarios 10–11
Source composite curve 92–93, 203, 203f
developing 92f
for formic acid example 96f
sliding to the left 94–98, 95f
Source-prioritization rule 92, 102–103, 210, 223
Sources
described 133–134
selection of 90–91
splitting of 288f
Source-sink allocation 287f
Source-sink mapping diagram 98
allocating sources to sinks 103
for AN example 139f
for formic acid problem 104f
multicomponent 105–106
Source-sink representation 287
Soy to biodiesel, ecological cycle for 389f
Species perspective, process from 134f
Species viewpoint, process flow sheet from 133–134
Species-interception network (SPIN) 134–143
Split fractions, mixing of 288f
Spray column 404f
Spray towers 404
Steam, medium for capturing and delivering heat 194
Steam power plant, schematic representation of 168f
Steam tables, modeling performance of a steam power plant 172–173
Steam turbine
power plant with 167–174
purchased cost of 25f
topping system 171
Stick temperature, for recovery furnace 341
Stoichiometric, calculations 63
Stoichiometric targeting 63–64
Stoichiometric targets, levels of 63–64, 64f
Stoichiometric-economic stoichionomic targeting 64–66
Stoichiometry-based targeting 63–66
Stover-to-ethanol plant 20
Straight-line method 27
Strategies, for segregation, mixing, and direct recycle 336–339

Stream matching, for specialty chemical example 160f
Stream population 123, 157
Strictly convex function 267
Stripper, sources of heat added to 190
Stripping 401, 407f
Subnetworks, mass exchanged between any two streams 306
Summation 277
Superheated-steam enthalpy 171
SuperPro Designer 21–23
Superstructure
after prescreening 368f
generated 367f
Surplus composite line, development of 195–196
Sustainability
balancing three objectives 1
definition of 1
primary dimensions of 1f
Sustainable design
alternatives 5–12
defined 3
objectives of 3
pillars for 4
problems 10–11
Sustainable development, defined 1
Sustainable process design, pillars of 5f
Synthesis approach 318–322
Synthesis task, formulating 330–336
Systematic design methodology 201

T

Table of exchangeable heat loads (TEHL) 152
Table of exchangeable loads (TEL) 231–232
for dephenolization example 233t
for gaseous waste 321t
for lean streams 321t
for MSAs in dephenolization example 307t
for waste streams 299, 301t, 307t
TAC see Total annualized cost (TAC)
Targeted species, discharged out of the process 67–70
Targeting 3, 63
Targeting step, bypassing 395
Task identification 3
TCI see Total capital investment (TCI)
Techno-economic analysis 370, 370f
TEL see Table of exchangeable loads (TEL)
Temperature, conversion factors 398

Temperature interval diagram 156f
  ammonia removal case study 359f
  discretized streams over multiple periods 355f
  pharmaceutical example 348f
Temperature-entropy diagram 165–166, 168f
Temperature-interval diagram (TID) 152
  for pharmaceutical case study 153f
  for specialty chemical processing facility 159f
Terminal discharge, reducing for targeted species 70
Ternary (three-component) systems, source-sink diagram extended to represent 105–106
Ternary composition representation 106f
Ternary mixture, representation of 106f
Ternary property-cluster diagram
  for the absorber 217f
  for the degreaser 217f
Theoretical stoichiometric targets 63
Thermal efficiency, of the boiler 171–174
Thermal energy
  AR heat pumps driven primarily by 190
  integration through synthesis of heat-exchange networks (HENs) 165
Thermal equilibrium, requiring an infinitely large heat-transfer area 148
Thermal pinch analysis, integration with 174–178
Thermal pinch diagram 150f
  constructing 148
  minimum utility targets via 148–152
  for pharmaceutical process 151f
Thermal pinch point 149–150
Thermodynamic cycles 165
Thermodynamic feasibility, of heat exchange 152
Thermodynamic limitations, of mass exchange 113
Three-dimensional metrics 3
Three-property cluster source-sink mapping diagram 209–210
TH-shaftwork targeting model 198
TID see Temperature-interval diagram (TID)

Time-value, of money 33–40
Tire-to-fuel plant, reduction of discharge in 77–79, 261–262
Toluene, removal from wastewater 121
Total annualized cost (TAC) 27
  defined 55–56
  of mass-exchange systems 405
  using to compare alternatives 57
Total capital investment (TCI) 15
  components of 16f
  sum of FCI and WCI 26
Total direct cost 18
Total indirect cost 18
Total installed equipment cost 18
Toulene, removal from aqueous waste 407–408
Transesterification route, for producing biodiesel 370–372
Transport parameters, estimation of 247
Transportation problem
  for certain commodities from sources to destinations 265–266
  developing a set formulation for 280–281
Trial-and-error plot, determining DCR ROI for 49f
Tributaries
  inputs and outputs of typical 380f
  pollutant balance for 383
  of a watershed 379
Tricresyl phosphate process 342f
Trigeneration 198
T-S representation, of a Carnot heat pump 180f
Turbine, specific work produced by 170
Turbine hardware model, based on Willans line 198
Turbine power, estimating 195
Turnaround periods 31–32
Turnover ratios, for key industries 20–21, 22t
Two-dimensional data, entering 279
Two-dimensional metrics 2–3

U

Uncertainties, inappropriate accounting for 393–395
Underestimation piecewise linearization, of a function 272–277
United Nations Intergovernmental Panel on Climate Change (UN IPCC) 1–2

Updates using cost indices 17, 23
U.S. annual inflation rate, time-based variation in 36f
U.S. Consumer Price Index, time-based variation in 36f
U.S. Product Price Index for Chemicals and Allied Products, time-based variation in 36f
Useful life period 27
Utilities, prices of selected 26t
Utility cost, minimizing 239

V

Vapor-compression heat pumps
  most efficient 180
  schematic representation of 179f
  thermal pinch diagram and 185–186
Variable annual operating cost 31
Variable charges 31
Vatavuk cost index 23
Vector of optimization variables 255
Very high pressure (VHP) 194–195
Vinyl acetate case study 292–295
Vinyl acetate monomer (VAM)
  manufacturing 107–108
  plant flow diagram 292
  schematic flow sheet of 293f
Viscose rayon production, simplified flow sheet for 323f
Visualization tools, optimizing certain process objectives 209
VOC recycle, in a metal degreasing plant with multiple properties 213–218
VOC-condensation system, involving three units 237
VOC-recovery problem, converting into a heat-transfer task 238
VOCs (volatile organic compounds), recovering 237
Volatile organic compounds see VOCs (volatile organic compounds)
Volume, conversion factors 398

W

Warehouses, heat going through 345
Washing water, constraints imposed on quality and quantity 342
Waste cooking oil (WCO)
  to biodiesel process 372f

plant producing biodiesel 372
Waste sink 287
Waste-interception network (WIN), becoming mass-exchange network (MEN) 327
Wastewater
  with dilute pollutants at ambient temperature 398
  recycling to the boiler 5–8
  recycling to the scrubber 5–8
  segregation of 8f
  sources of 118
  toluene removal from 121
Water
  desalination 275–277
  modeling removal of an inorganic salt from 246
Water flux, predicting 245
Water intensity index 2
Water recovery, in a food process 99–100
Water recycle, difficulty involved in 341
Watersheds, material flow analysis and reverse problem formulation for 379–384
WCI (working capital investment) 15
WCO see Waste cooking oil (WCO)
WCO-to-biodiesel process, annual profit of 372f
Weak black liquor 319
Well-to-wheels GHG emissions, for alternate pathways 390f
White liquor 319, 341
Working capital investment (WCI) 15
World Business Council for Sustainable Development (WBCSD) 2–3

Y

Yield
  acetaldehyde from ethanol 79, 262
  ethanol from biomass 65, 68, 69
  flowchart for targeting 74f
  hydrogen from biomass 260
  maximization 72, 262
  of the process 72
  of the reactor 72, 263
  targeting 72, 262